Vitamins							Minerals						
Vitamin E RDA (mg/day)[e]	Vitamin K AI (μg/day)	Calcium AI (mg/day)	Phosphorus RDA (mg/day)	Magnesium RDA (mg/day)	Iron RDA (mg/day)	Zinc RDA (mg/day)	Iodine RDA (μg/day)	Selenium RDA (μg/day)	Copper RDA (μg/day)	Manganese AI (mg/day)	Fluoride AI (mg/day)	Chromium AI (μg/day)	Molybdenum RDA (μg/day)
4	2.0	210	100	30	0.27	2	110	15	200	0.003	0.01	0.2	2
5	2.5	270	275	75	11	3	130	20	220	0.6	0.5	5.5	3
6	30	500	460	80	7	3	90	20	340	1.2	0.7	11	17
7	55	800	500	130	10	5	90	30	440	1.5	1.0	15	22
11	60	1300	1250	240	8	8	120	40	700	1.9	2	25	34
15	75	1300	1250	410	11	11	150	55	890	2.2	3	35	43
15	120	1000	700	400	8	11	150	55	900	2.3	4	35	45
15	120	1000	700	420	8	11	150	55	900	2.3	4	35	45
15	120	1200	700	420	8	11	150	55	900	2.3	4	30	45
15	120	1200	700	420	8	11	150	55	900	2.3	4	30	45
11	60	1300	1250	240	8	8	120	40	700	1.6	2	21	34
15	75	1300	1250	360	15	9	150	55	890	1.6	3	24	43
15	90	1000	700	310	18	8	150	55	900	1.8	3	25	45
15	90	1000	700	320	18	8	150	55	900	1.8	3	25	45
15	90	1200	700	320	8	8	150	55	900	1.8	3	20	45
15	90	1200	700	320	8	8	150	55	900	1.8	3	20	45
15	75	1300	1250	400	27	13	220	60	1000	2.0	3	29	50
15	90	1000	700	350	27	11	220	60	1000	2.0	3	30	50
15	90	1000	700	360	27	11	220	60	1000	2.0	3	30	50
19	75	1300	1250	360	10	14	290	70	1300	2.6	3	44	50
19	90	1000	700	310	9	12	290	70	1300	2.6	3	45	50
19	90	1000	700	320	9	12	290	70	1300	2.6	3	45	50

[e] Vitamin E recommendations are expressed as α-tocopherol.

SOURCE: Adapted with permission from the *Dietary Reference Intakes* series, National Academy Press. Copyright 1997, 1998, 2000, 2001, by the National Academy of Sciences. Courtesy of the National Academy Press, Washington, D.C.

Minerals									
Zinc (mg/day)	Iodine (μg/day)	Selenium (μg/day)	Copper (μg/day)	Manganese (mg/day)	Fluoride (mg/day)	Molybdenum (μg/day)	Boron (mg/day)	Nickel (mg/day)	Vanadium (mg/day)
4	—	45	—	—	0.7	—	—	—	—
5	—	60	—	—	0.9	—	—	—	—
7	200	90	1000	2	1.3	300	3	0.2	—
12	300	150	3000	3	2.2	600	6	0.3	—
23	600	280	5000	6	10	1100	11	0.6	—
34	900	400	8000	9	10	1700	17	1.0	—
40	1100	400	10,000	11	10	2000	20	1.0	1.8
40	1100	400	10,000	11	10	2000	20	1.0	1.8
34	900	400	8000	9	10	1700	17	1.0	—
40	1100	400	10,000	11	10	2000	20	1.0	—
34	900	400	8000	9	10	1700	17	1.0	—
40	1100	400	10,000	11	10	2000	20	1.0	—

NOTE: An Upper Limit was not established for vitamins and minerals not listed and for those age groups listed with a dash (—) because of a lack of data, not because these nutrients are safe to consume at any level of intake. All nutrients can have adverse effects when intakes are excessive.

SOURCE: Adapted with permission from the *Dietary Reference Intakes* series, National Academy Press. Copyright 1997, 1998, 2000, 2001, by the National Academy of Sciences. Courtesy of the National Academy Press, Washington, D.C.

Nutrition through the Life Cycle

Judith E. Brown
Ph.D., M.P.H., R.D.
Public Health Nutrition
Division of Epidemiology, School of Public Health
University of Minnesota

Janet Sugarman Isaacs, PhD., R.D.
Department of Nutrition Science,
School of Health-Related Professions and
Civitan International Research Center, Sparks Clinics
University of Alabama at Birmingham

U. Beate Krinke, Ph.D., M.P.H., R.D.
Public Health Nutrition
Division of Epidemiology, School of Public Health
University of Minnesota

Maureen A. Murtaugh, Ph.D., R.D.
Postdoctoral Fellow
Division of Epidemiology, School of Public Health
University of Minnesota

Jamie Stang, Ph.D., M.P.H., R.D.
Public Health Nutrition
Division of Epidemiology, School of Public Health
University of Minnesota

Nancy H. Wooldridge, M.S., R.D., L.D.
Pediatric Pulmonary Center
Department of Pediatrics
University of Alabama at Birmingham

Contributors
Carolyn Sharbaugh, M.S., R.D.
Consultant
Denise Sofka, M.P.H., R.D.
DHHS Health Resources and Services Administration
Maternal and Child Health Bureau
Lori Roth-Yousey, M.P.H., R.D., L.N., Consultant

WADSWORTH
™
THOMSON LEARNING

Australia • Canada • Mexico • Singapore • Spain
United Kingdom • United States

WADSWORTH
THOMSON LEARNING

PUBLISHER: Peter Marshall
DEVELOPMENT EDITOR: Elizabeth Howe
ASSISTANT EDITOR: John Boyd
EDITORIAL ASSISTANT: Andrea Kesterke
MARKETING MANAGER: Jennifer Somerville
ADVERTISING PROJECT MANAGER: Stacey Purviance
PROJECT MANAGER: Sandra Craig
PRINT/MEDIA BUYER: Barbara Britton
PERMISSIONS EDITOR: Robert Kauser
DESIGN AND PRODUCTION: Ann Borman
COPY EDITORS: Ben Shriver, Cheryl Welms, Ann Whetstone
ILLUSTRATOR: Stan Maddock
INDEX: Terry Casey
COVER IMAGES: Photo Disc
COMPOSITOR: Parkwood Composition
PRINTER: Quebecor World, Iowa

Printed in the United States of America
1 2 3 4 5 6 7 05 04 03 02 01

For more information about our products, contact us at:
Thomson Learning Academic Resource Center
1-800-423-0563
For permission to use material from this text, contact us by:
Phone: 1-800-730-2214
Fax: 1-800-730-2215
Web: http://www.thomsonrights.com

Wadsworth/Thomson Learning
10 Davis Drive
Belmont, CA 94002-3098
USA

ASIA
Thomson Learning
60 Albert Street, #15-01
Albert Complex
Singapore 189969

AUSTRALIA
Nelson Thomson Learning
102 Dodds Street
South Melbourne, Victoria 3205
Australia

CANADA
Nelson Thomson Learning
1120 Birchmount Road
Toronto, Ontario M1K 5G4
Canada

EUROPE/MIDDLE EAST/AFRICA
Thomson Learning
Berkshire House
168-173 High Holborn
London WC1 V7AA
United Kingdom

LATIN AMERICA
Thomson Learning
Seneca, 53
Colonia Polanco
11560 Mexico D.F.
Mexico

SPAIN
Paraninfo Thomson Learning
Calle/Magallanes, 25
28015 Madrid, Spain

Library of Congress Cataloging-in-Publication Data
Brown, Judith E.
    Nutrition through the life cycle / by Judith E. Brown with Janet S. Isaacs . . . [et.al.].
    p. cm.
    Includes bibliographic references and index.
    ISBN 0-534-58986-3
    1. Nutrition. 2. Children—Nutrition. 3. Aged—Nutrition. I. Isaacs, Janet S. II. Title

QP141 .B8742002
612.3'9—dc21
                                                  2001046932

CONTENTS IN BRIEF

CONTENTS

Chapter 5
NUTRITION DURING PREGNANCY
Conditions and Interventions 107

Chapter 6
NUTRITION DURING LACTATION 135

Chapter 7
NUTRITION DURING LACTATION
Conditions and Interventions 169

Chapter 8
INFANT NUTRITION 191

PREFACE

On behalf of the expert authors represented in this text, welcome to the first edition of *Nutrition through the Life Cycle*. May you find the text to represent the innovative, comprehensive, and engaging approach to life cycle nutrition education it is intended to be.

Nutrition through the Life Cycle was developed with the needs of instructors teaching, and students taking, a two-to-four credit course in life cycle nutrition in mind. It is written at a level that assumes students have had an introductory nutrition course. Chapter 1 summarizes key elements of introductory nutrition and gives students who need it a chance to update or renew their knowledge. Subsequent chapters for Nutrition through the Life Cycle were developed by a cadre of expert authors who are actively engaged in clinical practice, teaching, or research related to nutrition during specific phases of the life cycle. All of us remained totally dedicated to the goals established for the text at its conception: to make the text comprehensive, logically organized, science-based, and realistic.

Coverage of the life cycle begins with preconceptional nutrition and continues with each major phase of the life cycle through adulthood and the special needs of the elderly. Each of these 18 chapters was developed from a common organizational framework that includes key nutrition concepts, public health statistics, physiological principles, nutritional needs and recommendations, model programs, case studies, and recommended practices. To meet the knowledge needs of students with the variety of career goals represented in many life cycle nutrition courses, we developed two chapters for each life cycle phase.

The first chapter for each life cycle phase covers normal nutrition topics and the second nutrition-related conditions and interventions. Every chapter focuses on scientifically based information and employs the most up-to-date resources and references available. Each chapter ends with a list of electronic and/or print resources that will lead students to reliable information on scientific and applied aspects of life cycle nutrition.

Overall, the text is intended to give instructors a tool they can happily use to enhance their teaching efforts, and to give students an engaging and rewarding educational experience they will carry with them throughout their lives.

Resources for Students and Instructors

An electronic *Instructor's Manual with Test Bank*, available via the Web, contains a test bank, classroom activities, chapter outlines, and more.

Acknowledgments

Development of *Nutrition through the Life Cycle* was made possible by Peter Marshall, the Publisher of Wadsworth Nutrition titles. His excitement for the project was contagious, and his vision for the text delighted us all. A thousand thanks go out to Beth Howe, the developmental editor on the project, who steadfastly guided us through problem areas while engendering the respect and admiration of all of us. Ann Borman, Senior Production Editor, designed the text and cover. Thanks to her, our unattractive manuscript pages turned miraculously into the text you are holding. We were fortunate to have many other capable Wadsworth professionals contribute to this text, including Jennifer Somerville, Marketing Manager.

Reviewers

Many thanks to the following reviewers whose careful reading and thoughtful comments helped shape this first edition.

Betty Alford
Texas Women's University

Leta Aljadir
University of Delaware

Dea Hanson Baxter
Georgia State University

Shelley Hancock
University of Alabama

Tay Seacord Kennedy
Oklahoma State University

Younghee Kim
Bowling Green State University

Barbara Kirks
California State University, Chico

Kaye Stanek Krogstrand
University of Nebraska

Sally Ann Lederman
Columbia University

Richard Lewis
University of Georgia

Harriet McCoy
University of Arkansas
Cherie Moore
Cuesta College

Sharon McWhinney
Prairie View A&M University

Robert Reynolds
University of Illinois at Chicago

Sharon Nickols-Richardson
Virginia Polytechnic Institute and
State University

Adria Sherman
Rutgers University

Carmen R. Roman-Shriver
Texas Tech University

Joanne Slavin
University of Minnesota

Joanne Spaide
Professor Emeritus University of
Northern Iowa

Diana-Marie Spillman
Miami University,
Oxford Ohio

Wendy Stuhldreher
Slippery Rock University of
Pennsylvania

Anne VanBeber
Texas Christian University

Phyllis Moser-Veillon
University of Maryland

Janelle Walter
Baylor University

Suzy Weems
Stephen F. Austin State University

Kay Wilder
Point Loma Nazarene College

Judith E. Brown, August 2001

CHAPTER 1

Photo Disc

To be surprised, to wonder, is to begin to understand.

José Ortega y Gasset

NUTRITION BASICS

Prepared by **Judith E. Brown**

CHAPTER OUTLINE

- Introduction
- Principles of the Science of Nutrition
- Nutritional Assessment
- Public Food and Nutrition Programs
- Nationwide Priorities for Improvements in Nutritional Health

KEY NUTRITION CONCEPTS

1 Nutrition is the study of foods, their nutrients and other chemical constituents, and the effects of food constituents on health.

2 Nutrition is an interdisciplinary science.

3 Nutrition recommendations for the public change as new knowledge about nutrition and health relationships is gained.

4 At the core of the science of nutrition are principles that represent basic truths and serve as the foundation of our understanding about nutrition.

INTRODUCTION

Need to freshen up your knowledge of nutrition? Or, do you need to get up to speed on basic nutrition for the course? This chapter presents information about nutrition that paves the way to greater understanding of specific needs and benefits related to nutrition by life cycle stage.

Nutrition is an interdisciplinary science focused on the study of foods, *nutrients,* and other food constituents and health. The body of knowledge about nutrition is large and is growing rapidly, changing views on what constitutes the best nutrition advice. You are encouraged to refer to basic nutrition texts and to use the online resources listed at the end of this chapter to fill in any knowledge gaps. You are also encouraged to stay informed in the future and to keep an open mind about the best nutrition advice for many health-related issues. Scientific evidence that drives decisions about nutrition and health changes with time.

This chapter centers on (1) the principles of the science of nutrition, (2) the nutrients, (3) nutritional assessment, (4) public food and nutrition programs, and (5) nationwide priorities for improvements in the public's nutritional health.

NUTRIENTS   Chemical substances in foods that are used by the body for growth and health.

FOOD SECURITY   Access at all times to a sufficient supply of safe, nutritious foods.

FOOD INSECURITY   Limited or uncertain availability of safe, nutritious foods.

CALORIE   A unit of measure of the amount of energy supplied by food. Also know as the "kilocalorie," or the "large Calorie."

PRINCIPLES OF THE SCIENCE OF NUTRITION

Every field of science is governed by a set of principles that provides the foundation for growth in knowledge. These principles change little with time. Knowledge of the principles of nutrition listed in Table 1.1 will serve as a springboard to greater understanding of the nutrition and health relationships explored in the chapters to come.

PRINCIPLE #1   Food is a basic need of humans.

Humans need enough food to live and the right assortment of foods for optimal health (Illustration 1.1). People who have enough food to meet their needs at all times experience *food security.* They are able to acquire food in socially acceptable ways—without having to scavenge or steal food. *Food insecurity* exists when the availability of safe, nutritious foods, or the ability to acquire them in socially acceptable ways, is limited or uncertain.[1]

Table 1.1   Principles of human nutrition.

PRINCIPLE #1   Food is a basic need of humans.

PRINCIPLE #2   Foods provide energy (calories), nutrients, and other substances needed for growth and health.

PRINCIPLE #3   Health problems related to nutrition originate within cells.

PRINCIPLE #4   Poor nutrition can result from both inadequate and excessive levels of nutrient intake.

PRINCIPLE #5   Humans have adaptive mechanisms for managing fluctuations in food intake.

PRINCIPLE #6   Malnutrition can result from poor diets and from disease states, genetic factors, or combinations of these causes.

PRINCIPLE #7   Some groups of people are at higher risk of becoming inadequately nourished than others.

PRINCIPLE #8   Poor nutrition can influence the development of certain chronic diseases.

PRINCIPLE #9   Adequacy and balance are key characteristics of a healthy diet.

PRINCIPLE #10   There are no "good" or "bad" foods.

PRINCIPLE #2   Foods provide energy (calories), nutrients, and other substances needed for growth and health.

People eat foods for many different reasons. The most compelling reason is the requirement for *calories* (energy), nutrients, and other substances supplied by foods for growth and health.

A calorie is a measure of the amount of energy transferred from food to the body. Because calories

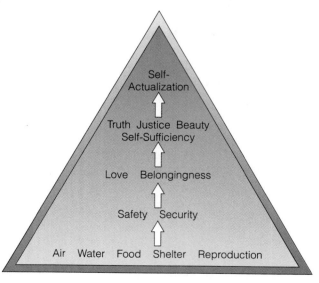

Illustration 1.1   The need for food is part of Maslow's hierarchy of needs.

are a unit of measure and not a substance actually present in food, they are not considered to be nutrients. Nutrients are chemical substances in food that are used by the body for a variety of functions that support growth, tissue maintenance and repair, and ongoing health. Essentially, every part of our body was once a nutrient consumed in food.

There are six categories of nutrients (Table 1.2), and each category except water consists of a number of different substances.

Essential and Nonessential Nutrients

Of the many nutrients required for growth and health, some must be provided by the diet while others can be made by the body.

ESSENTIAL NUTRIENTS   Nutrients the body cannot manufacture, or produce in sufficient amounts, are referred to as *essential nutrients*. Here *essential* means "required in the diet." The following nutrients are considered essential:

- Carbohydrates,
- Certain amino acids (the essential amino acids: histidine, isoleucine, leucine, lysine, methionine, phenylalanine, threonine, tryptophan, and valine),
- Linoleic acid and alpha-linolenic acid (essential fatty acids),
- Vitamins,
- Minerals, and
- Water.

NONESSENTIAL NUTRIENTS   Cholesterol, creatine, and glucose are examples of *nonessential nutrients*. Nonessential nutrients are present in food and used by the body, but they do not have to be part of our diet. Many of the beneficial chemical substances in plants are not considered essential, for example, yet they play important roles in maintaining health.

REQUIREMENTS FOR ESSENTIAL NUTRIENTS   All humans require the same set of essential nutrients, but the amount of nutrients needed varies based on:

- Age
- Gender
- Growth
- Genetic traits
- Pregnancy and lactation
- Body size
- Lifestyle habits (e.g., smoking, alcohol intake)
- Illness
- Medication use

Amounts of essential nutrients required each day vary a great deal, from cups (for water) to micrograms (for example, for folate and vitamin B_{12}).

Dietary Intake Standards

Dietary intake standards developed for the public cannot take into account all of the factors that influence nutrient needs, but they do account for the major ones of age, gender, growth, and pregnancy and lactation. Intake standards are called Dietary Reference Intakes (DRIs). DRIs have been developed for most of the essential nutrients and are updated every 5 to 10 years. (These are listed on the inside front covers of this text.) The DRIs are levels of nutrient intake intended for use as reference values for planning and assessing diets for healthy people. They consist of the Recommended Dietary Allowances (RDAs), which specify intake levels that meet the nutrient needs of over 98% of healthy people, and the other categories of intake standards described in Illustration 1.2. It is recommended that individuals aim for nutrient intakes that approximate the RDAs or Adequate Intake (AI) levels. Estimated Average Requirements (EARs) should be used to examine the possibility of inadequate intakes in individuals and within groups. Additional tests are required to confirm inadequate nutrient intakes and status.[2]

ESSENTIAL NUTRIENTS   Substances required for growth and health that cannot be produced, or produced in sufficient amounts, by the body. They must be obtained from the diet.

NONESSENTIAL NUTRIENTS   Nutrients required for growth and health that can be produced by the body from other components of the diet.

Table 1.2   The six categories of nutrients.

1. *Carbohydrates*   Chemical substances in foods that consist of a single sugar molecule or multiples of sugar molecules in various forms. Sugar and fruit, starchy vegetables, and whole grain products are good dietary sources.

2. *Protein*   Chemical substances in foods that are made up of chains of amino acids. Animal products and dried beans are examples of protein sources.

3. *Fat*   Components of food that are soluble in fat but not in water. They are more properly referred to as "lipids." Lipids are composed of glycerol attached to one to three fatty acids. Oil, butter, sausage, and avocado are examples of rich sources of dietary lipids.

4. *Vitamins*   Thirteen specific chemical substances that perform specific functions in the body. Vitamins are present in many foods and are essential components of the diet. Vegetables, fruits, and grains are good sources of vitamins.

5. *Minerals*   In the context of nutrition, minerals consist of 15 elements found in foods that perform particular functions in the body. Milk, dark, leafy vegetables, and meat are good sources of minerals.

6. *Water*   An essential component of the diet provided by food and fluid.

TOLERABLE UPPER LEVELS OF INTAKE (ULs) The DRIs include a table indicating levels of daily nutrient intakes from foods, fortified products, and supplements that should not be exceeded. They can be used to assess the safety of high intakes of nutrients, particularly from supplements.

STANDARDS OF NUTRIENT INTAKE FOR NUTRITION LABELS The Nutrition Facts panel on packaged foods uses standard levels of nutrient intakes based on an earlier edition of recommended dietary intake levels. The levels are known as *Daily Values (DVs)* and are used to identify the amount of a nutrient provided in a serving of food compared to the standard level. The "% DV" listed on nutrition labels represents the percentages of the standards obtained from one serving of the food product. Table 1.3 lists DV standard amounts for nutrients that are mandatory or voluntary components of nutrition labels.

> **DAILY VALUES (DVs)** Scientifically agreed-upon standards for daily intakes of nutrients from the diet developed for use on nutrition labels.

Carbohydrates

Carbohydrates are used by the body mainly as a source of readily available energy. They consist of the simple sugars (monosaccharides and disaccharides), complex carbohydrates (the polysaccharides), and alcohol sugars. Alcohol is a close chemical relative of carbohydrates and is usually considered to be part of this nutrient category. The most basic forms of carbohydrates are single molecules called monosaccharides. They are most often present in foods in the form of glucose (also called "blood sugar" and "dextrose"), fructose ("fruit sugar"), and galactose. When two simple sugars combine, the result is a disaccharide. The most common disaccharides are:

- Sucrose (glucose + fructose, or common table sugar),
- Maltose (glucose + glucose, or maltose sugar), and
- Lactose (glucose + galactose, or milk sugar).

Complex carbohydrates (also called polysaccharides) are considered "complex" because they have more elaborate chemical structures than the simple sugars. They include:

- Starches (the plant storage form of carbohydrate),
- Glycogen (the animal form of stored carbohydrate), and
- Dietary fiber.

Each type of simple and complex carbohydrate, except dietary fiber, provides four calories per gram. Dietary fibers are not considered a source of energy because they cannot be broken down by digestive enzymes. The main function of dietary fiber is to provide "bulk" for normal elimination. It has other beneficial properties, however. High-fiber diets reduce the rate of glucose absorption (a benefit for people with diabetes) and may help prevent cardiovascular disease and some types of cancer.[3,4]

Nonalcoholic in the beverage sense, alcohol sugars are like simple sugars, except that they include a chemical component of alcohol. Xylitol, mannitol, and sorbitol are common forms of alcohol sugars. Some are very sweet, and only small amounts are needed to sweeten commercial beverages, gums, yogurt, and other products. Unlike the simple sugars, alcohol sugars do not promote tooth decay.

Illustration 1.2 **Theoretical framework, terms, and abbreviations used in the Dietary Reference Intakes.**

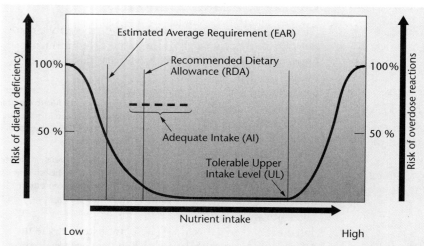

- **Dietary Reference Intakes (DRIs).** This is the general term used for the new nutrient intake standards for healthy people.
- **Recommended Dietary Allowances (RDAs).** These are levels of essential nutrient intake judged to be adequate to meet the known nutrient needs of practically all healthy persons while decreasing the risk of certain chronic diseases.
- **Adequate Intakes (AIs).** These are "tentative" RDAs. AIs are based on less conclusive scientific information than are the RDAs.
- **Estimated Average Requirements (EARs).** These are nutrient intake values that are estimated to meet the requirements of half the healthy individuals in a group. The EARs are used to assess adequacy of intakes of population groups.
- **Tolerable Upper Levels of Intake (ULs).** These are upper limits of nutrient intake compatible with health. The ULs do not reflect desired levels of intake. Rather, they represent total, daily levels of nutrient intake from food, fortified foods, and supplements that should not be exceeded.

Table 1.3 Daily Values (DVs) for nutrition labeling based on intakes of 2000 calories per day in adults and children aged four years and above.

MANDATORY COMPONENTS OF THE NUTRITION LABEL

Food Component	Daily Value (DV)
Total fat	65 g[a]
Saturated fat	20 g
Cholesterol	300 mg[a]
Sodium	2400 mg
Total carbohydrate	300 g
Dietary fiber	25 g
Vitamin A	5000 IU[a]
Vitamin C	60 mg
Calcium	1000 mg
Iron	18 mg

VOLUNTARY COMPONENTS OF THE NUTRITION LABEL

Food Component	Daily Value (DV)
Vitamin D	400 IU
Vitamin E	30 IU
Vitamin K	80 mcg[a]
Thiamin	1.5 mg
Riboflavin	1.7 mg
Niacin	20 mg
Vitamin B_6	2 mg
Folate	400 mcg
Vitamin B_{12}	6 mcg
Biotin	300 mcg
Pantothenic acid	10 mg
Phosphorus	1000 mg
Iodine	150 mcg
Magnesium	400 mg
Zinc	15 mg
Selenium	70 mcg
Copper	2 mg
Manganese	2 mg
Chromium	120 mcg
Molybdenum	75 mcg
Chloride	3400 mg
Potassium	3500 mg

[a]g = grams; mg = milligrams; IU = International Units; mcg = micrograms.

Alcohol (consumed as ethanol) is considered to be part of the carbohydrate family because its chemical structure is similar to that of glucose. It is a product of the fermentation of sugar with yeast. With seven calories per gram, alcohol has more calories per gram than do other carbohydrates.

RECOMMENDED INTAKE LEVEL Recommended intake of carbohydrate is based on their contribution to total energy intake. It is recommended that 55–60% of calories come from carbohydrates, especially from complex carbohydrates such as whole grain products,

pasta, and breads. Only 10% of total energy intake should be from simple sugars. It is recommended that adults consume between 25 and 30 grams of dietary fiber daily.

FOOD SOURCES OF CARBOHYDRATES Carbohydrates are widely distributed in plant foods, while milk is the only important animal source of carbohydrates (lactose). Table 1.4 lists selected food sources by type of carbohydrate. Additional information about total carbohydrate and dietary fiber content of foods can be found in the food composition table in Appendix A at the back of this book and on nutrition information labels on food packages.

Protein

Protein in foods provides the body with *amino acids* used to build and maintain tissues such as muscle, bone, enzymes, and red blood cells. Protein can also be used by the body as a source of energy—it provides four calories per gram. However, this is not a primary function of protein. Of the common types of amino acids, nine must be provided by the diet and are classified as "essential amino acids." Many other amino acids obtained from food perform important functions, but since the body can manufacture these from other amino acids, they are classified as "nonessential amino acids."

Food sources of protein differ in quality, based on the types of amino acids they contain. Foods of high protein quality include all of the essential amino acids. Protein from milk, cheese, meat, eggs, and other animal products is considered high quality. Plant sources of protein, with the exception of soybeans, do not provide all nine essential amino acids. Combinations of plant foods, such as grains or seeds with dried beans, however, yield high-quality protein. Amino acids found in these individual foods "complement" each other, thus providing a source of high-quality protein.

AMINO ACIDS The "building blocks" of protein. They are nitrogen-containing constituents of food.

KWASHIORKOR A disease syndrome in children caused by protein deficiency. It is characterized by edema (or swelling), loss of muscle mass, fatty liver, rough skin, reddening of the hair, growth retardation, and apathy.

RECOMMENDED PROTEIN INTAKE DRIs for protein are shown on the inside front cover of this text. In general, proteins should contribute approximately 12% of total energy intake. Protein deficiency, although rare in economically developed countries, leads to loss of muscle tissue, weakness, reduced resistance to disease, kidney and heart problems, and *kwashiorkor* in children.

Table 1.4 **Food sources of carbohydrates.**

A. SIMPLE SUGARS (MONO- AND DISACCHARIDES)

	Portion Size	Grams of Carbohydrate		Portion Size	Grams of Carbohydrate
Breakfast Cereals			Apple	1 med	16
Raisin Bran	1 c	18	Orange	1 med	14
Corn Pops	1 c	14	Peach	1 med	8
Frosted Cheerios	1 c	13	**Vegetables**		
Bran Flakes	¾ c	5	Corn	½ c	3
Grape-Nuts	½ c	3	Broccoli	½ c	2
Special K	1 c	3	Potato	1 med	1
Wheat Chex	1 c	2	**Beverages**		
Corn Flakes	1 c	2	Soft drinks	12 oz	38
Sweeteners			Fruit drinks	1 c	29
Honey	1 tsp	6	Skim milk	1 c	12
Corn syrup	1 tsp	5	Whole milk	1 c	11
Maple syrup	1 tsp	4	**Candy**		
Table sugar	1 tsp	4	Hard candy	1 oz	28
Fruits			Gumdrops	1 oz	25
Watermelon	1 pc (4" × 8")	25	Caramels	1 oz	21
Banana	1 med	21	Milk chocolate	1 oz	16

B. COMPLEX CARBOHYDRATES (STARCHES)

	Portion Size	Grams of Carbohydrate		Portion Size	Grams of Carbohydrate
Grain Products			**Dried Beans**		
Rice, white, cooked	½ c	21	White beans, cooked	½ c	13
Pasta, cooked	½ c	15	Kidney beans, cooked	½ c	12
Oatmeal, cooked	½ c	12	Lima beans, cooked	½ c	11
Cheerios	1 c	11	**Vegetables**		
Corn Flakes	1 c	11	Potato	1 med	30
Bread, whole wheat	1 slice	7	Corn	½ c	10
			Broccoli	½ c	2

C. DIETARY FIBER

	Portion Size	Grams of Carbohydrate		Portion Size	Grams of Carbohydrate
Grain Products			Green peas	½ c	4
Bran Buds	½ c	12	Carrots	½ c	3
All Bran	½ c	10	Potato, with skin	1 med	4
Raisin Bran	1 c	7	Collard greens	½ c	3
Bran Flakes	¾ c	5	Corn	½ c	3
Oatmeal, cooked	1 c	4	Cauliflower	½ c	2
Bread, whole wheat	1 sl	2	**Nuts**		
Fruits			Almonds	½ c	5
Avocado	½ med	7	Peanuts	½ c	3
Raspberries	1 c	5	Peanut butter	2 tbs	2
Mango	1 med	4	**Dried Beans**		
Pear, with skin	1 med	4	Pinto beans, cooked	½ c	10
Orange	1 med	3	Black beans, cooked	½ c	8
Banana	1 med	2	Black-eyed peas, cooked	½ c	8
Vegetables			Navy beans, cooked	½ c	6
Lima beans	½ c	5	Lentils, cooked	½ c	5

FOOD SOURCES OF PROTEIN   Animal products and dried beans are particularly good sources of protein. These and other food sources of protein are listed in Table 1.5.

Fats

Fats in food share the property of being soluble in fats but not water. They are actually a subcategory of *lipids*. Lipids include fats and oil, the distinction being that fats are generally solid at room temperature whereas oils are not. Both forms of lipids are made of *glycerol* linked to one to three *fatty acids.*

Fats and oils are a concentrated source of energy, providing nine calories per gram. Fats perform a number of important functions in the body. They are precursors for cholesterol and sex hormone synthesis, components of cell membranes, vehicles for carrying certain vitamins that are soluble in fats only, and suppliers of the *essential fatty acids* required for growth and health.

ESSENTIAL FATTY ACIDS   There are two known essential fatty acids: linoleic acid and alpha-linolenic acid. Because these fatty acids are essential, they must be supplied through dietary intake. The central nervous system is particularly rich in derivatives of these two fatty acids. They are found in phospholipids, which along with cholesterol, are the primary lipids in the brain and other nervous system tissue.

Linoleic acid   Linoleic acid is the parent of the n-6 fatty acid family. One of the major derivatives of linoleic acid is arachidonic acid. Arachidonic acid is used to form prostaglandins (hormones with multiple forms and functions in the body), and it also serves as a primary structural component of the central nervous system. Most vegetable oils, meats, and human milk are good sources of linoleic acid. American diets tend to provide sufficient levels of linoleic acid, and considerable amounts are stored in body fat.

Alpha-linolenic acid   Alpha-linolenic acid is the parent of the n-3 fatty acid family. It is present in many types of dark green vegetables, vegetable oils, and flaxseed. The most important derivatives of this essential fatty acid are eicosapentaenoic acid (EPA) and docosahexaenoic acid (DHA). EPA and DHA also enter the body through intake of fatty, cold water fish and shellfish and human milk. EPA is the precursor of some prostaglandins, and DHA is found in large amounts in the central nervous system, the retina of the eye, and the testes. The body stores only small amounts of alpha-linolenic acid, EPA, and DHA.[5]

GLYCEROL   A component of fats that is soluble in water. It is similar to glucose in chemical structure.

FATTY ACIDS   The fat-soluble components of fats in foods.

ESSENTIAL FATTY ACIDS   Components of fat that are a required part of the diet, i.e., linoleic and alpha-linolenic acids. Both contain unsaturated fatty acids.

SATURATED FATS   Fats in which adjacent carbons in the fatty acid component are linked by single bonds only, e.g., -C-C-C-C-.

UNSATURATED FATS   Fats in which adjacent carbons in one or more fatty acids are linked by one or more double bonds, e.g., -C-C=C-C=C-.

MONOUNSATURATED FATS   Fats in which only one pair of adjacent carbons in one or more of its fatty acids is linked by a double bond, e.g., -C-C=C-C-.

POLYUNSATURATED FATS   Fats in which more than one pair of adjacent carbons in one or more of its fatty acids are linked by two or more double bonds, e.g., -C-C=C-C=C-.

SATURATED AND UNSATURATED   Fats come in two basic types: *saturated* and *unsaturated.* Whether a fat is saturated or not depends on whether it has one or more double bonds between the carbon atoms of the fatty acid part of the fat. If one double bond is present in one or more of the fatty acids components, the fat is considered *monounsaturated;* if two or more are present, the fat is *polyunsaturated.*

Some unsaturated fatty acids (the n-3-fatty acids) are highly unsaturated. Alpha-linolenic acid, for example, contains three double bonds, EPA five, and DHA six. These fatty acids are more unstable than fatty acids with fewer double bonds because double bonds between atoms are weaker than single bonds.

Saturated fats contain no double bonds between carbons and tend to be solid at room temperature. Animal products such as butter, cheese, and meats and two plant oils (coconut and palm) are rich sources of saturated fats. Fat we consume in our diets, whether it contains primarily saturated or unsaturated fatty acids, is generally in the triglyceride (or triacylglycerol) form. This type of fat consists of a glycerol backbone and three fatty acids:

Table 1.5   Food sources of protein.

	Portion Size	Grams of Protein
Meats		
Beef, lean	3 oz	26
Tuna, in water	3 oz	24
Hamburger, lean	3 oz	24
Chicken, no skin	3 oz	24
Lamb	3 oz	22
Pork chop, lean	3 oz	20
Haddock, broiled	3 oz	19
Egg	1 med	6
Dairy Products		
Cottage cheese, low fat	½ c	14
Yogurt, low fat	1 c	13
Milk, skim	1 c	9
Milk, whole	1 c	8
Swiss cheese	1 oz	8
Cheddar cheese	1 oz	7
Grain Products		
Oatmeal, cooked	½ c	4
Pasta, cooked	½ c	4
Bread	1 slice	2
Rice, white or brown	½ c	2

Fats also come in the form of:

- Monoglycerides (or monoacylglycerols), consisting of glycerol plus one fatty acid and
- Diglycerides (or diacylglycerols), consisting of glycerol and two fatty acids.

Although most foods contain both saturated and unsaturated fats, animal foods tend to contain more saturated and less unsaturated fat than plant foods.

Hydrogenation and Trans Fats   Oils can be made solid by adding hydrogen to the double bonds of their fatty acids. This process, called hydrogenation, makes some of the fatty acids in oils saturated and enhances storage life and baking qualities. Hydrogenation may alter the structure of the fatty acids, however, producing *trans fats*. This type of fat is naturally present in a few foods, but the primary dietary sources are products made from hydrogenated fats. Diets high in trans fats are related to an increased risk of heart disease, as are diets high in saturated fat.[6,7]

CHOLESTEROL   Dietary *cholesterol* is a fatlike, clear liquid substance found in lean and fat components of animal products. Cholesterol is a component of all animal cell membranes, the brain, and the nerves; it is the precursor of estrogen, testosterone, and vitamin D, which is manufactured in the skin upon exposure to sunlight. The body produces about one-third of the cholesterol our bodies use, and the remainder comes from the diet. The extent to which dietary cholesterol intake modifies blood cholesterol level appears to vary a good deal based on genetic tendencies as well as on dietary intake of cholesterol and saturated fat.[8] Dietary cholesterol intake affects blood cholesterol level substantially less than does saturated fat intake, however.[9] Leading sources of dietary cholesterol are egg yolks, meat, milk and milk products, and fats such as butter.

CHOLESTEROL   A fat-soluble, colorless liquid found in animals but not plants.

RECOMMENDED INTAKE OF FATS   It is recommended that 30% or less of total caloric intake among adults be from fats, and less than 10% of calories be from saturated fats. Dietary cholesterol intake should be less than 300 milligrams (mg) a day.

FOOD SOURCES OF FAT   The fat content of many foods can be identified by reading the nutrition information labels on food packages. The amount of total fat, saturated fat, and cholesterol in a serving of food is listed on the labels. The amount of unsaturated fat in a serving equals total fat minus saturated fat.

Table 1.6 lists the total fat, saturated fat, unsaturated fat, trans fat, cholesterol, and n-3-fatty acid contents of selected foods. The food composition table in Appendix A lists the fact content of many other foods.

Vitamins

Vitamins are chemical substances in foods that perform specific functions in the body. Thirteen have been discovered so far. They are classified as either fat soluble or water soluble (Table 1.7).

The B-complex vitamins and vitamin C are soluble in water and found dissolved in water in foods. The fat-soluble vitamins consist of vitamins A, D, E, and K and are present in the fat portions of foods. (To remember the fat-soluble vitamins, think of "DEKA" for vitamins D, E, K, and A.) Only these chemical substances are truly vitamins. Substances such as coenzyme Q_{10}, inositol, provitamin B_5 complex, and pangamic acid (vitamin B_{15}) may be called vitamins, but they are not. With the exception of vitamin B_{12}, water-soluble vitamin stores in the body are limited and run out within a few weeks to a few months after intake becomes inadequate. Fat-soluble vitamins are stored in the body's fat tissues and the liver. These stores can be sizable and last a period of months to years when intake is low.

Connections
Why aren't vitamins named A through F?

Vitamins as we know them today don't follow alphabetical order because errors were made when vitamins were discovered. They were named in alphabetical sequence, but some substances turned out not really to be vitamins. The letters assigned to these substances were dropped, so the sequence is interrupted.

Excessive consumption of the fat-soluble vitamins from supplements, especially of vitamins A and D, produces various symptoms of toxicity. High intake of the water-soluble vitamins from supplements can also produce adverse health effects. Toxicity symptoms from water-soluble vitamins, however, tend to last a shorter time and are more quickly remedied. Vitamin overdoses are very rarely related to food intake.

Vitamins do not provide energy or serve as structural components of the body. They play critical roles

Table 1.6   Food sources of fats.

A. TOTAL FAT

Fats and Oils	Portion Size	Grams of Fat		Portion Size	Grams of Fat
Mayonnaise	1 tbs	11.0	Veggie pita	1	17.0
Salad dressing	1 tbs	6.0	Subway meatball sandwich	1	16.0
Vegetable oils	1 tsp	4.7	Subway turkey sandwich	1	4.0
Butter	1 tsp	4.0	**Milk and Milk Products**		
Margarine	1 tsp	4.0	Cheddar cheese	1 oz	9.5
Meats			Milk, whole	1 c	8.5
Sausage	4 links	18.0	American cheese	1 oz	6.0
Hot dog	2 oz	17.0	Cottage cheese, regular	½ c	5.1
Hamburger, 21% fat	3 oz	15.0	Milk, 2%	1 c	5.0
Hamburger, 16% fat	3 oz	13.5	Milk, 1%	1 c	2.9
Steak, rib eye	3 oz	9.9	Milk, skim	1 c	0.4
Bacon	3 strips	9.0	Yogurt, frozen	1 c	0.3
Steak, round	3 oz	5.2	**Other Foods**		
Chicken, baked no skin	3 oz	4.0	Avocado	½	15.0
Flounder, baked	3 oz	1.0	Almonds	1 oz	15.0
Shrimp, boiled	3 oz	1.0	Cashews	1 oz	13.2
Fast Foods			French fries, small serving	1	10.0
Whopper	8.9 oz	32.0	Taco chips	1 oz (10 chips)	10.0
Big Mac	6.6 oz	31.4	Potato chips	1 oz (14 chips)	7.0
Quarter Pounder with Cheese	6.8 oz	28.6	Peanut butter	1 tbs	6.1
			Egg	1	6.0

B. SATURATED FATS

Fats and Oils	Portion Size	Grams of Fat		Portion Size	Grams of Fat
Margarine	1 tsp	2.9	Haddock, breaded, fried	3 oz	3.0
Butter	1 tsp	2.4	Rabbit	3 oz	3.0
Salad dressing, Ranch	1 tbs	1.2	Pork chop, lean	3 oz	2.7
Peanut oil	1 tsp	0.9	Steak, round, lean	3 oz	2.0
Olive oil	1 tsp	0.7	Turkey, roasted	3 oz	2.0
Salad dressing, 1000 Island	1 tbs	0.5	Chicken, baked no skin	3 oz	1.7
Canola oil	1 tsp	0.3	Prime rib, lean	3 oz	1.3
Milk and Milk Products			Venison	3 oz	1.1
Cheddar cheese	1 oz	5.9	Tuna, in water	3 oz	0.4
American cheese	1 oz	5.5	**Fast Foods**		
Milk, whole	1 c	5.1	Croissant w/egg, bacon, & cheese	1	16.0
Cottage cheese, regular	½ c	3.0	Sausage crescent	1	16.0
Milk, 2%	1 c	2.9	Whopper	1	11.0
Milk, 1%	1 c	1.5	Cheeseburger	1	9.0
Milk, skim	1 c	0.3	Bac'n Cheddar Deluxe	1	8.7
Meats			Taco, regular	1	4.0
Hamburger, 21% fat	3 oz	6.7	Chicken breast sandwich	1	3.0
Sausage, links	4	5.6	**Nuts and Seeds**		
Hot dog	1	4.9	Macadamia nuts	1 oz	3.2
Chicken, fried with skin	3 oz	3.8	Peanuts, dry roasted	1 oz	1.9
Salami	3 oz	3.6	Sunflower seeds	1 oz	1.6

C. UNSATURATED FATS

Fats and Oils	Portion Size	Grams of Fat		Portion Size	Grams of Fat
Canola oil	1 tsp	4.1	Margarine	1 tsp	2.9
Vegetable oils	1 tsp	3.6	Butter	1 tsp	1.3

continued

Table 1.6 Food sources of fats. (continued)

C. UNSATURATED FATS (CONTINUED)

	Portion Size	Grams of Fat		Portion Size	Grams of Fat
Milk and Milk Products			Pork chop, lean	3 oz	5.3
Cottage cheese, regular	½ c	3.0	Turkey, roasted	3 oz	4.5
Cheddar cheese	1 oz	2.9	Tuna, in water	3 oz	0.7
American cheese	1 oz	2.8	Egg	1	5.0
Milk, whole	1 c	2.8	**Nuts and Seeds**		
Meats			Sunflower seeds	1 oz	16.6
Hamburger, 21% fat	3 oz	10.9	Almonds	1 oz	12.6
Haddock, breaded, fried	3 oz	6.5	Peanuts	1 oz	11.3
Chicken, baked no skin	3 oz	6.0	Cashews	1 oz	10.2

D. TRANS FATS

	Portion Size	Grams of Fat		Portion Size	Grams of Fat
Fats and Oils			**Milk**		
Margarine, stick	1 tsp	0.6	Whole	1 c	0.2
Margarine, tub (soft)	1 tsp	0.3	**Other Foods**		
Shortening	1 tsp	0.3	Doughnut	1	3.2
Butter	1 tsp	0.1	Danish pastry	1	3.0
Margarine, "no-trans fat"	1 tsp	0	French fries, small serving	1	2.9
Meats			Cookies	2	1.8
Beef	3 oz	0.5	Corn chips	1 oz	1.4
Chicken	3 oz	0.1	Cake	1 slice	1.0
			Crackers	4 squares	0.5

E. CHOLESTEROL

	Portion Size	Milligrams of Fat		Portion Size	Milligrams of Fat
Fats and Oils			Ostrich, ground	3 oz	63
Butter	1 tsp	10.3	Pork chop, lean	3 oz	60
Vegetable oils, margarines	1 tsp	0	Hamburger, 10% fat	3 oz	60
Meats			Venison	3 oz	48
Brain	3 oz	1476	Wild pig	3 oz	33
Liver	3 oz	470	Goat, roasted	3 oz	32
Egg	1	212	Tuna, in water	3 oz	25
Veal	3 oz	128	**Milk and Milk Products**		
Shrimp	3 oz	107	Ice cream, regular	1 c	56
Prime rib	3 oz	80	Milk, whole	1 c	34
Chicken, baked no skin	3 oz	75	Milk, 2%	1 c	22
Salmon, broiled	3 oz	74	Yogurt, low fat	1 c	17
Turkey, baked no skin	3 oz	65	Milk, 1%	1 c	14
Hamburger, 20% fat	3 oz	64	Milk, skim	1 c	7

F. N-3 FATTY ACIDS

	Portion Size	Grams of Fat		Portion Size	Grams of Fat
Fish and Seafood			Halibut	3.5 oz	1.2
Mackerel	3.5 oz	2.6	Salmon, pink	3.5 oz	1.0
Lake trout	3.5 oz	2.0	Halibut	3.5 oz	0.9
Herring	3.5 oz	1.8	Bass, striped	3.5 oz	0.8
Salmon, sockeye	3.5 oz	1.5	Oysters	3.5 oz	0.6
Whitefish, lake	3.5 oz	1.5	Catfish	3.5 oz	0.5
Anchovies	3.5 oz	1.4	Pollock	3.5 oz	0.5
Salmon, Chinook	3.5 oz	1.4	Shrimp	3.5 oz	0.5
Bluefish	3.5 oz	1.2	Trout, rainbow	3.5 oz	0.5

Table 1.6 Food sources of fats. (continued)

F. N-3 FATTY ACIDS (CONTINUED)					
	Portion Size	Grams of Fat		Portion Size	Grams of Fat
Tuna	3.5 oz	0.5	Lobster	3.5 oz	0.2
Carp	3.5 oz	0.3	Salmon, red	3.5 oz	0.2
Crab	3.5 oz	0.3	Scallops	3.5 oz	0.2
Pike, walleye	3.5 oz	0.3	Snapper, red	3.5 oz	0.2
Flounder	3.5 oz	0.2	Swordfish	3.5 oz	0.2
Haddock	3.5 oz	0.2			

in the regulation of reactions in the body that convert carbohydrates and fats into energy and protein into muscle, enzymes, red blood cells, and other protein-containing components of the body. Vitamin A, for example, is needed to replace the cells that line the mouth and esophagus, thiamin for conversion of glucose to energy, and folate for the production of body proteins. Other vitamins (vitamins C and E, and beta-carotene) act as *antioxidants*. By preventing or repairing damage to cells due to oxidation, these vitamins help maintain body tissues and prevent disease.

Primary functions, consequences of deficiency and overdose, primary food sources, and comments about each vitamin are listed in Table 1.8 (pp. 12–17).

RECOMMENDED INTAKE OF VITAMINS Recommendations for levels of intake of vitamins are presented in the tables on the inside front covers of this text. Note that Tolerable Upper Levels of Intake for many vitamins are also given and represent levels of intake that should not be exceeded.

FOOD SOURCES OF VITAMINS Table 1.9 (pp. 16–19) lists food sources of each vitamin. Additional information about the vitamin content of foods can be found in Appendix A.

Table 1.7 Vitamin solubility.

Water-Soluble Vitamins	Fat-Soluble Vitamins
B-complex vitamins	Vitamin A (retinol, beta-carotene)
Thiamin (B$_1$)	Vitamin D (1,25 dihydroxy-cholecalciferol)
Riboflavin (B$_2$)	
Niacin (B$_3$)	Vitamin E (alpha-tocopherol)
Vitamin B$_6$	Vitamin K
Folate	
Vitamin B$_{12}$	
Biotin	
Pantothenic acid	
Vitamin C (ascorbic acid)	

Other Substances in Food

Things don't happen by accident in nature. If you observe it, it has a reason for being there.

Norman Krinsky, Tufts University

There are many substances in foods in addition to nutrients that affect health. Some foods contain naturally occurring toxins, such as poison in puffer fish and solanine in green sections near the skin of some potatoes. Consumption of the poison in puffer fish can kill; large doses of solanine can interfere with nerve impulses. Some plant pigments, hormones, and other naturally occurring substances that protect plants from insects, oxidization, and other damaging exposures also appear to benefit human health. These substances in plants are referred to as *phytochemicals* and knowledge about their effects on human health is advancing rapidly. Consumption of foods rich in specific pigments and other phytochemicals, rather than consumption of isolated phytochemicals, may help prevent certain types of cancer, infections, and heart disease.[9] High intakes of certain phytochemicals from vegetables, fruits, and whole grain products may partially account for lower rates of heart disease and cancer observed in people with high intakes of these foods.[10]

ANTIOXIDANTS Chemical substance that prevent or repair damage to cells caused by exposure to oxidizing agents such as oxygen, ozone, and smoke and to other oxidizing agents normally produced in the body. Many different antioxidants are found in foods; some are made by the body.

PHYTOCHEMICALS (*Phyto* = plants) Chemical substances in plants, some of which affect body processes in humans that may benefit health.

Minerals

Humans require the 15 minerals listed in Table 1.10 (p. 19). Minerals are unlike other nutrients in that they consist of single atoms and carry a charge in solution. The property of being charged (or having an unequal number of electrons and protons) is related to many of the functions of minerals. The charge carried by minerals allows them to combine with other minerals to form stable complexes in bone, teeth, cartilage,

(text continues on page 19)

Table 1.8   Summary of the vitamins.

THE WATER-SOLUBLE VITAMINS		
	Primary Functions	**Consequences of Deficiency**
Thiamin (vitamin B$_1$) AI[a] women: 1.1 mg men: 1.2 mg	• Helps body release energy from carbohydrates ingested • Facilitates growth and maintenance of nerve and muscle tissues • Promotes normal appetite	• Fatigue, weakness • Nerve disorders, mental confusion, apathy • Impaired growth • Swelling • Heart irregularity and failure
Riboflavin (vitamin B$_2$) AI women: 1.1 mg men: 1.3 mg	• Helps body capture and use energy released from carbohydrates, proteins, and fats • Aids in cell division • Promotes growth and tissue repair • Promotes normal vision	• Reddened lips, cracks at both corners of the mouth. • Fatigue
Niacin (vitamin B$_3$) RDA women: 14 mg men: 16 mg UL: 35 mg (from supplements and fortified foods)	• Helps body capture and use energy released from carbohydrates, proteins, and fats • Assists in the manufacture of body fats • Helps maintain normal nervous system functions	• Skin disorders • Nervous and mental disorders • Diarrhea, indigestion • Fatigue

Vitamin B$_6$ (pyridoxine)
 AI   women: 1.3 mg
        men: 1.3 mg
        UL: 100 mg
	• Needed for reactions that build proteins and protein tissues
	 • Assists in the conversion of tryptophan to niacin
	 • Needed for normal red blood cell formation
	 • Promotes normal functioning of the nervous system
		• Irritability, depression
		 • Convulsions, twitching
		 • Muscular weakness
		 • Dermatitis near the eyes
		 • Anemia
		 • Kidney stones

Folate (folacin, folic acid)
 RDA:
      women: 400 mcg
      men: 400 mcg
      UL: 1000 mcg (from supplements and fortified foods)
	• Needed for reactions that utilize amino acids (the building blocks of protein) for protein tissue formation
	 • Promotes the normal formation of red blood cells
		• Anemia
		 • Diarrhea
		 • Red, sore tongue
		 • Neural tube defects, low birthweight (in pregnancy); increased risk of heart disease and stroke

Vitamin B$_{12}$ (cyanocobalamin) AI women: 2.4 mcg men: 2.4 mcg	• Helps maintain nerve tissues • Aids in reactions that build up protein tissues • Needed for normal red blood cell development	• Neurological disorders (nervousness, tingling sensations, brain degeneration) • Pernicious anemia • Fatigue • Deficiency reported in 39% of adults in one study
Biotin AI women: 30 mcg men: 30 mcg	• Needed for the body's manufacture of fats, proteins, and glycogen	• Depression, fatigue, nausea • Hair loss, dry and scaly skin • Muscular pain

[a]AI (Adequate Intakes) and RDAs (Recommended Dietary Allowances) are for 19–30 year olds; UL (Upper Limits) are for 19–70 year olds, 1997–2001.

Table 1.8   Summary of the vitamins. (continued)

THE WATER-SOLUBLE VITAMINS

Consequences of Overdose	Primary Food Sources	Highlights and Comments
• High intakes of thiamin are rapidly excreted by the kidneys. Oral doses of 500 mg/day or less are considered safe.	• Grains and grain products (cereals, rice, pasta, bread) • Ready-to-eat cereals • Pork and ham, liver • Milk, cheese, yogurt • Dried beans and nuts	• Need increases with carbohydrate intake. • There is no "e" on the end of thiamin! • Deficiency rare in the U.S.; may occur in people with alcoholism. • Enriched grains and cereals prevent thiamin deficiency.
• None known. High doses are rapidly excreted by the kidneys.	• Milk, yogurt, cheese • Grains and grain products (cereals, rice, pasta, bread) • Liver, poultry, fish, beef • Eggs	• Destroyed by exposure to light
• Flushing, headache, cramps, rapid heartbeat, nausea, diarrhea, decreased liver function with doses above 0.5 g per day	• Meats (all types) • Grains and grain products (cereals, rice, pasta, bread) • Dried beans and nuts • Milk, cheese, yogurt • Ready-to-eat cereals • Coffee • Potatoes	• Niacin has a precursor—tryptophan. Tryptophan, an amino acid, is converted to niacin by the body. Much of our niacin intake comes from tryptophan.
• Bone pain, loss of feeling in fingers and toes, muscular weakness, numbness, loss of balance (mimicking multiple sclerosis)	• Oatmeal, bread, breakfast cereals • Bananas, avocados, prunes, tomatoes, potatoes • Chicken, liver • Dried beans • Meats (all types), milk • Green and leafy vegetables	• Vitamins go from B_3 to B_6 because B_4 and B_5 were found to be duplicates of vitamins already identified.
• May cover up signs of vitamin B_{12} deficiency (pernicious anemia)	• Fortified, refined grain products (bread, flour, pasta) • Ready-to-eat cereals • Dark green, leafy vegetables (spinach, collards, romaine) • Broccoli, brussels sprouts • Oranges, bananas, grapefruit • Milk, cheese, yogurt • Dried beans	• Folate means "foliage." It was first discovered in leafy green vegetables. • This vitamin is easily destroyed by heat. • Synthetic form added to fortified grain products is better absorbed than naturally occurring folates.
• None known. Excess vitamin B_{12} is rapidly excreted by the kidneys or is not absorbed into the bloodstream. • Vitamin B_{12} injections may cause a temporary feeling of heightened energy.	• Animal products: beef, lamb, liver, clams, crab, fish, poultry, eggs • Milk and milk products • Ready-to-eat cereals	• Older people and vegans are at risk for vitamin B_{12} deficiency. • Some people become vitamin B_{12} deficient because they are genetically unable to absorb it. • Vitamin B_{12} is found in animal products and microorganisms only.
• None known. Excesses are rapidly excreted.	• Grain and cereal products • Meats, dried beans, cooked eggs • Vegetables	• Deficiency is extremely rare. May be induced by the overconsumption of raw eggs.

continued

Table 1.8   Summary of the vitamins. (continued)

THE WATER-SOLUBLE VITAMINS		
	Primary Functions	**Consequences of Deficiency**
Pantothenic acid (pantothenate) AI women: 5 mg men: 5 mg	• Needed for the release of energy from fat and carbohydrates	• Fatigue, sleep disturbances, impaired coordination • Vomiting, nausea

Vitamin C (ascorbic acid)
RDA women: 75 mg
       men: 90 mg
       UL: 2000 mg
	• Needed for the manufacture of collagen
	• Helps the body fight infections, repair wounds
	• Acts as an antioxidant
	• Enhances iron absorption
		• Bleeding and bruising easily due to weakened blood vessels, cartilage, and other tissues containing collagen
		• Slow recovery from infections and poor wound healing
		• Fatigue, depression
		• Deficiency reported in 9–24% of adults in one study

Vitamin A
I. Retinol
RDA women: 700 mcg
       men: 900 mcg
       UL: 3000 mcg
	• Needed for the formation and maintenance of mucous membranes, skin, bone
	• Needed for vision in dim light
		• Increased susceptibility to infection, increased incidence and severity of infection (including measles and HIV)
		• Impaired vision, blindness
		• Inability to see in dim light

2. Beta-carotene (a vitamin A precursor or "provitamin") No RDA; suggested intake: 6 mg	• Acts as an antioxidant; prevents damage to cell membranes and the contents of cells by repairing damage caused by free radicals	• Deficiency disease related only to lack of vitamin A

Vitamin E (alpha-tocopherol)
RDA women: 15 mg
      men: 15 mg
      UL: 1000 mg
	• Acts as an antioxidant, prevents damage to cell membranes in blood cells, lungs, and other tissues by repairing damage caused by free radicals
	• Reduces the ability of LDL cholesterol (the "bad" cholesterol) to form plaque in arteries
		• Muscle loss, nerve damage
		• Anemia
		• Weakness
		• Many adults may have non-optimal blood levels.

Vitamin D
(1,25 dihydroxy-cholecalciferol)
AI   women: 5 mcg
       (200 IU)
     men: 5 mcg
       (200 IU)
     UL: 50 mcg
       (2000 IU)
	• Needed for the absorption of calcium and phosphorus, and for their utilization in bone formation, nerve and muscle activity.	• Weak, deformed bones (children)
		• Loss of calcium from bones (adults), osteoporosis

Table 1.8   Summary of the vitamins. (continued)

THE WATER-SOLUBLE VITAMINS		
Consequences of Overdose	**Primary Food Sources**	**Highlights and Comments**
• None known. Excesses are rapidly excreted.	• Many foods, including meats, grains, vegetables, fruits, and milk	• Deficiency is very rare.
• Intakes of 1 g or more per day can cause nausea, cramps, and diarrhea and may increase the risk of kidney stones.	• Fruits: oranges, lemons, limes, strawberries, cantaloupe, honeydew melon, grapefruit, kiwi fruit, mango, papaya • Vegetables: broccoli, green and red peppers, collards, cabbage, tomato, asparagus, potatoes • Ready-to-eat cereals	• Need increases among smokers (to 110–125 mg per day). • Is fragile; easily destroyed by heat and exposure to air • Supplements may decrease duration and symptoms of colds. • Deficiency may develop within 3 weeks of very low intake.
• Vitamin A toxicity (hypervitaminosis A) with acute doses of 500,000 IU, or long-term intake of 50,000 IU per day. Limit retinol use in pregnancy to 5000 IU daily. • Nausea, irritability, blurred vision, weakness • Increased pressure in the skull, headache • Liver damage • Hair loss, dry skin • Birth defects	• Vitamin A is found in animal products only. • Liver, butter, margarine, milk, cheese, eggs • Ready-to-eat cereals	• Symptoms of vitamin A toxicity may mimic those of brain tumors and liver disease. Vitamin A toxicity is sometimes misdiagnosed because of the similarities in symptoms. • 1 mcg retinol equivalent = 5 IU vitamin A or 6 mcg beta-carotene
• High intakes from supplements may increase lung damage. • With high intakes and supplemental doses (over 12 mg/day for months), skin may turn yellow-orange. • Possibly related to reversible loss of fertility in women	• Deep orange, yellow, and green vegetables and fruits are often good sources. • Carrots, sweet potatoes, pumpkin, spinach, collards, red peppers, broccoli, cantaloupe, apricots, tomatoes, vegetable juice	• The body converts beta-carotene to vitamin A. Other carotenes are also present in food, and some are converted to vitamin A. Beta-carotene and vitamin A perform different roles in the body, however. • May decrease sunburn.
• Intakes of up to 800 IU per day are unrelated to toxic side effects; over 800 IU per day may increase bleeding (blood-clotting time). • Avoid supplement use if aspirin, anticoagulants, or fish oil supplements are taken regularly.	• Oils and fats • Salad dressings, mayonnaise, margarine, shortening, butter • Whole grains, wheat germ • Leafy, green vegetables, tomatoes • Nuts and seeds • Eggs	• Vitamin E is destroyed by exposure to oxygen and heat. • Oils naturally contain vitamin E. It's there to protect the fat from breakdown due to free radicals. • Eight forms of vitamin E exist, and each has different antioxidant strength. • Natural form is better absorbed than synthetic form: 15 IU alpha-tocopherol = 22 IU d-alpha tocopherol (natural form) and 33 IU synthetic vitamin E.
• Mental retardation in young children • Abnormal bone growth and formation • Nausea, diarrhea, irritability, weight loss • Deposition of calcium in organs such as the kidneys, liver, and heart	• Vitamin D–fortified milk and margarine • Butter • Fish • Eggs • Mushrooms • Milk products such as cheese, yogurt, and ice cream are generally not fortified with vitamin D.	• Vitamin D is manufactured from cholesterol in cells beneath the surface of the skin upon exposure of the skin to sunlight. • Deficiency may be common in ill, homebound, and elderly and hospitalized adults. • Breastfed infants with little sun exposure may benefit from vitamin D supplements.

continued

Table 1.8   Summary of the vitamins. (continued)

THE WATER-SOLUBLE VITAMINS

	Primary Functions	**Consequences of Deficiency**
Vitamin K (phylloquinone, menaquinone) AI women: 90 mcg men: 120 mcg	• Is an essential component of mechanisms that cause blood to clot when bleeding occurs • Aids in the incorporation of calcium into bones	• Bleeding, bruises • Decreased calcium in bones • Deficiency is rare. May be induced by the long-term use (months or more) of antibiotics.

Table 1.9   Food sources of vitamins.

THIAMIN

Food	Amount	Thiamin (Milligrams)	Food	Amount	Thiamin (Milligrams)
Meats			Rice	½ c	0.1
Pork roast	3 oz	0.8	Bread	1 slice	0.1
Beef	3 oz	0.4	Vegetables		
Ham	3 oz	0.4	Peas	½ c	0.3
Liver	3 oz	0.2	Lima beans	½ c	0.2
Nuts and Seeds			Corn	½ c	0.1
Sunflower seeds	¼ c	0.7	Broccoli	½ c	0.1
Peanuts	¼ c	0.1	Potato	1 med	0.1
Almonds	¼ c	0.1	Fruits		
Grains			Orange juice	1 c	0.2
Bran flakes	1 c (1 oz)	0.6	Orange	1	0.1
Macaroni	½ c	0.1	Avocado	½	0.1

RIBOFLAVIN

Food	Amount	Riboflavin (Milligrams)	Food	Amount	Riboflavin (Milligrams)
Milk and Milk Products			Beef	3 oz	0.2
Milk	1 c	0.5	Tuna	3 oz	0.1
2% milk	1 c	0.5	Vegetables		
Yogurt, low fat	1 c	0.5	Collard greens	½ c	0.3
Skim milk	1 c	0.4	Broccoli	½ c	0.2
Yogurt	1 c	0.1	Spinach, cooked	½ c	0.1
American cheese	1 oz	0.1	Eggs		
Cheddar cheese	1 oz	0.1	Egg	1	0.2
Meats			Grains		
Liver	3 oz	3.6	Macaroni	½ c	0.1
Pork chop	3 oz	0.3	Bread	1 slice	0.1

NIACIN

Food	Amount	Niacin (Milligrams)	Food	Amount	Niacin (Milligrams)
Meats			Pork	3 oz	4.5
Liver	3 oz	14.0	Haddock	3 oz	2.7
Tuna	3 oz	10.3	Scallops	3 oz	1.1
Turkey	3 oz	9.5	Nuts and Seeds		
Chicken	3 oz	7.9	Peanuts	1 oz	4.9
Salmon	3 oz	6.9	Vegetables		
Veal	3 oz	5.2	Asparagus	½ c	1.5
Beef (round steak)	3 oz	5.1			

Table 1.8 Summary of the vitamins. (continued)

THE WATER-SOLUBLE VITAMINS

Consequences of Overdose	Primary Food Sources	Highlights and Comments
• Toxicity is only a problem when synthetic forms of vitamin K are taken in excessive amounts. That may cause liver disease.	• Leafy, green vegetables • Grain products	• Vitamin K is produced by bacteria in the gut. Part of our vitamin K supply comes from these bacteria. • Newborns are given a vitamin K injection because they have "sterile" guts and consequently no vitamin K–producing bacteria.

Table 1.9 Food sources of vitamins. (continued)

NIACIN

Food	Amount	Niacin (Milligrams)	Food	Amount	Niacin (Milligrams)
Grains			Bread, enriched	1 slice	0.7
Wheat germ	1 oz	1.5	Milk and Milk Products		
Brown rice	½ c	1.2	Cottage cheese	½ c	2.6
Noodles, enriched	½ c	1.0	Milk	1 c	1.9
Rice, white, enriched	½ c	1.0			

VITAMIN B$_6$

Food	Amount	Vitamin B$_6$ (Milligrams)	Food	Amount	Vitamin B$_6$ (Milligrams)
Meats			Dried beans, cooked	½ c	0.4
Liver	3 oz	0.8	Fruits		
Salmon	3 oz	0.7	Banana	1	0.6
Other fish	3 oz	0.6	Avocado	½	0.4
Chicken	3 oz	0.4	Watermelon	1 c	0.3
Ham	3 oz	0.4	Vegetables		
Hamburger	3 oz	0.4	Turnip greens	½ c	0.7
Veal	3 oz	0.4	Brussels sprouts	½ c	0.4
Pork	3 oz	0.3	Potato	1	0.2
Beef	3 oz	0.2	Sweet potato	½ c	0.2
Eggs			Carrots	½ c	0.2
Egg	1	0.3	Peas	½ c	0.1
Legumes					
Split peas	½ c	0.6			

FOLATE

Food	Amount	Folate (Micrograms)	Food	Amount	Folate (Micrograms)
Vegetables			Broccoli	½ c	43
Garbanzo beans	½ c	141	Fruits		
Navy beans	½ c	128	Cantaloupe	¼ whole	100
Asparagus	½ c	120	Orange juice	1 c	87
Brussels sprouts	½ c	116	Orange	1	59
Black-eyed peas	½ c	102	Grains[a]		
Spinach, cooked	½ c	99	Ready-to-eat cereals	1 c/1 oz	100–400
Romaine lettuce	1 c	86	Oatmeal	½ c	97
Lima beans	½ c	71	Noodles	½ c	45
Peas	½ c	70	Wheat germ	2 tbs	40
Collard greens, cooked	½ c	56	Wild rice	½ c	37
Sweet potato	½ c	43			

[a]Fortified, refined grain products such as bread, rice, pasta, and crackers provide approximately 40 to 60 mcg of folic acid per standard serving.

continued

Table 1.9   Food sources of vitamins. (continued)

VITAMIN B₁₂

Food	Amount	Vitamin B$_{12}$ (Micrograms)	Food	Amount	Vitamin B$_{12}$ (Micrograms)
Meats			Milk and Milk Products		
Liver	3 oz	6.8	Skim milk	1 c	1.0
Trout	3 oz	3.6	Milk	1 c	0.9
Beef	3 oz	2.2	Yogurt	1 c	0.8
Clams	3 oz	2.0	Cottage cheese	½ c	0.7
Crab	3 oz	1.8	American cheese	1 oz	0.2
Lamb	3 oz	1.8	Cheddar cheese	1 oz	0.2
Tuna	3 oz	1.8	Eggs		
Veal	3 oz	1.7	Egg	1	0.6
Hamburger, regular	3 oz	1.5			

VITAMIN C

Food	Amount	Vitamin C (Milligrams)	Food	Amount	Vitamin C (Milligrams)
Fruits			Watermelon	1 c	15
Orange juice, vitamin C–fortified	1 c	108	Vegetables		
Kiwi fruit	1 or ½ c	108	Green peppers	½ c	95
Grapefruit juice, fresh	1 c	94	Cauliflower, raw	½ c	75
Cranberry juice cocktail	1 c	90	Broccoli	½ c	70
Orange	1	85	Brussels sprouts	½ c	65
Strawberries, fresh	1 c	84	Collard greens	½ c	48
Orange juice, fresh	1 c	82	Vegetable (V-8) juice	¾ c	45
Cantaloupe	¼ whole	63	Tomato juice	¾ c	33
Grapefruit	1 med	51	Cauliflower, cooked	½ c	30
Raspberries, fresh	1 c	31	Potato	1 med	29
			Tomato	1 med	23

VITAMIN A (RETINOL)

Food Sources of Vitamin A (Retinol)	Amount	Vitamin A Food (Micrograms RE)[b]	Sources of Vitamin A (Retinol)	Amount	Vitamin A (Micrograms RE)[b]
Meats			2% milk	1 c	139
Liver	3 oz	9124	American cheese	1 oz	82
Salmon	3 oz	53	Whole milk	1 c	76
Tuna	3 oz	14	Swiss cheese	1 oz	65
Eggs			Fats		
Egg	1 med	84	Margarine, fortified	1 tsp	46
Milk and Milk Products			Butter	1 tsp	38
Skim milk, fortified	1 c	149			

VITAMIN A AND BETA-CAROTENE

Food Sources of Beta-Carotene	Amount	Vitamin A Value (Micrograms RE)[b]	Food Sources of Beta-Carotene	Amount	Vitamin A Value (Micrograms RE)[b]
Vegetables			Green peppers	½ c	40
Pumpkin, canned	½ c	2712	Fruits		
Sweet potato, canned	½ c	1935	Cantaloupe	¼ whole	430
Carrots, raw	½ c	1913	Apricots, canned	½ c	210
Spinach, cooked	½ c	739	Nectarine	1 med	101
Collard greens, cooked	½ c	175	Watermelon	1 c	59
Broccoli, cooked	½ c	109	Peaches, canned	½ c	47
Winter squash	½ c	53	Papaya	½ c	20

[b]RE (retinol equivalent) = 3.33 IU.

Table 1.9 Food sources of vitamins. (continued)

VITAMIN E

Food	Amount	Vitamin E (IU)c	Food	Amount	Vitamin E (IU)c
Oils			Collard greens	½ c	3.1
Oil	1 tbs	6.7	Asparagus	½ c	2.1
Mayonnaise	1 tbs	3.4	Spinach, raw	1 c	1.5
Margarine	1 tbs	2.7	Grains		
Salad dressing	1 tbs	2.2	Wheat germ	2 tbs	4.2
Nuts and Seeds			Bread, whole wheat	1 slice	2.5
Sunflower seeds	¼ c	27.1	Bread, white	1 slice	1.2
Almonds	¼ c	12.7	Seafood		
Peanuts	¼ c	4.9	Crab	3 oz	4.5
Cashews	¼ c	0.7	Shrimp	3 oz	3.7
Vegetables			Fish	3 oz	2.4
Sweet potato	½ c	6.9			

VITAMIN D

Food	Amount	Vitamin D (IU)d	Food	Amount	Vitamin D (IU)d
Milk			Organ meats		
Milk, whole, low fat, or skim	1 c	100	Beef liver	3 oz	42
Fish and seafoods			Chicken liver	3 oz	40
Salmon	3 oz	340	Eggs		
Tuna	3 oz	150	Egg yolk	1	27
Shrimp	3 oz	127			

c15 mg alpha-tocopherol = 22 IU d-alpha tocopherol (natural form) and 33 IU synthetic vitamin E.

d40 IU = 1 mcg.

and other tissues. In body fluids, charged minerals serve as a source of electrical power that stimulates muscles to contract (e.g., the heart to beat) and nerves to react. They also help the body maintain an adequate amount of water in tissues and control how acidic or basic body fluids remain.

The tendency of minerals to form complexes has implications for the absorption of minerals from food. Calcium and zinc, for example, may combine with other minerals in supplements or with dietary fiber and form complexes that cannot be absorbed. Therefore, in general, the proportion of total mineral intake that is absorbed is less than for vitamins.

Functions, consequences of deficiency and overdose, primary food sources, and comments about the 15 minerals needed by humans are summarized in Table 1.11 (pp. 20–23).

RECOMMENDED INTAKE OF MINERALS Recommendations for intake of minerals are presented in the tables on the inside front covers of this text. Note that Tolerable Upper Levels of Intake for many minerals are also given in a separate table.

FOOD SOURCES OF MINERALS Table 1.12 (pp. 24–26) lists food sources of each mineral. Additional information about the mineral content of foods can be found in Appendix A.

Water

Water is the last, but not least, nutrient category. Adults are about 60% water by weight. Water is the medium in which most chemical reactions take place in the body. It plays a role in energy transformation, the excretion of wastes, and temperature regulation.

People need enough water to replace daily losses from perspiration, urination, and exhalation. In normal weather conditions and with normal physical activity levels, adult males need about 12 cups of water each day, and females 9 cups.[11] The need for water is greater in hot and humid climates, and when physical activity levels are high. People generally consume around 6 cups of water from water and other fluids and 4 cups from foods each day.

Table 1.10 Minerals required by humans.

Calcium	Fluoride	Chromium
Phosphorus	Iodine	Molybdenum
Magnesium	Selenium	Sodium
Iron	Copper	Potassium
Zinc	Manganese	Chloride

Table 1.11   Summary of minerals.

	Primary Functions	Consequences of Deficiency
Calcium AI* women: 1000 mg men: 1000 mg UL: 2500 mg	• Component of bones and teeth • Needed for muscle and nerve activity, blood clotting	• Poorly mineralized, weak bones (osteoporosis) • Rickets in children • Osteomalacia (rickets in adults) • Stunted growth in children • Convulsions, muscle spasms
Phosphorus RDA women: 700 mg men: 700 mg UL: 4000 mg	• Component of bones and teeth • Component of certain enzymes and other substances involved in energy formation • Needed to maintain the right acid-base balance of body fluids	• Loss of appetite • Nausea, vomiting • Weakness • Confusion • Loss of calcium from bones
Magnesium RDA women: 310 mg men: 400 mg UL: 350 mg (from supplements only)	• Component of bones and teeth • Needed for nerve activity • Activates enzymes involved in energy and protein formation	• Stunted growth in children • Weakness • Muscle spasms • Personality changes
Iron RDA women: 18 mg men: 8 mg UL: 45 mg	• Transports oxygen as a component of hemoglobin in red blood cells • Component of myoglobin (a muscle protein) • Needed for certain reactions involving energy formation	• Iron deficiency • Iron-deficiency anemia • Weakness, fatigue • Pale appearance • Reduced attention span and resistance to infection • Mental retardation, developmental delay in children
Zinc RDA women: 8 mg men: 11 mg UL: 40 mg	• Required for the activation of many enzymes involved in the reproduction of proteins • Component of insulin, many enzymes	• Growth failure • Delayed sexual maturation • Slow wound healing • Loss of taste and appetite • In pregnancy, low-birth-weight infants and preterm delivery
Fluoride AI women: 3 mg men: 4 mg UL: 10 mg	• Component of bones and teeth (enamel)	• Tooth decay and other dental diseases
Iodine RDA women: 150 mcg men: 150 mcg UL: 1100 mcg	• Component of thyroid hormones that help regulate energy production and growth	• Goiter • Cretinism (mental retardation, hearing loss, growth failure)

*AIs and RDA are for women and men 19–30 years of age, ULs are for males and females 19 to 70 years of age, 1997–2001.

Table 1.11   Summary of minerals. (continued)

Consequences of Overdose	Primary Food Sources	Highlights and Comments
• Drowsiness • Calcium deposits in kidneys, liver, and other tissues • Suppression of bone remodeling • Decreased zinc absorption	• Milk and milk products (cheese, yogurt) • Broccoli • Dried beans • Calcium-fortified foods (some juices, breakfast cereals, bread, for example)	• The average intake of calcium among U.S. women is approximately 60% of the DRI. • One in four women and one in eight men in the U.S. develop osteoporosis. • Adequate calcium and vitamin D status must be maintained to prevent bone loss.
• Muscle spasms	• Milk and milk products (cheese, yogurt) • Meats • Seeds, nuts • Phosphates added to foods	• Deficiency is generally related to disease processes.
• Diarrhea • Dehydration • Impaired nerve activity due to disrupted utilization of calcium	• Plant foods (dried beans, tofu, peanuts, potatoes, green vegetables) • Milk • Bread • Ready-to-eat cereals • Coffee	• Magnesium is primarily found in plant foods where it is attached to chlorophyll. • Average intake among U.S. adults is below the RDA.
• Hemochromatosis ("iron poisoning") • Vomiting, abdominal pain • Blue coloration of skin • Liver and heart damage, diabetes • Decreased zinc absorption • Atherosclerosis (plaque buildup) in older adults	• Liver, beef, pork • Dried beans • Iron-fortified cereals • Prunes, apricots, raisins • Spinach • Bread • Pasta	• Cooking foods in iron and stainless steel pans increases the iron content of the foods. • Vitamin C, meat, and alcohol increase iron absorption. • Iron deficiency is the most common nutritional deficiency in the world. • Average iron intake of young children and women in the U.S. is low.
• Over 25 mg/day is associated with nausea, vomiting, weakness, fatigue, susceptibility to infection, copper deficiency, and metallic taste in mouth. • Increased blood lipids	• Meats (all kinds) • Grains • Nuts • Milk and milk products (cheese, yogurt) • Ready-to-eat cereals • Bread	• Like iron, zinc is better absorbed from meats than from plants. • Marginal zinc deficiency may be common, especially in children. • Zinc supplements may decrease duration and severity of the common cold.
• Fluorosis • Brittle bones • Mottled teeth • Nerve abnormalities	• Fluoridated water and foods and beverages made with it • Tea • Shrimp, crab	• Toothpastes, mouth rinses, and other dental care products may provide fluoride. • Fluoride overdose has been caused by ingestion of fluoridated toothpaste.
• Over 1 mg/day may produce pimples, goiter, and decreased thyroid function.	• Iodized salt • Milk and milk products • Seaweed, seafoods • Bread from commercial bakeries	• Iodine deficiency was a major problem in the U.S. in the 1920s and 1930s. Now intakes are marginally excessive. Deficiency remains a major health problem in some developing countries. • Amount of iodine in plants depends on iodine content of soil. • Most of the iodine in our diet comes from the incidental addition of iodine to foods from cleaning compounds used by food manufacturers.

continued

Table 1.11   Summary of minerals. (continued)

	Primary Functions	Consequences of Deficiency
Selenium RDA women: 55 mcg men: 55 mcg UL: 400 mcg	• Acts as an antioxidant in conjunction with vitamin E (protects cells from damage due to exposure to oxygen) • Needed for thyroid hormone production	• Anemia • Muscle pain and tenderness • Keshan disease (heart failure), Kashin-Beck disease (joint disease)
Copper RDA women: 900 mcg men: 900 mcg UL: 10,000 mcg	• Component of enzymes involved in the body's utilization of iron and oxygen • Functions in growth, immunity, cholesterol and glucose utilization, brain development	• Anemia • Seizures • Nerve and bone abnormalities in children • Growth retardation
Manganese AI women: 2.3 mg men: 1.8 mg	• Needed for the formation of body fat and bone	• Weight loss • Rash • Nausea and vomiting
Chromium AI women: 35 mcg men: 25 mcg	• Required for the normal utilization of glucose and fat	• Elevated blood glucose and triglyceride levels • Weight loss
Molybdenum RDA women: 45 mcg men: 45 mcg UL: 2000 mcg	• Component of enzymes involved in the transfer of oxygen from one molecule to another	• Rapid heartbeat and breathing • Nausea, vomiting • Coma
Sodium No DRI. Recommended intake: 2400 mg	• Needed to maintain the right acid-base balance in body fluids • Helps maintain an appropriate amount of water in blood and body tissues • Needed for muscle and nerve activity	• Weakness • Apathy • Poor appetite • Muscle cramps • Headache • Swelling
Potassium No DRI	• Same as for sodium	• Weakness • Irritability, mental confusion • Irregular heartbeat • Paralysis
Chloride No DRI	• Component of hydrochloric acid secreted by the stomach (used in digestion) • Needed to maintain the right acid-base balance of body fluids • Helps maintain an appropriate water balance in the body	• Muscle cramps • Apathy • Poor appetite • Long-term mental retardation in infants

Table 1.11   Summary of minerals. (continued)

Consequences of Overdose	Primary Food Sources	Highlights and Comments
• "Selenosis." Symptoms of selenosis are hair and fingernail loss, weakness, liver damage, irritability, and "garlic" or "metallic" breath.	• Meats and seafoods • Eggs • Whole grains	• Content of foods depends on amount of selenium in soil, water, and animal feeds. • May play a role in the prevention of some types of cancer.
• Wilson's disease (excessive accumulation of copper in the liver and kidneys) • Vomiting, diarrhea • Tremors • Liver disease	• Bread • Potatoes • Grains • Dried beans • Nuts and seeds • Seafood • Ready-to-eat cereals	• Toxicity can result from copper pipes and cooking pans. • Average intake in the U.S. is below the RDA.
• Infertility in men • Disruptions in the nervous system (psychotic symptoms) • Muscle spasms	• Whole grains • Coffee, tea • Dried beans • Nuts	• Toxicity is related to overexposure to manganese dust in miners.
• Kidney and skin damage	• Whole grains • Wheat germ • Liver, meat • Beer, wine • Oysters	• Toxicity usually results from exposure in chrome-making industries or overuse of supplements. • Supplements do not build muscle mass or increase endurance.
• Loss of copper from the body • Joint pain • Growth failure • Anemia • Gout	• Dried beans • Grains • Dark green vegetables • Liver • Milk and milk products	• Deficiency is extraordinarily rare.
• High blood pressure in susceptible people • Kidney disease • Heart problems	• Foods processed with salt • Cured foods (corned beef, ham, bacon, pickles, sauerkraut) • Table and sea salt • Bread • Milk, cheese • Salad dressing	• Very few foods naturally contain much sodium. • Processed foods are the leading source of dietary sodium. • High-sodium diets are associated with the development of hypertension in "salt-sensitive" people.
• Irregular heartbeat, heart attack	• Plant foods (potatoes, squash, lima beans, tomatoes, plantains, bananas, oranges, avocados) • Meats • Milk and milk products • Coffee	• Content of vegetables is often reduced in processed foods. • Diuretics (water pills) and other anti-hypertension drugs may deplete potassium. • Salt substitutes often contain potassium.
• Vomiting	• Same as for sodium. (Most of the chloride in our diets comes from salt.)	• Excessive vomiting and diarrhea may cause chloride deficiency. • Legislation regulating the composition of infant formulas was enacted in response to formula-related chloride deficiency and subsequent mental retardation in infants.

Table 1.12 Food sources of minerals.

MAGNESIUM

Food	Amount	Magnesium (mg)	Food	Amount	Magnesium (mg)
Legumes			**Vegetables**		
Lentils, cooked	½ c	134	Bean sprouts	½ c	98
Split peas, cooked	½ c	134	Black-eyed peas	½ c	58
Tofu	½ c	130	Spinach, cooked	½ c	48
Nuts			Lima beans	½ c	32
Peanuts	¼ c	247	**Milk and Milk Products**		
Cashews	¼ c	93	Milk	1 c	30
Almonds	¼ c	80	Cheddar cheese	1 oz	8
Grains			American cheese	1 oz	6
Bran buds	1 c	240	**Meats**		
Wild rice, cooked	½ c	119	Chicken	3 oz	25
Breakfast cereal, fortified	1 c	85	Beef	3 oz	20
Wheat germ	2 tbs	45	Pork	3 oz	20

CALCIUM*

Food	Amount	Calcium (mg)	Food	Amount	Calcium (mg)
Milk and Milk Products			Ice milk	1 c	180
Yogurt, low fat	1 c	413	American cheese	1 oz	175
Milk shake			Custard	½ c	150
(low-fat frozen yogurt)	1¼ c	352	Cottage cheese	1½ c	70
Yogurt with fruit, low fat	1 c	315	Cottage cheese, low fat	½ c	69
Skim milk	1 c	301	**Vegetables**		
1% milk	1 c	300	Spinach, cooked	½ c	122
2% milk	1 c	298	Kale	1½ c	47
3.25% milk (whole)	1 c	288	Broccoli	½ c	36
Swiss cheese	1 oz	270	**Legumes**		
Milk shake (whole milk)	1¼ c	250	Tofu	½ c	260
Frozen yogurt, low fat	1 c	248	Dried beans, cooked	½ c	60
Frappuccino	1 c	220	**Foods Fortified with Calcium**		
Cheddar cheese	1 oz	204	Orange juice	1 c	350
Frozen yogurt	1 c	200	Frozen waffles	2	300
Cream soup	1 c	186	Soy milk	1 c	200–400
Pudding	½ c	185	Breakfast cereals	1 c	150–1000
Ice cream	1 c	180			

SELENIUM

Food	Amount	Selenium (mcg)	Food	Amount	Selenium (mcg)
Seafood			Ham	3 oz	29
Lobster	3 oz	66	Beef	3 oz	22
Tuna	3 oz	60	Bacon	3 oz	21
Shrimp	3 oz	54	Chicken	3 oz	18
Oysters	3 oz	48	Lamb	3 oz	14
Fish	3 oz	40	Veal	3 oz	10
Meats			**Eggs**		
Liver	3 oz	56	Egg	1 med	37

*Actually, the richest source of calcium is alligator meat; 3½ ounces contain about 1231 milligrams of calcium. But just try to find it on your grocer's shelf!

Table 1.12   Food sources of minerals. (continued)

SODIUM*

Food	Amount	Sodium (mg)	Food	Amount	Sodium (mg)
Miscellaneous			Meat loaf	3 oz	555
Salt	1 tsp	2132	Sausage	3 oz	483
Dill pickle	1 (4½ oz)	1930	Hot dog	1	477
Sea salt	1 tsp	1716	Fish, smoked	3 oz	444
Chicken broth	1 c	1571	Bologna	1 oz	370
Ravioli, canned	1 c	1065	**Milk and Milk Products**		
Spaghetti with sauce, canned	1 c	955	Cream soup	1 c	1070
			Cottage cheese	½ c	455
Baking soda	1 tsp	821	American cheese	1 oz	405
Beef broth	1 c	782	Cheese spread	1 oz	274
Gravy	¼ c	720	Parmesan cheese	1 oz	247
Italian dressing	2 tbs	720	Gouda cheese	1 oz	232
Pretzels	5 (1 oz)	500	Cheddar cheese	1 oz	175
Green olives	5	465	Skim milk	1 c	125
Pizza with cheese	1 wedge	455	Whole milk	1 c	120
Soy sauce	1 tsp	444	**Grains**		
Cheese twists	1 c	329	Bran flakes	1 c	363
Bacon	3 slices	303	Cornflakes	1 c	325
French dressing	2 tbs	220	Croissant	1 med	270
Potato chips	1 oz (10 pieces)	200	Bagel	1	260
Catsup	1 tbs	155	English muffin	1	203
Meats			White bread	1 slice	130
Corned beef	3 oz	808	Whole wheat bread	1 slice	130
Ham	3 oz	800	Saltine crackers	4 squares	125
Fish, canned	3 oz	735			

IRON

Food	Amount	Iron (mg)	Food	Amount	Iron (mg)
Meat and Meat Alternates			English muffin	1	1.6
Liver	3 oz	7.5	Rye bread	1 slice	1.0
Round steak	3 oz	3.0	Whole wheat bread	1 slice	0.8
Hamburger, lean	3 oz	3.0	White bread	1 slice	0.6
Baked beans	½ c	3.0	**Fruits**		
Pork	3 oz	2.7	Prune juice	1 c	9.0
White beans	½ c	2.7	Apricots, dried	½ c	2.5
Soybeans	½ c	2.5	Prunes	5 med	2.0
Pork and beans	½ c	2.3	Raisins	¼ c	1.3
Fish	3 oz	1.0	Plums	3 med	1.1
Chicken	3 oz	1.0	**Vegetables**		
Grains			Spinach, cooked	½ c	2.3
Breakfast cereal, iron-fortified	1 c	8.0 (4–18)	Lima beans	½ c	2.2
Oatmeal, fortified, cooked	1 c	8.0	Black-eyed peas	½ c	1.7
			Peas	½ c	1.6
Bagel	1	1.7	Asparagus	½ c	1.5

continued

Table 1.12   Food sources of minerals. (continued)

ZINC

Food	Amount	Zinc (mg)	Food	Amount	Zinc (mg)
Meats			Oatmeal, cooked	1 c	1.2
Liver	3 oz	4.6	Bran flakes	1 c	1.0
Beef	3 oz	4.0	Brown rice, cooked	½ c	0.6
Crab	½ c	3.5	White rice	½ c	0.4
Lamb	3 oz	3.5	**Nuts and Seeds**		
Turkey ham	3 oz	2.5	Pecans	¼ c	2.0
Pork	3 oz	2.4	Cashews	¼ c	1.8
Chicken	3 oz	2.0	Sunflower seeds	¼ c	1.7
Legumes			Peanut butter	2 tbs	0.9
Dried beans, cooked	½ c	1.0	**Milk and Milk Products**		
Split peas, cooked	½ c	0.9	Cheddar cheese	1 oz	1.1
Grains			Whole milk	1 c	0.9
Breakfast cereal, fortified	1 c	1.5–4.0	American cheese	1 oz	0.8
Wheat germ	2 tbs	2.4			

PHOSPHORUS

Food	Amount	Phosphorus (mg)	Food	Amount	Phosphorus (mg)
Milk and Milk Products			**Grains**		
Yogurt	1 c	327	Bran flakes	1 c	180
Skim milk	1 c	250	Shredded wheat	2 large biscuits	81
Whole milk	1 c	250	Whole wheat bread	1 slice	52
Cottage cheese	½ c	150	Noodles, cooked	½ c	47
American cheese	1 oz	130	Rice, cooked	½ c	29
Meats			White bread	1 slice	24
Pork	3 oz	275	**Vegetables**		
Hamburger	3 oz	165	Potato	1 med	101
Tuna	3 oz	162	Corn	½ c	73
Lobster	3 oz	125	Peas	½ c	70
Chicken	3 oz	120	French fries	½ c	61
Nuts and Seeds			Broccoli	½ c	54
Sunflower seeds	¼ c	319	**Other**		
Peanuts	¼ c	141	Milk chocolate	1 oz	66
Pine nuts	¼ c	106	Cola	12 oz	51
Peanut butter	1 tbs	61	Diet cola	12 oz	45

POTASSIUM

Food	Amount	Potassium (mg)	Food	Amount	Potassium (mg)
Vegetables			Hamburger	3 oz	480
Potato	1 med	780	Lamb	3 oz	382
Winter squash	½ c	327	Pork	3 oz	335
Tomato	1 med	300	Chicken	3 oz	208
Celery	1 stalk	270	Grains		
Carrots	1 med	245	Bran buds	1 c	1080
Broccoli	½ c	205	Bran flakes	1 c	248
Fruits			Raisin bran	1 c	242
Avocado	½ med	680	Wheat flakes	1 c	96
Orange juice	1 c	469	Milk and Milk Products		
Banana	1 med	440	Yogurt	1 c	531
Raisins	¼ c	370	Skim milk	1 c	400
Prunes	4 large	300	Whole milk	1 c	370
Watermelon	1 c	158	Other		
Meats			Salt substitutes	1 tsp	1300–2378
Fish	3 oz	500			

DIETARY SOURCES OF WATER   The best sources of water are tap and bottled water; nonalcoholic beverages such as fruit juice, milk, and vegetables juice; and brothy soups. Alcohol tends to increase water loss though urine; so beverages such as beer and wine are not as "hydrating" as water is. Caffeinated beverages provide some hydration, but less than water.

> **PRINCIPLE # 3**   Health problems related to nutrition originate within cells.

The functions of each cell are maintained by the nutrients it receives. Problems arise when a cell's need for nutrients differs from the amounts that are available. Cells (Illustration 1.3) are the building blocks of tissues (such as bones and muscles), organs (the heart, kidney, and liver, for example), and systems (such as the circulatory and respiratory systems). Normal cell health and functions are maintained when a nutritional and environmental utopia exists within and around cells. This state of optimal, cellular nutrient conditions supports *homeostasis* in the body.

Disruptions in the availability of nutrients, or the presence of harmful substances in the cell's environment, initiate diseases and disorders that eventually affect tissues, organs, and systems. For example, too little folate reduces the conversion of the amino acid methionine to cysteine. This causes a buildup of an intermediary product called homocysteine. High levels of homocysteine disrupt normal cell processes and enhance the deposition of cholesterol and other materials into artery walls.

> **PRINCIPLE #4**   Poor nutrition can result from both inadequate and excessive levels of nutrient intake.

Each nutrient has a range of intake levels that corresponds to optimum functioning of that nutrient (Illustration 1.4. Intake levels below and above this range are associated with impaired functions. Inadequate intake of an essential nutrient, if prolonged, results in obvious deficiency diseases. Marginally deficient diets produce subtle changes in behavior or physical condition. If the optimal intake range is exceeded (usually by overdoses of supplements), mild to severe changes in mental and physical functions occur, depending on the amount of the excess and the nutrient involved. Overt vitamin C deficiency, for example, produces irritability, bleeding gums, pain upon being touched, and failure of bone growth. Marginal deficiency may cause delayed wound healing. The length of time a deficiency or toxicity takes to develop depends on the type and amount of the nutrient consumed and the extent of body nutrient reserves. Intakes of 32 mg/day of vitamin C, or about a third of the RDA for adults (75 mg and 90 mg per day for women and men, respectively), lower blood vitamin C levels to the deficient state within three weeks.[12] On the excessive side, too much supplemental vitamin C causes diarrhea. For nutrients, enough is as good as a feast.

> **HOMEOSTASIS**   Constancy of the internal environment. The balance of fluids, nutrients, gases, temperature, and other conditions needed to ensure ongoing, proper functioning of cells and, therefore, all parts of the body.

STEPS IN THE DEVELOPMENT OF NUTRIENT DEFICIENCIES AND TOXICITIES   Poor nutrition due to inadequate diets generally develops in the stages outlined in Illustration 1.5. After a period of deficient intake of an essential nutrient, tissue reserves become depleted, and subsequently blood levels of the nutrient decline. When the blood level can no longer supply cells with optimal amounts of nutrients, cell processes change. These changes have a negative effect on the cell's ability to form proteins appropriately, regulate energy formation and use, protect itself from

Illustration 1.3   **Schematic representation of the structure of a human cell.**

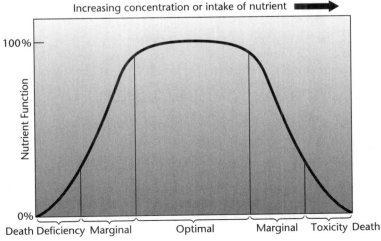

Illustration 1.4   **Nutrient function and consequences by level of intake.**

Deficiency	Toxicity
Inadequate nutrient intake	Excessive nutrient intake
⬇	⬇
Depletion of tissue reserves of the nutrient	Saturation of tissue reserves of the nutrient
⬇	⬇
Decreased blood nutrient level	Increased blood nutrient level
⬇	⬇
Insufficient nutrient available to cells	Excessive nutrient available to cells
⬇	⬇
Impaired cellular functions	Impaired cellular functions
⬇	⬇
Physical signs and symptoms of deficiency	Physical signs and symptoms of toxicity
⬇	⬇
Long-term impairment of health	Long-term impairment of health

Illustration 1.5 Usual steps in the development of nutrient deficiencies and toxicities.

oxidation, or carry out other normal functions. If the deficiency continues, groups of cells malfunction, which leads to problems related to tissue and organ functions. Physical signs of the deficiency may then develop, such as growth failure with protein deficiency or an inability to walk as a result of beriberi (thiamin deficiency). Eventually, some problems produced by the deficiency can no longer be reversed by increased nutrient intake. Blindness that results from serious vitamin A deficiency, for example, is irreversible.

Excessively high intakes of many essential nutrients produce toxicity diseases. Excessive vitamin A, for example, produces hypervitaminosis A, and selenium overdose leads to selenosis. Signs of toxicity stem from an increased level of the nutrient in the blood and the subsequent oversupply of the nutrient to cells. The high nutrient load upsets the balance needed for optimal cell function. These changes in cell function lead to the signs and symptoms of a toxicity disease.

MALNUTRITION   Poor nutrition resulting from an excess or lack of calories or nutrients.

For both deficiency and toxicity diseases, the best way to correct the problem is at the level of intake. Identifying and fixing intake problems prevents related health problems from developing.

NUTRIENT DEFICIENCIES ARE USUALLY MULTIPLE

Most foods contain many nutrients, so poor diets are generally inadequate in many nutrients. Calcium and vitamin D, for example, are present in milk.

Deficiencies of both of these nutrients may develop from a low milk intake and an otherwise poor diet.

The "Ripple Effect"   Dietary changes affect the level of intake of many nutrients. Switching from a high-fat to a low-fat diet, for instance, generally results in a lower intake of calories and higher intake of dietary fiber and vitamins. Consequently, dietary changes introduced for the purpose of improving intake of a particular nutrient produce a "ripple effect" on the intake of other nutrients.

> **PRINCIPLE #5**   Humans have adaptive mechanisms for managing fluctuations in food intake.

Healthy humans have adaptive mechanisms that partially protect the body from poor health due to fluctuations in nutrient intake. These mechanisms act to conserve nutrients when dietary supply is low and to eliminate them when excessively high amounts are present. Dietary surpluses of nutrients such as iron, calcium, vitamin A, and vitamin B_{12}, are stored within tissues for later use. In the case of iron and calcium, absorption is also regulated so that the amount absorbed changes in response to the body's need for them. The body has a low storage capacity for other nutrients, such as vitamin C and water, and excesses are eliminated through urine or stools. Fluctuations in energy intake are primarily regulated by changes in appetite. If too few calories are consumed, however, the body will obtain energy from its glycogen and fat stores. If caloric intakes remain low and a significant amount of body weight is lost, the body downregulates its need for energy by lowering body temperature and the capacity for physical work. When energy intake exceeds need, the extra is converted to glycogen, fat, or both and stored for later use.

Although they provide an important buffer, these built-in mechanisms do not protect humans from all of the consequences of poor diets. An excessive vitamin A or selenium intake over time, for example, results in toxicity disease; excessive energy intake creates health problems related to obesity; and deficient intakes of other vitamins and minerals compromise health in many ways.

> **PRINCIPLE #6**   Malnutrition can result from poor diets and from disease states, genetic factors, or combinations of these causes.

Malnutrition means "poor nutrition" and results from either inadequate or excessive availability of energy and nutrients. Niacin toxicity, obesity, iron deficiency, and kwashiorkor (protein deficiency in children) are examples of malnutrition.

Malnutrition can result from poor diets and also from diseases that interfere with the body's ability to use the nutrients consumed. *Primary malnutrition* results when a poor nutritional state is dietary in origin. *Secondary malnutrition,* on the other hand, is precipitated by a disease state, surgical procedure, or medication. Diarrhea, alcoholism, AIDS, and gastrointestinal tract bleeding are examples of conditions that may cause malnutrition.

A portion of the population is susceptible to malnutrition due to genetic factors or, more commonly, to a combination of genetic factors and adverse environmental exposures. People may be born with a genetic tendency to produce too much cholesterol, absorb high levels of iron, or utilize folate poorly. Some cases of obesity are related to the combined influences of genetic traits and underactivity.

> **PRINCIPLE #7**   Some groups of people are at higher risk of becoming inadequately nourished than others.

Women who are pregnant or breastfeeding, infants, children, people who are ill, and frail elderly persons have a greater need for nutrients than healthy adults and elderly people. As a result, they are at higher risk of becoming inadequately nourished than others. Within these groups those at highest risk of nutritional insults are the poor. In cases of widespread food shortages, such as those induced by war or natural disaster, the health of these nutritionally vulnerable groups is compromised the soonest and the most.

> **PRINCIPLE #8**   Poor nutrition can influence the development of certain chronic diseases.

Poor nutrition not only results in deficiency or toxicity diseases, it also plays an important role in the development of heart disease, hypertension, cancer, osteoporosis, and other chronic diseases. Diets high in animal fat, for example, are related to the development of heart disease, diets low in vegetables and fruits to cancer, diets low in calcium to osteoporosis, and diets high in sugar to tooth decay.[13] The effects of habitually poor diets on chronic disease development often take years to become apparent.

> **PRINCIPLE #9**   Adequacy and balance are key characteristics of a healthy diet.

Healthy diets contain many different foods that together provide calories and nutrients in amounts that promote the optimal functioning of cells and of the body. A variety of food is required to obtain all the nutrients needed because no one food contains them all, but many different combinations of food can make up a healthy diet.

Adequate diets are most easily obtained by consuming foods that are good sources of a number of nutrients but not packed with calories. Such foods are considered *nutrient dense.* Those that provide calories and low amounts of nutrients are considered *"empty-calorie"* foods. Vegetables, fruits, lean meats, dried beans, breads, and cereals are nutrient dense. Foods such as beer, chips, candy, pastries, sodas, and fruit drinks lead the list of empty-calorie foods.

> **PRINCIPLE #10**   There are no "good" or "bad" foods.

All things in nutriment are good or bad relatively.

Hippocrates

Unless you are talking about spoiled food or poisonous mushrooms, there is no such thing as a bad food. There are, however, combinations of foods that add up to an unhealthy diet. Occasional consumption of a hot dog, fried chicken, or a chocolate sundae isn't going to shorten a person's life span. Making them dietary staples, along with other empty-calorie foods, will take its toll, however.

NUTRITIONAL ASSESSMENT

Nutritional assessment of groups and individuals is a prerequisite to planning for the prevention or solution of nutrition-related health problems. It represents a broad area within the field of nutrition and is only highlighted here. Resources related to the selection of appropriate nutritional assessment techniques and their implementation are listed at the end of this chapter.

Nutritional status may be assessed for a population group or for an individual. Community-level assessment identifies a population's status using broad nutrition and health indicators, whereas individual assessment provides the baseline for anticipatory guidance and nutrition intervention.

> **PRIMARY MALNUTRITION**   Malnutrition that results directly from inadequate or excessive dietary intake of energy or nutrients.
>
> **SECONDARY MALNUTRITION**   Malnutrition that results from a condition (e.g., disease, surgical procedure, medication use) rather than primarily from dietary intake.
>
> **NUTRIENT-DENSE FOODS**   Foods that contain relatively high amounts of nutrients compared to their caloric value.
>
> **EMPTY-CALORIE FOODS**   Foods that provide an excess of calories relative to their nutrient content.

Community-Level Assessment

A target community's "state of nutritional health" can generally be estimated using existing vital statistics data, seeking the opinions of target group members and local health experts, and making observations. Knowledge of average household incomes; the proportion of families participating in the Food Stamp Program, soup kitchens, school breakfast programs, or food banks; and the age distribution of the group can help identify key nutrition concerns and issues. In large communities, rates of infant mortality, heart diseases, and cancer can reveal whether the incidence of these problems is unusually high.

Information gathered from community-level nutritional assessment can be used to develop community-wide programs addressing specific problem areas, such as childhood obesity or iron-deficiency anemia. Nutrition programs should be integrated into community-based health programs.[14]

Individual-Level Nutritional Assessment

Nutritional assessment of individuals has four major components:

- Clinical/physical assessment
- Dietary assessment
- Anthropometric assessment
- Biochemical assessment

Data from all of these areas are needed to describe a person's nutritional status. Data on height and weight provide information on weight status, for example, and knowledge of blood iron levels tells you something about iron status. It cannot be concluded (although it sometimes is) that people who are normal weight or have good iron status are "well nourished." Single measures do not describe a person's nutritional status.

REGISTERED DIETITIAN An individual who has acquired knowledge and skills necessary to pass a national registration examination and who participates in continuing, professional education.

CLINICAL/PHYSICAL ASSESSMENT A clinical/physical assessment involves visual inspection of a person by a trained *registered dietitian* or other qualified professional to note features that may be related to malnutrition. Excessive or inadequate body fat, paleness, bruises, and brittle hair are examples of features that may suggest nutrition-related problems. Physical characteristics are nonspecific indicators, but they can support other findings related to nutritional status. They cannot be used as the sole criterion upon which to base a decision about the presence or absence of a particular nutrition problem.

Dietary Assessment

Garbage in, garbage out.

Many methods are used for assessing dietary intake. For clinical purposes, 24-hour dietary recalls and food records analyzed by computer programs are most common. Single, 24-hour recalls are most useful

for estimating dietary intakes for groups, whereas multiple recalls and food records are generally used for assessments of individual diets.

Becoming proficient at administering 24-hours recalls takes training and practice. Food records, on the other hand, are completed by clients themselves. These are more accurate if the client has also received some training. Generally, the purpose of assessing of an individuals's diet is to estimate the person's overall diet quality so that strengths and weaknesses can be identified or to assess intake of specific nutrients that may be involved in disease states.

All methods of dietary assessment and all computer programs used to analyze food intake have limitations. People often underestimate their food intake (especially if they have a weight problem), and some people have trouble remembering what they ate. Computer programs may not be updated frequently enough to contain current information on the composition of foods and may not include all of the various foods people eat. Specific foods, ingredients, and portion sizes must be carefully recorded and entered if reliable results are to be produced.

Information on at least three days of dietary intake (preferably two usual weekdays and one weekend day) is needed to obtain a reliable estimate of intake by food group, calories, and nutrients. A good approach is to have a trained dietary interviewer administer a 24-hour recall and then have the client record her/his own diet on two other days. The experience of thinking about what was eaten, portion sizes, ingredients, and recipes helps train people to complete their food records accurately. Completed records should be reviewed with the client during a telephone call or clinic session to make sure they are accurate.

Several high-quality computer programs and Internet resources are available for dietary assessment. The Healthy Eating Index (HEI), which was developed and is maintained by the U.S. Department of Agriculture (USDA), is an example of a high-quality Internet resource. This interactive program provides an analysis of food intake based on the "Healthy Eating Index" and evaluates dietary intake based on the Food Guide Pyramid recommendations and guidelines related to intake of fat, cholesterol, sodium, and other nutrients. Web addresses for the HEI and other dietary assessment resources are given at the end of this chapter.

Anthropometric Assessment

Individual measures of body size (height, weight, percent body fat, bone density, and head and waist circumferences, for example) are useful in the assessment of nutritional status—if done correctly. Each measure requires use of standard techniques and calibrated instruments by trained personnel. Unfortunately, *anthropometric* measurements are frequently performed and recorded incorrectly in clinical practice. Training on anthropometric measures is often available through public health agencies and programs such as WIC (Special Supplemental Nutrition Program for Women, Infants, and Children), and courses and training sessions are sometimes presented at universities.

Biochemical Assessment

Nutrient and enzyme levels, DNA characteristics, and other biological markers are components of a biochemical assessment of nutritional status. Which biological markers are measured depends on what problems are suspected, based on other evidence. For example, a young child who tires easily, has a short attention span, and does not appear to be consuming sufficient iron based on dietary assessment results may have blood taken for analyses of hemoglobin and serum ferritin (markers of iron status). Suspected inborn errors of *metabolism* that may underlie nutrient malabsorption may be identified through DNA or other tests. Such results provide specific information on a component of a person's nutritional status and are very helpful in diagnosing a particular condition.

After the fact-finding phase of nutritional assessment, the nutritionist or other professionals must "apply their brains to their clients' problems." There is no "one size fits all" approach to solving nutrition problems—each has to be figured out individually.

PUBLIC FOOD AND NUTRITION PROGRAMS

A variety of federal, state, and local programs are available to provide food and nutrition services to families and individuals. Many communities have nutrition coalitions or partnership groups that collaborate on meeting the food and nutritional needs of members of the community. Programs representing church-based feeding sites, food shelves, Second Harvest Programs, the Salvation Army, missions, and others are usually a part of local coalitions. These central resources can be identified by contacting the local public health or cooperative extension agency. State-level programs are generally part of large national programs.

Government-sponsored food and nutrition programs are widely available throughout the country. Some programs, such as the School Lunch Program, benefit many children. Other programs are targeted to families and individuals in need. "Need" is generally defined as individual and household incomes below the poverty line. Some programs, such as WIC, have eligibility standards of 185% of the poverty line (Table 1.13). Income guidelines change every year and are sev-

eral thousand dollars higher for people in Alaska and Hawaii. Federal poverty guidelines can be identified on the Web by searching for the *Federal Register* and then the USDA income eligibility guidelines.

ANTHROPOMETRY The science of measuring the human body and its parts.

METABOLISM The chemical changes that take place in the body.

Table 1.14 presents summary information on existing federal food and nutrition programs. You can get more information on these programs through the Internet at the address: www.nutrition.gov.

NATIONWIDE PRIORITIES FOR IMPROVEMENTS IN NUTRITIONAL HEALTH

Public health initiatives involving population-based improvements in food safety, food availability, and nutritional status have led to major gains in the health status of the country's population. Among the important components of this success story are programs that have expanded the availability of housing; safe food and water; foods fortified with iodine, iron, vitamin D, or folate; fluoridated public water; food assistance; and nutrition education. Since 1900, the average life span of persons in the United States has lengthened by more than 30 years, and 25 years of this gain are estimated to be due to the quiet revolutions that have taken place in public health.[15]

Although the United States spends more on health care than any other country, statistics from the World Health Organization indicate that it ranks 26th among developed countries of the world in life expectancy. Today's priorities for improvements in the public's health and longevity center on reducing obesity, infant mortality, smoking, excessive alcohol consumption, accidents, violence, and physical inactivity. Goals for dietary changes are a central part of the nation's overall plan for health improvements. For the two out of three Americans who do not smoke or

Table 1.13 **Income eligibility standards for the WIC program (≤185% of poverty income through June 30, 2001).**

Household Size	Household Income per Year
1	$15,448
2	20,813
3	26,178
4	31,543
5	36,908
6	42,273
7	47,638
8	53,003
Each additional member	+5,365

Table 1.14   Summary of federal food and nutrition programs.

Program	Activity
Child and Adult Care Food Program (CACFP)	Reimburses child- and adult-care organizations in low-income areas for provision of nutritious foods.
Summer Food Service Program	Provides foods to children in low-income areas during summers.
School Breakfast and Lunch Programs	Provide free breakfasts and reduced-cost or no-cost lunches to children from families who cannot afford to buy them.
Food Stamp Program	Subsidizes food purchases of low-income families and individuals.
WIC (Special Supplemental Nutrition Program for Women, Infants, and Children)	Serves low-income, high-risk, and pregnant and breastfeeding women and children up to five years of age. Provides supplemental, nutritious foods and nutrition education as an adjunct to health care.
Head Start Program	Includes nutrition education for children and parents and supplies meals for children in the program.
Team Nutrition	A USDA nutrition education program that brings active learning about nutrition to schoolchildren.

drink excessively, dietary intakes represent the major environmental influence on long-term health.[13]

Goals for improving the nutritional health of the nation are summarized in the document "Healthy People 2010: Objectives for the Nation" (Table 1.15). As the seeds of many chronic disease are planted during pregnancy and childhood, major emphasis is placed on dietary habits early in life.

Food Intake Recommendations

Two major guides offer recommendations for food intake: the Food Guide Pyramid and the "Dietary Guidelines for Americans." The Food Guide Pyramid guides people in the selection of foods from five food groups (Illustration 1.6). Each food group has a recommended number of daily servings, and serving numbers vary based on age. Foods within each group

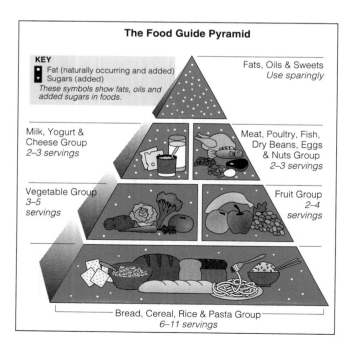

The Food Guide Pyramid

KEY
■ Fat (naturally occurring and added)
▽ Sugars (added)
These symbols show fats, oils and added sugars in foods.

Fats, Oils & Sweets
Use sparingly

Milk, Yogurt & Cheese Group
2–3 servings

Meat, Poultry, Fish, Dry Beans, Eggs & Nuts Group
2–3 servings

Vegetable Group
3–5 servings

Fruit Group
2–4 servings

Bread, Cereal, Rice & Pasta Group
6–11 servings

WHAT COUNTS AS ONE SERVING?

The amount you eat may be more than one serving. For example, a dinner portion of spaghetti would count as two or three servings.

Bread, Cereal, Rice, and Pasta Group
1 slice of bread
½ cup of cooked rice or pasta
½ cup of cooked cereal
1 ounce of ready-to-eat cereal

Milk, Yogurt, and Cheese Group
1 cup of milk or yogurt
1½ ounces of natural cheese
2 ounces of process cheese

Vegetable Group
½ cup of chopped raw or cooked vegetables
1 cup of leafy raw vegetables

Meat, Poultry, Fish, Dry Beans, Eggs, and Nuts Group
2½ to 3 ounces of cooked lean meat, poultry, or fish
Count ½ cup of cooked beans, or 1 egg, or 2 tablespoons of peanut butter as 1 ounce of lean meat

Fruit Group
1 piece of fruit or melon wedge
¾ cup of juice
½ cup of canned fruit
¼ cup of dried fruit

Fats and Sweets
LIMIT CALORIES FROM THESE especially if you need to lose weight.

Illustration 1.6   The Food Guide Pyramid.

Table 1.15 2010 Nutrition Objectives for the Nation.

√ Increase the proportion of adults who are at a healthy weight from 42 to 60%.

√ Reduce the proportion of adults who are obese from 23 to 15%.

√ Reduce the proportion of children and adolescents who are overweight or obese from 11 to 5%.

√ Reduce growth retardation among low-income children under age 5 years from 8 to 5%.

√ Increase the proportion of persons aged 2 years and older who:

- consume at least two daily servings of fruit from 28 to 75%.
- consume at least three daily servings of vegetables, with at least one-third being dark green or deep yellow vegetables from 3 to 50%.
- consume at least six daily servings of grain products, with at least three being whole grains from 7 to 50%.
- consume less than 10% of calories from saturated fat from 36 to 75%.
- consume no more than 30% of total calories from fat from 33 to 75%.
- consume 2400 mg or less of sodium daily from 21 to 65%.
- meet dietary recommendations for calcium from 46 to 75%.

√ Reduce iron deficiency among young children and females of childbearing age from 4–11% to 1–7%.

√ Reduce anemia among low-income pregnant females in their third trimester from 29 to 20%.

√ Increase the proportion of worksites that offer nutrition or weight management classes or counseling from 55 to 85%.

√ Increase the proportion of physician office visits made by patients with the diagnosis of cardiovascular disease, diabetes, osteoporosis, or hyperlipidemia that include counseling or education related to diet and nutrition from 42 to 75%.

√ Increase food security among U.S. households and in so doing reduce hunger rates.

Other Objectives Related to Nutrition

√ Reduce infant mortality from 7.6 to no more than 5 per 1000 live births.

√ Reduce the incidence of spina bifida and other neural tube defects from 7 to 3 per 10,000 live births.

√ Reduce the incidence of birth defects from 1.7 to 1.2 per 1000 live births.

√ Increase the proportion of women who receive preconceptional counseling.

√ Increases the proportion of pregnant women who begin prenatal care in the first trimester from 81 to 90% or more.

√ Reduce low birthweight (<2500 grams) from 7.3 to 5%.

√ Reduce preterm births (<37 weeks) from 9.1 to 7.6%.

√ Increase abstinence from alcohol use by pregnant women from 79 to 95%.

√ Reduce the incidence of fetal alcohol syndrome.

√ Increase the proportion of women who gain weight appropriately during pregnancy.

√ Increase from 60 to 75% the proportion of women who exclusively breastfeed after delivery.

are assigned specific portion sizes that correspond to a serving. This food guide is periodically updated.

The "Dietary Guidelines for Americans" which is updated every five years, recommends nutrition behaviors as well as the types and amounts of food that promote health (Table 1.16). The complete version of the 2000 guidelines can be found at the Web site www.usda.gov.cnpp.

In the Dietary Guidelines, under the recommendation to "Let the Pyramid guide your food choices," is an informative description of how to read a nutrition information label on food packages. Nutrition labels provide a wealth of facts on the nutrient profile of foods. Refer to this page at the "Dietary Guidelines for Americans" Web site if you need to update your knowledge about nutrition labels.

Table 1.16
Dietary Guidelines for Americans, 2000.

- Aim for a healthy weight.
- Be physically active each day.
- Let the Pyramid guide your food choices.
- Choose a variety of grains daily, especially whole grains.
- Choose a variety of fruits and vegetables daily.
- Keep foods safe to eat.
- Choose a diet that is low in saturated fat and cholesterol and moderate in total fat.
- Choose beverages and foods to moderate your intake of sugars.
- Choose and prepare foods with less salt.
- If you drink alcoholic beverages, do so in moderation.

Resources

Start here if you want to identify government resources related to nutrition information, services, and programs.
Web site: www.nutrition.gov

USDA's Center for Nutrition Policy and Promotion

Select Interactive Healthy Eating Index to assess one day's dietary intake. This site also provides the 2000 Dietary Guidelines.
Web site: www.usda.gov/cnpp
One-stop shopping for information on U.S. government programs.
Web site: www.firstgov.gov

USDA's Food and Nutrition Resource Center.

Link to WIC and other federal programs.
Web site: www.nal.usda.gov

American Dietetic Association

Leads to a wide variety of resources on nutrition and health.
Web site: www.eatright.org

Dietary Guidelines Alliance

Provides suggestions on how to meet the Dietary Guidelines and other information.
Web site: www.ificinfo.health.org/iaay/index.htm
Want more information on the Dietary Reference Intakes? You can get full reports here.
Web site: www.iom.edu/fnb

National Academy of Sciences

Provides the summary report on applications of the DRIs for dietary assessment.
Web site: www.nap.edu/openbook/0309065542/html/1.html
Leads to high-quality sites in many areas of nutrition.
Web site: www.navigator.tufts.edu

Medline

Get access to scientific journal articles on a variety of nutrition topics.
Web site: www.ncbi.nlm.nih.gov/PubMed

U.S. Department of Health and Human Services

Federal poverty guidelines are available at this site.
Web site: http://.aspe.hhs.gov/poverty/01poverty.htm
Provides international dietary guidelines.
Web site: www.fcs.uga.edu~selbon/apple/guides/choose.html

Healthy People 2010: Objectives for the Nation

The home page for "Healthy People 2010."
Web site: web/health.gov/healthypeople
Read about the health status of Canadians and nutrition and other goals for Canada.
Web site: www.healthcanada.com

References

1. National data on household food security. Nutrition Today 2000;35:4.

2. Summary report on application of the DRIs for dietary assessment. Washington, DC: National Academy of Sciences, 2000.

3. Anderson JW, Allgood LD, Turner J, et al. Effects of psyllium on glucose and serum lipid responses in men with type 2 diabetes and hypercholesterolemia. Am J Clin Nutr 1999;70:466–73.

4. Cohen LA. Dietary fiber and breast cancer. Anticancer Res 1999;19:3685–8.

5. Butte NF. Carbohydrate and lipid metabolism in pregnancy: normal compared with gestational diabetes mellitus. Am J Clin Nutr 2000;71:1256S–61S.

6. Aro A. Epidemiology of trans fatty acids and coronary heart disease in Europe. Nutr Meta & Cardiovasc Dis 1998;8:402.

7. Hu FB, Stampfer MJ, Manson JE, et al. Dietary saturated fats and their food sources in relation to the risk of coronary heart disease in women. Am J Clin Nutr 1999;70:1001–8.

8. McNamara DJ. Cholesterol intake and plasma cholesterol: an update. J Am Coll Nutr 1997;16:530–4.

9. Milner JA. Functional foods: the US perspective. Am J Clin Nutr 2000;71:1654S–9S.

10. Craig WJ. Phytochemicals: guardians of our health. J Am Diet Assoc 1997;97:S199–S204.

11. Kleiner SM. Water: an essential but overlooked nutrient. J Am Diet Assoc 1999;99:200–6.

12. Johnston CS, Corte C. People with marginal vitamin C status are at high risk of developing vitamin C deficiency. J Am Diet Assoc 1999;99:854–6.

13. Deckelbaum RJ, Fisher EA, Winston M, et al. Summary of a scientific conference on preventive nutrition: pediatrics to geriatrics. Circulation 1999;100:450–6.

14. Making the future: developing community-based nutrition services. Association of Public Health Nutrition State and Territorial Nutrition Directors, and the Maternal and Child Health Bureau, HRSA, PHS, 1996.

15. Public health achievements. Nutrition Today 1999 May/June:2.

CHAPTER 2

Corbis

> *To bring on the menses, recover the flesh by giving her puddings, roast meats, a good wine, fresh air, and sun.*
>
> Amusing advice on the treatment of infertility from 1847

PRECONCEPTION NUTRITION

Prepared by **Judith E. Brown**

CHAPTER OUTLINE

KEY NUTRITION CONCEPTS

1 Fertility is achieved and maintained by carefully orchestrated, complex processes that can be disrupted by a number of factors related to body composition and dietary intake.

2 Oral contraceptives and contraceptive implants can adversely affect some aspects of nutritional status.

3 Optimal nutritional status prior to pregnancy enhances the likelihood of conception and helps ensure a healthy pregnancy and robust newborn.

INTRODUCTION

Human reproduction is the result of a superb orchestration of complex and interrelated genetic, biological, environmental, and behavioral processes. Given favorable states of health, these processes occur smoothly in females and males and set the stage for successful reproduction. However, less than optimal states of health, brought about by conditions such as acute undernutrition or high levels of alcohol intake, can disrupt these finely tuned processes and diminish reproductive capacity.

INFERTILITY   Absence of production of children. Commonly used to mean a biological inability to bear children.

INFECUNDITY   Biological inability to bear children after one year of unprotected intercourse.

FERTILITY   Actual production of children. Word best applies to specific vital statistic rates, but is commonly taken to mean the ability to bear children.

FECUNDITY   Biological ability to bear children.

MISCARRIAGE   Generally defined as the loss of a conceptus in the first 20 weeks of pregnancy. Also called *spontaneous abortion*.

ENDOCRINE   A system of ductless glands, such as the thyroid, adrenal glands, ovaries, and testes, that produces secretions that affect body functions.

IMMUNOLOGICAL   Having to do with the immune system and its functions in protecting the body from bacterial, viral, fungal, or other infections and from foreign proteins, i.e., those proteins that differ from proteins normally found in the body.

SUBFERTILITY   Reduced level of fertility characterized by unusually long time to conception (over 12 months) or repeated, early pregnancy losses.

Sometimes conception occurs in the presence of poor nutritional or health status. Such events increase the likelihood that fetal growth and development, and the health of the mother during pregnancy, will be compromised.

This chapter first highlights vital statistics related to the preconception period and presents background information on reproductive physiology. Then the focus is on (1) nutrition and the development and maintenance of the biological capacity to reproduce, (2) nutritional effects of contraceptives, (3) preconceptional nutritional status and the course and outcome of pregnancy, and (4) model programs that promote preconceptional nutritional health. The following chapter addresses the role of nutrition in specific conditions, such as premenstrual syndrome, eating disorders, and polycystic ovary syndrome, that affect preconceptional health or very early pregnancy outcomes.

PRECONCEPTION OVERVIEW

Approximately 15% of all couples in the Western world are involuntarily childless. They are generally considered to be *infertile,* or more correctly *infecund.*

Fertility refers to the actual production of children, whereas *fecundity* addresses the biological capacity to bear children. *Fertility* best applies to vital statistics on fertility rates, or the number of births per 1000 women of childbearing age (15–44 years in most statistical reports). For example, in 1998 the U.S. fertility rate was 65.6 births per 1000 women aged 15 to 44 years.[1] Even scientists and clinicians rarely use these terms correctly, however, and to do so in this chapter would cause undue confusion. Hence, we will use the familiar meaning of *infertility.*

Infertility is generally defined as the lack of conception after one year of unprotected intercourse. This definition leads to a "yes/no" answer about fertility that is misleading. Approximately 40% of couples diagnosed as infertile will conceive a child in three years without the help of technology.[2] Given these chances, fertility procedures can be considered successful only if they increase conceptions by over 40%, and most currently do not.[3] Chances of conceiving decrease the longer infertility lasts and as women and men age beyond 35 years.[4]

Healthy couples having regular, unprotected intercourse have a 25 to 30% chance of a diagnosed pregnancy within a given menstrual cycle. However, many more conceptions probably occurred. Studies show that 25–30% of conceptions are lost by resorption into the uterine wall within the first six weeks after conception.[5] Another 9% are lost by *miscarriage* in the first 20 weeks of pregnancy.[6]

The most common cause of miscarriage is the presence of a severe defect in the fetus. Miscarriages can also be caused by maternal infection, structural abnormalities of the uterus, *endocrine* or *immunological* disturbances, and unknown, random events.[7]

Women who experience multiple miscarriages (sometimes defined as three or more), men who have sperm abnormalities (such as low sperm count or density, malformed or immobile sperm), and women who ovulate infrequently are considered *subfertile.* It is estimated that 18% of married couples in the United States are subfertile due to delayed time to conception (over 12 months) or repeated, early pregnancy losses.[2] A silver lining to subfertility is that the reproductive capacity of one individual can compensate for diminished potential in the other.

2010 Nutrition Objectives for the Nation Related to the Preconceptional Period

National priorities for improvement in health status prior to pregnancy include five related to nutrition (see Table 2.1). Objectives for improving intake of vegetables, calcium, whole grains, and other dietary

Table 2.1 2010 nutrition objectives for the nation related to preconception.

√ Increase the proportion of adults who are at a healthy weight from 42 to 60%.

√ Reduce the proportion of adults who are obese from 23 to 15%.

√ Reduce iron deficiency among young children and females of childbearing age from 4–11% to 1–7%.

√ Reduce the incidence of spina bifida and other neural tube defects from 7 to 3 per 10,000 live births and birth defects from 1.7 to 1.2 per 1000 live births.

√ Increase the proportion of women who receive preconceptional counseling.

components apply to men as well as women prior to pregnancy.

REPRODUCTIVE PHYSIOLOGY

The reproductive systems of males and females (Illustration 2.1) begin developing in the first months of pregnancy and continue to grow in size and complexity of function through *puberty.* Females are born with their full complement of immature *ova* and males with sperm-producing capabilities. The capacity for reproduction is established during puberty when hormonal changes cause the maturation of the reproductive system over the course of three to five years.

Approximately seven million immature ova, or *primordial follicles,* are formed during early fetal development, but only about one-half million per ovary remain by the onset of puberty. During a woman's fertile years, some 400–500 ova will mature and be released for possible fertilization. Very few ova remain by *menopause.* Sperm numbers and viability decrease somewhat after approximately 35 years of age, but are still produced from puberty onward.[4] Since females are born with their lifetime supply of ova, the number with chromosomes damaged by oxidation, radioactive particle exposure, and aging increases with time. Consequently, children born to women older than roughly 35 years of age are more likely to have disorders related to defects in chromosomes than are children born to younger women.[2]

Female Reproductive System

During puberty females develop monthly *menstrual cycles,* the purpose of which is to prepare an ovum for fertilization by sperm and the uterus for implantation of a fertilized egg. Menstrual cycles result from complex interactions among hormones secreted by the hypothalamus, the pituitary gland, and the ovary.

Knowledge of hormonal changes during the menstrual cycle is expanding, and the process is more complex than this presentation, which focuses on nutritional effects on hormonal changes in the menstrual cycle and on fertility.

Menstrual cycles are 28 days long on average, but it is not uncommon for cycles to be several days shorter or longer. The first day of the cycle is when menses, or blood flow, begins. The first half of the cycle is called the *follicular phase,* the last 14 days is the *luteal phase.* Hormonal changes during these phases of the menstrual cycle are shown in Illustration 2.2. (Note: Content on hormonal effects during the menstrual cycle may not be a required part of all courses using this text. Check with your instructor. Refer to this section for definitions of hormones and their abbreviations.)

PUBERTY The stage in life during which humans become biologically capable of reproduction. In girls, puberty generally occurs between the ages of 9 and 16, and in boys, between the ages of 11 and 17.

OVA Eggs of the female produced and stored within the ovaries (singular = *ovum*).

MENOPAUSE Cessation of the menstrual cycle and reproductive capacity in females.

MENSTRUAL CYCLE An approximately four-week interval in which hormones direct a buildup of blood and nutrient stores within the wall of the uterus and ovum maturation and release. If the ovum is fertilized by a sperm, the stored blood and nutrients are used to support the growth of the fertilized ovum. If fertilization does not occur, they are released from the uterine wall over a period of three to seven days. The period of blood flow is called the *menses,* or the menstrual period.

HORMONAL EFFECTS DURING THE MENSTRUAL CYCLE
At the beginning of the follicular phase, estrogen stimulates the hypothalamus to secrete *gonadotropin releasing hormone (GnRH),* which causes the pituitary gland to release the *follicle stimulating hormone (FSH)* and *luteinizing hormone (LH).* (See Table 2.2 for definitions of these hormones.) FSH prompts the growth and maturation of 6–20 follicles, or capsules in the surface of the ovary, in which ova mature. The presence of FSH stimulates the production of *estrogen* by cells within the follicles. Estrogen and FSH further stimulate the growth and maturation of the follicle while rising LH levels cause cells within the follicles to secrete *progesterone.* Estrogen and progesterone also prompt the uterine wall (or endometrium) to store glycogen and other nutrients and to expand the growth of blood vessels and connective tissue. These changes prepare the uterus for nourishing a conceptus after implantation. Just prior to ovulation on day 14 of a 28-day menstrual cycle, blood levels of FSH and LH peak. The surge in LH level results in the release of an ovum from a follicle and Voilà! Ovulation occurs.

The luteal phase of the menstrual cycle begins after ovulation. Much of the hormonal activity that

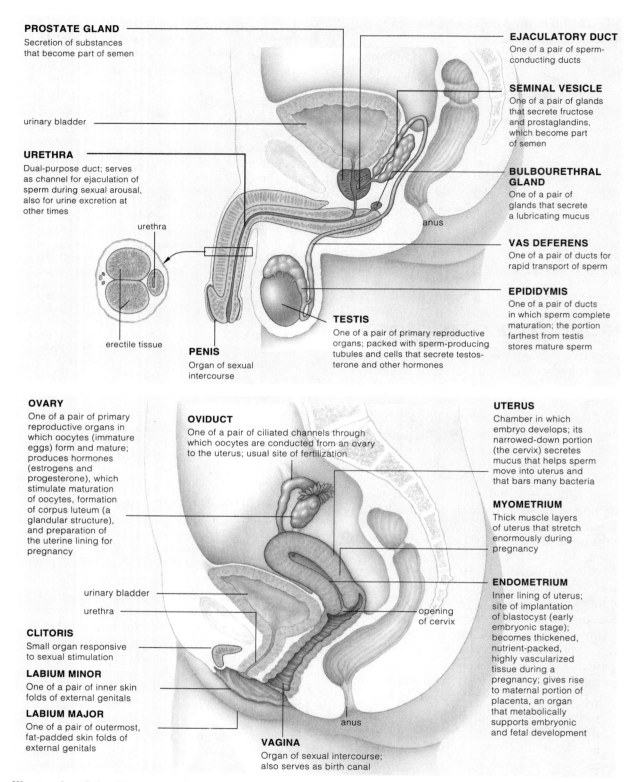

PROSTATE GLAND
Secretion of substances that become part of semen

urinary bladder

URETHRA
Dual-purpose duct; serves as channel for ejaculation of sperm during sexual arousal, also for urine excretion at other times

urethra

erectile tissue

PENIS
Organ of sexual intercourse

EJACULATORY DUCT
One of a pair of sperm-conducting ducts

SEMINAL VESICLE
One of a pair of glands that secrete fructose and prostaglandins, which become part of semen

BULBOURETHRAL GLAND
One of a pair of glands that secrete a lubricating mucus

anus

VAS DEFERENS
One of a pair of ducts for rapid transport of sperm

EPIDIDYMIS
One of a pair of ducts in which sperm complete maturation; the portion farthest from testis stores mature sperm

TESTIS
One of a pair of primary reproductive organs; packed with sperm-producing tubules and cells that secrete testosterone and other hormones

OVARY
One of a pair of primary reproductive organs in which oocytes (immature eggs) form and mature; produces hormones (estrogens and progesterone), which stimulate maturation of oocytes, formation of corpus luteum (a glandular structure), and preparation of the uterine lining for pregnancy

OVIDUCT
One of a pair of ciliated channels through which oocytes are conducted from an ovary to the uterus; usual site of fertilization

urinary bladder

urethra

CLITORIS
Small organ responsive to sexual stimulation

LABIUM MINOR
One of a pair of inner skin folds of external genitals

LABIUM MAJOR
One of a pair of outermost, fat-padded skin folds of external genitals

opening of cervix

anus

VAGINA
Organ of sexual intercourse; also serves as birth canal

UTERUS
Chamber in which embryo develops; its narrowed-down portion (the cervix) secretes mucus that helps sperm move into uterus and that bars many bacteria

MYOMETRIUM
Thick muscle layers of uterus that stretch enormously during pregnancy

ENDOMETRIUM
Inner lining of uterus; site of implantation of blastocyst (early embryonic stage); becomes thickened, nutrient-packed, highly vascularized tissue during a pregnancy; gives rise to maternal portion of placenta, an organ that metabolically supports embryonic and fetal development

Illustration 2.1   Mature male and female reproductive systems.

regulates biological processes during this half of the cycle is initiated by the cells in the follicle left behind when the egg was released. These cells grow in number and size and form the *corpus luteum* from the original follicle. The corpus luteum secretes large amounts of progesterone and some estrogen. These hormones now inhibit the production of GnRH, and thus the secretion of FSH and LH. Without sufficient

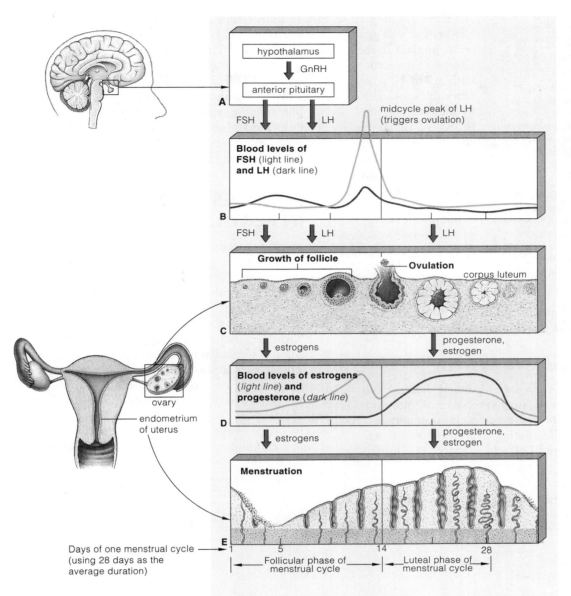

Illustration 2.2

Changes in the ovary and uterus, correlated with changing hormone levels during the follicular and luteal phases of the menstrual cycle.

FSH and LH, ova within follicles do not mature and are not released. (This is also how estrogen and progesterone in some birth control pills inhibit ova maturation and release.) Estrogen and progesterone secreted by the corpus luteum further stimulate the development of the endometrium. If the ovum is not fertilized, the production of hormones by the corpus luteum declines, and blood levels of progesterone and estrogen fall. This decline removes their inhibitory effect on GnRH release, and GnRH is again able to stimulate release of FSH for the next cycle of follicle development, and of LH for the stimulation of progesterone and estrogen production. Decreased levels of progesterone and estrogen also cause blood vessels in the uterine wall to constrict and to release its outer layer in the *menstrual flow.* Cramps and other side-effects of menstruation can be traced to the production of *prostaglandins* by the uterus. These prostaglandins cause the uterus to contract and release its contents.

CORPUS LUTEUM (*corpus* = body, *luteum* = yellow)   A tissue about ½ inch in diameter formed from the follicle that contained the ovum prior to its release. It produces estrogen and progesterone. The "yellow body" derivation comes from the accumulation of lipid precursors of these hormones in the corpus luteum.

PROSTAGLANDINS   A group of physiologically active substances made from the fatty acid arachidonic acid. They are present in many tissues and perform such functions as the constriction or dilation of blood vessels and the stimulation of smooth muscles and the uterus. (They are not produced solely in the prostate gland as once was thought.)

TESTES   Male reproductive glands located in the scrotum. Also called testicles.

EPIDIDYMIS   Tissues on top of the testes that store sperm.

PELVIC INFLAMMATORY DISEASE (PID)
A general term applied to infections of the cervix, uterus, fallopian tubes, or ovaries. Occurs predominantly in young women and is generally caused by infection with a sexually transmitted disease, such as gonorrhea or chlamydia, or with intrauterine device (IUD) use.

ENDOMETRIOSIS   A disease characterized by the presence of endometrial tissue in abnormal locations, such as deep within the uterine wall, in the ovary, or in other sites within the body. The condition is quite painful and is associated with abnormal menstrual cycles and infertility in 30–40% of affected women.

If the ovum is fertilized, it will generally implant in the lining of the uterus within 8 to 10 days. Hormones secreted by the dividing, fertilized egg signal the corpus luteum to increase in size and to continue to produce enough estrogen and progesterone to maintain the nutrient and blood vessel supply in the endometrium. The corpus luteum ceases to function within the first few months of pregnancy, when it is no longer needed for hormone production.

Male Reproductive System

Reproductive capacity in males is established by complex interactions among the hypothalamus, pituitary gland, and *testes*. The process in males is ongoing rather than cyclic. Fluctuating levels of GnRH signal the release of FSH and LH, which trigger the production of testosterone (Table 2.2) by the testes. Testosterone stimulates the maturation of sperm, which takes 70–80 days. When mature, sperm are transported to the *epididymis* for storage. Upon ejaculation, sperm mix with fluids from various glands on their way to the urethra for release.

Just as some aspects of female reproductive processes remain unclear, scientists have yet to fully elucidate hormonal and other processes involved in male reproduction.

SOURCES OF DISRUPTIONS IN FERTILITY

The intricate mechanisms that regulate fertility can be disrupted by many factors, including adverse nutritional exposures, contraceptive use, crowding, severe stress, infection, tubal damage and other structural problems, and chromosomal abnormalities (Table 2.3).[8] Conditions that modify fertility appear to affect hormones that regulate ovulation, the presence or length of the luteal phase, sperm production, or the tubular passageways that ova and sperm must travel for conception to occur. Sexually transmitted infections, for example, can result in *pelvic inflammatory disease (PID)*, which may lead to scarring and blockage of the fallopian tubes.[9] *Endometriosis* is also a common cause of reduced fertility. It devel-

Table 2.2 Hormones that affect reproduction.

HORMONE	ABBREVIATION	SOURCE	ACTION
Gonadotropin releasing hormone	GnRH	Hypothalamus	Stimulates release of FSH and LH
Follicle stimulating hormone	FSH	Pituitary	Stimulates the maturation of ova and sperm
Luteinizing hormone	LH	Pituitary	Stimulates secretion of estrogen, progesterone, and testosterone and growth of the corpus luteum
Estrogen (most abundant form is estradiol)		Ovaries, testes, fat cells, corpus luteum, and placenta (during pregnancy)	Stimulates release of GnRH in follicular phase and inhibits in luteal; stimulates thickening of uterine wall during menstrual cycle
Progesterone (progestin, progestogen, and gestagon are similar)		Ovaries and placenta	"Pro-gestational": prepares uterus for fertilized ovum and to maintain a pregnancy; stimulates uterine lining buildup during menstrual cycle; helps stimulate cell division of fertilized ova; inhibits action of testosterone
Testosterone		Mostly by testes	Stimulates maturation of male sex organs and sperm, formation of muscle tissue and other functions

Table 2.3 Factors related to impaired fertility in women and men.

Females and Males	Females	Males
• Weight loss >15% of normal weight	• Recent oral contraceptive use (within 2 months)	• Inadequate zinc status
• Negative energy balance	• Anorexia nervosa, bulimia nervosa	• Inadequate antioxidant status (selenium, vitamins C and E, carotinoids)
• Inadequate body fat,	• High caffeine intake (?)[a]	• Heavy metal exposure (lead, mercury, cadmium, manganese)
• Excessive body fat, especially central fat	• High fiber intake	• Halogen (in some pesticides) and glycol exposure (in antifreeze, de-icers)
• Extreme levels of exercise	• Vegetarian diets (?)	• Estrogen exposure (in DDT, PCBs)
• High alcohol intake	• Carotenemia (?)	• Chromosomal abnormalities in sperm
• Endocrine disorders (e.g., hypothyroidism, Cushing's disease)	• Age >35 years	• Sperm defects
• Structural abnormalities of the reproductive tract	• Pelvic inflammatory disease (PID)	• Excessive heat to testes
• Celiac disease	• Endometriosis	• Steroid abuse
• Crowding		
• Severe stress		
• Infection (sexually transmitted diseases)		
• Diabetes		

[a]The "?" indicates that the relationship to impaired fertility is controversial.

ops when portions of the endometrial wall that build up during menstrual cycles become embedded within other body tissues. Endocrine abnormalities that modify hormonal regulation of fertility are the leading diagnoses related to infertility. "Unknown cause" is the second leading diagnosis, however, and is applied to about one-half of all cases of male and female infertility.[2]

NUTRITION-RELATED DISRUPTIONS IN FERTILITY

Disruptions in fertility related to nutritional status include undernutrition, weight loss, obesity, high activity levels, and intake of specific foods and food components. Nutritional factors generally exert only temporary influence on fertility; normal fertility returns once the problem is corrected.

Undernutrition and Fertility

Does undernutrition decrease fertility in populations? The answer depends on whether the undernutrition is long term (chronic) or short term (acute). Chronic undernutrition appears to reduce fertility by only a small amount.[10] Acute undernutrition due to famine or deliberate weight loss in normal-weight women clearly decreases fertility.

CHRONIC UNDERNUTRITION The primary effect of chronic undernutrition on reproduction in women is the birth of small and frail infants who have a high likelihood of death in the first year of life.[11] Infant death rates in developing countries where malnutrition is common often exceed 50 per 1000 live births. By contrast, infant death rates are less than 5 per 1000 live births in developed countries such as Sweden and Japan.[12]

The effect of chronic undernutrition on fertility is difficult to study accurately, and conclusions about relationships will change as more is learned. Investigations of relationships between chronic undernutrition and fertility are complicated by differences in the use of contraception, age of puberty and marriage, breastfeeding duration (longer periods of breastfeeding increase the time to the next pregnancy), access to induced abortions, and social and economic incentives or constraints on family size.[13,14] In less developed countries with poor access to contraception, births per woman average 6 to 8, whereas in developed countries they average at or below the replacement level of 2.1 births per woman.[15] Without careful study, it might appear fertility is *lower* in better nourished women in developed countries than in poorly nourished women in less developed countries.

Of the environmental factors that influence fertility, education and child survival appear to be the most important. Fertility rates in poor countries decline substantially as women become educated and as child survival increases.[16,17]

ACUTE UNDERNUTRITION Undernutrition among previously well-nourished women is associated with a dramatic decline in fertility that recovers when food intake does.[11] Periods of feast and famine in the nomadic !Kung tribe of Botswana and the Turkana people of Kenya, for example, are associated with major shifts in fertility.[18,19] These groups of hunter-gatherers (although relatively few survive now), experience sharp, seasonal fluctuations in body weight depending on the success of their hunting and foraging for plant foods. Birthrates decline substantially during periods of famine and increase with food availability. When these hunter-gatherers become farmers and food supply is more dependable, body weight increases, activity levels decrease, and pregnancy rates go up.

Other evidence also suggests a connection. Food shortages in Europe in the seventeenth and eighteenth centuries were accompanied by dramatic declines in birthrates. Famine in Holland during World War II led to calorie intakes of about 1000 per day among women. One out of two women in famine-affected areas stopped menstruating, and the birthrate dropped by 53%. Fertility status improved within four months after the end of the famine, but for many women it took as long as a year for their menstrual cycles to return to normal.[20] Similarly, the 1974–1975 famine in Bangladesh resulted in a 40% decline in births.[10]

Famines are associated with more than disruptions in the food supply. They are usually accompanied by low availability of fuel for heating and cooking, poor living conditions, anxiety, fear, and despair. These factors also probably contribute to the declines in fertility observed with famine.

Acute reduction in food intake appears to reduce reproductive capacity by modifying hormonal signals that regulate menstrual cycles in females. It also appears to impair sperm maturation in males.[10]

Body Fat, Weight, and Fertility

A number of theories on the critical levels of body fat required to sustain normal fertility have been advanced.[11] However, the critical levels of body fat proposed do not uniformly correspond to lowered fertility in different population groups. Consequently, it is not clear that specific levels of body fat provide meaningful cutpoints for identifying risk for infertility.[21] Fertility in general is lower in women who have a **body mass index** of less than 20 (underweight) or greater than 30 kg/m² (obese). Thus, differences in body fat content may play a role in the fertility decline.

Estrogen is produced by fat tissue, and availability of estrogen changes with varying amounts of body fat. Therefore, it is difficult to dismiss the likelihood that body fat plays a role in the development of fertility problems. Underfat women may experience impaired fertility related to low levels of estrogen, and overly fat women due to excessive estrogen.[21]

> **BODY MASS INDEX (BMI)**   Weight in kg/height in m². BMIs <19.8 are considered underweight, 19.8–25 normal weight, 25–30 overweight, and BMIs of 30 and higher obesity.
>
> **AMENORRHEA**   Absence of menstrual cycle.
>
> **ANOVULATORY CYCLES**   Menstrual cycles in which ovulation does not occur.

> **Connections**
> **The Yam-Twin Link**
>
> The Yoruba in western Nigeria consume large quantities of yams, and they are also reported to have the highest incidence of twin births in the world. It is thought that the yams contain a sufficient amount of *phytoestrogen*, plant estrogen, to affect ovulation.[22]

Weight Loss and Fertility

In normal-weight women, weight loss that exceeds approximately 10–15% of usual weight decreases estrogen. Consequences of these hormonal changes include **amenorrhea, anovulatory cycles,** and short or absent luteal phases.[11,23] It is estimated that about 30% of cases of impaired fertility are related to simple weight loss, or "weight-related amenorrhea" as it is called.[11] Hormone levels tend to return to normal when weight is restored to within 95% of previous weight.[23] Case Study 2.1 provides an example of the effect of weight loss on fertility.

Weight gain is the recommended, first-line treatment for weight-related amenorrhea. In many cases, however, the advice is more easily given than applied. About 10% of underweight women will not consider weight gain and may change health care providers in search of a different solution to infertility.[23] Treatment of underweight women with Clomid (clomiphene citrate) generally does not improve fertility until weight is regained. Fertility may be improved through the use of GnRH and several other hormones.[24] However, twice as many infants born to underweight women receiving such therapy are small-for-gestational age compared to infants born to underweight women who gain weight and experience unassisted conception.[11]

The eating disorder anorexia nervosa is associated with similar, but more severe, changes in endocrine

Case Study 2.1
Cyclic Infertility with Weight Loss and Gain

After four years of amenorrhea, Tonya sought medical care to help her become pregnant. She was convinced that her lack of menstrual periods was the cause of her infertility. Tonya is 5 feet 4 inches and had weighed 121 pounds before dropping to 107 pounds four years ago. Her FSH and LH levels were both abnormally low, and she was not ovulating. When the importance and methods of weight gain were explained, Tonya agreed to gain some weight. After she had regained 7 pounds, her LH level was normal, but her FSH level was still low and the luteal phase of her cycles was abnormally short. When her weight reached 119 pounds, her LH and FSH levels and her menstrual cycles were normal.

This case is not the only clinical picture observed in women experiencing amenorrhea after weight loss. In some cases FSH is low and LH release and levels normal; other cases are characterized by elevated estrogen levels, and so on. Each case must be considered individually.

and hypothalamic functions than those seen with weight loss in normal-weight women. (This topic is covered in the chapter on nutrition-related conditions prior to conception.)

Weight Loss in Males and Infertility

Weight loss decreases fertility in men just as it does in women. In the classic starvation experiments by Keys during World War II,[25] men experiencing a 50% reduction in caloric intake reported substantially reduced sexual drive early in the study. Sperm viability and motility decreased as weight reached 10 to 15% below normal, and sperm production ceased entirely when weight loss exceeded 25% of normal weight. Sperm production and libido returned to normal after weight was regained.

Exercise and Infertility

The adverse effects of intense levels of physical activity on fertility were observed over 40 years ago in female competitive athletes. Since then, a number of studies have shown that young female athletes may experience delayed age at puberty and lack menstrual cycles.[26,27] Average age of menarche for competitive female athletes and ballet dancers is often delayed by two to four years. The delay in menarche increases if females begin training for events that require thinness (such as gymnastics) before menarche normally would begin. Very high levels of exercise can also interrupt previously established, normal menstrual cycles.[27] The presence of abnormal cycles reportedly ranges from

about 21% in top Norwegian athletes to 40% among Olympic competitors.[26]

Delays and interruptions in normal menstrual cycles appear to result from hormonal changes related to weight loss and to high levels of physical activity. Very high levels of physical activity are related to modified estrogen, LH, FSH, and other hormone levels.[26, 28, 29] Weight and hormonal status generally revert to normal after high levels of training diminish or end.

Some of the hormones involved in fertility impairments perform other important functions in the body, which may also be disrupted. Reduced levels of estrogen that accompany low levels of body fat and amenorrhea, for example, may decrease bone density and increase the risk of shortness and osteoporosis.[11]

Diet and Fertility

Certain dietary components appear to influence fertility by modifying estrogen, LH, and other hormone levels in women. There may be effects in males, too, but these have not been reported. Vegetarian diets, low fat intake, and high intakes of dietary fiber, caffeine, and alcohol are related to impaired fertility. The effects of several of these factors on fertility status are probably interrelated, but none directly causes infertility. They impair fertility largely by increasing the time it takes to conceive, due to menstrual cycle irregularities.[30]

Vegetarian Diets and Fertility

Women who regularly consume plant-based, low-fat, high-fiber diets (>25 g per day) and no meat appear to have lower circulating levels of estrogen and may be more likely to have irregular menstrual cycles than omnivores.[28,31,32] These results apply to vegetarians who are thin, normal weight, or overweight.[28] Irregularity makes ovulation sporadic and may increase the time it takes for conception.

Diets providing less than 20% of calories from fat appear to lengthen menstrual cycles among women in general,[33] and high-fiber diets may reduce estrogen levels.[28,34]

Since the mechanisms underlying these effects are not yet known, there is ample room for more study on the effects of dietary components on fertility in men and women.

Carotenemia and Fertility

Several studies indicate that women with *carotenemia* may experience amenorrhea and menstrual dysfunction.[35,36] Women who consume 12 mg or more of beta-carotene daily for over six weeks by constant munching on foods such as dried green pepper flakes or carrots

CAROTENEMIA   A condition, caused by ingestion of high amounts of carotenoids (or carotenes) from plant foods, in which the skin turns yellowish orange.

or by taking "tanning pills" tend to develop carotenemia. Skin color and fertility return to normal within two to six weeks after high levels of intake are discontinued. Why carotenemia may cause infertility is not known, and the relationship must be considered speculative until additional studies are undertaken. It is not clear that the carotene content of the diet, rather than other substances in plentiful supply in plant foods, accounts for the relationship observed.

Caffeine and Fertility

Should women concerned about infertility consume coffee and other foods with caffeine? It appears that high intakes of caffeine may prolong the time it takes to become pregnant. In a study of European women, researchers found that the chance of a 10-month interval from attempting conception to achieving conception was twice as likely among women who consumed over 4 cups of coffee, or approximately >500 mg caffeine, per day versus women who consumed little coffee.[37] Another study reported that intake of over 300 mg of caffeine daily from coffee, sodas, and tea decreased the chance of conceiving by 27% per cycle compared to negligible caffeine intake.[38] In both studies, the effect of caffeine on time to conception was stronger in women who smoked.

It is not known why caffeine appears to prolong time to conception. Nor is it clear if it is caffeine, one

Table 2.4
Caffeine content of foods and beverages.

FOODS AND BEVERAGES	CAFFEINE (mg)
Coffee, 1 c	
Drip	137–153
Percolated	97–125
Instant	61–70
Decaffeinated	0.5–4.0
Tea, 1 c	
Brewed 5 minutes	32–176
Instant	40–80
Soft Drinks, 12 oz	
Mountain Dew	54
Coca-Cola	46
Diet Coca-Cola	46
Dr. Pepper	40
Pepsi-Cola	38
Diet Pepsi-Cola	37
Gingerale	0
7-Up	0
Chocolate Products	
Cocoa, chocolate milk, 1 c	10–17
Milk chocolate, 1 oz	1–15
Chocolate syrup, 2 tbl	4

or more of the hundreds of other chemical substances in coffee, or an unexamined characteristic of women who consume lots of caffeine that accounts for the apparent effects of caffeine on fertility. The prudent course of action is to advise concerned women who consume more than several cups of regular coffee each day to cut down on coffee and other products high in caffeine. Table 2.4 shows the caffeine content of selected beverages and foods. Women who cut out coffee altogether should be advised to do so gradually, to reduce "caffeine withdrawal headaches" and fatigue.

Alcohol and Fertility

Alcohol may influence fertility by decreasing estrogen and testosterone levels and disrupting normal menstrual cycles. In a study of 430 Danish couples attempting pregnancy for six months, consumption of one to five alcohol-containing drinks per week by women was related to a 39% lower chance of conception. Alcoholic beverage consumption of over 10 drinks per week was related to a 66% reduction in the probability of conception during the six-month period.[39] Not all studies show alcohol intake affecting fertility, however, so the effect, if real, may be weak. Nevertheless, since alcohol should not be consumed during pregnancy due to the risk of fetal malformations and mental retardation, it makes sense that women who are attempting conception should not consume alcohol.

Other Factors Contributing to Infertility in Males

The genesis of infertility in males and females is not well understood, but less is known about factors that affect males' reproductive capacity than about females'. Examinations of fertility status in males tend to focus on sperm quality, assessed as sperm number (concentration), motility (movement), and morphology (shape).[2] A number of chromosomal abnormalities (*Klinefelter's syndrome,* for example) and environmental toxins have been related to modification of sperm. Certain drugs used for hypertension and cancer, disease processes such as diabetes and atherosclerosis (hardening of the arteries), and endocrine disorders that disrupt testosterone production affect male fertility primarily in ways that involve sperm production or erectile function.

Sperm production appears to be sensitive to a male's exposure to a number of nutritional and other environmental factors (see Table 2.3). Since sperm development occurs over 70 to 80 days, however, there is a delay of up to three months between onset of the exposure (such as use of a medication or nutrient toxicity) and the appearance of sperm abnormalities.

LOW ZINC STATUS Zinc intakes of 5 mg per day or less are associated with decreased *semen* volume and testosterone levels in experimental studies. Zinc plays a critical role in male reproduction potential, and the concentration of zinc is semen is high. Because zinc is concentrated in semen, it is lost via ejaculated semen. Zinc in seminal fluid is an essential cofactor for enzymes involved in testosterone production, DNA replication, protein synthesis, and cell division. It appears to protect sperm from bacteria and chromosomal damage.

Zinc also plays important roles in sexual organ development in males. Deficiency of zinc prior to and during early adolescence is associated with *hypogonadism* and lack of sexual development such as reduced growth of the penis, failure of the voice to deepen, and lack of muscle enlargement. Hypogonadism has been observed in some Middle Eastern males due to diets based on whole grains with infrequent consumption of animal products. Zinc in whole grains is poorly absorbed, whereas animal products generally contain good amounts of absorbable zinc.

Supplementation with 10 mg of zinc or more per day reverses zinc-related fertility and sexual maturational problems in males over time.[2,40]

ANTIOXIDANT NUTRIENTS Sperm are particularly susceptible to oxidative damage because of their high concentration of polyunsaturated fatty acids. Antioxidants such as selenium, vitamins C and E, beta-carotene, and other carotenoids protect sperm DNA from oxidative damage and promote normal sperm motility and function.[2]

ALCOHOL INTAKE Impaired fertility is common in alcoholics and is related to direct toxic effects of alcohol on the testes. Light to moderate alcohol consumption does not appear to affect fertility in males.[41]

HEAVY METALS Exposure to high levels of lead and mercury is related to decreased sperm production. Inhaled or ingested lead is transported to the pituitary gland where it appears to disrupt hormonal communications with the testes. The result is lowered testosterone levels and decreased sperm production and motility. The men most likely to be exposed to excess lead tend to be workers in smelting and battery factories.[41] Mercury can build up in fish living in contaminated waters. Ingestion of fish from waters contaminated with mercury in Hong Kong has been associated with decreased sperm count.[42] Consumption of fish from the U.S. Great Lakes does not appear to pose a similar problem.[43]

Exposure to excess levels of cadmium, manganese, boron, cobalt, copper, nickel, silver, or tin may also affect male fertility, but evidence from human studies is sparse. These metals may build up in male reproductive systems through the inhalation of fumes or dust containing particles or through contaminated water.[41]

HALOGENS Occupational exposure to pesticides made from halogen compounds (such as dibromochloropropane) has been observed to cause sperm count reduction and male infertility.[41]

GLYCOLS Glycols are widely used in antifreeze, solvents, and de-icers for airplanes. Occupational exposures to these compounds can decrease fertility in males threefold.[41]

KLINEFELTER'S SYNDROME A congenital abnormality in which testes are small and firm, legs abnormally long, and intelligence generally subnormal.

SCROTUM A muscular sac containing the testes.

SEMEN The penile ejaculate containing a mixture of sperm and secretions from the testes, prostate, and other glands. It is rich is zinc, fructose, and other nutrients. Also called seminal fluid.

HYPOGONADISM Atrophy or reduced development of testes or ovaries. Results in immature development of secondary sexual characteristics.

HORMONES   Exposure to synthetic estrogens in pharmaceutical factories and to the estrogen-like substances of DDT (dichlorodiphenlytrichloroethane) and PCBs (polychlorinated biphenyls) has been found to

VENOUS THROMBOEMBOLISM   A blood clot in a vein.

reduce libido and increase impotence and breast size in males.[41] It has been speculated that increased exposure to estrogenic compounds and other pollutants may be responsible for the reported decrease in sperm count in men over the last 50 years.[41] This observation begs a closer look at changes in sperm count over recent decades.

Is sperm count declining in men? The issue is not settled. But the answer to this question appears to be changing from "yes" to "no." Approximately 17% of males have low sperm counts, and several studies have noted declines in the number of sperm produced of about 1 to 2% per year over recent decades.[44] Recent studies, however, have not found such a decline and conclude that popular press headlines about declines in sperm counts are unjustified.[44,45]

HEAT   Elevating the temperature of the scrotum and testes can reduce sperm count. Long-haul truck driving, welding, and foundry work may cause increased temperatures and decreased fertility in some men.[41] Prolonged and frequent exposure to hot tub water is also thought to reduce sperm count in men.

STEROID ABUSE   Doses of steroids (e.g., testosterone) used by some body builders are up to 40 times higher than therapeutic doses. Side effects of steroid abuse are multiple and include atrophy (shrinking) of the testicles, absence of sperm, and decreased libido. Fertility generally returns to normal after steroid use ends, but other disease risks may linger.[46]

NUTRITION AND CONTRACEPTIVES

The contraceptive revolution emerged in full force in the 1960s when use of pills with heavy doses of estradiol, the most biologically active form of estrogen, and progesterone became widespread. The adverse side effects of these oral contraceptives were plentiful and included increased risk of heart attack and stroke, elevated blood lipids, glucose intolerance, weight gain, and folate and vitamin B$_6$ deficiencies. New generations of oral contraceptives employed increasingly lower hormone doses, and side effects diminished substantially. Nevertheless, some remain. Those related to use of the newer contraceptives are listed in Table 2.5. The current generation of oral contraceptives elevates

some blood lipids and carries a slight risk of **venous thromboembolism**. Their continuing effect on the formation of blood clots appears to be related to the progesterone content of the pills.[48]

The newest generation of fertility control products for females includes contraceptive implants, patches, and injections. These, too, have a number of nutritional side effects. Development of hormonal contraceptive methods for control of fertility in males lags far behind advances in female contraception. Oral contraceptives that suppress sperm production are being tested, however, and products may emerge on the market by 2005.

Oral Contraceptives and Nutritional Status

Many types of birth control pills containing low doses of different forms of estrogen and progesterone are currently available. Use of these pills by healthy, nonsmoking women is not associated with an increased risk of heart attack, glucose intolerance, or nutrient deficiencies, nor is use related to weight gain. Their use is, however, associated with a three-to-fourfold increased risk of blood clot formation in veins of the lungs and other parts of the body.[49] The new generation of oral contraceptives increases blood levels of triglycerides by about 30% and total cholesterol levels by approximately 6% on average. HDL cholesterol—the "good" blood cholesterol fraction—is increased slightly by these contraceptives.[50]

Oral contraceptives still decrease blood levels of certain nutrients. Blood levels of vitamin B$_{12}$ were found to be an average of 33% lower in oral contraceptive users versus nonusers in a Canadian study of 14- to 20-year-old females.[51] Serum copper levels

Connections
Better Nutrition and Higher Fertility in Animals

• Animal studies, which may not apply to humans, show that "flushing," or providing pigs with plentiful food during the follicular phase, can increase the likelihood of an above-average number of eggs being released and bigger litters in gilts (young female pigs).

• Animals fed on a higher plane of nutrition (i.e., increased feeding level) prior to mating mature sooner, have more body fat, and ovulate more often than animals fed less.

• Low-fat diets (10% fat) in sheep and goats decrease egg release to below an average of one per cycle, whereas higher-fat diets (20–30% fat) increase the number of eggs released at one time to two to three on average.

• It is suggested that feeding gilts and sheep on a high plane of nutrition increases glucose and nutrient availability to the ovary and that stimulates the maturation of more than one egg at a time.[47]

Whether these intriguing findings parallel biological effects on human fertility awaits additional research.

Table 2.5   Nutrition-related side effects of contraceptives.

Oral Contraceptives
Increased blood levels of HDL cholesterol (the "good" cholesterol)
Increased blood levels of triglycerides and LDL cholesterol
Increased risk of venous thromboembolism (blood clots)
Decreased blood levels of vitamin B_{12}
Increased blood levels of copper

Contraceptive Injections (Depo-Provera)
Weight gain
Increased blood levels of LDL cholesterol and insulin
Decreased blood levels of HDL cholesterol
Decreased bone density

Contraceptive Implants (Norplant)
Weight gain

were 34 to 55% higher and may be related to the increased risk of blood clot formation observed among oral contraceptive users.[52]

It is recommended that females who are obese, are over the age of 35 years and smoke, have cardiovascular disease, or are immobilized use nonhormone methods of contraception because of their increased risk of venous thromboembolism.[50] It is generally also recommended that women stop using oral contraceptive pills about three months prior to attempting pregnancy.

Contraceptive Injections

DMPA (depot medroxyprogesterone acetate), or Depo-Provera, is the primary type of injectable contraception used in females. Injections are given six to eight weeks apart and then every three months. Although highly effective, Depo-Provera has discontinuation rates that average over 50% within the first year of use. Weight gain is a leading reason for discontinuation (27%); irregular periods (24%), fatigue (23%), headache (25%), and abdominal pain (18%) are also commonly reported reasons for discontinuation.[53,54] Weight gain averages 12 pounds during the first year of Depo-Provera use,[54] but not all studies report weight gain.[55] Long-term use of this contraceptive is also related to decreased bone density and blood levels of HDL cholesterol and increased levels of LDL cholesterol (the "bad" cholesterol) and insulin.[56,57]

Contraceptive Implants

Norplant (levonorgestral) is the leading contraceptive implant and prevents contraception for up to seven years.[58] This contraceptive method is highly effective for normal-weight and underweight women, as evi-

denced by a 1.9% pregnancy rate.[59] Pregnancy rates are 4.2% in obese women, however.[60] High rates of side effects, especially erratic bleeding (69%), weight gain (41%), and headaches (30%), lead to early removal of the implant in about half of users.[61,62] Average weight gain one year after the implant has been reported to be 9 pounds.[63]

Contraceptive Patches

The newest form of female contraception is a patch that releases a type of estrogen and progesterone. Tests on humans indicate it is highly effective and easy to use. The patch is placed on the abdomen for three weeks and then taken off for a week.[64] Side effects are yet to be reported, but may mimic those for oral contraceptive pills.

OTHER PRECONCEPTUAL NUTRITION CONCERNS

Approximately eight to ten days after an ovum is fertilized, it implants into the uterine wall. Within the first month after conception, the developing *embryo* will have grown from a single cell to millions of cells, basic structures of organs will have formed, and the blueprint for future growth and development will have been established. All this often happens before women know they are pregnant or attend a prenatal clinic. The time to establish a state of optimal health and nutritional status is before conception.

EMBRYO   The developing organism from conception through eight weeks.

FETUS   The developing organism from eight weeks after conception to the moment of birth.

Very Early Pregnancy Nutrition Exposures

Table 2.6 summarizes major nutritional exposures that adversely affect the growth and development of the embryo and *fetus*.[65] It is important to be aware that any of these conditions, if present preconceptionally, may impair embryonic and fetal growth and development. This chapter touches on the importance of adequate folate intake prior to conception. Chapter 4, on nutrition and pregnancy, expands the folate discussion and returns to the other topics listed in Table 2.6.

Folate Status Prior to Conception and Neural Tube Defects

Folate status prior to conception is an important concern because inadequate folate very early in pregnancy can cause *neural tube defects*. These defects develop within 21 days after conception—or before many women even know they are pregnant, and well before prenatal care begins.[66] Knowledge of the folate–neural tube defect relationship, and awareness that folate intake is inadequate in many women of childbearing age, prompted efforts to increase folate intake. In particular, efforts are focused on encouraging women to consume more folic acid, a highly absorbable form of this B vitamin. In 1998, that task was made easier when the Food and Drug Administration mandated that refined grains products such as white bread, grits, crackers, rice, and pastas be fortified with folic acid.

NEURAL TUBE DEFECTS   Spina bifida and other malformation of the neural tube. Defects result from incomplete formation of the neural tube during the first month after conception.

Although intake of folic acid from fortified foods is increasing, there are still concerns that some women are unaware of the importance of consuming these foods or other sources of folate prior to pregnancy. A 2000 Gallup Poll commissioned by the March of Dimes found that only one in seven women was aware that folate decreases the risk of neural tube defects, and only one in ten knew it was needed in adequate amounts prior to pregnancy.

Women can get enough folate by consuming a good basic diet and a fully fortified breakfast cereal (Smart Start, Total, Product 19, for example) or a regular breakfast cereal (Cheerios, Corn Flakes, Raisin Bran, etc.) and six to eight servings of refined grain products each day. Folic acid supplements (400 mcg per day) can also provide folic acid.

Recommended Dietary Intakes for Preconceptional Women

Recommendations for food and nutrient intakes for women who may become pregnant differ from those for adult women in general in several ways. It is recommended that women who may become pregnant consume 400 mcg of folic acid from fortified foods or supplements, take no more than 5000 IU of vitamin A (retinol or retinoic acid) from supplements daily, and limit or omit alcohol-containing beverages.[65] Recommendations for nutrient intake given in the Dietary Reference Dietary Intakes (DRIs), listed on the inside front covers of this text, should be applied, with careful attention paid to the Tolerable Upper Intake Levels listed in a separate table.

Food selection prior to pregnancy should be based on the Food Guide Pyramid recommendations for adults (see Illustration 2.3). The number of servings from each food group within the range of ser-

Table 2.6   Nutritional exposures before and very early in pregnancy that disrupt fetal growth and development.

Weight Status
- Underweight increases the risk of maternal complications during pregnancy and the delivery of small and early newborns.
- Obesity increases the risk of clinical complications during pregnancy and delivery of newborns with neural tube defects or excessive body fat.

Nutrient Status
- Insufficient folate intake increases the risk of embryonic development of neural tube defects.
- Excessive vitamin A intake (retinol, retinoic acid) increases the risk the fetus will develop facial and heart abnormalities.
- High maternal blood levels of lead increase the risk of mental retardation in the offspring.
- Iodine deficiency early in pregnancy increases the risk that children will experience impaired mental and physical development.
- Iron deficiency increases the risk of early delivery and development of iron deficiency in the child within the first few years of life.

Alcohol
- Regular intake of alcohol increases the risk of "fetal alcohol syndrome" and "fetal alcohol effects," both which include impaired mental and physical development.

Diabetes
- Poorly controlled blood glucose levels early in pregnancy increase the risk of fetal malformations, excessive infant size at birth, and the development of diabetes in the offspring later in life.

vings given will be determined primarily by a woman's physical activity level and body size.

Herbal Remedies for Fertility-Related Problems

Herbal remedies are increasingly being used to treat fertility-related problems, but much remains to be learned about their safety and effectiveness. Studies have not yet assessed the safety of herbal remedies used for menstrual disorders, such as black cohosh, evening primrose oil, and natural estrogens and progesterone. We do not know with certainty that these remedies are effective or that they and other herbal remedies do not adversely affect embryonic and fetal development.

Women who choose to use these products for fertility-related problems should inform their health care provider and avoid using the herbal remedies if conception is possible.

MODEL PRECONCEPTIONAL NUTRITION PROGRAMS

Two model programs related to nutrition during the preconceptional period, one in the United States and one in Indonesia, are highlighted in this section.

Preconceptional Benefits of WIC

WIC (Special Supplemental Nutrition Program for Women, Infants, and Children) is a nationwide, supplemental food and nutrition education program for pregnant and breastfeeding women and children up to age five. Women normally enter the program while pregnant, but in a model program in California, women were experimentally provided WIC food supplements

Illustration 2.3 Food Guide Pyramid recommendations for preconceptional women.

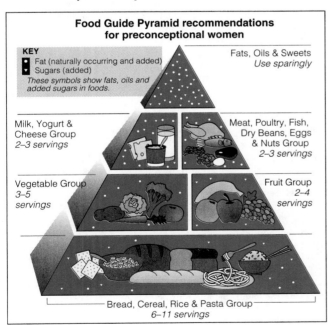

Food Guide Pyramid recommendations for preconceptional women

KEY
- Fat (naturally occurring and added)
- Sugars (added)

These symbols show fats, oils and added sugars in foods.

Fats, Oils & Sweets
Use sparingly

Milk, Yogurt & Cheese Group
2–3 servings

Meat, Poultry, Fish, Dry Beans, Eggs & Nuts Group
2–3 servings

Vegetable Group
3–5 servings

Fruit Group
2–4 servings

Bread, Cereal, Rice & Pasta Group
6–11 servings

FOOD GROUP	RECOMMENDED NUMBER OF SERVINGS
Bread, cereal, rice, and pasta	6–11
Vegetables	3–5
Fruits	2–4
Milk, yogurt, and cheese	2–3
Meat, poultry, fish, dry beans, eggs, and nuts	2–3

and nutrition education during *and* between consecutive pregnancies. Women who received WIC benefits during one pregnancy through to the first two months of the next pregnancy had better iron status and delivered newborns with higher birthweights and greater lengths than women who received WIC benefits during pregnancy only.[67] The study demonstrates that low-income women at nutritional risk benefit from WIC services before, as well as during, pregnancy.

Decreasing Iron Deficiency in Preconceptional Women in Indonesia

Approximately one in every two women in Indonesia experiences iron-deficiency anemia during pregnancy. In a unique effort to prevent this problem, the Ministry of Health initiated regulations that require a couple applying for a marriage license to receive advice on iron status from those dispensing the license. All women are now advised to take 30–60 mg of iron along with folic acid in a supplement. Of 344 women studied after the program was initiated, 98% reported that they had purchased and taken iron-folate tablets; 56% had taken at least 30 tablets. The incidence of iron deficiency in this group of women dropped by almost half.[68]

Preconception Care

Increasingly, routine health care visits and educational sessions are being recommended and introduced into health care organizations. Services focus on risk assessment of behaviors such as weight status, dietary intake, folate and iron status, and vitamin, mineral, and herbal supplement use, and on the presence of diseases such as diabetes, hypertension, infection, and genetic traits that may be transmitted to offspring. Psycho-social needs should also be addressed as part of preconceptional care, and referrals made to appropriate services for issues such as eating disorders, abuse, violence, or lack of food or shelter.[69] The desire of couples planning for pregnancy to have a healthy newborn makes the preconceptional period a prime time for positive behavioral changes. It presents opportunities to make lasting improvements in the health and well-being of individuals and families.

Starting pregnancy in the best health status possible can make an important difference to reproductive outcomes. It should be recognized, however, that even in ideal conditions, continued infertility, early pregnancy loss, fetal malformations, and maternal complications will sometimes occur.

Resources

WWW RESOURCES

The National Center for Health Statistics provides information on fertility and birth rates, and other vital statistics data. *Web site:* www.cdc.gov/nchs

Merck Manual of Diagnosis and Therapy

This ever-popular and useful guide is available online and includes information on infertility and related conditions. *Web site:* www.merckmanual.org

Medscape Women's Health Journal

Receive automatic updates on women's health, fertility, and contraception topics by subscribing to this free online journal. *Web site:* www.medscape.com/medscape/WomensHealth/journal

The Babycenter Company

A popular source of consumer advice on preconception planning, nutrition, and related topics, which is, however, heavily infiltrated with advertisements. *Web site:* www.babycenter.com

Public Health Service

A gateway to consumer and services information on fertility clinics and women's health from the U.S. government. *Web site:* www.healthfinder.org/justforyou/women/default.htm

References

1. Ventura SJ, Martin JA, Curtin SC, et al. Births: final data for 1998. Natl Vital Stat Rep 2000;48:1–100.

2. Wong WY, Thomas CM, Merkus JM, et al. Male factor subfertility: possible causes and the impact of nutritional factors. Fertil Steril 2000;73:435–42.

3. te Velde ER, Cohlen BJ. The management of infertility [editorial]. N Engl J Med 1999;340:224–6.

4. Ford WC, North K, Taylor H, et al. Increasing paternal age is associated with delayed conception in a large population of fertile couples: evidence for declining fecundity in older men. The ALSPAC Study Team (Avon Longitudinal Study of Pregnancy and Childhood). Hum Reprod 2000;15:1703–8.

5. Wilcox AJ, Baird DD, Weinberg CR. Time of implantation of the conceptus and loss of pregnancy. N Engl J Med 1999;340:1796–9.

6. Wilcox AJ, Weinberg CR, O'Connor JF, et al. Incidence of early loss of pregnancy. N Engl J Med 1988;319:189–94.

7. Kallen B. Epidemiology of human reproduction. Boca Raton, FL: CRC Press; 1988.

8. Warren MP. Effects of undernutrition on reproductive function in the human. Endocrine Reviews 1983;4:363–77.

9. Wiley AS. The ecology of low natural fertility in Ladakh. J Biosoc Sci 1998;30:457.

10. Bongaarts J. Does malnutrition affect fecundity? A summary of evidence. Science 1980;208:564–9.

11. van der Spuy ZM. Nutrition and reproduction. Clin Obstet Gynecol 1985;12:579–604.

12. Wegman ME. Infant mortality: some international comparisons. Pediatrics 1996;98:1020–7.

13. Rao VG, Sugunan AP, Sehgal SC. Nutritional deficiency disorders and high mortality among children of the Great Andamanese tribe. Natl Med J India 1998;11:65–8.

14. Wood JW. Maternal nutrition and reproduction: why demographers and physiologists disagree about a fundamental relationship. Ann NY Acad Sci 1994;709:101–16.

15. Hirschman C. Why fertility changes. Annu Rev Sociol 1994;20:203–33.

16. Ramakrishnan U, Barnhart H, Schroeder DG, et al. Early childhood nutrition, education and fertility milestones in Guatemala. J Nutr 1999;129:2196–202.

17. Pena R, Liljestrand J, Zelaya E, et al. Fertility and infant mortality trends in Nicaragua 1964–1993. The role of women's education. J Epidemiol Community Health 1999;53:132–7.

18. Kolata GB. Kung hunter-gatherers: feminism, diet, and birth control. Science 1974;185:932–4.

19. Leslie PW, Campbell KL, Little MA, et al. Evaluation of reproductive function in Turkana women with enzyme immunoassays of urinary hormones in the field. Hum Biol 1996;68:95–117.

20. Stein Z, Susser M, Saenger G, et al. Famine and human development: the Dutch hunger winter of 1944–1945. New York: Oxford University Press, 1975: pp 119–48.

21. Zaadstra BM, Seidell JC, Van Noord PA, et al. Fat and female fecundity: prospective study of effect of body fat distribution on conception rates. BMJ 1993;306:484–7.

22. Nylander PPS. The phenomenon of twinning. In: Baron SL, Thomson AM, eds. Obstetrical epidemiology. London: Academic Press; 1983: pp 143–65.

23. Bates GW. Body weight control practice as a cause of infertility. Clin Obstet Gynecol 1985;28:632–44.

24. Thomas AK, Mander J, Hale J, et al. Induction of ovulation with subcutaneous pulsatile gonadotropin-releasing hormone: correlation with body weight and other parameters. Fertil Steril 1989;51:786–90.

25. Keys A, Brozek J, Henschel A, et al. The biology of human starvation. Minneapolis: University of Minnesota Press; 1950.

26. Oian P, Augestad LB, Molne K, et al. Menstrual dysfunction in Norwegian top athletes. Acta Obstet Gynecol Scand 1984;63:693–7.

27. Cumming DC, Wheeler GD, Harber VJ. Physical activity, nutrition, and reproduction. Ann NY Acad Sci 1994;709:55–76.

28. Feng W, Marshall R, Lewis-Barned NJ, et al. Low follicular oestrogen levels in New Zealand women consuming high fibre diets: a risk factor for osteopenia? NZ Med J 1993;106:319–22.

29. Kraemer RR, Heleniak RJ, Tryniecki JL, et al. Follicular and luteal phase hormonal responses to low-volume resistive exercise. Med Sci Sports Exer 1995;27:809–17.

30. Brown JE. Preconceptional nutrition and reproductive outcomes. Ann NY Acad Sci 1993;678:286–92.

31. Hill PB, Garbaczewski L, Daynes G, et al. Gonadotropin release and meat consumption in vegetarian women. Am J Clin Nutr 1986; 43:37–41.

32. Wyshak G, Snow RC. Fiber consumption and menstrual regularity in young women. J Women's Health 1993;2:295–9.

33. Reichman ME, Judd JT, Taylor PR, et al. Effect of dietary fat on length of the follicular phase of the menstrual cycle in a

controlled diet setting. J Clin Endocrinol Metab 1992;74:1171–5.

34. de Ridder CM, Thijssen JHH, Van 't Veer P, et al. Dietary habits, sexual maturation, and plasma hormones in pubertal girls: a longitudinal study. Am J Clin Nutr 1991;54:805–13.

35. Kemmann E, Pasquale SA, Skaf R. Amenorrhea associated with carotenemia. JAMA 1983;249:926–9.

36. Wentz AC. Body weight and amenorrhea. Obstet Gynecol 1980;56:482–7.

37. Bolumar F, Olsen J, Rebagliato M, et al. Caffeine intake and delayed conception: a European multicenter study on infertility and subfecundity. European study group on infertility subfecundity. Am J Epidemiol 1997;145:324–34.

38. Hatch EE, Bracken MB. Association of delayed conception with caffeine consumption. Am J Epidemiol 1993;138:1082–92.

39. Jensen J, Lendorf A, Stimpel H, et al. The prevalence and etiology of impotence in 101 male hypertensive outpatients. Am J Hypertens 1999;12:271–5.

40. Hunt CD, Johnson PE, Herbel J, et al. Effects of dietary zinc depletion on seminal volume and zinc loss, serum testosterone concentrations, and sperm morphology in young men. American J Clin Nutr 1992; 56:148–57.

41. Lamb EJ, Bennett S. Epidemiologic studies of male factors in infertility. Ann NY Acad Sci 1994;709:165–78.

42. Dickman MD, Leung KM. Mercury and organochlorine exposure from fish consumption in Hong Kong. Chemosphere 1998;37:991–1015.

43. Buck GM, Sever LE, Mendola P, et al. Consumption of contaminated sport fish from Lake Ontario and time-to-pregnancy. New York State Angler Cohort. Am J Epidemiol 1997;146:949–54.

44. Sokol RZ, Shulman P, Paulson RJ. Comparison of two methods for the measurement of sperm concentration. Fertil Steril 2000;73:591–4.

45. Joffe M. Time trends in biological fertility in Britain. Lancet 2000;355:1961–5.

46. Lloyd FH, Powell P, Murdoch AP. Anabolic steroid abuse by body builders and male subfertility. BMJ 1996;313: 100–1.

47. Robinson JJ. Nutrition in the reproduction of farm animals. Nutr Res Rev 1990;3:253–76.

48. Vandenbroucke JP, Rosing J, Bloemenkamp KW, et al. Oral contraceptives and the risk of venous thrombosis. N Engl J Med 2001;344:1527–35.

49. Skegg D. Oral contraceptives and venous thromboembolism. NZ Med J 2000;113:385.

50. Crook D. The metabolic consequences of treating postmenopausal women with non-oral hormone replacement therapy. Br J Obstet Gynaecol 1997;104Suppl16:4–13.

51. Green TJ, Houghton LA, Donovan U, et al. Oral contraceptives did not affect biochemical folate indexes and homocysteine concentrations in adolescent females. J Am Diet Assoc 1998;98:49–55.

52. Berg G, Kohlmeier L, Brenner H. Use of oral contraceptives and serum beta-carotene. Eur J Clin Nutr 1997;51:181–7.

53. Polaneczky M, Guarnaccia M, Alon J, et al. Early experience with the contraceptive use of depot medroxyprogesterone acetate in an inner-city clinic population. Fam Plann Perspect 1996;28:174–8.

54. Matson SC, Henderson KA, McGrath GJ. Physical findings and symptoms of depot medroxyprogesterone acetate use in adolescent females. J Pediatr Adolesc Gynecol 1997;10:18–23.

55. Pelkman CL, Chow M, Heinbach RA, et al. Short-term effects of a progestational contraceptive drug on food intake, resting energy expenditure, and body weight in young women. Am J Clin Nutr 2001;73:19–26.

56. Archer B, Irwin D, Jensen K, et al. Depot medroxyprogesterone. Management of side-effects commonly associated with its contraceptive use. J Nurse Midwifery 1997;42:104–11.

57. Westhoff C. Depot medroxyprogesterone acetate contraception. Metabolic parameters and mood changes. J Reprod Med 1996;41:401–6.

58. Sivin I, Mishell DR, Jr., Diaz S, et al. Prolonged effectiveness of Norplant(R) capsule implants: a 7-year study. Contraception 2000;61:187–94.

59. Meirik O, Farley TM, Sivin I. Safety and efficacy of levonorgestrel implant, intrauterine device, and sterilization. Obstet Gynecol 2001;97:539–47.

60. Kaunitz AM. Long-acting hormonal contraception: assessing impact on bone density, weight, and mood. Int J Fertil Womens Med 1999;44:110–7.

61. Haugen MM, Evans CB, Kim MH. Patient satisfaction with a levonorgestrel-releasing contraceptive implant. Reasons for and patterns of removal. J Reprod Med 1996;41:849–54.

62. Harel Z, Biro FM, Kollar LM, et al. Adolescents' reasons for and experience after discontinuation of the long-acting contraceptives Depo-Provera and Norplant. J Adolesc Health 1996;19:118–23.

63. Berenson AB, Wiemann CM, Rickerr VI, et al. Contraceptive outcomes among adolescents prescribed Norplant implants versus oral contraceptives after one year of use. Am J Obstet Gynecol 1997;176:586–92.

64. Audet MC et al. Evaluation of contraceptive efficacy and cycle control of transdermal contraceptive patch vs. an oral contraceptive. JAMA 2001;285:2347–54. Shangold G. Phase II/III trials of the 17-deacetylnorgestimate, ethinyl estradiol patch. Am J Obstet Gynecol 2000.

65. National Academy of Sciences. Nutrition during pregnancy. I. Weight gain. II. Nutrient supplements. Washington, DC: National Academy Press; 1990.

66. Eskes TK. Open or closed? A world of difference: a history of homocysteine research. Nutr Rev 1998;56:236–44.

67. Caan B, Horgen DM, Margen S, et al. Benefits associated with WIC supplemental feeding during the interpregnancy interval. Am J Clin Nutr 1987;45:29–41.

68. Jus'at I, Achadi EL, Galloway R, et al. Reaching young Indonesian women through marriage registries: an innovative approach for anemia control. J Nutr 2000;130:456S–8S.

69. Dixon D, et al. The first prenatal visit. Clin Rv 2000;10:53–74.

CHAPTER 3

Photo Disc

The importance of nutrition to the course and outcome of pregnancy doesn't begin at conception. It is part of a continuum of physiologic events. . . . Women's nutritional status before conception influences physiologic events during pregnancy, and nutrition during pregnancy sets the stage for meeting nutritional needs during lactation.

J. King[1]

PRECONCEPTION NUTRITION:

Conditions and Interventions

Prepared by **Judith E. Brown**

CHAPTER OUTLINE

- Introduction
- Premenstral Syndrome
- Obesity and Fertility
- Eating Disorders and Fertility
- Diabetes Mellitus Prior to Pregnancy
- Polycystic Ovary Syndrome
- Inborn Errors of Metabolism

KEY NUTRITION CONCEPTS

1 Nutritional and health status directly before and during the first two months after conception influences embryonic development and the risk of complications during pregnancy.

2 Nutrition therapy plays an important role in the management of a number of conditions that affect preconceptional health and reproductive outcomes.

INTRODUCTION

This chapter addresses specific nutrition-related conditions of women before conception and during the *periconceptional period,* and the interventions that address them. Conditions presented here are relatively rare but have important implications for health and well-being and for reproductive outcomes. We begin with a discussion of premenstrual syndrome and progress to obesity, eating disorders, diabetes, polycystic ovary syndrome, and inborn errors of metabolism. Then we address the role of complementary and alternative medical therapy in the treatment of disorders of menstruation.

PREMENSTRUAL SYNDROME

It wasn't until 1987 that PMS, or *premenstrual syndrome,* moved from the psychogenic disorder section of medical textbooks to chapters on physiologically based problems. It is diagnosed according to criteria stipulated in the fourth version of the Diagnostic and Statistical Manual of Mental Disorders (DSM IV). A standard questionnaire, rather than physical examination or laboratory tests, is used to diagnose PMS.

PMS is characterized by severe, life-disrupting physiological and psychological changes that begin in the luteal phase of the menstrual cycle and end with menses (menstrual bleeding). Common physical signs and psychological symptoms of PMS are listed in Table 3.1. For a diagnosis, at least five signs or symptoms of PMS intense enough to disrupt work or social life must be present in three consecutive luteal phases.

Some symptoms of PMS occur in about 40% of women of childbearing age, and they are severe enough to be considered PMS in about 5 to 10% of these women.[2,3] Although up to 70% of women suffer through monthly bouts of lower abdominal cramps, bloating, back pain, headache, food cravings, or irritability, these conditions do not qualify as PMS. They are considered to represent *dysmenorrhea,* a condition related to prostaglandin release near to and during menses.[4]

PERICONCEPTIONAL PERIOD   Around the time of conception, generally defined as the month before and the month after conception.

PREMENSTRUAL SYNDROME   (*Premenstrual* = the period of time preceding menstrual bleeding; *syndrome* = a constellation of symptoms) A condition occurring among women of reproductive age that includes a group of physical, psychological, and behavioral symptoms with onset in the luteal phase and subsiding with menstrual bleeding. Also called premenstrual dysphoric disorder (PMDD).

DYSMENORRHEA   Painful menstruation due to abdominal cramps, back pain, headache, and/or other symptoms.

Table 3.1   Common signs and symptoms of PMS.[1,2]

Physical Signs	Psychological Symptoms
• Fatigue	• Craving for sweet or salty foods
• Abdominal bloating	• Depression
• Swelling of the hands or feet	• Irritability
• Headache	• Mood swings
• Tender breasts	• Anxiety
• Nausea	• Social withdrawal

The cause of PMS in unknown but is thought to be related to abnormal serotonin activity following ovulation.[2] Almost all remedies tested for PMS show about a 30% decline in symptoms with placebo, so an effective treatment must bring relief to an even higher proportion of women receiving it in experimental studies. Serotonin re-uptake inhibitors, which are the active ingredient in some types of antidepressants, effectively reduce PMS symptoms.[2] Decreased caffeine intake, exercise and stress reduction; magnesium, calcium, or vitamin B$_6$ supplements; and a number of herbal remedies are also used to treat PMS.

Caffeine Intake and PMS

To decrease PMS symptoms, it is commonly recommended that women reduce their intake of coffee and other beverages high in caffeine. This recommendation is holding up with time. A study at the University of Oregon demonstrated that PMS symptoms in college women increased in severity as coffee intake increased from one cup to 8 to 10 cups a day.[5] Risk of severe symptoms were eight times higher in women consuming the highest average daily amount of coffee (8 to 10 cups) compared to non-coffee drinkers.

Exercise and Stress Reduction

Increasing daily physical activity and reducing daily stressors appear to diminish PMS symptoms in many women.[3,6] Regular physical activity tends to improve energy level, mood, and feeling of well-being in women with PMS. Relieving stress, by such techniques as sitting comfortably and quietly with eyes closed while relaxing deep muscles, breathing through the nose, and exhaling while silently saying a word such as "one," appears to decrease symptoms. When done for 15 to 20 minutes twice daily over five months, this exercise was associated with a 58% improvement in PMS symptoms.[6]

Magnesium, Calcium, and Vitamin B$_6$ Supplements and PMS Symptoms

Magnesium and calcium, but not vitamin B$_6$, supplements appear to decrease specific symptoms of PMS in many women. Mechanisms underlying improvements are incompletely understood.

MAGNESIUM Magnesium supplements of 200 mg per day given during two cycles have been shown to decrease swelling, breast tenderness, and abdominal bloating symptoms of PMS. The beneficial response to magnesium was seen during the second month of treatment.[7] The 200 mg daily dose of magnesium is below the Tolerable Upper Intake Level (UL) for magnesium of 350 mg daily and is therefore considered safe.

CALCIUM Calcium supplements of 1200 mg per day for three cycles were found to reduce the PMS symptoms of irritability, depression, anxiety, headaches, and cramps by 48%, versus a reduction of 30% in the placebo group. The effect of calcium increased with duration of supplement use.[8] The UL for calcium is 2500 mg per day.

VITAMIN B$_6$ Initial studies of the effectiveness of supplemental vitamin B$_6$ for PMS treatment appeared promising. Subsequent studies, however, indicate that B$_6$ supplementation improves symptoms to about the same extent as do placebos.[9] It is no longer considered an effective treatment, and warnings have been issued about the toxic effects of supplemental levels of B$_6$ of 200 mg/day or more. These doses of vitamin B$_6$ are related to the development of neurological changes such as numbness in the hands and feet, and a burning and tingling sensation in the fingers and skin.[5] The UL for vitamin B$_6$ is 100 mg per day in adults.

HERBAL REMEDIES FOR PMS "Vitex," or Chastetree (named after the plant monks were said to have chewed to inspire chastity), shows promise for safely relieving PMS symptoms. Side effects include headache, cramps, and skin itching and rash. It is not yet considered safe for women who may become or who are pregnant, or who are taking oral contraceptive pills.[10] Evening primrose oil, which contains high amounts of the essential fatty acids linoleic and alpha-linolenic acid, does not appear to beat a placebo in term of PMS relief.[11] Black cohosh extract appears to improve PMS symptoms through effects on hormone activity. It causes few side effects at levels of 40 mg per day, but should not be taken by women who may become pregnant.[12]

OBESITY AND FERTILITY

Obesity in men and women impairs fertility. Obesity in men is associated with reduced *sex hormone binding globulin* (SHBG) levels and sperm count.[14] Many obese women experience highly irregular, anovulatory, or no menstrual cycles.[15,16] Obesity in women is related to reduced SHBG level, also, and to increased estrogen, blood glucose, or insulin levels.[17,18] Weight gain in women may trigger polycystic ovary syndrome, a condition characterized by low SHBG, anovulation, and elevated insulin level. (This condition is described later in this chapter.)

Central Body Fat and Fertility

The presence of central body obesity, indicated by a waist circumference of 38 inches or greater, is a strong risk factor for impaired fertility. In a study of women attending an artificial insemination clinic due to infertility in their partners, conception occurred in just half as many women with high central body fat as in other women.[19] In general, it takes women with high central body stores longer to become pregnant than it does women with normal levels of central fat.[20]

> **SEX HORMONE BINDING GLOBULIN** A protein that binds with the "sex hormones" testosterone and estrogen. Also called steroid hormone binding globulin, as testosterone and estrogen are produced from cholesterol and are therefore considered to be "steroid hormones." These hormones are inactive when bound to SHBG, but are available for use when needed. Low levels of SHBG are related to increased availability of testosterone and estrogen in the body.[13]

Weight loss should be the first therapy option for men and women who are obese and infertile.[16] (Read about one woman's experience with weight loss and fertility in Case Study 3.1.)

Studies of both women and men have demonstrated that weight loss of 7 to 22 pounds in women with BMIs over 25 kg/m^2, and of 100 pounds in massively obese men are related to a return of fertility in a majority of study participants.[13,15–18] Weight loss in the four studies of women was accomplished by diet and exercise; the study of massively obese men used gastric bypass surgery. Weight loss produced a drop in testosterone and increase in SHBG in males, and increased SHBG and decreased estrogen, glucose, and insulin levels in women.

Weight loss is considered the first therapeutic option for infertility in obese people in part because it is less costly than medications and has many health

Photo Disc

Case Study 3.1
Anna Marie's Tale

Exercise can be bad for you—or at least it was for Anna Marie. She and her husband Mark already had two delightful children, full time jobs, and hectic schedules. Mark wanted more children, but Anna Marie was dead-set against it. Mark refused to use contraception and made Anna Marie promise not to use any, either. Anna Marie had made the promise because she thought she could avoid pregnancy by staying at her weight of 210 pounds. At this weight, Anna Marie seldom had a menstrual period and figured the odds of conception were slim. For two years Anna Maria's plan for avoiding conception worked.

As the children grew older Anna Maria found she had a bit of free time, and she used it to indulge her love of swimming. Within months of beginning her program of swimming regularly, however, Anna Marie abandoned it. Her menstrual periods had become regular and her contraception method was lost.

Anna Marie's weight at 210 pounds had remained stable during the months she had swum. It appeared that her improved level of physical fitness and body fat improved her fertility status, too.

benefits, and also because hormone therapy often does not work in the presence of obesity.[16,21]

EATING DISORDERS AND FERTILITY

Both *anorexia nervosa* and *bulimia nervosa* are related to menstrual irregularities and infertility.[22,23] These disorders affect about 3 to 5% of young women, and likely twice that many have clinically important symptoms related to these eating disorders.[24] Amenorrhea is a cardinal manifestation of anorexia nervosa, and little bleeding during menses (oligomenorrhea) or amenorrhea may occur in women with bulimia nervosa. Amenorrhea in anorexia nervosa is related to irregular release of GnRH (defined in Chapter 2) and very low levels of estrogen. Menses generally returns upon weight gain, but some cases of infertility persist even after normal weight is attained. This effect may be related to continued low levels of body fat, low dietary fat intake, excessive exercise, or other factors.[24]

Interventions for Women with Anorexia Nervosa or Bulimia Nervosa

The primary therapeutic goal for anorexia nervosa is normalization of body weight, and for bulimia nervosa, normalization of eating behaviors.

Recommended treatment for anorexia nervosa involves individual, family, or group therapy. Hospitalization may be required in severe cases. Certain psychotherapeutic medications are moderately effective for treating bulimia nervosa, but cognitive-behavioral therapy is the best, established approach.

DIABETES MELLITUS PRIOR TO PREGNANCY

Many women with diabetes mellitus are unaware that the disorder increases the risk of maternal and fetal complications and fail to get blood glucose under excellent control prior to conception.[26] High blood glucose levels during the first two months of pregnancy are *teratogenic;* they are associated with a two to threefold increase in *congenital abnormalities* in newborns. Exposure to high blood glucose during the first two months in utero is related to malformations of the pelvis, central nervous system, and heart in newborns, as well as to higher rates of miscarriage.[27]

Management approaches to blood glucose control in diabetes depends, in part, on the type of diabetes present. Women may have "type 1" diabetes characterized by onset before the adult years, or "type 2," which most often occurs in adults. People with type 1 diabetes do not produce any, or enough, insulin and must take insulin. People with type 2, on the other hand, produce insulin but do not utilize it well due to *insulin resistance.*

What Is Insulin Resistance?

Insulin is a hormone released by the pancreas that increases glucose passage into cells by attaching to and stimulating receptors on cell membranes. When insulin is bound to these receptors, enzymes are activated that open cell membrane "doors" to the passage of glucose. With insulin resistance, cells "resist" this process and that lowers the amount of glucose that can be transported into the cells. In response to high levels of blood glucose, more insulin is released by the pancreas. If the capacity of the pancreas to produce increasingly large quantities of insulin needed for glucose absorption into cells is exceeded, blood glucose levels become and remain abnormally high.

The incidence of insulin resistance in adults and children is increasing in the United States and many other countries and is now considered to be a major public health problem.[28] It is estimated that about half of cases of insulin resistance are related primarily to genetic factors and half to behaviors. Most cases results from an interaction of genetic and behavioral factors. Obesity (especially large stores of central body fat), high-fat diets, smoking, physical inactivity, and small size at birth have been related to insulin resistance. The condition is also related to the development of polycystic ovary syndrome, syndrome X (a metabolic syndrome characterized by high blood triglyceride and insulin levels, low HDL and LDL cholesterol levels, and high blood pressure), and increased risk of hypertension and heart disease.[29,30]

The risk of developing insulin resistance is reduced by:

- Weight loss if overweight or obese,
- Regular physical activity,
- Diets in which fats are primarily monounsaturated and carbohydrates are primarily complex rather than simple, (see Chapter 1 for lists of food sources)
- Smoking cessation, and
- Normal weight and length in newborns.

Management of Type 2 Diabetes

Some people with this type of diabetes can manage their glucose levels with diet and exercise, whereas others will need an oral medication that increases insulin production or sensitivity to further boost glucose absorption into cells. These oral medications cannot be used during pregnancy because they also increase insulin in the fetus and cause excessive fetal growth and fat gain.[27] For both type 1 and 2 diabetes, diet and exercise are key components of blood glucose management.

Individualized diet and exercise recommendations and an educational and follow-up program developed and implemented by registered dietitians, certified diabetes educators (CDE), physicians, and nurses are preferred for diabetes management. Carefully planned and monitored dietary recommendations are a major component of blood glucose control. Individual blood glucose levels vary a good deal in response to diet composition so dietary prescriptions must be tailor-made for every person. Diets that are relatively low in total carbohydrates (40–45% of total calories) and moderate in protein (20–25%) and fat (35%) are often employed to achieve the best blood glucose control possible. Diets developed for people with diabetes emphasize:

- Complex carbohydrates featuring whole grains and high-fiber foods,
- Avoidance of simple sugars and carbohydrate foods that raise blood glucose levels the most (these are addressed in Chapter 5),
- Monounsaturated fats,
- Ample vegetables high in antioxidants (they help decrease cellular oxidative damage and malformations), and
- Three regular meals and snacks daily.[31–33]

Daily physical activity and frequent monitoring of blood glucose levels are also recommended.[31,33]

Reducing the Risk of Type 2 Diabetes

Women who are overweight and have borderline-high blood glucose levels (or "impaired glucose tolerance") may be able to reduce their risk of developing type 2

ANOREXIA NERVOSA (*anorexia* = poor appetite; *nervosa* = mental disorder) A primarily psychological disorder characterized by extreme underweight, malnutrition, amenorrhea, low bone density, irrational fear of weight gain, restricted food intake, hyperactivity, and disturbances in body image.

BULIMIA NERVOSA (*bulimia* = ox hunger) A disorder characterized by repeated bouts of uncontrolled, rapid ingestion of large quantities of food (binge eating) followed by self-induced vomiting, laxatives or diuretic use, fasting, or vigorous exercise in order to prevent weight gain. Binge eating is often followed by feelings of disgust and guilt. Menstrual cycle abnormalities may accompany this disorder.

TERATOGENIC Exposures that produce malformations in embryos or fetuses.

CONGENITAL ABNORMALITY A structural, functional, or metabolic abnormality present at birth. They may be caused by environmental or genetic factors, or by a combination of the two. Also called birth defects and congenital anomalies. Structural abnormalities are generally referred to as congenital malformations, and metabolic abnormalities as inborn errors of metabolism.

INSULIN RESISTANCE A condition in which cell membranes have reduced sensitivity to insulin so that more insulin than normal is required to transport a given amount of glucose into cells. It is characterized by elevated levels of serum insulin and triglycerides, low HDL cholesterol levels, and increased blood pressure. Nearly everyone with type 2 diabetes is insulin resistant.

diabetes during pregnancy. Modest levels of weight loss (about 10 pounds or 4.5 kg), reduction in fat intake (especially trans fats), increased fiber intake, and regular physical activity prevent or delay the onset of type 2 diabetes in 50% or more of adults.[34] Whether such lifestyle changes before pregnancy in women at risk of type 2 diabetes reduce the chances that the disorder will develop during pregnancy is not known. However, improvements in body weight, dietary intake, and physical activity level would confer health benefits and are low-risk interventions. They should be considered for women who are overweight and have impaired glucose tolerance prior to pregnancy.

POLYCYSTIC OVARY SYNDROME

Case scenario: Lupe, a 28 year old woman who is 5 feet 3 inches tall and weighs 208 pounds, and her husband want to start having a family. Her irregular periods and a failure to become pregnant as soon as desired brought her to see her nurse practitioner. She had undesired hair growth on her upper lip and wanted that investigated, too. The nurse practitioner determined that Lupe's blood pressure was elevated and her BMI was 37 kg/m². Laboratory tests showed that her blood level of insulin was high.

Lupe was diagnosed as having *polycystic ovary syndrome*, or PCOS. About 6% of women of reproductive age have PCOS.[35] Core elements of this complex disorder include menstrual dysfunction such as amenorrhea, infertility, insulin resistance, elevated insulin and testosterone blood levels, and obesity. It is a progressive disease if not resolved. Insulin resistance and elevated insulin levels worsen with time and increase the risk of hypertension and gestational diabetes during pregnancy,[36,37] and of hypertension, heart disease, and diabetes later in life.[38,39]

> **Polycystic Ovary Syndrome (PCOS)** (*polycysts* = many cysts, i.e., abnormal sacs with membranous linings) A condition in females characterized by insulin resistance, high blood insulin and testosterone levels, obesity, menstrual dysfunction, amenorrhea, infertility, hirsutism (excess body hair), and acne.
>
> **PKU (Phenylketonuria)** An inherited error in phenylalanine metabolism most commonly caused by a deficiency of phenylalanine hydroxylase, which converts the essential amino acid phenylalanine to the nonessential amino acid tyrosine.
>
> **Celiac Disease** Malabsorption with fatty stools (steatorrhea) due to an inherited sensitivity to the gliadin portion of gluten in wheat, rye, barley, and oats. It is often responsible for iron, folate, and zinc deficiencies. Also called celiac sprue and nontropical sprue.

Nutrition Interventions for Women with PCOS

The first-line therapy for PCOS is weight loss (if required) through reduced caloric intake and increased physical activity. Weight loss decreases insulin resistance, blood insulin, testosterone levels, and blood pressure. Menstrual function and fertility also improve with weight loss.[40] As Pastorek has noted, it is better to prevent or reduce obesity and avoid damage due to the progression of PCOS than to treat it with medications, surgery, or both.[41]

INBORN ERRORS OF METABOLISM

Two inborn errors of metabolism that effect embryonic development or fertility are covered here: *PKU (phenylketonuria)* and *celiac disease.*

PKU (Phenylketonuria)

Phenylketonuria derives its name from the characteristic presence of phenylalanine in the urine of people with this condition. PKU is an inherited condition that causes elevation in blood phenylalanine levels due to low levels or lack of the enzyme phenylalanine hydroxylase. Lack of this enzyme diminishes the conversion of the essential amino acid phenylalanine to tyrosine, a nonessential amino acid, and causes phenylalanine to accumulate in blood. If present during very early pregnancy, high blood levels of phenylalanine impair normal central nervous system development of the embryo. Untreated women with PKU have a 92% chance of delivering a newborn with mental retardation, and a 73% chance that the infant will be born with an abnormally small head (microcephaly).[42,43]

Nutrition Intervention for Women with PKU

PKU can be successfully managed by a low phenylalanine diet instituted and monitored with the help of an experienced registered dietitian. This diet should be followed throughout life, but it is critical that it be adhered to prior to conception and maintained throughout pregnancy. Women who establish normal blood phenylalanine levels before pregnancy and maintain it throughout pregnancy tend to deliver infants with almost normal intelligence.[44] The later in pregnancy that control is achieved, the more severe mental retardation and other problems become.[42,43]

Many women with PKU are mentally retarded and may not know that they should follow a special diet prior to pregnancy. To help them start the diet before pregnancy, health departments keep records of the addresses and ages of many women with PKU so they can be contacted. Women with PKU may be provided dietary counseling and infants are given low phenylalanine formula and foods.

Celiac Disease

People with this inherited condition are sensitive to the protein gliadin, found in gluten in wheat and rye, and to a lesser extent in barley and oats. The sensitivity causes malabsorption of fats and other food components when gluten is consumed, and a flattening of the intestinal lining. Malabsorption can lead to a number of nutrient deficiencies and reduced growth. Diagnosis of celiac disease is often delayed, so many people do not know they have it and do not make the required dietary changes. Symptoms of celiac disease tend to be present an average of 11 years before the condition is diagnosed.[45]

It has become clear that celiac disease is associated with infertility in females and males.[46] In men it can cause delayed sexual maturation and abnormal testosterone utilization. Infertility in females with the condition appears to be related to amenorrhea. Women with untreated celiac disease are at increased risk for miscarriage and fetal growth restriction during pregnancy.[47] Vitamin and mineral deficiencies that accompany untreated celiac disease may contribute to the risk of infertility and compromised pregnancy outcomes.[46]

Nutrition Management of Women with Celiac Disease

Not all people with celiac disease have overt symptoms, so it may be missed as an underlying cause of infertility. It should be considered in unexplained cases of infertility. If celiac disease is identified, fertility generally returns once gluten is removed from the diet and nutrient deficiencies are corrected.[46] Counseling by a registered dietitian facilitates adoption of a diet that restores health to people with celiac disease. A number of gluten-free grain products are available in specialty food stores and over the Internet.

Resources

U.S. Public Health Service

High quality information on eating disorder, PMS, and other conditions as presented in this chapter can be found here.
Available from: www.healthfinder.org/justforyou/women/default.htm

National Women's Health Information Center

U.S. Public Health Service site represents the pooled resources of several government agencies and provides updated reports on a wide assortment of nutrition and women's health topics.
Available from: www.4women.gov

Medscape.com

Individuals with fertility-related disorders may be able to find out if clinical trials of new drugs or procedures are being tested in which they may participate.
Available from: www.medscape.com/misc/cwatchindex.cfm

National Institutes of Health

Check out the safety and effectiveness of complementary and alternative medical treatments for fertility-related problems at these sites.
Available from: www.almed.od.nih.gov and http://dietarysupplements.info.nih.gov

USDA and Nonprofit Organizations

Each of the following sites provides information about eating disorders.
Available from: www.laureate.com/aboutned.html
Available from:
www.medpatients.com/Health%20Resources/NAANAD.htm
Available from: www.nal.usda.gov/fnic/etext/fnic.html

References

1. Mortola JF, Girton L, Beck L, et al. Diagnosis of premenstrual syndrome by a simple, prospective, and reliable instrument: the calendar of premenstrual experiences. Obstet Gynecol 1990;76:302–7.

2. Freeman EW, Rickels K, Sondheimer SJ, et al. Differential response to antidepressants in women with premenstrual syndrome/premenstrual dysphoric disorder: a randomized controlled trial. Arch Gen Psychiatry 1999;56:932–9.

3. Ugarriza DN, Klingner S, O'Brien S. Premenstrual syndrome: diagnosis and intervention. Nurse Pract 1998;23:40, 45,49–52.

4. Vaitukaitis JL. Premenstrual syndrome (editorial). N Engl J Med 1984;311: 1371–3.

5. Daugherty JE. Treatment strategies for premenstrual syndrome. Am Fam Physician 1998;58:183–92,197–8.

6. Goodale IL, Domar AD, Benson H. Alleviation of premenstrual syndrome symptoms with the relaxation response. Obstet Gynecol 1990;75:649–55.

7. Walker AF, De Souza MC, Vickers MF, et al. Magnesium supplementation alleviates premenstrual symptoms of fluid retention. J Womens Health 1998;7:1157–65.

8. Thys-Jacobs S, Starkey P, Bernstein D, et al. Calcium carbonate and the premenstrual syndrome: effects on premenstrual and menstrual symptoms. Premenstrual Syndrome Study Group. Am J Obstet Gynecol 1998;179:444–52.

9. Berman MK, Taylor ML, Freeman E. Vitamin B-6 in premenstrual syndrome. J Am Diet Assoc 1990;90:859–61.

10. Vitex or chaste tree, it helps balance women's cycles. Environ Nutr 1999;22:8.

11. Collins A, et al. Essential fatty acids in the treatment of premenstrual syndrome. Obstet Gynecol 1993;81:93–8.

12. Pettit JL. Alternative medicine: black cohosh. Clin Rev 2000;10:117.

13. Pasquali R, Vicennati V, Scopinaro N, et al. Achievement of near-normal body weight as the prerequisite to normalize sex hormone-binding globulin concentrations in massively obese men. Int J Obes Relat Metab Disord 1997;21:1–5.

14. Kyung NH, Barkan A, Klibanski A, et al. Effect of carbohydrate supplementation on reproductive hormones during fasting in men. J Clin Endocrinol Metab 1985;60: 827–35.

15. Mitchell GW, Rogers J. The influence of weight reduction on amenorrhea in obese women. N Engl J Med 1953;249: 835–7.

16. Clark AM, Thornley B, Tomlinson L, et al. Weight loss in obese infertile women results in improvement in reproductive outcome for all forms of fertility treatment. Hum Reprod 1998;13:1502–5.

17. Clark AM, Ledger W, Galletly C, et al. Weight loss results in significant improvement in pregnancy and ovulation rates in anovulatory obese women. Hum Reprod 1995;10:2705–12.

18. Hollmann M, Runnebaum B, I. G. Effects of weight loss on the hormonal profile in obese, infertile women. Hum Reprod 1996;11:1884–91.

19. Zaadstra BM, Seidell JC, Van Noord PA, et al. Fat and female fecundity: prospective study of effect of body fat distribution on conception rates. BMJ 1993;306:484–7.

20. Kaye SA, Folsom AF, Soler JT, et al. Associations of body mass and fat distribution with sex hormone concentrations in postmenopausal women. International J Epidemiol 1990;131:794–803.

21. Norman RJ, Clark AM. Obesity and reproductive disorders: a review. Reprod Fertil Dev 1998;10:55–63.

22. Robinson PH. Review article: recognition and treatment of eating disorders in primary and secondary care. Aliment Pharmacol Ther 2000;14:367–77.

23. Warren MP. Effects of undernutrition on reproductive function in the human. Endocrine Rev 1983;4:363–77.

24. Becker AE, Grinspoon SK, Klibanski A, et al. Eating disorders. N Engl J Med 1999;340:1092–8.

25. Agras WS, Walsh T, Fairburn CG, et al. A multicenter comparison of cognitive-behavioral therapy and interpersonal psychotherapy for bulimia nervosa. Arch Gen Psychiatry 2000;57:459–66.

26. Lorber DL. Preconception counseling of women with diabetes. Pract Diabetol 1995:14;12–16.

27. Carr SR. Effect of maternal hyperglycemia on fetal development, American Dietetic Association Annual Meeting and Exhibition, Kansas City, MO, October 10, 1998.

28. Type 2 diabetes in children and adolescents. American Diabetes Association. Pediatrics 2000;105:671–80.

29. Brunzell JD, Hokanson JE. Dyslipidemia of central obesity and insulin resistance. Diabetes Care 1999;22Suppl3: C10–3.

30. The ins and outs of insulin resistance. Food Insight 2000;1:1,4.

31. Brand-Miller J, Foster-Powell K. Diets with a low glycemic index: from theory to practice. Nutr Today1999;34:64–72.

32. Reece EA, Homko CJ. Diabetes mellitus in pregnancy. What are the best treatment options? Drug Saf 1998;18:209–20.

33. Thomas-Dobersen D. Nutritional management of gestational diabetes and nutritional management of women with a history of gestational diabetes: two different therapies or the same? Clin Diab 1999;17:170–6.

34. Tuomilehto J, Lindstrom J, Eriksson JG, et al. Prevention of type 2 diabetes mellitus by changes in lifestyle among subjects with impaired glucose tolerance. N Engl J Med 2001;344:1343–50.

35. Nestler JE. Role of hyperinsulinemia in the pathogenesis of the polycystic ovary syndrome, and its clinical implications. Semin Reprod Endocrinol 1997;15:111–22.

36. Kashyap S, Claman P. Polycystic ovary disease and the risk of pregnancy-induced hypertension. J Reprod Med 2000;45: 991–4.

37. Radon PA, McMahon MJ, Meyer WR. Impaired glucose tolerance in pregnant women with polycystic ovary syndrome. Obstet Gynecol 1999;94:194–7.

38. Gupta S. Polycystic ovarian syndrome: is community care appropriate? Int J Clin Pract 1999;53:359–62.

39. Pasquali R, Gambineri A, Anconetani B, et al. The natural history of the metabolic syndrome in young women with the polycystic ovary syndrome and the effect of long-term oestrogen-progestagen treatment. Clin Endocrinol (Oxf) 1999;50:517–27.

40. Pasquali R, Casimirri F, Vicennati V. Weight control and its beneficial effect on fertility in women with obesity and polycystic ovary syndrome. Hum Reprod 1997;12Suppl1:82–7.

41. Pastorek JG. Addressing obesity in medical practice: An interactive series. Medscape Women's Health 2000;5.

42. Waisbren SE, Hanley W, Levy HL, et al. Outcome at age 4 years in offspring of women with maternal phenylketonuria: the Maternal PKU Collaborative Study. Jama 2000;283:756–62.

43. Friedman EG, Koch R, Azen C, et al. The International Collaborative Study on maternal phenylketonuria: organization, study design and description of the sample. Eur J Pediatr 1996;155Suppl1:S158–61.

44. Acosta PB, Matalon K, Castiglioni L, et al. Intake of major nutrients by women in the Maternal Phenylketonuria (MPKU) Study and effects on plasma phenylalanine concentrations. Am J Clin Nutr 2001;73: 792–6.

45. Green PHR, Stavropoulos SN, Panagi SG, et al. Characteristics of adult celiac disease in the USA: results of a national survey. Am J Gastroenterol 2001;96:126–31.

46. Sher KS, Jayanthi V, Probert CS, et al. Infertility, obstetric and gynaecological problems in coeliac sprue. Dig Dis 1994; 12:186–90.

47. Eliakim R, Sherer DM. Celiac disease: fertility and pregnancy. Gynecol Obstet Invest 2001;51:3–7.

CHAPTER 4

Photo Disc

Everyone is kneaded out of the same dough but not baked in the same oven.

Yiddish proverb

NUTRITION DURING PREGNANCY

Prepared by **Judith E. Brown**

CHAPTER OUTLINE

- Introduction
- The Status of Pregnancy Outcomes
- Physiology of Pregnancy
- Embryonic and Fetal Growth and Development
- Pregnancy Weight Gain
- Nutrition and the Course and Outcome of Pregnancy
- Model Nutrition Programs for Risk Reduction in Pregnancy

KEY NUTRITION CONCEPTS

1 Many aspects of nutritional status, such as dietary intake, supplement use, and weight change, influence the course and outcome of pregnancy.

2 The fetus is not a parasite; it depends on the mother's nutrient intake to meet its nutritional needs.

3 Periods of rapid growth and development of fetal organs and tissues occur during specific times during pregnancy. Essential nutrients must be available in required amounts during these times for fetal growth and development to proceed optimally.

4 The risk of heart disease, diabetes, hypertension, and other health problems during adulthood may be influenced by maternal nutrition during pregnancy.

INTRODUCTION

The nine months of pregnancy represent the most intense period of growth and development humans ever experience. How well these processes go depends on many factors, most of which are modifiable. Of the factors affecting fetal growth and development that are within our control to change, nutritional status stands out. At no other time in life are the benefits of optimal nutritional status more obvious than during pregnancy.

This chapter addresses the status of pregnancy outcomes in the United States and other countries. It covers physiological changes that take place to accommodate pregnancy, and the impact of these changes on maternal nutritional needs. The chapter presents the roles of nutrition in fostering fetal growth, development, and long term health, and covers dietary supplement use and weight-gain recommendations. The discussion goes on to consider common problems during pregnancy that can be addressed with nutritional remedies. We begin by highlighting vital statistic reports that clearly show a need for improving pregnancy outcomes in the United States.

THE STATUS OF PREGNANCY OUTCOMES

The status of reproductive outcomes in the United States and other economically developed countries is routinely assessed through examination of a particular set of vital statistics data called "natality statistics." (*Natality* means related to birth.) Natality statistics summarize important information about the occurrence of pregnancy complications and harmful behaviors, in addition to infant mortality (death) and morbidity (illness) rates within a specific population. These data are used to identify problems in need of resolution and to identify progress in meeting national goals for improvement in the course and outcome of pregnancy.

Illustration 4.1 presents a timeline of key intervals and events before, during, and after pregnancy. Specific time points and periods are labeled. Table 4.1 shows terms characterizing and changing rates of different natality statistics. They are frequently referred to in this chapter. Also referred to often in this chapter are weights in grams (g) and kilograms (kg), as well as in pounds and ounces. There are 448 grams in a pound and 2.2 pounds in a kilogram.

Infant Mortality

Infant mortality is a mirror of a population's health status.

Infant mortality reflects the general health status of a population to a considerable degree because so many of the environmental factors that affect the health of pregnant women and newborns also affect in the rest of the population.[3] The graph presented in Illustration 4.2 demonstrates this point. This historical view of infant mortality rates indicates that population-wide improvements in social circumstances, infectious disease control, and availability of safe and nutritious foods have corresponded to greater

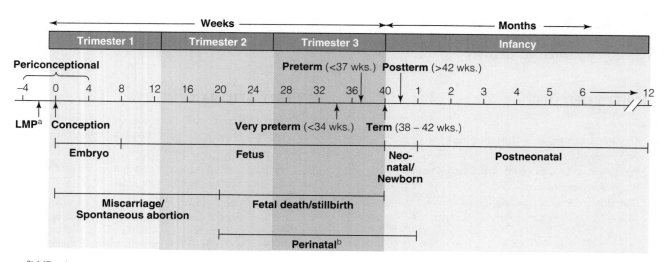

aLMP = last menstrual period

bPerinatal definition varies from 20 to 24 weeks gestation to 7 to 28 days after birth.

Illustration 4.1   Time-related terms before, during, and after pregnancy.

Table 4.1   Natality statistics: Rates, definitions, and trends in the rates in the United States[2]

RATES	1995	1998	Definition
Maternal mortality	7.1	7.1	Deaths/100,000 live births
Fetal deaths (stillbirths)	7.0	6.8	Deaths/1000 pregnancies over 20 weeks gestation
Perinatal mortality	7.6	7.3	Deaths/1000 deliveries over 20 weeks gestation to 7 days after birth
Neonatal mortality	4.9	4.8	Deaths from delivery to 28 days/1000 live births
Postneonatal mortality	2.7	2.4	Deaths from 28 days after birth to 1 year/1000 live births
Infant mortality	7.6	7.2	Deaths from birth to age 1 year/1000 live births
Preterm	11.0	11.6	Births <37 weeks gestation/100 live births
Very preterm	1.9	1.9	Births <34 weeks gestation/100 live births
Low birthweight	7.3	7.6	Newborn weights <2,500 g (5 lb 8 oz)/100 live births
Very low birthweight	1.4	1.5	Newborn weights <1500 g (3 lb 4 oz)/100 live births
Multifetal pregnancies			
Twins	1 in 40	1 in 36	Number of twin births/total live births
Triplets+	1 in 784	1 in 500	Number of triplets plus higher order multiple births/total live births
Adolescent pregnancies	56.8	51.1	Births /1000 females aged 15 to 19 years

reductions in infant mortality than have technological advances in medical care.[3] Small improvements in infant mortality in the past few decades in the United States are largely due to technological advances in medical care that save ill newborns. High-level medical care has not favorably impacted the need for such care, however.[3] Rates of low birthweight and preterm infants have changed only slightly or increased since the advent of technology-driven NICUs (neonatal intensive care units) in high-level U.S. hospitals.

Infant mortality reflects deaths during the first year of life. However, two-thirds of deaths to *liveborn infants* occur within the first month after birth, or during the neonatal period. Deaths in the first month after birth and

LIVEBORN INFANT   The World Health Organization developed a standard definition of liveborn to be used by all countries when assessing an infant's status at birth. By this definition, a liveborn infant is the outcome of delivery when a completely expelled or extracted fetus breathes, shows any sign of life such as beating of the heart, pulsation of the umbilical cord, or definite movement of voluntary muscles, whether or not the cord has been cut or the placenta is still attached.

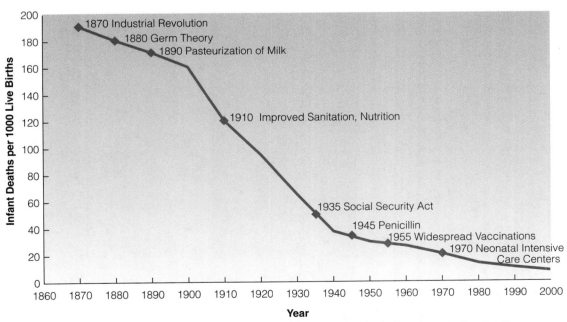

Illustration 4.2   Chronology of events related to declines in infant mortality in the United States.

SOURCE: Judith E. Brown, 2001

fetal deaths (or stillbirths) result largely from health problems that develop in the mother or fetus during pregnancy. Together they constitute the 11th leading cause of death in the United States.[2]

IMPROVEMENT NEEDED IN THE U.S. INFANT MORTALITY RATE   The United States ranks 26th in infant mortality rate among economically developed countries with populations over 2.5 million (Table 4.2).

This relatively poor state of infant outcomes exists despite the fact that the United States spends more money on medical care per person than any other country. Rather than advancing in the international rankings, the nation is falling further behind. In 1967, for example, the United States ranked 13th in infant mortality.[4]

High infant mortality rates take a toll on life expectancy. Average length of life of people within a population decreases sharply as the number of infants

Table 4.2   Infant Mortality Rates (IMR), deaths/1000 live births in the first year, and international ranking in 1996; life expectancy at birth and international ranking for 1995; and per capita health care expenditures in dollars, 1996.

Country	IMR	Rank	Life Expectancy, Years	Rank	Per Capita Health Care Expenditure
Japan	3.8	1	79.7	1	1760
Singapore	3.8	1	76.2	15	—
Finland	4.0	3	76.6	14	1525
Norway	4.0	3	77.8	8	2017
Sweden	4.0	3	78.9	2	1762
Hong Kong	4.1	6	—	—	—
Spain	4.7	7	77.9	7	1762
Switzerland	4.7	7	78.5	3	2611
France	4.9	9	78.4	4	2047
Germany	5.0	10	76.6	14	2364
Austria	5.1	11	76.8	13	1905
Ireland	5.5	12	75.3	20	1293
Belgium	5.6	13	76.4	14	1768
Denmark	5.7	14	75.4	19	2402
Netherlands	5.7	14	77.5	10	1933
Australia	5.8	16	78.0	6	1909
Northern Ireland	5.8	16	75.8	17	—
Czech Republic	6.0	18	73.2	24	943
Italy	6.0	18	77.6	9	1613
Canada	6.1	20	78.2	5	2175
England and Wales	6.1	20	77.0	12	1391
Scotland	6.2	22	74.9	22	—
Israel	6.3	23	77.3	11	—
New Zealand	6.7	24	76.1	16	1357
Portugal	6.9	25	74.9	22	1148
United States	7.3	26	75.7	18	3912
Cuba[a]	8.0	27	75.0	21	—
Greece	8.1	28	78.0	5	1196
Slovakia	9.9	29	72.3	25	—
Puerto Rico	10.4	30	74.3	24	—
Hungary	10.9	31	70.0	28	642
Kuwait[a]	11.5	32	—	—	—
Chile[a]	11.7	33	74.8	23	—
Costa Rica[a]	11.8	34	75.4	19	—
Poland	12.2	35	72.0	26	386
Bulgaria	14.8	36	71.0	27	—
Russia	18.2	37	65.0	30	—
Romania	22.3	38	69.5	29	—

SOURCE: Table developed from information in Health, United States, 2000, available at www.cdc.gov/nchs
[a]Developing country.

Table 4.3 **Range of birthweights by gestational age, U.S. 1996–1998.**

BIRTHWEIGHT		WEEKS GESTATION
Pounds (lb) and Ounces (oz)	**grams**	
<1 lb 2 oz	<500	<22
1 lb 2 oz–2 lb 3 oz	500–999	22–27
2 lb 3 oz–3 lb 5 oz	1000–1499	27–29
3 lb 5 oz–4 lb 6 oz	1500–1999	29–31
4 lb 6 oz–5 lb 8 oz	2000–2499	31–33
5 lb 8 oz–6 lb 10 oz	2500–2999	33–36
6 lb 10 oz–7 lb 11 oz	3000–3499	36–40
7 lb 11 oz–8 lb 13 oz	3500–3999	40+
8 lb 13 oz–9 lb 14 oz	4000–4499	40+
9 lb 14 oz–11 lb	4500–4999	40+
>11 lb	5000+	40+

SOURCE: Data obtained from National Center for Health Statistics, 1996–2000.

Illustration 4.3 **Percentage of infants born low birthweight by race.**

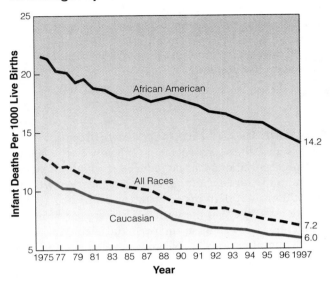

Infants born low birthweight or preterm are at substantially higher risk of dying in the first year of life than are larger and older newborns. Low birthweight infants, for example, make up 7.6% of all births, yet comprise 64% of all infant deaths.[5] The 11.6% of newborns delivered prior to 37 weeks of pregnancy similarly account for a disproportionately large number of infant deaths.[3] Low birthweight and preterm infant outcomes are intertwined in that the shorter the pregnancy, the less newborns tend to weigh. Table 4.3 shows increases in birthweight with the duration of pregnancy.

with short lives increases. As shown in Table 4.2, 17 countries with lower infant mortality rates than the United States also have greater life expectancy.

LOW BIRTHWEIGHT, PRETERM DELIVERY, AND INFANT MORTALITY Infants born low birthweight or preterm are at substantially higher risk of dying in the first year of life than are larger and older newborns. Low birthweight infants, for example, make up 7.6% of all births, yet comprise 64% of all infant deaths.[5] The 11.6% of newborns delivered prior to 37 weeks of pregnancy similarly account for a disproportionately large number of infant deaths.[3] Low birthweight and preterm infant outcomes are intertwined in that the shorter the pregnancy, the less newborns tend to weigh. Table 4.3 shows increases in birthweight with the duration of pregnancy.

Rates of low birthweight in the United States have trended slowly upward since 1983 and have remained approximately two times higher in African American infants than in other infants (Illustration 4.3).

Infant mortality is also a social mirror.

LOW BIRTHWEIGHT AMONG AFRICAN AMERICAN NEWBORNS The high rate of low birthweight in African American infants clearly represents a problem in need of resolution. It has been suggested that racial, or genetic, differences are responsible. This view, however, is not supported by research results. David and colleagues[6], for example, implemented a study in which they compared average birthweights and rates of low birthweight among African Americans born in Africa or the United States, and Caucasian Americans in Illinois. Table 4.4 shows the results of these comparisons. It is apparent from the results that factors other than race or genetics are related to differences in low birthweight rates. These authors

Table 4.4 **Comparisons of average birthweight and low birthweight rates among infants born to African Americans and Caucasians.[6]**

	African Americans		Caucasians
	Africa Born	**U.S. Born**	**U.S. Born**
Average Birthweight	3333 g	3089 g	3446 g
Low Birthweight	7.1%	13.3%	4.3%
Low Birthweight among Well Educated Women with No History of Fetal Loss and Early Prenatal Care	3.6%	7.5%	2.4%

concluded that discrimination and its effects on education and job opportunities, stress levels, and quality of health care may take a toll on maternal health and compromise the course and outcome of pregnancy.

Birthweight is the most powerful predictor of newborn outcome.

R. Williams, 1982[7]

REDUCING INFANT MORTALITY AND MORBIDITY

Deaths and illnesses associated with low birthweight and preterm infants can be reduced through improvements in the birthweight of newborns. Infants weighing 3500 to 4500 grams at birth (or 7 lb 12 oz to 10 lb) are least likely to die within the first year of life (Table 4.5), as well as in the perinatal, neonatal, and postneonatal periods.[7,8] Newborns weighing 3500 to 4500 grams are also at an advantage as a group in terms of overall health status and subsequent mental development.[9] They are less likely to develop heart disease, diabetes, lung disease, and hypertension later in life.[10] Reducing the proportion of infants born small or early would clearly decrease infant mortality. How can that be achieved? The answer is through reduction in the prevalence of risk factors for low weight at birth and preterm delivery.

Low birthweight and preterm delivery are affected by a number of factors, some of which are within our control to change and some of which are not. Modifiable risk factors include:

- Underweight prior to pregnancy,
- Low pregnancy weight gain,
- Smoking during pregnancy,
- Certain maternal infections, and
- Iron deficiency anemia early in pregnancy.[11]

Inadequate attention is focused on these factors in prenatal care settings, and services that address these risk factors are not always reimbursed by health insurers.

It is suggested that preventive efforts include improvements in the health of women prior to conception. Interventions that:

- Improve the quality of dietary intake,
- Normalize body weight,
- Increase exercise levels,
- Curb tobacco, alcohol, and drug use,
- Reduce stress levels, and
- Increase access to high quality medical care

may play important roles in reducing infant mortality.[12] Social change toward the elimination of discriminatory practices against groups of people would also likely play an important role.[6] The types of interventions that would reduce infant mortality call primarily for behavioral and social changes.[13]

Health Objectives for the Year 2010

National health objectives for pregnant women and newborns focus on the reduction of low birthweight, preterm delivery, and infant mortality. A number of the objectives are related to improvements in nutritional status (Table 4.6).

PHYSIOLOGY OF PREGNANCY

Conception triggers thousands of complex and sequenced biological changes that transform two united cells into a member of the next generation of human beings. The rapidity with which structures and func-

Table 4.5
Birthweight-specific infant mortality rates.

Birthweight	Infant Deaths/ 1000 Live Births
Less than 1500 (3.3 lb)	890
1500–1999 g (3.3–4.4 lb)	40
2000–2499 g (4.4–5.5 lb)	17
2500–2999 g (5.5–6.6 lb)	6.7
3000–3499 g (6.6–7.7 lb)	3.5
3500–3999 g (7.7–8.8 lb)	2.5
4000–4499 g (8.8–9.9 lb)	2.2
4500–4999 g (9.9–11 lb)	3.0
5000 g and more (11 lb+)	8.2

SOURCE: Health, United States, 2000; www.cdc.gov/nchs, 2000.[2]

Table 4.6 Objectives for the nation related to pregnant women and infants.

- Reduce anemia among low-income pregnant females in their third trimester from 29 to 20%.
- Reduce infant mortality from 7.6 to no more than 5 per 1000 live births.
- Reduce the incidence of spina bifida and other neural tube defects from 7 to 3 per 10,000 live births.
- Reduce low birthweight (<2500 grams) from 7.3 to 5%.
- Reduce preterm births (<37 weeks) from 9.1 to 7.6%.
- Increase abstinence from alcohol use by pregnant women from 79 to 95%.
- Reduce the incidence of fetal alcohol syndrome.
- Increase the proportion of women who gain weight appropriately during pregnancy.

tions develop in mother and fetus and the time-critical nature of energy and nutrient needs make maternal nutritional status a key element of successful reproduction.

Pregnancy begins at conception; that occurs approximately 14 days before a woman's next menstrual period is scheduled to begin. Assessed from conception, pregnancy averages 38 weeks, or 266 days, in length. Most commonly, however, pregnancy duration is given as 40 weeks (280 days) because it is measured from the date of the first day of the last menstrual period (LMP). Consequently, the common way of measuring pregnancy duration includes two nonpregnant weeks at the beginning. The anticipated date of delivery is denoted by the ancient terminology of "estimated date of confinement," or EDC. Assessment of duration of pregnancy as weeks from conception is correctly termed "gestational age," whereas time in pregnancy estimated from LMP reflects "menstrual age." It is particularly important to get these terms straight during early fetal development when a two week error in duration of pregnancy may mean miscalculating the timing of nutrient-related events in pregnancy.

Maternal Physiology

Changes in maternal physiology during pregnancy are so profound that they were previously considered abnormal and in need of correction. Low-sodium diets were routinely advised to reduce fluid retention; weight gain and dietary intake were restricted to prevent complications at delivery; and excessive levels of iron and other supplements were given to bring blood nutrient levels back up to "normal." We now know that what is considered normal physiological status of nonpregnant women cannot be considered normal for women who are pregnant. Fortunately, it is now understood that attempts to bring maternal physiological changes back to nonpregnant levels may cause more harm to the pregnancy than good.

Changes in maternal body composition and functions occur in a specific sequence during pregnancy.

The order of the sequence is absolute because the successful completion of each change depends on the one before it.[14] Since maternal physiological changes set the stage for fetal growth and development, they begin in earnest within a week after conception.[14]

The sequence of physiological changes taking place during pregnancy are listed in Table 4.7. The table indicates the timing of maximal rates of change in maternal tissues, the *placenta,* and fetal weight across pregnancy. To provide the fetus with sufficient energy, nutrients, and oxygen for growth, the mother must first expand the volume of plasma that can be circulated. Maternal nutrient stores are accumulated next. These stores are established in advance

> **PLACENTA** A disk-shaped organ of nutrient and gas interchange between mother and fetus. At term, the placenta weighs about 15% of the weight of the fetus.

of the time they will be needed to support large gains in fetal weight. Similarly, the maximal rate of placental growth is timed to precede that of fetal weight gain. This sequence of events ensures that the placenta is fully prepared for the high level of functioning that will be needed as fetal weight increases most rapidly. Fetuses depend on the functioning of multiple systems established well in advance of their maximal rates of growth and development. Abnormalities in the development of any of these physiological systems can modify fetal growth and development.

Normal Physiological Changes During Pregnancy

Physiological changes in pregnancy can be divided into two basic groups: those occurring in the first and those in the second half of pregnancy. In general, physiological changes in the first half are considered "maternal anabolic" changes because they build the capacity of the mother's body to deliver relatively large quantities of blood, oxygen, and nutrients to the fetus in the second half of pregnancy. The second half is a time of "maternal catabolic" changes in which energy and nutrient stores, and the heightened capacity to deliver stored energy and nutrients to the fetus, predominate (Table 4.8). Approximately 10% of fetal growth is accomplished in the first half of pregnancy, and the remaining 90% in the second half.[15]

The list of physiological changes that normally occur during pregnancy is extensive (Table 4.9), and such changes affect every maternal organ and system. Changes that are most

Table 4.7 Sequence of tissue development and approximate gestational week of maximal rates of change in maternal systems, the placenta, and fetus during pregnancy.[14]

Tissue	Sequence of Development	Gestational Week of Maximal Rate of Growth
Maternal plasma volume	1	20
Maternal nutrient stores	2	20
Placental weight	3	31
Uterine blood flow	4	37
Fetal weight	5	37

Table 4.8   Summary of maternal anabolic and catabolic phases of pregnancy.[15–17]

Maternal Anabolic Phase 0–20 Weeks	Maternal Catabolic Phase 20+ Weeks
Blood volume expansion, increased cardiac output	Mobilization of fat and nutrient stores
Buildup of fat, nutrient, and liver glycogen stores	Increased production and blood levels of glucose, triglycerides, and fatty acids; decreased liver glycogen stores
Growth of some maternal organs	Accelerated fasting metabolism
Increased appetite, food intake (positive caloric balance)	Increased appetite and food intake decline somewhat near term
Decreased exercise tolerance	Increased exercise tolerance
Increased levels of anabolic hormones	Increased levels of catabolic hormones

directly related to maternal energy and nutrient needs are discussed further.

BODY WATER CHANGES   A woman's body gains a good deal of water during pregnancy, primarily due to increased volumes of plasma and extracellular fluid, as well as amniotic fluid.[18] Total body water increases in pregnancy range from 7 to 10 liters (or approximately 7 to 10 quarts or about two to two-and-a-half gallons). About two-thirds of the expansion is intracellular (blood and body tissues) and one-third is extracellular (fluid in spaces between cells).[14] Plasma volume begins to increase within a few weeks after conception and reaches a maximum at approximately 34 weeks. Early pregnancy surges in plasma volume appear to be the primary reason that pregnant women feel tired and become exhausted easily when undertaking exercise performed routinely prior to pregnancy. Fatigue associated with plasma volume increases in the second and third months of pregnancy declines as other compensatory physiological adjustments are made.

Table 4.9   Normal changes in maternal physiology during pregnancy.[16,18]

Blood Volume Expansion
- Blood volume increases 20%
- Plasma volume increases 50%
- Edema (occurs in 60–75% of women)

Hemodilution
- Concentrations of most vitamins and minerals in blood decrease

Blood Lipid Levels
- Increased concentrations of cholesterol, LDL cholesterol, triglycerides, HDL cholesterol

Blood Glucose Levels
- Increased insulin resistance (increased plasma levels of glucose and insulin)

Maternal Organ and Tissue Enlargement
- Heart, thyroid, liver, kidneys, uterus, breasts, adipose tissue

Circulatory System
- Increased cardiac output through increased heart rate and stroke volume (30–50%)
- Increased heart rate (16% or 6 beats/min)
- Decreased blood pressure in the first half of pregnancy (−9%), followed by a return to nonpregnancy levels in the second half

Respiratory System
- Increased tidal volume, or the amount of air inhaled and exhaled (30–40%)
- Increased oxygen consumption (10%)

Food Intake
- Increased appetite and food intake; weight gain
- Taste and odor changes, modification in preference for some foods
- Increased thirst

Gastrointestinal Changes
- Relaxed gastrointestinal tract muscle tone
- Increased gastric and intestinal transit time
- Nausea (70%), vomiting (40%)
- Heartburn
- Constipation

Kidney Changes
- Increased glomerular filtration rate (50–60%)
- Increased sodium conservation
- Increased nutrient spillage into urine; protein is conserved
- Increased risk of urinary tract infection

Immune System
- Suppressed immunity
- Increased risk of urinary and reproductive tract infection

Basal metabolism
- Increased basal metabolic rate in second half of pregnancy
- Increased body temperature

Hormones
- Placental secretions of large amounts of hormones needed to support physiological changes of pregnancy

Gains in body water vary a good deal among women during normal pregnancy. High gains are associated with increasing degrees of *edema* and weight gain. If not accompanied by hypertension, edema generally reflects a healthy expansion of plasma volume. Birthweight is strongly related to plasma volume: generally, the greater the expansion, the greater the newborn size.[14]

The increased volume of water in blood is responsible for the "dilution effect" of pregnancy on blood concentrations of most vitamins and minerals. Although total amounts of nutrients in the blood increase during pregnancy, their concentration in blood decreases due to the disproportionate expansion of plasma volume.[14]

HORMONAL CHANGES Many physiological changes in pregnancy are modulated by hormones produced by the placenta (Table 4.10 and Illustration 4.4).[16] The placenta serves many roles, but a key one is the production of *steroid hormones,* such as progesterone and estrogen. The placenta is also the main supplier of many other hormones needed to support the physiological changes of pregnancy.

MATERNAL NUTRIENT METABOLISM Adjustments in maternal nutrient metabolism are apparent within the first few weeks after conception and progress throughout pregnancy.[15] The changes are sequenced and interrelated in complex ways that are appreciated but not yet fully understood.

Illustration 4.4 Changes in maternal plasma concentration of hormones during pregnancy.

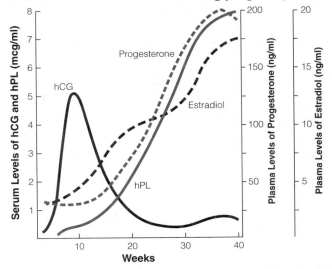

SOURCE: Rosso, P. Nutrition and metabolism in pregnancy: mother and fetus. Oxford University Press, 1990. Used by permission.

Carbohydrate Metabolism Many adjustments in carbohydrate metabolism are made during pregnancy that promote the availability of glucose to the fetus. Glucose is the fetus's preferred fuel, even though fats can be utilized for energy. Continued availability of a fetal supply of glucose is accomplished primarily through metabolic changes that promote maternal insulin resistance. These changes, sometimes referred to as the "diabetogenic effect of pregnancy" make normal pregnant women slightly carbohydrate intolerant in the third trimester of pregnancy.[19] Illustration 4.5 provides an example of the normal levels of plasma glucose and insulin during late pregnancy compared to prepregnancy levels.

EDEMA Swelling (usually of the legs and feet but can also extend throughout the body) due to an accumulation of extracellular fluid.

STEROID HORMONES Hormones such as progesterone and estrogen produced primarily from cholesterol.

Carbohydrate metabolism in the first half of pregnancy is characterized by estrogen- and progesterone-stimulated increases in insulin production and conversion of glucose to glycogen and fat. In the second half, rising levels of hCS and prolactin inhibit the conversion of glucose to glycogen and fat.[16] At the same time, insulin resistance builds in the mother, increasing her reliance on fats for energy. Decreased conversion of glucose to glycogen and fat, lowered maternal utilization of glucose, and increased liver production of glucose help to ensure that a constant supply of glucose for fetal growth and development is available in the second half of pregnancy.

Table 4.10 Key placental hormones and examples of their roles in pregnancy.[15,16]

Human chorionic gonadotropin (hCG): maintains the corpus luteum in early pregnancy. Referred to as "the hormone of pregnancy."

Progesterone: maintains the implant; relaxes smooth muscles of the blood vessels and gastrointestinal and urinary tracts; promotes lipid deposition.

Estrogen: increases lipid formation and storage, protein synthesis, and uterine blood flow; prompts breast development; promotes ligament flexibility.

Human placental lactogen (hPL, also called human chorionic somatotropin or hCS): increases maternal insulin resistance to maintain glucose availability for fetal use, promotes protein synthesis and the breakdown of fat for energy for maternal use.

Prolactin: promotes insulin resistance and fat breakdown and utilization.

Leptin: may participate in the regulation of appetite and lipid metabolism, and utilization of fat stores.

Fasting maternal blood glucose levels decline in the third trimester due to increased utilization of glucose by the rapidly growing fetus. However, postmeal blood glucose concentrations are elevated and remain higher longer than before pregnancy.[19]

Accelerated Fasting Metabolism   Maternal metabolism is rapidly converted toward *glucogenic amino acid* utilization, fat oxidation, and increased production of *ketones* with fasts that last longer than 12 hours. Decreased levels of plasma glucose and insulin, and increased levels of triglycerides, free fatty acids, and ketones are seen hours before they occur in nonpregnant fasting women. The rapid conversion to fasting metabolism allows pregnant women to use primarily stored fat for energy while sparing glucose and amino acids for fetal use.[19]

GLUCOGENIC AMINO ACIDS   Amino acids such as alanine and glutamate that can be converted to glucose.

KETONES   Metabolic by-products of the breakdown of fatty acids in energy formation. β-hydroxybutyric acid, acetoacetic acid, and acetone are the major ketones, or "ketone bodies."

Although these metabolic adaptations help ensure a constant fetal supply of glucose, fasting eventually increases the dependence of the fetus on ketone bodies for energy. Prolonged fetal utilization of ketones, such as occurs in women with poorly controlled diabetes or in those who lose weight during part or all of pregnancy, is associated with reduced growth and impaired intellectual development of the offspring.[20]

Protein Metabolism   Nitrogen and protein are needed in increased amounts during pregnancy for synthesis of new maternal and fetal tissues. To some extent the increased need for protein is met through reduced levels of nitrogen excretion and the conservation of amino acids for protein tissue synthesis. There is no evidence, however, that the mother's body stores protein early in pregnancy in order to meet fetal needs for protein later in pregnancy. Maternal and fetal needs for protein are primarily fulfilled by the mother's intake of protein during pregnancy.[15]

Fat Metabolism   Multiple changes occur in the body's utilization of fats during pregnancy. Overall, changes in lipid metabolism promote the accumulation of maternal fat stores in the first half of pregnancy and enhance fat mobilization in the second half.[19] In addition to seeing increasing maternal reliance on fat stores for energy as pregnancy progresses, we see blood levels of many lipoproteins increase dramatically (Table 4.11). Plasma triglyceride levels increase first and most dramatically, reaching three times nonpregnant levels by term.[15,21] Cholesterol-containing lipoproteins, phospholipids, and fatty acids also increase, but to a lesser extent than do triglycerides. The increased cholesterol supply is used by the placenta for steroid hormone synthesis and by the fetus for nerve and cell membrane formation.[19] Small increases in HDL cholesterol in pregnancy appear to decline within a year postpartum and remain lower than prepregnancy levels. It is speculated that declines in HDL cholesterol after pregnancy may contribute to an increased risk of heart disease in women.[22] Other changes in serum lipids appear to revert to prepregnancy levels postpartum.[22]

By the third trimester of pregnancy most women have a lipid profile that would be considered atherogenic, if not for pregnancy. These blood lipid changes

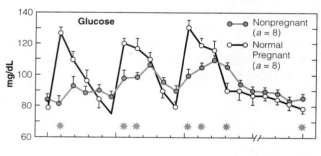

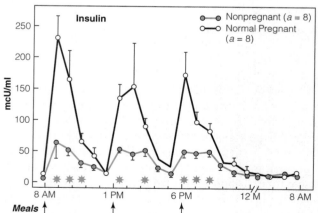

Illustration 4.5 Plasma glucose and insulin levels in nonpregnant women and in women near term.

SOURCE: Used with permission from Mosby, Inc. Phelps et al, Am J Obstet Gynecol 1981 140:730–36. Fig. 2-3.

Table 4.11   Changes in cholesterol and triglyceride levels during pregnancy.[21,22]

Trimester	Cholesterol mmol/L	Triglycerides mmol/L
1	5.78	1.19
2	6.88	1.32
3	8.14	2.58
Nonpregnant	5.11	0.80

are normal, however, which is the reason it is not recommended that blood lipid screening be undertaken during pregnancy.[22] Changes in blood lipid levels during pregnancy appear to be unrelated to maternal dietary intake.[23]

Mineral Metabolism   Impressive changes in mineral metabolism occur during pregnancy. Calcium metabolism is characterized by an increased rate of bone turnover and reformation.[24] Elevated levels of body water and tissue synthesis during pregnancy are accompanied by increased requirements for sodium and other minerals. Sodium metabolism is delicately balanced during pregnancy to promote an accumulation of sodium by the mother, placenta, and fetus. This is accomplished by changes in the kidneys that increase aldosterone secretion and the retention of sodium. This normal change in pregnancy renders attempts to prevent and treat high blood pressure in pregnancy by reducing sodium intake ineffective and potentially harmful. Sodium restriction may overstress mechanisms that act to conserve sodium and lead to functional and growth impairments due to sodium depletion.[25]

Common Health Problems During Pregnancy

Some of the physiological changes that occur in pregnancy are accompanied by side effects that can dull the bliss of expecting a child by making women feel physically miserable. Common ailments of pregnancy, such as nausea and vomiting, heartburn, and constipation, are generally more amenable to prevention than to treatment, but often can be relieved through dietary measures.

Nausea and Vomiting   Nausea occurs in about 7 in 10 pregnancies, and vomiting in 4 of 10. The conditions are so common that they are considered a normal part of pregnancy. Unless severe or prolonged, nausea and vomiting during pregnancy are associated with a reduction in risk of miscarriage of greater than 60% and with healthy newborn outcomes.[26] The cause of nausea and vomiting is not yet clear, but they are thought to be related to increased levels of human chorionic gonadotropin, progesterone, estrogen, or other hormones early in pregnancy.[27]

In the past, nausea and vomiting of pregnancy was called "morning sickness" because it was thought to occur mostly after waking up. It occurs at all times of day and tends to begin during the fourth week after conception. In many women, nausea and vomiting suddenly disappear, and for most (but not all) women the symptoms end around the tenth week.

Iron supplements may aggravate nausea and vomiting when given in the first trimester of pregnancy.[26]

Hyperemesis gravidarum   Between 1 and 2% of pregnant women with nausea and vomiting develop "hyperemesis gravidarum" (more commonly called hyperemesis).[28] Hyperemesis is characterized by severe nausea and vomiting that last throughout much of pregnancy. It can be debilitating. In addition to the mother feeling very sick, frequent vomiting can lead to weight loss, electrolyte imbalances, and dehydration. Women with hyperemesis who gain weight normally during pregnancy (about 30 pounds total) are not at increased risk of delivering small infants, but women who gain less (21–22 pounds) are.[29]

Management of Nausea and Vomiting   Many approaches to the treatment of nausea and vomiting are used in clinical practice, but only a few are considered safe and effective. Dietary interventions represent the safest method, primarily because the short- and long-term safety of many drugs and herbal remedies early in pregnancy is unclear.[30] It is generally recommended that women with nausea and vomiting:

- Continue to gain weight,
- Separate liquid and solid food intake,
- Avoid odors and foods that trigger nausea, and
- Select foods that are well tolerated.

Many women find that hard-boiled eggs, potato chips, popcorn, yogurt, crackers, and other high-carbohydrate foods are well tolerated. Personal support and understanding are important components of counseling women with nausea and vomiting. Care should be taken to individualize dietary advice based on each woman's food preferences and tolerances. Women with hyperemesis may require rehydration therapy to restore fluids and electrolyte balance. (See the Connections feature.)

Periodically, articles will appear in the popular press claiming that nausea and vomiting are caused by certain foods and that women should avoid them to protect their fetus from harmful substances in the foods. Not too long ago it was claimed that bitter tasting vegetables, for example, should be avoided. When put to the test, this notion was found to be groundless.[31] Theoretical claims that certain foods elicit nausea and vomiting in order to protect the fetus from harmful effects of the food

Connections
Heavenly Relief

Said one woman with hyperemesis after rehydration therapy "I thought I died and went to heaven." Women with hyperemesis can feel absolutely awful.

should be considered unreliable until proven in scientific studies.

One drug containing an antihistamine and vitamin B_6 and vitamin B_6 supplements (10 mg three times a day) was found to reduce nausea and vomiting in some women.[30] (The Tolerable Upper Intake Level for vitamin B_6 in pregnancy is 100 mg per day.) The long-term safety of this drug is not known, but in the short term it appears to be safe.[30]

HEARTBURN   Pregnancy has little effect on gastrointestinal secretions or nutrient absorption but has major effects on relaxation of gastrointestinal tract muscles. This effect is attributed primarily to progesterone. Relaxation of the muscular valve known as the cardiac or lower esophageal sphincter at the top of the stomach is thought to be the principal reason for the 30–50% incidence of heartburn in women during pregnancy. The loose upper valve may allow stomach contents to be pushed back into the esophagus.[32]

Management of Heartburn   Dietary advice for the prevention and management of heartburn include:

- Ingestion of small meals frequently,
- Not going to bed with a full stomach, and
- Avoiding foods that seem to make heartburn worse.

Elevating the upper body during sleep and not bending the upper body below the waist also reduces gastric reflux. Antacid tablets which act locally in the stomach are often recommended, but heartburn pills are not.[32]

CONSTIPATION   Relaxed gastrointestinal muscle tone is also thought to be primarily responsible for the increased incidence of constipation and hemorrhoids in pregnancy. The best way to prevent these maladies is

to consume approximately 30 grams of dietary fiber daily.[33] (Food sources of fiber are listed in Table 1.4 in Chapter 1.) Laxative pills are not recommended for use by pregnant women, but bulk-forming fiber in products such as Metamucil, Citrucel, and Perdiem are considered safe and effective for the prevention and treatment of constipation.[32] Women should drink a cup or more of water along with the fiber supplement.

The Placenta

The word *placenta* is derived from the Latin word for cake. The placenta, with its round, disk-like shape, looks somewhat like a cake—the more so the more active the imagination. It is genetically part of fetal tissues and larger than the fetus for most of pregnancy. Development of the placenta precedes fetal development.

Functions of the placenta include hormone and enzyme production, nutrient and gas exchange between the mother and fetus, and removal of waste products from the fetus. Its structure, including a double lining of cells separating maternal and fetal blood, acts as a barrier to some harmful compounds, and it governs the rate of passage of nutrients and other substances into and out of the fetal circulation (Illustration 4.6). The barrier role of the placenta is better described as a fence than as a filter that guards the fetus against all things harmful. Many potentially harmful substances (alcohol, excessive levels of some vitamins, drugs, and certain viruses, for example) do pass through the placenta to the fetus. The placenta is a barrier to the passage of maternal red blood cells, bacteria, and many large proteins. The placenta also prevents the mixing of fetal and maternal blood until delivery, when ruptures in blood vessels may occur.

NUTRIENT TRANSFER   Nutrient transfer across the placenta depends on a number of factors including:

- The size of molecules available for transport,
- Lipid solubility of the particles being transported, and
- The concentration of nutrients in maternal and fetal blood.

Small molecules with little or no charge (water, for example) and lipids (cholesterol and ketones, for instance) pass through the placenta most easily, while large molecules (e.g., insulin and enzymes), aren't transferred at all.

Nutrient exchange between the mother and fetus is unregulated for some nutrients, oxygen, and carbon dioxide; it is highly regulated for other nutrients. Nutrient transfer based on concentration gradients determined by the levels of the nutrient in the maternal and the fetal blood are unregulated. In these cases,

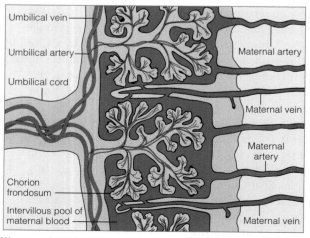

Illustration 4.6   Structure of the placenta.

nutrients cross placenta membranes by simple diffusion from blood with high concentration of the nutrient to blood with the lower concentration.

Three primary mechanisms regulate nutrient transfer: facilitated diffusion, active transport, and endocytosis (or pinocytosis). Table 4.12 summarizes mechanisms of nutrient transfer across the placenta and provides examples of nutrients transported by each specific mechanism as they are now known.

The fetus receives small amounts of water and other nutrients from ingestion of *amniotic fluid.* By the second half of pregnancy the fetus is able to swallow and absorb water, minerals, nitrogenous waste products, and other substances in amniotic fluid.

THE FETUS IS NOT A PARASITE The fetus is not a "parasite"—it cannot take whatever nutrients it needs from the mother's body at the mother's expense. When maternal nutrient intakes fall below optimum levels or adjustment thresholds, fetal growth and development are compromised more than maternal health.[15] In general, nutrients will first be used to support maternal nutrient needs for her health and physiological changes, and next for placental development, before they become available at optimal levels to the fetus. For example:

- Underweight women gaining the same amount of weight as normal weight women tend to deliver smaller infants and to retain more of the weight gained during pregnancy at the expense of fetal growth.[16]

- Fetal growth tends to be reduced in pregnant teenagers who gain height during pregnancy compared to fetal growth in teens who do not grow during pregnancy.[35]

- Vitamin and mineral deficiencies and toxicities in newborns have been observed in women who showed no signs of deficiency or toxicity diseases during pregnancy.[16]

> **AMNIOTIC FLUID** The fluid contained in the amniotic sac that surrounds the fetus in the uterus.
>
> **GROWTH** Increase in an organism's size through cell multiplication (hyperplasia) and enlargement of cell size (hypertrophy).
>
> **DEVELOPMENT** Progression of the physical and mental capabilities of an organism through growth and differentiation of organs and tissues, and integration of functions.

If the fetus did act as a parasite it would harm the mother for its own benefit. Rather, the fetus is generally harmed more by poor maternal nutritional status than is the mother.[34]

EMBRYONIC AND FETAL GROWTH AND DEVELOPMENT

The rate of human *growth* and *development* is higher during gestation than at any time thereafter. If the rate of weight gain achieved in the nine months of gestation continued after delivery, infants would weigh about 160 pounds at their first birthday and be 20

Table 4.12 Mechanisms of nutrient transport across the placenta.[16, 34]

MECHANISM	EXAMPLES OF NUTRIENTS
Passive diffusion (also called simple diffusion) Nutrients transferred from blood with higher concentration levels to blood with lower concentration levels	Water, some amino acids and glucose, free fatty acids, ketones, vitamins E and K[a], some minerals (sodium, chloride), gases
Facilitated diffusion Receptors ("carriers") on cell membranes increase the rate of nutrient transfer	Some glucose, iron, vitamins A and D
Active transport Energy (from ATP) and cell membrane receptors	Water soluble vitamins, some minerals (calcium, zinc, iron, potassium) and amino acids required for transfer
Endocytosis (also called pinocytosis) Nutrients and other molecules are engulfed by placenta membrane and released into fetal blood supply	Immunoglobulins, albumin

[a]Vitamin K crosses the placenta slowly and to a limited degree.

feet tall by age 20![36] Table 4.13 shows fetuses of various age during pregnancy and overviews some of the amazing transformations that take place during embryonic and fetal growth and development.

Critical Periods of Growth and Development

Fetal growth and development proceed along genetically determined pathways in which cells are programmed to multiply, *differentiate,* and establish long-term functional levels during set time intervals. Such time intervals are known as *critical periods* and are most intense during the first two months after conception, when most organs and tissues form. On the whole, critical periods represent a "one way street" because it is not possible to reverse directions and correct errors in growth or development that occurred during a previous, critical period. Consequently, adverse effects of nutritional and other insults occurring during critical periods of growth and development persist throughout life.[37]

> **DIFFERENTIATION**   Cellular acquisition of one or more characteristics or functions different from that of the original cells.
>
> **CRITICAL PERIODS**   Preprogrammed time periods during embryonic and fetal development when specific cells, organs, and tissues are formed and integrated, or functional levels established. Also called *sensitive periods.*

HYPERPLASIA   Critical periods of growth and development are characterized by hyperplasia, or an increase in cell multiplication. Since every human cell has a specific amount of DNA, periods of hyperplasia can be determined by noting times during gestation when the DNA content of specific organs and tissues increases sharply. The critical period of rapid cell multiplication of the forebrain, for example, is between 10 and 20 weeks of gestation (Illustration 4.7).

The brain is the first organ that develops in humans, and along with the rest of the central nervous system, it is given priority access to energy and nutrient supplies. Thus, in conditions of low energy and nutrient availability, the needs of the central nervous system will be met before those of other fetal tissues such as the liver or muscles.

Table 4.13   Notes on normal embryonic and fetal growth and development.[16,36]

Day 1	Conception, one cell	**Week 9**	Embryo now considered a fetus
Day 3	Hollow ball of cells has formed (the blastocyst) and cells begin to differentiate	**Month 3**	Weighs an ounce; primitive egg and sperm cells developed, hard palate fuses, breathes in amniotic fluid
Day 6–8	Blastocyst comprised of approximately 250 cells implants into uterine wall	**Month 4**	Weighs about 6 ounces; placenta diameter is 3 inches
Day 12	Embryo is composed of thousands of cells, differentiation well underway	**Month 5**	Weighs about a pound, 11 inches long; skeleton begins to calcify, hair grows
Week 4 (21–28 days)	¼ inch long; rudimentary head, trunk, arms; heart "practices" beating; spinal cord and two major brain lobes present	**Month 6**	14 inches long; fat accumulation begins, permanent teeth buds form; lungs, gastrointestinal tract, and kidneys formed but are not fully functional
Week 5 (28–35 days)	Rudimentary kidney, liver, gastrointestinal tract, circulatory system, eyes, ears, mouth, hands, arms, and gastrointestinal tract; heart beats 65 times per minute circulating its own, newly formed blood	**Month 7**	Gains ½ to 1 ounce per day
Week 7 (49–56 days)	½ inch long, weighs 2–3 grams; brain sends impulses, gastrointestinal tract produces enzymes, kidney eliminates some waste products, liver produces red blood cells, muscles work. (Approximately 25% of blastocysts and embryos will be lost before 7 weeks.)	**Months 8 and 9**	Gains about an ounce per day; stores fat, glycogen, iron, folate, B_6 and B_{12}, riboflavin, calcium, magnesium, vitamins A, E, D; functions of organs continue to develop. Growth rate declines near term. Placenta weighs 500–650 g (1–1½ lbs) at term

Illustration 4.7   The critical period of cell multiplication of the forebrain. Growth in cell numbers is indicated by increases in DNA content of a given amount of tissue.

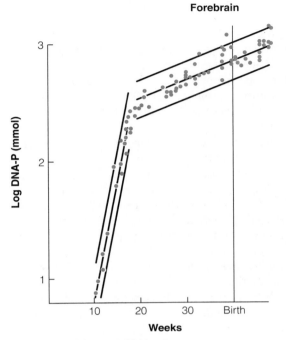

Used by permission of the BMJ Publishing Group.

Deficits or excesses in nutrients supplied to the embryo and fetus during critical periods of cell multiplication can produce lifelong defects in organ and tissue structure and function. The organ or tissue undergoing critical periods of growth at the time of the adverse exposure will be affected most.[38] For example, the neural tube develops into the brain and spinal cord during weeks three and four after conception. If folate supplies are inadequate during this critical period of growth, permanent defects in brain or spinal cord formation occur, regardless of folate availability at other times. Other tissues, such as the pancreas, which does not undergo rapid cell multiplication until the third trimester of pregnancy, do not appear to be affected by the early shortage of folate.

Some degree of hyperplasia takes place in a number of organs and tissues in the first year or two after birth and during the adolescent growth spurt. Cells of the central nervous system, for instance, continue to multiply for about two years after birth, but at a much slower pace than early in pregnancy. Skeletal and muscle cells increase in number during the adolescent growth spurt.[39]

In utero and early life changes in DNA content of the brain have been investigated in fetuses, infants, and young children dying from non-nutritional causes and

from undernutrition. Illustration 4.8 present results from one such study that show deficits in DNA content (or cell number) in the brains of children dying of protein-energy malnutrition versus those dying from accidents. Deficits in DNA were apparent a few months after birth, indicating that severe malnutrition early in pregnancy reduced brain cell number in utero.[40]

HYPERPLASIA AND HYPERTROPHY   Cell multiplication continues at a lower rate after critical periods of cell multiplication and is accompanied by increases in the size of cells. This phase of growth can be seen in Illustration 4.7, where it begins around 20 weeks in the forebrain when the rate of increase in DNA content slows. Cell size increases mainly due to an accumulation of protein and lipids inside of cells. Consequently, increases in cell size can be determined by measuring the protein or lipid content of cells. Specialized functions of cells, such as production of digestive enzymes by cells within the small intestine or neurotransmitters by nerve cells, occur along with increases in cell number and size.[16]

HYPERTROPHY   Periods of hyperplasia-hypertrophy are followed by hypertrophy only. During this phase cells continue to accumulate protein and lipids and functional levels continue to grow in sophistication, but cells no longer multiply. Reductions in cell size caused by unfavorable nutrient environments or other conditions are associated with deficits in organ and tissue functions, such as reduced mental capabilities or declines in muscular coordination. Such functional changes can often be reduced or reversed later if deficits are corrected.[37]

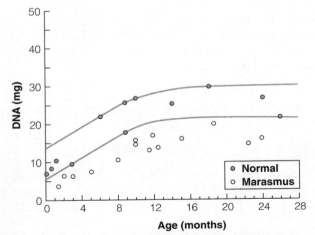

Illustration 4.8   DNA content of the cerebellum of the human brain in young children dying from non-nutritional causes and from undernutrition.

SOURCE: Used by permission of Winick M. Malnutrition and brain development. J Pediatr 1969;74:667–79.

MATURATION   The last phase of growth and development is maturation—or the stabilization of cell number and size. This phase occurs after tissues and organs are fully developed later in life.

Fetal Body Composition

The fetus undergoes marked changes in body composition during pregnancy (Table 4.14). The general trend is toward progressive increases in fat, protein, and mineral content. Some of the most drastic changes take place in the last five weeks of pregnancy, when fat and mineral content increase substantially.

Variation in Fetal Growth

Given a healthy mother and fetal access to needed amounts of energy, nutrients, and oxygen and freedom from toxins, fetal genetic growth potential is achieved.[42] However, as evidenced by the relatively high rate of low birthweight in the United States, optimal conditions required for achievement of genetic growth potential often do not exist during pregnancy. Although some of the conditions that limit fetal growth are yet to be identified, it is clear that factors such as prepregnancy underweight and shortness, low weight gain during pregnancy, poor dietary intakes, smoking, drug abuse, genetic abnormalities, and certain clinical complications of pregnancy are associated with reduced fetal growth.[11]

SMALL FOR GESTATIONAL AGE (SGA) Newborn weight is <10th percentile for gestational age. Also called *small for date* (SFD).

DISPROPORTIONATELY SMALL FOR GESTATIONAL AGE (dSGA)   Newborn weight is <10th percentile of weight for gestational age, length and head circumference are normal. Also called *asymmetrical SGA.*

PROPORTIONATELY SMALL FOR GESTATIONAL AGE (pSGA)   Newborn weight, length, and head circumference are <10th percentile for gestational age. Also called *symmetrical SGA.*

Risk of illness and death varies substantially with size at birth and is particularly high for newborns experiencing intrauterine growth retardation (IUGR).[38,43] For a portion of newborns, smallness at birth is normal and may reflect familial genetic traits.

Because IUGR is complicated to determine, it is usually approximated by assessment of size for gestational age using a reference standard (Table 4.15).[44] Infants are generally considered likely to have experienced intrauterine growth retardation if their weight for gestational age or length is low. Newborns whose weight is less than the tenth percentile for gestational age are considered *small for gestational age,* or SGA. This determination is further categorized into *disproportionately small for gestational age* (dSGA) and *proportionately small for gestational age* (pSGA). Newborns who weigh less than the tenth percentile of weight for gestational age but have normal length and head circumference for age are considered dSGA. If weight, length, and head circumference are less than the tenth percentile for gestational age, than the newborn is considered pSGA.[42] Approximately two-thirds of SGA newborns in the U.S. are disproportionately small, and one-third are proportionately small.[42]

dSGA   Infants who are disproportionately small for gestational age look skinny, wasted, and wrinkly. They tend to have small abdominal circumferences, reflecting a lack of glycogen stores in the liver, and little body fat. It appears that these infants have experienced in utero malnutrition in the third trimester of pregnancy and that it compromised liver glycogen and fat storage. Short-term episodes of malnutrition, such as maternal weight loss or low weight gain late in pregnancy, are thought to be related to dSGA. These infants generally have smaller organ sizes but the normal number of cells in organs and tissues. Although dSGA infants are at risk of developing the "hypos" after birth (hypoglycemia, hypocalcemia, hypomagnesiumenia, and hypothermia), with nutritional rehabilitation they generally experience good catch-up growth.[42] Unfortunately, disproportionately small infants appear to be at greater risk for heart disease, hypertension, and type 2 diabetes in the adult years.[38] (We return to this issue later in this chapter.)

pSGA   Proportionately SGA newborns look small but well proportioned. It is believed that these

Table 4.14   Estimated changes in body composition of the fetus by time in pregnancy.[16,41]

Component	10 Weeks	20 Weeks	30 Weeks	40 Weeks
Body weight, g	10	300	1667	3450
Water, g	9	263	1364	700
Protein, g	<1	22	134	446
Fat, g	<1	26	66	525
Sodium, meq	<1	32	136	243
Potassium, meq	<1	12	75	170
Calcium, mg	<1	1	10	28
Magnesium, mg	<1	5	31	76
Iron, mg	<1	17	104	278
Zinc, mg	<1	6	26	53

Table 4.15 Percentiles of weight for newborn gestational age.

Gestational Age (wk)	5th Pctl	10th Pctl	50th Pctl	90th Pctl	95th Pctl
20	249	275	412	772	912
21	280	314	433	790	957
22	330	376	496	826	1023
23	385	440	582	882	1107
24	435	498	674	977	1223
25	480	558	779	1138	1397
26	529	625	899	1362	1640
27	591	702	1035	1635	1927
28	670	798	1196	1977	2237
29	772	925	1394	2361	2553
30	910	1085	1637	2710	2847
31	1088	1278	1918	2986	3108
32	1294	1495	2203	3200	3338
33	1513	1725	2458	3370	3536
34	1735	1950	2667	3502	3697
35	1950	2159	2831	3596	3812
36	2156	2354	2974	3668	3888
37	2357	2541	3117	3755	3956
38	2543	2714	3263	3867	4027
39	2685	2852	3400	3980	4107
40	2761	2929	3495	4060	4185
41	2777	2948	3527	4094	4217
42	2764	2935	3522	4098	4213
43	2741	2907	3505	4096	4178
44	2724	2885	3491	4096	4122

Pctl = percentile

SOURCE: Alexander, G et al.[44] Reprinted with permission from the American College of Obstetricians and Gynecologists (Obstetrics and Gynecology, 1996, 87(2), pp 163–168).

The goal of nutritional rehabilitation for pSGA infants should be catch-up in weight and length, and not just weight. This goal appears to be easier to reach if pSGA infants are breastfed. Excessive weight gain by pSGA infants appears to increase the risk of obesity and insulin resistance-related disorders, such as hypertension and type 2 diabetes, later in life.[47]

LGA Newborns with weights greater than the 90th percentile for gestational age are considered to be large for gestational age. About 1–2% of U.S. newborns are LGA. Although it is difficult to predict LGA, it appears to be related to prepregnancy obesity, poorly controlled diabetes in pregnancy, excessive weight gain in pregnancy (over 44 pounds), and possibly other factors. Except for infants born to women with poorly controlled diabetes during pregnancy or other health problems, LGA newborns experience far lower illness and death rates than do SGA infants, and tend to be taller later in life.[7,48] Delivery and postpartum complications in mothers, however, tend to be higher with LGA newborns, and include increased rates of operative delivery, *shoulder dystocia,* and postpartum hemorrhage.[49]

PONDERAL INDEX Increasingly, Ponderal Index (PI) is being used to assess the appropriateness of newborn size. Like body mass index, it is a measure of weight-for-length, but it is calculated by dividing weight by the cube of length (PI = weight in grams/cm^3 × 1000). Values between approximately 23 and 25 reflect normal weight-for-length, lower values represent thinness and higher values heaviness at birth. Wider adoption of PI to assess newborn size in practice awaits development of standard values indicative of increased risk of thinness and heaviness at birth.

APPROPRIATE FOR GESTATIONAL AGE (AGA) Weight, length, and head circumference are between the 10th and 90th percentiles for gestational age.

LARGE FOR GESTATIONAL AGE (LGA) Weight for gestational age exceeds the 90th percentile for gestational age. Also defined as birthweight greater than 4500 g (>10 lb) and referred to as *excessively sized for gestational age,* or *macrosomic.*

SHOULDER DYSTOCIA Blockage or difficulty of delivery due to obstruction of the birth canal by the infant's shoulders.

infants have experienced long-term malnutrition in utero, due to factors such as prepregnancy underweight, consistently low rates of maternal weight gain in pregnancy and the corresponding inadequate dietary intake, or chronic exposure to alcohol.[16, 45] Because nutritional insults existed during critical periods of growth early in pregnancy, pSGA infants generally have a reduced number of cells in organs and tissues. These babies tend to exhibit fewer health problems at birth than do pSGA infants, but catch-up growth is poorer, even with nutritional rehabilitation.[42] On average, pSGA infants remain shorter and lighter, and have smaller head circumferences throughout life than do infants born *appropriate for gestational age* (AGA) or *large for gestational age* (LGA).[46]

Nutrition, Miscarriages, and Preterm Delivery

Several other pregnancy outcomes are related in part to maternal nutrition. Highlighted here are the roles played by nutrition in miscarriage and preterm delivery.

MISCARRIAGES   Approximately 31% of conceptions are lost by reabsorption into the uterus or expulsion before 20 weeks of conception, and roughly a third of these are recognized as a miscarriage.[50] Such early losses of embryos and fetuses are thought to be primarily caused by genetic, uterine, or hormonal abnormalities, reproductive tract infections, or to tissue rejection due to immune system disorders.[51] Intakes of caffeine over 500 mg per day (or the equivalent of about four cups of brewed coffee), especially in women with nausea and vomiting of pregnancy, sharply increase the risk of miscarriage.[52] The presence of nausea and vomiting is otherwise related to a very low risk of miscarriage.[26]

PRETERM DELIVERY   Infants born preterm are at greater risk than other infants of death, neurological problems reflected later in low IQ scores, congenital malformations, and chronic health problems such as *cerebral palsy.* The risk for these outcomes increases rapidly as gestational age at birth decreases. Infants born very preterm (<34 weeks) commonly have problems related to growth, digestion, respiration, and other conditions due to immaturity.[53] Low stores of fat, essential fatty acids, glycogen, and other nutrients in very preterm infants may also interfere with growth and health after delivery.

Although preterm delivery is a major health problem in the United States, its etiology remains unclear, and the search for effective prevention programs continues. A portion of preterm deliveries appears to be related to genital tract infections, uterine abnormalities, prepregnancy underweight, and low weight gain in pregnancy. It is also fairly common in women who have previously delivered preterm.[54] Improvements in prenatal care for women at risk of preterm delivery, such as close supervision of the pregnancy, inclusion of nutritional counseling, encouragement of adequate weight gain in underweight and normal weight women, and home visits, appear to decrease the risk of preterm delivery somewhat. Calcium, essential fatty acids, and fish oil supplements have been used to prevent preterm labor, but results of studies on their effectiveness are mixed.[54]

> **CEREBRAL PALSY**   A group of disorders characterized by impaired muscle activity and coordination present at birth or developed during early childhood.
>
> **FETAL ORIGINS HYPOTHESIS**   The theory that exposures to adverse nutritional and other conditions during critical or sensitive periods of growth and development can permanently affect body structures and functions. Such changes may predispose individuals to cardiovascular diseases, type 2 diabetes, hypertension, and other disorders later in life. Also called *metabolic programming* and the *Barker Hypothesis.*

The Fetal Origins Hypothesis of Chronic Disease Risk

The implications of the associations between fetal nutrition and adult disease are immense, and if substantiated, demand intense scrutiny of current prenatal nutrition policies.

J. King, 2000[1]

In the last decade, thinking about chronic disease risk has changed substantially. In contrast to earlier thinking that disease risk begins during childhood or in the adult years, studies testing the *fetal origins hypothesis* indicate that risks may begin in utero. The concept that chronic disease risk may be established in utero is strongly supported by animal studies and by epidemiological investigations in humans.[10] Much of the evidence that relates in utero exposures to later disease risk in humans comes from studies showing increased risk for diseases such as heart disease, hypertension, type 2 diabetes, gestational diabetes, and chronic bronchitis in small, short, and thin newborns (Table 4.16). Because newborn size is a primary indicator, it is generally thought that maternal nutrition may underlie the development of later disease risk.[38]

Relatively small reductions in newborn weight or disproportions in size have been related to increased later disease risk. Risk of cardiovascular disease (heart disease and stroke) for instance, is associated with birthweights below 7.5 pounds (3360 g)—weights that are often considered "normal." Results of the U.S. Nurses Study, which compared newborn birth-

Table 4.16

Diseases and other conditions in adults related to smallness or thinness at birth.[10,38]

Allergies	Obesity
Autoimmune diseases	Ovarian cancer
Bronchitis	Polycystic ovary syndrome
Cardiovascular disease	Schizophrenia
Decreased bone mineral content	Short stature
Gestational diabetes	Stroke
Hypertension	Subfertility in males
Kidney disease	Suicide
Metabolic syndrome	Type 2 diabetes

Table 4.17 Association of birthweight with the risk of cardiovascular disease in the U.S. Nurses Study.[55]

BIRTHWEIGHT	RELATIVE RISK OF:	
	Heart Disease	Stroke
<5 lb (2240 g)	1.5	2.3
5–5½ lb (2240–2500 g)	1.3	1.4
5½–7 lb (2500–3136 g)	1.1	1.3
7–8½ lb (3136–3808 g)	1.0	1.0
8½–10 lb (3808–4480 g)	1.0	1.0
>10 lb (>4480 g)	0.7	0.7

weight to risk of cardiovascular disease in adults aged 46 to 71 years, are provided in Table 4.17 to illustrate this point. Additionally, disease risk in people born small, short, or thin may be exacerbated by later, excessive weight gain.[10,56]

How is Disease Risk Increased In Utero?

Given less than optimal growing conditions during gestation, fetal tissues make adaptations to cope with energy and nutrient shortages and excesses, and with harmful substances. Although such adaptations may support in utero survival, they may increase the risk of disease later in life. Adaptations in structures and functions that occur during critical periods of fetal growth and development may result in *programming,* or permanent changes in the structure and function of organs and tissues. Altered newborn size appears to be an outward sign that such programming has occurred.

The Why and How of Programming The ability to modify structures and functions in response to available energy or nutrient supply is advantageous to the fetus in several ways. Adaptations could help the fetus survive by changing its requirements for energy or nutrients, and they may also biologically prepare the fetus for similar nutritional circumstances after birth. Fetal adaptations that lower energy or nutrient needs in response to a low supply, for example, may promote survival in an environment of limited food after birth.

Example of a Potential Mechanism Take the example of a fetus exposed to a low supply of glucose early in pregnancy. This circumstance would hinder central nervous system development and threaten fetal survival. Mechanisms are set in place, however, that triage available glucose to the central nervous system. This change represents an adaptation to how glucose utilization is programmed to operate by the body.

What adaptations are made to ensure the CNS gets priority access to glucose? Mechanisms behind adaptations have been mostly investigated in animals, so their application to humans is unclear. However, animal studies indicate that the expression of genes that produce insulin receptors on muscle cell membranes may be suppressed in response to a low availability of glucose. This would decrease uptake of glucose by muscle cells and reduce their growth. It would also increase the availability of glucose for central nervous system development.

> **Programming** The process by which exposure to adverse nutritional or other conditions during sensitive periods of growth and development produces long-term effects on body structures, functions, and disease risk.

Adaptations that decrease muscle utilization of glucose and reduce muscle size may serve the offspring well later in life if food availability and intake are limited. If food is abundant and food intake is high, however, such adaptations may lead to elevated blood levels of glucose and insulin. These changes may increase the risk of obesity, type 2 diabetes, gestational diabetes, and other disorders associated with elevated blood glucose and insulin levels.[56]

Other In Utero Exposures That May Be Related To Later Disease Risk

Some studies have shown a link between maternal nutritional exposures during pregnancy and later disease risk in infants with a wide range of birthweights. Low weight gain around mid-pregnancy, for example, has been associated with higher blood pressure in children. Children born to women receiving a calcium supplement during pregnancy have been found to have lower blood pressure than children born to women given a placebo.[57] Obviously, not all programmed changes that affect later disease risk impair fetal growth. In addition, chronic disease risk appears to be lower among infants born small due to preterm delivery alone than for infants born small, thin, or short for gestational age.[10]

Non-nutritional factors, such as hypertension during pregnancy and genetic abnormalities may also interfere with delivery of an optimal supply of energy and nutrients to the fetus during critical periods of growth and development. Conditions that alter the hormonal milieu of the placenta and fetus may also program later disease risk.[10]

Limitations of the Fetal Origins Hypothesis

The hypothesis that maternal and fetal nutrition exposures influence later disease risk is gaining support and recognition. Many questions are unanswered, however. Which specific nutritional exposures are responsible for increased disease risk? When do the vulnerable periods of fetal sensitivity to poor nutrition occur? To what extent are genetic factors responsible for the relationships observed? The implications of the associations between maternal and fetal nutrition and adult disease risk are immense, and if substantiated, demand intense scrutiny of current prenatal nutrition policies and recommendations.[1]

PREGNANCY WEIGHT GAIN

Any obstetrician who allows a woman to lose her attractiveness (i.e., gain too much weight) is depriving her of many things that make for her mental well-being, her husband's contentment, and her own personal satisfaction.

Loughran, American Journal of Obstetrics and Gynecology, 1946

Weight gain during pregnancy is an important consideration because newborn weight and health status tend to increase as weight gain increases. Birthweights of infants born to women with weight gains of 15 pounds (7 kg) for example, average 3100 grams (6 lb 14 oz). This weight is about 500 grams less than the average birthweight of 3600 grams (8 lb) in women gaining 30 pounds (13.6 kg). Rates of low birthweight are higher in women gaining too little weight during pregnancy.[34]

GRAVIDA Number of pregnancies a woman has experienced.

PARITY The number of previous deliveries experienced by a woman; *nulliparous* = no previous deliveries, *primiparous* = one previous delivery, *multiparous* = 2 or more previous deliveries. Women who have delivered infants are considered to be "parous."

Weight gain during pregnancy is an indicator of plasma volume expansion and positive calorie balance, and provides a rough index of dietary adequacy.[58] Multiple studies show broad agreement on amounts of weight gain that are related to the birth of infants with weights that place them within the lowest category of risk for death or health problems.[11] Yet, how much weight should be gained during pregnancy remains a hotly debated topic. Earlier in the last century, when gains were routinely restricted to 15 or 20 pounds, weight gain in pregnancy was seen as the cause of pregnancy hypertension, difficult deliveries, and obesity in women. Pregnant women would be placed on low-calorie diets and given diuretics and amphetamines and urged to use saccharin to limit weight gain.[59]

Although none of these notions have been shown to be true, weight gain during pregnancy still represents a prickly issue. Weight gain and body weight are not only a matter of health, but are also closely linked to some people's view of what is socially acceptable.

Psychological and sociological biases related to body weight and shape in women are an important reason why recommendations for weight gain in pregnancy based on scientific studies and consensus must be applied.

Pregnancy Weight Gain Recommendations

Current recommendations for weight gain in pregnancy are based primarily on gains associated with the birth of healthy-sized newborns (approximately 3500–4500 g or 7 lb 13 oz to 10 lb).[11] As shown in Illustration 4.9, however, prepregnancy weight status influences the relationship between weight gain and birthweight. The higher the weight before pregnancy, the lower the weight gain needed to produce healthy-sized infants. Recommended weight gains for women entering pregnancy underweight, normal weight, overweight, and obese are displayed in Table 4.18.[11] Because underweight women tend to retain some of the weight gained in pregnancy for their own needs, they need to gain more weight in pregnancy than do other women. Overweight and obese women, on the other hand, are able to use a portion of their energy stores to support fetal growth, so they need to gain less. A separate and higher pregnancy weight gain recommendation is given for women expecting twins. Young adolescents and women who are at the low end of the prepregnancy weight ranges are encouraged to gain near the upper end of their weight gain ranges, and short women and those at the higher end of the weight status range ought to gain near the lower end of the range. Because factors such as duration of gestation, smoking, maternal health status, *gravida,* and *parity,* also influence birthweight, gaining a certain

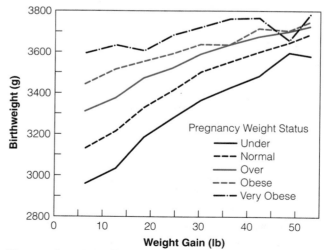

Illustration 4.9 Pregnancy weight gain by prepregnancy weight status and birthweight.

SOURCE: Used by permission of Harcourt Publishers Ltd. Brown JE. Clin Nutr 1988;7:181–90.

Table 4.18   Pregnancy weight gain recommendations.*

Prepregnancy Weight Status	Recommended Gain
Underweight, BMI <19.8	28–40 lb (12.7–18.2 kg)
Normal weight, BMI 19.8–26.0	25–35 lb (11.4–15.9 kg)
Overweight, BMI >26.0–29.0	15–25 lb (6.8–11.4 kg)
Obese, BMI >29.0	15 lb (6.8 kg) at least
Twin pregnancy	35–45 lb (15.9–20.5 kg)

*Young adolescents should achieve gains at the upper end of ranges, short women at the lower end.
SOURCE: Reprinted with permission. Institute of Medicine, National Academy of Sciences, 1990.

amount of weight during pregnancy does not guarantee that newborns will be a healthy size. It does improve the chances that that will happen, however.

Approximately one-third of U.S. women gain within the recommended weight ranges during pregnancy.[60] For all except the obese, women who gain within the recommended ranges are approximately half as likely to deliver low birthweight or SGA newborns as women who gain less. Rates of LGA newborns, Caesarean section deliveries, and postpartum weight retention tend to be higher when pregnancy weight gain exceeds that recommended.[60]

Restriction of pregnancy weight gain to levels below the recommended ranges is not recommended.[25] It does not decrease the risk of pregnancy-related hypertension and is associated with increased infant death and low birthweight, and poorer offspring growth and development.[9,25] In addition, low weight gain in pregnancy may increase the risk that infants will develop heart disease, type 2 diabetes, hypertension, and other types of chronic disease later in life.[38]

RATE OF PREGNANCY WEIGHT GAIN   Rates at which weight is gained during pregnancy appear to be as important to newborn outcomes as is total weight gain. Low rates of gain in the first trimester of pregnancy may down-regulate fetal growth and result in reduced birthweight and thinness.[61] Rates of gain of less than 0.5 pound (0.25 kg) per week in the second half of pregnancy, and of less than 0.75 pound (0.37 kg) per week in the third trimester of pregnancy by underweight and normal-weight women double the risk of preterm delivery and SGA newborns. Rates of gain of less than 0.5 pound (0.25 kg) per week in the third trimester by overweight and obese women also double the risk of preterm birth.[62] Third trimester weight gains exceeding approximately 1.5 pounds a week (0.7 kg), however, add little to birthweight in normal-weight and heavier women, and may increase postpartum weight retention.[63]

Rate of weight gain is generally highest around mid-pregnancy—which is prior to the time the fetus gains most of its weight (Illustration 4.10). In general, the pattern of gain should be within a few pounds of that represented by the weight gain curves shown in Illustration 4.11.[11] Some weight (4 to 5 pounds) should be gained in the first trimester, followed by gradual and consistent gains thereafter. The rate of weight gain often slows a bit a few weeks prior to delivery, but as is the case for the rest of pregnancy, weight should not be lost until after delivery.[61]

Composition of Weight Gain in Pregnancy

A question often asked by pregnant women is "Where does the weight gain go?" Where the weight gain generally goes by time in pregnancy is shown in Table 4.19. The fetus actually comprises only about a third of the total weight gained during pregnancy in women who enter pregnancy normal weight or underweight. Most of the rest of the weight is accounted for by the increased weight of maternal tissues.

BODY FAT CHANGES   Pregnant women store a significant amount of body fat in normal pregnancy in order to meet their own and their fetus's energy needs, and quite likely to prepare for the energy demands of lactation. Body fat stores increase the most between 10 and 20 weeks of pregnancy, or before fetal energy requirements are highest. Levels of stored fat tend to decrease before the end of pregnancy. Only 0.5 kg of the approximately 3.5 kg of fat stored during pregnancy is deposited in the fetus.[15,19]

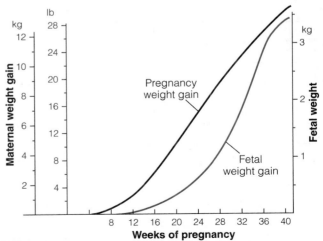

Illustration 4.10   Rates of maternal and fetal weight gain during pregnancy.

SOURCE: Curves drawn by Judith E. Brown, 2001[61]

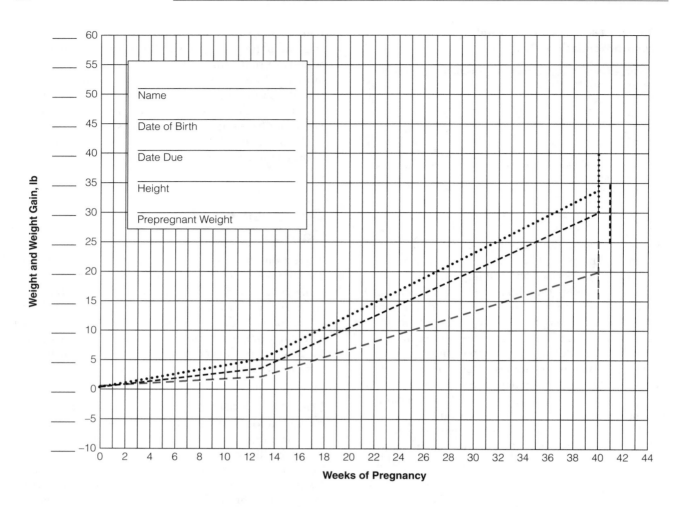

Name

Date of Birth

Date Due

Height

Prepregnant Weight

Weights and Weight Gain, lb

Weeks of Pregnancy

Weight Record			
Date	Weeks of Gestation	Weight	Notes

Please bring this chart with you to each prenatal visit.

Illustration 4.11   The Institute of Medicine's prenatal weight gain graph.[11]

SOURCE: Reprinted with permission of the National Academy of Sciences.

Table 4.19 Components of weight gain during pregnancy for healthy, normal-weight women delivering a 3500 gram (about 8 lb) infant at term.[14–16, 41]

	WEIGHT GAIN, G			
Component	10 Weeks	20 Weeks	30 Weeks	40 Weeks
Fetus	5	300	1500	3550
Placenta	20	170	430	670
Uterus	140	320	600	1120
Amniotic fluid	30	350	750	896
Breasts	45	180	360	448
Blood supply	100	600	1300	1344
Extracellular fluid	0	265	803	3200
Maternal fat stores	315	2135	3640	3500
Total weight gain at term = 14.7 kg or 32 lb				

Postpartum Weight Retention

Concern about the role of pregnancy weight gain in fostering long-term maternal obesity has increased in the United States along with the rising incidence of obesity in adults. Increased weight after pregnancy appears to be related to a variety of factors, including excessively high weight gain in pregnancy (over 45 lb, or 20.5 kg), weight gain after delivery, and low activity levels.[64]

Women tend to lose about 15 pounds the day of delivery, but subsequent weight loss is highly variable.[64] On average, however, women who gain within the recommended ranges of weight gain are 2.0 pounds (0.9 kg) heavier one year after delivery than they were before pregnancy.[65] This gain is slightly above the amount of weight women tend to gain with age.[66] Postpartum weight retention tends to be slightly less in women who breastfeed for at least six months after pregnancy.[67] Women who gain less than the recommended amount of weight gain in pregnancy do not retain less weight on average after pregnancy than do women who gain within the ranges.[65]

NUTRITION AND THE COURSE AND OUTCOME OF PREGNANCY

A mother who wishes her child to have black eyes should frequently eat mice.

Prenatal diet folklore from Ancient Rome[68]

The history of beliefs about the effects of maternal diet on the course and outcome of pregnancy is rife with superstition, ill founded and hazardous conclusions, and unhelpful suggestions. (The Connections feature provides some examples.) Societies have shared a belief in the importance of "eating right" during pregnancy for the sake of the child, but actual knowledge about

maternal nutrition and the course and outcome of pregnancy has been acquired only relatively recently.

Much of the scientific interest in the effects of maternal nutrition on the course and outcome of pregnancy comes from studies done in the first half of the twentieth century. Ecological studies on effects of famines in Europe and Japan during World War II on the course and outcome of pregnancy demonstrated potential negative, as well as positive, effects of food intake on fertility and newborn outcomes.

Connections
Diet and Pregnancy Folklore

To produce milk, mix fragrant bread with ground poppy plant and require the pregnant women to sit cross-legged while eating it. **Egypt, circa 1500 bc**

If pregnant women eat suckling pigs, the fetus will have weak joints and will be subject to diseases of the joints. **Hippocrates**

If a pregnant woman's desire for food is kindled and unsatisfied, there is a danger that the uterus may sink down. **Constantine**

To deny a pregnant woman any food she may desire might cause the unborn child to die of weakness. **Ancient Rome**

It is better for women with child to eat too much than too little lest her child should want nourishment. **Culpepper, 1651, in a Directory for Midwives**

(It was a common belief during this time that labor was brought about by a woman's inability to supply enough food for the baby. If it wanted more, it would have to come out to get it!)

The diet of the pregnant women should be restricted to lessen the size of the fetus and ease delivery. **Prochownick**

Famine and Pregnancy Outcome

THE DUTCH HUNGER WINTER, 1943–1944 As mentioned briefly in Chapter 2, people in many parts of Holland experienced severe food shortages for an eight-month period during World War II due to enemy occupation of major cities. Although people in Holland were generally well nourished and had a reasonable standard of living before the disaster, conditions rapidly deteriorated during the famine. In addition to intakes that averaged only about 1100 cal and 34 g of protein per day, fuel was in low supply and the winter harsh.

Carefully kept records by health officials showed a sharp decline in pregnancy rates of over 50% during the famine, an effect attributed to absent and irregular menstrual periods. Average birthweight declined by

372 g (13 oz), delivery of low birthweight infants increased by 50%, and rates of infant deaths increased. Birthweight did not fully "catch up" in infants born to women exposed to famine early in pregnancy even if they received enough food later in pregnancy. This result supports the notion that the fetal growth trajectory may be established early in pregnancy and that early nutritional deprivations limit fetal growth regardless of food intake later in pregnancy.[69]

Although the famine in Holland was associated with major declines in fertility and newborn health and survival, the rather good nutritional status of women prior to the famine likely protected pregnant women and their infants from more severe disruptions in health. Normal fertility status and newborn outcomes returned within a year after the famine ended.[70]

Studies undertaken in the last 30 years on adults who were born to women during the hunger winter (the Dutch famine cohort) show relationships between the timing of famine during pregnancy and adult offspring health outcomes. Examples of relationships identified are shown in Table 4.20.

THE SEIGE OF LENINGRAD, 1942 Unlike people in Holland, the population in Leningrad had experienced moderate deprivations in nutritional status and quality of life prior to the famine. As was the case for pregnant women in Holland, the famine resulted in average intakes of approximately 1100 cal per day. Infertility and low birthweight rates increased over 50%, infant death rates rose, and birthweights dropped by an average of 535 g (1.2 lb) during the famine.[73] Rates of pSGA newborns also increased, suggesting that the poor nutritional status of women coming into pregnancy and persistent undernutrition during pregnancy interfered with critical periods of fetal growth.

FOOD SHORTAGES IN JAPAN Effects of World War II–associated food shortages on reproductive outcomes in Japan were similar to those observed in Holland. Japanese women tended to be well nourished prior to the shortages. Lack of food before and during pregnancy was reflected in decreased fertility status among women and to reductions in birthweight that averaged 200 g.

Social and economic improvements occurring in Japan after the war led to increased availability of many foods, including animal products. This higher plane of nutrition achieved during the postwar years in Japan was accompanied by major increases in newborn size and the "growing up" of Japanese children. In a trend that continues today, subsequent generations of Japanese adults averaged two inches taller than the previous generation.[74] Infant mortality in Japan, which ranked among the highest for industrialized nations prior to World War II, declined incredibly after the war and remains well below rates in the United States and in a number of other developed countries.[5]

Food shortages continue to occur in various parts of the world and to adversely affect fertility and the course and outcome of pregnancy. Effects have become predictable, such that declines in fertility and newborn size and vitality are viewed to be part of the consequence of such disasters. The siege of Sarajevo, which decreased food availability during 1993–1994, for example, led to reduced caloric and nutrient intakes during pregnancy, reduced weight gain, and increased proportions of poor pregnancy outcomes. During these two years perinatal mortality increased from 16 to 36 per 1000 deliveries, and the proportion of infants born with congenital abnormalities increased from 0.4 to 3.0%.[75]

Contemporary Prenatal Nutrition Research Results

Faulty nutrition, not just malnutrition, can influence fetal health.

Bertha Burke, 1948[76]

Carefully conducted studies of diet and pregnancy outcome in the first half of the twentieth century began the era of scientifically based recommendations on nutrition and pregnancy. The now classic studies conducted by Bertha Burke at Harvard in the 1940s were particularly influential.[14,76,77] These studies showed that diet quality during pregnancy, assessed using diet histories, was strongly related to newborn health status. Newborns assessed as having optimal physical condition by pediatricians were found to be much

Table 4.20   Exposure to the Dutch World War II famine by time in pregnancy and adult offspring health risks.[71,72]

PERIOD OF FAMINE		
First Trimester	**First and/or Second Trimester**	**Second Half of Pregnancy**
Schizophrenia	Antisocial personality disorder	Decreased glucose tolerance
High LDL and low HDL cholesterol		
High body weight and central body fat		
Infertility		
Neural tube defects		

more common among women consuming high quality diets, whereas those with the poorest physical condition were born to women with the poorest quality diets. Average birthweight of newborns assessed as being in optimal physical condition was 7 pounds, 15 ounces in females, and 8 pounds, 8 ounces in males.[77] Although Burke's studies did not show that high quality pregnancy diets by themselves were responsible for robust newborn health, they provided some of the first evidence that prenatal diet quality may strongly influence pregnancy outcome.

Thousands of other studies on the effects of nutrition on the course and outcome of pregnancy are now available. The following sections highlight research results and recommendations related to calories, key nutrients, and other substances in food that influence the course and outcome of pregnancy.

Energy Requirement in Pregnancy

Energy requirements increase during pregnancy mainly due to increased maternal body mass and fetal growth. The additional requirements can be allocated to different maternal and fetal tissues by estimating the amount of oxygen used (or "consumed") by the various tissues. Illustration 4.12 shows the results of work on oxygen consumption during pregnancy undertaken by Hytten and Leitch.[78] Approximately one-third of the increased calorie need in pregnancy is related to increased work of the heart, and another third to increased energy needs for respiration and accretion of breast tissue, uterine muscles, and the placenta. The fetus accounts for about a third of the increased energy needs of pregnancy.

The increased need for energy in pregnancy averages 300 cal a day, or a total of 80,000 cal.[14] This is a rough estimate that by no means applies to all women. Additional energy requirements of women have been found by different studies to range from 210–570 cal a day.[15] The need for additional calories during pregnancy may be a good deal lower in women who perform little exercise, and higher in women who are very active. Low levels of energy expenditure from physical activity is common in the first trimester of pregnancy, and the energy savings may produce a positive caloric balance even though a woman's caloric intake hasn't changed much. Contrary to a previous belief, energy needs of pregnant women do not appear to be affected by "metabolic efficiencies" of pregnancy that decrease caloric need.[79]

Illustration 4.13 shows the difference between caloric (kcal) intake and estimated caloric balance throughout pregnancy in a group of women served by a health maintenance organization.[79] The graph indicates that estimated caloric balance is higher than caloric intake throughout pregnancy and becomes negative postpartum. The positive caloric balance during pregnancy observed is due to the fact that women consumed more calories than they expended in physical activity and basal metabolism.

ASSESSMENT OF CALORIC INTAKE   Adequacy of calorie intake is most easily assessed in practice by

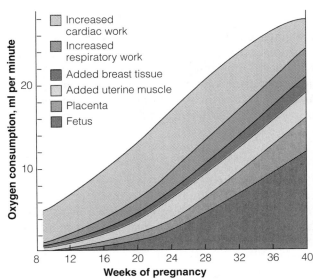

Illustration 4.12   Components of increased oxygen consumption in normal pregnancy.

SOURCE: Reprinted with permission. Hytten and Leitch, 1980.[78]

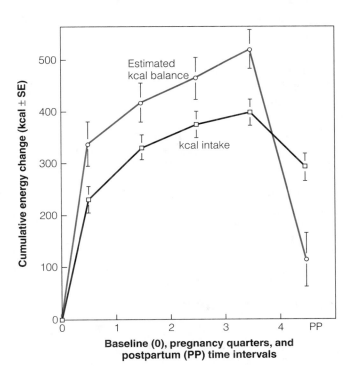

Illustration 4.13   Estimated caloric balance in pregnancy through 6–8 weeks postpartum.

SOURCE: Reprinted with permission. Brown and Kahn, 1997.[79]

Photo Disc

Case Study 4.1
Fasting in Labor: Yes or No?

The answer to this question is not known due to the lack of sufficient research, but opinions about it tend to be strong.

As in every area of medicine, some obstetrical practices are based more on ritual and intuition than scientific evidence. The practice of restricting fluid and food intake during labor appears to be one such practice, as there is no evidence that it improves outcomes of mothers or infants. Labor is a high energy, Olympic-caliber event. Yet many women are made to fast their way through it, dependent on intravenous fluid and glucose, and occasional sips of water or ice chips. This is the practice in most U.S. hospitals, but the routine in hospitals in other countries allows unrestricted fluids or fluids and certain solid foods.[80]

The motivation behind fluid and food restriction during labor is based on the concern that anesthesia (if needed) will delay stomach emptying and increase the risk of aspiration—the regurgitation and inhalation of stomach contents. This situation is very rare but can lead to the development of pneumonia. The odds of dying from aspiration in labor, however, are less than those for being struck by lightning, or less than 1 in 600,000. In addition, an empty stomach does not guarantee that aspiration will not occur. Hydrochloric acid and other digestive juices normally present in an empty stomach can enter the lungs if aspiration occurs. There is also evidence that intravenous glucose solutions given to laboring women may produce high blood glucose levels in the fetus, which then drop rapidly after delivery.[80]

It is possible that allowing women to eat high-carbohydrate, pre-event-type foods as athletes do and to drink hydrating fluids would help them maintain their strength throughout this physically demanding event. This and other considerations must be tested before it can be confidently concluded that current practice represents best practice.

pregnancy weight gain. Rates of gain in women who do not have noticeable edema is a good indicator of caloric balance. Women who are losing weight are in negative balance, and those gaining weight in positive caloric balance. Adjustments in rate of weight gain can be approached through modifying physical activity level, calorie intake, or both. Women should maintain a positive caloric balance and rate of weight gain throughout pregnancy. Meal skipping or fasting during pregnancy are not advised. Whether women should restrict their food and fluid intake during labor, however, is hotly debated (Case Study 4.1).

Carbohydrate Intake During Pregnancy

Approximately 55% of total caloric intake during pregnancy should come from carbohydrate and consist mainly of complex and unrefined sources of carbohydrate such as starchy vegetables, whole and fortified grain products, and fruits. Inclusion of foods that provide 25–35 g of fiber a day adds to the diet a variety of beneficial phytochemicals and a measure of protection against constipation.[33]

ARTIFICIAL SWEETENERS There is no evidence that consumption of aspartame (Nutrasweet) or acesulfame K (Sunette) is harmful in pregnancy.[81] Diet soft drinks and other artificially sweetened beverages and foods are often poor sources of nutrients, however, and may displace other, more nutrient-dense foods in the diet.

Alcohol and Pregnancy Outcome

Prenatal exposure to alcohol is a leading, preventable cause of birth defects, mental retardation, and developmental disorders.[82] Approximately 1 in 12 women consume alcohol during pregnancy, and 1 in 30 consume five or more drinks on one occasion at least monthly.[83] It is estimated that approximately 50,000 infants are born each year in the United States with some degree of alcohol-related damage.[84]

Alcohol consumed by a woman easily crosses the placenta to the fetus. Because the fetus has yet to fully develop enzymes that break it down, alcohol lingers in the fetal circulation. This situation, plus the fact that the fetus is smaller and has far less blood than the mother, increases the harmful effects of alcohol on the fetus compared to the mother. Alcohol exposure during critical periods of growth and development can permanently impair organ and tissue formation, growth, health, and mental development.

Poor dietary intakes of some women who consume alcohol regularly in pregnancy, as well as the negative effect of alcohol on the availability of certain nutrients, may also contribute to the harmful effects of alcohol exposure during pregnancy.[85]

Consumption of four or more drinks a day, or occasional episodes of consumption of five or more drinks in a row is considered to represent "heavy" alcohol intake during pregnancy. Heavy drinking during pregnancy increases the risk of miscarriage, stillbirth, and infant death within the first month after delivery.[84] Approximately 40% of the fetuses born to women who drink heavily early in pregnancy will develop fetal alcohol syndrome (FAS). The likelihood that the fetus will be affected by FAS increases as the number of drinks consumed early in pregnancy increases (Table 4.21).

FAS FAS was first described in 1973 and consists of pSGA, mental retardation, and a set of common malformations (Illustration 4.14). Diagnosis of FAS is difficult, however, because many of the facial features are not unique to the syndrome. Short noses, flat nasal bridges, and thin upper lips, for example, are also sometimes normal facial features.

Children with FAS tend to have poor coordination, short attention span, behavioral problems, and to remain small for their age. Adults with FAS often find it difficult to hold a job or live independently.[84]

FETAL ALCOHOL EFFECTS (FAE) Approximately 10 times the number of infants born with FAS have lesser degrees of alcohol-related damage known as

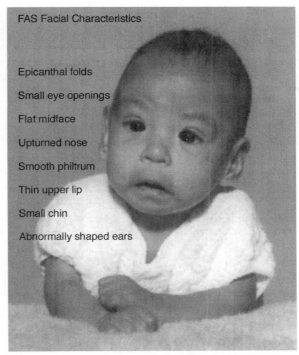

Illustration 4.14 Features of FAS in children.[85]

FAE. It is characterized by mental and behavioral abnormalities but not malformations.[82]

Alcohol Related Birth Defects and Neurodevelopmental Disorders Fetal exposure to alcohol has been linked to many other neurodevelopmental problems since it was first described in 1973. To reflect advances in knowledge, it has been recommended that the terms "alcohol related birth defects" and "alcohol related neurodevelopmental disorders" replace the term FAE. The changes (summarized in Illustration 4.15) emphasize the wide range of neurodevelopmental problems such as mental retardation, aggressiveness, destructiveness, nervousness, and short attention span associated with intrauterine exposure to excess levels of alcohol.[82]

Protein Requirement

The recommended protein intake for pregnancy is 60 g per day, an amount that is generally exceeded by nonpregnant women consuming a Western diet.[87] Protein content of diets can be simply estimated by evaluating women's usual daily intake of major sources of protein. A tool for estimating protein intake is shown in Table 4.22.

VEGAN PREGNANT WOMEN Protein needs of women who do not consume animal products are somewhat higher than those for women who do, because of the lower quality of plant proteins.

Table 4.21 Approximate incidence of structural abnormalities of the brain in 5 to 14 week old fetuses exposed to alcohol.

Maternal alcohol intake, drinks	Abnormal brain structure
13 to 31/day	100%
6.3 to 13/day	83%
2 to 6.3 occasionally	29%
≤2/day	0%

SOURCE: Table is based on formation presented by Konovalov et al, 1997.[86]

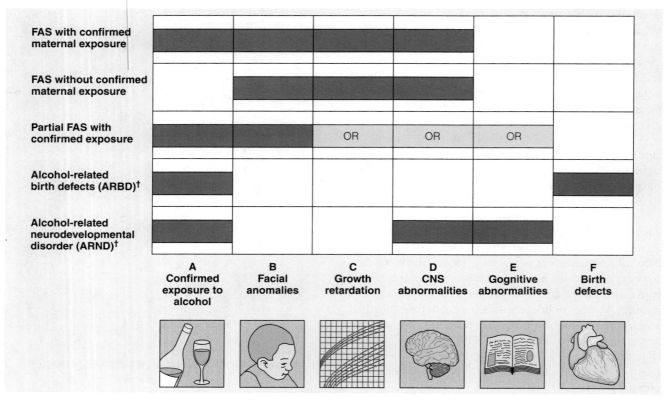

	A Confirmed exposure to alcohol	B Facial anomalies	C Growth retardation	D CNS abnormalities	E Cognitive abnormalities	F Birth defects
FAS with confirmed maternal exposure	■	■	■	■		
FAS without confirmed maternal exposure		■	■	■	■	
Partial FAS with confirmed exposure	■	■	OR	OR	OR	
Alcohol-related birth defects (ARBD)[†]	■					■
Alcohol-related neurodevelopmental disorder (ARND)[†]	■			■	■	

Illustration 4.15   Diagnostic classification of FAS, Alcohol Related Birth Defects, and Neurodevelopmental Disorders.[82]

SOURCE: Reprinted with permission of the American Academy of Pediatrics.[82]

Combinations of different plant sources of protein, such as grains and legumes (e.g., corn and lima beans, rice and beans, bread and humus), consumed daily can provide adequate, high-quality protein for pregnancy.

Vegetarian Diets in Pregnancy

The topic of vegetarian dietary practices often brings with it a variety of images and attitudes regarding those who follow such practices. Those attitudes may have limited, if any, basis in actual fact.

Patricia Johnston, 1988[88]

Nutrient needs in pregnancy may be met by many different types of diets, including those that do not include animal products. It is the type and amount of food consumed that determines the appropriateness of dietary intake during pregnancy, not the label placed on it.

The quality of dietary intakes during pregnancy of vegetarian women can be estimated by comparing number of servings from the Food Guide Pyramid groups with that recommended for pregnancy.

Nonanimal-product alternatives to meat, eggs, milk, and milk products such as legumes, nuts, seeds, and fortified soymilk should be included in the assessment. Careful attention should be focused on the adequacy of highly restrictive vegetarian diets to ensure that intakes of energy, vitamin D and B_{12}, calcium, or zinc are sufficient. Computerized nutrient analysis of

Table 4.22

Tool for estimating protein intake.

Food	Protein, grams	How much protein is there in this usual day's diet?	
Milk, 1 c	8		
Cheese, 1 oz	7	2 sl toast	6
Egg, 1	7	1 c milk	8
Meat, 1 oz	7	3 oz tuna	21
Dried beans, 1 c	13	2 sl bread	6
Bread, 1 sl or oz	3	2 oz chicken	14
		1 oz cheese	7
		2 tortillas	6
		½ c refried beans	7
		Total g protein =	75

several days of usual food intake may be especially helpful in assessing such diets.[88] Evaluation of rate of weight gain in pregnancy is generally a good way to assess the adequacy of energy intake.

Maternal Intake of Essential Fatty Acids and Pregnancy Outcome

The essential fatty acids, linoleic acid and alpha-linolenic acid, were presented in Chapter 1. They are referred to again here because inadequate maternal intake of these fatty acids and their derivatives may impair fetal growth and development. Recent research has focused on the importance of maternal intake of alpha-linolenic acid and its derivative docosahexaenoic acid (DHA) in fetal vision development. Adequacy of maternal intake of DHA from foods is being emphasized because it does not appear that sufficient amounts of DHA can be produced by the body from alpha-linolenic acid to meet fetal needs.

DHA AND VISION DEVELOPMENT Knowledge of the effects of DHA on vision development come primarily from studies involving preterm infants but may apply to term infants as well. Preterm infants are likely to be born with inadequate DHA in the retina, the part of the eye that receives light rays. DHA is incorporated into the retina at a high rate during retinal development in the final months of pregnancy and the first six months after birth.[19] Lack of DHA during retinal development appears to reduce the sharpness of vision and slows visual recognition of objects. Neurological development measured by IQ score and problem-solving skills may also benefit from the presence of adequate levels of DHA during fetal and infant development.[89]

Human milk tends to be a good source of DHA, whereas infant formulas are not. (Few formulas sold in the United States are fortified with DHA, and recommendations regarding its addition to all infant formulas are pending.) It is suspected that the DHA content of human milk is related to the superior intellectual performance of preterm infants given human milk versus those fed formula. Women consuming vegetarian diets that exclude fish and other seafood are likely to have low levels of DHA available for fetal growth and development.[19]

DIETARY INTAKE RECOMMENDATIONS RELATED TO ESSENTIAL FATTY ACIDS Physiological effects of the essential fatty acids are interrelated, and high intakes of one combined with low intakes of the other can be detrimental to health. Consequently, a balanced intake of the essential fatty acids is required for their optimal

functioning. Although much debated, the ideal ratio of intake of linoleic and alpha-linolenic acid remains unclear. For most North Americans, it appears that intake of linoleic acid exceeds that of alpha-linolenic acid by too wide a margin. Americans are being urged to increase their intake of food sources of alpha-linolenic acid and DHA (Table 4.23). Although under consideration, there is currently no consistent recommendation regarding the use of DHA supplements during pregnancy or infancy.[19]

The Need for Water During Pregnancy

The large increase in water need during pregnancy is generally met by increased levels of thirst. On average, women consume about 9 cups of fluid daily

Table 4.23 **Food sources of the essential fatty acids and their derivatives.**

Linoleic Acid (n-6)	Alpha-linolenic Acid (n-3)
• Safflower oil	• Linseed oil
• Walnut oil	• Spinach
• Soybean oil	• Brussels sprouts
• Sunflower oil	• Kale
• Black current oil	• Human milk
• Borage oil	• Canola oil
• Walnut oil	• Soybean oil
• Evening primrose oil	• Flax seed, flaxseed oil
• Peanut oil	• Walnuts
• Cottonseed oil	**DHA**
• Corn oil	• Mackerel
• Sesame oil	• Atlantic salmon
• Canola oil	• Cod, cod liver oil
Arachidonic acid[a]	• Bluefish
• Beef	• Halibut
• Egg yolk	• Sockeye salmon
• Liver	• Striped bass
• Pork	• Tuna
• Chicken	• Flounder
• Human milk	• Anchovies
	• Sardines
	• Scallops
	• Haddock
	• Human milk

[a]These foods contain small amounts of arachidonic acid, or ≤3% of total fatty acid content of the food.

during pregnancy.[90] In hot and humid climates, however, and among women who engage in physical activity in such climates, thirst mechanisms may not keep up with need. Women in such climates should drink enough to keep urine light colored and remember to drink plenty of fluids when it is hot or they are exercising. Water, diluted fruit juice, iced tea, and other unsweetened beverages are good choices for staying hydrated, because they are low in sugar.

Folate and Pregnancy Outcome

Inadequate folate during pregnancy has long been associated with anemia in pregnancy and reduced fetal growth.[91] Only during the last two decades, however, has the broad spectrum of effects of folate been recognized. Discoveries of the multiple effects of inadequate folate intake on the development of congenital abnormalities and clinical complications of pregnancy represent some of the most important advances in our knowledge about nutrition and pregnancy.

FOLATE BACKGROUNDER   The term *folate* encompasses all compounds that have the properties of folic acid and includes monoglutamate and polyglutamate forms of the vitamin. The monoglutamate form of folate is represented primarily by folic acid, a synthetic form of folate used in fortified foods and supplements. A similar monoglutamate form of folate naturally occurs in a few foods. Food sources of folate contain primarily the polyglutamate form of folate. The two major types of folates are often distinguished by referring to the monoglutamates as folic acid and the polyglutamates as dietary folate.

Bioavailability of folic acid and dietary folate differ substantially. Folic acid is nearly 100% bioavailable if taken in a supplement on an empty stomach, and 85% bioavailable if consumed with food or in fortified foods. Naturally occurring folates are 50% bioavailable on average.[92]

Folate requirements increase dramatically during pregnancy due to the extensive organ and tissue growth that takes place.

FUNCTIONS OF FOLATE   Folate is a methyl group (CH_3) donor and enzyme cofactor in metabolic reactions involved in the synthesis of DNA, gene expression, and gene regulation. Deficiency of folate impairs these processes, leading to abnormal cell division and tissue formation.[93] Folate serves as a methyl donor in the conversion of homocysteine to the amino acid methionine. The conversion of homocysteine to methionine depends primarily on three enzymes and folate, vitamin B_{12}, and vitamin B_6 cofactors. Lack of folate in particular, and less commonly a lack of vitamin B_{12}, as well as genetic

abnormalities in the enzymes can lead to an accumulation of homocysteine.[94] This may result in methionine shortage at a crucial stage of fetal development. High cellular and plasma levels of homocysteine may also contribute to the development of some of the adverse effects of folate inadequacy.[94] A common genetic defect has been identified in the enzyme 5,10-methylene tetrahydrofolate reductase (MTHFR). This variant of the normal enzyme reduces the level of activity of MTHFR by about half.[94] Variant forms of MTHFR and defects in methionine synthase are thought to be present in approximately 30% of the population.[94]

Folate and Congenital Abnormalities

There is a widespread belief that congenital malformations are always the result of defective genes. Perhaps a nutritional deficiency resulting in a defective gene leads to the same congenital abnormality.

R.D. Mussey, 1949[68]

It has been known since the 1950s that low and high intakes of certain vitamins and minerals caused congenital abnormalities in laboratory animals. It was also known that neural tube defects, brain and heart defects, and cleft palate could be caused by feeding pregnant rats folate-deficient diets.[95] Firmly held beliefs that only severe malnutrition affected fetal growth and that genetic errors were the sole cause of congenital abnormalities delayed recognition of the importance of folate to human pregnancy.[96]

Neural tube defects (NTDs) are malformations of the spinal cord and brain. There are three major types of NTDs:

- Spina bifida is marked by the spinal cord failing to close, leaving a gap where spinal fluid collects during pregnancy (see Illustration 4.16). Paralysis below the gap in the spinal cord occurs in severe cases.
- Anencephaly is the absence of the brain or spinal cord.
- Encephalocele is characterized by the protrusion of the brain through the skull.

It is now well accepted that inadequate availability of folate between 21 and 27 days after conception (when the embryo is only 2–3 mm in length) can interrupt normal cell differentiation and cause NTDs.[97] Neural tube defects are among the most common types of congenital abnormalities identified in infants, with approximately 4000 pregnancies affected each year in the United States.[98] They are among the most preventable types of congenital abnormalities that exist.[99] Approximately 70% of cases of NTDs can be

Illustration 4.16   A newborn child with spina bifida.

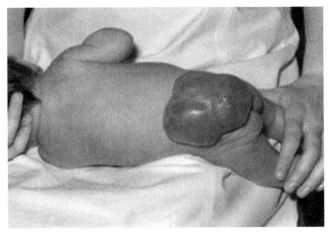

SOURCE: Photo Researchers, Inc.

prevented by consumption of adequate folate before and during very early pregnancy.[91]

FOLATE STATUS OF WOMEN IN THE UNITED STATES

Folate status is assessed by serum and red cell folate levels. Of the two measures, red cell folate levels are the preferred indicator because they represent long-term folate intake whereas serum folate levels reflect only recent intake. Levels of red cell folate of over 300 ng/mL (or 680 nmol/L) are associated with very low risk of NTDs.[100] These levels of red cell folate can generally be achieved by folic acid intakes that average 400 mcg daily.[101] As shown in Illustration 4.17, red cell folate levels are higher among women who consume folic acid fortified cereals or supplements compared to women consuming folate from food only.[101]

Folate status in women of child-bearing age has improved since the advent of folic acid fortification of refined grain products in 1998. Average levels of red cell folate in U.S. women have increased from 181 to 315 ng/mL since fortification began.[99] Low levels of intake of folic acid fortified grain products and breakfast cereals still leave some women with too little folate, however.[102]

DIETARY SOURCES OF FOLATE

Many vegetables and fruits are good sources of folate (See Table 1.9 in Chapter 1) but only a few foods contain the highly bioavailable form of folate. Table 4.24 lists some foods that naturally contain the highly bioavailable, monoglutamate form of folate and foods that provide folic acid through fortification.

Adequacy of folic acid intake before and during pregnancy can be estimated by adding up the amount of folic acid in foods typically consumed in the daily

Illustration 4.17   Mean red cell folate level in pre-conceptional women by level of intake of various sources of folate.

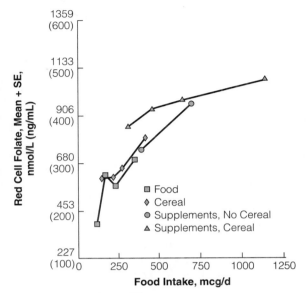

SOURCE: Brown, JE et al, 1997.[101]

diet using the data in the table. Whole grain products including breads and pastas, brown rice, oatmeal, shredded wheat, and organic grain products are often not fortified with folic acid.

RECOMMENDED INTAKE OF FOLATE

Due to variation in folate bioavailability, the DRI for folate takes into consideration a measure called "dietary folate equivalents," or DFE. One DFE equals

- 1 mcg food folate,
- 0.06 mcg folic acid consumed in fortified foods or a supplement taken with food, and

Table 4.24   Food sources of folic acid.

Amount		Folic Acid, mcg
A. Foods		
Orange	1	40
Orange juice	6 oz	82
Pineapple juice	6 oz	44
Papaya juice	6 oz	40
Dried beans	½ c	50
B. Fortified Foods		
Highly fortified breakfast cereals[a]	1 c or 1 oz	400
Breakfast cereals	1 c or 1 oz	100
Bread, roll	1 sl or 1 oz	40
Pasta	½ c	30
Rice	½ c	30

[a]Includes All Bran, Complete, Crispix, Healthy Choice, Just Right, Product 19, Smart Start, Special K, Total, and Life cereals.

- 0.5 mcg of folic acid taken as a supplement on an empty stomach.

Folic acid taken in a supplement without food provides twice the dietary folate equivalents as does an equivalent amount of folate from food.

It is recommended that women consume 600 mcg DFE of folate per day during pregnancy and include 400 mcg folic acid from fortified foods or supplements.[92] The remaining 200 mcg DFE should be obtained from vegetables and fruits. These nutrient-dense foods provide an average of 40 mcg of folate per serving.[101] Because NTDs develop before women may realize they are pregnant, adequate folate should be consumed several months prior to, as well as throughout, pregnancy.

Women who have previously delivered an infant affected by a NTD are being urged to take 4000 mcg (4.0 mg) of folic acid in a supplement to reduce the risk of recurrence.[98] This dose, however, may be much higher than needed based on results of clinical trials.[103] The upper limit for intake of folic acid from fortified foods and supplements is set at 1000 mcg per day. There is no upper limit for folate consumed in its naturally occurring form in foods. The 1000 mcg level represents an amount of folic acid that may mask the neurological signs of vitamin B_{12} deficiency. If left untreated, vitamin B_{12} deficiency leads to irreversible neurological damage.[98]

Vitamin A and Pregnancy Outcome

Vitamin A is a key nutrient in pregnancy because it plays important roles in reactions involved in cell differentiation. Deficiency of this vitamin is rare in pregnant women in developed countries, but it is a major problem in many developing countries. When vitamin A deficiency does occur it can produce malformations of fetal lungs, urinary tract, and heart.[104]

Of more concern than vitamin A deficiency in the United States are problems associated with excessive intakes of vitamin A in the form of retinol or retinoic acid (but not beta-carotene). Intakes of these forms of vitamin A of over 10,000 IU per day, and the use of medications such as Accutane and Retin A for acne and wrinkle treatment, increase the risk of fetal abnormalities. Effects are particularly striking in infants born to women using Accutane or Retin A early in pregnancy. Fetal exposure to the high doses of retinoic acid in these drugs tends to develop "retinoic acid syndrome." Features of this syndrome include small ears or no ears, abnormal or missing ear canals, brain malformation, and heart defects.[16]

Due to its potential toxicity, it is recommended that women take no more than 5000 IU of vitamin A as retinol from supplements during pregnancy.[105]

Most supplements made today contain beta-carotene rather than retinol. Although strong warnings not to take Retin A or Accutane if pregnancy is possible are issued to women, ill-timed use continues to occur to some extent, as does the retinoic acid syndrome.

Vitamin D Requirement

Several groups of women are at risk for vitamin D deficiency in pregnancy. Included are women:

- Who consume small amounts of vitamin D fortified milk, or who consume raw milk,
- Whose skin is rarely exposed directly to the sun,
- Who consistently use sun block,
- Who have dark skin, and
- Who are vegans.

The primary effect of insufficient vitamin D during pregnancy is poor fetal bone formation due to poor utilization of calcium.[16] Infants born to women with inadequate vitamin D status in pregnancy tend to be small, to have poorly calcified bones and abnormal enamel, and to show low blood levels of calcium after birth.[106]

The need for vitamin D (5 mcg/day) can be met by including three cups of vitamin D-fortified milk in the daily diet, or generally by exposing the face, hands, and arms to sunlight for one-half to two hours per week. Winter sunlight is too weak in northern climates to permit vitamin D formation in the skin, however.[106]

Calcium Requirements in Pregnancy

Calcium metabolism changes meaningfully during pregnancy. Absorption of calcium from food increases, excretion of calcium in urine likewise increases, and bone mineral turnover takes place at a higher rate.[107] The additional requirement for calcium in the last quarter of pregnancy is approximately 300 mg per day and may be obtained by increased absorption and by release of calcium from bone.[15] (Calcium is not taken from the teeth, however.) Calcium lost from bones appears to be replaced after pregnancy in women with adequate intakes of calcium and vitamin D.[1] Inadequate calcium intake has been related to increased blood pressure during pregnancy, decreased subsequent bone remineralization, increased blood pressure of infants, and decreased breast milk concentration of calcium.[107]

CALCIUM AND THE RELEASE OF LEAD FROM BONES
Lead is transferred from the mother to the fetus and can disrupt normal development of the central ner-

vous system.[108] Pregnant women who don't consume enough calcium show greater increases in blood lead levels than women who consume 1000 mg (the DRI for calcium) or more per day. Bone tissues contain about 95% of the body's lead content, and the lead is released into the bloodstream when bones demineralize. Bone tissues demineralize to a greater extent in pregnant women who fail to consume adequate calcium.[108]

Calcium needs during pregnancy can be met through the consumption of three cups of milk or calcium-fortified soy milk, two cups of calcium fortified orange juice plus a cup of milk, or by choosing a sufficient number of other good sources of calcium daily. (Table 1.12 in Chapter 1 lists food sources of calcium.)

FLUORIDE   Teeth develop in utero, so why isn't it recommended that pregnant women consume sufficient fluoride so that the fetus builds cavity-resistant teeth? This is a logical question, but it has the answer "because only trace amounts of fluoride pass through the placenta to the fetus." Children of pregnant women given fluoride supplements during pregnancy have the same rates of dental caries as do children of women who did not receive supplements.[109]

Iron Status and the Course and Outcome of Pregnancy

Iron status is a leading topic of discussions in prenatal nutrition because the need for iron increases substantially. Women require about 1000 mg of additional iron during pregnancy:

- 300 mg is used by the fetus and placenta,
- 250 mg is lost at delivery, and
- 450 mg is used to increase red blood cell mass.

Maternal iron stores get a boost after delivery when iron liberated during the breakdown of surplus red blood cells is recycled.

Many women enter pregnancy with little stored iron and consequently are at risk of developing *iron deficiency anemia* in pregnancy.[11]

IRON-DEFICIENCY ANEMIA IN PREGNANCY   Over recent decades rates of iron deficiency anemia in pregnancy have remained high in women in both developing and developed countries (Table 4.25). Iron-deficiency anemia at the beginning of pregnancy increases the risk of preterm delivery and low birthweight infants by two-to-three times.[110] The mechanisms underlying these effects are unknown, but they may be related to deceased oxygen delivery to the placenta and fetus and to increased rates of infection

in iron-deficient women. *Iron deficiency* often occurs toward the end of pregnancy even among women who enter pregnancy with some iron stores. It is far more common than iron-deficiency anemia. However, neither iron deficiency nor iron-deficiency anemia in the second half of pregnancy appears to be related to the risk of preterm delivery or low birthweight infants.[111]

ASSESSMENT OF IRON STATUS   Red cell mass increases substantially (30%) in pregnancy. However, plasma volume expands more (by about 50%). The higher increase in plasma volume compared to red cell mass makes it appear that amounts of hemoglobin, ferritin, and packed red blood cells have decreased.[18] They have not decreased but rather have become diluted by the large increase in plasma volume. Hemoglobin concentration normally decreases until the middle of the second trimester and then rises somewhat in the third. It is not necessary to prevent normal declines in hemoglobin level during pregnancy.[112]

> **IRON DEFICIENCY ANEMIA**   A condition often marked by low hemoglobin level. It is characterized by the signs of iron deficiency plus paleness, exhaustion, and a rapid heart rate.
>
> **IRON DEFICIENCY**   A condition marked by depleted iron stores. It is characterized by weakness, fatigue, short attention span, poor appetite, increased susceptibility to infection, and irritability.

Due to the dilution effects of increased plasma volume, changes in hemoglobin levels tend to be more indicative of plasma volume expansion than of iron status.[110] Low levels of hemoglobin or serum ferritin may be associated with high plasma volume expansion (hypervolemia), and high hemoglobin levels are related to low plasma volume expansion (hypovolemia). Low levels of plasma volume expansion are associated with reduced fetal growth, whereas newborns tend to be larger in women with higher levels of plasma volume expansion.[111]

The Centers for Disease Control have developed standard hemoglobin levels to be used in the identification of iron-deficiency anemia in pregnant women. These standards (shown in Table 4.26) represent

Table 4.25   Estimates of the incidence of iron-deficiency anemia in women in developing and developed countries.[2,111]

	% WITH IRON-DEFICIENCY ANEMIA	
	Developing Countries	Developed Countries
Nonpregnant	43	12
Pregnant	56	18

Table 4.26 CDC's Gestational age-specific cutoffs for anemia in pregnancy.[113]

Gestational Weeks	Hemoglobin (g/dL) Indicating Anemia[a]
12	<11.0
16	<10.6
20	<10.5
24	<10.5
28	<10.7
32	<11.0
36	<11.4
40	<11.9

[a]For women living in high attitudes, hemoglobin values should be increased by 0.2 g/dL for every 1000 feet above 3000 and by 0.3 g/dL for every 1000 feet above 7000. Hemoglobin values should be adjusted upwards for cigarette smokers by 0.3 g/dL.

levels below the fifth percentile of hemoglobin values in pregnancy.[113]

By trimester, hemoglobin levels indicative of iron-deficiency anemia are:

- <11.0 g/dL in the first and third, and
- <10.5 g/dL in the second trimester.

Serum ferritin cut-points indicative of iron deficiency anemia in pregnancy have also been developed:[114]

	SERUM FERRITIN, ng/mL
Normal	>35
Depleted Stores	<20
Iron Deficiency	≤15

Hemoglobin and serum ferritin are the most commonly employed measures of iron status in pregnant women.[11]

The diagnosis of iron-deficiency anemia is more complicated than often thought. No single test of iron status is totally accurate because many factors, including infection and inflammatory disease, affect iron status and because each test measures a different aspect of iron status. It is best to base the diagnosis of iron-deficiency anemia on results of several tests.[113]

Women entering pregnancy with adequate iron stores tend to absorb about 10% of total iron ingested; those with low stores absorb more—about 20% of the iron consumed. The largest percentage of iron absorption, 40%, occurs in women who enter pregnancy with iron-deficiency anemia.

Iron absorption from foods and supplements is enhanced in women with low iron stores during pregnancy, and absorption increases as pregnancy progresses.[11] It is highest after the thirtieth week of pregnancy when the greatest amount of iron transfer to the fetus occurs. The amount of iron the placenta

can transfer to the fetus, however, is limited in women who are iron deficient. Maternal iron depletion in pregnancy decreases fetal iron stores and increases the risk that infants will develop iron-deficiency anemia.[115]

ABSORPTION OF IRON FROM SUPPLEMENTS Absorption of iron from multimineral supplements is substantially lower than is iron absorption from supplements containing iron only. For example, women given a multimineral supplement containing iron, calcium, and magnesium absorb less than 5% of the iron, whereas women given a similar dose of iron in a supplement containing iron only absorb over twice that much.[116]

The amount of iron absorbed from supplements depends primarily on women's need for iron and the amount of iron in the supplement. As can be seen in Illustration 4.18, the amount of iron absorbed from supplements decreases substantially as the dose of iron increases. Although controversial, it is thought that the small level of improvement in iron absorption that occurs with doses of iron over 30 or 60 mg may be too little to justify their use.[112] In addition, the acceptance of high levels of iron supplementation by women is often poor due to side effects related to the ingestion of high amounts of iron. Side effects such as nausea, cramps, gas, and constipation are associated with the presence of free iron in the intestines, and they increase as doses of supplemental iron increase (Table 4.27).[117,118] Side effects experienced in iron supplements users is a major reason why women fail to take them.[110,112]

Amounts of elemental iron in supplements vary depending on the form of the iron compound in the supplement (Table 4.28). The proportion of iron absorbed from a constant amount of iron from each type of supplement listed in Table 4.28 is approximately equal.[119]

Since 1997 the FDA has required that the iron content of multivitamin and mineral supplements not exceed 30 mg. This action limits the excess iron that will remain in the gut after ingestion of multivitamin and mineral supplements with iron. Higher doses of iron have to be provided as single-ingredient iron supplements.

RECOMMENDATIONS RELATED TO IRON SUPPLEMENTATION IN PREGNANCY It is recommended that all pregnant women in the United States take a 30 mg iron supplement daily after 12 weeks of pregnancy. Women with iron deficiency anemia should take 120–180 mg of iron per day.[20] Iron supplementation during pregnancy and usual clinical practices related to it are subjects of heated debate, however.

Illustration 4.18 Effect of dose of supplemental iron on iron absorption in women during pregnancy.

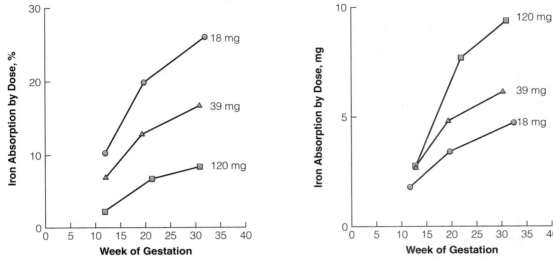

SOURCE: Used by permission of Oxford University Press, Inc. Rosso P. Nutrition and metabolism in pregnancy: mother and fetus. Oxford University Press; 1990.

Although iron supplements increase iron stores and help prevent anemia in many women who take them, some pregnant women do not benefit from iron supplements because they do not need them.[112] Unused iron supplements, when stored and later found by young children, pose a risk of iron poisoning. In addition, iron doses over 30 mg per day decrease zinc absorption and lower zinc status.[11] Since iron status early rather than late in pregnancy is associated with adverse newborn outcomes, iron supplements may be routinely given too late in pregnancy to improve these outcomes.[112]

Proposed Alternatives to Routine Iron Supplementation
It has been suggested that women's iron status be assessed at the first prenatal visit to determine if there is a need for iron supplements. A 30 mg iron supplement would be indicated when hemoglobin levels are <11 g/dL, or if serum ferritin levels are <30 mcg/L. Women with higher values would be monitored for iron status but not given a supplement.[119]

RECOMMENDED INTAKE OF IRON DURING PREGNANCY The increased need for iron can be met by intakes that lead to an additional 3.7 mg absorbed iron per day on average throughout pregnancy. This is a large increase, especially considering that nonpregnant women consuming the DRI for iron (18 mg) absorb only around 1.8 mg of iron daily. Given an ongoing need for 1.8 mg of absorbed iron a day, and the additional need of 3.7 mg of iron daily for pregnancy, the total need for absorbed iron during pregnancy is 5.5 mg daily. Assuming 20% of iron consumed is absorbed, average iron consumption of 27 mg per day (the DRI for iron for pregnancy) will meet the iron needs of pregnancy. The Upper Limit for iron intake during pregnancy is set at 45 mg per day.

Zinc Requirement in Pregnancy

Zinc functions as a cofactor for many enzymes, including those involved in protein synthesis. Like iron, the bioavailability of zinc is higher in meats and low in plants, especially whole grains. Low levels of zinc intake, reliance on whole grain cereals for dietary

Table 4.27 Increased occurrence of side effects in women by supplemental iron dose.[117,118]

Dose of Iron, mg/day	Side Effects
60	32%
120	40%
240	72%

Table 4.28 Percent of elemental iron by weight in various types of iron supplements.

Supplement Type	Iron Content
Ferrous sulfate	20%
Ferrous gluconate	12%
Ferrous fumarate	32%

zinc, and doses of iron supplements over 30 mg daily all decrease zinc status.[120]

A firm consensus on effects of zinc deficiency in pregnancy has not been reached, although animal studies clearly show that it is related to growth retardation and malformations.[121] Levels of serum zinc representative of marginal deficiency have been associated with preterm delivery, intrapartum hemorrhage, infections, and prolonged labor in human studies. It is difficult to know if these results reflect true relationships, however, because serum zinc does not appear to be a very good marker of zinc status.[120]

Iodine and Pregnancy Outcome

Iodine is needed for the synthesis of thyroid hormones, and deficiency of it early in pregnancy can lead to *hypothyroidism* in the offspring. Hypothyroidism in infants is endemic in parts of southern and eastern Europe, Asia, Africa, and Latin America.[122] The incidence of infant hypothyroidism has been found to decrease by over 70% when at-risk women in developing countries are given iodine supplements before or in the first half of pregnancy. Rates of infant deaths are also substantially improved, as is the psychomotor development of the offspring.[123] Iodine supplementation in the second half of pregnancy does not improve infant outcomes.[122]

HYPOTHYROIDISM A condition characterized by growth impairment and mental retardation and deafness when caused by inadequate maternal intake of iodine during pregnancy. Used to be called *cretinism*.

Women in developed countries tend to consume ample amounts of iodine due to the availability of iodine-fortified salt and the use of iodine solutions to clean commercial bread-making vats.

The Need for Sodium During Pregnancy

Sodium plays a critical role in maintaining the body's water balance. Requirements for it increase markedly during pregnancy due to plasma volume expansion. But the need for increased amounts of sodium in pregnancy hasn't always been appreciated. Thirty years ago in the United States it was accepted practice to put all pregnant women on low-sodium diets. (Routine sodium restriction is still practiced in some European countries.) It was then thought that sodium increased water retention and blood pressure, and that sodium restriction would prevent edema and high blood pressure. We now know this isn't accurate and that inadequate sodium intake can complicate the course and outcome of pregnancy.[124] Sodium restriction during pregnancy may exhaust sodium conservation mechanisms and lead to excessive sodium loss.[125]

Sodium restriction is not indicated in normal pregnancy or for the control of edema or high blood pressure that develops in pregnancy. Women should be either given no advice on modifying sodium intake, or if asked, advised to consume salt "to taste."[126]

Caffeine Use in Pregnancy

Caffeine has long been suspected of causing adverse effects in pregnant women because it increases heart rate, acts as a diuretic, and stimulates the central nervous system. It easily passes from maternal to fetal blood and lingers in the fetus longer than in maternal blood because the fetus excretes it more slowly.[127]

Due to its high caffeine level, coffee has generally been the target of investigations on the effects of caffeine intake on pregnancy outcome. Coffee, however, contains hundreds of substances, and some of these have effects similar to those of caffeine. So, although conclusions about caffeine's effects on pregnancy are largely based on coffee intake, it is possible that other components of coffee are responsible for effects observed. Coffee is by far the largest contributor to caffeine intake in most people,[128] and pregnant women consume on average 170 mg caffeine from coffee per day in pregnancy.[129] (Table 2.4 in Chapter 2 provides a list of the caffeine content of beverages and foods.)

Despite the possibilities, caffeine and coffee intake during pregnancy have not been related to an increased risk of fetal malformations, reduced fetal growth, labor or delivery complications, preterm delivery, or developmental problems in offspring.[127,128] No long-term consequences of coffee intake during pregnancy have appeared on children seven years later. Children of women who drank coffee during pregnancy have been found to have similar levels of intellectual and neuromotor development when compared to non-coffee drinkers.[129] High levels of coffee and caffeine intake (over approximately 500 mg per day) have, however, been related to miscarriage.[53] It is generally concluded that intake of four or fewer cups of coffee per day during pregnancy is safe.

Healthy Diets for Pregnancy

Healthy diets for women during pregnancy are described in terms of calories and nutrient intake, and by food choices. Such diets have a number of characteristics in common (Table 4.29).

Adequacy of caloric intake during pregnancy is generally based on rate of weight gain, but for nutrients it is based on the DRIs (Table 4.30). Nutrient intakes during pregnancy should approximate those given in the DRI table, and food intake should correspond to the recommended types and quantity of food

Table 4.29　Basics of a good diet for normal pregnancy.

Good pregnancy diets:

1. Provide sufficient calories to support appropriate rates of weight gain.
2. Follow the Food Guide Pyramid food groups recommendations.
3. Provide all essential nutrients at recommended levels of intake from the diet (with the possible exception of iron).
4. Include 400 mcg of folic acid daily.
5. Provide ample dietary fiber (25–35 g/day)
6. Include 10 cups of fluid daily.
7. Include salt "to taste."
8. Exclude alcohol and limit coffee intake to ≤4 cups per day.
9. Are satisfying and enjoyable.

Table 4.30　Dietary Reference Intakes (DRIs) for pregnant and nonpregnant women aged 19–30 years.*

Nutrient	Pregnant	Nonpregnant	Upper Limit (UL)
Vitamin A, mcg	770	700	3000
Vitamin C, mg	85	75	2000
Vitamin D, mcg[a]	5	5	50
Vitamin E, mg	15	15	1000[c]
Vitamin K, mcg	90	90	—
Thiamin, mg	1.4	1.1	—
Riboflavin, mg	1.4	1.1	—
Niacin, mg	18	14	35[c]
Vitamin B_6	1.9	1.3	100
Folate, mcg[b]	600	400	1000[c,d]
Vitamin B_{12}, mcg	2.6	2.4	—
Pantothenic acid, mcg	6	5	—
Biotin, mcg	30	30	—
Choline, g	450	425	3.5
Calcium, mg	1000	1000	2500
Chromium, mcg	30	25	—
Copper, mcg	1000	900	10,000
Fluoride, mg	3	3	10
Iodine, mcg	220	150	1100
Iron, mg	27	18	45
Magnesium, mg	350	310	350[c]
Manganese, mg	2	1.8	11
Molybdenum, mcg	50	45	2000
Phosphorus, mg	700	700	3500
Selenium, mcg	60	55	400
Zinc, mg	11	8	40

*DRIs for females <19 and >30 years are listed on the inside, front covers of this book.
[a] 1 mcg = 40 IU vitamin D, DRI applies in the absence of adequate sunlight.
[b] As Dietary Folate Equivalent (DFE). 1 DFE = 1 mcg food folate = 0.6 mcg folic acid from fortified food or supplement consumed with food = 0.5 mcg of a supplement taken on an empty stomach.
[c] UL applies to intake from supplements or synthetic form only.
[d] Applies to intake of folic acid.

recommended in the Food Guide Pyramid. (The Food Guide Pyramid recommendations are presented in an upcoming section on dietary assessment in pregnancy.)

Effect of Taste and Smell Changes During Pregnancy on Dietary Intake

No inner voice directs women to consume foods that provide needed nutrients during pregnancy. Pregnant women may, however, develop food preferences and aversions due to changes in the sense of taste and smell; and they may experience *pica.*

Changes in the way certain foods taste and the odor of foods and other substances affect two out of three women during pregnancy. Asked to recall, many previously pregnant women could tell you what foods tasted really good to them, and which odors made them feel queasy to even think about. Increased preference for foods such as sweets, fruits, salty foods, and dairy products are common.[15] The odor of meat being cooked, coffee, perfume, cigarette smoke, and gasoline are common nasal offenders and may stimulate episodes of nausea.[130] The biological bases for such changes are not known, but they are suspected of being related to hormonal changes of pregnancy.

Pica

Pica permits the mind no rest until it is satisfied.

　F.W. Craig, 1935

Classified as an eating disorder, pica affects over half of pregnant women in some locations of the southern part of the United States. It is more common in African Americans than other ethnic groups, and is common enough to be considered a normal behavior in some countries.[131]

> **PICA**　An eating disorder characterized by the compulsion to eat substances that are not food.
>
> **GEOPHAGIA**　Compulsive consumption of clay or dirt.

Historically, one type of pica—*geophagia,* was thought to provide women with additional minerals and to ease gastrointestinal upsets. The cause of pica remains a mystery.[131]

Nonfood items most commonly craved and consumed by pregnant women with pica include ice or freezer frost (*pagophagia*), laundry starch or corn-starch (*amylophagia*), baking soda and powder, and clay or dirt (geophagia). Women experiencing pica are more likely to be iron deficient than those who don't, and iron-deficiency anemia is especially common among pregnant women who compulsively consume ice or freezer frost.[132] It is not clear, however, whether iron deficiency leads to pica or if pica leads to iron deficiency.

PAGOPHAGIA   Compulsive consumption of ice or freezer frost.

AMYLOPHAGIA   Compulsive consumption of laundry starch or cornstarch.

Pica does not appear to be related to newborn weight or preterm delivery. It can, however, complicate control of gestational diabetes if starch is eaten, and has caused lead poisoning, intestinal obstruction, and parasitic infestation of the gastrointestinal tract.[132] Powdered milk is sometimes accepted by women as an alternative to laundry starch or cornstarch, and treating anemia often stops the craving for ice or freezer frost.

Assessment of Dietary Intake During Pregnancy

Routine assessment of dietary practices is recommended for all pregnant women to determine the need for an improved diet or vitamin and mineral supplements.[11] Dietary assessment in pregnancy should cover usual dietary intake, dietary supplement use, and weight gain progress. For best results, several days of accurately recorded, usual intake should be used.

There are several levels of dietary assessment that can be undertaken. (Internet resources for dietary assessment are listed at the end of the chapter.) Which is best primarily depends on the skill level of the health professional responsible for interpreting the results. Results of food-based assessments are rather straightforward to interpret, whereas computerized assessments of levels of nutrient intake are more complex.

The Food Guide Pyramid provides a good way to assess overall quality of usual dietary intake. Table 4.31 presents an example of a recording form that can be used to assess dietary quality based on the Food Guide Pyramid. Computerized analysis, given accurate records and entry of dietary intake and a high-quality nutrient data base, provides results useful for estimating the quantity of calories and nutrients consumed. Detailed knowledge of dietary intake is particularly useful for women at risk of nutrient inadequacies or excesses, and for women with conditions such as gestational diabetes, food in-

tolerances, multiple fetuses, and other special dietary needs.

Evaluation of dietary supplement use includes an examination of types and amounts of supplements consumed. Levels of intake that exceed the Upper Limits (ULs) for pregnancy, as well as the use of herbs and other dietary supplements not known to be safe for pregnancy, should be identified.

CULTURAL CONSIDERATIONS   People tend to be attached to existing food preferences, many of which may have deep cultural roots. Dietary recommendations will differ for Native Alaskans accustomed to a diet based on wild game; for Cambodians, Vietnamese, and Somalians who may think no meal is complete without rice; and for lactose intolerant individuals.

The belief that consumption of certain foods "marks" the baby is common in many cultures. It may be thought, for example, that a woman who loves mangos and eats lots of them during pregnancy may have a baby born with a "mango-shaped" birthmark. Some cultures would hold that the baby will also have learned to love mangos because mom ate them often while pregnant.

Dietary recommendations that are not consistent with a person's usual dietary practices and beliefs, or that are not viewed as acceptable or even preferred by the woman, are least likely to be effective. For best results, dietary adjustments recommended for individual pregnant women should take into account her usual practices and preferences.

Vitamin and Mineral Supplementation During Pregnancy

The only supplement recommended for all pregnant women is iron.[11] Vitamin and mineral supplements may be indicated for specific individuals, however. An appropriate vitamin or mineral supplement is recommended when dietary and clinical assessments determine that a woman is deficient in a particular nutrient. It is also recommended that certain groups of women at high risk for poor dietary intake or increased nutrient need be given a multivitamin and mineral supplement in place of the iron supplement. Individuals in these groups include women who:

- Have a poor-quality, unchangeable diet,
- Are pregnant with multiple fetuses,
- Are vegan,
- Smoke cigarettes,
- Have iron-deficiency anemia, or
- Use illicit drugs or abuse alcohol.

Table 4.31   Recording and evaluation form for dietary intake in pregnancy using the Food Guide Pyramid recommendations.

Food Group	Serving Sizes	Recommended Servings	Actual Servings	Difference Consumed
Breads, Cereals, Rice, and Pasta		6–11		
• Bread	1 slice or 1 oz			
• Roll, biscuit,				
• Muffin, bagel,				
• Tortilla	1 oz			
• Ready-to-eat cereal	1 cup or 1 oz			
• Pasta, rice, cooked cereal	½ cup			
Vegetables		3–5		
• Raw	1 cup			
• Cooked	½ cup			
• Juice	¾ cup or 6 oz			
Fruits		2–4		
• Fresh	1 piece			
• Canned or cooked	½ cup			
• Juice	¾ cup or 6 oz			
Milk, Yogurt, Cheese		3		
• Milk, fortified soymilk	1 cup			
• Yogurt	1 cup			
• Cheese	1½ oz			
• Cottage cheese	½ cup			
Meat, Poultry, Fish, Dried Beans, Eggs, Nuts		3		
• Meat, fish, poultry	2–3 oz			
• Dried beans, cooked	½ cup			
• Tofu	½ cup			
• Eggs	2			
• Peanut butter	4 tbs			
• Nuts and seeds	½ cup or 3 oz			
Fats, Oils, and Sweets				
Based on caloric need				

The moderate-dose-level, multivitamin and mineral supplement recommended for use during pregnancy has the following composition:

Vitamins
Vitamin B$_6$, 2 mg
Folate, 200 mcg
Vitamin C, 50 mg
Vitamin D, 5 mcg

Minerals
Iron, 30 mg
Zinc, 15 mg
Copper, 2 mg
Calcium, 250 mg

Herbal Remedies and Pregnancy

Herbal remedies are becoming commonly used during pregnancy, although very little is known about their safety and effectiveness. Little is known because herbs are rarely tested in pregnant women due to concerns about potential damage to the fetus.

About one-third of commonly used herbal remedies have been deemed unsafe for use by pregnant women.[133] Table 4.32 provides a list of some of these herbs.

Advice to use herbal remedies during pregnancy appears to be based primarily on their traditional use in different societies. This strategy for assessing the safety of herbs doesn't always work. Some herbs considered safe based on traditional use have been found to produce malformations in animal studies.[133] Others, such as blue cohosh, which was previously thought to safely induce uterine contractions, may increase the risk of heart failure in the baby.[135] Peppermint tea and ginger root for nausea appear to be safe.[133]

Manufacturers of herbal remedies do not have to prove they are safe for use by pregnant women. However, the FDA does advise that claims related to pregnancy not be made for herbal supplements.

Table 4.32 Herbs to avoid in pregnancy.[134,135]

Aloe vera	Ergot
Anise	Feverfew
Black cohosh	Ginkgo
Black haw	Ginseng
Blue cohosh	Juniper
Borage	Kava
Buckthorn	Pennyroyal
Comfrey	Raspberry leaf
Cotton root	Saw palmetto
Dandelion leaf	Senna
Ephedra, Ma Huang	

Exercise and Pregnancy Outcome

Exercise during pregnancy has important implications for a women's nutritional status during pregnancy. Because exercise during pregnancy influences caloric need, nutrient utilization, and weight gain, it may potentially interfere with fetal growth and development. Energy costs for weight-bearing exercise increase during pregnancy due to maternal weight gain, and unless accompanied by increased caloric intake, could produce a negative caloric balance and reduced fetal growth. Vigorous exercise during pregnancy could reduce glucose and oxygen availability to the placenta and fetus, and in that way compromise fetal growth.[15]

With that said, there is no evidence that moderate or vigorous exercise undertaken by healthy women consuming high-quality diets and gaining appropriate amounts of weight is harmful to mother or fetus. The bulk of evidence indicates that exercise during pregnancy benefits both mother and her fetus. Women who exercise regularly during pregnancy feel healthier and have an enhanced sense of well-being; furthermore, their labors appear to be somewhat shorter than is the case for women who do not exercise.[136] Exercise during pregnancy, though, can reduce fetal growth in women who are poorly nourished and gain little weight in pregnancy.[135] It is also important for women to avoid dehydration by drinking plenty of fluids while exercising and not to become overheated during physical activity. Activities such as contact sports and scuba diving should not be undertaken due to their potential effects on the fetus.[136]

Is it safe to begin an exercise program during pregnancy? Although the basis of the concern isn't clear, it has been thought that out-of-shape women should not start exercising in pregnancy. This doesn't appear to be the case. In fact, beginning a weight-bearing exercise program may improve fetal growth. This effect was shown in a study involving non-exercising pregnant women who began to exercise at eight weeks of pregnancy. Women participated in three to five weight-bearing exercise sessions a week until delivery. Placenta function was better and newborn weight and length greater in exercising women compared to women who did not exercise.[137] No downside appears to beginning a program of moderate exercise during pregnancy in healthy women.

T. GONDII, OR LOXOPLASMOSIS A parasitic infection that can impair fetal brain development. The source of the infection is often hands contaminated with soil or cat litter box contents, or raw or partially cooked pork, lamb, or venison.

L. MONOCYTOGENES, OR LISTERIA A foodborne bacterial infection that can lead to preterm delivery and stillbirth in pregnant women. Listeria infection is commonly associated with the ingestion of soft cheeses, unpasteurized milk, ready-to-eat deli meats, and hot dogs.

Food Safety Issues During Pregnancy

Foodborne illness can be devastating during pregnancy. Increased progesterone levels that normally occur decrease pregnant women's ability to resist infectious diseases, so they are more susceptible to the effects of foodborne infections. The most common types of foodborne infection in U.S. women are due to *T. gondii* and *L. monocytogenes*.[138] To prevent these and other foodborne infections, pregnant women should not eat raw fish, oysters, soft cheese (brie and camembert, for example), raw or undercooked meat, or unpasteurized milk.

MERCURY CONTAMINATION Seafoods have come under fire as a potential source of mercury overload due to contamination of waters and fish by fungicides, fossil fuel exhaust, and products used in smelting plants, pulp and paper mills, leather tanning facilities, electroplating companies, and chemical manufacturing plants. Mercury is known to pass from the mother's blood to the fetus. It is a fetal neurotoxin that can produce mild to severe effects on fetal brain development. Fetuses exposed to high amounts of mercury can develop mental retardation, hearing loss, numbness, and seizures.[139] Pregnant women are generally only slightly affected by the mercury overload. It accumulates in the mother's tissues and may increase fetal exposure to mercury during pregnancy and lactation.[139]

High levels of mercury are most likely to be present in the muscles of large, long-lived predatory fish such as shark, swordfish, tile fish, tuna, trout, walleye, pickerel, orange roughy, and bass. Mercury content of bottom feeders, such as carp, channel catfish, and white sucker is generally less than half the amount found in predatory fish. The FDA advises that pregnant women and women who may become pregnant not eat shark, swordfish, king mackerel, and tile fish.[139] Weekly consumption of 12 ounces or less during pregnancy of a variety of other types of cooked fish less than 20 inches long appears to be safe.[139] The World Wide Web address for local fish advisories is listed at the end of this chapter.

MODEL NUTRITION PROGRAMS FOR RISK REDUCTION IN PREGNANCY

... pregnancy may be the most sensitive period of the lifecycle in which intervention may reap the greatest benefits.

A. Prentice[108]

Two programs that have been shown to substantially improve pregnancy outcomes are highlighted in this section. The first is the intervention program offered by the Montreal Diet Dispensary and the second the Special Supplemental Nutrition Program for Women, Infants, and Children (WIC).

The Montreal Diet Dispensary

The Montreal Diet Dispensary (MDD) has served low-income, high-risk pregnant females with nutritional assessment and intervention services since the early 1900s. Part of the rationale for the WIC program in the United States was based on the successes of the MDD program. The program is located in the large, comfortable house (see Illustration 4.19) in urban Montreal. Clients are warmly welcomed into a nonthreatening, relaxed setting.

Developed as an adjunct to routine prenatal care, the MDD's intervention strategy has four major components:

1. Assessment of usual dietary intake and the risk profile of each pregnant woman, including calories, protein and selected vitamin and mineral adequacy; and of stress level,
2. Determination of individual nutritional rehabilitation needs based on results of the assessment,
3. Teaching clients the importance of optimal nutrition and about changes that should be made through practical examples, and
4. Regular follow-up and supervision.

The MDD dietitians are carefully trained and hold the interests of their clients first in their hearts. Clients are treated with respect, openness, and affection; in addition, client needs, such as transportation or emergency food or housing, are also addressed by staff. Staff interactions with clients are nonjudgmental in nature and include positive feedback and praise for dietary changes and other successes of clients.

Illustration 4.19 The Montreal Diet Dispensary.

The initial client visit to the MDD takes about 75 minutes and two-week follow-up visits are scheduled for 40 minutes. Women are identified as undernourished if their protein intake falls below that recommended for pregnancy, and an additional protein allowance is added to the diet. Women who are underweight are given an additional allowance of 20 grams protein and 200 calories per day for each additional pound of weight gain needed up to a two-pound gain per week. Women identified as being under excessive stress (such as having a partner in jail, being homeless, or abused) are given an additional allowance of protein and calories and lots of positive attention. Food supplements in the form of milk and eggs and vitamin supplements are provided to women who need them.

IMPACT OF MDD SERVICES Multiple studies have shown that women receiving MDD services have higher birthweight infants (+107 grams), fewer low birthweight infants (−50%), and infants with lower rates of perinatal mortality than is the case for similar women not receiving MDD services.[140,141]

The program is cost effective in terms of savings on newborn critical care, and programs based on MDD services have grown across Canada. Expenditures per client average $450. The program is primarily supported by Centraide of Greater Montreal, provincial and federal programs, and other contributions.[142]

The WIC Program

WIC represents an outstanding example of a successful public program intended to serve the nutritional needs of low-income women and families. It is cited as a model program in several other chapters and highlighted here.

In operation since 1974, WIC provides nutritional assessment, education and counseling, food supplements, and access to health services to over 6 million participants. WIC serves low-income pregnant, postpartum, and breastfeeding women; and children up to five years of age at nutritional risk. Supplemental food provided to women includes milk, ready-to-eat cereals, dried beans, fruit juice, and cheese; some programs offer vouchers for farmer's markets.

Participation in WIC is related to reduced rates of iron-deficiency anemia in pregnancy, higher birthweight infants, decreased low birthweight infants, and lower rates of iron-deficiency anemia in women after delivery. For each dollar invested in WIC, approximately three dollars are saved in health care costs. Internet addresses that will lead you to additional information about WIC are listed among the Resources at the end of this chapter.

Resources

Environmental Protection Agency

This site links to local fresh water fish advisories.
Available from: www.epa.gov/ost/fish

Health Canada, Office of Nutrition Policy and Promotion

Nutrition Web Site that provides access to their policy statement on nutrient needs of pregnant women.
Available from: www.hc-sc.gc.ca/nutrition

National Agricultural Library

A site that provides links to vast resources related to education materials for WIC, vegetarian diets, health fraud, and other topics.
Available from: www.nal.usda.gov/fnic

Nutritionquest

Free assessments of fat, vegetable and fruit, and fiber intake can be performed here. More detailed dietary assessments are available for a fee.
Available from: www.nutritionquest.com/freescreen.html

Public Health Service and USDA

The U.S. Government's site that provides information on food and nutrition programs and eligibility, links to scientific references, and information about the nutrition needs of infants, children, adults, and seniors.
Available from: www.nutrition.gov

University of Illinois Food Science and Nutrition Department

The site provides the "Nutrition Analysis Tool" that is free for noncommercial use. In-depth food constituent results are provided.
Available from: www.nat.uiuc.edu

University of Washington

Provides information on ethnic diets, cultural and health beliefs of a wide population groups immigrating to the United States and Canada.
Available from: healthlinks.washington.edu/clinical/ethnomed

USDA

Provides access to information about the WIC program, the WIC Works Food Safety Resource List, and other resources.
Available from: www.nal.usda.gov/fnic

USDA

Select "Health Eating Index" and run dietary intake records one day at a time. Provides analysis of diet by food groups and selected nutrients.
Available from: www.usda.gov/cnpp

USDA

This is the best site for food composition data.
Available from: www.nal.usda.gov/fnic/foodcomp

References

1. King JC. Preface. Am J Clin Nutr 2000;71:1217S.

2. Health United States, 2000. Vol. 2001; www.cdc.gov/nchs

3. Wegman ME. Infant mortality in the 20th century, dramatic but uneven progress. J Nutr 2001;131:401S–8S.

4. Chase HC. Ranking countries by infant mortality rates. Public Health Rep 1969;84:19–27.

5. Guyer B, MacDorman MF, Martin JA, et al. Annual summary of vital statistics—1997. Pediatrics 1998;102:1333–49.

6. David RJ, Collins JW, Jr. Differing birth weight among infants of US-born blacks, African-born blacks, and US-born whites. N Eng J Med 1997;337:1209–14.

7. Williams RL, Creasy RK, Cunningham GC, et al. Fetal growth and perinatal viability in California. Obstet Gynecol 1982;59:624–32.

8. Brown JE. Weight gain during pregnancy: what is optimal? Clin Nutr 1988;7:181–90.

9. Singer JE, Westphal M, Niswander K. Relationships of weight gain during pregnancy to birth weight and infant growth and development in the first year of life. A report from the collaborative study of cerebral palsy. Obstet Gynecol 1968;31:417–23.

10. Godfrey KM. Maternal regulation of fetal development and health in adult life.

Eur J Obstet Gynecol Repro Biol 1998;78:141–50.

11. National Academy of Sciences. Nutrition during pregnancy. I. Weight gain. II. Nutrient supplements. Washington, DC: National Academy Press; 1990.

12. Bloom BS. Changing infant mortality: the need to spend more while getting less. Pediatrics 1984;73:862–6.

13. Paneth N. Technology at birth. Am J Public Health 1990;80:791–2.

14. Hytten FE, Leitch I. The physiology of human pregnancy. Oxford: Blackwell Scientific Publications; 1971.

15. King JC. Physiology of pregnancy and nutrient metabolism. Am J Clin Nutr 2000;71:1218S–25S.

16. Rosso P. Nutrition and metabolism in pregnancy. New York: Oxford University Press; 1990: pp 117–118, 125, 150–151.

17. Winick M. Nutrition, pregnancy, and early infancy. Baltimore: Williams & Wilkins; 1989: p 182.

18. Cruikshank DP, et al. Maternal physiology in pregnancy. In: Gabbe SG, et al., eds. Obstetrics: normal and problem pregnancies. New York: Churchill Livingstone; 1996: pp 91–109.

19. Butte NF. Carbohydrate and lipid metabolism in pregnancy: normal compared with gestational diabetes mellitus. Am J Clin Nutr 2000;71:1256S–61S.

20. Naeye RL, Chez RA. Effects of maternal acetonuria and low pregnancy weight gain on children's psychomotor development. Am J Obstet Gynecol 1981;139:189–93.

21. van Stiphout WAHJ, Hofman A, de Bruijn AM. Serum lipids in young women before, during, and after pregnancy. Am J Epidemiol 1987;126:922–8.

22. Martin U, Davies C, Hayavi S, et al. Is normal pregnancy atherogenic? Clin Sci (Colch) 1999;96:421–5.

23. Ortega RM, Gaspar MJ, Cantero M. Influence of maternal serum lipids and maternal diet during the third trimester of pregnancy on umbilical cord blood lipids in two populations of Spanish newborns. Int J Vitam Nutr Res 1996;66:250–7.

24. King JC. Effect of reproduction on the bioavailability of calcium, zinc and selenium. J Nutr 2001;131:1355S–8S.

25. American College of Obstetricians and Gynecologists. Nutrition during pregnancy. ACOG Technical Bulletin No. 179 1993:1–7.

26. Weigel RM, Weigel MM. Nausea and vomiting of early pregnancy and pregnancy outcome. A meta-analytical review. Br J Obstet Gynecol 1989;96:1312–18.

27. Erick M. Hyperolfaction and hyperemesis gravidarum: what is the relationship? Nutr Rev 1995;53:289–95.

28. Tsang IS, Katz VL, Wells SD. Maternal and fetal outcomes in hyperemesis gravidarum. Int J Gynaecol Obstet 1996;55:231–5.

29. Gross S, Librach C, Cecutti A. Maternal weight loss associated with hyperemesis gravidarum: a predictor of fetal outcome. Am J Obstet Gynecol 1989;160:906–9.

30. Scialli A. Revived focus on treatment for morning sickness. Medscape Wire 2000.

31. Brown JE, Kahn ES, Hartman TJ. Profet, profits and proof: do nausea and vomiting of early pregnancy protect women from "harmful" vegetables? Am J Obstet Gynecol 1997;176:179–81.

32. Baron TH, Ramirez B, Richter JE. Gastrointestinal motility disorders during pregnancy. Ann Intern Med 1993;118:366–75.

33. Anderson JW. Health implications of wheat fiber. Am J Clin Nutr 1985;41:1103–12.

34. Rosso P, Cramoy C. Nutrition and pregnancy. In: Winick M, ed. Human nutrition: pre- and post-natal development. New York: Plenum Press; 1979: pp 133–228.

35. Scholl TO, Stein TP, Smith WK. Leptin and maternal growth during adolescent pregnancy. Am J Clin Nutr 2000; 72:1542–7.

36. Flanigan G. The first nine months of life. New York: Simon and Schuster; 1962: pp 17–66.

37. Rozovski SJ, Winick M. Nutrition and cellular growth. In: Winick M, ed. Human nutrition: pre- and postnatal development. New York: Plenum Press; 1979: pp 61–102.

38. Godfrey KM, Barker DJ. Fetal nutrition and adult disease. Am J Clin Nutr 2000; 71:1344S–52S.

39. Smith DW. Growth and its disorders: basics and standards, approach and classifications, growth deficiency disorders, growth excess disorders, obesity. In: Schaeffer AJ, Markowtiz M, eds. Major problems in clinical pediatrics. Vol. 15. Philadelphia: W. B. Saunders Company; 1979: pp 134–169.

40. Winick M. Malnutrition and brain development. J Pediatr 1969;74:667–79.

41. Ziegler EE, O'Donnell AM, Nelson SE, Fomon SJ. Body composition of the reference fetus. Growth 1976;40:329–41.

42. Rosso P, Winick M. Intrauterine growth retardation. A new systematic approach based on the clinical and biochemical characteristics of this condition. J Perinat Med 1974;2:147–60.

43. Hediger ML, Overpeck MD, Kuczmarski RJ, et al. Muscularity and fatness of infants and young children born small- or large-for-gestational-age. Pediatrics 1998;102:E601–7.

44. Alexander GR, Himes JH, Kaufman RB, et al. A United States national reference for fetal growth. Obstet Gynecol 1996;87:163–8.

45. Day NL, Zuo Y, Richardson GA, et al. Prenatal alcohol use and offspring size at 10 years of age. Alcohol Clin Exp Res 1999;23:863–9.

46. Hediger ML, Overpeck MD, Maurer KR, et al. Growth of infants and young children born small or large for gestational age: findings from the Third National Health and Nutrition Examination Survey. Arch Pediatr Adolesc Med 1998;152:1225–31.

47. Lucas A, Fewtrell MS, Davies PSW, et al. Breastfeeding and catch-up growth in infants born small for gestational age. Acta Paediatr 1997;86:564–9.

48. Luo ZC, Albertsson-Wikland K, Karlberg J. Length and body mass index at birth and target height influences on patterns of postnatal growth in children born small for gestational age. Pediatrics 1998; 102:E72.

49. Lazer S, Biale Y, Mazor M, et al. Complications associated with the macrosomic fetus. J Repro Med 1986;31:501–5.

50. Kallen B. Endpoints studied in the epidemiology of reproduction. Epidemiology of human reproduction. Boca Raton, FL: CRC Press, Inc.; 1988: pp 142–9

51. Tempfer C, et al. The etiology of habitual abortion: a review. Geburtshilfe und Frauenheilkunde 2000;60:604–8.

52. Wen W, Shu XO, Jacobs DR, Jr., et al. The associations of maternal caffeine consumption and nausea with spontaneous abortion. Epidemiology 2001;12:38–42.

53. Kramer MS, Demissie K, Yang H, et al. The contribution of mild and moderate preterm birth to infant mortality. Fetal and Infant Health Study Group of the Canadian Perinatal Surveillance System. JAMA 2000;284:843–9.

54. Goldenberg RL, Rouse DJ. Prevention of premature birth. N Eng J Med 1998;339:313–20.

55. Rich-Edwards JW, Stampfer MJ, Manson JE, et al. Birth weight and risk of cardiovascular disease in a cohort of women followed up since 1976. BMJ 1997;315:396–400.

56. Lucas A, Fewtrell MS, Cole TJ. Fetal origins of adult disease—the hypothesis revisited. BMJ 1999;319:245–9.

57. Belizan JM, Villar J, Bergel E, et al. Long-term effect of calcium supplementation during pregnancy on the blood pressure of offspring; follow up of a randomised controlled trial. BMJ 1997;315:281–5.

58. Villar J, Cossio TG. Nutritional factors associated with low birth weight and short gestational age. Clin Nutr 1986;5:78–85.

59. Eastman NJ. Expectant motherhood. Boston: Little, Brown and Company; 1947: p 198.

60. Parker JD, Abrams B. Prenatal weight gain advice: an examination of the recent prenatal weight gain recommendations of the Institute of Medicine. Obstet Gynecol 1992;79:664–9.

61. Brown JE, Murtaugh MA, Jacobs DR, et al. Variation in newborn size by trimester weight change in pregnancy. Am J Clin Nutr 2001 (in press).

62. Carmichael SL, Abrams B. A critical review of the relationship between gestational weight gain and preterm delivery. Obstet Gynecol 1997;89:865–73.

63. To WW, Cheung W. The relationship between weight gain in pregnancy, birthweight and postpartum weight retention. Aust N Z J Obstet Gynaecol 1998;38: 176–9.

64. Abrams B. Prenatal weight gain and postpartum weight retention: a delicate balance (editorial). Am J Public Health 1993;83:1082–4.

65. Keppel KG, Taffel SM. Pregnancy-related weight gain and retention: implications of the 1990 Institute of Medicine guidelines. Am J Public Health 1993;83: 1100–3.

66. Brown JE, Kaye SA, Folsom AR. Parity-related weight change in women. Int J Obes 1992;16:627–31.

67. Gigante DP, Victora CG, Barros FC. Breast-feeding has a limited long-term effect on anthropometry and body composition of Brazilian mothers. J Nutr 2001;131:78–84.

68. Mussey RD. Nutrition and human reproduction: an historical review. Am J Obstet Gynecol 1949;58:1037–48.

69. Stein AD, Ravelli AC, Lumey LH. Famine, third-trimester pregnancy weight gain, and intrauterine growth: the Dutch Famine Birth Cohort Study. Hum Biol 1995;67:135–50.

70. Smith CA. Effects of maternal undernutrition upon the newborn infant in Holland (1944–1945). J Pediatr 1947;30: 229–43.

71. Susser M, Stein Z. Timing in prenatal nutrition: a reprise of the Dutch Famine Study. Nutr Rev 1994;52:84–94.

72. Roseboom TJ, van der Meulen JH, Osmond C, et al. Plasma lipid profiles in adults after prenatal exposure to the Dutch famine. Am J Clin Nutr 2000; 72:1101–6.

73. Antonov AN. Children born during the siege of Leningrad in 1942. J Pediatr 1947;30:250–9.

74. Gruenwald P, Funakawa H, Mitani S, et al. Influence of environmental factors on fetal growth in man. Lancet 1967;I:1026–8.

75. Simic S, Idrizbegovic S, Jaganjac N, et al. Nutritional effects of the siege on newborn babies in Sarajevo. Eur J Clin Nutr 1995;49 Suppl 2:S33–6.

76. Burke BS. Nutritional needs in pregnancy in relation to nutritional intakes as shown by dietary histories. Obstetr Gynecol Survey 1948;3:716–30.

77. Burke BS, Harding VV, Stuart HC. Nutrition studies during pregnancy. IV. Relation of protein content of mother's diet during pregnancy to birth length, birth weight, and condition of infant at birth. J Pediatr 1943;23:506–15.

78. Hytten F, Chamberlain G. Clinical physiology in obstetrics. Oxford: Blackwell Scientific Publications; 1980: pp 92–130.

79. Brown JE, Kahn ESB. Maternal nutrition and the outcome of pregnancy: a renaissance in research. Clin Perinatol 1997;24:433–49.

80. Sleutel M, Golden SS. Fasting in labor: relic or requirement. J Obstet Gynecol Neonatal Nurs 1999;28:507–12.

81. Duffy VB, Anderson GH. Position of The American Dietetic Association: Use of nutritive and nonnutritive sweeteners. J Amer Diet Assoc 1998;98:580–7.

82. American Academy of Pediatrics. Committee on Substance Abuse and Committee on Children With Disabilities. Fetal alcohol syndrome and alcohol-related neurodevelopmental disorders. Pediatrics 2000;106:358–61.

83. Ventura SJ, Martin JA, Curtin SC, et al. Births: final data for 1998. Natl Vital Stat Rep 2000;48:1-100.

84. March of Dimes Birth Defects Foundation. Public health education information sheet: drinking alcohol during pregnancy, 1997.

85. Polygenis D, Wharton S, Malmberg C, et al. Moderate alcohol consumption during pregnancy and the incidence of fetal malformations: a meta-analysis. Neurotoxicol Teratol 1998;20:61–7.

86. Konovalov HV, Kovetsky NS, Bobryshev YV, et al. Disorders of brain development in the progeny of mothers who used alcohol during pregnancy. Early Hum Dev 1997;48:153–66.

87. Food and Nutrition Board. National Academy of Sciences (Institute of Medicine) NRC, Subcommittee on the 10th Edition of the RDAs, Commission on Life Sciences. Recommended dietary allowances. Washington, DC: National Academy Press, 1989.

88. Johnston PK. Counseling the pregnant vegetarian. Am J Clin Nutr 1988;48:901–5.

89. Conner WE. Importance of n-3 fatty acids in health and disease. Am J Clin Nutr 2000;71:171S–5S.

90. Ershow AG, Brown LM, Cantor KP. Intake of tapwater and total water by pregnant and lactating women. Am J Public Health 1991;81:328–34.

91. Lumley J, Watson L, Watson M, et al. Periconceptional supplementation with folate and/or multivitamins for preventing neural tube defects. Cochrane Database Syst Rev 2000;2.

92. Suitor CW, Bailey LB. Dietary folate equivalents: interpretation and application. J Amer Diet Assoc 2000;100:88–94.

93. Bailey LB. Evaluation of a new Recommended Dietary Allowance for folate. J Amer Diet Assoc 1992;92:463–71.

94. Fodinger M, Wagner OF, Horl WH, et al. Recent insights into the molecular genetics of the homocysteine metabolism. Kidney Int 2001;59Suppl78:S238–42.

95. Warkany J. Production of congenital malformations by dietary measures. JAMA 1958;168:2020–3.

96. Smithells RW. Availability of folic acid. Lancet 1984;1:508.

97. Eskes TKAB. From birth to conception. Open or closed. Eur J Obstet Gynecol Reprod Biol 1998;78:169–77.

98. Folic acid. MMWR Morb Mortal Wkly Rep 2001;50:185–9.

99. Folate status in women of childbearing age—United States, 1999. MMWR Morb Mortal Wkly Rep 2000;49:962–5.

100. Daly LE, Kirke PN, Molloy A, et al. Folate levels and neural tube defects. JAMA 1995;274:1698–702.

101. Brown JE, Jacobs DR, Jr., Hartman TJ, et al. Predictors of red cell folate level in women attempting pregnancy. JAMA 1997;277:548–52.

102. Cuskelly GJ, McNulty H, Scott JM. Fortification with low amounts of folic acid makes a significant difference in folate status in young women: implications for the prevention of neural tube defects. Am J Clin Nutr 1999;70:234–9.

103. Shaw GM, Schaffer D, Velie EM, et al. Periconceptional vitamin use, dietary folate, and the occurrence of neural tube defects. Epidemiology 1995;6:219–26.

104. Azais-Braesco V, Pascal G. Vitamin A in pregnancy: requirements and safety limits. Am J Clin Nutr 2000;71:1325S–33S.

105. Smithells D. Vitamins in early pregnancy. BMJ 1996;313:128–9.

106. Specker BL. Do North American women need supplemental vitamin D during pregnancy or lactation? Am J Clin Nutr 1994;59:484S–91S.

107. Prentice A. Maternal calcium metabolism and bone mineral status. Am J Clin Nutr 2000;71:1312S–6S.

108. Hertz-Picciotto I, Schramm M, Watt-Morse M, et al. Patterns and determinants of blood lead during pregnancy. Am J Epidemiol 2000;152:829–37.

109. Leverett DH, Adair SM, Vaughan BW, et al. Randomized clinical trial of the effect of prenatal fluoride supplements in preventing dental caries. Caries Res 1997;31: 174–9.

110. Allen LH. Biological mechanisms that might underlie iron's effects on fetal growth and preterm birth. J Nutr 2001;131:581S–9S.

111. Allen LH. Anemia and iron deficiency: effects on pregnancy outcome. Am J Clin Nutr 2000;71:1280S–4S.

112. Beard JL. Iron requirements in adolescent females. J Nutr 2000;130:440S–2S.

113. Recommendations to prevent and control iron deficiency in the United States.

Centers for Disease Control and Prevention. MMWR Morb Mortal Wkly Rep 1998;47:1–29.

114. Godel JC, Pabst HF, Hodges PE, et al. Iron status and pregnancy in a northern Canadian population: relationship to diet and iron supplementation. Can J Public Health 1992;83:339–43.

115. Sweet DG, Savage G, Tubman TR, et al. Study of maternal influences on fetal iron status at term using cord blood transferrin receptors. Arch Dis Child Fetal Neonatal Ed 2001;84:F40–3.

116. Seligman PA, Caskey JH, Frazier JL, et al. Measurements of iron absorption from prenatal multivitamin—mineral supplements. Obstet Gynecol 1983;61:356–62.

117. Dawson EB, Dawson R, Behrens J, et al. Iron in prenatal multivitamin/multimineral supplements. Bioavailability. J Reprod Med 1998;43:133–40.

118. Singh K, Fong YF, Kuperan P. A comparison between intravenous iron polymaltose complex (Ferrum Hausmann) and oral ferrous fumarate in the treatment of iron deficiency anaemia in pregnancy. Eur J Haematol 1998;60:119–24.

119. Iron deficiency anemia: recommended guidelines for prevention, detection, and management among US children and women of childbearing age. Washington, DC: National Academy Press, 1993.

120. King JC. Determinants of maternal zinc status during pregnancy. Am J Clin Nutr 2000;71:1334S–43S.

121. Tamura T, Goldenberg RL, Johnston KE, et al. Maternal plasma zinc concentrations and pregnancy outcome. Am J Clin Nutr 2000;71:109–13.

122. Xue-Yi C, Xin-Min J, Zhi-Hong D, et al. Timing of vulnerability of the brain to iodine deficiency in endemic cretinism. N Eng J Med 1994;331:1739–44.

123. Mahomed K, Gulmezoglu AM. Maternal iodine supplements in areas of deficiency (Cochrane Review). The Cochrane Library. Issue 3. Oxford: Update Software, 1999.

124. Delemarre FM, Steegers EA, Berendes JN, et al. Eclampsia despite strict dietary sodium restriction. Gynecol Obstet Invest 2001;51:64–5.

125. Pike RL, Gursky DS. Further evidence of deleterious effects produced by sodium restriction during pregnancy. Am J Clin Nutr 1970;23:883–9.

126. Duley L, Henderson-Smart D. Reduced salt intake compared to normal dietary salt, or high intake, in pregnancy. Cochrane Database Syst Rev 2000;2.

127. Nehlig A, Debry G. Consequences on the newborn of chronic maternal consumption of coffee during gestation and lactation: a review. J Am Coll Nutr 1994;13:6–21.

128. Hinds TS, West WL, Knight EM, et al. The effect of caffeine on pregnancy outcome variables. Nutr Rev 1996;54:203–7.

129. Barr HM, Streissguth AP. Caffeine use during pregnancy and child outcome: a 7-year prospective study. Neurotoxicol Teratol 1991;13:441–8.

130. Patten-Hitt E. Nausea during pregnancy linked to keen sense of smell. Reuters Health 2000.

131. Johns T, Duquette M. Detoxification and mineral supplementation as functions of geophagy. Am J Clin Nutr 1991;53:448–56.

132. Rainville AJ. Pica practices of pregnant women are associated with lower maternal hemoglobin level at delivery. J Amer Diet Assoc 1998;98:293–6.

133. FDA statement concerning structure/function rule and pregnancy claims. HHS Statement, U.S. Department of Health and Human Services 2000; 65 FR 1000; January 6, 2000; Docket No. 98N-0044.

134. Belew C. Herbs and the childbearing woman. Guidelines for midwives. J Nurse Midwifery 1999;44:231–52.

135. Jones FA. Herbs: useful plants. J R Soc Med 1996;89:717–9.

136. Artal R, Sherman C. Exercise during pregnancy—safe and beneficial for most. Physician and Sportsmedicine 1999;27:51–56.

137. Clapp JF, III, Kim H, Burciu B, et al. Beginning regular exercise in early pregnancy: effect on fetoplacental growth. Am J Obstet Gynecol 2000;183:1484–8.

138. Smith JL. Foodborne infections during pregnancy. J Food Prot 1999;62:818–29.

139. EPA fish advisory update. Vol. 2001, 2001.

140. Higgins AC, Moxley JE, Pencharz PB, et al. Impact of the Higgins Nutrition Intervention Program on birth weight: a within-mother analysis. J Amer Diet Assoc 1989;89:1097–103.

141. Dubois S, Coulombe C, Pencharz P, et al. Ability of the Higgins Nutrition Intervention Program to improve adolescent pregnancy outcome. J Amer Diet Assoc 1997;97:871–8.

142. Duquette MP, Director of the Montreal Diet Dispensary. Personal communication, 2000.

CHAPTER 5

Eyewire

Women at greater risk of adverse birth outcomes benefit the most from educational health care messages.

M. D. Kogan et al. 1994

NUTRITION DURING PREGNANCY:

Conditions and Interventions

Prepared by **Judith E. Brown**

CHAPTER OUTLINE

- Introduction
- Hypertensive Disorders of Pregnancy
- Diabetes in Pregnancy
- Multifetal Pregnancies
- HIV/AIDS During Pregnancy
- Eating Disorders in Pregnancy
- Nutrition and Adolescent Pregnancy
- Evidence-Based Practice

KEY NUTRITION CONCEPTS

1 Some complications of pregnancy are related to women's nutritional status.

2 Nutritional interventions for a number of complications of pregnancy can benefit maternal and infant health outcomes.

3 Nutritional interventions during pregnancy should be based on scientific evidence that supports their safety, effectiveness, and affordability.

INTRODUCTION

Almost all healthy women expect that their pregnancies will proceed normally and that they will be rewarded at delivery with a healthy newborn. For the vast majority of pregnancies, this expectation is fulfilled. For other women, however, the path to a healthy newborn is strewn with obstacles in the form of health problems that women bring into or develop during pregnancy. This chapter addresses a number of these health conditions and the role of nutrition in their etiology and management. The specific health conditions presented are hypertensive disorders of pregnancy, preexisting and gestational diabetes, obesity, multifetal pregnancy, HIV/AIDS, and eating disorders. Nutritional considerations for adolescent pregnancy are also presented.

HYPERTENSIVE DISORDERS OF PREGNANCY

Hypertensive disorders of pregnancy are the second leading cause of maternal mortality in the United States. They affect 6 to 8% of pregnancies and contribute significantly to stillbirths, fetal and newborn deaths, and other adverse outcomes of pregnancy. The causes of most cases of hypertension during pregnancy remain unknown, and cures for these disorders remain elusive.[1]

Several types of hypertensive disorders in pregnancy have been identified (Table 5.1). In the past, the major types of hypertensive disorders in pregnancy were grouped under the heading "pregnancy-induced hypertension," or PIH. This terminology is being phased out in favor of the classification scheme for hypertensive disorders of pregnancy presented in Table 5.1. Use of edema as an indicator of hypertensive disorders is fading because edema is not predictive of adverse pregnancy outcomes.[1] In addition, new recommendations for the diagnosis of hypertensive disorders of pregnancy no longer include specific levels of increase in blood pressure, such as 30 mm Hg systolic and 15 mm Hg diastolic, as have been used in the past.[1]

Chronic Hypertension

The incidence of chronic hypertension, or that diagnosed before 20 weeks after conception as opposed to hypertension that develops later in pregnancy, ranges from 1 to 5% depending on the population studied. The condition is more likely to occur in African Americans, obese women, women over 35 years of age, and women who experienced high blood pressure in a previous pregnancy.[2]

Table 5.1   Definitions and features of hypertensive disorders of pregnancy.*[1]

Chronic Hypertension

Hypertension that is present before pregnancy or diagnosed before 20 weeks of pregnancy. Hypertension is defined as blood pressure ≥140 mm Hg systolic or ≥90 mm Hg diastolic blood pressure.

Hypertension first diagnosed during pregnancy that does not resolve after pregnancy is also classified as chronic hypertension.

Gestational Hypertension

This condition exists when elevated blood pressure levels are detected for the first time after midpregnancy. It is not accompanied by proteinuria. If blood pressure returns to normal by 12 weeks postpartum, the condition is considered to be transient hypertension of pregnancy. If it remains elevated, then the woman is considered to have chronic hypertension.

Women with gestational hypertension are at lower risk for poor pregnancy outcomes than are women with preeclampsia.

Preeclampsia-Eclampsia

A pregnancy-specific syndrome that usually occurs after 20 weeks gestation (but that may occur earlier) in previously normotensive women. It is determined by increased blood pressure during pregnancy to ≥140 mm Hg systolic or ≥90 mm Hg diastolic and is accompanied by proteinuria. In the absence of proteinuria, the disease is highly suspected when increased blood pressure is accompanied by headache, blurred vision, abdominal pain, low platelet count, and abnormal liver enzyme values.

- Proteinuria is defined as the urinary excretion of ≥0.3 grams of protein in a 24-hour urine specimen. This usually correlates well with readings of ≥30 mg/dL protein, or ≥2 on dipstick readings taken in samples from women free of urinary tract infection. In the absence of urinary tract infection, proteinuria is a manifestation of kidney damage.

- Eclampsia is defined as the occurrence of seizures that cannot be attributed to other causes in women with preeclampsia.

Preeclampsia Superimposed on Chronic Hypertension

This disorder is characterized by the development of proteinuria during pregnancy in women with chronic hypertension. In women with hypertension and proteinuria before 20 weeks of pregnancy, it is indicated by a sudden increase in proteinuria, blood pressure, or abnormal platelet or liver enzyme levels.

*Blood pressure values used to determine status should be based on two or more measurements of blood pressure in relaxed settings.

Women with mild hypertension may be taken off anti-hypertension medications preconceptionally or early in pregnancy, as the drugs do not appear to improve the course or outcome of pregnancy.[3] Mild hypertension in healthy women that does not become worse during pregnancy appears to pose few risks to maternal and newborn health. Pregnancies among women with blood pressures ≥160/110 mm Hg—either or both values—are associated with an increased risk of fetal death, preterm delivery, and fetal growth retardation. Selection of the proper anti-hypertension medicines for women during pregnancy reduces these risks somewhat. Some anti-hypertension medicines, though, reduce maternal blood sodium levels and limit plasma volume expansion. This, in turn, decreases fetal growth.

NUTRITIONAL CONSIDERATIONS FOR WOMEN WITH CHRONIC HYPERTENSION IN PREGNANCY

Preconceptional and pregnancy diets of women with hypertension should be carefully monitored with the aim of achieving adequate and balanced diets for pregnancy. Weight gain recommendations are the same as for other pregnant women.

Women with salt-sensitive hypertension, or hypertension that responds to dietary sodium intake, must be managed along a fine line between consuming too much sodium for good blood pressure control, and consuming too little at the potential cost of impaired fetal growth.[3] It is generally recommended that women with hypertension that was managed successfully in part by a low-sodium diet prior to pregnancy continue that dietary approach.[1]

Although exercise is beneficial to pregnant and nonpregnant women alike, it is not clear whether it is safe for women with chronic hypertension. Therefore, aerobic exercise is discouraged for women with chronic hypertension until more is known about its safety.[1]

Gestational Hypertension

Gestational hypertension is generally diagnosed after 20 weeks of pregnancy and represents a less serious condition than preeclampsia. Unlike those suffering from preeclampsia, women with gestational diabetes have neither elevated serum insulin levels nor proteinuria.[4] Women with this disorder are at greater risk for hypertension later in life.[5]

Preeclampsia-Eclampsia

Preeclampsia-eclampsia represents a syndrome characterized by:

- Deficits in *prostacyclin* relative to *thromboxane,*
- Blood vessel spasms and constriction,

- Increased blood pressure,
- Adverse maternal immune system responses to the placenta,
- Platelet aggregation and blood coagulation,
- Alterations of hormonal and other systems related to blood volume and pressure control,
- Oxidative tissue damage and inflammation,
- Alterations in calcium regulatory hormones, and
- Reduced calcium excretion.[1,6]

Organs most affected by the disease are the placenta and the mother's kidney, liver, and brain.

Eclampsia can be a life-threatening condition and one that is difficult to predict. Eclamptic seizures appear to be related to hypertension, the tendency of blood to clot, and spasms of and damage to blood vessels in the brain. It complicates about 1 in 2000 pregnancies.[7]

Signs and symptoms of preeclampsia range from mild to severe (Table 5.2), as do the health consequences (Table 5.3). The cause of preeclampsia is unknown and the only cure is delivery.[1] Signs and symptoms of preeclampsia generally disappear rapidly after delivery, but eclampsia may occur within 12 days following delivery.[5]

PROSTACYCLIN A potent inhibitor of platelet aggregation and a powerful vasodilator and blood pressure reducer derived from n-3 fatty acids.

THROMBOXANE The parent of a group of thromboxanes derived from the n-6 fatty acid arachidonic acid. Thromboxane increases platelet aggregation and constricts blood vessels causing blood pressure to increase.

PREECLAMPSIA AND INSULIN RESISTANCE Growing evidence indicates that the onset and effects of

Table 5.2
Signs and symptoms of preeclampsia.[8–10]

- Hypertension
- Increased urinary protein (albumin)
- Decreased plasma volume expansion (hemoglobin levels >13 g/dL)
- Low urine output
- Persistent and severe headaches
- Sensitivity of the eyes to bright light
- Blurred vision
- Abdominal pain
- Nausea
- Increased platelet aggregation, vasoconstriction related to increased thromboxane levels and decreased levels of prostacyclin

Table 5.3   Outcomes related to the existence of preeclampsia during pregnancy.[2,6,11]

Mother
- Early delivery by cesarean section
- Acute renal dysfunction
- Increased risk of gestational diabetes, hypertension, and type 2 diabetes later in life
- Abruptio placenta (rupture of the placenta)

Newborn
- Growth restriction
- Respiratory distress syndrome

Table 5.4   A sampling of factors that may influence the development of preeclampsia in women with insulin resistance.[4,6,12]

Prepregnancy
- Elevated serum insulin, free fatty acids, triglycerides
- Disrupted platelet functions

Pregnancy
- Genetic predisposition to hypertension
- Hormonal and metabolic changes
- Renal disease
- Nutritional status

preeclampsia are at least partially mediated by insulin resistance.[12] (For background information on insulin resistance see Chapter 3, in the section Diabetes Melitus Prior to Pregnancy.) The evidence is based on the presence of features of the insulin resistance syndrome in women with preeclampsia and the fact that women with insulin resistance are at increased risk of developing preeclampsia.[6]

Shared features of preeclampsia and insulin resistance include elevated serum insulin, hypertension, dyslipidemia (reduced HDL cholesterol and elevated free fatty acids and trigylcerides), relative glucose intolerance, disruption of platelet functions and coagulation abnormalities, atherosclerotic changes, and obesity.[13] A number of these conditions may be "clinically silent" or unnoticed in women with preeclampsia but represent persistent alterations in body functions. Abnormalities associated with insulin resistance may worsen during pregnancy due to the profound hormonal and other metabolic changes that occur. The likelihood that this situation will arise appears to be increased in women with insulin resistance who are genetically predisposed to developing hypertension (Table 5.4).[6]

Elevated fasting insulin of twice the level of normal values and postprandial levels that are four times higher than normal for pregnancy are found in preeclampsia and may be accompanied by normal glucose levels. The simultaneous presence of normal glucose and high insulin levels indicates that women can still produce enough insulin to compensate for low cell membrane sensitivity to the action of insulin. The ability of pancreatic beta cells to continue to produce high enough levels of insulin to overcome the effects of insulin resistance may eventually end, however. Women with preeclampsia are at increased risk for de-

ENDOTHELIUM   The layer of flat cells lining blood and lymph vessels. These cells produce a variety of proteins that play a role in blood pressure regulation and body fluid distribution.

veloping diabetes during pregnancy and type 2 diabetes later in life.[6] In addition, about 15% of women with gestational diabetes and 30% of women with type 2 diabetes before pregnancy will develop preeclampsia.[6] High insulin levels found in women with insulin resistance favor the development of hypertension, atherosclerosis, heart disease, stroke, and increased levels of body fat in the years following pregnancy.[6]

High blood pressure associated with insulin resistance may participate in physiological mechanisms that impair functions of the *endothelium* and modify the balance between various types of prostaglandins such as prostacyclin and thromboxane.[13] As presented in Chapter 2, prostaglandins are physiologically active substances made from the fatty acid arachidonic acid. Found in many tissues, they constrict or dilate blood vessels, among other functions.

RISK FACTORS FOR THE DEVELOPMENT OF PREECLAMPSIA   The roots of preeclampsia lie very early in pregnancy, but as of yet there is no reliable means of identifying women who will develop it before the condition is established.[5] However, women with insulin resistance, obesity, or other characteristics listed in Table 5.5 are at increased risk for developing the disease. Characteristics listed confer at least a two-fold increase in risk.

Increased rates of preterm delivery and low birthweight in infants born to women with preeclampsia are partly related to clinical decisions to deliver fetuses early in order to treat the disease. Most infants born to women with this disorder are normal weight, however, and some newborns are large for gestational age. Variations in birthweight associated with preeclampsia appear be related to the severity of the disease in individual women.[16]

The risk of developing hypertension associated with preeclampsia is higher in women who were born SGA.

Table 5.5 Risk factors for preeclampsia.[2,5,14,15]

- First pregnancy (nulliparous)
- Obesity, especially high levels of central body fat
- Underweight
- Mother's smallness at birth
- African Americans, American Indians
- History of preeclampsia
- Preexisting diabetes mellitus
- Age over 35 years
- Multifetal pregnancy
- Insulin resistance
- Chronic hypertension
- Renal disease
- History of preeclampsia in the mother's or father's mother
- High blood levels of homocysteine
- Inadequate diet (possibly related to inadequate vitamins C and E, calcium, zinc, and the n-3 fatty acids)

Table 5.6 Status of effectiveness of preventive measures for preeclampsia.[1,3,17]

- Antihypertensive medications: Controversial, may not improve pregnancy outcomes
- Low dose aspirin: Unproven to be effective in prevention
- Calcium supplementation: Appears to reduce the incidence of preeclampsia in high-risk women
- Magnesium supplementation: Does not appear to prevent preeclampsia
- Fish oils (n-3 fatty acids): No reduction in preeclampsia in high-risk women
- Vitamins C and E: May prevent preeclampsia

It appears that growth restriction in utero may impair mechanisms involved in the regulation of blood pressure and increase the probability that high blood pressure will develop with the physiological stresses of pregnancy.[15] Interestingly, women or men whose mothers had preeclampsia while pregnant with them are more likely to have children born from pregnancies complicated by preeclampsia.[14]

Several drugs (metformin and troglitazone, for example) enhance insulin sensitivity, but their safety for treating insulin resistance in pregnancy is not known. Insulin sensitivity can be improved by weight loss after pregnancy, exercise, and diet modification.[6]

NUTRIENT INTAKE AND PREECLAMPSIA Inadequate nutrient status has been investigated as a potential cause of preeclampsia, and various nutrients have been given to women with the aim of reducing its occurrence. Some of the results are promising. Table 5.6 provides an overview of nutritional and other remedies that have been tested for their usefulness in preventing preeclampsia.

Calcium and Preeclampsia Calcium plays a role in the maintenance of normal blood vessel tone and has been implicated in the etiology of hypertension in adults. Whether it plays a role in the development of preeclampsia in pregnant women continues to be investigated.

Randomized, placebo-controlled clinical trials have found that intakes around the DRI for calcium in pregnancy (1000 mg per day), or one to two grams of supplemental calcium daily (1000–2000 mg) decrease blood pressure and preeclampsia in pregnant women. (The UL for calcium in pregnant women is 2.5 grams.) Results of various trials indicate that calcium supplements decrease blood pressure in pregnant women in general by about 20%. Women at high risk of preeclampsia, however, experience a 78% reduction in occurrence, and women with low, baseline intakes of calcium a 68% reduction in preeclampsia. Calcium does not appear to affect the risk of preterm delivery in women in general, but a 56% reduction is seen in women at high risk of developing preeclampsia.[18] In a randomized, placebo-controlled trial, women at risk of preeclampsia given 450 mg of linoleic acid and 600 mg of calcium per day in the third trimester experienced a 75% reduction of preeclampsia. Infants born to women taking the supplements weighed an average of 124 grams more than those of women who received the placebo.[19]

Studies have also found that increased calcium intakes from food or supplements ranging from 375–2000 mg per day during pregnancy may decrease blood pressure levels in children.[20,21]

Vitamin C and E and the Prevention of Preeclampsia Supplementation with vitamin C (1000 mg) and E (400 IU) from 16 to 22 weeks of gestation to delivery show promise in reducing oxidative damage caused by preeclampsia,[22] and have been found to reduce its occurrence by over half.[23] Decisions on the use of antioxidant supplementation to prevent preeclampsia awaits the results of additional clinical trials.

Effects of calcium and antioxidant nutrients on the risk of preeclampsia may depend on when the increased intakes begin. It is possible that protective effects are diminished by initiating these changes too late in pregnancy, or after about 20 weeks.[2] Such options are being seriously considered because the effectiveness of current therapies for treating preeclampsia is controversial.[16]

Folic Acid, Hyperhomocysteinemia and Preeclampsia
Women with preeclampsia tend to experience many of the same clinical features as do women at risk of heart disease. Such conditions include insulin resistance and atherosclerotic changes in blood vessels (of the placenta in pregnant women). High blood levels of homocysteine are related to these same conditions, giving rise to the theory that folic acid and hyperhomocysteinemia may be related to preeclampsia.[24] Women with elevated blood levels of homocysteine are over four times more likely to have preeclampsia or eclampsia than are women with low homocysteine levels.[24] Although it is not known whether adequate intake of folic acid reduces preeclampsia or its symptoms, it does appear to normalize plasma homocysteine levels in pregnant women with preeclampsia.[25]

High concentrations of homocysteine in pregnancy are also associated with repeated spontaneous abortion and placenta abruptio.[26]

n-3 Fatty Acids and the Risk of Preeclampsia   Only a few trials of the effectiveness of n-3 fatty acid supplements or fish consumption on the occurrence of preeclampsia have been reported. It is speculated that prostacyclin levels would be increased, and thromboxane levels decreased by these fatty acids, and that vasoconstriction would be lowered as a result. Results of the trials are mixed, however. It appears that n-3 fatty acids may increase birthweight and gestational age at delivery somewhat but may not affect the occurrence of preeclampsia.[21,27]

Low Sodium Diets Do Not Prevent Hypertension in Pregnancy   Salt restriction is standard policy in the Netherlands for the prevention of preeclampsia and other hypertensive disorders of pregnancy—but may not continue to be. Results of a randomized, controlled clinical trial of the effectiveness of sodium restriction in preventing preeclampsia undertaken in the Netherlands revealed that it did not work. The authors concluded that sodium restriction during pregnancy is not warranted.[28] Sodium restriction for the prevention of hypertension fell out of favor in most parts of the United States and Canada decades ago.

Other Nutritional Theories Related to Preeclampsia   Through the years a number of nutritional causes of preeclampsia have been proposed and corresponding nutritional remedies implemented in practice—even in the absence of supportive studies. These include the belief that low-protein diets, high and rapid weight gain in pregnancy, and high salt intakes prompted the development of preeclampsia.[1,29] In one situation, belief in the safety and potential bene-

fits of a rice and fruit diet motivated a physician to recommend it for the prevention of preeclampsia. At least one woman who followed the advice ended up winning a malpractice law suit. The grossly inadequate diet was found to be related to her daughter's mental retardation.[30] The bottom line is that it is unwise to test unsubstantiated theories about nutrition and diseases such as preeclampsia on women and their fetuses.

Preeclampsia Case Presentation

Signs, symptoms, severity, and causes of preeclampsia vary from woman to woman. Therefore, appropriate interventions for women presenting differing aspects of the syndrome are best designed on a case-by-case basis.[16] Case Study 5.1 describes the course of preeclampsia in one woman having the condition by way of example.

Nutritional Recommendations and Interventions for Preeclampsia

In the best of circumstances, dietary interventions for preeclampsia would begin prior to pregnancy. This may provide the opportunity for women to decrease body weight and stores of central body fat, and become physically fit, as well as allow them to consume a diet that reduces the need for insulin. Short of those circumstances, dietary recommendations and interventions should begin as early in pregnancy as possible and target at-risk women.

Nutritional and physical activity recommendations that may benefit women at risk of preeclampsia are within the normal scope of practice. They include:

- 1000 mg per day of dietary calcium
- 600 DFE of folate, which includes 400 mcg of folic acid daily
- Four or more servings of vegetables and fruits daily
- No restriction of sodium intake (with the possible exception of some women with chronic hypertension)
- Consumption of the assortment of other basic foods recommended in the Food Guide Pyramid
- Moderate exercise (for example, walking, swimming, noncompetitive tennis, or dancing for 30 minutes at least three times a week) unless medically contraindicated
- Weight gain that follows recommendations based on prepregnancy weight status
- Three regular meals and snacks daily
- Consumption of low-glycemic-index rather than high-glycemic-index carbohydrate foods.[1,28,31–34]

Case Study 5.1
A Case of Preeclampsia

Susan became pregnant when she was 31 years old and weighed 209 pounds (95 kg). She had experienced hypertension and renal disease in her early 20s, but had been free of these disorders for five years before pregnancy. Six months before the current pregnancy Susan spontaneously aborted an 11 week old fetus affected by congenital malformations. She had a strong family history of heart disease; both her mother and 44 year old brother had died of a heart attack.

At 17 weeks gestation, Susan was identified as having proteinuria and was excreting 2300 mg protein in her urine in a 24-hour period. It was reasoned that the protein loss was related to scarring in tubules of the kidney due to her earlier bout with renal disease. By 21 weeks of pregnancy her blood pressure had increased to 152/100 mm Hg. Ultrasound revealed an active fetus who was at the 25th percentile for weight by gestational age. Susan was given antihypertension medications, a multivitamin and mineral supplement, and aspirin, but at 23 weeks her blood pressure was still high.

Susan was admitted to a hospital at 27 weeks because the proteinuria and hypertension had worsened. Protein excretion had increased to 7 grams a day and blood pressure remained high but did not increase further. The level of protein excretion was now thought to be an affect of the preeclampsia, in addition to that related to previous kidney damage. Due to her family history of heart disease, homocysteine levels were tested and found to be elevated. Susan was not experiencing the symptoms of severe headache, blurred vision, abdominal pain, or nausea that are commonly associated with severe preeclampsia. She also was found to have normal platelet and liver enzyme values, and normal kidney function. A diagnosis of preeclampsia was made and confirmed by renal biopsy.

Susan developed a low platelet count and elevated liver enzymes during week 28 and a cesarean section delivery was performed. A severely growth retarded 655-gram fetus and 133-gram placenta were delivered. The infant had a long hospital stay but was eventually able to go home.

It was decided at a case conference held after Susan's delivery that her elevated homocysteine levels may have been related to an inadequate folate intake, and that the high homocysteine level may have contributed to the severity of preeclampsia she experienced. Many other possible scenarios that would account for the severity of her disease were discussed as well.

Food sources of low-glycemic-index carbohydrates include whole grain products, other carbohydrates high in fiber, milk, apples, peaches, and dried beans. Foods such as white bread, corn flakes, hard candy, watermelon, and rice have high glycemic indices. (A table of the glycemic index of foods is presented in this chapter in the section on diabetes in pregnancy.) Such diets also limit the consumption of simple sugars and foods and beverages that contain them.

Whether supplemental calcium or vitamins C and E should be recommended to women at risk of preeclampsia isn't clear, but doses below the UL for pregnancy are likely safe. Additional information is needed on the benefits of these supplements to outcomes such as intrauterine growth retardation and perinatal death associated with preeclampsia before they can be confidently recommended or ignored. The multivitamin and mineral supplement recommended by the Institute of Medicine[35] and discussed previously should be given to women considered to be at risk of nutrient shortages.

DIABETES IN PREGNANCY

Diabetes is the second leading complication in pregnancy[36] and has several forms:

- Gestational diabetes
- Type 2 diabetes
- Type 1 diabetes
- Other specific types[37]

Gestational diabetes and type 2 diabetes are part of the same disease. This chapter focuses on gestational diabetes, which develops during pregnancy. Information related to type 1 diabetes, or insulin-dependent diabetes, is presented as well.

Gestational Diabetes

> *The difference between gestational and type 2 diabetes may be the moment of detection.*
>
> Branchtein[38]

Approximately 1 to 3% of pregnant women develop *gestational diabetes* during pregnancy.[39] The cause of gestational diabetes is controversial but it is considered to be a form of non–insulin dependent diabetes, or type 2.[40] It is not clear whether the disorder occurs primarily due to increasing insulin resistance and exaggerated liver production of glucose late in pregnancy, or to limits on the ability of beta cells to produce enough insulin to overcome the effects of increased insulin resistance in pregnancy.[6,41] Gestational diabetes in underweight and normal-weight women appears to be related to insulin resistance combined with a deficit in insulin production, whereas insulin resistance and not inadequate insulin production may underlie it in obese women.[42]

GESTATIONAL DIABETES Carbohydrate intolerance with onset or first recognition in pregnancy.

Women who develop gestational diabetes appear to enter pregnancy with a predisposition to insulin resistance and type 2 diabetes that is expressed due to physiological changes that occur during pregnancy. The insulin resistance brought into pregnancy, or the tendency to develop it, may be clinically silent in that glucose levels may not be elevated and blood pressure may be normal. However, high blood levels of glucose and other signs related to increased insulin resistance develop as pregnancy progresses. Women with gestational diabetes not only develop elevated levels of blood glucose but also of triglycerides, fatty acids, and sometimes blood pressure. In some cases, gestational diabetes appears to be related to exaggerated metabolic changes favoring elevated blood glucose levels[41] or reduced insulin output.[43]

High maternal blood glucose levels reach the fetus (Illustration 5.1) and cause the fetus to increase insulin production to lower it. The higher the level of blood glucose received, the larger the fetal output of insulin.

Consequences of Poorly Controlled Gestational Diabetes

Potential consequences associated with gestational diabetes are summarized in Table 5.7. Exposure to high

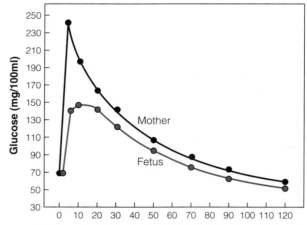

Illustration 5.1   Concentrations of fetal blood glucose following an intravenous dose of glucose to the mother.

SOURCE: In: Hytten F and Chamberland G. Clinical Physiology in Obstetrics, Oxford:Blackwell Scientific Publishers, 1980 and based on data from Coultart et al. Reproduced with permission.[44]

insulin levels in utero leads to increased glucose uptake into cells and the conversion of glucose to triglycerides. These changes increase fetal formation of fat and muscle tissue and may program metabolic adaptations that increase the likelihood that insulin resistance, type 2 diabetes, high blood pressure, and obesity will develop later in life.[41] The chances that these disorders will occur increase with higher maternal levels of glucose and triglycerides during pregnancy.[45]

Effects of high maternal levels of glucose and triglycerides are particularly striking in the Pima

Table 5.7   Adverse outcomes associated with gestational diabetes.[6,41,45]

Mother
- Cesarean delivery to prevent shoulder dystocia
- Increased risk for preeclampsia during pregnancy
- Increased risk of type 2 diabetes, hypertension, and obesity later in life
- Increased risk for gestational diabetes in a subsequent pregnancy

Offspring
- Stillbirth
- Spontaneous abortion
- Macrosomia (>10 pounds or >4500 grams)
- Neonatal hypoglycemia
- Increased risk of resistance, type 2 diabetes, high blood pressure, and obesity later in life

Indians of Arizona. Fetal exposure to poorly controlled maternal diabetes increases the risk that children will develop type 2 diabetes by 10-fold. Offspring of diabetic, Pima mothers are heavier at birth, have higher BMIs throughout childhood, and a seven to 20 times greater incidence of type 2 diabetes in early adulthood. Although risks of these conditions increase in offspring of women with poorly controlled diabetes in general, the pronounced effect in Pima Indians is likely due to a strong genetic tendency toward insulin resistance and obesity.[46]

The end of pregnancy initially restores insulin sensitivity in most women with gestational diabetes. However, a degree of insulin resistance often remains.[6] Close to half of women with gestational diabetes in a previous pregnancy will develop it in a subsequent pregnancy.[34,47] The cumulative incidence of type 2 diabetes in women who experienced gestational diabetes is approximately 50% after five years. Women with weight gain after pregnancy and repeated pregnancies continue to experience insulin insufficiency and resistance; this group is even at higher risk of developing type 2 diabetes later in life. Among women with gestational diabetes, those requiring insulin therapy have higher blood pressure than women whose gestational diabetes was controlled with diet and exercise.[6]

Risk Factors for Gestational Diabetes

Both type 2 and gestational diabetes are linked to multiple inherited predispositions and their environmental triggers, such as excess body fat and low physical activity levels.[6] About half of women who develop gestational diabetes have no identified risk for the disease, however.[8] Risk factors for gestational diabetes are outlined in Table 5.8.

Diagnosis of Gestational Diabetes

The diagnosis of gestational diabetes is principally based on blood glucose levels, and disagreement continues about what levels of blood glucose are most likely to predict which women are at significant risk. The disagreement exists because the relationship between maternal blood glucose levels and adverse perinatal outcomes is continuous. Consequently, cutpoints selected for blood glucose levels diagnostic of gestational diabetes are arbitrary and subject to change.[51] Current standards for blood glucose levels indicative of risk for gestational diabetes, as well as characteristics of women which place them at potential risk, are outlined next.

Table 5.8   Risk factors for gestational diabetes.[47–50]

- Obesity, especially high levels of central body fat
- Weight gain between pregnancies
- Underweight
- Age over 35 years
- Native American, Hispanic, African American, South or East Asian, Pacific Islander, Indigenous Australian ancestry
- Family history of gestational diabetes
- History of delivery of a macrosomic newborn (>4500 g or 10 lbs)
- Chronic hypertension
- Mother was SGA at birth
- History of gestational diabetes in a previous pregnancy
- Diabetes in pregnant women's mothers during their pregnancy with them and LGA at birth

CRITERIA FOR THE DIAGNOSIS OF GESTATIONAL DIABETES   Glucose screening is recommended for women at high risk of gestational diabetes at the initial visit or as soon as possible thereafter. High risk is identified in women who have one or more of the following:

- Marked obesity
- Diabetes in a mother, father, sister, or brother
- History of glucose intolerance
- Previous macrosomic infant
- Current glucosuria

A 50 g oral glucose challenge test is generally used for blood glucose screening. This test can be done without fasting. Blood is collected one hour after the glucose load is consumed and tested for glucose content. This test should be followed by an oral glucose tolerance test if glucose level is high, or ≥ 130 mg/dL (7.2 mmol/L). (You can convert mg/dL to millimoles per liter, or mmol/L, by multiplying mg/dL by 0.05551.)

The oral glucose tolerance test (OGTT) is the basis for the diagnosis of most cases of gestational diabetes. It can be bypassed among women with very high glucose screening results and treatment started. A 100-gram glucose, three-hour test is used for the OGTT. (The practice of "loading women up" with a high carbohydrate diet three days in advance of the test is no longer recommended. It does not alter results of the OGTT nor reduce the proportion of women incorrectly diagnosed with gestational diabetes.)[52] The beverage provided for this test is very sweet and it takes some women a bit of time to drink the full amount. A diagnosis of gestational diabetes is made when two or more values for venous serum or plasma glucose concentrations exceed these levels:

Overnight fast	95 mg/dL
One hour after glucose load	180 mg/dL
Two hours after glucose load	155 mg/dL
Three hours after glucose load	140 mg/dL

A 75 gram oral glucose load can also be used, but the three-hour glucose level will likely differ from results using the 100-gram glucose load test. Two or more abnormal glucose values using a 75-gram load is diagnostic of gestational diabetes. Because of their increased risk for preeclampsia, women with gestational diabetes should be closely monitored for preeclampsia.[50]

A plasma glucose screening between 24 and 28 weeks of pregnancy is recommended for women at "average risk" and for high-risk women not determined by glucose screen to have elevated glucose levels earlier. Average risk is defined as women who neither fit the low nor the high-risk profile.

Glucose screens are not recommended for women at low risk as defined by:

- Age <25 years
- Member of a low-risk ethnic group (those other than Hispanic, African American, South or East Asian, Pacific Islander, Native American, or Indigenous Australian.)
- No diabetes in first-degree relatives
- Normal prepregnancy weight and weight gain during pregnancy
- No history of glucose intolerance
- No prior poor obstetrical outcomes.[50]

Women with gestational diabetes may notice an increased level of thirst (especially in the morning), an increased volume of urine, and other signs related to high blood glucose levels.[53]

Urinary glucose cannot be used to diagnose nor monitor gestational diabetes because the results do not correspond to blood glucose levels.[54]

Treatment of Gestational Diabetes

A team approach to caring for women with diabetes in pregnancy is advised. Such teams often consist of an obstetrician, a registered dietitian who is also a certified diabetes educator, a nurse educator, and an endocrinologist.[55] The primary approach to treatment is medical nutrition therapy that begins with attempts to normalize blood glucose levels with diet and exercise. If postprandial glucose levels remain high two weeks after institution of the diet and exercise plan, then insulin injections will be added. Postprandial blood glucose rather than fasting glucose levels are re-

lated to fetal overgrowth and are the main indicators of adequacy of blood glucose control.[41]

Medical nutrition therapy has been shown to effectively normalize blood glucose levels and to decrease the risk of adverse perinatal outcomes. Results shown in Table 5.9 demonstrate the effect and the usefulness of identifying and intervening upon women with gestational diabetes. It can also be noted from the results that a higher proportion of large newborns occurs even with medical nutrition therapy, but that the incidence is substantially less than in women with untreated gestational diabetes.

Blood glucose levels can be brought down by low caloric intakes. However, accelerated rates of starvation metabolism during pregnancy, as well as potentially deleterious effects of resulting ketonemia on fetal development, exclude this approach to blood glucose control.[41,57] Correspondingly, restriction of pregnancy weight gain to below recommended amounts is not advised.[58] Aggressive treatment of gestational diabetes that excessively limits caloric intake and weight gain increases the risk of SGA newborns.[59] On the other hand, excessively high caloric balances and weight gains are of concern because they increase the risk of macrosomia.[60]

Type 2 diabetes in nonpregnant individuals is often treated with sulfonylurea oral medications. These drugs cannot be used in pregnancy because they cross the placenta and stimulate fetal insulin production. Other types of oral medications are being tested for use among women with gestational diabetes.[37]

Presentation of a Case Study

No two individual women with gestational diabetes share the same history, risks, needs, and response to treatment. The case presented next represents an individual's experience with the disorder.

Table 5.9 Comparison of outcomes of unrecognized and diet-treated gestational diabetes.

GESTATIONAL DIABETES			
Outcome	**Unrecognized**	**Diet-Treated**	**Controls**
LGA (>90th percentile)	44%	9%	5%
Macrosomia (>4500 g)	44%	15%	8%
Shoulder dystocia	25%	3%	3%
Birth trauma	25%	0%	0%

SOURCE: Data from Adams, 1998.[56]

Case Study 5.2
Elizabeth's Story: Gestational Diabetes

Elizabeth is a 36 year old Hispanic woman in her second pregnancy. The first pregnancy resulted in the birth of a 3450 gram (7.7 lb) healthy infant. Before this pregnancy Elizabeth was of normal weight status with a BMI of 24.5 kg/m². Pregnancy progressed normally through the first 26 weeks, and her weight gain of 20 pounds was within the normal limits.

Since Elizabeth was considered to be at "average risk," she was given a 50 g oral glucose screening test at her 26-week appointment. She was referred for a 100 g oral glucose tolerance test (OGTT) because the screening result was 147 mg/dL glucose. The next day's OGTT results identified the following glucose levels:

Fasting: 90
1 hour: 195
2 hour: 163
3 hour: 135 mg/dL

Based on elevated one- and two-hour results, a diagnosis of gestational diabetes was made. Elizabeth's certified nurse midwife saw her the next day to answer her questions and provide education on gestational diabetes and its implications to her care. The midwife also advised Elizabeth on her newborn's risk for hypoglycemia and the need for her to have her glucose levels tested again after pregnancy. She stressed the importance of keeping good control of blood glucose levels, of monitoring her blood glucose levels regularly, and of carefully following the diet and exercise plan that would be worked out. Elizabeth was advised that her glucose would be tested again after she had been on the diet and exercise plan for a week and that if her blood glucose levels remained high after two weeks, she would be put on insulin injections. Elizabeth was given an appointment with the clinic's dietitian and certified diabetes educator.

After performing a dietary assessment and learning about her food and exercise preferences, the dietitian estimated Elizabeth's caloric need. Elizabeth's current body weight of 69 kg (152 lb) was multiplied by 30 calories to yield an estimate of a need for 2,070 calories per day. A food plan including 48% of calories from carbohydrate in 3 meals and 2 snacks daily was negotiated with Elizabeth. It was concluded that she would maintain her current, moderate level of physical activity. Elizabeth was instructed on blood glucose assessment using a glucometer that stored the results in its built-in memory, and fasting urinary ketone testing procedures. She was asked to call the dietitian or midwife if questions or problems arose.

The thought of insulin injections if the diet plan didn't work was a major motivation for Elizabeth. She carefully followed her diet plan, recording her food intake, testing her blood levels of glucose four times a day. She also performed a dipstick test for urinary ketones after she awoke in the morning. When she returned to the clinic the next week she was rewarded by normal glucose values—but she had lost a half pound and had elevated urine levels of ketones twice during the past week. Her diet plan was re-evaluated, a bedtime snack added, and the caloric level of the plan was increased by 400 calories a day to allow for weight gain.

Elizabeth showed no signs of developing hypertension or other complications, and fetal growth appeared to be normal based on ultrasound results. Consequently, the consulting endocrinologist and obstetrician decided there was no need for them to see Elizabeth. A week after she started the revised diet plan she had gained a pound and continued to gain about a pound a week for the next 12 weeks while maintaining acceptable blood glucose and urinary ketone levels for the vast majority of the time. *continued*

Case Study 5.2
Elizabeth's Story: Gestational Diabetes, continued

Elizabeth went into labor at the beginning of the 39th week of pregnancy. After six hours of labor Elizabeth delivered a healthy, 3550 gram (7.9 lb) girl whose blood glucose levels remained within the normal range through two hours after birth. Elizabeth returned to the clinic for a follow-up glucose test two months after delivery and the two-hour OGTT test results were normal. She was concerned, however, that she wasn't losing weight quickly enough, so she asked for an appointment with the dietitian.

After following the diet and exercise plan negotiated with the dietitian for five months, Elizabeth was two pounds below her prepregnancy weight and physically fit. She felt great and was determined to keep her weight under control and to stay physically active. Type 2 diabetes or hypertension was not going to be in her future if she had anything to do with it!

Exercise Benefits and Recommendations

Insulin resistance is decreased and blood glucose control enhanced by regular aerobic exercise such as walking, jogging, biking, golfing, hiking, swimming, and moderate weight lifting. This appears to be the case as well in women with gestational diabetes. Weight lifting with the arms three days a week for 20 minutes per session for six weeks, and exercising on a recumbent bicycle at 50% VO_2 max for 45 minutes three times a week, have been found to normalize blood glucose levels in some women.[61]

Levels of exercise that approximate 50% of VO_2 max, or maximal oxygen uptake, are most often recommended for women with gestational diabetes. These levels are estimated in practice using a formula for heart rates associated with various levels of VO_2 max. The formula is: $220 - age \times 0.50$ (for 50% of VO_2 max) = heart beats per minute. If we took the example of a 29 year old, the estimated heart rate at 50% of VO_2 max would be $220 - 29 \times 0.50$, or 96 beats per minute. This would be the maximum heart rate she should experience while exercising. Levels of exercise should make women become slightly sweaty but not overheated, dehydrated, or exhausted.[58]

Exercise and certain types of physical activity are not safe for all women, for example some women with threatened preterm labor or hypertension. Consequently, intentions to exercise must first be approved by the primary health care provider. Physical activities undertaken should not threaten the fetus or mother as can happen if the exercise causes women to lose their balance, bounce, or put pressure on the abdomen. Exercise that is safe to the fetus and does not raise blood pressure is generally considered safe.[58]

Nutritional Management of Women with Gestational Diabetes

Primary outcome goals for women with gestational diabetes are well-controlled blood glucose levels and a healthy newborn. Other goals include the normalization of carbohydrate metabolism and a reduction in the mother's and offspring's subsequent risk of diabetes, hypertension, and obesity.[58] For the majority of women, diet and exercise changes will be the primary way to achieve these goals. In other women supplemental insulin will also help achieve these goals.

Except in cases of very high glucose levels, dietary changes are given a two-week trial before insulin is used. A two-week trial of diet and exercise changes are generally undertaken due to the general impression that blood glucose control not achieved during a two-week trial is unlikely to be achieved given a longer period of time.[62] Blood glucose control is more likely to require insulin in obese women than in normal-weight or underweight women.[8]

Components of the nutritional management of women with gestational diabetes include:

- Dietary and exercise assessments
- Development of an individualized diet and exercise plan
- Monitoring weight gain
- Interpretation of blood glucose and urinary ketone results
- Follow-up during pregnancy and postpartum[34,61]

Women with type 2 diabetes coming into pregnancy are managed in a very similar fashion as are women with gestational diabetes, only nutritional care begins earlier. Ideally, normal blood glucose levels

Table 5.10   Estimating levels of caloric need in women with gestational diabetes.[61]

Current Weight Status	Definition	kcal/kg Body Weight
Underweight	<80% of average weight	35–40
Normal weight	80–120% of average weight	30
Overweight	120–150% of average weight	24
Obese	>150% of average weight	12–15

should be established prior to conception and then maintained in good control through pregnancy. Diet and exercise plans for women with type 2 diabetes can often be based on what has worked in the past, thus simplifying planning to needs associated with pregnancy.

THE DIET PLAN   In general, diets developed for women with gestational diabetes emphasize:

- Complex carbohydrates including whole grain breads and cereals, vegetables, fruits, and high-fiber foods
- Limited intake of simple sugars and foods and beverages that contain them
- Low-glycemic-index foods, or carbohydrate foods that don't raise glucose levels greatly
- Unsaturated fats
- Three regular meals and snacks daily

Dietary planning is based around a calculated level of caloric need. These initial estimates of caloric need are intended to meet both maternal and fetal demand for energy while limiting increases in blood glucose levels. They are based on the pregnant woman's current weight and her need to gain weight during pregnancy. Estimated levels of caloric need according to women's current weight status are shown in Table 5.10.

Calories are generally distributed among the meals and three snacks, with lunch serving as the largest meal and breakfast and snacks being limited to 10 to 15% of total calories.[58]

Caloric levels and meal and snack plans are considered to be starting points and often require modifications after results of blood glucose home monitoring tests are known. Caloric levels for overweight and obese women are particularly likely to initially fall short of need,[63] and morning blood glucose levels are most likely to be high.[64] Reduction of calories from carbohydrate is indicated at breakfast for women with high morning glucose levels.

Specific recommendations for the distribution of calories from carbohydrate and fat primarily depend on individual eating habits and the effect of carbohydrate and fat on blood glucose levels. The distribution of calories utilized in practice, however, commonly range from:

- 40–50% of calories from carbohydrate
- 30–40% from fat
- 20% from protein

The relatively low-carbohydrate, high-fat diet decreases the need for insulin by lowering the amount of glucose absorbed from food, and blunts postprandial increases in blood glucose and insulin levels (Table 5.11). The addition of high-fiber foods to diet plans may also enhance blood glucose control. These changes in turn reduce fetal overgrowth and other adverse effects of insulin resistance and high levels of glucose and insulin.[61]

Table 5.11   Effects of six weeks of low- and high-carbohydrate diets on maternal and newborn outcomes in women with gestational diabetes.[65]

| OUTCOME | DIET | |
	Low-carbohydrate (<42% of Total Calories)	High-carbohydrate (>45% of Total Calories)
Postprandial glucose values	110 mg/dL	132 mg/dL
Fasting glucose	92 mg/dl	94 mg/dL
Insulin requirement	5%	33%
Urinary ketones	10%	0%
Birthweight, g	3694	3890
LGA	9%	42%
Cesarean delivery due to LGA fetus	3%	48%

Consumption of Low-Glycemic-Index Foods

The topic of whether low-glycemic-index (GI) foods benefit women with diabetes in pregnancy has been much debated. After due consideration, however, it appears that the benefits are real enough to support a recommendation that they should be the carbohydrate foods of choice. Low-GI foods help women sustain modest improvements in blood glucose levels and decrease insulin requirements.[32,66] A remaining concern centers on the complexity of incorporating glycemic index values into individual diet plans.[67]

WHAT IS THE GLYCEMIC INDEX?   The extent to which blood glucose levels increase after carbohydrates are consumed varies depending on the food source of carbohydrate. Difference in the blood glucose raising property of food sources of carbohydrates has led to the development of the "glycemic index," or the GI value of carbohydrate foods.[68] The glycemic index of foods is calculated by the formula:

$$GI = \frac{\text{Blood glucose response to 50 g of a food}}{\text{Blood glucose response to 50 g of a reference food (eg. glucose)}} \times 100$$

A food that elicits a high increase in blood glucose compared to that of a reference food is considered a high-GI food. Foods that cause lower elevations in blood glucose level are moderate- or low-GI foods. Table 5.12 lists the GI rating of a number of food sources of carbohydrate. Values for different foods shown clearly demonstrate that GI value cannot be predicted solely based on the simple sugar content of foods. Glycemic index values do vary somewhat from study to study, so absolute GI values are not given in the table.

A concern related to the glycemic index is whether values differ if carbohydrate foods are eaten with a meal. This concern is not justified based on test results of carbohydrate foods eaten alone or with meals.[32] Results of one such study are shown in Illustration 5.2. In this study 50 grams of white bread (a high-GI food) or spaghetti (a lower-GI food) were given in a standard, 600-calorie meal, and blood glucose responses were noted. Results show that GI values for white bread and spaghetti differed little from those obtained when the foods were eaten alone.[68] Blood glucose response to high-carbohydrate foods does not appear to change during pregnancy.[66]

EXAMPLE MEAL PLANS   Individualized diet plans for women with gestational diabetes include a variety of different foods that correspond to the preferences

Table 5.12   Glycemic index of foods.[*] (High-glycemic-index foods raise blood glucose levels more than low-glycemic-index foods.)[32,69,70]

HIGH	MODERATE	LOW
Grain Products		
White bread	Cream of Wheat	Heavy, mixed grain
Wheat bread	Mueslix	bread
French bread	Kudos bar, whole	Pumpernickel
Flour tortilla	grain	bread
Crackers	Bran Chex	All Bran
Muffins	Sourdough bread	Mueslix, toasted
Cake	Rye bread	
Rice (white or	Basmati rice	
brown)	Oatmeal	
Corn flakes	Pasta	
Rice Krispies		
Cheerios		
Vegetables		
Potatoes	Sweet potatoes	Dried beans
Pumpkin		Lentils
		Green peas
		Tomato, raw
		Corn
		Carrots, raw
Fruits		
Watermelon	Raisins	Apple
	Banana	Orange
	Pineapple	Peach
		Grapefruit
		Plum
Candy, Other		
Jelly beans	Potato chips	Snickers bar
Lifesavers	Mars bar	Fructose
Skittles	Ice cream, full-fat	Yogurt, low-fat
Honey		Skim milk
		Whole milk
		Chocolate

[*]Classification of foods varies somewhat depending on GI test results.

and needs of women. Two examples of such diet plans are shown in Table 5.13. One menu provides approximately 2200, and the other 2400 calories. Both menus include low-glycemic-index food sources of carbohydrate and the nutrients needed by women during pregnancy.

URINARY KETONE TESTING   Women with gestational diabetes are generally instructed to monitor urinary ketone levels using dipsticks. In the past all dipsticks used were insensitive to ß-hydroxybutyrate, the primary ketone spilled into urine. Sticks that de-

Illustration 5.2 **Blood glucose response in people with type 2 diabetes to meals containing white bread or spaghetti.**

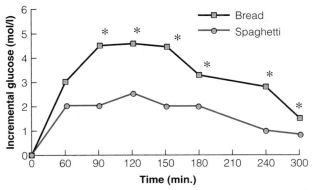

SOURCE: Riccardi G and Rivellese AA. Diabetes: nutrition in prevention and management. Nutr Metab Cardiovasc Dis 1999;9(suppl):33–6.[68] Reproduced with permission.

tect ß-hydroxybutyrate are becoming increasingly available and should be used. When interpreting results of urinary ketone tests keep in mind that 10–20% of pregnant women spill ketones after an overnight fast.[58] This means the severity and consistency of positive findings for urinary ketones should be considered.

Postpartum Follow-up

The relatively low-carbohydrate, high-fat diet instituted during pregnancy is generally changed to a higher-carbohydrate, lower-fat diet after delivery. Emphasis on low-glycemic-index and high-fiber foods, along with gradual weight loss may also be components of postpartum nutrition counseling.

Prevention of Gestational Diabetes During Pregnancy

Reducing overweight and obesity, increasing physical activity, and decreasing insulin resistance prior to pregnancy are important components of reducing the risk of gestational diabetes.[63] Screening programs that identify women with insulin resistance or glucose intolerance prior to pregnancy have also been advocated for risk reduction.[40]

Type 1 Diabetes During Pregnancy

Women with type 1 diabetes have deficient insulin output and must rely on insulin injections or an insulin pump to meet their need for insulin. Type 1 diabetes represents a potentially more hazardous condition to mother and fetus than do most cases of gestational di-

abetes. This condition places women at risk of kidney disease, hypertension, and other complications of pregnancy. Newborns of women with this type of diabetes are at increased risk of mortality, being SGA or LGA, and of experiencing hypoglycemia and other problems within 12 hours after birth. Hypoglycemia occurs in about half of macrosomic infants.[71] Coming into pregnancy with this type of diabetes also increases the risk of congenital malformations of the pelvis, central nervous system, and heart in offspring three-fold (from 2–3% to 6–9%).[37,72] Good control of blood glucose and diets rich in antioxidants reduce the risk of malformations. Maintenance of normal glucose levels from the start of pregnancy decreases the risk of fetal malformations and macrosomia.[72]

Blood glucose control from the beginning of pregnancy is also important because fetal growth trajectory may be largely determined in the first half of pregnancy. Exposure to high amounts of glucose and insulin when fetal growth trajectory is being established may set the "metabolic stage" for fetal accumulation of fat and lean tissue later in pregnancy.[73] Even relatively low elevations in blood glucose levels can meaningfully increase birthweight.[55] Unfortunately, only 10% of women with type 1 diabetes receive preconceptional care.[72]

NUTRITIONAL MANAGEMENT OF TYPE 1 DIABETES IN PREGNANCY Primary goals for the nutritional management of type 1 diabetes in pregnancy are continual control of blood glucose levels, nutritional adequacy of dietary intake, achievement of recommended amounts of weight gain, and a healthy mother and newborn. Close home monitoring of glucose levels and adjustments in dietary intake, exercise, and insulin dose based on the results are key events that increase the likelihood of reaching these goals. Monitoring urinary ketones is particularly important in women with type 1 diabetes because they are more prone to developing ketosis than are women with gestational diabetes.[37] Inclusion of ample amounts of dietary fiber (25–35 g per day) reduces insulin requirements in many women with type 1 diabetes in pregnancy.[64]

MULTIFETAL PREGNANCIES

Rates of multifetal pregnancy in the United States have increased remarkably since 1980. Twin births, which accounted for 1 in 56 births in 1980, constituted 1 in 34 births in 1999. Rates of triplet and higher order multiple births (referred to as "triplet+" births) have increased from 1 in 2941 to 1 in 541 births. Increases in the rates of multifetal pregnancy during the past decade are outlined in Table 5.14.

Table 5.13   Examples of three-meal, three-snack one-day menus at two caloric and carbohydrate levels for women with gestational diabetes.

1. 2,220 CALORIE DIET	Carbohydrates, g	Calories	2. 2,400 CALORIES	Carbohydrates, g	Calories
Breakfast			**Breakfast**		
All Bran, ½ c	22	80	Complete Wheat		
2% milk, ½ c	6	61	Bran Flakes, ¾ c	23	90
Mozzarella cheese			2% milk, ½ c	6	61
stick, 1 oz	1	78	Egg, 1	1	74
Black coffee, tea			Black coffee, tea		
Morning Snack			**Morning Snack**		
Oat bran bagel, ½	19	98	Peanuts, 2 oz	10	326
Sugar-free, low-fat			Carrot, 1	7	31
yogurt, 1 c	17	155	Graham crackers,		
			4 small or 1 sheet	11	59
Lunch					
Tuna salad, ½ c	19	192	**Lunch**		
Whole grain			Beef or chicken		
bread, 2 sl	24	130	burrito, 1	33	255
Carrot and celery			Salsa, ½ c	7	33
sticks, 1 c	7	31	Black beans, 1 c	40	228
Potato salad, ½ c	14	179	Apple, 1	21	81
Orange, 1	15	62	Black coffee, tea,		
Black coffee, tea,			water, or diet soda		
water, or diet soda					
			Midday Snack		
Midday Snack			Banana, ½	28	55
Peaches canned in			2% milk, 1 c	12	121
juice, ½ c	29	109			
2% milk, 1 c	12	121	**Dinner**		
			Lean pork chop, 4 oz	0	263
Dinner			Pinto beans, 1 c	22	116
Lean roast beef, 3 oz	0	188	Corn bread, 1 oz	12	92
Broccoli, 1 c with	10	50	Margarine, 1 tsp	1	33
Cheese (melted), 1 oz	1	105	Garden salad, 2 cup	0	10
Roll, 2 oz	30	167	Feta cheese, 1 oz	1	74
Margarine, 2 tsp	0	67	Salad dressing, 2 tbs	3	104
Grapes, 15	14	53	Black coffee, tea,		
Black coffee, tea, water,			water, or diet soda		
or diet soda					
			Bedtime Snack		
Bedtime Snack			Peanut butter, 2 tbs	12	190
Hard-boiled egg, 1	1	74	Rice cake, 1	8	35
Saltine crackers, 4	8	50	2% milk, 1 c	12	121
2% milk, 1 c	12	121			
			Total:	270 g	2442
Total:	261 g	2,171		or 44% of total calories	
	or 48% of calories				

SOURCE: Brown, 2001[61]

The extraordinary rise in multiple births over the past two decades, especially in triplet+ births, is primarily related to the increased availability and use of assisted reproduction techniques such as ovulation-inducing drugs and in vitro fertilization. Only one in five triplet+ pregnancies are spontaneously conceived. Rates of twin birth are highest by far in women 45–49 years old, the age group most likely to receive medical and technological interventions to achieve pregnancy.[74,75]

Progressively older ages at which childbearing is occurring in the United States also contributes to rising rates of multifetal pregnancies. The chances of a spontaneous multifetal pregnancy increase with age

Table 5.14 **Numbers of births to women with multifetal pregnancy.**[74]

NUMBER OF BIRTHS		
	1989	**1999**
Twin	90,118	114,307
Triplets	2529	6742
Quadruplets	229	512
Quintuplets or higher	40	67

after about 35 years. Rates of spontaneous multifetal pregnancy also increase with increasing weight status. For example, the rate of twin pregnancy is about two times higher in obese than in underweight women.[76] Rates of triplet+ pregnancies appear to be headed downward due to improved assisted reproductive technologies that reduce higher order, multifetal pregnancies.[74]

Upward trends in low birthweight and preterm delivery in the United States over recent years has been strongly influenced by the upsurge in multiple births. Only 3% of newborns are from multifetal pregnancies, yet they account for 21% of all low birthweight newborns, 14% of preterm births, and 13% of infant deaths.[75]

Background Information about Multiple Fetuses

The most common type of multifetal pregnancy, those with twin fetuses, come in several types and levels of risk. Twins are dizygotic (DZ) if two eggs were fertilized, and monozygotic (MZ) if one egg was. Monozygotic twins result when the fertilized and rapidly dividing egg splits in two within days after conception. The term "identical" is often used to describe MZ twins, and "fraternal" to denote DZ twins. These terms are misleading, so the preferred terms are "monozygotic" and "dizygotic."[77] About 70% of twins are DZ, and 30% MZ.

Monozygotic twins are always the same sex, whereas DZ twins are the same sex half the time and different sexes half the time. Monozygotic twins are genetically identical in almost all ways, but are seldom absolutely identical. Genetic differences in pairs of MZ twins can result from chromosome abnormalities in one twin, unequal genetic expression of maternally and paternally derived genes, and environmental effects on gene expression.[78] Rates of MZ twins are remarkably stable across population groups and do not appear to be influenced by heredity.[78]

Dizygotic twins represent individuals with differing genetic "fingerprints." The incidence of DZ twin pregnancies is influenced both by inherited and environmental factors. Rates of DZ twins vary among racial groups and by country. Rates tend to decrease in populations during famine and to increase when food availability and nutritional status improve.[79] Periconceptional vitamin and mineral supplement use has also been related to an increased incidence of DZ twin pregnancy.[80]

Twins also vary in the number of placentas; some are born having used the same placenta, whereas more commonly each fetus has its own. Twins may share a common amniotic sac and one of the membranes around the sac (the chorion), or have separate amniotic sacs and membranes (Illustration 5.3). Twins at

Illustration 5.3 **Twins (a) with two amniotic sacs, two chorions, and two placentas, (b) with one amniotic sac, chorion, and placenta, and (c) with two amniotic sacs, one chorion, and fused placentas.**

SOURCE: Reprinted with permission from George Thiene Verlag, Stuttgart, 1983

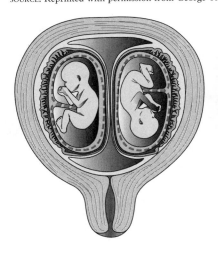
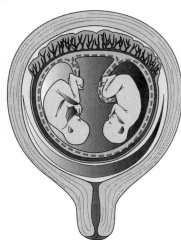
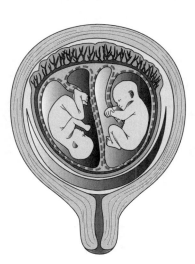

highest risk of death, malformations, growth retardation, short gestation, and other serious problems are those that share the same amniotic sac and chorion, and to a lesser degree, MZ twins in general.[78, 82]

Determining twin type is not always an easy task during or after pregnancy. Definitive diagnoses of tough cases can be made through DNA fingerprinting.[78]

In Utero Growth of Twins and Triplets

Fetal growth patterns of twins and triplets compared to singleton fetuses are shown in Illustration 5.4. Rates of weight gain for each group of fetuses is the same until about 28 weeks of gestation. Rates of weight gain begin to decline in twin and triplet fetuses after that point, however, and remain lower until delivery. Variations in birthweight of twin and triplet newborns appear to be related to factors that affect fetal growth after 28 weeks of pregnancy.[83]

The Vanishing Twin Phenomenon   The disappearance of embryos within 13 weeks of conception is not unusual. It has been estimated that 6 to 12% of pregnancies begin as twins, but that only about 3% result in the birth of twins. Most fetal losses silently occur by absorption into the uterus within the first eight weeks after conception.[84] The prognosis for continued viability of a pregnancy associated with a vanishing twin tends to be good.[84]

Risks Associated with Multifetal Pregnancy

Singleton pregnancy is the biological norm for humans, so it may be expected that multifetal pregnancy would

be accompanied by increasing health risks (Table 5.15).[85] Multifetal pregnancies present substantial risks to both mother and fetuses, and the risks increase as the number of fetuses increase (Table 5.16). Newborns from twin pregnancies at lowest risk of death in the perinatal period weigh between 3000 and 3500 grams (6.7 to 7.8 lb) at birth and are born between 37–39 weeks gestation. Triplets tend to do best when they weigh over 2000 grams (4.5 lb) and are born between 34 and 35 weeks gestation.[85,86] Unfortunately, these outcomes do not represent the usual. Data presented in Table 5.17 show that median weights of twins born at 37, 38, and 39 weeks gestation fall below the 3000 to 3500 gram range. The 3000 to 3500 gram birthweight range for twins, and the >2000 gram mark for triplet newborns can, however, serve as goals for the provision of nutrition services.

Interventions and Services for Risk Reduction

Special multidisciplinary programs that offer women with multifetal pregnancy a consistent, main provider of care, preterm prevention education, increased attention to nutritional needs, and intensive follow-up achieve better pregnancy outcomes than does routine prenatal care.[85,88] Rates of very low birthweight (<1500 g or <3.3 lb) have been reported to be substantially lower (6 versus 26%), neonatal intensive care admissions three times lower (13 versus 38%),

Illustration 5.4   Rates of fetal weight gain in singleton, twin, and triplet fetuses.

SOURCE: Reprinted with permission from Maclennan, 1994[83]

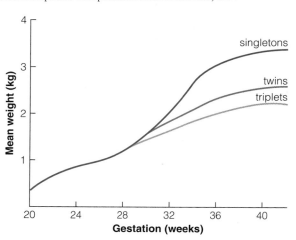

Table 5.15   Risks to mother and fetuses associated with multifetal pregnancy.[85]

Pregnant Women
- Preeclampsia
- Iron-deficiency anemia
- Gestational diabetes
- Hyperemesis gravidarum
- Placenta previa
- Kidney disease
- Fetal loss
- Preterm delivery
- Cesarean delivery

Newborns
- Neonatal death
- Congenital abnormalities
- Respiratory distress syndrome
- Intraventricular hemorrhage
- Cerebral palsy

Table 5.16   Average birthweight and gestational age at delivery, and low birthweight rates of singleton, twin, and triplet newborns.[74,85]

	Mean Birthweight	Mean Gestational Age	Low Birthweight Rate
Singletons	3440 g (7.7 lb)	39–40 weeks	6%
Twins	2400 g (5.4 lb)	37 weeks	54%
Triplets	1800 g (4.0 lb)	33–34 weeks	90%

and perinatal mortality strikingly lower (1 versus 8%) among women who receive such services.[89] Interventions offered by the Montreal Diet Dispensary, which focuses on improving the nutritional status and well-being of the pregnant women served, have been shown to substantially reduce poor outcomes compared to similar women not receiving the services. Improvements include a 27% reduction in the rate of low birthweight, 47% decline in very low birthweight, 32% lower rate of preterm delivery, and a 79% drop in mortality during the first seven days after birth.[90]

Nutrition and the Outcome of Multifetal Pregnancy

Nutritional factors are suspected of playing a major role in the course and outcome of multifetal pregnancy, but much remains to be learned. Of the nutritional factors that influence multifetal pregnancy, weight gain during twin pregnancy has been studied most.

WEIGHT GAIN IN MULTIFETAL PREGNANCY   As with singleton pregnancy, weight gain is linearly related to birthweight, and weight gains associated with newborn weight vary based on prepregnancy weight status (Table 5.18).[85] The Institute of Medicine recommends that women with twins gain 35 to 45 pounds (15.9 to 20.5 kg). It is advised that underweight women gain toward the upper end, and overweight and obese women near the lower end of this range.[35]

RATE OF WEIGHT GAIN IN TWIN PREGNANCY   A positive rate of weight gain in the first half of twin pregnancy is strongly associated with increased birthweight.[91] On the other hand, weight loss after 28 weeks of pregnancy increases the risk of preterm delivery by three fold.[92]

Recommended rates of weight gain for women with twin pregnancy are:

- 0.5 pounds (0.2 kg) per week in the first trimester, and
- 1.5 pounds (0.7 kg) per week in the second and third trimesters.

Table 5.17   Median birthweight for gestational age at delivery of twins.

Gestational Age, Weeks	Birthweight
28	995 g (2.2 lb)
29	1145 g (2.6 lb)
30	1300 g (2.9 lb)
31	1445 g (3.2 lb)
32	1580 g (3.5 lb)
33	1750 g (3.9 lb)
34	1905 g (4.3 lb)
35	2165 g (4.8 lb)
36	2275 g (5.1 lb)
37	2430 g (5.4 lb)
38	2565 g (5.7 lb)
39	2680 g (6.0 lb)
40	2810 g (6.3 lb)
41	2685 g (6.0 lb)

SOURCE: Data from Cohen SB, 1997.[87]

Table 5.18   Prepregnancy weight status and weight gain relationships in twin pregnancy.[76]

Prepregnancy weight status	Weight Gain Related to Birthweights of ≥2,500 g (5.5 lb)
Underweight	44.2 lb (20.1 kg)
Normal weight	40.9 lb (18.6 kg)
Overweight	37.8 lb (17.2 kg)
Obese	37.2 lb (16.9 kg)
Very obese	29.2 lb (13.3 kg)

WEIGHT GAIN IN TRIPLET PREGNANCY   Several studies have examined the relationship between weight gain and birthweight in women with triplets. The general result is that weight gains of about 50 pounds (22.7 kg) correspond to healthy-sized triplets. Rates of gain related to a total weight gain of 50 pounds in women who will average 33 to 34 weeks of gestation are 1.5 pounds (0.7 kg) per week or more starting as early in pregnancy as possible.[85]

Dietary Intake in Twin Pregnancy

Ensuring "adequate nutrition" is widely acknowledged to be a key component of prenatal care for women with multifetal pregnancy. However, it is not clear what constitutes adequate nutrition. Energy and nutrient needs clearly increase during multifetal pregnancy due to increased levels of maternal blood volume, extracellular fluid, and uterine, placental, and fetal growth. These increases place higher energy and nutrient demands on the mother in terms of the nutritional costs of building and maintaining these tissues. Although their newborns are smaller, women with twins still produce around 5000 g (11.2 lb) of fetal weight, and women with triplets 5400 g (13.4 lb) or more.

Evidence of higher caloric need for tissue maintenance and growth in multifetal than singleton pregnancy comes from studies that show increased weight gain and a quicker onset of starvation metabolism in women expecting more than one newborn.[85] Reduced rates of twin deliveries, as well as the higher incidence of twins in overweight and obese women, imply that energy status is an important factor in multifetal pregnancy.[85] Whereas it is obvious that energy and nutrient needs are higher in multifetal than singleton pregnancy, levels of energy balance and nutrient intake associated with optimal outcomes of multifetal pregnancy have not been quantitated.

One prospective study of dietary intakes of women pregnant with twins has been undertaken.[93] Results from this study indicate that women with twins enter pregnancy with higher average caloric intakes (2030 versus 1789 cal per day), and consume an average of 265 cal more per day during pregnancy than women with singleton pregnancy. Nutrient intakes during pregnancy are also higher in women bearing twins than with singleton pregnancy.

Several studies have concluded that the need for specific nutrients is increased during multifetal pregnancy. The need for essential fatty acids (linoleic and alpha-linolenic acid) appears to be increased in multifetal pregnancy. Poor essential fatty acid status is related to neurologic abnormalities and vision impairments in twin offspring.[94] Requirements for iron and calcium have also been found to be increased based on the magnitude of physiological changes that take place in multifetal pregnancy. Levels of essential fatty acids, iron, or calcium required by women to meet these increased needs is unknown, however.[85]

VITAMIN AND MINERAL SUPPLEMENTS AND MULTIFETAL PREGNANCY   Benefits and hazards of multivitamin and mineral supplement use in multifetal pregnancy have not been reported. Consequently, whether they should be provided and if so in what dose levels is unknown. Levels of nutrient intake exceeding the DRIs Tolerable Upper Intake Levels should be avoided.

Nutritional Recommendations for Women with Multifetal Pregnancy

Due to the lack of study results, nutritional recommendations for women with multifetal pregnancy are largely based on logical assumptions and theories (Table 5.19). It is reasoned, for example, that caloric needs for twin pregnancy can be extrapolated from weight gain. Theoretically, to achieve a 40-pound (18.2 kg) weight gain, or 10 pounds (4.5 kg) more than in singleton pregnancy, women with twins would need to consume approximately 35,000 cal more during pregnancy than do women with singleton pregnancies. This increase would amount to about 150 cal per day above the level for singleton pregnancy, or an average of 450 more cal per day than prepregnancy. To achieve higher rates of gain, underweight women may need a higher level, and overweight and obese women lower levels, of energy intake. Energy needs will also vary by energy expenditure levels. As for singleton pregnancy, adequacy of caloric intake can be estimated by weight gain progress.

Food intake recommendations for women with multifetal pregnancy are primarily estimated based on assumptions related to caloric and nutrient needs. Women with multifetal pregnancy likely benefit from diets selected from the Food Guide Pyramid groups and nutrient intakes that meet or slightly exceed the DRIs.

RECOMMENDATIONS FROM THE POPULAR PRESS
Web sites, books, and pamphlets are available that provide ample amounts of scientifically unsupported "guesses" about food and nutrient requirements of women with multifetal pregnancy. Even if presented with steely resolution, any advice that strays from current scientifically based wisdom about nutritional needs of women during pregnancy should be side-stepped.

Table 5.19 "Best practice" recommendations for nutrition during multifetal pregnancy.[85]

Weight Gain

Twin pregnancy: Overall gain of 35–45 pounds (15.9–20.5 kg). Underweight women should gain at the upper, and overweight and obese women at the lower end of this range.

- First trimester: 4–6 pounds (1.8–2.7 kg)
- Second and third trimesters: 1.5 pounds (0.7 kg) per week

Triplet pregnancy: Overall gain of approximately 50 pounds (22.7 kg)

- Gain of 1.5 pounds (0.7 kg) per week through pregnancy

Daily Food Intake

Twin pregnancy

- Breads, cereal, rice, pasta: >6 servings
- Vegetables: 3+ servings
- Fruits: 2+ servings
- Meat, poultry, fish, dry beans, eggs, nuts: 3+ servings
- Milk, yogurt, cheese: 3+ servings
- Fats and sugars: use oils rich in essential fatty acids (safflower oil, sunflower, walnut oil, and canola oil, for example)

Triplet pregnancy

- Foods intake from the Food Guide Pyramid groups should be consumed at a level that promotes targeted weight gain

Caloric Intake

Twin pregnancy

- 450 calories above prepregnancy intake, or the amount consistent with targeted weight gain progress.

Triplet pregnancy

- Caloric intake levels should promote targeted weight gain progress

Nutrient Intake

Twin and triplet pregnancy

- DRI levels or somewhat more than these levels
- Intakes should be lower than ULs.

Vitamin and Mineral Supplements

Twin pregnancy

- Use a supplement providing:

iron: 30 mg	vitamin B_6: 2 mg
zinc: 15 mg	folate: 300 mcg
copper: 2 mg	vitamin C: 50 mg
calcium: 250 mg	vitamin D: 200 IU

Triplet pregnancy

- Provide a supplement containing at least the above levels for twin pregnancy while avoiding excessive amounts

Case Example

Selecting one case example of multifetal pregnancy is difficult because women's experiences vary a good deal. Some women feel exhausted throughout most of pregnancy, have poor appetites, and find it difficult to get a full night's sleep. Others experience very few discomforts, have great appetites, and are able to stay physically active up to delivery. Some women will deliver growth-retarded, very preterm infants, while other women will deliver three alert and hungry newborns weighing over seven pounds each. It is often difficult to predict which women will have remarkably healthy or complication-ridden multifetal pregnancies.

As you read the situation described in Case Study 5.3, try to predict what the outcome will be based on the health status and risk factors that are part of the case. The exercise may give you a taste of the difficulty involved.

HIV/AIDS DURING PREGNANCY

The world first became aware of acquired immunodeficiency syndrome (AIDS) in the summer of 1981. It was caused by a newly recognized microbe, the human immunodeficiency virus (HIV). Since then, over 100,000 women of childbearing age in the United States have been diagnosed with AIDS[95] and 15 million worldwide. Transmission of the virus during pregnancy and delivery are major routes to the spread of the infection. Approximately 20% of children with HIV/AIDS are infected during pregnancy or delivery, and 14–21% during breastfeeding.[96]

Treatment of HIV/AIDS

The primary focus of care for pregnant women with HIV/AIDS is the prevention of transmission of the virus to the fetus and infant. The transmission of the AIDS virus from mother to child is negligible if treatment is provided before, during, and after pregnancy.[96] In developing countries where there is not enough money to purchase the drugs, transmission rates can be substantially reduced by giving the mother a short course of a specific anti-HIV drug before delivery.[97]

Consequences of HIV/AIDS During Pregnancy

Disease processes such as compromised immune system functions related to HIV/AIDS progress during pregnancy, but it does not appear that the infection

Photo Disc

Case Study 5.3
Rebecca's Twin Pregnancy

Rebecca is a 39 year old corporate comptroller who had remarried last year. It was the second marriage for both Rebecca and her husband. Rebecca had not had children, but her husband has a daughter. Both Rebecca and her husband were looking forward to having children as soon as possible, but Rebecca's periods were irregular and she had not conceived within the past year. Knowing Rebecca's biological clock was ticking, the couple made an appointment with a fertility specialist.

Tests undertaken indicated that Rebecca had low levels of gonadotropin-releasing hormone and of ferritin. The doctor suggested that her low levels of this hormone may be related to her BMI of 18.2 kg/m² and that weight gain should be attempted before hormone therapy. Although she was very attached to her slim figure, Rebecca decided it was worth it and gave up skipping meals and her soup and salad dinners for three square meals daily. At the doctor's advice, she added a folic acid fortified cereal and several more servings of grain products to her daily diet. She also bought and faithfully took the daily dose of 30 mg of supplemental iron recommended by her doctor. Within four months her weight had increased from 120 to 132, and conception occurred. Three months later she found out that she was pregnant with twins.

The diagnosis of a twin pregnancy came as a shock—and as a relief. Although the doctor had said her blood pressure and glucose values were fine and that pregnancy was progressing normally, she had been feeling terrible. She had not spent a day without feeling nauseous since the second month of pregnancy, was tired all the time, and for the last two weeks had trouble keeping down any foods other than hard-boiled eggs, crackers, potato chips, and water. Most of all, Rebecca was concerned that she hadn't gained any weight or eaten a good diet for the past few weeks. Her phone call to the doctor lead to a referral to a nutritionist and an appointment the next evening.

Discussions with the nutritionist led to several changes. Rebecca added new foods to her diet (yogurt, canned peaches, applesauce, and pumpkin pie) based on discussions with the nutritionist. She now ate and drank by the clock, alternating fluids with solid foods, and stayed away from odors that triggered her feeling of nausea. She began to feel better within a week. Her appetite had picked up substantially and she found herself gaining around two pounds a week over the next several months. Aside from bothersome problems such as heartburn and constipation, and a slightly elevated blood pressure and edema in her hands and feet, both mother and babies were doing well.

By the end of the eighth month of pregnancy Rebecca felt like she was as big as a house, was uncomfortable, and decided to start maternity leave. She had never weighed 172 pounds before and was already looking forward to getting back to her formerly slim self. She didn't have as long to wait as she imagined. Rebecca lost 25 of the 42 pounds she had gained during pregnancy within 24 hours of delivering at 38 weeks of pregnancy. Nearly 15 of those pounds were accounted for by healthy twin babies weighing 3250 and 3432 grams.

itself is related to adverse pregnancy outcomes. Although adverse pregnancy outcomes such as preterm delivery, fetal growth retardation, and low birthweight tend to be higher in women with HIV/AIDS, differences are most closely related to poverty, poor food availability, compromised health status, and the coexistence of other infections.[98]

Nutritional Factors and HIV/AIDS During Pregnancy

HIV/AIDS is related to poor nutritional status that further compromises the body's ability to fight infections. The disease can lead to nutrient losses and fat malabsorption due to diarrhea, and inflammatory responses to the infection cause the loss of lean muscle mass. Risks of inadequate nutrient status of a wide variety of vitamins and minerals increases as the disease progresses. Nutritional needs increase the most during the later stages of HIV/AIDS as diarrhea, wasting, and reductions in CD4 counts (a measure of white blood cells that help the body fight infection) increase. New drugs used to treat HIV/AIDS are associated with increased insulin resistance and the accumulation of central body fat.[99]

Results of a number of studies have indicated that vitamin A and multiple vitamin supplementation during pregnancy may reduce the transmission of HIV/AIDS to offspring and improve pregnancy outcomes.[96,100] The accuracy of these results, however, have recently been questioned by other studies showing no such effects.[101] Multivitamin supplementation does, however, appear to increase birthweight in developing countries without the disease.[101] Early and appropriate nutrition intervention may retard progression of the disease during pregnancy. Although nutrition intervention to correct nutritional problems is recommended for women with HIV/AIDS, routine supplementation may not decrease transmission or improve the course and outcome of pregnancy in well-nourished women.

The compromised immune status of women with HIV/AIDS, and further decreases in immune response during pregnancy, mean that women with the disease are at high risk of developing foodborne infections during pregnancy. Risk of infection originating from foods can be decreased if raw or uncooked meats and seafoods and unpasteurized milk products and honey are not consumed. Safe food handling practices at home can also reduce the risk of foodborne infection.

Nutritional Management of Women with HIV/AIDS During Pregnancy

Goals for the nutritional management of women with HIV/AIDS include:

- Maintenance of a positive nitrogen balance and preservation of lean muscle mass,
- Adequate intake of energy and nutrients to support maternal physiological changes and fetal growth and development,
- Correction of elements of poor nutritional status identified by nutritional assessment,
- Avoidance of foodborne infections, and
- Delivery of a healthy newborn.

Nutrient requirements of HIV-infected pregnant women are not known, but it is suspected that energy and nutrient needs will be somewhat higher due to the effects of the virus on the body.[99,102]

Insufficient information exists to provide specific standards for nutritional care for women with HIV/AIDS during pregnancy. Consequently, nutritional recommendations for women with HIV/AIDS are consistent with recommendations for pregnant women in general. As is the case for other pregnant women, foods should be the primary source of nutrients in women with HIV/AIDS.[99,102]

EATING DISORDERS IN PREGNANCY

Eating disorders represent relatively rare conditions in pregnancy because many women with such disorders are subfertile or infertile. They can have far reaching effects on both mother and fetus, however, when they do occur.[103]

The type of eating disorder most commonly observed among pregnant women in the United States is bulimia nervosa, or a condition marked by both severe food restriction and bingeing and purging.[103] It is estimated that 1–3% of adolescents and young women in the United States have this condition.[104] Women with this condition exhibit poorly controlled eating patterns marked by recurrent episodes of binge eating. To prevent weight gain, they will induce vomiting, use laxatives, exercise intensly, or fast after binges. Self-worth in women with bulimia nervosa is usually closely tied to their weight and shape. A history of sexual abuse is common among women with this eating disorder, as well as in women with anorexia nervosa.[105]

Pregnancy is rarely suspected in women with anorexia nervosa because amenorrhea is a diagnostic criteria for the disorder. Nonetheless, women with anorexia nervosa may occasionally ovulate and become at risk for conception. To women with anorexia nervosa, body weight is of utmost importance and they are generally fully dedicated to achieving extreme thinness. Adolescents and women with this condition will refuse to eat, even when ravenously hungry, limit their food choice to low-calorie foods only, and exercise excessively.[105] The example presented in Case Study 5.4 relates to a woman with anorexia nervosa during pregnancy.

Eating disorder symptoms often subside during the second and third trimesters of pregnancy, but they rarely vanish altogether. Symptoms tend to return after delivery, sometimes to a more severe extent than was the case prior to pregnancy.[106]

Consequences of Eating Disorders in Pregnancy

Women with eating disorders during pregnancy are at higher risk for spontaneous abortion, hypertension,

Photo Disc

Case Study 5.4
A Young Pregnant Woman with
Symptoms of Anorexia Nervosa and Bulimia

Ms. Schwartz, a markedly underweight, 21 year old physical trainer, was first seen by her family practice nurse with complaints about breast tenderness and fluid retention. Physical exam results and answers to questions about her history of unprotected intercourse indicated that Ms. Schwartz was pregnant. Since she was amenorrheic, Ms. Schwartz assumed she could not get pregnant and never used contraception. For the past five years she had maintained her low level of weight by alternating days of eating very little with days of bingeing and purging.

Ultrasound results confirmed the pregnancy and estimated that she was about 24 weeks pregnant. Upon confirmation of the news, Ms. Schwartz urgently pleaded for an abortion. She was terribly afraid of pregnancy and the weight she would gain. When told it was too late to legally perform an abortion, Ms. Schwartz did not seek subsequent care for her pregnancy and could not be reached for follow-up. She was admitted to a community hospital in labor during her 35th week of pregnancy and delivered a 3 pound 1 ounce baby. The baby died of SIDS four months after leaving the hospital.

and difficult deliveries than women without an eating disorder. Pregnancy weight gain is generally below the recommended amounts, and newborns tend to be smaller and to experience higher rates of neonatal complications.[106]

Treatment of Women with Eating Disorders During Pregnancy

It is recommended that pregnant women with eating disorders be referred to an eating disorders clinic or specialist. Most large communities have special clinics and programs for women with eating disorders, and they commonly use a team approach to problem solving around the eating disorder. Nutritionists or dietitians often participate in these services because they are knowledgeable of the woman's individual nutritional needs and those of pregnancy.

Health professionals serving women with eating disorders in pregnancy can facilitate open communication and behavioral change by gently encouraging women to talk about their eating disorder, fears, and concerns.[106]

Nutritional Interventions for Women with Eating Disorders

Behavioral changes required for improvements in nutritional status and weight gain in women with eating disorders are most likely to work when the changes

are considered acceptable to the women with the disorder. Frequently, the health professional will present the types of changes that need to be made and why, and then will work with the woman to develop specific plans accomplishing these changes.

NUTRITION AND ADOLESCENT PREGNANCY

Since 1991 rates of pregnancy among teens have fallen from about 62 to 50 per 1000 females aged 15 to 19 years. Although the rates are declining, the United States continues to have one of the highest rates of adolescent pregnancy of all developed countries.[74]

Adolescents are at higher risk for a number of clinical complications and other unfavorable outcomes compared to adult women (Table 5.20). The downward shift in birthweight observed in newborns of adolescent mothers is related to increased rates of perinatal mortality, and suggests the existence of unfavorable nutrition and health conditions during pregnancy.[29]

The extent to which increased rates of poor outcomes in pregnant teens are associated with biological immaturity or with lifestyle factors such as drug use, smoking, and poor dietary intakes (that influence health status) is unclear. Age-related differences in outcome diminish substantially when potentially harmful lifestyle factors are taken into account, diminishing the theory that biological immaturity accounts for the dif-

Table 5.20 Risks associated with adolescent pregnancy.[29,107]

- Low birthweight
- Perinatal death
- Cesarean delivery
- Cephalopelvic disproportion (head too large for birth canal)
- Preeclampsia
- Iron-deficiency anemia
- Delayed, reduced educational achievement
- Low income

ferences. Very young adolescents becoming pregnant within a few years after the onset of menstruation may be at risk due to biological immaturity. They tend to have shorter gestations and a higher likelihood of cephalopelvic disproportion.[29] Poorly nourished, growing adolescent mothers may compete with the fetus for calories and nutrients—and win.[107]

Dietary Recommendations for Pregnant Adolescents

Recommendations for pregnant adolescents are basically the same as for older pregnant women with a few exceptions. Adolescents within two years of the start of menses are likely still growing and may need additional calories to support their own and their fetus's growth.[29] Caloric need should be met by a nutrient-dense diet and lead to weight gains that follow those recommended. Pregnant adolescents have a higher requirement for calcium. The DRI for pregnant teens for calcium is 1300 mg per day, or 300 mg higher than for adult pregnant women. This increased need can be met by the consumption of four daily servings of milk and milk products combined with a varied, basic diet.

The importance of lifestyle and other environmental factors to pregnancy outcome in teens emphasizes the need for special, comprehensive teen pregnancy health care programs. Because most pregnant adolescents are low-income, referral to appropriate food and nutrition programs and other assistance related to health care, housing, and education should be core components of services.

EVIDENCE-BASED PRACTICE

The clinical, nutritional management of the conditions covered in this chapter, as well as other complications during pregnancy, is not all evidence-based. Such practices are a problem when they burden women and families with costs, call for dietary changes not known to work, or potentially cause harm. Practices not based on evidence, that likely pose little risk or burden, and that may potentially be of help, should nonetheless be carefully evaluated. It is not uncommon that outdated practices linger far too long at the expense of missed opportunities for real improvements. Use of practices not supported by scientific evidence should always be questioned and confirmed to represent "best practice" insofar as that can be determined. To know what best practice is requires vigilant attention to scientific developments related to the nutritional management of clinical conditions during pregnancy.

Resources

Canadian Diabetes Association

The section "About Diabetes" clearly presents facts on gestational diabetes.
Available from: www.diabetes.ca

National Institutes of Health

This site provides answers to a comprehensive set of questions related to gestational diabetes.
Available from: www.nichd.gov/publications/pubs/gesttoc.htm
Story M and Stang J, eds. Maternal and Child Health Bureau,

HRSA, U.S. Public Health Service, 2000. Ordering information: Contact Dr. Jamie Stang, stang@epi.umn.edu.
On adolescent pregnancy: nutrition and the pregnant adolescent.
Berkowitz KM. Insulin resistance and preeclampsia. Clinics in Perinatology 1998;25:873-885.
On insulin resistance, preeclampsia, and gestational diabetes.

References

1. Report of the National High Blood Pressure Education Program Working Group on High Blood Pressure in Pregnancy. Am J Obstet Gynecol 2000;183:S1–S22.

2. Sibai BM, Lindheimer M, Hauth J, et al. Risk factors for preeclampsia, abruptio placentae, and adverse neonatal outcomes among women with chronic hypertension. National Institute of Child Health and Human Development Network of Maternal-Fetal Medicine Units. N Engl J Med 1998;339:667–71.

3. Haddad B, Sibai BM. Chronic hypertension in pregnancy. Ann Med 1999;31:246–52.

4. Solomon CG, Carroll JS, Okamura K, et al. Higher cholesterol and insulin levels in pregnancy are associated with increased risk for pregnancy-induced hypertension. Am J Hypertens 1999;12:276–82.

5. Pipkin FB. Risk factors for preeclampsia (editorial). N Engl J Med 2001;344:925–6.

6. Berkowitz KM. Insulin resistance and preeclampsia. Clin Perinatol 1998;25:873–85.

7. Prevent recurrent eclamptic seizures with magnesium sulfate, an unconventional anticonvulsant. Drug Ther Perspect 2000;16:6–8.

8. Galtier-Dereure F, Boegner C, Bringer J. Obesity and pregnancy: complications and cost. Am J Clin Nutr 2000;71:1242S–8S.

9. Ritchie LD, King JC. Dietary calcium and pregnancy-induced hypertension: is there a relation? Am J Clin Nutr 2000;71:1371S–4S.

10. Velzing-Aarts FV, van der Klis FR, van der Dijs FP, et al. Umbilical vessels of preeclamptic women have low contents of both n-3 and n-6 long-chain polyunsaturated fatty acids. Am J Clin Nutr 1999;69:293–8.

11. Hauth JC, Ewell MG, Levine RJ, et al. Pregnancy outcomes in healthy nulliparas who developed hypertension. Calcium for Preeclampsia Prevention Study Group. Obstet Gynecol 2000;95:24–8.

12. Innes KE, Wimsatt JH. Pregnancy-induced hypertension and insulin resistance: evidence for a connection. Acta Obstet Gynecol Scand 1999;78:263–84.

13. Kaaja R, Laivuori H, Laakso M, et al. Evidence of a state of increased insulin resistance in preeclampsia. Metabolism 1999;48:892–6.

14. Esplin MS, Fausett MB, Fraser A, et al. Paternal and maternal components of the predisposition to preeclampsia. N Engl J Med 2001;344:867–72.

15. Klebanoff MA, Secher NJ, Mednick BR, et al. Maternal size at birth and the development of hypertension during pregnancy: a test of the Barker hypothesis. Arch Intern Med 1999;159:1607–12.

16. Xiong X, Demianczuk NN, Buekens P, et al. Association of preeclampsia with high birth weight for age. Am J Obstet Gynecol 2000;183:148–55.

17. Duley L, Henderson-Smart D, Knight M, et al. Antiplatelet drugs for prevention of pre-eclampsia and its consequences: systematic review. BMJ 2001;322:329–33.

18. Atallah AN, Hofmeyr GJ, Duley L. Calcium supplementation during pregnancy for preventing hypertensive disorders and related problems. Cochrane Database Syst Rev 2000;2.

19. Herrera JA, Arevalo-Herrera M, Herrera S. Prevention of preeclampsia by linoleic acid and calcium supplementation: a randomized controlled trial. Obstet Gynecol 1998;91:585–90.

20. Belizan JM, Villar J, Bergel E, et al. Long-term effect of calcium supplementation during pregnancy on the blood pressure of offspring; follow up of a randomised controlled trial. BMJ 1997;315:281–5.

21. Wallenburg HC. Prevention of pre-eclampsia: status and perspectives 2000. Eur J Obstet Gynecol Reprod Biol 2001;94:13–22.

22. Chambers JC, Fusi L, Malik IS, et al. Association of maternal endothelial dysfunction with preeclampsia. JAMA 2001;285:1607–12.

23. Chappell LC, Seed PT, Briley AL, et al. Effect of antioxidants on the occurrence of pre-eclampsia in women at increased risk: a randomised trial. Lancet 1999;354:810–6.

24. Rajkovic A, Mahomed K, Malinow MR, et al. Plasma homocyst(e)ine concentrations in eclamptic and preeclamptic African women postpartum. Obstet Gynecol 1999;94:355–60.

25. Leeda M, Riyazi N, de Vries JI, et al. Effects of folic acid and vitamin B6 supplementation on women with hyperhomocysteinemia and a history of preeclampsia or fetal growth restriction. Am J Obstet Gynecol 1998;179:135–9.

26. Scholl TO, Johnson WG. Folic acid: influence on the outcome of pregnancy. Am J Clin Nutr 2000;71:1295S–303S.

27. Makrides M, Gibson RA. Long-chain polyunsaturated fatty acid requirements during pregnancy and lactation. Am J Clin Nutr 2000;71:307S–11S.

28. Knuist M, Bonsel GJ, Zondervan HA, et al. Low sodium diet and pregnancy-induced hypertension: a multi-centre randomised controlled trial. Br J Obstet Gynaecol 1998;105:430–4.

29. Rosso P. Nutrition and metabolism in pregnancy. New York: Oxford University Press; 1990: 117–118, 125, 150–151.

30. Woman wins lawsuit on nutritional malpractice: Community Nutrition Institute Weekly Report, 1977:5.

31. Yeo S, Steele NM, Chang MC, et al. Effect of exercise on blood pressure in pregnant women with a high risk of gestational hypertensive disorders. J Reprod Med 2000;45:293–8.

32. Brand-Miller J, Foster-Powell K. Diets with a low glycemic index: from theory to practice. Nutrition Today 1999;34:64–72.

33. Reece EA, Homko CJ. Diabetes mellitus in pregnancy. What are the best treatment options? Drug Saf 1998;18:209–20.

34. Thomas-Dobersen D. Nutritional management of gestational diabetes and nutritional management of women with a history of gestational diabetes: two different therapies or the same? Clin Diab 1999;17:170–6.

35. National Academy of Sciences. Nutrition during pregnancy. I. Weight gain. II. Nutrient supplements. Washington, D.C.: National Academy Press, 1990.

36. Ventura SJ, Martin JA, Curtin SC, et al. Births: final data for 1998. Natl Vital Stat Rep 2000;48:1–100.

37. Carr SR. Effect of maternal hyperglycemia on fetal development, American Dietetic Association Annual Meeting and Exhibition, Kansas City, MO, October 10, 1998.

38. Branchtein L, Schmidt MI, Mengue SS, et al. Waist circumference and waist-to-hip ratio are related to gestational glucose tolerance. Diabetes Care 1997;20:509–11.

39. Damm P. Gestational diabetes mellitus and subsequent development of overt diabetes mellitus. Dan Med Bull 1998;45:495–509.

40. Jovanovic L. Aberrations in diabetes metabolism throughout pregnancy, 60th Scientific Sessions of the American Diabetes Association, San Antonio, TX, June 11, 2000.

41. Butte NF. Carbohydrate and lipid metabolism in pregnancy: normal compared with gestational diabetes mellitus. Am J Clin Nutr 2000;71:1256S–61S.

42. Carducci AA, Corrado F, Sobbrio G, et al. Glucose tolerance and insulin secretion in pregnancy. Diabetes, Nutrition & Metabolism—Clinical & Experimental 1999;12:264–70.

43. Homko CJ, Sivan E, Reece EA, et al. Fuel metabolism during pregnancy. Semin Reprod Endocrinol 1999;17:119–25.

44. Hytten F, Chamberlain G. Clinical physiology in obstetrics. Oxford: Blackwell Scientific Publications; 1980.

45. Silverman BL, Metzger BE, Cho NH, et al. Impaired glucose tolerance in adolescent offspring of diabetic mothers. Relationship to fetal hyperinsulinism. Diabetes Care 1995;18:611–7.

46. Lindsay RS, Hanson RL, Bennett PH, et al. Secular trends in birth weight, BMI, and diabetes in the offspring of diabetic mothers. Diabetes Care 2000;23:1249–54.

47. MacNeill S, Dodds L, Hamilton DC, et al. Rates and risk factors for recurrence of gestational diabetes. Diabetes Care 2001;24:659–62.

48. McMahon MJ, Ananth CV, Liston RM. Gestational diabetes mellitus. Risk factors, obstetric complications and infant outcomes. J Reprod Med 1998;43:372–8.

49. Pole JD, Dodds LA. Maternal outcomes associated with weight change between pregnancies. Can J Public Health 1999;90:233–236.

50. Kjos SL, Buchanan TA. Gestational diabetes mellitus. N Engl J Med 1999;341:1749–56.

51. Coustan DR, Carpenter MW. The diagnosis of gestational diabetes. Diabetes Care 1998;21Suppl2:B5–8.

52. Crowe SM, Mastrobattista JM, Monga M. Oral glucose tolerance test and the

preparatory diet. Am J Obstet Gynecol 2000;182:1052–4.

53. Coleman T. Patient centred care of diabetes in general practice. Study failed to measure patient centredness of GPs' consulting behavior. BMJ1999;318:1621–2.

54. Cruikshank DP, et al. Maternal physiology in pregnancy. In: Gabbe SG, et al., eds. Obstetrics: normal and problem pregnancies. New York: Churchill Livingstone; 1996: pp 91–109.

55. Mello G, Parretti E, Mecacci F, et al. What degree of maternal metabolic control in women with type 1 diabetes is associated with normal body size and proportions in full-term infants? Diabetes Care 2000;23:1494–8.

56. Adams KM, Li H, Nelson RL, et al. Sequelae of unrecognized gestational diabetes. Am J Obstet Gynecol 1998;178:1321–32.

57. Jovanovic L. Time to reassess the optimal dietary prescription for women with gestational diabetes [editorial]. Am J Clin Nutr 1999;70:3–4.

58. Jovanovic L. Management of diet and exercise in gestational diabetes mellitus. Prenat Neonat Med 1998;3:534–541.

59. Ecker JL, Mascola MA, Riley LE. Gestational diabetes. N Engl J Med 2000;342:896–7.

60. Pezzarossa A, Orlandi N, Baggi V, et al. Effects of maternal weight variations and gestational diabetes mellitus on neonatal birth weight. J Diabetes Complications 1996;10:78–83.

61. Fagen C, King JD, Erick M. Nutrition management in women with gestational diabetes mellitus: a review by ADA's Diabetes Care and Education Dietetic Practice Group. J Am Diet Assoc 1995;95:460–7.

62. McFarland MB, Langer O, Conway DL, et al. Dietary therapy for gestational diabetes: how long is long enough? Obstet Gynecol 1999;93:978–82.

63. Butte NF. Dieting and exercise in overweight, lactating women [editorial]. N Engl J Med 2000;342:502–3.

64. Kalkwarf HJ, Bell RC, Khoury JC, et al. Dietary fiber intakes and insulin requirements in pregnant women with type 1 diabetes. J Am Diet Assoc 2001;101:305–10.

65. Major CA, Henry MJ, De Veciana M, et al. The effects of carbohydrate restriction in patients with diet-controlled gestational diabetes. Obstet Gynecol 1998;91:600–4.

66. Clapp JF, III Effect of dietary carbohydrate on the glucose and insulin response to mixed caloric intake and exercise in both nonpregnant and pregnant women. Diabetes Care 1998;21Suppl2:B107–12.

67. Franz MJ. In defense of the American Diabetes Association's recommendations on the glycemic index. Nutrition Today 1999;34:78–81.

68. Riccardi G, Rivellese AA. Diabetes: nutrition in prevention and management. Nutr Metab Cardiovasc Dis 1999;9:33–6.

69. Nishimune T, Yakushiji T, Sumimoto T, et al. Glycemic response and fiber content of some foods. Am J Clin Nutr 1991;54:414–9.

70. Soh NL, Brand-Miller J. The glycaemic index of potatoes: the effect of variety, cooking method and maturity. Eur J Clin Nutr 1999;53:249–54.

71. Landon MB. Diabetes mellitus and other endocrine disorders. In: Gabbe SG, et al., eds. Obstetrics, normal and problem pregnancies. New York: Churchill Livingstone; 1996: pp 1037–81.

72. Reece EA, Homko CJ, Wu YK, et al. The role of free radicals and membrane lipids in diabetes-induced congenital malformations. J Soc Gynecol Investig 1998;5:178–87.

73. Raychaudhuri K, Maresh MJ. Glycemic control throughout pregnancy and fetal growth in insulin-dependent diabetes. Obstet Gynecol 2000;95:190–4.

74. Ventura SJ, et al. Births: final data for 1999. Vol. 2001: www.cdc.gov/nchs, 2001.

75. Guyer B, Hoyert DL, Martin JA, et al. Annual summary of vital statistics—1998. Pediatrics 1999;104:1229–46.

76. Brown JE, Schloesser PT. Prepregnancy weight status, prenatal weight gain, and the outcome of term twin gestations. Am J Obst Gynecol 1990;162:182–6.

77. Keith LG. Multiple pregnancy: Epidemiology, gestation and perinatal outcome. New York: Parthenon Publishing Group, 1995.

78. Keith L, Machin G. Zygosity testing. Current status and evolving issues. J Reprod Med 1997;42:699–707.

79. Imaizumi Y. A comparative study of twinning and triplet rates in 17 countries, 1971–1996. Acta Genet Med Gemellol 1998;47:101–14.

80. Czeizel AE, Métneki J, Dudás I. Higher rate of multiple births after periconceptional vitamin supplementation. New Engl J Med 1994;330:1687–8.

81. Lavery JP. The human placenta. Gaithersburg, MD: Aspen Publishers, Inc.; 1987: p 69.

82. Hoskins RE. Zygosity as a risk factor for complications and outcomes of twin pregnancy. Acta Genet Med Gemellol 1995;44:11–23.

83. Maclennan AH. Multiple gestation. Clinical characteristics and management. In: Creasy RK, Resnick R, eds. Maternal-fetal medicine: principles and practice. Philadelphia: W. B. Saunders; 1994: pp 589–601.

84. Landy HJ, Keith LG. The vanishing twin: a review. Hum Reprod Update 1998;4:177–83.

85. Brown JE, Carlson M. Nutrition and multifetal pregnancy. J Am Diet Assoc 2000;100:343–8.

86. Hartley RS, Emanuel I, Hitti J. Perinatal mortality and neonatal morbidity rates among twin pairs at different gestational ages: optimal delivery timing at 37 to 38 weeks' gestation. Am J Obstet Gynecol 2001;184:451–8.

87. Cohen SB, Dulitzky M, Lipitz S, et al. New birth weight nomograms for twin gestation on the basis of accurate gestational age. Am J Obstet Gynecol 1997;177:1101–4.

88. Kogan MD, Alexander GR, Kotelchuck M, et al. Trends in twin birth outcomes and prenatal care utilization in the United States, 1981–1997. JAMA 2000;284:335–41.

89. Ellings JM, Newman RB, Hulsey TC, et al. Reduction in very low birth weight deliveries and perinatal mortality in a specialized, multidisciplinary twin clinic. Obstet Gynecol 1993;81:387–91.

90. Dubois S, Dougherty C, Duquette M-P, et al. Twin pregnancy: the impact of the Higgins Nutrition Intervention Program on maternal and neonatal outcomes. Am J Clin Nutr 1991;53:1397–403.

91. Luke B, Gillespie B, Min SJ, et al. Critical periods of maternal weight gain: effect on twin birth weight. Am J Obstet Gynecol 1997;177:1055–62.

92. Konwinski T, Gerard C, Hult AM, et al. Maternal pregestational weight and multiple pregnancy duration. Acta Genet Med Gemellol 1973;22:44–7.

93. Brown JE, Murtaugh MA, Jacobs DR, et al. Early pregnancy weight change and newborn size (Abstract), FASEB, Orlando, FL, March 31–April 4, 2000.

94. Zeijdner EE, van Houwelingen AC, Kester AD, et al. Essential fatty acid status in plasma phospholipids of mother and neonate after multiple pregnancy. Prostaglandins Leukotrienes and Essential Fatty Acids 1997;56:395–401.

95. Health United States, 2000. Vol. 2001.

96. Fawzi WW, Msamanga GI, Spiegelman D, et al. Randomised trial of effects of vitamin supplements on pregnancy outcomes and T cell counts in HIV-1-infected women in Tanzania. Lancet 1998;351:1477–82.

97. French M, Lenzo N, John M, et al. Highly active antiretroviral therapy. Lancet 1998;351:1056–7.

98. Rich KC, Fowler MG, Mofenson LM, et al. Maternal and infant factors predicting disease progression in human immunodeficiency virus type 1-infected infants. Women and Infants Transmission Study Group. Pediatrics 2000;105:e8.

99. Fields-Gardner C, Ayoob KT. Position of the American Dietetic Association and Dietitians of Canada: nutrition intervention in the care of persons with human immunodeficiency virus infection. J Am Diet Assoc 2000;100:708–17.

100. Azais-Braesco V, Pascal G. Vitamin A in pregnancy: requirements and safety limits. Am J Clin Nutr 2000;71:1325S–33S.

101. Fawzi WW, Msamanga G, Hunter D, et al. Randomized trial of vitamin supplements in relation to vertical transmission of HIV-1 in Tanzania. J Acquir Immune Defic Syndr 2000;23:246–54.

102. Wunderlich SM. Nutritional assessment and support of HIV-positive pregnant women and their children: perspectives from an industrialized country. Nutrition Today 2000;35:107–112.

103. Bonne OB, Rubinoff B, Berry EM. Delayed detection of pregnancy in patients with anorexia nervosa: two case reports. Int J Eat Disord 1996;20:423–25.

104. Herzog DB, Greenwood DN, Dorer DJ, et al. Mortality in eating disorders: a descriptive study. Int J Eat Disord 2000;28:20–6.

105. Becker AE, Grinspoon SK, Klibanski A, et al. Eating disorders. N Engl J Med 1999;340:1092–8.

106. Little L, Lowkes E. Critical issues in the care of pregnant women with eating disorders and the impact on their children. J Midwifery Womens Health 2000;45:301–7.

107. Lenders CM, McElrath TF, Scholl TO. Nutrition in adolescent pregnancy. Curr Opin Pediatr 2000;12:291–6.

CHAPTER 6

Arthur Tress, Photonica

Thinking that baby formula is as good as breast milk is believing that thirty years of technology is superior to three million years of nature's evolution.

Christine Northrup, M.D.

NUTRITION DURING LACTATION

Prepared by **Maureen A. Murtaugh**

CHAPTER OUTLINE

- Introduction
- Benefits of Breastfeeding
- Breastfeeding Goals for the United States
- Lactation Physiology
- Breast Milk Supply and Demand
- Human Milk Composition
- The Breastfeeding Infant
- Maternal Diet
- Maternal Energy Balance and Milk Composition
- Factors Influencing Breastfeeding Initiation and Duration
- Breastfeeding Promotion, Facilitation, and Support
- Public Food and Nutrition Programs
- Model Breastfeeding Promotion Programs

KEY NUTRITION CONCEPTS

1 Human milk is the best food for newborn infants for the first year of life or longer.

2 Maternal diet does not significantly alter the protein, carbohydrate, fat, and major mineral composition of breast milk, but it does affect the fatty acid profile and the amounts of some vitamins and trace minerals.

3 When maternal diet is inadequate, the quality of the milk is preserved over the quantity for the majority of nutrients.

4 Health care policies and procedures and the knowledge and attitudes of health care providers affect community breastfeeding rates.

INTRODUCTION

The benefits of breastfeeding for both mothers and infants are well established. Federal efforts to promote breastfeeding and greater understanding of its advantages have contributed to a resurgence of breastfeeding in the United States since the 1970s. Nevertheless, racial and ethnic disparities in breastfeeding rates persist, and despite the knowledge that the benefits increase with longer duration, there has been little increase in the duration of lactation.

The health care system, workplace, and community can either hinder or facilitate the initiation and continuation of breastfeeding. Health programs can play a significant role in increasing breastfeeding rates. Health care professionals who want to promote breastfeeding should know the physiology of lactation, the composition of human milk, and the benefits to mothers and infants. Helping women achieve the appropriate nutritional status for breastfeeding requires an understanding of energy needs, weight goals, the effects of exercise during breastfeeding, vitamin and mineral requirements, and the use of herbal preparations.

BENEFITS OF BREASTFEEDING

Breastfeeding Benefits for Mothers

Women who breastfeed experience hormonal, physical, and psychosocial benefits.[1] Breastfeeding immediately after birth increases levels of the hormone oxytocin, which stimulates contractions of the uterus, minimizes maternal postpartum blood loss, and helps the uterus to return to nonpregnant size.[2]

After the birth, the return of fertility (through monthly ovulation) is delayed in most women during breastfeeding, particularly with exclusive breastfeeding.[3] This delay in ovulation results in longer intervals between pregnancies. Breastfeeding alone, however, is not as effective as other available birth control methods. Consequently, many health care professionals in the United States do not suggest breastfeeding as an option for birth control.

Many women experience psychological benefits including increased self-confidence and bonding with their infants.[4] Although some women still consider faster return to pre-pregnancy weight a benefit of breastfeeding, women may lose or gain weight while nursing. The impact of breastfeeding on maternal weight is discussed in more detail later in this chapter. In addition to these short-term benefits, women who nurse at a younger age and for longer duration have a lower risk of breast and ovarian cancer.[5,6]

Breastfeeding Benefits for Infants

Human milk is uniquely superior for infant feeding and is species-specific; all substitute feeding options differ markedly from it. The breastfed infant is the reference or normative model against which all alternative feeding methods must be measured with regard to growth, health, development, and all other outcomes.

American Academy of Pediatrics Policy Statement on Breastfeeding, 1997[7]

NUTRITIONAL BENEFITS The superiority of the composition of human milk is widely recognized. Although companies that make human milk substitutes (HMSs) use human milk as the standard, human milk is superior to HMSs in a variety of ways:

- With its dynamic composition and appropriate balance of nutrients, human milk provides optimal nutrition to the infant.[8,9]
- The balance of nutrients in human milk matches human infant requirements for growth and development closely;[10] no other animal milk or HMS meets infant needs as well.
- Human milk is isosmotic (of similar ion concentration; in this case human milk and plasma are of similar ion concentration) and therefore meets the requirements for infants without other forms of food or water.
- The relatively low protein content of breast milk compared to cow's milk meets the infant's needs without overloading the immature kidneys with nitrogen.
- Whey protein in human milk forms a soft, easily digestible curd.
- Human milk provides generous amounts of lipids in the form of essential fatty acids, saturated fatty acids, medium-chain triglycerides, and cholesterol.
- Long-chain polyunsaturated fatty acids, especially docosahexaenoic acid (DHA), which promotes optimal development of the central nervous system, are present in human milk but not in the HMSs marketed in the United States.
- Minerals in breast milk are largely protein bound and balanced to enhance their availability and meet infant needs with minimal demand on maternal reserves.

IMMUNOLOGICAL BENEFITS   One of the more important realizations about breastfeeding in the last decade is the ability of human milk to protect against infection. Many components of human milk are active against infection.[8,9] Cellular components (*T-* and *B-lymphocytes, neutrophils, macrophages,* and *epithelial cells*) are especially high in *colostrum* but are also present for months in mature human milk in lower concentrations.

Secretory immunoglobulin A, the predominant *immunoglobulin* in human milk, helps to protect the infant's gastrointestinal tract. These proteins in human milk bind iron and vitamin B_{12}, making the nutrients unavailable for the growth of pathogens in the infant's gastrointestinal tract. Such factors are also responsible for the different kinds of intestinal flora (natural bacteria of the gastrointestinal tract) found in breastfed infants.

Growth factors and hormones in human milk, such as insulin, enhance the maturation of the infant's gastrointestinal tract. These substances also help to protect the infant, especially neonates, against viral and bacterial pathogens.

In addition to protecting against viral and bacterial pathogens, breastfeeding enhances the immune response to immunizations including polio, tetanus, diphtheria, and Haemophilus influenza. Breastfeeding also enhances the immune response to respiratory syncytial virus (RSV) infection, a common infant respiratory infection.[1] Protection against infection is strongest during the first several months of life for infants who are exclusively breastfed and continues throughout the duration of breastfeeding.

COGNITIVE BENEFITS   Several reports have linked breastfeeding, and especially duration of breastfeeding, with cognitive benefits, although the data can be difficult to interpret because of differences in education, socioeconomics status, and other factors between women who choose to breastfeed and women who choose not to. Recognition that the fatty acid composition of milk plays an important role in neuropsychological development bolsters the credibility of psychological or cognitive benefits from breastfeeding. The differences in *cognitive function* are greater between premature infants fed human milk versus HMS than for term infants. The differences increase with the duration of breastfeeding.[11]

REDUCED MORBIDITY   Not surprisingly, in countries with high infant *morbidity* and *mortality rates,* poor sanitation, and questionable water supplies, breastfed infants experience reduced rates of illness. Even in the United States, where modern health care systems, safe water, and proper sanitation are commonplace, there is a clear relationship between breastfeeding and reduced rates of illness in infants. In U.S. samples, the incidence of diarrhea is estimated to be 50% lower in exclusively breastfed infants.[12,13] Ear infections are 19% lower, and the number of prolonged episodes of ear infection was 80% lower among breastfed infants than among infants fed HMS. In a U.S. population study, breastfed infants experienced 17% less coughing and wheezing, and 29% less vomiting than infants fed HMS.[13]

In addition to the lower rate of acute illnesses in breastfed children, breastfeeding also seems to protect against chronic childhood diseases. Breastfeeding may reduce the risk of celiac disease,[14] inflammatory bowel disease,[15] and childhood cancer.[16] HMS feeding results in an increase in the risk of allergy (30%) and asthmatic disease (25%).[17]

These reductions in acute and chronic infant illness increase with greater use of human milk.[18] For example, infants who receive some human milk and some HMS are at 60% greater risk of ear infection than those fed exclusively human milk.[18] The risks, particularly for allergy and asthmatic disease, are reduced for the duration of breastfeeding and for months to years after weaning.[17]

Breastfeeding may also play a role in reducing the risk of sudden infant death syndrome (SIDS), but this is still under investigation. Researchers disagree about whether breastfeeding has a primary effect in reducing risk of SIDS. A recent analysis of available studies found that bottle-feeding increases the risk of SIDS, but confounding variables may be responsible for this finding.[19]

Great attention has been paid to the role of breastfeeding in preventing obesity, but this

---

T-LYMPHOCYTES   White blood cells that are active in fighting infection may also be called *T-cells;* the *T* in T-cell stands for thymus). These cells coordinate the immune system by secreting hormones that act on other cells.

B-LYMPHOCYTES   White blood cells that are responsible for producing immunoglobulins.

NEUTROPHILS   A class of white blood cells that are involved in protecting against infection.

MACROPHAGES   A white blood cell that acts mainly through phagocytosis.

EPITHELIAL CELLS   Cells that line the surface of the body.

COLOSTRUM   The milk produced in the first 2–3 days after the baby is born. Colostrum is higher in protein and lower in lactose than milk produced after the milk supply is established.

SECRETORY IMMUNOGLOBULIN A   A protein found in secretions that protect the body's mucosal surfaces from infections. It may act by reducing the binding of a microorganism with cells lining the digestive tract. It is present in human colostrum but not transferred across the placenta.

IMMUNOGLOBULIN   A specific protein that is produced by blood cells to fight infection.

COGNITIVE FUNCTION   The process of thinking.

MORBIDITY   The rate of illnesses in a population.

MORTALITY RATE   Rate of death.

relationship has been difficult to quantify. Breastfed infants typically gain less weight and are leaner than HMS-fed infants at one year of age without any difference in activity level or development.[20] A recent study suggests that breastfeeding is not as strong as the mother's weight in predicting childhood risk for overweight.[21] Diet and physical activity may be more effective in reducing overweight in children.

SOCIOECONOMIC BENEFITS OF BREASTFEEDING

A decrease in medical care for breastfed infants is the primary socioeconomic benefit of breastfeeding. Medicaid costs for breastfed WIC infants in Colorado were $175 lower than for infants who were fed HMS.[22] Never-breastfed infants required more care for lower respiratory tract illness, otitis media, and gastrointestinal disease than infants breastfed for at least three months.[23] Each 1000 never-breastfed infants had 2033 more sick care visits, 212 more days of hospitalization, and 609 more prescriptions. In addition, in one study of two companies with established lactation programs, the one-day maternal absenteeism from work due to infant illness was approximately two-thirds lower in breastfeeding women than in nonbreastfeeding women.[24] Thus, companies benefit through lower medical costs and greater employee productivity.

BREASTFEEDING GOALS FOR THE UNITED STATES

During the twentieth century, infant feeding practices have undergone dramatic changes that reflect shifts in values and attitudes in the U.S. society as a whole. They have tended to occur first among those women at the forefront of changes in dominant social values and among those with resources (whether it is time, energy or money) to permit adoption of new feeding practices.

Institute of Medicine, Subcommittee on Nutrition during Lactation, 1991[25]

In the early 1900s, almost all infants in the United States were breastfed. As HMSs became available, breastfeeding rates steadily declined, reaching levels below 30% in the 1950s and 1960s and then rising dramatically in the 1970s (Illustration 6.1).[26] In the early 1980s, levels peaked above 60% at hospital discharge and then steadily declined until the early 1990s. For the past decade, breastfeeding rates have been steadily increasing among all sociodemographic groups, but they still remain below the Healthy People 2000 breastfeeding goals for the nation.[27]

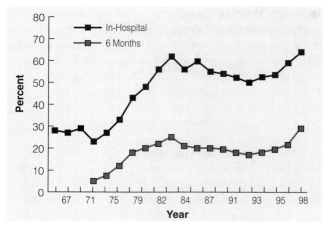

Illustration 6.1   U.S. breastfeeding rates: 1965–1998.[28,29]

Despite the well-documented advantages of breastfeeding, in 1998 only 64% of all mothers breastfed in the early postpartum period and less than one-third (29%) breastfed at six months postpartum.[28] The ethnic and racial disparities in breastfeeding remain disturbingly wide; the rates of breastfeeding at 6 and 12 months among African American women (19% and 9%, respectively) are approximately half the rates among white women (31% and 17%). Rates of breastfeeding are highest among college-educated women and women aged 35 years and older. The rates of breastfeeding are lowest among woman whose infants are at greatest risk of poor health and development—women aged 21 years and under and those with low educational levels.[28]

In recognition of the health and economic benefits of breastfeeding, national goals for breastfeeding rates have been established and revised since 1980.[27,28] Healthy People 2010 contains broad-reaching national health goals for the new decade, focusing on two major themes: (1) increasing the quality and years of healthy life and (2) eliminating racial and ethnic disparities in health status.[28] In addition to adding specific breastfeeding goals for black or African Americans, and Hispanic or Latina Americans, Healthy People 2010 places increased emphasis on the duration of breastfeeding. By the year 2010, the goal is for at least 75% of American women to breastfeed their infants in the early postpartum period, at least 50% at six months, and 25% at one year (Table 6.1). Increasing the rates of breastfeeding is a compelling public health goal, particularly among the racial and ethnic groups who are less likely to initiate and sustain breastfeeding throughout the infant's first year.[1]

If the United States is to reach these breastfeeding goals, health professionals must take an active role in promoting and supporting lactation. Becoming a

Table 6.1 Healthy People 2010: Breastfeeding Objectives for the Nation.[1]

Objective: Increase the proportion of mothers who breastfeed their babies.

	1998 BASELINE	2010 GOAL
In Early Postpartum Period		
All women	64%	75%
Black or African American	45	75
Hispanic or Latina	66	75
White	68	75
At Six Months		
All women	29	50
Black or African American	19	50
Hispanic or Latina	28	50
White	31	50
At One Year		
All women	16	25
Black or African American	9	25
Hispanic or Latina	17	25
White	19	25

SOURCE: U.S. Department of Health and Human Services. Healthy People 2010: Conference Edition—volumes I and II. Washington, DC: U.S. Department of Health and Human Services, Public Health Service, Office of the Assistant Secretary for Health, January 2000.[1]

breastfeeding advocate requires a thorough understanding of lactation physiology.

LACTATION PHYSIOLOGY

Mammary Gland Anatomy

The *mammary gland* lies on the front of the chest. The breast extends from approximately the second to the sixth rib and lies on top of the pectoralis muscle and between the sternum and the anterior axillary line.

Functional Unit of the Mammary gland

The functional units of the mammary gland are the *alveoli* (Illustration 6.2). Each alveolus is composed of a cluster of *secretory cells* with a duct in the center. *Myoepithelial cells* surround the secretory cells. The myopethelial cells can contract under the influence of *oxytocin* and cause milk to be ejected into the ducts. The ducts are arranged like branches of a tree, with each smaller duct leading to a larger duct. These branchlike ducts lead to the nipple. Dilations in the *lactiferous sinuses* behind the nipple allow for storage of milk.

Mammary Gland Development

During puberty, the ovaries mature, and the release of estrogen and progesterone increases (Table 6.2). The cyclic release of these two hormones governs pubertal breast development (Illustration 6.3). The mammary gland develops its lobular structure (*lobes*) under the cyclic production of progesterone and is usually complete within 12 to 18 months after menarche. As the ductal system matures, cells that can secrete milk develop, and the nipple grows and its pigmentation changes. Fibrous and fatty tissues increase around the ducts.

In pregnancy, the luteal and placental hormones (placental lactogen and chorionic gonadotropin) allow further preparation for breastfeeding (Illustration 6.3). Estrogen stimulates development of

MAMMARY GLAND The source of milk for offspring, also commonly called the breast. The presence of mammary glands is a characteristic of mammals.

ALVEOLI Rounded or oblong cavities present in the breast (singular = *alveolus*).

SECRETORY CELLS Cells in the acinus (milk gland) that are responsible for secreting milk components into the ducts.

MYOEPITHELIAL CELLS Specialized cells that line the alveoli and can contract to cause milk to be secreted into the duct.

OXYTOCIN A hormone produced during letdown that causes milk to be ejected into the ducts

LACTIFEROUS SINUSES Larger ducts for storage of milk behind the nipple.

LOBES Rounded structures of the mammary gland.

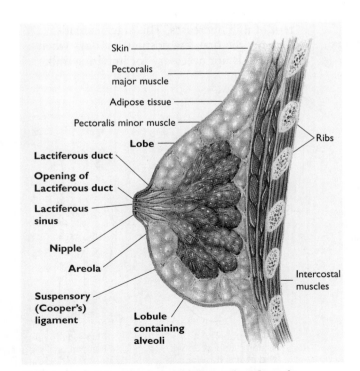

Illustration 6.2 Breast of a lactating female.
This cutaway view shows the mammary glands and ducts.

Table 6.2   Hormones contributing to breast development and lactation.

Hormone	Role in Lactation	Stage of Lactation
Estrogen	Ductal growth	Mammary gland differentiation with menstruation
Progesterone	Alveolar development	After onset of menses and during pregnancy
Human growth hormone	Development of terminal end buds	Mammary gland development
Human placental lactogen	Alveolar development	Pregnancy
Prolactin	Alveolar development and milk production	Pregnancy and breastfeeding (from the third trimester of pregnancy to weaning)
Oxytocin	Letdown: ejection of milk from myoepithelial cells	From the onset of milk secretion to weaning.

the glands that will make milk. Progesterone allows the ducts to elongate and the cells that line them (epithelial cells) to duplicate. Production of milk is possible as early as 16 weeks of gestation.

Lactogenesis

Breast milk production, or *lactogenesis,* occurs in three stages.[30] The first stage, or lactogenesis I, begins during the last trimester of pregnancy; the second and third stages, lactogenesis II and III, occur after birth.

> **LACTOGENESIS**   Another term for human milk production.
>
> **PROLACTIN**   A hormone that stimulates milk production.

- *Lactogenesis I.* During lactogenesis I, milk begins to form, and the lactose and protein content of milk increases. This stage extends through the first few postpartum days when suckling is not necessary for initiating milk production.

- *Lactogenesis II.* This stage begins two to five days postpartum and is marked by increased blood flow to the mammary gland. Clinically, it is considered the onset of copious milk secretion, or "when milk comes in." Significant changes in both the milk composition and the quantity of milk that can be produced occur over the first 10 days of the baby's life.
- *Lactogenesis III.* This stage begins about 10 days after birth and is the stage in which the milk supply is maintained.[31]

Hormonal Control of Lactation

Prolactin and oxytocin are necessary for establishing and maintaining a milk supply. *Prolactin* is a hormone that stimulates milk production. Suckling is a major stimulator of prolactin secretion: prolactin levels double with suckling.[8] Stress, sleep, and sexual intercourse also stimulate prolactin levels. Prolactin activity is suppressed in the last three months of pregnancy by prolactin inhibiting factor that is released by the hypothalamus. This inhibition of prolactin al-

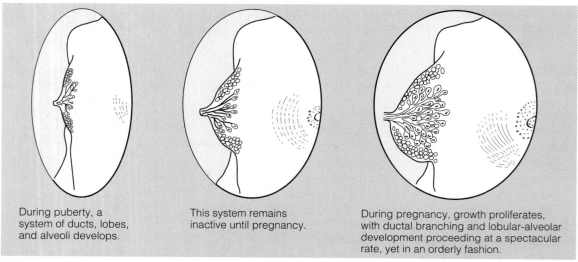

During puberty, a system of ducts, lobes, and alveoli develops.

This system remains inactive until pregnancy.

During pregnancy, growth proliferates, with ductal branching and lobular-alveolar development proceeding at a spectacular rate, yet in an orderly fashion.

Illustration 6.3   Breast development from puberty to lactation.

lows the body to prepare for milk production during pregnancy while preventing milk production until the baby is born. The actual level of prolactin in the blood is not related to the amount of milk made, but prolactin is necessary for milk synthesis to occur.[32]

Oxytocin release is also stimulated by suckling or nipple stimulation. Its main role is in letdown, or the ejection of milk from the milk gland (acinus) into the milk ducts and lactiferous sinuses. Women may experience tingling or sometimes sharp shooting pain that lasts about a minute and corresponds with contractions in the milk ducts. Oxytocin also acts on the uterus, causing it to contract, seal blood vessels, and shrink in size.

Secretion of Milk

Understanding the mechanisms of milk secretion is important to understanding how factors such as nutritional status, supplementation, medications, and disease may affect breastfeeding. As described by Neville et al.,[33] the secretory cell in the breast uses five pathways for milk secretion (Illustration 6.4). These transcellular (through the cell) and paracellular (between the cells) pathways include:

I. Exocytosis of milk protein and lactose in secretory vesicles derived from *golgi*
II. Milk fat secretion via the milk fat globule
III. Secretion of ions and water across the cell membrane
IV. Pinocytosis-exocytosis of immunoglobulins
V. Paracellular pathway for plasma components and leukocytes

PATHWAY I *Exocytosis* allows lactose, calcium, phosphate, and citrate and some proteins to be secreted into the ducts. Lactose, the major carbohydrate of human milk and the milk of many other species, is synthesized from glucose and galactose in the secretory cells of the breast using glucose derived from the maternal blood supply. An enzyme, lactalbumin, catalyzes the conversion to lactose. During pregnancy, progesterone levels inhibit lactalbumin production. At delivery, a drop in progesterone and an increase in prolactin allow for a large increase in lactose production over the first weeks of lactation.

> **GOLGI** Groups of intracellular vesicles where the processing and packaging of secreted proteins take place.
>
> **EXOCYTOSIS** Release of the cellular contents of a vesicle when the membrane of that vesicle merges with the plasma membrane.
>
> **VESICLE** A small liquid-containing sac.

Milk proteins (casein and whey, immunoglobulins, milk fat globule proteins, free amino acids) are secreted into milk using pathways I, II, IV, and V. Prolactin, cortisol, and insulin stimulate production of milk proteins including caseins and whey in the secretory cells. Glucose or phosphates are added to proteins for stability. Some of the proteins combine with minerals (i.e., calcium) to increase their solubility. The processed milk proteins are stored in *vesicles* in the secretory cells with lactose and other components until they are secreted along with lactose by exocytosis (pathway I). As milk fat globules form in the secretory cells, proteins from the secretory cells are incorporated into the milk fat globules. These proteins in milk fat globule are secreted through pathway II. Immunoglobulins are transported by pathway IV; they are engulfed into the cell by endocytosis and then secreted into milk by exocytosis. Free amino acids pass between secretory cells (pathway V) from the mother's blood supply into the milk.

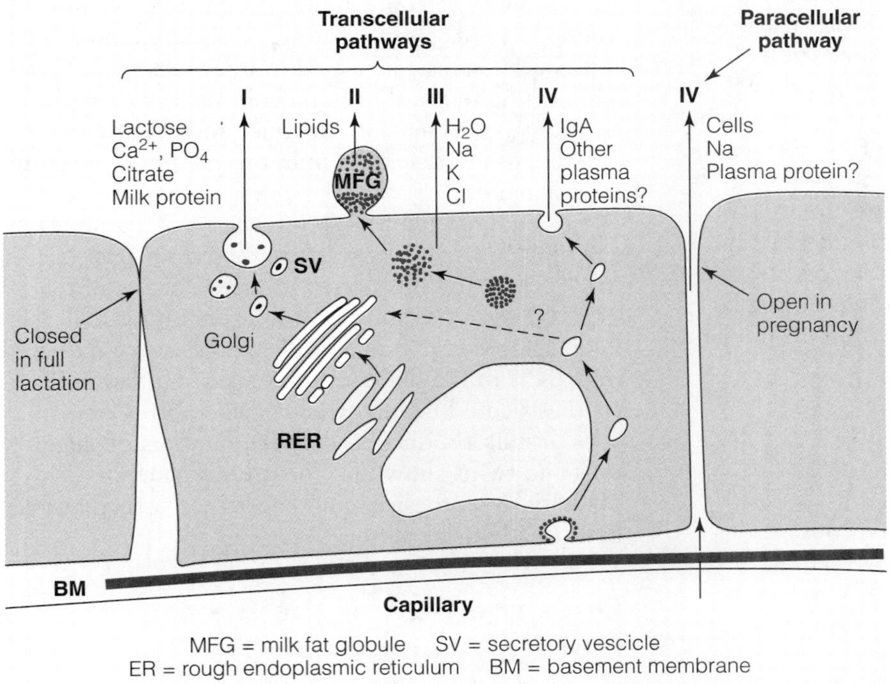

Illustration 6.4   The pathways for secretion of milk components.

MFG = milk fat globule   SV = secretory vesicle
ER = rough endoplasmic reticulum   BM = basement membrane

From LACTATION: Physiology, Nutrition and Breast Feeding, M.C. Neville & M.R. Neifert. Used by permission of Plenum Publishers.

PATHWAY II The milk fat globules are secreted by pathway

II, as shown in Illustration 6.4. Human milk fat is provided by maternal serum triglycerides and by synthesis of short-chain fatty acids. When fatty acids are made in the breast, glucose and acetate from the mother's blood are the major fuels. Because fat is not soluble in water, fat is secreted as a human milk globule. Fat globules appear first as small triglyceride-rich droplets. These droplets fuse and contain some of the secretory cell membrane, some of the secretory cell's fluids, and some plasma enzymes. The secretory cell membrane in the fat globule allows the emulsification of fat and stabilizes the milk fat globules.[34]

PATHWAY III   Water, sodium, potassium, and chloride are able to pass through alveolar cell membranes in either direction (passive diffusion). Water flow follows the concentration of glucose (needed for lactose production).

PATHWAY IV   The pinocytosis-excocytosis pathway is used to transfer immunoglobulin A and other plasma proteins from the mother's blood to the milk. These proteins are taken into the alveolar cell from the mother's plasma by *pinocytosis* and then secreted by exocytosis into the milk space.

PATHWAY V   The between-cell, or paracellular, path for cells is used for the flow of fluid, sodium and other solutes, and small plasma proteins into milk. This pathway is thought to be open, or leaky, during pregnancy. These paracellular spaces allow fluid and solutes to flow between the ducts and the fluids in the tissue of the breast.[35] Therefore, when the pathways are open, more sodium and cells from the mother's blood can pass into the milk space. The paracellular spaces close after delivery, allowing milk to be stored during lactogenesis II and III by preventing the milk from being reabsorbed into the breast tissues.

PINOCYTOSIS   The process of taking fluid-filled vesicles into cells.

The Letdown Reflex

I have a strong let-down—milk jets out 3 or 4 feet when I express a squirt or two before nursing.

C. Martin and N. F. Krebs[36]

The letdown reflex allows milk to be released from the breast. An infant suckling at the nipple usually causes letdown. The stimuli from the infant suckling are passed through nerves to the hypothalamus, which responds by promoting oxytocin release from the posterior pituitary gland (Illustration 6.5). The oxytocin causes contraction of the myoepithelial cells surrounding the secretory cells. As a result, milk is released through the ducts to the lactiferous sinus, making it available to the infant. Other stimuli, such as hearing a baby cry, sexual arousal, and thinking about nursing, can also cause letdown, and milk will leak from the breasts.

BREAST MILK SUPPLY AND DEMAND

Can Women Make Enough Milk?

Typical milk production begins at approximately 600 ml (240 ml = 1 cup) in the first month postpartum and increases to approximately 750–800 ml per day by 4–5 months postpartum.[25] Milk production can range from 450–1200 ml per day in women who are nursing one infant.[37] Infant weight, the caloric density of milk, and the infant's age contribute significantly to the infant's demand for milk. Milk production increases to meet the demand of twins, triplets, or infants and toddlers suckling simultaneously; it can also be increased by pumping the milk.

Traditionally, factors such as how vigorously an infant nurses, how much time the infant is at the breast, and how many times he or she nurses in a day were thought to control milk production. We now know that milk synthesis (rate of accumulation of milk in the breast) is related to infant demand.[38] That is, the removal of milk from the breast seems to be the signal to make more milk, and most women are able to increase their milk production to meet infant demand.[39]

A study of infant milk intake found that an average of 24% of milk was left in the breast after feeding.[40] Thus, the short-term storage capacity of the breast does not seem to be a limit on infant milk intake. The average rate of milk synthesis in a day was only 64% of the highest rate, suggesting that milk synthesis could be increased considerably. Comparisons of milk production between mothers of singletons and twins show that the breasts have the capacity to synthesize much more than a singleton infant usually drinks.[41]

Does the Size of the Breast Limit a Woman's Ability to Nurse Her Infant?

The size of a woman's breast does not determine the amount of milk production tissue (clusters of alveoli

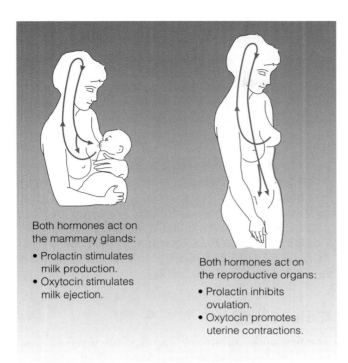

Both hormones act on the mammary glands:
- Prolactin stimulates milk production.
- Oxytocin stimulates milk ejection.

Both hormones act on the reproductive organs:
- Prolactin inhibits ovulation.
- Oxytocin promotes uterine contractions.

Illustration 6.5   The letdown reflex.
An infant suckling at the breast stimulates the pituitary to release the hormones prolactin and oxytocin.

containing secretory cells that produce the milk).[42] Much of the variation in breast size is due to the amount of fat in the breast. The size of the breast does limit *storage* because it puts limits on the expansion of the ducts and lactiferous sinuses. Daily milk production is not related to the total milk storage capacity within the breast, however.[40] This means that a women with small breasts can produce the same amount of milk as a woman with large breast, although the latter may be able to feed her infant less frequently to deliver the same volume of milk.

Is Feeding Frequency Related to the Amount of Milk a Woman Can Make?

Feeding frequency is not consistently related to milk production. The rate of milk synthesis is highly variable between breasts and between feedings.[43] The amount of milk produced in 24 hours and the total milk withdrawn in that 24-hour period are highly related, however.[40] Milk synthesis is able to quickly respond to infant demand.

The breast responds to the degree of emptying during a feeding, and this response is a link between maternal milk supply and infant demand. Daly [38–43] proposed that the breast responds to the infant's need by measuring how completely the infant empties the breast. For example, if a lot of milk is left in the breast, then milk synthesis will be low to prevent engorgement, whereas if the breast is fully emptied, synthesis will be high to replenish the milk supply. Thus, to prevent overfullness, the frequency of feeding has to be greater than the 24-hour demand divided by the total storage capacity. The implication is that women with smaller breasts have less storage capacity and may need to feed their infant more frequently than women with a larger storage capacity.

The exact mechanism of milk supply and demand, or autocrine control, is not well understood, but seems to be related to a protein called Feedback Inhibitor of Lactation that has been identified in the milk of animals and humans.[43,44] Feedback Inhibitor of Lactation is an active whey protein that inhibits milk secretion. The protein inhibits all milk components equally according to their concentration in milk. Therefore, this protein seems to affect milk quantity only, not milk composition.

The physiology of milk production is an active area for research. Recently, researchers proposed a role for nitric oxide in increasing blood supply to the breast, the erection of the nipple, and the enlargement of the mammary ducts.[45] Nitric oxide appears to be synthesized in the breast and seems to peak at approximately four days postpartum, just before the volume of milk production increases rapidly. Women with higher milk production (>350 ml or more of milk) by day 5 had significantly higher levels of nitric oxide than women who made less milk (<350 ml per day). The fact that nitric oxide is produced in the capillaries of the mammary gland would support a role in the local control of lactation, but further work is needed to understand its contribution, if any, to regulation of milk synthesis.

Pumping or Expressing Milk

A woman may need to pump or express milk for many reasons including maternal or infant illness or separation. Women can express milk by several different methods: manually or by using a hand pump, a commercial electric pump, or a hospital-grade electric pump. A pump that allows a mother to pump both breasts at the same time (10 minutes per session) can save time over single pumping (20 minutes per session).[46] Electric pumps are efficient and may increase prolactin more than hand expression or hand pumping.[46,47]

Insufficient milk production is a common problem among women who express milk. Researchers working with women who pumped their breasts report that 8 to 12 or more milk expressions per day were necessary to stimulate adequate production of milk.[46] The optimal number of expressions in a 24-hour period is likely to differ according to how well women empty the breast and the storage capacity of the breast. Women who are able to establish an adequate volume of milk (>500 ml per day) in the first two weeks postpartum are more likely to still have enough milk for their infant at four to five weeks postpartum.[46] Only about half of a small group of women (*n* = 39) who were able to pump less milk (<250 ml per day) at two weeks postpartum achieved an adequate milk supply for their premature infant by four to five weeks. According to Meier,[48] establishing an abundant supply of milk means 750 to 1000 ml per day by about 10 days postpartum. This recommendation is consistent with the advice to nurse the infant (or pump) early and often to build a good milk supply.

HUMAN MILK COMPOSITION

Thus, the complexity of milk as a system designed to deliver nutrients and nonnutritive messages to the neonate has increased.

R. G. Jensen[9]

Human milk is an elegantly designed natural resource. It is the only food needed by the majority of healthy infants for approximately six months. Human milk is designed not only to nurture, but also to protect infants from infectious disease. Its composition is changeable over a single feeding, over a day, according to the age of the infant or gestation at delivery, the presence of infection in the breast, with menses, and with maternal nutritional status.

As our ability to measure and identify its components increases, we recognize that the composition of human milk is complex. Hundreds of components of human milk have been identified, and their nutritive and nonnutritive roles are under investigation. The basic nutrient composition of colostrum and mature milk is provided in Table 6.3. *The Handbook of Milk Composition*[9] and *Breastfeeding: A Guide for the Medical Professional*[8] provide more complete descriptions of human and other milks.

Colostrum

The first milk, colostrum, is a thick, often yellow fluid produced during lactogensis II (days 1–3 after infant birth). Infants may drink only 2 to 10 ml (½ to 2 teaspoons) of colostrum per feeding in the first two to three days. Colostrum provides about 58–70 cal/100 ml and is higher in protein, and lower in carbohydrate and fat than mature milk (produced two weeks after infant birth). Secretory immunogolulin A and lactoferrin are the primary proteins in colostrum, but other proteins found in mature milk are not present. The concentration of mononuclear cells (a specific type of white blood cell) from the mother that provides immune protection is highest in colostrum. Colostrum has higher concentrations of sodium, potassium, and chloride than are found in more mature milk.

Water

Milk is isotonic with maternal plasma. This biological design means that babies do not need water or other fluids to maintain hydration, even in hot climates.[54] As a major component of human milk, water allows suspension of milk sugars, proteins, immunoglobulin A, sodium, potassium, citrate, magnesium, calcium, chloride, and water-soluble vitamins.

Energy

Human milk provides approximately 0.65 cal/ml, although the energy content varies with its fat (and, to a lesser degree, protein and carbohydrate) composition. Breastfed infants consume fewer calories than those fed HMS.[55,56] Whether this difference in energy intake has to do with the composition of human milk, the ability to see the volume of breastfeeding,

Table 6.3
Human milk composition (per 100 ml).[a]

Milk Component	Mature Milk
Lactose	7.2 g
Protein	1.05 g
Fat	3.9 g
Calories	65 cal
Vitamin A	48.5 mcg
Thiamin	.4 mg
Riboflavin	.035 mg
Niacin	.18 mg
Vitamin B_6	.013 mg
Pantothenic acid	.22–.25 mg
Biotin	.038 mcg
Folate	.085 mcg
Vitamin B_{12}	.042mcg
Vitamin C	4 mg
Vitamin D	51 IU biological activity
Vitamin E	0.23 mg
Vitamin K	.25 g
Calcium	26.4 mg
Phosphorus	12.4 mg
Magnesium	3.4 mg
Chromium	.025 mcg
Copper	25 mcg
Iron	.035 mg
Zinc[b]	.12 mg

[a]Data for vitamins and minerals are Dietary Reference Intakes.[49–52] Protein, calorie, and carbohydrate values are from the Recommended Dietary Allowances, 1989.[53]
[b]Zinc content of milk declines very rapidly from colostrum: 0.4 mg/dL to 0.2 mg/dL at two months and 0.12 mg/dL at six months.

suckling at the breast versus an artificial nipple, or other factors is not known. Infants who are breastfed are thinner for their weight at 8–11 months than HMS-fed infants, but these differences disappear by 12–23 months of age. Few differences are notable by five years of age.[21]

Lipids

Human milk lipids are packaged in a human milk globule membrane that allows the fat to be emulsified in milk. Lipids are the second greatest component by concentration (3–5% in mature milk) as well as the most variable. Lipids[9] provide half of the energy of human milk. Human milk fat is low at the beginning of a feeding (foremilk) and higher at the end (hindmilk).

EFFECT OF MATERNAL DIET ON FAT COMPOSITION
The fatty acid profile, but not the fat content, of human milk varies with the diet of the mother.[57]

When diets rich in polyunsaturated fats are consumed, more polyunsaturated fatty acids are present in the milk. When very low fat diets, with adequate calories from carbohydrate and protein, are consumed, more medium-chain fatty acids are synthesized in the breast. When a mother loses weight, the fatty acid profile of her fat stores is reflected in her milk.[58]

DHA Recent interest in lipids in human milk arises from the developmental advantages provided by docosahexaenoic acid (DHA) and cholesterol.[59] DHA, essential for retinal development, has its greatest accumulation in the last month of pregnancy. Therefore, the advantages of human milk are particularly great for premature infants born before 36–38 weeks. Both maternal diet and supplements can change the level of DHA in human milk.[60]

CHOLESTEROL Cholesterol, an essential component of all cell membranes, is needed for growth and replication of cells. Cholesterol concentration ranges from 10–20 mg/dL changes with the time of day.[9] Breastfed infants have higher intakes of cholesterol and higher levels of serum cholesterol than infants fed HMS,[61] which has little cholesterol and relatively high levels of polyunsaturated fatty acids. Whether early exposure to cholesterol through breast milk alters lifetime cholesterol metabolism in humans remains to be seen.

Protein

The protein content of mature human milk is relatively low (0.8–1.0%) compared to other mammalian milks. The concentration of proteins synthesized in the breast is more affected by the age of the infant (time since delivery) than by the proteins that arise from maternal serum. The proteins synthesized by the breast are more variable because hormones that change in the weeks and months after birth regulate the gene expression for their production.[62] Despite the relatively low concentration, human milk proteins have important nutritive and nonnutritive value. Proteins exhibit a variety of antiviral and antimicrobial effects. Enzymes present in human milk may also protect by generating components with biological activity such as the platelet-activating factor acetylhydrolase that prevents inflammatory reactions.

CASEIN Casein is the major class of protein in mature milk from women who deliver either at term or preterm.[63] Casein, calcium phosphate, and other ions such as magnesium and citrate appear as an aggregate and are the source of milk's white appearance.[64] Casein's digestive products, casein phosphopeptides, keep calcium in soluble form and facilitate its intestinal absorption.

WHEY PROTEINS   Whey proteins are the proteins that remain soluble after casein is precipitated from milk. Whey proteins include milk and serum proteins, enzymes, and immunoglobulins, among others. Several mineral-, hormone- or vitamin-binding proteins have also been identified as components of whey proteins. These include lactoferrin, which is understood to carry iron in a bioavailable form and to have bacteriostatic activity. Finally, enzymes that aid in digestion and protect against bacteria round out the diversity of whey proteins.

NONPROTEIN NITROGEN   Nonprotein nitrogen provides 20–25% of the nitrogen in milk.[64] Urea accounts for 30–50% of the nonprotein nitrogen and nucleotides for 20%, depending on the stage of lactation and the diet of the mother. Some of this nitrogen is available for the infant to use for producing nonessential amino acids. Some is used to produce other proteins with biological roles such as hormones, growth factors, free amino acids, nucleic acids, nucleotides, and carnitine. The role of individual nucleotides in human milk is under investigation; however, growth and immune advantage are believed to be the roles of nucleotides as a whole.

Milk Carbohydrates

Although lactose is the dominant carbohydrate in human milk, other monosaccharides (single sugars), neutral and acid oligosaccharides (chains of simple sugars), and protein-bound carbohydrates are also present.[65] Lactose enhances calcium absorption and contributes galactose. As the second largest carbohydrate component, oligosaccharides contribute calories at low osmolality, stimulate the growth of bifidus bacteria in the gut, and inhibit bacterial adhesion to epithelial surfaces.

Fat-Soluble Vitamins

VITAMIN A   The vitamin A content of colostrum is approximately twice that of mature milk. The high content of carotenes in colostrum is responsible for its characteristic yellow color. In mature milk, vitamin A is present at 75 mcg/dL or 280 IU/dL.[53] These levels are adequate to meet infant needs.[66]

VITAMIN D   Vitamin D is present in both lipid and aqueous (water) compartments of human milk. The majority of vitamin D is in the form of 25-OH2 vitamin D. Vitamin D levels in human milk vary with maternal diet and exposure to sunshine.[67] Maternal exposure to sunlight reportedly increases the vitamin D level 10-fold.[68] An increasing incidence of vitamin D–deficiency rickets among infants who are exclusively breastfed indicates that the vitamin D content of human milk is vulnerable. Hence, efforts should be made to maintain an adequate vitamin D content of human milk through sunlight, maternal supplementation, consumption of vitamin D–fortified dairy products, or direct supplementation of infants.

VITAMIN E   The level of total tocopherols in human milk is related to the milk's fat content. Human milk contains 40 mcg of vitamin E per gram of lipid in the milk.[69] Levels of alpha-tocopherol decrease from colostrum to transitional milk to mature milk, whereas beta- and gamma-tocopherols remain stable throughout each stage of lactation. The level of vitamin E present in human milk is adequate to meet the needs of full-term infants for muscle integrity and resistance of red blood cells to hemolysis (breaking of red blood cells). The levels of vitamin E in preterm milk have been reported to be the same[70] and higher[71] than term milk. In both reports, however, the levels present were not considered adequate to meet the needs of preterm infants.

VITAMIN K   Vitamin K is present in human milk at levels of 2.3 mcg/dL.[72] Approximately 5% of breastfed infants are at risk for vitamin K deficiency, based on vitamin K–dependent clotting factors. Cases of vitamin K deficiency among exclusively breastfed infants who did not receive vitamin K at birth have been reported. Infants fed either human milk or HMS are vulnerable to hemorrhagic disease of the newborn, however. All infants in the United States receive a vitamin K supplement (by injection) at birth.[73]

Water-Soluble Vitamins

Water-soluble vitamins in human milk are generally responsive to the content of the maternal diet or supplements (vitamin C, riboflavin, niacin, B6, and biotin). Problems with levels of these nutrients in human milk are related to their deficiency in the mother's diet. Clinical problems relating to water-soluble vitamins are rare in infants nursed by a mother with an inadequate diet.[8] Vitamin B_6 is the most likely to be deficient in human milk; however, levels of B_6 in human milk directly reflect maternal intake.[74]

VITAMIN B_{12} AND FOLIC ACID   Vitamin B_{12} and folic acid are bound to whey proteins in human milk;

therefore, their content in milk is less influenced by maternal intake than that of the other water-soluble vitamins. Factors that influence protein secretion (hormones, the age of the infant, or time since delivery) are more likely to alter the human milk levels of B_{12} and folate.[25,75] Infant illness associated with low folate levels in milk has not been reported. Folate levels increase with the duration of lactation despite a decrease in maternal serum and red blood cell folate levels.[75,76] Low levels of B_{12} have been found in the milk of women who follow vegan diets, have latent pernicious anemia (B_{12} deficiency) caused by hypothyroidism, or are generally malnourished.[76,77]

Minerals in Human Milk

The minerals in human milk contribute substantially to its osmolality. *Monovalent ion* secretion is managed closely by the alveolar cells, in balance with lactose, to maintain the isosmotic composition of human milk.

The mineral content of milk is related to the growth rate of the infant. The mineral content of human milk is much lower than the concentration in cow's milk and the milks of other animals whose offspring reach adult size in one to two years. The concentration of minerals decreases over the first four months postpartum, with the exception of magnesium. This decline during the period of rapid growth is not what one would expect, but infant growth is well supported.[78] The lower mineral concentration of human milk is easier for the kidneys to handle and is considered a significant benefit of human milk.

BIOAVAILABILITY An important feature of magnesium, calcium, iron, and zinc in human milk is the packaging that makes them more available (bioavailable) to the infant.[79] Packaging minerals so that the infant can use them efficiently also reduces the burden on the mother because less of the mineral is needed in the milk. For example, iron is 49% available from human milk, but only 10% available from cow's milk and cow's milk–based HMS. Exclusively breastfed infants have little risk of anemia,[80] despite the seemingly low concentration of iron in human milk. One study suggests that infants who are exclusively breastfed for six and a half months are less likely to be anemic than those nursed exclusively for five and a half months.[81] The exact mechanisms that make iron from human milk available and allow breastfed infants to maintain intact iron stores are not well understood.

ZINC The importance of zinc to human growth is well established. Human milk zinc is bound to protein and is highly available compared to zinc in cow's milk and cow's milk–based HMS. Zinc content of colostrum is approximately 0.4 mg/dL regardless of the time of day.[82] By six months the content of mature human milk is 0.12 mg/dL.[52] Both the zinc intake (per kilogram) and the zinc requirements of infants decline after the first few months.[83] Rare cases of zinc deficiency, which appears as intractable diaper rash, occur in exclusively breastfed infants.[84] A defect in the mammary gland uptake of zinc has been described as the cause of low milk concentration when maternal serum zinc concentrations are normal.[84] It is not clear how well serum zinc represents the whole-body or cellular levels of zinc, but in these cases, infants seem to respond to zinc supplementation.

MONOVALENT ION An atom with an electrical charge of +1 or −1.

TRACE MINERALS Trace minerals (copper, selenium, chromium, manganese, molybdenum, nickel, and fluorine) are present in the human body in small concentrations and are essential for growth and development. Less is known about trace minerals and infant health than about their nutrients. In general, however, the levels of trace minerals in human milk are not altered by the mother's diet or supplement use.

Taste of Human Milk

> . . . *too full o' th' milk of human kindness to catch the nearest way.*
>
> William Shakespeare, *MacBeth* Act I, Scene V

The centuries-old belief that a breastfeeding woman's diet influences the composition of her milk and has a long-lasting influence on the child is reflected in this line from Shakespeare. The flavor of human milk is an important taste experience for newborn infants, but is often ignored in discussions of the benefits of human milk or its composition. Human milk tastes slightly sweet,[85] and it carries the flavors of compounds ingested such as mint, garlic, vanilla, and alcohol.[86]

Infant responses to flavors in milk seem to depend on the length of time since the mother consumed the food, the amount of the flavor that the mother consumed, and the frequency with which the flavor is consumed (new versus repeated exposure). Infants seem more interested in their mother's milk when flavors are new to them. Researchers found that infants nursed at the breast longer if a flavor (garlic) was new to them than if the mother had taken garlic tablets for several days.[87] Infants who were exposed to carrot juice flavor in their mother's milk ate less of a carrot-flavored cereal and spent less time feeding at the breast than infants who had not been exposed to the carrot flavor. Thus, exposing infants to a variety of flavors in human milk may contribute to their interest in and consumption of human milk as well as their acceptance of new flavors in solid foods.[88]

THE BREASTFEEDING INFANT

Optimal Duration of Breastfeeding

The wide gap between the "best breastfeeding practice" and the norms for breastfeeding in the United States makes study of the optimal duration particularly important for health care professionals. The health benefits to the mother-child pair should be the primary criteria used to determine the optimal duration of breastfeeding, rather than whether the cultural environment makes such duration practical. The American Academy of Pediatrics (AAP)[7] has taken a clear stand on this issue, saying that breastfeeding should continue for a year or longer.

Breastfeeding can prevent intestinal blood loss in infants—a factor that should be considered when determining its optimal duration. Infants fed cow's milk before the age of six months suffer nutritionally significant losses of iron via intestinal blood loss (30% increase in the presence of blood in the infants fed cow's milk)[89]—an observation that supports the AAP's recommendation.[7] Through one year, breastfed infants suffer fewer acute infections than do formula-fed infants, a finding that supports continued breastfeeding beyond the introduction of solids.

Reflexes

Healthy term infants are born with several reflexes that enable them to nourish themselves. Observations show that 18-week-old fetuses start sucking. By 34 weeks gestation, the suck has adequate pace and rhythm to be nutritive. The gag reflex prevents the fetus from taking food and fluids into the lungs and is developed by 28 weeks gestation. These reflexes allow term infants to suck and swallow in a coordinated pattern that protects the airways.

Two other reflexes describe infants' ability to position themselves to breastfeed. With oral search reflex, the infant opens her mouth wide in proximity to the breast while thrusting her tongue forward. With the rooting reflex, the infant turns to the side if stimulated on the side of the upper or lower lip. The infant comes forward, opens her mouth and extends her tongue when the center of the upper or lower lip is stimulated.

The presence of these reflexes is important to the success of breastfeeding. Successful nursing also requires appropriate positioning of the infant and adequate maternal letdown and milk production. Appropriate positioning and maternal assessment of infant nursing behaviors must be learned. Support from lactation consultants and/or other health care professionals trained in lactation may be necessary.

Breastfeeding Positioning

Proper positioning of the infant at the breast is important to breastfeeding success. Mothers need to learn from health professionals experienced in optimal positioning because improper positioning causes pain and possible damage to nipple and breast tissue. The mother may need to use cushions, pillows, or a footstool to be comfortable and positioned well to nurse the infant (see Illustration 6.6). Once the mother is comfortable, she should stimulate the oral search reflex by touching the baby's bottom lip with her nipple. The infant will open his mouth wide and should be brought to the breast with the nipple centered in his mouth (Illustration 6.7). This process is called latching on. An infant who is properly latched at the breast has all or most of the areola (the dark pigmented skin around the nipple) in his mouth. The mother should hear swallowing, but not smacking, clicking, or slurping. The baby's nose should be close to the breast with his breathing unrestricted.

Identifying Hunger and Satiety

Early hunger behavior occurs in a period of increased alertness. During this time, infants begin to bring their hands to their mouths, suck on them, and start moving their head from side to side with their mouth open in the rooting reflex. Infants should be fed when these signs of hunger are displayed, rather than waiting for crying, a late sign. Recognizing early hunger behaviors and initiating feeding before the infant becomes very upset can be particularly helpful for mothers and infants who have difficulty nursing.

Nutritive and nonnutritive sucks are different. Feedings begin with nonnutritive sucking. The infant sucks quickly and not particularly rhythmically. Nutritive sucking is slower and more rhythmic as the infant begins to suck and swallow. In a quiet room, a mother can hear the infant suck.

Infants should be allowed to nurse as long as they want at one breast. Infants who are fed for shorter periods from both breasts can get larger amounts of foremilk, which can cause diarrhea because of the high lactose content.[90] Allowing the infant to nurse at one breast until satisfied creates a

a. Cradle hold

b. Football or clutch hold

c. Cross-cuddle hold

Illustration 6.6   Positions for breastfeeding.

pattern that assures that the infant gets both foremilk and hindmilk. The higher fat content of hindmilk may help in signaling satiety. The infant will stop nursing when she is full. If the infant is still hungry, after burping, she can be offered the other breast.

Feeding Frequency

Stomach emptying occurs in about one and a half hours for breastfed infants. Ten to 12 feedings per day are normal for newborn infants. As infants develop, feeding frequency will depend, in part, on maternal storage capacity (see "Breast Milk Supply and Demand" earlier in this chapter). Different feeding patterns can meet infant needs. In one study, infants who did not feed from midnight through early morning consumed more in the other feedings, particularly in the morning. The milk intake and weight gain of those infants in the first four months of life were similar to those of infants whose feedings were distributed over 24 hours.[91]

Identifying Breastfeeding Malnutrition

A normal newborn weight loss of up to 7% can occur in the first week postpartum. A loss of 10% should trigger an evaluation of milk transfer to the infant by a lactation consultant or other trained professional. This evaluation should result in support for maintaining breastfeeding. Malnourished infants become sleepy, are nonresponsive, have a weak cry, and wet few diapers. The clinician can use the diagnostic flowchart in Illustration 6.8 to help diagnose failure

to thrive in breastfed infants. By the fifth to seventh day postpartum, infants who are getting adequate nourishment have wet diapers approximately six times a day and three to four soft yellowish stools per day.[7] Infants who are slow gainers are alert, bright, responsive, and developing normally.[8] Their urine is pale yellow and dilute while stools are loose and seedy (some small particles are present in the stool). In contrast, infants who are failing to thrive are apathetic and hard to arouse and have a weak cry. They have few wet diapers, and their urine is concentrated. Their stools are infrequent.

Mothers of slow-gaining infants, in particular, should be advised to let the infant nurse at one breast until it is empty or the infant stops nursing, rather than switching breasts after a specific amount of time. This regimen assures that the infant gets hindmilk with its higher fat and calorie content. Lawrence[8] recommends evaluation of slow-gaining infants when the slow-gaining pattern is recognized (Illustration 6.8).

Infant Supplements

Breastfed infants need few supplements, except in specialized conditions. As discussed above, infants should be given vitamin D supplements if exposure to sunlight is inadequate or if the mother's diet does not include vitamin D–fortified dairy products. Adequate exposure to sunlight is defined at 30 minute per week in just diapers or two hours per week in clothes but no hat.[92] A recommendation for routine vitamin D supplementation for exclusively breastfed infants is being considered in order to prevent further cases of rickets.

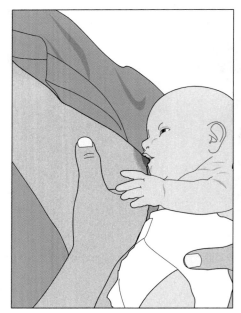

a. Touching the baby's bottom lip with the nipple stimulates the oral search reflex.

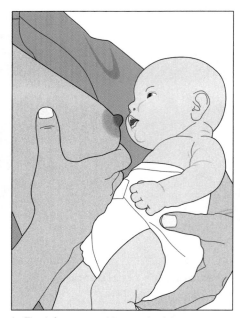

b. The infant opens his mouth wide.

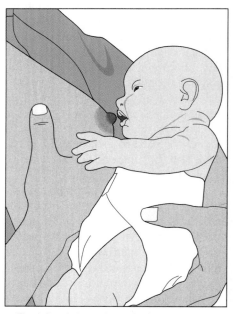

c. The infant is brought to the breast with the nipple centered in his mouth.

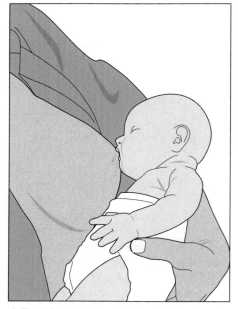

d. The infant is properly latched at the breast and has all of the areola in his mouth.

Illustration 6.7   Attachment.

Infants should not be supplemented with fluoride until six months of age. Thereafter, infants may need a fluoride supplement if the water has less than 0.3 parts per million (ppm) of fluoride.[7] Breastfed infants do not need iron-fortified HMS,[8] as they rarely experience iron deficiency. The excess iron in HMS might bind with lactoferrin in human milk, resulting in a loss of the lactoferrin's protective activity.

Tooth Decay

There is no evidence that human milk is particularly prone to causing dental caries. Nevertheless, caries due to feeding can occur in children who are breastfed.[93] Frequent nursing at night after one year of age should be considered a risk factor for dental caries. Children who are nursed for long stretches may develop dental

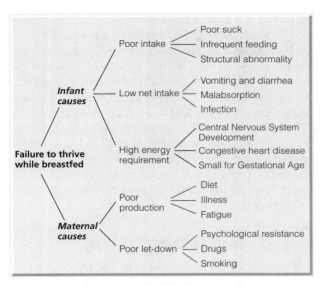

Illustration 6.8 **Diagnostic flowchart for failure to thrive.**[8]

From Lawrence, R.A. and Lawrence, R.M., BREASTFEEDING: A Guide for the Medical Profession 5/E. Used by permission of Mosby, Inc.

caries between ages two and three years. Since susceptibility to caries is genetically programmed, fluoride supplements are indicated for infants in families with a history of dental enamel problems or caries.

Breastfed babies have straighter teeth due to the development of a well-rounded dental arch.[94] A well-rounded arch may also help prevent sleep apnea later in life.

MATERNAL DIET

Both the Food Guide Pyramid[95] and the "Dietary Guidelines for Americans"[96] are appropriate guidelines for breastfeeding women (see Chapter 1). Breastfeeding women should follow the Food Guide Pyramid when choosing their foods. Four servings of dairy products fortified with vitamins A and D each day help to achieve adequate calorie, protein, vitamin D, and calcium intake.

Although it is widely believed that components of the maternal diet are related to infant colic, few well-designed studies have examined this relationship. One retrospective cross-sectional study of the diets of breastfeeding women suggests that cow's milk, onions, cruciferous vegetables, and chocolate are associated with an increased risk of colic in infants.[97] The results may have been biased, however, by the participating women's belief that these foods were related to their infant's colic.

Energy and Nutrient Needs

Until recently, when safe methods of studying energy metabolism became available, determining the energy needs of lactating women was very difficult. As a result, the energy needs of lactation are estimated using the factorial method, which adds the estimated requirements of lactation to the requirements of non-lactating women. The 1989 RDAs for lactating women assume that milk production is 80% efficient, that human milk contains 70 cal/dL, and that the average amount of milk secreted in the first six months is 750 ml.[53] The 1989 RDA of 500 additional calories per day also assumes that 250 calories per day are mobilized from maternal fat stores. It is important to understand that public health policy supports estimating the energy and nutrient needs of breastfeeding women conservatively to ensure adequate nutrition for women and their children. An overestimation of needs is possible using this approach.

Vitamin and mineral intakes that do not meet recommended levels have been reported for lactating women.[49–53] Intakes of folate, thiamin, vitamin A, calcium, iron, and zinc have been reported to be inadequate. Ten percent of lactating women have thiamin intakes below the recommended levels,[49] whereas fewer than 5% of nonlactating women have intakes below the 1998 DRI. These reports of inadequate intake have not been followed by reports of deficits in the nutritional status of the mother-infant pair, however. Concern over inadequate maternal dietary intake must be carefully balanced against the possibility that women will be discouraged from breastfeeding because they believe their diet is not optimal.

Studies of dietary intake of lactating women in developed countries[98,99] have rarely documented energy intakes that meet the recommended levels.[53] In contrast, studies that measure components of energy expenditure and intake[100] or adequacy of milk production and infant growth[37] have simultaneously documented adequate lactation (and infant growth) with dietary intakes below the 1989 RDA.[53]

We now understand that women use several mechanisms to meet the needs of lactation. Adjustments in energy intake and energy expenditure must be balanced to meet those needs. Goldberg et al.[101] found that women increased food intake (56% of the energy need for milk production) and decreased physical activity (44% of the energy need for milk production) to meet the increase in energy needs for lactation. Doubly labeled water studies suggest that the components of energy expenditure vary greatly and that measurements of dietary intake can be unreliable.[102] Therefore, a single recommendation for lactating women could never address all of the individual ways that women meet their energy needs.

Assessment of adequacy of energy intake of breast-feeding women should always be made within the context of the mother's overall nutritional status and weight changes and the adequacy of the infant's growth.

MATERNAL ENERGY BALANCE AND MILK COMPOSITION

The composition of breast milk depends on maternal nutritional status. Protein-calorie malnutrition results in an energy deficit that reduces the volume of milk produced, but usually does not compromise the composition of the milk. Several studies show that milk production is maintained when there is a modest level of negative calorie balance. Animal models first identified a potential threshold effect of energy restriction. Baboons fed 60% of their voluntary intake significantly reduced milk production.[103] Yet baboons fed 80% of their usual intake maintained milk production. Randomized studies such as the one performed on baboons cannot ethically be done with humans. Nevertheless, a series of human studies on weight loss during breastfeeding discussed below support a threshold effect of energy limitations on lactation.

Weight Loss during Breastfeeding

Because breastfeeding has an average estimated cost of 750 calories per day, it could enhance maternal weight loss.[53] Studies of maternal energy intake rarely report intakes that reach the recommended levels of the RDA,[25,53] also indicating that weight loss should result. In addition, mechanisms that favor use of maternal fat stores and delivery of nutrients to the breast seem to occur during lactation.[104]

Despite these mechanisms that should favor weight loss in breastfeeding women, on average, postpartum weight changes are relatively small and inconsistent.[105] Even more surprising, postpartum weight changes (−0.1 kg/mo) are smaller in developing countries than in industrialized nations (−0.8 kg/mo) (Table 6.4). Possibly changes in postpartum weight are small and inconsistent because changes in energy intake, energy expenditure, and fat mobilization easily meet the demands.

Strode et al.[106] first observed women who voluntarily reduced their energy intake to 68% of their estimated needs for seven days. No differences in infant intake or milk composition were observed. In the week following the diet, women who consumed fewer than 1500 cal tended to experience a decrease in milk volume. Women who consumed over 1500 cal per day experienced no decrease. Despite its voluntary

nature and the lack of milk samples throughout the day, this study provides important support for the idea that modest and/or short-term reductions in energy are not associated with decreases in milk production. In addition, 22 healthy postpartum women who participated in a 10-week weight-loss program, which reduced energy intake by 23%, maintained milk production.[107] The women lost an average of about a pound a week during the 10 weeks. A number of study design limitations, including small numbers and lack of randomization and/or control groups, limit generalizations based on this information, however.

Exercise and Breastfeeding

Two studies[108,109] have examined the effect of increasing energy expenditure on weight and lactation. In a cross-sectional study, vigorous exercise increased energy expenditure, but these women also increased their energy intake, so the calorie deficit was similar in the two groups.[108] There were no significant differences in milk volume between the groups, although the exercising women tended to have higher volumes. A later 12-week exercise intervention trial studied women who were six weeks postpartum and exclusively breastfeeding to examine the effects of exercise on body composition and energy expenditure during breastfeeding.[109] The women were randomly assigned to two groups. One group followed an aerobic exercise regimen for 45 minutes, five days a week. The other group did no exercise. The exercising women increased energy expenditure by 400 cal per day and increased energy intake to compensate for this use. Both groups experienced similar weight changes, milk volume, milk composition, and infant weight gain and serum prolactin levels. These studies suggest that lactating women efficiently balance their energy intake to support energy expenditure.

The available evidence suggests that modest energy restriction combined with increases in activity may be effective in helping women to lose weight, while improving their metabolic profile and increasing fat losses. Although only a small number of women have been studied, the consistent lack of effect on milk production (infant intake), milk energy output, and infant growth is encouraging.

The mechanisms responsible for maintaining milk production may be different. Aerobic activity appears to enhance fatty acid mobilization to meet the needs of milk production, whereas restricted energy intake requires increases in prolactin levels to promote use of dietary fatty acids or to promote mobilization from fat. These prolactin increases explain how milk production can be maintained despite a negative energy balance.

Table 6.4 **Longitudinal studies of weight change during lactation.**[105]

AFFLUENT POPULATIONS

Reference	Country	No. in Study	Month Postpartum	Mean Weight Change (kg/mo)
Naismith and Ritchie 1995	UK	22	0.3–3	−0.87
			3–6	−0.90
Manning–Dalton and Allen 1983	US	27	0.5–3	−0.78
Butte et al. 1984	US	45	1–4	0.67
Sadurskis et al. 1989	Sweden	23	0.25–6	−0.77
Brewer et al. 1989	US	21	0–3	−2.25
			3–6	−0.43
Goldberg et al. 1991	UK	10	1–3	−0.13
Van Raaij et al. 1991	Netherlands	16	1–2	+0.20
			2–3	−0.5
			3–6	−0.73
Dewey et al. 1993	US	46	1–3	−0.60
			3–6	−0.73

UNDERPRIVILEGED POPULATIONS

Reference	Country	No. in Study	Month Postpartum	Mean Weight Change (kg/mo)
Prentice et al. 1981	Gambia	143	0–3	+0.70 dry season
				−0.50 wet season
			3–6	+0.45 dry season
				−0.80 wet season
Adair et al. 1983	Taiwan	225	0–1	+0.79
			1–3	−0.14
			3–6	−0.28
Mbofung and Atinmo 1985	Nigeria	43	1–3	−0.45
Novotny and Haas 1987	Bolivia	18	3–6	−0.22
Guillermmo-Tuazon et al. 1992	Philippines	40	1.5–6	−0.27
Allen et al. 1994	Egypt	38	0–6	+0.03
	Kenya	109	0–6	−0.18
	Mexico	74	0–6	−0.15
Fornes and Dorea 1995	Brazil	39	0.5–3	−0.73
Babosa et al. 1997	Mexico	40	0.5–6	−0.14

Diet Supplements

A 1990 Institute of Medicine report, *Nutrition during Lactation*,[75] stated that well-nourished breastfeeding women do not need routine vitamin or mineral supplementation. Instead supplementation should target specific nutritional needs of individual women. Supplementation strategies should take into account how nutrients are secreted into human milk and the potential for nutrient-nutrient interactions in mothers and their infants.

Breastfed infants can suffer from deficiencies of vitamins D and B_{12} because human milk can have inadequate amounts of these nutrients even when mothers do not exhibit deficiency symptoms. Either the mother or her infant should take a vitamin D supplement if the mother does not use vitamin D–fortified dairy products or is not exposed to sunlight. Risk factors for nutritional rickets include exclusively breastfed infants or infants with highly pigmented skin fed human milk and weaning foods.[110] In addition, mothers who live in northern or cloudy climates, who are practicing Muslims (little skin exposure), or who follow a vegan diet, as well as mothers whose infants have little exposure to sunlight, should take special precautions to be sure that their infants do not suffer from nutritional rickets. In North Carolina alone, at least 30 cases of nutritional rickets have been identified in exclusively breastfed infants.[110] B_{12} deficiency has affected infants of women with gastric bypass and vegan diets.[111,112] Women with intestinal disease, such as Crohn's disease, should also be considered at risk and supplemented.

Fluids

There is no evidence that increasing fluid intake will increase milk production or that a short-term fluid deficit results in a decrease in milk production. Fluid demands rise during breastfeeding, however, so women should drink fluids to thirst. Once a mother and her infant have the nursing routine down, she may find it convenient to have something to drink while she nurses. Although many women want to know how many glasses to drink per day, the amount needed varies depending on climate, milk production, body size, and other factors. Therefore, women are advised to drink enough fluids to keep their urine pale yellow.

MEDICINAL HERBS   Plants used to prevent or remedy illness.[114]

Vegetarian Diets

Breastfeeding women can follow vegetarian eating patterns and be well nourished. The goal is to adequately nourish the mother and child, not to force women to use supplements and/or products that are not part of their normal eating patterns. Vegetarians who do not consume dairy products and/or eggs, however, may have an increased risk of inadequate intakes of calories, protein, calcium, vitamin D, vitamin B_{12}, iron, and zinc. Vegetarians' intakes of protein are generally adequate as long as they achieve adequate energy intake. Breastfeeding women who consume no animal products should use plant foods with bioavailable B_{12} from plant sources such as yeasts, seaweed, and some soy products. Women who are unable or unwilling to get adequate B_{12} from foods should take a vitamin B_{12} supplement. The 1991 Institute of Medicine[25] report recommends a multiple vitamin-mineral supplement for vegetarians because human milk may be low in vitamin B_{12} even when mothers do not exhibit deficiency symptoms.[112]

Herbal Preparations

Although numerous herbs have traditional uses dating back thousands of years, scientific information about herb use during lactation, particularly recent studies, is sparse. The limited pertinent safety data are based on traditional use, animal studies, and knowledge of the pharmacologic activities of the products constituents.[113] A mother may perceive herbs as natural and, therefore, safe and even preferable to conventional beverages, over-the-counter medicines, or prescription drugs. Many herbs are far from benign, however, and many are contraindicated during lactation (Table 6.5). Because little is known about the amount secreted in human milk or the effects on preterm or term infants, herbs that are central nervous system stimulants, cathartic laxatives, hepatotoxic, carcinogenic, cytotoxic, or mutogenic, or that contain potentially toxic essential oils, are not recommended during lactation.[113]

The toxic effects of herbs are often not the fault of the herb itself, but are caused by products containing misidentified plants or contaminants such as heavy metals, synthetic drugs, microbial toxins, and toxic botanicals.[115] *Medicinal herbs* in the United States are regulated as dietary supplements and are not tested for safety or efficacy.

Herbal teas that are safe for the infant and mother during lactation are presented in Table 6.6. Lawrence[8] recommends using only herbal teas "that are prepared carefully, using only herbs for essence (e.g., Celestial Seasonings brand tea) and avoiding heavy doses of herbs with active principles." Careful attention should also be given to preparation, avoiding long steeping times.

Some culinary herbs may lead to problems when used extensively. In lactation and herbal texts, sage has a folk reputation for lowering milk supply,[113] as do parsley and peppermint, especially if the oil is taken internally in therapeutic doses.[116] Consumed on occasion, however, in small amounts as part of a reasonably varied diet, peppermint, parsley, sage, and other culinary herbs currently have no documented negative effect on lactation.

If a mother is consuming a large amount of any herbal product, its contents should be checked. Important information to obtain on the product includes its name, a list of all ingredients, the names of the plants or other components (include the plants' Latin names if possible), details of the preparation, and the amount consumed.[116] Reliable sources of herbal information[117–119] or the regional poison control center may be able to identify potentially harmful pharmacological and toxicological ingredients.

Although various herbal gels, ointments, or creams are often suggested for use on the nipples, any substance applied to the breast or nipples could easily be ingested by the nursing child. The use of herbal oils is not recommended.[116] In one infant, severe breathing difficulties were documented after the mother used menthol, a significant component of peppermint oil, on her nipples.[120]

FACTORS INFLUENCING BREASTFEEDING INITIATION AND DURATION

All new mothers, both low income and more affluent, need support for breastfeeding. Low-income women,

Table 6.5 Medicinal herbs considered not appropriate for use during pregnancy or lactation.[115]

Agnus castus	Aloes	Angelica	Apricot kernal
Aristolchia	Asafoetida	Avens	Blue flag
Bogbean	Boldo	Bonese	Borage
Broom	Buchu	Buckthorn	Burdock
Calamus	Calendula	Cascara	Chamonile, German
Chamomile, Roman	Chaparral	Cohosh, black	Cohush, blue
Coltsfoot	Comfrey	Cornsilk	Cottonroot
Crotalaria	Darniana	Devil's claw	Dogbane
Dong Quai	Ephedra	Eucalyptus	Eupatorium
Euphorbia	Fenugreek	Ferverfew	Foxglove
Frangula	Fucus	Gentian	Germander
Ginseng, eleuthero	Ginseng, panax	Golden seal	Ground ivy
Groundsel	Guarana	Hawthorne	Heliotropium
Hops	Horehound, back	Horehound, white	Horsetail
Hydrocotlye	Jamaica dogwood	Juniper	Liferoot
Lobelia	Male fern	Mandrake	Mate
Meadowsweet	Melilot	Mistletoe	Motherwort
Myrrh	Nettle	Osha	Passionflower
Pennyroyal	Petasites	Plantain	Pleurisy root
Pokeroot	Poplar	Prickly ash	Pulsatilla
Queen's delight	Ragwort	Red clover	Rue
Sassafras	Scullcap	Senna	Shepard's purse
Skunk cabbage	Squill	St. John's wort	Stephania
Tansy	Tonka bean	Uva-ursi	Vervain
Wild carrot	Willow	Wormwood	Yarrow
Yellow dock	Yohimbine		

Note: Exclusion from this list should not be a recommendation for safety.

however, often lack the education, support, and confidence to interpret the abundant and pervasive mixed messages on infant feeding practices.[120] Consider the strikingly different context for pregnancy, birth, and parenting for low-income women and their more affluent counterparts:

Profile of a low-income pregnant woman: "She says she wants to do what is best for her baby and, in fact, knows the breast is best. However, she is afraid breastfeeding will cause her baby to be too "clingy." She feels extremely uncomfortable about nursing around family, much less in public. She is certainly not up for the pain she has heard breastfeeding causes. To make things more difficult, in the hospital, she is separated from her baby soon after delivery and is given little assistance for getting started. She is sent home from the hospital with samples of free formula."

Profile of an affluent pregnant women: "The affluent expectant mother has friends who have breastfed and have helped build her confidence that she can breastfeed successfully. She may have been able to choose her birth setting and select a hospital with knowledgeable staff allowing mother and baby to stay together around the clock. Because she knows there may be bumps in the road getting started, she seeks out support from friends or the doctor after discharge. At home, she has a supportive husband who is proud of

her for offering the best for their baby. If she returns to work, she knows she can still breastfeed to keep that special closeness with her baby even after returning to work."

Common barriers to breastfeeding initiation expressed by expectant mothers[121–123] include the following:

- Embarrassment.
- Time and social constraints, and concerns about loss of freedom (particularly issues for working moms).
- Lack of support from family and friends.
- Lack of confidence.

Table 6.6 Herbal teas considered safe during lactation.[8]

Tea	Origin/Use
Chicory	Root/caffeine-free coffee substitute
Orange spice	Mixture/flavoring
Peppermint	Leaves/flavoring
Raspberry	Fruit/flavoring
Red bush tea	Leaves, fine twigs/beverage
Rose hips	Fruits/vitamin C

Reprinted with permission from Lawrence, RA. Breastfeeding: a guide for the medical professions. St. Louis: Mosby; 1999.

- Concerns about diet and health practices.
- In adolescents, fear of pain.

Additional obstacles to the initiation and continuation of breastfeeding[7,124–126] include:

- Insufficient prenatal breastfeeding education.
- Health care provider apathy and misinformation.
- Inadequate health care provider lactation management training.
- Disruptive hospital policies.
- Early hospital discharge.
- Lack of routine follow-up care and postpartum home health visits.
- Maternal employment, especially in the absence of workplace facilities and support for breastfeeding.
- Lack of broad societal support.
- Media portrayal of bottle-feeding as the norm.
- Commercial promotion of infant formula through distribution of hospital discharge packs, coupons for free or discounted formula, and television and general magazine advertising.

BREASTFEEDING PROMOTION, FACILITATION, AND SUPPORT

Significant steps must be taken to increase breastfeeding rates in the United States and to close the wide racial and ethnic gaps in breastfeeding. This goal can only be achieved by supporting breastfeeding in the family, community, workplace, health care sector, and society.

U.S. Department of Health and Human Services Blueprint for Action on Breastfeeding, 2000[1]

The support a woman receives from those around her has a direct effect on her capability to breastfeed optimally.[1] The health care system, her workplace, and her community can all work to facilitate the initiation and continuation of breastfeeding.

Role of the Health Care System in Supporting Breastfeeding

Health care providers and facilities can exert tremendous influence over the mother-infant dyad by promoting and modeling optimal breastfeeding practices during prenatal care, at delivery, and after discharge.

PRENATAL BREASTFEEDING EDUCATION AND SUPPORT Culturally competent prenatal breastfeeding education that is given frequently in person can have a significant positive influence on breastfeeding rates.[122,127] Best Start Social Marketing[121] has developed an effective three-step counseling strategy (Table 6.7) that quickly identifies a woman's particular barriers to choosing breastfeeding and provides targeted education while affirming the woman's ability to breastfeed.

The Best Start approach avoids questions that force the woman to choose, such as "Are you going to breast or bottle-feed?" and instead uses open-ended questions, such as "What have you heard about breastfeeding?" or "What questions do you have about breastfeeding?" Such questions give the woman an opportunity to begin a dialogue with her provider about the infant feeding decision. By using follow-up probes as necessary, the counselor can identify the woman's specific concerns. Counselors should avoid overwhelming the woman with too much information, which can give the impression that breastfeeding is difficult.[121]

Another effective strategy utilizes peer counselors and peer group discussions with at least one or two women who have successfully breastfed.[129] Exposure to mothers nursing their babies increases a woman's level of comfort with breastfeeding and provides a forum for informal discussion with family and friends. Hearing about someone else's personal experience can help a woman realize that other women share her concerns.[123]

Toward the end of pregnancy, women need information on what to expect in the hospital or birthing

---

Table 6.7   The best start three step breastfeeding counseling strategy.[121]

1. Ask open-ended questions to identify the woman's concerns:
   - Dietitian: "What have you heard about breast-feeding?"
   - Client: "I hear it's best for my baby, but all my friends say it really hurts!"

2. Affirm her feelings by reassuring her that these feelings are normal:
   - Dietitian: "You know, most women worry about whether it will hurt."

3. Educate by explaining how other women like her have dealt with her concerns. Avoid overeducating or giving the impression that breastfeeding is hard to master:
   - Dietitian: "Did you know that it is not supposed to be painful, and if you are having discomfort, there are people who can help make it better?"

center and practical tips for initiating breastfeeding.[123] Several key points are shown in Table 6.8. Since fathers,[130] grandmothers,[130] *doulas,* friends, and social networks[131] all play a powerful role in infant feeding decisions, these influential people should be included as often as possible in breastfeeding promotion efforts.[123,130,132]

The environment for the delivery of prenatal care can inhibit or facilitate breastfeeding. It should provide positive messages about breastfeeding, such as posters on the walls and magazines and literature in the waiting room that promote breastfeeding; there should be no advertisements or promotions of formula. Nevertheless, patient education materials that include formula advertising, samples, and business reply cards for free formula[133,134] are often available in U.S. and Canadian prenatal care settings, in direct violation of the World Health Organization's (WHO's) International Code on the Marketing of Breast Milk Substitutes (Table 6.9). Women who have been exposed to materials and products from formula companies prenatally are more likely to cease breastfeeding in the first two weeks.[135] Use of these materials conveys a subtle message that infant formula is equivalent to breast milk.

Although not all women will choose to breastfeed, the goal of prenatal breastfeeding education is to empower all women with sufficient knowledge to make an informed decision about infant feeding methods. Although some professionals are concerned that breastfeeding promotional efforts may cause women who choose formula-feeding to feel guilty, experience indicates that women want information to make the best possible decision for themselves and their infants. Some women who formula-feed report feeling angry about not getting enough breastfeeding information during their pregnancy.[122] No studies support the avoidance of guilt as a reason not to promote breastfeeding. In recognition of the benefits of breastfeeding and the important role of health professionals in promoting and supporting it, the leading health and professional organizations in the United States that provide perinatal care have established policies supporting breastfeeding as the preferred infant feeding method.[7,124,137–139]

> **DOULA** An individual who gives psychological encouragement and physical assistance to a mother during the pregnancy, birth, and lactation; may be a relative, friend, or neighbor and is usually but not necessarily female.[8]

LACTATION SUPPORT IN HOSPITALS AND BIRTHING CENTERS Hospital policies and routines can have effects on a woman's critical early experience with breastfeeding that extend far beyond her short stay at the facility.[140] Illustration 6.9 provides examples of hospital practices that influence this pivotal initiation experience. Like prenatal care settings, hospitals should not distribute free samples of infant formula or coupons in their discharge packs because these can have detrimental effects on breastfeeding success, particularly among vulnerable groups such as new mothers and low-income women.[140]

In an effort to promote, protect, and support breastfeeding in hospitals and birthing centers worldwide, the WHO and UNICEF established the Baby Friendly Hospital Initiative in 1991. This initiative

---

Table 6.8 Key teaching points prior to birth.[121,128]

In the hospital or birthing center, the mother should:
- Request early first feeding.
- Practice frequent, exclusive breast-feeding.
- Ask to be taught swallowing indicators.
- Learn indicators of sufficient intake.
- Ask for help if breast-feeding hurts.
- Know sources for help.
- Understand postpartum rest and recovery needs.
- Avoid supplements unless medically indicated.

Table 6.9 World Health Organization's International/UNICEF Code on the Marketing of Breast Milk Substitutes.

- No advertising of breast-milk substitutes.
- No free samples or supplies.
- No promotion of products through health care facilities.
- No company sales representative to advise mothers.
- No gifts or personal samples to health workers.
- No gifts or pictures idealizing formula feeding, including pictures of infants, on the labels of the infant milk containers.
- Information to health workers should be scientific and factual.
- All information on artificial feeding, including labels, should explain the benefits of breast-feeding and the costs and hazards associated with formula feeding.
- Unsuitable products should not be promoted for babies.
- Manufacturers and distributors should comply with the Code's provisions even if countries have not adopted laws or other measures.

Hospital Practices that Influence Breastfeeding Initiation

	← Strongly Encouraging →	← Encouraging →	← Discouraging →	← Strongly Discouraging →
Physical Contact	• Baby put to breast immediately in delivery room • Baby not taken from mother after delivery • Woman helped by staff to suckle baby in recovery room • Rooming-in; staff help with baby care in room, not only in nursery	• Staff sensitivity to cultural norms and expectations of woman	• Scheduled feedings regardless of mother's breastfeeding wishes	• Mother-infant separation at birth • Mother-infant housed on separate floors in postpartum period • Mother separated from baby due to bilirubin problem • No rooming-in policy
Verbal Communication	• Staff initiates discussion re: woman's intention to breastfeed pre- and intrapartum • Staff encourages and reinforces breastfeeding immediately on labor and delivery • Staff discusses use of breast pump and realities of separation from baby, re: breastfeeding	• Appropriate language skills of staff, teaching how to handle breast engorgement and nipple problem • Staff's own skills and comfort re: art of breastfeeding and time to teach woman on one-to-one basis	• Staff instructs woman "to get good night's rest and miss the feed" • Strict times allotted for breastfeeding regardless of mother/baby's feeding "cycle"	• Woman told to "take it easy," "get your rest" … impression that breastfeeding is effortful/tiring • Woman told she doesn't "do it right," staff interrupts her efforts, corrects her re: positions, etc.
Non-verbal Communication	• Pictures of woman breastfeeding • Staff (doctors as well as nurses) give reinforcement for breastfeeding (respect, smiles, affirmation) • Nurse (or any attendant) making mother comfortable and helping to arrange baby at breast for nursing • Woman sees others breastfeeding in hospital • Closed circuit TV show in hospital on breastfeeding	• Literature on breastfeeding in understandable terms	• Pictures of woman bottle-feeding • Staff interrupts her breastfeeding session for lab tests, etc. • Woman doesn't see others breastfeeding	• Woman gives infant formula kit and infant food literature • Woman sees official-looking nurses authoritatively caring for babies by bottle-feeding (leads to woman's insecurities re: own capability of care)
Experiential	• If breastfeeding not immediately successful, staff continues to be supportive • Previous success with breastfeeding experience in hospital			• Previous failure with breastfeeding experience in hospital

Illustration 6.9   Hospital practices that influence breastfeeding initiation.[1,141]

focuses on 10 evidence-based components of hospital care that influence on breastfeeding success (Table 6.10).[142] In early 2001, 30 hospitals and birthing centers in the United States had met all of the criteria in Table 6.10 and were designated as Baby Friendly.

Lactation Support after Discharge

Breastfeeding support is essential in the first few weeks after delivery when lactation is being established.[1] A study of inner-city Baltimore WIC program participants[125] provides strong evidence that 7 to 10 days postpartum is the critical window for providing breastfeeding support; 35% of mothers who initiated breastfeeding in the hospital had stopped by 7 to 10 days.

A pediatrician, nurse, or other knowledgeable health care practitioner should see all breastfeeding mothers and their newborns for a home visit or in the office when the newborn is two to four days old.[7] During the visit, the practitioner should observe the mother breastfeeding to ensure that she is successful, revisit the major concerns she identified during pregnancy, and discuss any new concerns. The practitioner should also arm the mother with information on sources of trained help available in the commu-

nity, such as *lactation consultants,* peer counselors, the WIC program, and *La Leche League,* should questions or complications arise.[7,123] Follow-up telephone calls, as necessary, provide additional support to mothers who are not fully confident in their ability to breastfeed successfully.[123]

The Workplace

The increase in the proportion of women working outside the home that began after World War II has been one of the most significant social and economic trends in modern U.S. history.[144] In 1940, one in four U.S. workers was female; by 1998, almost one in two workers was female.[144] About 70% of employed mothers with children under three years of age work full-time;[1] about one-third of these mothers return to work within three months and about two-thirds within six months after childbirth.

Planning to return to work full-time does not appear to affect breastfeeding initiation rates.[26] Breastfeeding duration, however, is adversely affected by employment. At five to six months 25% of nonemployed women are still breastfeeding, compared with 23.4% of mothers working part-time, and only 14.3% of those employed full-time.[26] Occupation influences duration of breastfeeding. Women in professional occupations breastfeed significantly longer than women in sales, clerical, or technical occupations.[145]

Studies indicate that women who continue to breastfeed after returning to work miss less time from work because of baby-related illnesses and have shorter absences when they do miss work than women who do not breastfeed.[24] Worksite programs that support breastfeeding facilitate the continuation of breastfeeding after mothers return to their jobs and offer additional advantages to employers: companies with such programs find that employee morale and

LACTATION CONSULTANT A health care professional who provides education and management to prevent and solve breastfeeding problems and to encourage a social environment that effectively supports the breastfeeding mother-infant dyad. Those who successfully complete the International Board of Lactation Consultant Examiners (IBLCE) certification process are entitled to use IBCLC (International Board Certified Lactation Consultant) after their names (http://www.iblce.org/).

LE LECHE LEAGUE An international, nonprofit, nonsectarian organization dedicated to providing education, information, support, and encouragement to women who want to breastfeed. It was founded in 1956 by seven women who had learned about successful breastfeeding while nursing their own babies. Currently, approximately 7100 accredited lay leaders facilitate more than 3000 monthly mother-to-mother breastfeeding support group meetings around the world (http://www.lalecheleague.org).

Table 6.10 The Baby-Friendly Hospital Initiative: Ten Steps to Successful Breastfeeding.[143]

1. Have a written breastfeeding policy that is routinely communicated to all health care staff.

2. Train all health care staff in skills necessary to implement this policy.

3. Inform all pregnant women about the benefits and management of breastfeeding.

4. Help mothers initiate breastfeeding within a half-hour of birth.

5. Show mothers how to breastfeed and how to maintain lactation even if they should be separated from their infants.

6. Give newborn infants no food or drink other than breast milk, unless medically indicated.

7. Practice rooming-in—allow mothers and infants to remain together—24 hours a day.

8. Encourage breastfeeding on demand.

9. Give no artificial teats or pacifiers (also called dummies or soothers) to breastfeeding infants.

10. Foster the establishment of breastfeeding support groups and refer mothers to them on discharge from the hospital and clinic.

loyalty, the company's image as family-friendly, recruiting for personnel, and the retention rate of employees after childbirth all improve.[146] Sanvita, a worksite lactation support program, has helped companies achieve a $1.50 to $4.50 return for each dollar invested.[147] Aetna, Inc., achieved a $2.18 return for every $1.00 invested in its lactation support program.[146] Key elements of worksite lactation support programs are presented in Table 6.11.

Mothers planning to return to work have several choices. Breast milk can be expressed during the day into sterile containers, refrigerated or frozen, and then used for subsequent bottle-feedings when they are at work. With onsite child care, a woman can breastfeed during breaks and lunch hours. Another possibility is to train the body to produce milk only when the mother is home during the evenings and at night. To train her body, a woman should omit one feeding at a time during the periods of the day when she will not be feeding or expressing milk. Doing this will help her reduce her milk supply without experiencing engorgement. Then she can gradually wean to the feedings at the appropriate time of the day. This method works because removal of milk is the stimulius for milk production. Generally, at least two feedings per day are needed for women to continue

Table 6.11   Important elements of worksite lactation support programs.[1]

- Prenatal lactation education tailored for working women.
- Corporate policies providing information for all employees on the benefits of breastfeeding and on why their breastfeeding co-workers need support.
- Education for personnel about the services available to support breastfeeding women.
- Adequate breaks, flexible work hours, job sharing, and part-time work.
- Private "Mother's Rooms" for expressing milk in a secure and relaxing environment.
- Access to hospital-grade, autocycling breast pumps at the workplace.
- Small refrigerators for the safe storage of breast milk.
- Subsidization or purchase of individually owned portable breast pumps for employees.
- Access to lactation professional on-site or by phone to give breastfeeding education, counseling, and support during pregnancy, after delivery, and when the mother returns to work.
- Coordination with on-site or near-site child-care programs so the infant can be breastfed during the day.
- Support groups for working mothers with children.

making milk. No evidence indicates that it is necessary to introduce a bottle sooner than 10 days before returning to work.[36] Unless a mother is returning to work immediately after delivery, a bottle should not be introduced before lactation is well established, which is usually at least four weeks.

Women should seek child-care providers or facilities that are supportive of breastfeeding. Supportive facilities provide accommodations for mothers who wish to breastfeed their children at the facility or have their children fed expressed milk. Facilities should also follow established child-care standards, including standards for the storage and handling of expressed breast milk.

The Community

To increase breastfeeding rates in a community, it is important to identify community attitudes and obstacles to breastfeeding and to solicit support for breastfeeding from community leaders. A multidisciplinary task force with representatives from physicians, hospitals and birthing centers, public health, home visitors, La Leche League, government, industry, school boards, and journalists can be an effective vehicle for assessing community breastfeeding support needs and sponsoring collaborative efforts to overcoming obstacles to breastfeeding.[123] Barriers to breastfeeding may include lack of access to reliable and culturally appropriate sources of information and social support, cultural perception of bottle-feeding as the norm, aggressive marketing of breast milk substitutes, and laws that prohibit breastfeeding in public. In the past decade, legislative efforts have been made to protect a woman's right to breastfeed in public and on federal property and to prevent workplace discrimination.

PUBLIC FOOD AND NUTRITION PROGRAMS

National Breastfeeding Policy

The U.S. Department of Health and Human Services (DHHS) is as the leading federal agency in the effort to promote, protect, and support breastfeeding for U.S. families. Over the past 15 years, the Office of the Surgeon General and the Maternal and Child Health Bureau have highlighted the public health importance of breastfeeding though numerous workshops and publications (Table 6.12). In 2000, the DHHS Blueprint for Action on Breastfeeding outlined a comprehensive framework to increase breastfeeding rates in the United States and to promote optimal breastfeeding practices. The action plan is based on education, training, awareness, support, and research.[1]

In the 1990s, the DHHS, through its Maternal and Child Health Bureau and the Centers for Disease Control, supported the establishment of the U.S. Breastfeeding Committee (USBC) to fulfill one of the goals identified in the *Innocenti Declaration:* each nation should establish "a multisectoral national breastfeeding committee composed of representatives from relevant government departments, nongovernmental organizations, and health professional associations." The USBC is a collaborative partnership of about 30 major organizations.

Many DHHS agencies have breastfeeding initiatives. The Title V Maternal and Child Health programs of the Health Resources and Services Administration provide substantial support services, training, and research for breastfeeding (http://www.mchb.hrsa.gov). The Centers for Disease Control and Prevention plays a major role in supporting breastfeeding nationally through applied research, program evaluation, and surveillance (http:www.cdc.gov). The National Institutes of Health provides substantial support for breastfeeding research, including $13 million in fiscal year 1998.[1]

Table 6.12   Landmark U.S. breastfeeding policy statements and conferences.

- Report of the Surgeon General's Workshop on Breastfeeding and Human Lactation[148]

- Follow-up Report: Surgeon General's Workshop on Breastfeeding and Human Lactation[149]

- Healthy People 2000 Breastfeeding Goals for the Nation[27]

- DHHS Maternal and Child Health Bureau National Workshop "Call to Action: Better Nutrition for Mothers, Children, and Families"[150]

- Second Follow-up Report: Surgeon General's Workshop on Breastfeeding and Human Lactation[151]

- 1998 National Breastfeeding Policy Conference, presented by the UCLA Center for Healthier Children, Families and Communities, Breastfeeding Resource Program in cooperation with the U.S. Department of Health and Human Services, Health Resources, and Services Administration, Maternal and Child Health Bureau, and the Centers for Disease Control and Prevention[152]

- Healthy People 2010 Breastfeeding Goals for the Nation[28]

- DHHS Blueprint for Action on Breastfeeding[1]

- U.S. Breastfeeding Committee Strategic Plan

USDA WIC Program

WIC (Special Supplemental Nutrition Program for Women, Infants, and Children) is a federal program operated by the U.S. Department of Agriculture (USDA) Food and Nutrition Service in partnership with state and local health departments. Created in 1972, WIC is designed to provide nutrition education, supplementary foods, and referrals for health and social services to economically disadvantaged women who are pregnant, postpartum, or caring for infants and children under five. WIC operates through a network of 50 state health departments, 32 Indian tribal organizations, five U.S. territories and 2000 local agencies providing services to more than 7.3 million program participants each month (http://www.fns.usda.gov/wic/).

In 1989, Congress mandated (Public Law 101-147) that a specific portion of each state's WIC budget allocation to be used exclusively for the promotion and support of breastfeeding among its participants and authorized the use of WIC administrative funds to purchase breastfeeding aids such as breast pumps. Through this legislation, each state

> **INNOCENTI DECLARATION**   The Innocenti Declaration on the Protection, Promotion, and Support of breastfeeding was produced and adopted by participants at the WHO/UNICEF policymakers' meeting on "Breastfeeding in the 1990s: A Global Initiative," held at the Spedale degli Innocenti, Florence, Italy on 1 August 1990. The Declaration established exclusive breastfeeding from birth to 4–6 months of age as a global goal for optimal maternal and child health.

has a breastfeeding coordinator and a plan to coordinate operations with local agency programs for breastfeeding promotion. Reauthorization legislation, the Healthy Meals for Healthy Americans Act of 1994 (Public Law 103-448), increased the budget allocation for breastfeeding promotion to $21 for each pregnant and breastfeeding woman. WIC's 1999 budget appropriation increased this amount further to $23.53. In 1998, WIC state agencies spent over $50 million for breastfeeding promotion.

The USDA hosts semiannual meetings of the U.S. Breastfeeding Consortium to exchange ideas on how the USDA, other federal agencies, and private health interests can work together to promote breastfeeding, especially in the WIC program. More than 25 organizations participate, including health professional associations, breastfeeding advocacy groups, and other federal agencies.

The USDA Food and Nutrition Information Center supports a WIC Works Web site to serve health and nutrition professionals working in the WIC program. The WIC Works site (http://www.nal.usda.gov/wicworks) includes an e-mail discussion group, links to training materials on breastfeeding promotion, and information on how to share resources and recommendations.

MODEL BREASTFEEDING PROMOTION PROGRAMS

WIC National Breastfeeding Promotion Project—Loving Support Makes Breastfeeding Work

Although breastfeeding initiation and duration rates among WIC participants increased in the 1990s, the rates remained considerably lower than among women in higher socioeconomic levels. In 1995, the USDA Food and Nutrition Service entered into a co-operative agreement with Best Start Social Marketing (a nonprofit organization assisting public health, education, and social service organizations with *social marketing* services) to develop a national WIC breastfeeding promotion project to be implemented at the state level. The project has four goals: (1) to encourage WIC participants to begin and to continue breastfeeding, (2) to increase referrals to WIC clinics for breastfeeding support, (3) to increase general public acceptance of and support for breastfeeding, and (4) to provide support and technical assistance to WIC professionals in promoting breastfeeding.[154]

SOCIAL MARKETING A marketing effort that combines the principles of commercial marketing with health education to promote a socially beneficial idea, practice or product.[122]

Best start collected data in 10 pilot states through observations, personal and telephone interviews, and focus groups with WIC participants and individuals who might influence their infant feeding decisions such as mothers, boyfriends and husbands, health care providers, and WIC staff.[155] The researchers looked for motivations and perceived barriers to breastfeeding as well as social network influences on feeding choice.

Results from the research were then used to develop a marketing plan and program material. In contrast to the traditional public health approach of addressing breastfeeding as a medical health decision, the marketing plan repositioned breastfeeding as a way for a family to establish a special relationship with their child from the very onset of its life.[121,155] The campaign slogan, "Loving Support Makes Breastfeeding Work," capitalizes on the concept that everyone is important to a woman's breastfeeding success—her family, friends, doctors, and community. Campaign materials and a counseling program were developed to help mothers work through individual barriers and constraints to breastfeeding.

In 1997, the campaign was implemented at the state level. Participating states could implement all or part of the campaign with technical assistance from the federal level. The states had the opportunity to purchase campaign materials including pamphlets, posters, and radio and television public service announcements that address barriers and encourage breastfeeding. Training was provided on coalition building, utilizing local media, promoting effective breastfeeding counseling strategies and techniques (Table 6.7), and managing peer counseling programs.[121] Since 1997, the Loving Support Makes Breastfeeding work campaign has expanded to other states; currently, 72 state agencies, Indian tribal organizations, and territories are participating at various levels.[156,157]

A preliminary evaluation of the program's impact in Mississippi found improvement in both breastfeeding rates and attitudes toward and awareness of breastfeeding. Prior to initiation of the Loving Support campaign in 1997, Mississippi ranked 50th in the nation in breastfeeding initiation and duration rates. Mississippi's campaign included a public awareness component with television and radio spots, newspaper ads, and billboards in high-traffic areas; a patient and family education program in health departments and WIC clinics; extensive outreach with health providers in the community and 25 hospitals; and community outreach including training at childcare centers and worksites. In 1999, the state moved into 48th place nationally in breastfeeding initiation rates at hospital discharge. Duration rates at six months, which climbed from 7.0% to 15.4% from 1998 to 1999, moved Mississippi WIC from 50th to 33rd in the country. Preliminary data from focus groups conducted across the state in 1999, with support from the USDA, revealed that the campaign had a significant impact on WIC program participants' breastfeeding knowledge and attitudes.[156]

Other states are currently assessing their individual campaigns. Based on data from the Ross Breastfeeding Survey, breastfeeding initiation rates in Iowa increased from 57.8% to 65.1% after a year of the campaign.[155] The rates of women still nursing six months after birth also increased; before the start of the campaign, 20.4% of women were still breastfeeding after six months compared with 32.2% a year after the program's start. Increased breastfeeding support from relatives and friends was also documented from data collected in a mail survey.

Wellstart International

Wellstart International is an independent, nonprofit organization headquartered in San Diego, California, that is dedicated to supporting the health and nutrition of mothers and infants worldwide through the promotion of breastfeeding. Wellstart provides education and technical assistance to perinatal health care providers and educators around the world, enabling them to promote maternal and child health in their own settings by supporting breastfeeding. Wellstart faculty and staff offer in-depth clinical and program expertise and both domestic and international experience to hospitals, clinics, and university schools of medicine, nursing, and nutrition, as well as to a wide variety of governmental and nongovernmental health and population agencies. Wellstart is a designated WHO Collaborating Center on Breastfeeding Promotion.

Since 1977 Wellstart has helped health care professionals establish self-sustaining breastfeeding promotion programs worldwide. One example is the National Training and Technical Support Center for Breastfeeding in Cairo, Egypt, established in collaboration with the Egyptian Ministry of Health and Population. The Support Center trains health workers and provides technical support for other related program areas. It is part of the countrywide Technical Support Collaboration involving applied research, evaluation of university curricula, countrywide training of health workers, efforts to achieve Mother-Baby Friendly Hospital status, community outreach, behavior change activities, and monitoring and evaluation. The Support Center also creates documents such as Counseling Guidelines on Infant Feeding for Use in Egypt and the Final Report on the Technical Support Collaboration on Infant Feeding and disseminates them to service providers and community workers.

By combining community outreach activities with the use of Information, Education, and Communication (IEC) materials targeting key behaviors, the Support Center in Cairo has achieved significant behavior changes at the local, regional, and national levels:[157,158]

- *Initiation of breastfeeding.* Behaviors related to initiation of the first breastfeeding improved (p <0.001), with the optimal behavior of initiating breastfeeding within the first hour increasing from 46% to 72% within the demonstration areas.

- *Exclusive breastfeeding.* Exclusive breastfeeding for all age groups increased (p <0.001), with levels increasing from 24% to 74% for the 0–3 month age group, from 8% to 55% for the 4–6 month age group, and from 15.5% to 65% for the combined 0–6 month age group.
- *Infants receiving water and herbal teas.* The use of water or herbal teas within the 0–3 and 4–6 month age groups declined (p <0.001), as would be expected if exclusive breastfeeding was increasing.
- *Bottle and Pacifier Use.* For infants 0–12 months of age, bottle use decreased from 49% to 12%; pacifier use decreased from 42% to 13.6% (p <0.001).
- *Complementary feeding.* Data related to timely introduction of complementary foods, as well as types of complementary foods received by infants, also showed substantial improvements (p <0.001). The percentage of infants 7–9 months of age receiving no complementary food decreased from 16% to 3%.

Conclusion

On a global level, we must do everything we can "to increase women's confidence in their ability to breastfeed. Such empowerment involves the removal of constraints and influence that manipulate perceptions and behavior towards breastfeeding, often by subtle and indirect means. Furthermore, obstacles to breastfeeding within the health care system, the workplace, and the community must be eliminated." These words are from the WHO/UNICEF Innocenti Declaration on the Protection, Promotion, and Support of Breastfeeding, Florence, Italy, 1990. As we have seen, breastfeeding is best for the vast majority of infants and is physiologically possible for the vast majority of women. The challenge is to overcome social and cultural barriers and to provide support systems at the local, national, and international level so that the initiation and duration of breastfeeding will continue to increase.

Resources

American Academy of Pediatrics
    Breastfeeding Promotion in Pediatric Office Practices Program
    Telephone: (847) 228-5005, extension 4779
    Web site: http://www.aap.org/visit/brpromo.htm

Baby-Friendly USA Hospital Initiative
    Telephone: (508) 888-8044
    Web site: http://www.aboutus.com/a100/bfusa/

Best Start Social Marketing
    Telephone: (813) 971-2119
    Web site: beststart@beststartinc.org

Best Start, Inc.
    Loving Support Campaign materials
    Best Start, Inc., Tampa, FL.
    Beststart@mindspring.com

Case Western Reserve University School of Medicine
Online Breastfeeding Basics Course
Web site: www.breastfeedingbasics.org

DHHS HRSA Maternal and Child Health Bureau
Telephone: (301) 443-0205
Web site: http://www.mchb.hrsa.gov

International Lactation Consultants Association
Telephone: (919) 787-5181
Web site: http://www. ilca.org

La Leche League
Telephone: (847) 519-7730
Web site: http://www. lalecheleague.org

National Center for Education in Maternal and Child Health
Telephone: (703) 524-7802
Web site: http://www.ncemch.org

USDA Women, Infants, and Children (WIC) Program
Telephone: (703) 305-2736
Web site: http://www.fns.usda.gov/wic

Wellstart International
Telephone: (619) 295-5192

References

1. U.S. Department of Health and Human Services. HHS blueprint for action on breastfeeding. Washington, DC: U.S. Department of Health and Human Services, Office on Women's Health; 2000.

2. Heinig MJ, Dewey KG. Health effects of breast feeding for mothers: a critical review. Nutrition Reviews 1997;10:59–73.

3. McNeilly AS. Lactational amenorrhea. Endocrinol Metab Clin North Am 1993; 10:35–56.

4. Kuzela AL, Stifter CA, Worobey J. W. Breastfeeding and mother-infant interactions. J Reprod Infant Psychol 1990;8: 185–94.

5. Newcomb PA, Egan KM, Titus-Ernstoff L, et al. Lactation in relation to postmenopausal breast cancer. Am J Epidemiol 1999;150:174–82.

6. Rosenblatt KA, Thomas DB. Lactation and the risk of epithelial ovarian cancer. The WHO Collaborative Study of Neoplasia and Steroid Contraceptives. Int J Epidemiol 1993;22:192–7.

7. American Academy of Pediatrics. Breastfeeding and the use of human milk. Pediatrics 1997;100:1035–9.

8. Lawrence RA, Lawrence RM. Breastfeeding: a guide for the medical profession. St. Louis: Mosby;1999.

9. Jensen RG. Handbook of milk composition. New York: Academic Press; 1995.

10. Picciano MF. Human milk: nutritional aspects of a dynamic food. Biol Neonate 1998;74:84–93.

11. Rogan WJ, Gladen BC. Breast-feeding and cognitive development. Early Hum Dev 1993;31:181–93.

12. Dewey KG, Heinig MJ, Nommsen-Rivers LA. Differences in morbidity between breast-fed and formula-fed infants. J Pediatr 1995;126:696–702.

13. Raisler J, Alexander C, O'Campo P. Breastfeeding: a dose-response relationship? Am J Pub Health 1999;89:25–30.

14. Ivarsson A, Persson LA, Nystrom L, et al. Epidemic of coeliac disease in Swedish children. Acta Paediatr 2000;89:165–71.

15. Koletzko S, Sherman P, Corey M, et al. Role of infant feeding practices in development of Crohn's disease in childhood. Bmj 1989;298:1617–8.

16. Smulevich VB, Solionova LG, Belyakova SV. Parental occupation and other factors and cancer risk in children: I. Study methodology and non-occupational factors. Int J Cancer 1999;83:712–7.

17. Oddy WH, Holt PG, Sly PD, et al. Association between breast feeding and asthma in 6 year old children: findings of a prospective birth cohort study. Brit Meds Bmj 1999;319:815–9.

18. Scariati PD, Grummer-Strawn LM, Fein SB. A longitudinal analysis of infant morbidity and the extent of breastfeeding in the United States. Pediatrics 1997;99:E5.

19. McVea KL, Turner PD, Peppler DK. The role of breastfeeding in sudden infant death syndrome. J Hum Lact 2000;16: 13–20.

20. Dewey KG. Growth characteristics of breast-fed compared to formula-fed infants. Biol Neonate 1998;74:94–105.

21. Hediger ML, Overpeck MD, Ruan WJ, et al. Early infant feeding and growth status of US-born infants and children aged 4–71 mo: analyses from the third National Health and Nutrition Examination Survey, 1988–1994. Am J Clin Nutr 2000;72: 159–67.

22. Splett PL, Montgomery DL. The economic benefits of breastfeeding an infant in the WIC program twelve month follow-up study. Washington, DC: Food and Consumer Service, US Department of Agriculture; 1998.

23. Ball TM, Wright AL. Health care costs of formula-feeding in the first year of life. Pediatrics 1999;103:870–6.

24. Cohen RJ, Brown KH, Canahuati J, et al. Determinants of growth from birth to

12 months among breast-fed Honduran infants in relation to age of introduction of complementary foods. Pediatrics 1995;96:504–10.

25. Medicine Io. Nutrition during lactation. Washington, DC: National Academy Press; 1991.

26. Division RP. Mothers survey: updated breastfeeding trend through 1996. Columbus, OH: Abbot Laboratories, 1998.

27. U.S. Department of Health and Human Services. Healthy People 2000: National Health Promotion and Disease Prevention Objectives. Washington, DC: U.S. Department of Health and Human Services, Public Health Service, Office of the Assistant Secretary for Health; 2000.

28. U.S. Department of Health and Human Services. Healthy People 2010: Conference Edition—vols. 1 and 2. Washington, DC: U.S. Department of Health and Human Services, Public Health Service, Office of the Assistant Secretary for Health; 2000.

29. Ryan AS. The resurgence of breastfeeding in the United States. Pediatrics 1997; 99:e12.

30. Hartmann PE. Changes in the composition and yield of the mammary secretion of cows during the initiation of lactation. J Endocrinol 1973;59:231–47.

31. Vorherr H. Human lactation and breastfeeding. In: Larson BL, ed. The mammary gland/human lactation/milk synthesis. Vol. 4. New York: Academic Press; 1978.

32. Cox DB, Owens RA, Hartmann PE. Blood and milk prolactin and the rate of milk synthesis in women. Exp Physiol 1996;81:1007–20.

33. Neville MC. The physiological basis of milk secretion. Part I. Basic physiology. Ann NY Acad Sci 1990;5:861–8.

34. Jensen RG. The lipids of human milk. Boca Raton, FL: CRC Press; 1989.

35. Neville MC. Lactogensis in women: a cascade of events revealed by milk composition. In: Jensen RG, ed. Handbook of milk

composition. New York: Academic Press; 1995.

36. Martin C, Krebs NF. The nursing mother's problem solver. New York: Fireside Publishing; 2000.

37. Butte NF, Garza C, Smith EO, et al. Human milk intake and growth in exclusively breast-fed infants. J Pediatr 1984;104:187–95.

38. Daly SE, Hartmann PE. Infant demand and milk supply. Part 1. Infant demand and milk production in lactating women. J Hum Lact 1995;11:21–6.

39. Dewey KG, Lonnerdal B. Infant self-regulation of breast milk intake. Acta Paediatr Scand 1986;75:893–8.

40. Daly SE, Owens RA, Hartmann PE. The short-term synthesis and infant-regulated removal of milk in lactating women. Exp Physiol 1993;78:209–20.

41. Saint L, Maggiore P, Hartmann PE. Yield and nutrient content of milk in eight women breastfeeding twins and one woman breastfeeding triplets. Br J Nutr 1986;56:49–58.

42. Newton M. Human lactation. In: Kon SK, ed. Milk: the mammary gland and it's secretion. New York: Academic Press; 1961.

43. Daly SEJ, Hartmann PE. Infant demand and milk supply. Part 2. The short-term control of milk synthesis in lactating women. J Hum Lact 1995;11:27–37.

44. Linzell JL, Peaker M. The effects of oxytocin and milk removal on milk secretion in the goat. J Physiol 1971;216: 717–34.

45. Lizuka T, Sasaki MO, Uemura S, et al. Nitric oxide may trigger lactation in humans. J Pediatr 1997;131: 839–43.

46. Hill PD, Aldag JC, Chatterton RTJ. Breastfeeding experience and milk weight in lactating mothers pumping for preterm infants. Birth 1999;26.

47. Niefert M, Seacat JM. Practical aspects of breastfeeding the premature infant. Perinatol Neonatol. 1988;12:24.

48. Meier P. Breastfeeding the preterm infant: breastfeeding and human lactation. Boston: Jones & Bartlett; 1999.

49. Food and Nutrition Board, Institute of Medicine. Dietary reference intakes for thiamin riboflavin, niacin, vitamin B_6, folate vitamin B_{12}, pantothenic acid, biotin, and choline. Washington, DC: National Academy Press; 1998.

50. Food and Nutrition Board, Institute of Medicine. Dietary reference intakes for calcium, phosphorus, magnesium, vitamin D, and fluoride. Washington, DC: National Academy Press; 1999: p 448.

51. Food and Nutrition Board, Institute of Medicine. Dietary reference intakes for vitamin C, vitamin E, selenium, and carotenoids. Washington, DC: National Academy Press; 2000.

52. Food and Nutrition Board, Institute of Medicine. Dietary reference intakes for vitamin A, vitamin K, arsenic, boron, chromium, copper, iodine, iron, manganese, molybdenum, nickel, silicon, vanadium, and zinc. Washington, DC: National Academy Press; 2001.

53. Food and Nutrition Board NRC. Recommended dietary allowances. Washington, DC: National Academy Press; 1989.

54. Almroth SG. Water requirements of breastfed infants in a hot climate. Am J Clin Nutr 1978;31:1154–8.

55. Axelson I, Borulf S, Righard L, et al. Protein and energy intake during weaning: I. effects on growth. Acta Paediatr Scand 1987;76:321–7.

56. Butte NF, Garza C, Johnson CA, et al. Longitudinal changes in milk composition of mothers delivering preterm and term infants. Early Hum Dev 1984;9:153–62.

57. Connor WE, Lowensohn R, Hatcher L. Increased docosahexaenoic acid levels in human newborn infants by administration of sardines and fish oil during pregnancy. Lipids 1996;31suppl:S183–7.

58. Insull W, Ahrens EH. The fatty acids of human milk from mothers on diets taken ad libitum. Biochem J 1959;72:27.

59. Agostoni C, Riva E, Trojan S, et al. Docosahexaenoic acid status and developmental quotient of healthy term infants. Lancet 1995;346:638.

60. Makrides M, Neumann MA, Gibson RA. Effect of maternal docosahexaenoic acid (DHA) supplementation on breast milk composition. Eur J Clin Nutr 1996; 50:352–7.

61. Wong WW, Hachey DL, Insull W, et al. Effect of dietary cholesterol on cholesterol synthesis in breast-fed and formula-fed infants. J Lipid Res 1993;34:1403–11.

62. Rosen JM, Rodgers JR, Couch CH, et al. Multihormonal regulation of milk protein gene expression. Ann NY Acad Sci 1986;478:63–76.

63. Velona T, Abbiati L, Beretta B, et al. Protein profiles in breast milk from mothers delivering term and preterm babies. Pediatr Res 1999;45:658–63.

64. Lonnerdal B, Atkinson S. Nitrogenous components of milk. In: Jensen RG, ed. Handbook of milk composition. New York: Academic Press; 1995: pp 351–68.

65. Newburg DS, Neubauer SH. Carboyhdrates in milks: analysis, quantities, and signficance. In: Jensen RG, ed. Handbook of milk composition. New York: Academic Press, 1995: pp 273–349.

66. Canfield LM, Giuliano AR, Neilson EM, et al. Beta-carotene in breast milk and serum is increased after a single beta-carotene dose. Am J Clin Nutr 1997;66: 52–61.

67. Rothberg AD, Pettifor JM, Cohen DF, et al. Maternal-infant vitamin D relationships during breast-feeding. J Pediatr 1982; 101:500–3.

68. Ala-Houhala M, Koskinen T, Parviainen MT, et al. 25-Hydroxyvitamin D and vitamin D in human milk: effects of supplementation and season. Am J Clin Nutr 1988;48:1057–60.

69. Lammi-Keefe CJ, Jonas CR, Ferris AM, et al. Vitamin E in plasma and milk of lactating women with insulin-dependent diabetes mellitus. J Pediatr Gastroenterol Nutr 1995;20:305–9.

70. Haug M, Laubach C, Burke M, et al. Vitamin E in human milk from mothers of preterm and term infants. J Pediatr Gastroenterol Nutr 1987;6:605–9.

71. Chappell JE, Francis T, Clandinin MT. Vitamin A and E content of human milk at early stages of lactation. Early Hum Dev 1985;11:157–67.

72. Lammi-Keefe CJ, Jensen RG. Fat-soluble vitamins in human milk. Nutr Rev 1984;42:365–71.

73. American Academy of Pediatrics. Vitamin K supplementation for infants. Pediatrics 1971;48:483.

74. Andon MB, Reynolds RD, Moser-Veillon PB, et al. Dietary intake of total and glycosylated vitamin B-6 and the vitamin B-6 nutritional status of unsupplemented lactating women and their infants. Am J Clin Nutr 1989;50:1050–8.

75. Salmenpera L, Perheentupa J, Siimes MA. Folate nutrition is optimal in exclusively breast-fed infants but inadequate in some of their mothers and in formula-fed infants. J Pediatr Gastroenterol Nutr 1986;5:283–9.

76. Kuhne T, Bubl R, Baumgartner R. Maternal vegan diet causing a serious infantile neurological disorder due to vitamin B_{12} deficiency. Eur J Pediatr 1991;150:205–208

77. Specker BL, Black A, Allen L, et al. Vitamin B-12: low milk concentrations are related to low serum concentrations in vegetarian women and to methylmalonic aciduria in their infants. Am J Clin Nutr 1990;52:1073–6.

78. Butte NF, Garza C, Smith EO, et al. Macro- and trace-mineral intakes of exclusively breast-fed infants. Am J Clin Nutr 1987;45:42–8.

79. Fransson GB, Lonnerdal B. Zinc, copper, calcium, and magnesium in human milk. J Pediatr 1982;101:504–8.

80. Duncan B, Schifman RB, Corrigan JJ, Jr., et al. Iron and the exclusively breast-fed infant from birth to six months. J Pediatr Gastroenterol Nutr 1985;4:421–5.

81. Pisacane A, DeVizia B, Vallente A, et al. Iron status in breast-fed infants. J Pediatr 1995;127:429–31.

82. Picciano MF, Guthrie HA. Copper, iron, and zinc contents of mature human milk. Am J Clin Nutr 1976;29:242–54.

83. Krebs NF, Hambidge KM. Zinc requirements and zinc intakes of breast-fed infants. Am J Clin Nutr 1986;43:288–92.

84. Atkinson SA, Whelan D, Whyte RK, et al. Abnormal zinc content in human milk: risk for development of nutritional zinc deficiency in infants. Am J Dis Child 1989;143:608–11.

85. McDaniel MR, Barker E, Lederer CL. Sensory characterization of human milk. J Dairy Sci 1989;72:1149–58.

86. Menella JA. Mother's milk: a medium for early flavor experiences. J Hum Lact 1995;11.

87. Menella JA, Beauchamp GK. Smoking and the flavor of breastmilk. N Engl J Med 1998;339:149–56.

88. Gerrish CJ, Menella JA. Flavor variety enhances food acceptance in formula-fed infants. Am J Clin Nutr 2001;73:1080–5.

89. Ziegler EE, Foman SJ, Nelson SE, et al. Cow milk feeding in infancy: Further observations on blood loss from the gastrointestinal tract. J Pediatr 1990;116:11–18.

90. Wooldrige MS, Fischer C. Colic, "overfeeding" and symptoms of lactose malabsorption in the breast-fed baby. Lancet 1988;2:382–4.

91. Butte NF, Garza C, O'Brian, et al. Human milk intake and growth in exclusively breast-fed infants during the first four months of life. Early Hum Dev 1985;12:291–300.

92. Specker BL, Valanis B, Hertzberg V, et al. Sunshine exposure and serum 25-hydroxyvitamin D concentrations in exclusively breast-fed infants. J Pediatr 1985;107:372–6.

93. Brams M, Maloney J. "Nursing bottle caries" in breast-fed children. J Pediatr 1983;103:415–6.

94. Palmer B. The influence of breastfeeding on the development of the oral cavity: a commentary. J Hum Lact 1998;14:93–8.

95. U.S. Department of Agriculture. Food guide pyramid. Washington, DC: U.S. Department of Agriculture, U.S. Department of Health and Human Services; 1992.

96. U.S. Department of Agriculture. Nutrition and your health: Dietary guidelines for Americans. Washington, DC: U.S. Department of Agriculture, U.S. Department of Health and Human Services; 1995.

97. Lust KD, Brown JE, Thomas W. Maternal intake of cruciferous vegetables and other foods and colic symptoms in exclusively breast-fed infants. J Am Diet Assoc 1996;96:46–8.

98. Butte NF, Garza C, Stuff JE, et al. Effect of maternal diet and body composition on lactational performance. Am J Clin Nutr 1984;39:296–306.

99. Brewer MM, Bates MR, Vannoy LP. Postpartum changes in maternal weight and body fat depots in lactating vs nonlactating women. Am J Clin Nutr 1989;49:259–65.

100. Sadurskis A, Kabir N, Wager J, et al. Energy metabolism, body composition, and milk production in healthy Swedish women during lactation. Am J Clin Nutr 1988;48:44–49.

101. Goldberg GR, Prentice AM, Coward WA, et al. Longitudinal assessment of energy expenditure in pregnancy by the doubly labeled water method. Am J Clin Nutr 1993;57:494–505.

102. Prentice AM, Poppitt SD, Goldberg GR, et al. Energy balance in pregnancy and lactation. In: Allen LH, King J, Lonnerda B, eds. Nutrient regulation during pregnancy, lactation and infant growth. New York: Plenum Press; 1994.

103. Roberts S, Cole T, Coward W. Lactational performance in relation to energy intake in the baboon. Am J Clin Nutr 1985;41:1270–6.

104. Macnamara JP. Role and regulation of metabolism in adipose tissue during lactation. J Nutr Biochem 1995;6:120–9.

105. Butte NF, Hopkinson JM. Body composition changes during lactation are highly variable among women. J Nutr 1998;128:381S–5S.

106. Strode MA, Dewey KG, Lonnerdal B. Effects of short-term caloric restriction on lactational performance of well-nourished women. Acta Paediatr Scand 1986;75:222–9.

107. Dusdieker LB, Hemingway DL, Stumbo PJ. Is milk production impaired by dieting during lactation? Am J Clin Nutr 1994;59:833–40.

108. Lovelady CA, Lonnerdal B, Dewey KG. Lactation performance of exercising women. Am J Clin Nutr 1990;52:103–9.

109. Dewey KG, McCrory MA. Effects of dieting and physical activity on pregnancy and lactation. Am J Clin Nutr 1994;59:446S–52S; discussion 452S–3S.

110. Kreiter SR, Schwartz RP, Kirkman HNJ, et al. Nutritional rickets in African American breast-fed infants. J Pediatr 2000;137:143–5.

111. Wardinsky TD, Montes RG, Friederich RL, et al. Vitamin B12 deficiency associated with low breast-milk vitamin B12 concentration in an infant following maternal gastric bypass surgery [letter]. Arch Pediatr Adolescent Med 1995;149:1281–4.

112. Renault F, Verstrichel P, Plousard J, et al. Neuropathy in two cobalmin-deficient breast-fed infants of vegetarian mothers. Muscle Nerve 1999;22:252–4.

113. Hardy ML. Herbs of special interest to women. J Am Pharm Assoc 2000;40:234.

114. Humphrey SL, McKenna DJ. Herbs and breastfeeding. Breastfeeding Abstracts 1997;17:11–2.

115. Foote J, Rengers B. Medicinal herb use during pregnancy and lactation. Perinatal Nutr Report 1998;5:1–4.

116. Humphrey S. Sage advice on herbs and breastfeeding. LEAVEN 1998;34:43–7.

117. Tyler VE. Herbs of choice: the therapeutic use of phytomedicinals. New York: Pharmaceutical Products Press; 1994.

118. Tyler V. What pharmacists should know about herbal remedies. J Am Pharm Assoc 1996;36:29–37.

119. Cunningham E, Hansen K. Question of the month: where can I get information on evaluating herbal supplements? J Am Diet Assoc 1999;99:1240.

120. Leung A, Foster S. Encyclopedia of common natural ingredients used in foods, drugs, and cosmetics. New York: John Wiley & Sons; 1996.

121. Bryant C, Roy M. Best Start's three step counseling strategy. Tampa, FL: Best Start; 1997.

122. Bryant C, Coreil J, D'Angelo SL, et al. A strategy for promoting breastfeeding among economically disadvantaged women and adolescents. NAACOG's Clin Issu Perinat Women's Health Nurs 1992;3: 723–30.

123. Lazarov M, Evans A. Breastfeeding—Encouraging the best for low-income women. Zero to Three 2000:15–23.

124. American Dietetic Association. Position of the American Dietetic Association: promotion of breast-feeding. J Am Diet Assoc 1997;97:662–6.

125. Caulfield LE, Gross SM, Bentley ME, et al. WIC based interventions to promote breastfeeding among African-American women in Baltimore: effects on breastfeeding initiation and continuation. J Hum Lact 1998;14:15–22.

126. Freed GL, Clark SJ, Sorenson J, et al. National assessment of physicians' breastfeeding knowledge, attitudes, training, and experience. JAMA 1995;273:472–6.

127. Moreland JC, Lloyd L, Braun SB, et al. A new teaching model to prolong breastfeeding among Latinos. J Hum Lact 2000;16:337–41.

128. McCamman S, Page-Goetz S. Breastfeeding success: you can make the difference. Perinatal Nutr Report 1998;4:2–4.

129. Cadwell K. Reaching the goals of "Healthy People 2000" regarding breastfeeding. Clin Perinatol 1999;26:527–37.

130. Bentley ME, Caulfield LE, Gross SM, et al. Sources of influence on intention to breastfeed among African-American women at entry to WIC. J Hum Lact 1999;15:27–34.

131. Barron SP, Lane HW, Hannan TE, et al. Factors influencing duration of breast feeding among low-income women. J Am Diet Assoc 1988;88:1557–61.

132. Freed GL, Fraley JK, Schanler RJ. Attitudes of expectant fathers regarding breast-feeding. Pediatrics 1992;90:224–7.

133. Valaitis RK, Sheeshka JD, O'Brien MF. Do consumer infant feeding publications and products available in physicians' offices protect, promote, and support

breastfeeding? J Hum Lact 1997; 13:203–8.

134. Howard CR, Schaffer SJ, Lawrence RA. Attitudes, practices, and recommendations by obstetricians about infant feeding. Birth 1997;24:240–6.

135. Howard C, Howard F, Lawrence R, et al. Office prenatal formula advertising and its effect on breast-feeding patterns. Obstet Gynecol 2000;95:296–303.

136. World Health Organization. Contemporary patterns of breast-feeding. In: Report on the WHO collaborative study on breastfeeding. Geneva: World Health Organization; 1981.

137. American College of Obstetricians and Gynecologists. Breastfeeding: maternal and infant aspects. ACOG Educational Bulletin. Vol. 258. Washington, DC: American College of Obstetricians and Gynecologists; 2000.

138. American College of Nurse-Midwives. Clinical practices statement on breastfeeding. Washington, DC: American College of Nurse-Midwives; 1992.

139. American Academy of Family Physicians. Breastfeeding and infant Nutrition, 1994. AAFP reference manual 1998–1999. Washington, DC: American Academy of Family Physicians; 1999: p 51.

140. Perez-Escamilla R, Pollitt E, Lonnerdal B, et al. Infant feeding policies in maternity wards and their effect on breast-feeding success: an analytical overview. Am J Public Health 1994;84:89–97.

141. U.S. Department of Health, Education, and Welfare. Healthy people: The surgeon general's report on health promotion and disease prevention. Washington, DC: U.S. Department of Health, Education, and Welfare; 1979.

142. Saadeh R, Akre J. Ten steps to successful breastfeeding: a summary of the rationale and scientific evidence. Birth 1996;23:154–60.

143. World Health Organization. Protecting, promoting, and supporting breastfeeding: the special role of maternity services (a joint WHO/UNICEF statement). Geneva, Switzerland: WHO/UNICEF; 1989.

144. U.S. Department of Labor. Futurework: trends and challenges for work in the 21st century. Washington, DC: U.S. Department of Labor; 1999.

145. Kurinu N, Shiono, PH, Ezrine, SF, et al. Does maternal employment affect breast-feeding? Am J Public Health 1989; 79:1247–50.

146. National Healthy Mothers, Healthy Babies Coalition. Workplace models of excellence 2000: outstanding programs supporting working women that breastfeed. Alexandria, VA: National Healthy Mothers, Healthy Babies Coalition; 2000.

147. Medela I. Sanvita programs introductory pamphlet. McHenry, IL: 1994.

148. U.S. Department of Health and Human Services. Report of the surgeon general's workshop on breastfeeding and human lactation. Washington, DC: U.S. Department of Health and Human Services; 1984.

149. U.S. Department of Health and Human Services. Followup report: the surgeon general's workshop on breastfeeding and human lactation. Washington, DC: U.S. Department of Health and Human Services, Public Health Service, Health Resources and Services Administration; 1985.

150. Sharbaugh CS. Call to action: better nutrition for mothers, children, and families. Washington, DC: National Center for Education in Maternal and Child Health; 1990.

151. Spisak SG, Second followup report: the surgeon general's workshop on breastfeeding and human lactation. Washington, DC: National Center for Education in Maternal and Child Health; 1991.

152. Slusser WL, Thomas S. Report of the national breastfeeding policy conference. UCLA Center for Healthier Communities, Families and Children; 1999.

153. UNICEF/WHO. Innocenti Declaration on the Protection, Promotion and Support of Breastfeeding. Florence, Italy: UNICEF and WHO; 1990.

154. USDA Press Release No. 0283.97. Glickman kicks off USDA campaign to promote national breastfeeding week. 1997 Aug 6.

155, Andreason A. Marketing social change: changing behavior to promote health, social development, and the environment. San Francisco: Jossey-Bass, 1995.

156. Khoury A. Social Marketing Institute success stories: national WIC breastfeeding promotion project. http://www.social-marketing.org/success/cs-nationalwic.html: Mississippi WIC Program, 2001.

157. Carothers C. Social Marketing Institute success stories: national WIC breastfeeding promotion project.: Best Start Social Marketing, 2001.

158. Personal communication. Amal Khoury, PhD Mississippi WIC Program, 2001.

159. Personal communication. Cathy Carothers, Best Start Social Marketing, Inc, 2001.

CHAPTER 7

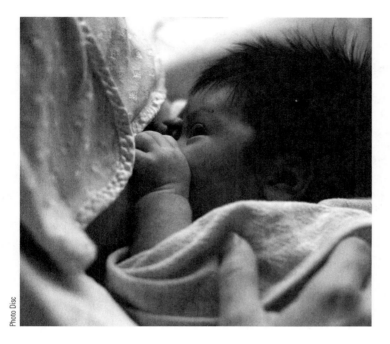

Photo Disc

The establishment of breastfeeding for at least six months, but optimally for at least a year, as a cultural norm supported by medical, social, and economic practices is a fundamental cornerstone of true wellness promotion.

American Dietetic Association, 2001

NUTRITION DURING LACTATION:

Conditions and Interventions

Prepared by **Maureen A. Murtaugh** with **Carolyn Sharbaugh** and **Denise Sofka**

CHAPTER OUTLINE

- Introduction
- Common Breastfeeding Conditions
- Maternal Medications
- Hyperbilirubinemia and Jaundice
- Breastfeeding Multiples
- Infant Allergies
- Human Milk and Preterm Infants
- Medical Contraindications to Breastfeeding
- Human Milk Collection and Storage
- Model Programs

KEY NUTRITION CONCEPTS

1 Human milk is the preferred feeding for all premature and sick newborns with rare exceptions.

2 Breastfeeding women need consistent, informed, and individualized care in the hospital and at home after discharge.

3 It is usually not necessary to discontinue breastfeeding to manage medical problems of the mother or infant; any medical decision to limit a mother's breastfeeding must be justified by the fact that the risk to her baby clearly outweighs the benefits of breastfeeding.

4 Feeding infants early in the postdelivery period whenever possible is important to successful breastfeeding. Early intervention to address questions or problems is equally important for maintaining breastfeeding.

5 Most medications, including over-the-counter as well as prescription drugs, drugs of abuse, alcohol, nicotine, and herbal remedies taken by nursing mothers are excreted in breast milk.

6 Twins and other multiples can be successfully breastfed without formula supplementation.

169

INTRODUCTION

The key to successful breastfeeding management is for the mother-infant breastfeeding dyad to receive support and informed, consistent, and individualized care from health care professionals both in the hospital and after discharge. The vast majority of women do not experience significant problems with breastfeeding, and many of the more common problems that do arise can be prevented through prenatal breastfeeding education and a positive, supportive breastfeeding initiation period.

This chapter discusses prevention and treatment of common breastfeeding conditions. Issues related to maternal use of medications, herbal remedies, and drugs of abuse are addressed. The chapter presents important considerations for breastfeeding multiples, preterm infants, and infants with medical problems. It provides information on the safe collection and storage of human milk and milk banks. The chapter concludes with case studies providing examples of management of challenging breastfeeding problems and with examples of model programs promoting support for breastfeeding in the health care system.

COMMON BREASTFEEDING CONDITIONS

Sore Nipples

Almost all women have some nipple pain when initiating nursing. Some women have pain when the infant first latches on, but pain should not be considered a normal part of breastfeeding. The first important step to prevent pain is proper positioning of the baby on the breast. The areola should be in the baby's mouth and the baby's tongue should be extended and against the lower lip. If a woman is experiencing pain, a lactation consultant or a health care professional well-trained in lactation should observe the mother nursing her baby. The lactation consultant can determine whether the pain is simply related to early breastfeeding, or if a problem exists. Sore nipples can result from poor positioning of the infant at the breast, infection (thrush or staphyloccus aureus), pumping with too much suction, or a problem with the infant's suck[1].

Women can take simple steps to prevent or manage nipple pain. Martin and Krebs[2] recommend that women let their breasts air dry after nursing, rub expressed milk and an all-purpose ointment (not petroleum based) on nipples, and use warm compresses on sore nipples. Use of a pump to express milk can help to maintain supply if the nipples are so sore that the mother cannot nurse. However, the suction on the breast pump should be adjusted carefully. High suction can make nipples sore and red.

Letdown Failure

After she latches on, take deep, long breaths—think yoga not Lamaze when you nurse. As you exhale, visualize the milk letting down through your breasts into the baby's mouth.

C Martin and NF Krebs[2]

Letdown failure is not common, but because letdown is necessary to successful breastfeeding, it is important to address the matter. Oxytocin nasal spray can be prescribed by a physician. The synthetic oxytocin is sprayed into the nose and stimulates letdown, but it can only be used for a few days to help women get through a tough time. Other methods should be used at the same time to stimulate letdown. Martin and Krebs recommend a number of techniques to help women relax and enhance letdown:[2]

- Play soothing music that the mother can focus on while nursing.
- Have the partner rub his knuckles down her spine.
- Try different nursing positions.
- Get out of the house. Most babies enjoy a walk.
- Arrange for some time alone (a few hours).
- Switch from caffeinated to decaffeinated beverages and water for a few weeks.

Overactive Letdown

Overactive letdown can also be a problem, especially among first-time mothers. When letdown is overactive, milk streams from the breast when feeding begins. Milk may also leak from the breast that the infant is not being nursed from. The milk streams quickly and the infant may be overwhelmed by the volume. The infant may choke or gulp to keep up with the flow. When the infant gulps, she takes in air and then may be fussy.

In order to manage overactive letdown, the mother should express milk until the flow slows and then put the infant to breast. (Expressed milk can be frozen for later use.) Expressing milk also allows infants to get hindmilk and prevents gas and colic that may result from a large volume of relatively low-fat milk with high lactose content.

Engorgement

Engorgement, when breasts are overfilled with milk, is common in first-time mothers. This occurs when the supply and demand process is not yet established and the milk is abundant. The best way to prevent engorgement is to nurse the infant frequently. (Newborn infants will often nurse every hour and a half.) If the infant is not available to nurse, expressing milk every

Case Study 7.1
Chronic Mastitis

This was the first and unremarkable pregnancy for BA (29 years old). The only pregnancy problem reported was a vaginal yeast infection during the eighth month, which was treated with an anti-fungal agent for seven days. BA reported experiencing "a little" breast enlargement during her pregnancy. She had a 17-hour labor, with a cesarean delivery after 3 hours of pushing.

Her infant was first put to the breast at two hours postpartum, and the infant latched well and suckled vigorously. The infant nursed every two hours over the first three to four days postpartum. BA's breasts became noticeably fuller during the third postpartum day and by the fourth postpartum day were painfully engorged. In addition, BA reported painful, burning, cracked nipples. The engorgement made it difficult for her baby to latch at the breast. The baby became irritable, and BA experienced a significant amount of pain. A lactation consultant provided BA with guidelines for engorgement management. Her advice included breast massage and hand expression prior to feeds to soften the areolar area, enabling the infant to achieve a deep latch. BA was advised to nurse frequently and to use cool compresses after feeds.

On day 5 the engorgement was still causing discomfort. Her nipples had become more cracked, abraded, and painful. The lactation consultant noted that the infant's latch had become shallow and tight, probably in an attempt to control the flow of milk. BA was advised to use varying positions to nurse to help relieve the sore areas on the nipple. A breast pump was provided to better facilitate softening the breasts prior to feedings. The infant showed all the signs of adequate intake, including 10 very wet and 3 soiled diapers during the 24 hours prior to the consultation.

By day 7 postpartum, BA had mastitis. She was treated with a 7-day course of dicloxacillin. The lactation consultant continued to advise BA to massage her breasts, apply warm compresses, and to express enough to soften the areolar area prior to feeding her infant. Further assistance was provided in achieving a proper latch.

By day 14 BA was feeling much better. The mastitis had resolved and her nipples were healing. She still had tenderness during feeds and a healing crack on the right side. Her breasts were still uncomfortably full and had occasional swelling and tenderness in some areas.

At three weeks postpartum BA developed an inflamed area on the right breast that remained red and tender despite applying warmth and massage to the area. The lactation consultant helped BA to position the infant in a way that allowed drainage of the inflamed area. She was treated with dicloxacillin. The crack on the right nipple had improved, but was still not completely healed. BA continued to show signs of oversupply such as breasts feeling uncomfortably full, even after feeding, and excessive milk leakage between feeds. The lactation consultant suggested that she nurse in a reclined position to reduce the milk flow to her baby. After 10 days of persistent burning pain in the nipple area, BA was seen and treated with fluconazole for her yeast infection. Seven days after starting the fluconazole, a topical nystatin ointment was prescribed for her nipples and an oral suspension for her infant.

At seven weeks postpartum, BA called the lactation consultant to report another flare-up of bacterial mastitis. Her health care provider prescribed a 10-day course of dicloxacillan. BA was still treating her nipples with nystatin ointment. At eight weeks postpartum her mastisis resolved, her nipple pain is still present, but improving. She nurses on one side only per feed and reports that the infant latches better when she is in a more reclined position.

Adapted from: Anonymous. Case management of a breastfeeding mother with persistent oversupply and recurrent breast infections.[114]

few hours will prevent engorgement while helping to build and maintain a milk supply.

Once engorgement occurs, there are several simple treatments to help ease the discomfort. It is important for the mother to express milk until her breasts are no longer hard before putting the infant to breast. This will make it more comfortable for her and easier for her baby to latch on. Women can use analgesics to reduce pain from engorgement. A warm shower, gently massaging the breast, and expressing milk will help to relieve pressure.

Plugged Duct

Milk stasis, or milk remaining in the ducts, is considered the cause of plugged ducts.[3] Treatment for plugged ducts is gentle massage, warm compresses, and complete emptying of the breast. Women should consider changing nursing positions to facilitate emptying the breast. When plugging occurs repeatedly, a gentle manual massage before nursing often results in the plug being expelled.

Infection

Mastitis is a bacterial infection of the breast most commonly found in breastfeeding women. Some women get mastitis after having cracked or sore nipples, and some get it without any noticeable problem on the surface of the breast, probably from a bloodborne source of infection. Missing a feeding or the infant sleeping through the night may precipitate engorgement, plugged ducts, and then mastitis.[4] Symptoms of mastitis are similar to those of a plugged duct (Table 7.1). In both conditions, there is a painful, enlarged, hard area in the breast, and often an area of redness on the surface of the breast. Cases of mastitis are usually accompanied by a fever and flu-like symptoms.

It is important for the mother to seek early treatment and to continue nursing through mastitis, unless it is too painful. The techniques used to minimize pain from engorgement may also be used for mastitis. Acetaminophen is commonly recommended to help with the pain. The pairing of antibiotics with emptying of the breast to treat bacterial mastitis is important.[5] In a randomized trial, half of 55 women treated only with antibiotics had breast abscess, recurrent mastitis, or symptoms lasting greater than two weeks compared to only 2 of 55 who also emptied their breasts (breasts can be emptied by feeding the baby or pumping). Significant delays in seeking treatment for mastitis are associated with the development of abscess and recurrent mastitis.[5,6] See also Case Study 7.1.

MATERNAL MEDICATIONS

It is equally inappropriate to discontinue breastfeeding when it is not medically necessary to do so as it is to continue breastfeeding while taking contraindicated drugs.

RA Lawrence[3]

The single most common medical issue physicians face in managing breastfeeding patients is maternal medication use.[3] Most medications—including over-the-counter as well as prescription drugs—taken by nursing mothers are excreted in breast milk, yet data on drug safety is meager and sometimes conflicting. Two key questions to address in the analysis of the risk of an infant's exposure to a drug excreted in breast milk are: How much of the drug is excreted in milk, and at this level of excretion, what is the risk of adverse effects?[7] Among the numerous variables to examine to answer these questions are: the pharmacokinetic properties of the drug; time-averaged breast

Table 7.1 Comparison of symptoms of engorgement, plugged duct and mastitis.

CHARACTERISTICS	ENGORGEMENT	PLUGGED DUCT	MASTITIS
Onset	Gradual most common early postpartum	Gradual after feedings	Sudden, after 10 days or more postpartum
Site	Both breasts	One breast	Usually one breast
Swelling and Heat	Generalized	May change positions, little or no heat	Red, hot, swollen, area on breast
Pain	Generalized	Mild but in a specific location	Intense in a specific location
Fever	No fever	No fever	Fever (>101° F)
Other Symptoms	None	None	Flu-like symptoms

Adapted from Lawrence and Lawrence 1999, Table 8-5.[3] Used by permission.

milk/plasma ratio; the drug *exposure index;* the infant's ability to absorb, detoxify, and excrete the agent; the dose, strength, and duration of dosing; and the infant's age, feeding pattern, total diet, and health.[3,7,8] Additional considerations are the well-established interethnic and racial differences in drug responsiveness, exposure of the infant to the drug during pregnancy, and whether the drug can be safely given to the infant directly.[3] With so many active variables, carefully controlled studies on large enough samples to validate the results are rare.

Fortunately, numerous resources (Table 7.2) based on a thorough evaluation of available evidence can assist the practitioner and mother in identifying which drugs are safe and which are not. The American Academy of Pediatrics (AAP) Committee on Drugs publishes guidelines for practitioners.[9] The guidelines provide a list of drugs divided into the following seven categories according to risk factors in relationship to breastfeeding:

1. Drugs contraindicated during breastfeeding
2. Drugs of abuse that are contraindicated during breastfeeding
3. Drugs that require temporary cessation of breastfeeding
4. Drugs whose effect on nursing infants is unknown, but may be of concern
5. Drugs that have been associated with significant effect on some nursing infants and should be given to nursing mothers with caution
6. Maternal medications usually compatible with breastfeeding
7. Food and environmental agents having no effect on breastfeeding

The list, which is updated periodically, includes only those drugs about which there is published information. Other useful monographs and review articles provide additional information on a wide array of medications.[7,10-13]

The Breastfeeding and Human Lactation Study Center at the University of Rochester (see Table 7.2) continually updates its database of more than 3000 references on drugs, medications, and contaminants in human milk and is a resource for complex questions on the risks to the breastfed infant. The Physician's Desk Reference (PDR) is not a good source for information about drugs and breastfeeding because the information is derived directly from pharmaceutical companies whose first concern is avoiding liability. When there are no studies that prove beyond a doubt that a drug is safe for nursing mothers, the drug companies must advise against use while breastfeeding—even if what is known about the drug suggests that there is little cause for concern.

Only a few medications are thought to be contraindications to breastfeeding: chemotherapeutic agents, radioactive isotopes, drugs of abuse, lithium, ergotamine, and drugs that suppress lactation.[8,9] In addition anticonvulsants, antihistamines, sulfa drugs, and salicylates may have effects on some breastfeeding infants.[8] Fortunately, safer alternative medications can be recommended as a substitute in most cases.

Many women have questions on the safety of oral contraceptive use during lactation. Preparations containing 2.5 mg or less of a 19-norprogestogen and 50 mcg or less of ethinyl estradial or 100 mcg or less of mestranol present no problem for the mother or infant.[3,8,9] Milk yield can be decreased with larger doses.

If a drug or surgery is elective, a mother may be able to delay it until the baby is weaned. If a breastfeeding mother needs a specific medication, and the hazards to the infant are minimal, she should be instructed to take the medication after breastfeeding, at the lowest effective dose, and for the shortest duration[11] Other important steps can be taken to further minimize the effects (Table 7.3). It is also sometimes possible to choose alternative routes for administration of a medication to reduce exposure. For example, prescribing an inhalant instead of a drug

> **MILK TO PLASMA DRUG CONCENTRATION RATIO (M/P RATIO)** The concentration of the drug in the milk versus the concentration in maternal plasma. Since the ratio varies over time, a time-averaged ratio provides more meaningful information than data obtained at a single time point.[7]
>
> **EXPOSURE INDEX** The average infant milk intake per kilogram body weight per day × (the milk to plasma ratio divided by the rate of drug clearance) × 100. It is indicative of the amount of the drug in the breast milk that the infant ingests and is expressed as a percentage of the therapeutic (or equivalent) dose for the infant.[7]

Table 7.2 Resources on drugs, medications, and contaminants in human milk.

- American Academy of Pediatrics, Committee on Drugs. The Transfer of Drugs and Other Chemicals Into Human Milk (RE9403). Pediatrics 93:137, 1994.[9]

- Briggs GG, Freeman RK, Yaffe SJ. 1998 *Drugs in Pregnancy and Lactation* (5th ed.). Baltimore, MD: Williams and Wilkins.[14]

- Hale TS. 1999. *Medications and Mothers' Milk* (8th ed.). Amarillo, TX: Pharmasoft Medial Publishing.[13]

- Finger Lakes Regional Poison and Drug Information Center (716-275-3232 or 800-333-0542). This service, available 24 hours a day, is staffed by physicians, nurses, pharmacists, and clinical toxicologists.

- The Breastfeeding and Human Lactation Center, University of Rochester. This service is available for complex medication questions (8 to 4 PM EST, Monday to Friday at 716-275-0088).

Table 7.3 Minimizing the effect of maternal medication.[3]

1. Avoid long-acting forms: Accumulation in the infant is a genuine concern because the infant may have more difficulty excreting a long-acting form of a drug, which usually requires detoxification in the liver.

2. Schedule doses carefully: Check usual absorption rates and peak blood levels of the drug and schedule the doses so that the least amount possible gets into the milk. In order to minimize milk levels of most drugs, the safest time for a mother to take the drug is usually immediately after her infant nurses.

3. Evaluate the infant: Watch for any unusual signs or symptoms, such as changes in feeding pattern or sleeping habits, fussiness, or rash.

4. Choose the drug that produces the least amount in the milk.

Table 7.4 Herbs traditionally used to affect milk production.

Herbs That Promote Milk Flow:
Caraway
Celery root and seed
Chaste tree berry
Fennel
Fenugreek
Goat's rue
Raspberry
Rauwolfia
Verbena

Herbs That Reduce Milk Flow:
Castor bean
Jasmine flower
Sage

Reprinted from Hardy, Mary L. Herbs of Special Interest to Women. Journal of the American Pharmaceutical Association, 40: 234, 2000.[16] Used by permission.

taken by mouth, or a topical application rather than oral dosing, reduces infant exposure. If a drug is to be taken for diagnostic testing (such as a radioactive agent) a mother may need to withhold breastfeeding for a short period of time, pumping and discarding her milk. Discontinuing breastfeeding due to maternal medications is a last resort but may be necessary for the health and well-being of the mother, for example if she needs chemotherapy or radioactive treatment.[15] Any decision to limit a mother's breastfeeding must be justified by the fact that the risk to her baby clearly outweighs the benefits of breastfeeding.

Herbal Remedies

Numerous herbs have been used in folk and traditional systems of healing to affect the flow of milk (Table 7.4), or treat mastitis, infant colic, and thrush.[16,17,18] However, as discussed in the previous chapter, pertinent safety data are limited, and the risks of using herbal remedies can outweigh the perceived benefits.[19,20,21] Medicinal herbs should be viewed as drugs, with evaluation of both their pharmacological and toxicological potential. In addition, mothers need to understand that the active ingredients, potency, and purity of herbal preparations sold in the United States are all unknown since they are sold as dietary supplements and not regulated as medications.[18]

Although health care practitioners may wish otherwise, some mothers may refuse prescription drugs and insist on using herbal alternatives. In balancing the risks and benefits in a given situation, consideration should be given to the benefits of continued breastfeeding to the baby and the mother. It is also important to consider the varied nature of lactation: newborns face different risks than older babies or tod-

dlers because of immaturity; infants consume varying amounts of human milk; mothers may be looking forward to many months of lactation yet need or desire the benefits of medicinal herbs. A few of the widely used herbs in the United States are discussed below. Information on other medicinal herbs can be found in several resources.[22–26]

ECHINACEA Echinacea is used for the common cold and to enhance the immune system. While no data exists about its entry into human milk, it has no known side effects.[3] In modest amounts, Echinacea may be even safer than many over-the-counter medicines for the common cold.[3]

GINSENG ROOT Ginseng root, widely believed to increase capacity for mental work and physical activity and to reduce stress, contains dozens of steroid-like glycosides, sterols, coumarins, flavonoids, and polysaccharides.[3] It is reported to have estrogen-like effects in some women, with mastalgia common with extended use and mammary nodularity also reported. While there has been considerable animal experimentation with ginseng root, human data is not extensive. The lack of standardized preparations, information on dosage, and accurate recording of side effects is a problem. According to Lawrence, it may be problematic during lactation because of the reported breast effects and occasional reports of vaginal bleeding.[3]

ST. JOHN'S WORT St. John's wort, widely used in the United States and Europe as an antidepressant and anxiolytic, has the potential to suppress lactation.

One of its active ingredients, hypericin, inhibits dopamine beta hydroxylase. This could lead to increased dopamine, increased prolactin inhibitory factor, and suppression of prolactin.[3] No clinical studies have investigated the effect of this herb on lactation. There is little conclusive evidence on the transfer of St. John's wort into breast milk or its effects on the newborn.[18] Antidepressants with an established safety profile during breastfeeding may be preferable alternatives.

CABBAGE LEAVES   Cabbage leaves (either cool or at room temperature) may be beneficial in relieving breast engorgement. They have been reported to reduce discomfort and swelling, although it is not known how the effects are mediated.[27] A randomized trial to evaluate the use of crushed cabbage leaves packed in the bra to prevent engorgement failed to find significant differences in engorgement, although the average duration of exclusive breastfeeding was longer.[28]

Alcohol

Avoid prescribing or proscribing it [alcohol] and . . . assist the mother in appropriately adjusting her alcohol consumption in both timing and volume.

R Lawrence[3]

The harmful effects of alcohol consumption during pregnancy are well documented, and drinking during pregnancy is clearly not recommended. Recommendations on alcohol consumption during lactation, however, are less clearcut and are controversial. Alcohol consumed by the mother passes quickly into her breast milk, and the effects on the breastfeeding baby are directly related to the amount the mother consumes.[29]

Research shows that lactating and nonlactating women process alcohol differently; lactating women's blood levels tend to peak lower after drinking and clear more quickly than nonlactating women. Alcohol passes quickly between the maternal blood and breast milk; the level of alcohol in breast milk matches the maternal plasma levels at the time of the infant feeding. Peak maternal plasma and breast milk levels are reached 30 to 60 minutes after alcohol consumption[30] and at 60 to 90 minutes when taken with food.[3] As the alcohol clears from a mother's blood, it clears from her milk. It takes a 120–pound woman about two to three hours to eliminate from her body the alcohol in one serving of beer or wine.[29]

While beer and wine have been recommended historically in many cultures to enhance lactation, there is evidence of a negative dose-related impact of alcohol on maternal oxytocin levels and the milk letdown reflex. At least a partial decrease in milk letdown is seen at 1.0 to 1.5 g/kg, and at higher levels all women have a partial to complete block in milk ejection.[3] The effect of alcohol on prolactin levels and milk production has not been studied in lactating women, although a study on males and nonlactating women showed the ingestion of beer increased the levels of prolactin in the blood within 30 minutes after it was ingested.[31,32] There is evidence that maternal alcohol consumption affects the odor of breastmilk and the volume consumed by the infant. Maternal alcohol consumption (0.3 g ethanol/kg) reduced infant breast milk intake by about 20%,[30] although infants appeared to suckle more vigorously.

Recent studies on the impact of maternal alcohol ingestion during lactation on infant sleep patterns and psychomotor development have raised concerns about regular consumption of alcohol while lactating. In one study, 11 of 13 breastfed infants had a reduction of more than 40 percent in active sleep after consuming their mother's expressed breast milk flavored with alcohol (32 mg) on one testing day and expressed breast milk alone on the other.[33] All infants spent significantly less total time sleeping after consumption of the breast milk with alcohol (56.8 minutes with alcohol compared to 78.2 minutes without). The investigators concluded that short-term exposure to small amounts of alcohol in breast milk produces distinctive changes in the infant's sleep-wake patterning.

In a study of 400 infants born to members of a health insurance plan, no differences in the infant's cognitive development scores were found at one year of age between infants whose mothers consumed alcohol while nursing and those that did not. However, Psychomotor Development Index scores at one year of age were slightly lower among infants who were exposed to alcohol through breastmilk than among those who were not exposed.[34] For example, if a mother consumed two drinks daily while nursing, the score decreased by nearly 0.5 SD. The results of this study are controversial because the infants were also exposed to alcohol in utero, binge drinking was not adequately taken into consideration, and the breastfeeding groups included infants who received up to 16 ounces of formula or supplemental milk per day.[35]

Since current research does not show that occasional use (one to two drinks) of alcohol is harmful to the baby, La Leche League continues to support the opinion that the occasional use of alcohol in limited amounts is compatible with breastfeeding.[35] The American Academy of Pediatrics places alcohol in the category "Maternal Medication Usually Compatible with Breastfeeding".[9] The Institute of Medicine Subcommittee on Nutrition During Lactation recommends that lactating women should be advised that if alcohol is consumed, intake should be limited to "no more than 0.5 grams of alcohol per kilogram of maternal body weight per day. For a 60-kilogram (132 pound) woman, 0.5 grams of alcohol per kilogram of body weight corresponds to approximately 2 to 2.5 ounces of liquor, 8 ounces of table wine, or 2 cans of beer." [36] There is concern that prohibiting alcohol may be too restrictive, especially when the research does not support any serious impact on the baby when a mother has an occasional drink.[35] Many feel that nursing mothers are already placed under too many restrictions and may be discouraged from initiating or continuing to breastfeed because they feel they will face too many limitations.

If a mother does choose to have a drink or two, she can wait for the alcohol to clear her system before nursing her baby, avoiding nursing for at least two hours. She can plan ahead and have alcohol-free expressed milk stored for the occasion. If she becomes engorged, she can pump her breasts as a means of comfort, and discard her alcohol-containing milk. Pumping and discarding breast milk does not hasten the removal of alcohol from the milk, as the alcohol content of milk matches the maternal plasma alcohol levels.

Nicotine (Smoking Cigarettes)

The AAP Committee on Drugs classifies nicotine in Category 2, Drugs of abuse that are contraindicated during lactation. Maternal smoking is associated with decreased milk volume,[37] inhibition of the milk ejection (letdown) reflex, and early weaning.[38] Well-documented data, however, provide clear evidence that children of smoking mothers do better if breastfed in regard to general health, respiratory illness, and risk of sudden infant death syndrome (SIDS).[3]

Although the mean 24-hour nicotine concentrations in breast milk of women who smoke rise as cigarette consumption increases, exposure of infants to nicotine in breastmilk is apparently limited. Dahlstrom et al.[39] estimated that the dose of nicotine in breastfeeding infants was 1 mcg per kilogram per feeding, based on data on nicotine concentrations in breast milk within 30 minutes after smoking. This intake is equivalent to 2% of the adult dose of nicotine chewing gum (30 to 60 mcg per kilogram per dose).[39] Lawrence reports that women who smoke 10 to 20 cigarettes per day have 0.4 to 0.5 mg of nicotine/L in their milk.[3] This calculates to an equivalent dose of 6 to 7.5 mg of nicotine in an adult. In an adult, 4 mg of nicotine may produce symptoms, and the adult lethal dosage is in the range of 40 to 60 mg. With gradual intake over a day's time, the infant can metabolize nicotine in the liver and excrete the chemical in the kidney.[3] Numerous studies of nicotine and cotinine concentrations in the nursing mother and her infant have confirmed that although bottle-fed infants born to smoking mothers and raised in a smoking environment have significant levels of nicotine and metabolites in their urine, breastfed infants have higher levels.[3]

Women should be counseled not to smoke while nursing or in the infant's presence. Mothers who are not willing to stop smoking should cut down, consider low-nicotine cigarettes, and delay feedings as long as possible after smoking. The half-life of nicotine is 95 minutes.

When used as directed, smoking cessation aids which replace nicotine do not appear to pose any more problems for the breastfeeding infant than maternal smoking does. According to the 1999 edition of "Medications and Mother's Milk" by Thomas W. Hale,[13] the blood level of nicotine in most smokers (20 cigarettes per day) approaches 44 ng/ml whereas levels in nicotine patch users average 17 ng/ml, depending on the dose in the patch. Dr. Hale writes, "Therefore nicotine levels in milk can be expected to be less in patch users than those found in smokers, assuming the patch is used correctly and the mother abstains from smoking. Individuals who both smoke and use the patch would have extremely high blood nicotine levels and could endanger the nursing infant."[13]

Also according to Dr. Hale, maternal serum nicotine levels with nicotine gum average 30 to 60% of those found in cigarette smokers. Dr. Hale cautions, "While patches (transdermal systems) produce a sustained and lower nicotine plasma level, nicotine gum may produce large variations in blood plasma levels when the gum is chewed rapidly, fluctuations similar to smoking itself. Mothers who choose to use nicotine gum and breastfeed should be counseled to refrain from breastfeeding for 2–3 hours after using the gum product."[13]

Marijuana

Delta-9-tetrhydrocannabinol (THC), an active ingredient in marijuana, transfers and concentrates in breast milk and is absorbed and metabolized by the nursing infant. There is evidence from animal studies of structural changes in the brain cells of newborn animals nursed by mothers whose milk contained THC. Impairment of deoxyribonucleic acid (DNA) and ribonucleic acid (RNA) formation and proteins essential for proper growth and development has been described.[3] In one study following breastfeeding mothers and their infants for 12 months, marijuana exposure via the mother's milk during the first month postpartum appeared to be associated with a decrease in infant motor development at one year of age.[40] Concerns about marijuana use during lactation include the amount of the drug the infant ingests while nursing and the amount inhaled from the environment. The possible effect on DNA and RNA metabolism should discourage any maternal use, especially since brain cell development is still taking place in the first months of life.

Caffeine

While caffeine ingestion is a frequent concern of breastfeeding mothers, moderate intake causes no problems for most breastfeeding mothers and babies. A dose of caffeine equivalent to a cup of coffee results in breast milk levels of 1% of the level in maternal plasma and, consequently, low levels in the infant.[3] However, since the infant's ability to metabolize caffeine does not fully develop until three to four months of age, caffeine does accumulate in the infant. Cases of caffeine excess in breastfed infants have been documented.[41] Symptoms, which include infants being wakeful, hyperactive, and fussy, did not require hospitalization and disappeared over a week's time after caffeine was removed from the maternal diet. No long-term effects of caffeine exposure during lactation have been documented.[42]

While most breastfed infants can tolerate a maternal caffeine intake equivalent to five or fewer 5-ounce cups of coffee per day, or less than 750 ml per day, some babies may be more sensitive than others. If a mother suspects her baby is reacting to caffeine, she may try avoiding caffeine from all sources (coffee, tea, soft drinks, over-the-counter medications, chocolate) for two to three weeks.[43]

Other Drugs of Abuse

Amphetamines, cocaine, heroin, and phencyclidine hydrochloride (angel dust, PCP) are classified by the AAP Committee on Drugs[9] as Category 2, Drugs of abuse that are contraindicated during lactation. The AAP guidelines strongly state that these compounds and all other drugs of abuse are hazardous not only to the nursing infant, but also to the mother's physical and emotional health.

HYPERBILIRUBINEMIA AND JAUNDICE

The AAP discourages the interruption of breastfeeding in healthy term newborns and encourages continued and frequent breastfeeding (at least eight to ten times every 24 hours).

American Academy of Pediatrics[44]

Infant jaundice is a yellow discoloring of the skin caused by too much bilirubin in the blood (*hyperbilirubinemia*). Each year, about 60% of newborns de-

velop jaundice within the first week of life. Since higher levels of hemoglobin are necessary in utero to carry the oxygen delivered to the fetus by the placenta, the normal full-term infant has a hematocrit of 50% to 60%. As soon as the infant is born and begins to breathe room air, the need for high levels of hemoglobin is gone, and excess red blood cells are destroyed. The released hemoglobin is broken down by the reticuloendothelial system; bilirubin, an insoluble byproduct of the breakdown of hemoglobin, is released into the circulation. It is removed from circulation by the liver, which converts bilirubin to a water-soluble form, and excretes it via the bile to the stool. The balance between liver cell uptake of bilirubin, the rate of bilirubin production, and the rate of bilirubin resorption through the intestines determines the total serum bilirubin (TSB) level. Excessive bilirubin is deposited in various tissues, including the skin, muscles, and mucous membranes of the body, causing the skin to take on a yellowish color. In healthy newborns, this condition is temporary and usually resolves within a few days without treatment.

> **HYPERBILIRUBINEMIA** Elevated blood levels of bilirubin, a yellow pigment that is a byproduct from the breakdown of fetal hemoglobin.
>
> **KERNICTERUS OR BILIRUBIN ENCEPHALOPATHY** The end result of very high untreated bilirubin levels. Excessive bilirubin in the system is deposited in the brain, permanently destroying brain cells.

Since bilirubin is a cell toxin, concern arises when TSB elevates to levels with the potential to cause permanent damage. The brain and brain cells, if destroyed by bilirubin deposits, do not regenerate.[45] *Bilirubin encephalophathy,* or *kernicterus,* has a mortality rate of 50%, and survivors usually are burdened with severe problems including cerebral palsy, high-frequency deafness, and mental retardation.[46] Fortunately, full-scale kernicterus rarely occurs today, but mild effects of bilirubin on the brain may be manifested clinically in later life with symptoms such as incoordination, hypertonicity, and mental retardation, or perhaps learning disabilities.[3,45]

Pathological Jaundice

High or rapidly rising levels of bilirubin evident at birth or in the first 24 hours are usually an indication of abnormal or pathological jaundice, which is caused by a pathological problem unrelated to feeding. Some causes of pathological jaundice include:[44]

- A disease condition resulting in increased red blood cell breakdown (e.g., Rh or ABO blood incompatibilities);
- A disease condition that interferes with the processing of the bilirubin by the liver (liver enzyme-deficiency diseases, infections, and

metabolic problems such as galactosemia and hypothyroidism);

- A disease condition that increases the resorption of bilirubin by the bowel (e.g., gastrointestinal obstruction).

In most cases, frequent breastfeeding (10 to 12 times every 24 hours) can continue during diagnosis and treatment of pathological jaundice.[44] An advantage of colostrum and mature human milk is stimulation of bowel movements, speeding elimination of bilirubin.[3] However, in jaundice caused by galactosemia, breastfeeding is contraindicated.

Normal Newborn Jaundice

After the first 24 hours, rising bilirubin levels in healthy term infants is reflective of the physiological breakdown of fetal hemoglobin, the increased resorption of the bilirubin from the intestines, and the limited ability of the newborn's immature liver to process large amounts of bilirubin as effectively as a mature liver. Bilirubin levels in the infant with normal newborn jaundice, often referred to as physiologic jaundice, usually peak between the third and fifth days after birth and are usually less than 12 mg of bilirubin per dL of blood. Infants of Asian ancestry, including Chinese, Japanese, and Korean, and Native Americans may have higher levels.[3]

MECONIUM   Dark green mucilaginous material in the intestine of the full-term fetus.

Treatment for Jaundice

Table 7.5 shows the American Academy of Pediatrics guidelines for the management of hyperbilirubinemia

in healthy term newborns.[44] Phototherapy involves placing the newborn under special fluorescent lights that, like sunlight, assist in removing jaundice from the skin. The light is absorbed by the bilirubin, changing it to a water-soluble product, which can then be eliminated without having to be conjugated by the liver.

Historically, treatment for jaundice in American hospitals involved phototherapy and discontinuing breastfeeding either permanently or until the bilirubin levels were acceptable.[47, 48] In addition, many health professions believed that newborn infants would become dehydrated if they were not supplemented with water or formula during the first days of breastfeeding. Recent research shows these practices are counterproductive. The benefits of early and frequent breastfeeding in the first days of life to prevention of hyperbilirubinemia through maintaining hydration and stimulating the passage of stool are now well documented.[3,47] The passage of stool in the newborn is important because there are 450 mg of bilirubin in the intestinal tract *meconium* of the average newborn.[3] To avoid reabsorption of bilirubin from the gut into the serum, passing meconium in the stool is critical. The current AAP hyperbilirubinemia management guidelines state: "The AAP discourages the interruption of breastfeeding in healthy term newborns and encourages continued and frequent breastfeeding (at least 8 to 10 times every 24 hours). Supplementing nursing with water or dextrose water does not lower the bilirubin level in jaundiced, healthy breastfeeding infants."[44]

Hyperbilirubinemia and Breastfeeding

Two clinical conditions, early and late jaundice, are associated with hyperbilirubinemia while breastfeeding

Table 7.5   Management of hyperbilirubinemia in the healthy term newborn.

INFANT AGE (HOURS)	TOTAL SERUM BILIRUBIN (TSB) LEVEL, mg/dL (pmol/L)			
	Consider Phototherapy*	Phototherapy	Exchange Transfusion if Intensive Phototherapy Fails**	Exchange Transfusion and Intensive Phototherapy
<=24++	......	....	......	....
25–48	≥12 (170)	≥15(260)	≥20 (340)	≥25 (430)
49–72	≥15 (260)	≥18 (310)	≥25 (430	≥30 (510)
>72	≥17 (290)	≥20 (340)	≥25 (430)	≥30 (510)

*Phototherapy at these TSB levels is a clinical option, meaning that the intervention is available and may be used on the bases of individual clinical judgment.
**Intensive phototherapy should produce a decline of TSB of 1 to 2 mg/dL within 4 to 6 hours and the TSB should continue to fall and remain below the threshold level for exchange transfusion. If this does not occur, it is considered a failure of phototherapy.
++Term infants who are clinically jaundiced at <=24 hours old are not considered healthy and require further evaluation.
SOURCE: American Academy of Pediatrics Provisional Committee for Quality Improvement and Subcommittee on Hyperbilirubinemia. Practice parameter: Management of hyperbilirubinemia in the healthy term newborn. Pediatrics 94(4):558-65, 1994.[44] Used with permission of the American Academy of Pediatrics.

(Table 7.6). The more common condition is called early breast milk jaundice, and is the exaggeration of the first phase of physiologic jaundice of the newborn in the breastfed infant. The second condition occurs less frequently and is variously called breast milk jaundice, late-onset jaundice, or breast milk jaundice syndrome.

Gartner provides evidence that early breast milk jaundice results from insufficient frequency of breast-feeding during the critical first few days of life.[47, 48] Lawrence outlines treatment guidelines (Table 7.7) aimed at treating the actual cause, that is, failed breast-feeding or inadequate stooling or underfeeding.[3] The goal is to evaluate the breastfeeding for frequency, length of suckling, and apparent supply of milk, and then adjust the breastfeeding to solve the problem.[48] If stooling is an issue, the infant should be stimulated to stool. If starvation is the problem, the infant should receive additional calories (formula) while the milk supply is being increased by better breastfeeding techniques.

Table 7.6 Comparison of early and late jaundice associated with hyperbilirubinemia while breastfeeding.

	Early Jaundice Occurs 2–5 Days of Age	Late Jaundice Occurs 5–10 Days of Age
Onset	Transient: 10 days	Persist >1 month
Incidence	More common with first child Approximately 60% of newborns in the United States become clinically jaundiced	All children of a given mother
Feeding	Infrequent feeds at breast Receiving water or dextrose water	Milk volume not a problem No supplements
Stools	Stools delayed and infrequent	Normal stooling
Total Serum Bilirubin	Peaks ≤15 mg/dL	May be >20 mg/dL
Treatment	None or phototherapy	Phototherapy Discontinue breast-feeding temporarily (4 to 12 hours) Rarely, exchange transfusion.
Associations	Low Apgar scores, water or dextrose water supplement, prematurity	None identified.

Adapted by permission from: Lawrence, R. A. and Lawrence, R. M. Breastfeeding: A Guide for the Medical Profession. 5th Ed. St. Louis: Mosby, 1999.[3]

Table 7.7 Management outline for early jaundice while breastfeeding.

1. Monitor all infants for initial stooling. Stimulate stool if no stool in 24 hours.

2. Initiate breastfeeding early and frequently. Frequent, short feeding is more effective than infrequent, pro-longed feeding, although total time may be the same.

3. Discourage water, dextrose water, or formula supplements.

4. Monitor weight, voidings, and stooling in association with breastfeeding pattern.

5. When bilirubin level approaches 15 mg/dL, stimulate stooling, augment feeds, stimulate breast milk produc-tion with pumping, and use phototherapy if this aggres-sive approach fails and bilirubin exceeds 20 mg/dL.

6. Be aware that no evidence suggests early jaundice is as-sociated with "an abnormality" of the breast milk, so withdrawing breast milk as a trial is only indicated if jaundice persists longer than 6 days or rises above 20 mg/dL or the mother has a history of a previously af-fected infant.

From: Lawrence, R. A. and Lawrence, R. M. Breastfeeding: A Guide for the Medical Profession. 5th edition. St. Louis: Mosby, 1999. [3]

Breast Milk Jaundice Syndrome

In contrast to normal newborn jaundice which peaks in the third day and then begins to drop, breast milk jaundice syndrome becomes apparent after the third day, and bilirubin levels may peak any time from the seventh to the tenth day, with untreated cases being reported to peak as late as the fifteenth day.[3] In breast milk jaundice syndrome no correlation exists with weight loss or gain, and stools are normal. Initially breast milk jaundice syndrome was thought to be an unusual and distinct type of newborn jaundice affect-ing 1 in 200 of all breastfed newborns.[3] More recent research reports a prolongation of the second or late stage of physiologic jaundice in approximately two-thirds of breastfed infants.[49] In the third week of life, a few (perhaps 1% or 2%) of healthy breastfed in-fants may have serum bilirubin concentrations that are as high as 15 or even 20 mg/dL.[3,47,50]

The undisputed cause of breast milk jaundice syn-drome is unresolved. It is believed to be caused by a combination of three factors: a substance in most mothers' milk that increases intestinal absorption of bilirubin, individual variations in the infant's ability to process bilirubin, and the inadequacy of feeding in the early days.[3,47,51] The AAP Hyperbilirubinemia Management Guidelines (Table 7.5) are also applied to breast milk jaundice syndrome, with the goal of lower-ing TSB bilirubin levels substantially over 20 mg/dL promptly.[3,44] To establish the diagnosis of breast milk jaundice firmly when the bilirubin level is above 16 mg/dL for more than 24 hours, Lawrence recommends a short, temporary interruption of breastfeeding (12 to 24 hours) while monitoring bilirubin levels.[3]

BREASTFEEDING MULTIPLES

The benefits of breastfeeding to mother and infant are multiplied with twins and higher-order multiples, who often are born at risk.[52] History and numerous case reports[53,54,55] provide ample evidence that an individual mother can provide adequate nourishment for more than one infant. In seventeenth century France, wet nurses in foundling homes were allowed to nurse up to six infants at one time.[3] Current breastfeeding initiation rates of nearly 70% have been reported by surveys of members of Mothers of Super Twins (MOST), Parents of Multiple Births Association (POMBA) of Canada, and Double Talk, a newsletter for parents of multiples.[52] Some mothers of triplets and quadruplets have fully breast-fed their babies.[54,56]

ALLERGY   Hypersensitivity to a physical or chemical agent.

FOOD ALLERGY (HYPERSENSITIVITY) Abnormal or exaggerated immunologic response, usually immunoglobulin E (IgE) mediated, to a specific food protein.

FOOD INTOLERANCE   An adverse reaction involving digestion or metabolism but not the immune system.

Frequency and effectiveness of breastfeeding are the keys to building a plentiful milk supply. The more often a baby nurses, the more milk there will be.[3,57] Mothers who exclusively breastfeed twins or triplets can produce 2 to 3 kg/day, although this involves nursing an average of 15 or more times per day.[58] The main obstacle to nursing multiples is not usually the milk supply, but time and fatigue of the mother.

Mothers of twins or higher order multiples often face special challenges in the intrapartum establishment of lactation.[52] Breastfeeding initiation may take place in the neonatal intensive care unit, usually because of prematurity and low birthweight. Mothers may be coping with the effects of a more physically demanding pregnancy and birth or complications of pregnancy. Mothers may experience exaggerated postpartum sleep deprivation related to round-the-clock care of two or more newborns, concern for sick newborns, or staggered infant discharge, which results in time divided between infants at home and in the hospital. In addition, every aspect of breastfeeding management is affected by the dynamics that multiple newborns create.

Health care professionals can help the mothers of multiples face the many challenges of breastfeeding by offering consistent, informed, individualized care and support in the hospital and after discharge.[57] Knowledgeable care providers can help parents distin-guish between multiples-specific issues, normal variations in individual infant's breastfeeding abilities and patterns, and actual breastfeeding problems.[52] Mothers need information on when and how to initiate simultaneous feedings, practical tips for managing nighttime nursing and fatigue, and on how to assure herself that her babies are getting ample nourishment.[57] Parents need to be informed of resources for parenting multiples and for receiving support for breastfeeding in their community.

A well-defined plan for the health care of the lactating woman that includes screening for nutritional problems and providing dietary guidance is also important.[36] Mothers should be encouraged to drink to satisfy their thirst, to eat nutritious foods, and to sleep when the babies sleep. As in singleton nursing, women nursing multiples should be encouraged to obtain their nutrients from a well-balanced, varied diet rather than from vitamin-mineral supplements.

INFANT ALLERGIES

Protection from *allergies* is one of the most important benefits of breastfeeding. Elimination of major food allergens from the diet of infants at high risk for atopic disease and their lactating mothers can delay or prevent some food allergy and atopic dermatitis.[3,8,59,60] Several mechanisms (Table 7.8) are thought to contribute to the protective effect of breastfeeding.

The development of infant *food allergy* is influenced by genetic risk for allergy, duration of breastfeeding, time for introduction of other foods, and by maternal diet and immune systems.[60]

Common pediatric food allergens include:

- Cow's milk
- Wheat
- Eggs
- Peanuts
- Soybeans
- Tree nuts (e.g., almond, Brazil nut, walnut, hazelnut)

Infants with a positive family history of allergies should be exclusively breastfed for six months with continuance of breastfeeding for as long as possible.[3,8] Breastfeeding mothers with a family history of allergies also should be counseled to avoid common allergens in their own diets. If there is no history of allergy to a specific food in the mother's or father's family, avoiding a food because it is a potential allergen is an unnecessary precaution. Only if there is a family history of allergies, or if a baby shows allergic symptoms should a mother consider avoiding certain foods. If a mother avoids certain foods, care must be taken to ensure that her diet remains nutritionally adequate.[61]

Table 7.8 Possible reasons for allergy preventive effects of breastfeeding.[60]

- Low content of allergens
- Transfer of maternal immunity
- Regulation of infant immunity
- Protection against infections
- Influence on gut microbial flora

SOURCE: Bjorksten, B and Kjellman, N-I M. Does breast-feeding prevent food allergy? Allergy Proc 12(4):233–237, 1991.[60] Used by permission.

Food Intolerance

Although infants may have sensitivities to certain foods, there is no scientific basis for the concern about gassy foods, such as cabbage or legumes, causing gas in the breastfed baby. In mothers, the normal intestinal flora produces gas from the action on fiber in the intestinal tract. Neither the fiber nor the gas is absorbed from the intestinal tract, and neither enters the mother's milk. Likewise, the acid content of the maternal diet does not affect the breast milk because it does not change the pH of the maternal plasma.

Characteristic essential oils in foods such as garlic and spices may pass into the milk, and an occasional infant objects to their presence. Studies by Mennella and Beauchamp confirm that the diet of lactating women alters the sensory qualities of her milk.[62, 63] Extensive clinical experience also suggests that some infants are sensitive to certain foods in the mother's diet. According to Lawrence, garlic, onions, cabbage, turnips, broccoli, beans, rhubarb, apricots, or prunes may be bothersome to some infants, making them colicky for 24 hours.[3] In the summer, a heavy diet of melon, peaches, and other fresh fruits may cause colic and diarrhea in the infant. Red pepper has been reported to cause dermatitis in the breastfed infant within an hour of milk ingestion.[64] Contrary to popular belief, chocolate rarely causes problems and can be consumed in moderation without causing colic, diarrhea, or constipation in most infants.[3]

If a mother suspects that her baby reacts to a specific food, it may be helpful for her to keep a record of foods eaten, along with notes on the baby's symptoms or behavior. If highly allergic or sensitive, infants may react to foods their mothers have eaten within minutes, although symptoms generally show up between 4 and 24 hours after exposure. While symptoms will improve in most infants after the offending food has been removed from the mother's diet for five to seven days, it may take two weeks to totally eliminate all traces of the offending substance from both the mother and baby.[65] See Case Study 7.2.

HUMAN MILK AND PRETERM INFANTS

Human milk is the preferred feeding for all infants, including premature and sick newborns, with rare exceptions. The ultimate decision on feeding of the infant is the mother's. Pediatricians should provide parents with complete, current information on the benefits and methods of breastfeeding to ensure that the feeding decision is a fully informed one. When direct breastfeeding is not possible, expressed human milk, fortified when necessary for the premature infant, should be provided.

AAP[66]

The benefits of breastfeeding may be most visible among preterm infants who are born immature and without adequate stores of nutrients. The nutritional benefits include ease of protein digestion, fat absorption, and improved lactose digestion.[66] The known health and developmental benefits include better visual acuity,[68] greater motor and mental development at 1.5 years of age,[69] greater verbal intelligence quotient at 7–8 years of age,[70] and a lower incidence of serious infectious disease including necrotizing enterocolitis and sepsis even among infants who also receive some human milk substitutes.[69–72] Lower weight and length gains and poorer bone mineralization has been reported for preterm infants fed human milk.[73,74]

The composition of milk from women who deliver preterm infants is higher in protein, slightly lower in lactose and higher in energy content (58–70 cal/oz) compared to the milk of women who deliver full–term infants (approximately 62 cal/oz). While preterm milk has higher calcium and phosphorus than term milk, study of human milk in the 1980s found that calcium absorption from human milk did not meet the level of fetal growth that would have been expected. Lawrence and Lawrence[3] suggest that human milk is not adequate to meet the needs for fat soluble vitamins (vitamins A, D, E and K) and vitamin C if the infant receives protein supplements. Concerns about bone mineralization and growth of infants fed unfortified human milk led to fortification of preterm human milk to prevent development of rickets, and increase the calorie and nutrient content, as recommended by the American Academy of Pediatrics.[66] Infants who receive fortified human milk do not need additional supplementation unless a specific nutritional problem is identified. The health benefits, particularly reduction in sepsis and necrotizing enterocolitis, outweigh the slightly lower rate of weight gain and length gain observed among preterm infants fed fortified human milk compared to those who receive a human milk substitute for premature infants.[73]

Case Study 7.2
Food Sensitivity

Q: My three-week old daughter has diarrhea that, according to my pediatrician, is the symptom of an allergic reaction to milk products I've eaten. I've eliminated milk from my diet, as well as my prenatal vitamins. But she's still having diarrhea. What am I doing wrong?

A: Diarrhea in a breastfed baby looks and smells different from a normal mustard-colored stool. Twelve or more poopy diapers within 24 hours, with watery foul-smelling stools, indicate diarrhea and require seeing a pediatrician immediately. Symptoms of food allergy—more accurately called a food sensitivity in babies—include:

- Constant fussiness
- Eczema (dry skin with red patches or a rash or hives)
- Diarrhea, foul-smelling gas and cramps, vomiting (especially with blood flecks)
- Bloating
- Canker sores
- Yeast infection (thrush)

A dairy sensitivity indicates an inability to digest the milk proteins in dairy products. Lactose and its various forms—milk sugar, glucose, galactose—are in many infant foods, pharmaceuticals, bakery products, sweets, and more.

So even though you've eliminated milk, it's possible that dairy products are sneaking into your diet. It may be in vitamin and herbal supplements, or a prescription drug you are taking. If that's a possibility, go to a pharmacist with a bag containing everything you are taking, over-the-counter or prescription and herbal supplements, for help in analyzing the contents.

You must examine product labels for casein, a phosphoprotein that's one of the chief components of milk and the basis of cheese used in many food and non-food products. A list of culprits:

- Milk solids ("curds")
- Whey
- Sodium caseinate (the most common form of caseinate) or any other ingredient that includes caseinate
- Sodium lactyulate
- Lactalbumin (and anything else that begins with "lact")
- "Nondairy" (a phrase that indicates less than 0.5% milk by weight and does not mean milk free)

And be cautious of generalizations like "natural ingredients" (a catchall that can include dairy products and by-products), or "hydrolyzed vegetable protein" (casein may be used in processing).

Are you keeping a food diary? Different symptoms can mean different food allergies, and babies who are allergic to one food can be sensitive to or allergic to another. A dairy sensitive baby may also develop soy allergies, especially if exposed very early to soy in formula. Go back to your pediatrician and discuss the problem with her.

If your baby is sensitive to dairy products, he may outgrow it by the time he's five or six months old, especially if you are cautious about your diet. By six months, an infant's bowel has matured enough to digest and absorb complex food and is at less risk for food sensitivities.

Early *enteral* feeding seems to be important for preterm infants (when medically appropriate). Early feeding may be important to the ability to digest[67] and to the development of the infant's digestive system.[75] Often, milk must be expressed and stored for preterm infants. Establishing a milk supply early seems to be important to the mother's ability to maintain a milk supply that will meet her infant's demands after several weeks (See Chapter 6, Can Women Make Enough Milk?). A woman who is pumping less than 750 mL of milk by 2 weeks may need additional support to establish a milk supply that will meet her infant's needs beyond the first month.

MEDICAL CONTRAINDICATIONS TO BREASTFEEDING

Few medical problems in the mother or baby are absolute contraindications to breastfeeding (Table 7.9). Even infants with metabolic disorders such as pheynylketonuria can continue to breastfeed in combination with a specialized formula to meet calorie and protein needs. When mothers or infants have medical or other problems that cause a poor suck or other feeding problems, early identification and appropriate support from a lactation consultant is necessary for successful breastfeeding. In some cases, pumping milk may be necessary to maintain a supply of milk while problems with the infant suck can be addressed and corrected.

Breastfeeding and HIV Infection

Breast milk transmission of HIV has been well documented. The first reports indicating the possibility of HIV 1 transmission through breastmilk were in breastfed infants of women who were infected postnatally through blood transfusion or through heterosexual exposure.[81–86] There were also reports of infants, with no other known exposure to HIV, who were infected through wetnursing and through pooled breast milk.[85,87] Generally, higher rates of mother-to-child transmission of HIV are observed where most infants are breastfed rather than where fewer infants are breastfed.[88] Factors associated with variability in transmission rates may include maternal nutritional status, breastfeeding practices (mixed versus exclusive breastfeeding), stage of HIV disease, and possibly differences in transmission of HIV subtypes.

Perinatal transmission of HIV can occur during pregnancy (intrauterine), during labor and delivery (intrapartum), or after delivery through breastfeeding (postpartum). Researchers estimate that in nonbreastfeeding settings, intrauterine transmission accounts for 25% to 40% of infection and that 60% to 75% of transmission occurs during labor and delivery.[89] In breastfeeding settings, about 20% to 25% of perinatal infections are believed to be associated with intrauterine transmission, 60% to 70% with intrapartum transmission or early breastfeeding, and 10% to 15% with later postpartum transmission through breastfeeding.[90] In a randomized trial of formula feeding versus breastfeeding in Kenya, as much as 44% of HIV infection was attributed to breastfeeding; by two years of age, formula-fed infants had a significantly higher rate of HIV-1-free survival (70% versus 58% of those breastfed).[91]

> **ENTERAL**   Fluid or food being delivered directly into the gastrointestinal system. The delivery can be by mouth or through a tube that is placed into the stomach or intestines.
>
> **HIV =**   human immunodeficiency virus
>
> **AIDS =**   acquired immunodeficiency syndrome

Results from prospective studies in developing countries have indicated that among infants born to HIV-infected mothers, those who are breastfed are more likely to be infected than those who are formula-fed, even allowing for other factors known to be associated with mother-to-child transmission of HIV.[91–98]

Insufficient information is available to estimate the exact association between duration of breastfeeding and the risk of transmission. However, strong evidence exists for a gradual and continued increase in transmission risk as long as the child is breastfed.[99,100] The risk of transmission through breast milk among women with recent infection (HIV infection acquired in the postpartum period) was nearly twice as high (29% as opposed to 16–42%).[88]

A recent study suggests that exclusive breastfeeding may pose less risk of HIV-1 transmission than the more common practice, namely, breastfeeding concurrent with the feeding of water, other fluids, and foods.[101,102] By contrast, mixed feeding was associated with increased transmission of HIV-1. Additional research is needed.

Newer research is focusing on the effect of breastfeeding on mortality among HIV-infected women. A recent paper by Ruth Nduati and colleagues reported a three-fold higher mortality rate in HIV-infected mothers who breastfed their infants compared with those who fed their infants with formula.[103] The authors suggest that the high energy demands of breastfeeding in HIV-infected mothers may accelerate the progression to HIV-related deaths.

In contrast, Coutsoudis and colleagues published an analysis of morbidity and mortality in mothers enrolled in a randomized study of Vitamin A supplementation conducted in Durban, South Africa, who were analyzed according to their chosen method of

Table 7.9   Summary of medical contraindications to breastfeeding in the United States.

PROBLEM	BREASTFEEDING	CONDITION
Infectious Diseases		
Acute infectious disease	Yes	Respiratory, reproductive, gastrointestional infection
HIV	No	HIV positive in developed countries
Active tuberculosis	Yes	After mother has received 2 or more weeks of treatment
Hepatitis		
A	Yes	As soon as mother receives gamma globulin
B	Yes	After infant receives HBIG, first dose of hepatitis B vaccine should be given before hospital discharge
C	Yes	If no co-infections (e.g., HIV)
Veneral warts	Yes	
Herpes viruses		
Cytomegalovirus	Yes	
Herpes simplex	Yes	Except if lesion on breast
Varicella-zoster (chickenpox)	Yes	As soon as mother becomes noninfectious
Epstein-Barr	Yes	
Toxoplasmosis	Yes	
Mastitis	Yes	
Lyme disease	Yes	As soon as mother initiates treatment
HTLV-I	No	
Over-the-Counter/Prescription Drugs and Street Drugs		
Antimetabolites	No	
Radiopharmaceuticals		
Diagnostic dose	Yes	After radioactive compound has cleared mother's plasma
Therapeutic dose	No	
Drugs of abuse	No	Exceptions: cigarettes, alcohol
Other medications	Yes	Drug-by-drug assessment
Environmental Contaminants		
Herbicides	Usually	Exposure unlikely (except workers heavily exposed to dioxins)
Pesticides		
DDT, DDE	Usually	Exposure unlikely
PCBx, PBBs	Usually	Levels in milk very low
Cyclodiene pesticides	Usually	Exposure unlikely
Heavy metals		
Lead	Yes	Unless maternal level >40 mg/dL
Mercury	Yes	Unless mother symptomatic and levels measurable in breast milk
Cadmium	Usually	Exposure unlikely
Radionuclides	Yes	Risk greater to bottle-fed infants
Metabolic Disorders		
Galactosemia	No	
Pheynylketonuria	Yes	Human milk supplemented with phyenylalanine-free formula

Modified from Lawrence, R. A. and Lawrence, R. M. Breastfeeding: A Guide for the Medical Profession. 5th ed. St. Louis: Mosby, 1999.[3] Used by permission.

infant feeding.[104] They showed no excess of any reported morbidity in mothers who breastfed compared with those who did not. According to the WHO Statement "neither of these studies provided detailed information on the mode, duration and quantity of breastfeeding and the associated mortality risks. In addition, the two groups of women enrolled in the trials are not directly comparable."[105] Because of limitations of the data any interpretation should be cautious, but the findings are important and additional research is needed.

Recommendations

Based on the state of our knowledge today, HIV-infected women in the United States should not breastfeed or provide their breastmilk for the nutrition of their own or other infants due to the risk of

HIV transmission to the child. This is the position of the U.S. Centers for Disease Control and Prevention (CDC) and the American Academy of Pediatrics. Women in the United States with or at risk for HIV in the postpartum period should be counseled by their provider about the risks of mother-to-child HIV transmission and the alternatives for infant feeding if they are considering ever breastfeeding. Internationally, the recommendations have recently changed. Originally, the World Health Organization (WHO) recommended that in developing countries or areas where the risk of infant mortality from infection is great, breastfeeding is recommend even if the mother has AIDS.[88] This policy was clarified at a meeting of the WHO, the Children's Charity UNICEF, and UNAIDS in Geneva in October 2000. The policy now states: "When replacement feeding is acceptable, feasible, affordable, sustainable and safe, avoidance of all breastfeeding by HIV infected mothers is recommended. Otherwise, exclusive breastfeeding is recommended during the first months of life."[106] This position is undergoing review, and investigations are underway which may support or change the current recommendations. Where the risk of mortality from other infections is not great, mothers with HIV should be counseled—if it is feasible—to obtain milk from a breastmilk bank, where screened mothers donate milk and the milk is pasteurized. The possibility of providing the mother's own treated expressed breastmilk to the baby at risk of HIV infection via breastfeeding is an alternative which needs to be fully explored.

HUMAN MILK COLLECTION AND STORAGE

Human milk is the most appropriate food for infants, and is also used as medical therapy for older children and adults with certain medical conditions. Human milk has a long history and proven track record both as nutrition and therapy.

Human Milk Banking Association of North America

The appropriate collection and storage of human milk is important whether the milk is for the mother's own infant or to be donated. All of the collection containers and tubing used should be cleaned by dishwasher or sterilized by boiling. Hand pumps, electric handheld pumps, hospital grade electric breast pumps and manual expression can be used to extract the milk.

Milk that is used for a mother's own baby is safe at room temperature for four hours and can be refrigerated at 0 to 4°C (32 to 39°F) for up to eight days without significant bacterial growth.[107,108] (Refrigeration inhibits bacterial growth, whereas freezing does not.) At home, milk can be frozen for 3 months in a self-defrosting refrigerator and up to 12 months in a freezer that maintains a stable temperature at 0°C.

Milk Banking

Human milk banks provide human milk to infants who cannot be breastfed by their mother. Premature and sick infants are most likely to receive banked milk. A woman can donate milk once, or on a continuing basis if her supply exceeds the demands of her infant. There is a long history of providing human milk to infants by persons other than the biological mother. Wet nurses were the main source of human milk until the early 1900s for infants not fed by their biological mothers. Milk banks began in Europe and followed in the United States. Some neonatal intensive care units had informal milk banks until the 1980s. As a result of the human immunodeficiency virus, the resurgence of tuberculosis, and risks related to donors who might abuse drugs, human milk banks are now scarce in North America. In 1999–2000 seven milk banks were operating.

Human milk donors are chosen by their health profile. Women are carefully screened before they can donate extra milk to milk banks. Milk banks that belong to the Milk Banking Association of North America require telephone screening, a written health and lifestyle history, and verification of the health of the mother and baby by the health care provider of each. Blood samples are tested for heptitis B, hepititis C, HIV, HTLV, and syphilis by the milk bank. Women are not accepted if they are acutely ill, have had a blood transfusion or an organ transplant within a year, drink more than 2 ounces of liquor daily, regularly use medications or megavitamins, smoke, or use street drugs. Additionally, women who eat no animal products must take vitamin supplements with B_{12} to be eligible to donate.

Human milk is carefully pasteurized to kill any potential pathogens while preserving the nutrients and active immune properties of the milk. The North American Human Milk Banking Association communicates closely with the Food and Drug Administration to follow guidelines for use of human tissues and fluids. Human milk for milk banks is stored frozen to preserve the immunologic and nutritional components. Rigid plastic (polypropylene) containers are recommended for keeping the milk composition stable. White blood cells stick to glass, but not to plastic containers.[109]

A prescription from a physician or a hospital is needed to order milk for an infant from one of the North American Milk Banking Association milk banks. Costs are approximately $2.50 per ounce before shipping charges, significantly more than the cost of human milk substitutes. Some insurance companies and Medicaid programs cover the fees when it is demonstrated that donor milk is the most appropriate therapy for a specific patient.

MODEL PROGRAMS

Loving Support for a Bright Future Breastfeeding Support Kits

The Maternal and Child Health Bureau working with Best Start Social Marketing, the U.S. Department of Agriculture, and many other organizations developed breastfeeding support kits for physicians and other health providers. The kits were developed in response to a need identified in a national survey of physician's knowledge and attitudes about breastfeeding.[110] The survey revealed that physicians were not prepared with sufficient knowledge and training to provide support and counseling to breastfeeding mothers.

Family physicians, pediatricians, and obstetricians are in a position to offer encouragement and advice to potential and current breastfeeding mothers. In fact, research has shown that accurate, appropriate counseling and encouragement by physicians can improve rates of breastfeeding initiation and duration.[110–113] The Loving Support for a Bright Future Breastfeeding Kits are designed to help health care providers develop an optimal environment for breastfeeding promotion within their practice or clinic settings. The kits contain helpful, easy-to-use references, such as a pocket guide for breastfeeding management, a resource guide, consumer materials, and community-based activities. The physician and health provider kits were distributed through professional organizations including the American Academy of Pediatrics, the American College of Obstetricians and Gynecologists, the American Academy of Family Physicians, Association of Women's Health, Obstetric and Neonatal Nurses, and the American College of Nurse Midwives. The kits are available through Best Start Social Marketing.

The Rush Mothers' Milk Club

The Rush Mothers' Milk Club at Rush Presbyterian–St. Luke's Medical Center provides support, education and encouragement to mothers who pump breast milk for their high-risk babies in the Special Care Nursery and the Pediatric Intensive Care Unit (http://www.rush.edu/patients/children/publications/notes/preemies.html). The club serves as a place for mothers to discuss their goals and concerns about breastfeeding. Mothers learn the value of their milk to their high-risk baby from the Special Care Nursery Staff and from the Rush Mothers' Milk Club. The mothers learn to measure the amount of fat and calories in their own milk and learn to capture the highest calorie portion of milk, which is usually produced during the last 10 minutes of pumping. To create a bond between the mother and baby, mothers are encouraged to use the breast pumps at the baby's bedside. Family members and friends are also encouraged to participate in the weekly Mothers' Milk Club meetings to learn the importance of breastfeeding to the high-risk infant.

The success of the Rush Mothers' Milk Club is measured by its breastfeeding initiation rates. Between 95 and 97% of all mothers who deliver high-risk infants at Rush Presbyterian–St. Luke's Medical Center begin to nurse, compared with national rates for high risk infants of only 30 to 40%. A group of low-income mothers who delivered babies below 1500 grams were able to provide enough milk for 95% of their baby's feedings in the first 3 days and 87% over the first 60 days. This initiation rate is approximately twice the national initiation rate for premature infants.

Resources

Best Start Social Marketing
 Telephone: 813-971-2119
 Web site: beststart@beststartinc.org

The Human Milk Banking Association of North America, Inc.
 Represents all of the North American human milk banks that collect, pasteurize, and distribute donated mother's milk.
 Available from: http://www.hmbana.org/index.html

International Lactation Consultants Association
 Telephone: 919-787-5181
 Web site: http://www.ilca.org

La Leche League
 Telephone: 847-519-7730
 Web site: http://www.lalecheleague.org

Martin C, and Krebs NF.
 The Nursing Mother's Problem solver. Fireside Publishing, New York, NY, 2000.

See Chapter 6 for additional resources.

References

1. Tait, P. Nipple pain in breastfeeding women: causes, treatment, and prevention strategies. J Midwifery Women's Health 2000;45(3):212–5.

2. Martin C, Krebs NF. The nursing mother's problem solver. New York: Simon and Schuster; 2000.

3. Lawrence RA, Lawrence RM. Breastfeeding: A guide for the medical profession. 5th ed. St. Louis: Mosby, Inc.; 1999.

4. Riordan J, Auerbach KG. Breastfeeding and human lactation. 2nd ed. Boston: Jones and Barlett; 1998.

5. Thomsen AC, Espersen T, Maigaard S. Course and treatment of milk stasis, noninfectious inflammation of the breast, and infectious mastitis in nursing women. Am J Obstet Gynecol 1984;149(5):492–5.

6. Matheson I, Aursnes I, Horgen M, et al. Bacteriological findings and clinical symptoms in relation to clinical outcome in puerperal mastitis. Acta Obstet et Gynecol Scand 1988;67(8):723–6.

7. Ito S. Drug therapy for breast-feeding women. N Engl J Med 2000;343(2):118–126.

8. Committee on Nutrition, American Academy of Pediatrics. Pediatric Nutrition Handbook, 4th ed. Kleinman, RE, editor. Elk Grove Village (ILL): American Academy of Pediatrics; 1998.

9. Committee on Drugs, American Academy of Pediatrics. The transfer of drugs and other chemicals into human milk. Pediatrics 1994;93:137.

10. Howard CR, Lawrence RA. Drugs and breastfeeding. Clin Perinatol 1999;26(2): 447–78.

11. Mitchell JL. Use of cough and cold preparations during breastfeeding. J Hum Lact 1999;15(4):347–9.

12. Kacew S. Adverse effects of drugs and chemicals in breast milk on the nursing infant. J Clin Pharmacol 1993;33(3):213–21.

13. Hale TW. Medications and mothers' milk. 8th ed. Amarillo (TX): Pharmasoft Medical Publishing; 1999.

14. Briggs GG, Freeman RK, Yaffe SJ. Drugs in pregnancy and lactation. 5th ed. Baltimore: Williams and Wilkins; 1998.

15. Gotsch G. Information please: Maternal medications and breastfeeding. New Beginnings 2000;17(2):55–6. (http://www.lalecheleague.org/NB/NBMarchApr00p55.html).

16. Hardy ML. Herbs of special interest to women. J Am Pharm Assoc 2000;40:234–42.

17. Belew C. Herbs and the childbearing women: Guidelines for midwives. J Nurse Midwifery 1999;44(3), 231–50.

18. Kopec K. Herbal medications and breastfeeding. J Hum Lact 1999;15(2):157–61.

19. Humphrey SL, McKenna DJ. Herbs and breastfeeding. Breastfeeding Abstracts 1997;17(2):11–2.

20. Humphrey S. Sage advice on herbs and breastfeeding. LEAVEN 1998;34(3):43–7 (http://www.lalecheleague.org).

21. Foote J, Bruce R. Medicinal herb use during pregnancy and lactation. The Perinatal Nutrition Report 1998;5(1):1–2.

22. Newall CA, Anderson LA, Phillipson JD. Herbal medicines: A guide for healthcare professionals. London: The Pharmaceutical Press; 1996.

23. DeSmet PA, Keller K, Hansel R, et al. Adverse effects of herbal drugs. New York: Springer-Verlag; 1992.

24. Ernst, E. Harmless herbs? A review of the recent literature. Am J Medicine 1998;104:170–78.

25. Cunningham E, Hansen K. Question of the month: Where can I get information on evaluating herbal supplements? J Am Diet Assoc 1999;99:1240.

26. Leung A, Foster S. Encyclopedia of common natural ingredients used in foods, drugs, and cosmetics. New York: John Wiley & Sons; 1996.

27. Roberts KL, Reiter M, Schuster D. A comparison of chilled and room temperature cabbage leaves in treating breast engorgement. J Hum Lact. 1995;11(3):191–4.

28. Nikodem VC, Danziger D, Gebka N, et al. Do cabbage leaves prevent breast engorgement? A randomized, controlled study. Birth. 1993;20(2):61–4.

29. Schulte P. Minimizing alcohol exposure of the breastfeeding infant. J Hum Lact. 1995;11(4):317–9.

30. Mennella JA, Beauchamp GK. The transfer of alcohol to human milk: Effect on flavor and the infant's behavior. N Engl J Med 1991;325:981–5.

31. De Rosa, C. Prolactin secretion after beer (letter). Lancet 1981;934.

32. DaSliva VA, Malheiros LR, Moraes-Santos AR, et al. Ethanol pharmacokinetics in lactating women. Braz. J Med Bio Res 1993; 26:1097.

33. Mennella JA, Gerrish CJ. Effects of exposure to alcohol in mother's milk on infant sleep. Pediatrics 1998;101(5):E2.

34. Little RE, Anderson KW, Ervin CH, et al. Maternal alcohol use during breastfeeding and infant mental and motor development at one year. N Eng J Med 1989;321:425.

35. Huotari, C. Alcohol and motherhood. LEAVEN. 1997;33(2):30–1 (http://www.lalecheleague.org/llleaderweb/LV/LVAprMay97p30.html).

36. Subcommittee on Nutrition During Lactation, Committee on Nutritional Status During Pregnancy and Lactation. Food and Nutrition Board, Institute of Medicine, National Academy of Sciences. Nutrition during lactation: Summary, conclusions, and recommendations. Washington: National Academy Press; 1991.

37. Vio F, Salazar G, Infante C. Smoking during pregnancy and lactation and its effects on breastmilk volume. Am J Clin Nutr. 1991;54:1011.

38. Schulte-Hobein B, Schwartz-Bickenbach D, Abt S, et al. Cigarette smoke exposure and development of infants throughout the first year of life: Influence of passive smoking and nursing on cotinine levels in breast milk and infant's urine. Acta Paediatr Scand. 1992;81:550.

39. Dahlstrom A, Lundell B, Curvall M, et al. Nicotine and cotinine concentrations in the nursing mother and her infant. Acta Paediatr Scand. 1990; 79:142.

40. Astley SJ, and Little RE. Maternal marijuana use during lactation and infant development at one year. Neurotoxicol Teratol. 1990;112(2):161–8.

41. Rivera-Calimlim. The significance of drugs in breast milk. Clin Perinatol. 1987; 14(1):51–70.

42. Nehlig A, and Debry G. Consequences on the newborn of chronic maternal consumption of coffee during gestation and lactation: A review. J Am Coll Nutr. 1994;13(1):6–21.

43. La Leche League International. Frequently Asked Questions on Caffeine and Breastfeeding. Http://www.lalecheleague.org/FAQ/Caffeine.html. November 8, 2000.

44. American Academy of Pediatrics Provisional Committee for Quality Improvement and Subcommittee on Hyperbilirubinemia. Practice parameter: Management of hyperbilirubinemia in the healthy term newborn. Pediatrics. 1994;(4):558–65.

45. Hansen TWR, Bratlid D. Bilirubin and brain toxicity. Acta Paediatr Scand. 1986; 75:513.

46. DeVries LS, Lary S, Whitelaw AG, et al. Relationship of serum bilirubin levels and hearing impairment in newborn infants. Early Hum Dev 1987;15:269.

47. Gartner LM, Herschel M. Jaundice and breastfeeding. Pediatr Clin North Am 2001 Apr;48(2):389–99.

48. Gartner LM, Lee KS. Jaundice in the breastfed infant. Clin Perinatol 1999;26(2): 431–45, vii.

49. Alonso EM, Whitington PF, Whitington SH, et al. Enterohepatic cirulation of nonconjugated bilirubin in rats fed with human milk. J Pediatr. 1991;118:425–30.

50. Garner L. On the question of the relationship between breastfeeding and jaundice in the first 5 days of life. Seminars in Perinatology. 1994;18(6):502–9.

51. Mohrbacher N, Stock J. The Breastfeeding Answer Book. Schaumburg (Ill): Le Leche League International; 1997.

52. Gromada KK, Spangler A. Breastfeeding twins and higher-order multiples.

J Obstet Gynecol Neonatal Nur 1998; 27(4): 441.

53. Hill PD, Brown LP, Harker TL. Initiation and frequency of breast expression in breastfeeding mothers of LBW and VLBW infants. Nurs. Res. 1995;44:352–5.

54. Mead LJ, Chuffo R, Lawlor-Klean P, et al. Breastfeeding success with preterm quadruplets. J Obstet Gynecol Neonatal Nurs 1992;21:221–7.

55. Biancuzzo M. Breastfeeding preterm twins: A case report. Birth 1994;21(2): 96–100.

56. Gromada KK. Breastfeeding more than one: Multiples and tandem breastfeeding. NAACOGS Clin Issu in Perinat Womens Health Nurs 1992;(3): 656–66.

57. Hattori R, Hattori H. Breastfeeding twins: Guidelines for success. Birth 1999;26(1):37–42.

58. Saint L, Maggiore P, Hartman PE. Yield and nutrient content of milk in eight women breastfeeding twins and one woman breastfeeding triplets. Br J Nutr 1986;56:49.

59. Chandra RK. Food allergy and nutrition in early life: implications for later health. Proc Nutr Soc 2000;59(2):272–7.

60. Bjorksten B, Kjellman NI. Does breastfeeding prevent food allergy? Allergy Proc 1991;12(4):233–7.

61. Holmberg-Marttila AT. Benefits and risks of elimination diets. Ann Med 1999;31(4):293–8.

62. Mennella JA, Beauchamp GK. Maternal diet alters the sensory qualities of human milk and nursling's behavior. Pediatrics 1991;88:737.

63. Mennella JA, Beauchamp GK. The effects of repeated exposure to garlic-flavored milk on the nursling's behavior. Pediatr Res 1993;34:805–8.

64. Cooper RL, Cooper MN: Red pepper-induced dermatitis in breast-fed infants. Dermatotolgy 1996;193:61.

65. Zeretzke K. Allergies and the breastfeeding family. New Beginnings 1998; 15(4):100 (http://www.lalecheleague.org/NB/NBJulAug98p100.html).

66. American Academy of Pediatrics. Breastfeeding and the use of human milk. Pediatrics 1997;100(6):1035–9.

67. Shulman RJ, Schanler RJ, Lau C, et al. Early feeding, feeding tolerance, and lactase activity in preterm infants. J Pediatr 1998;133(5):645–9.

68. Gibson RA, Makrides M. Polyunsaturated fatty acids and infant visual development: A critical appraisal of randomized clinical trials. Lipids 1999;34: 179–84.

69. Lucas A, Cole TJ. Breast milk and neonatal necrotizing enterocolitis. Lancet. 1990;1519–23.

70. Lucas A, Morley R, Cole TJ. Randomized trial of early diet in preterm babies and later intelligence quotient. BMJ 1998;317(7171):1481–7.

71. Hylander MA, Strobino DM, Dhanireddy R. Human milk feedings and infection among very low birth weight infants. Pediatrics 1998;102:3:e38.

72. Schanler RJ, Shulman RJ, Lau C, et al. Feeding strategies for premature infants: Randomized trial of gastrointestinal priming and tube-feeding method, Pediatrics 1999;103(2):434–9.

73. Schanler RJ, Schulman RJ, Lau C. Feeding strategies for premature infants: Beneficial outcomes of feeding fortified human milk versus preterm formula. Pediatr 1999;103:1150–7.

74. Schanler RJ, Abrams SA. Postnatal attainment of intrauterine macromineral accretion rates in low birth weight infants fed fortified human milk. J Pediatr 1995;126:441–7.

75. Schanler RJ. Bioactivity in milk and bacterial interactions in the developing immature intestine: The clinical perspective. J Nutr 2000;130:417S–19S.

76. Lucas A, Gibbs JA, Lyster RL, et al. Creamatocrit: simple clinical technique for estimating fat concentration and energy value of human milk. BMJ. 1978;1(6119): 1018–20.

77. Wang CD, Chu PS, Mellen BG, et al. Creamatocrit and the nutrient composition of human milk. J Perinatol 1999;19(5):343–6.

78. Polberger S, Lonnerdal B. Simple and rapid macronutrient analysis of human milk for individualized fortification: Basis for improved nutritional management of very-low-birth-weight infants? J Pediatr Gastroenterol Nutr 1993;17(3):283–90.

79. Griffin TL, Meier PP, Bradford LP, et al. Mothers' performing creamatocrit measures in the NICU: accuracy, reactions, and cost. J Obst Gynecol Neonatal Nurs 2000;29(3):249–57.

80. Smith L, Bickerton J, Pilcher G, et al. Creamatocrit, carbon content, and energy value of pooled banked human milk: implications for feeding preterm infants. Early Hum Dev 1985;11(1):75–80.

81. Palasanthiran P, Ziegler JB, Stewart GJ, et al. Breastfeeding during primary maternal immunodeficiency virus infection and risk of transmission from mother to infant. J Infect Dis 1993;167:441–4.

82. Van de Perre P, Simonon A, Msellati P, et al. Postnatal transmission of human immunodeficiency virus type 1 from mother to infant. A prospective cohort study in Kigali, Rwanda. N Engl J Med 1991; 325:593–8.

83. Stiehm R, Vink P. Transmission of human immunodeficiency virus infection by breastfeeding. J Pediatr 1991;118:410–2.

84. Hira SK, Mangrola UG, Mwale C, et al. Apparent vertical transmission of human immunodeficiency virus type 1 by breastfeeding in Zambia. J Pediatr 1990;117:421–4.

85. Colebunders R, Kapita B, Nekwet W, et al. Breastfeeding and transmission of HIV. Lancet 1988:2:1487.

86. Lepage P, Van de Perre P, Carael M, et al. Postnatal transmission of HIV from mother to child. Lancet Aug 15;2(8555):400.

87. Nduati R, John G, Kreiss J. Postnatal transmission of HIV-1 through pooled breast milk. Lancet 1994;344:1432.

88. World Health Organization. HIV and Infant Feeding. A review of HIV transmission through Brestfeeding. Geneva, Switzerland: World Health Organization; 1998.

89. Fowler MG, Simonds RJ, Roongpisuthipong A. Perinatal transmission of HIV-1. Pediatr Clin North Am 2000 Feb;47(1): 21–38.

90. Bertolli J, St. Louis, Simonds RJ, et al. Estimating the timing of mother-to-child transmission of human immunodeficiency virus in a breastfeeding population in Kinshasa, Zaire. J Infect Dis 1996 Oct;174(4):722–6.

91. Nduati R, John G, Mbori-Ngacha D, et al. Effect of breastfeeding and formula feeding on transmission of HIV-1: a randomized clinical trial. JAMA 2000 Mar 1;283(9):1167–74.

92. European Collaborative Study. Risk factors for mother-to-child transmission of HIV-1. Lancet 1992;339:1007–12.

93. Ryder RW, Manzila T, Baende E, et al. Evidence from Zaire that breastfeeding by HIV seropositive mother is not a major route for perinatal HIV-1 transmission but does decrease morbidity. AIDS 1991;5:709–14.

94. Blanche S, Rouzioux C, Moscato ML, et al. A prospective study of infants born to women seropositive for human immunodeficiency virus type I. N Engl J Med. 1989; 320:1643–8.

95. Tovo PA, De Martino M, Caramia G. Epidemiology, clinical features, and prognostic factors of pediatric HIV infection. Lancet. 1988;ii:1043–6.

96. Tess BH, Rodrigues LC, Newell ML, et al. Breastfeeding, genetic, obstetric and other risk factors associated with mother-to-child transmission of HIV-1 in Sao Paulo State, Brazil. AIDS. 1998;12:513–20.

97. Tess BH, Rodrigues CL, Newell ML, et al. Infant feeding and risk of mother-to-child transmission of HIV-1 in Sao Paulo State, Brazil. J Acquir Immune Defic Syndr Hum Retrovirol 1998 Oct 1;19(2):189–94.

98. Smith MM, Kuhn L. Infant-feeding patterns and HIV-1 transmission. Lancet 1999; 354:1903–4.

99. Leroy V, Newell ML, Dabis F, et al. for the Ghent International Working Group on Mother-to-Child Transmission of HIV. International multicenter provided analysis of late postnatal mother-to-child transmission of HIV-1 infection. Lancet. 1998;352: 597–600.

100. Miotti, Taha T, Kumwenda N, et al. HIV Transmission through breastfeeding, a study in Malawi. JAMA. 1999;282: 744–9.

101. Coutsoudis A, Pillay K, Spooner E, et al. Influence of infant feeding patterns on

early mother-to-child transmission of HIV-1 in Durban, South Africa. Lancet 1999; 354:471–6.

102. Smith MM, Kuhn L. Exclusive breast-feeding: does it have the potential to reduce breastfeeding transmission of HIV-1? Nutr Rev 2000; 58:333–40.

103. Nduati R, Richardson BA, John G, et al. Effect of breastfeeding on mortality among HIV-1 infected women; a random-ized trial. Lancet 2001; 357: 1651–5.

104. Coutsoudis A, Coovadia H, Pillay K, et al. Are HIV-infected women who breast-feed at increased risk of mortality? AIDS 2001;15: 653–5.

105. World Health Organization. Effect of Breastfeeding on Mortality among HIV-infected Women. Geneva, Switzerland: World Health Organization; 7 June 2001.

106. World Health Organization. New data on the prevention of mother-to-child transmission of hiv and their policy impli-cations: Conclusions and recommendations. WHO Technical Consultation on Behalf of the UNFPA/UNICEF/WHO/UNAIDS Inter-Agency Task Team on Mother-to-Child Transmission of HIV, Geneva, 11–13 October 2000.

107. Pardou A, Serruys E, Mascart-Lemone FM, et al. Human milk banking: Influence of storage processes and of bacterial con-tamination on some milk constituents. Bio Neonat 1994;65:302–9.

108. Tully, MR. Recommendations for handling mother's own milk. J Hum Lact 2000;16(2):149–51.

109. Paxson CI, JR, Cress CC. Survival of human milk leukocytes. J Pediatr 1979;94: 61–4.

110. Freed G, Clark S, Sorenson J, et al. National assessment of physicians' breast-feeding knowledge, attitudes, training and experience. JAMA. 1995;273(6):472–6.

111. Kristin N, Benton D, Rao S. Breastfeeding rates among black urban low-income women: effect of prenatal educa-tion. Pediatrics 1990;86:741–6.

112. Lawrence RA. Practices and attitudes toward breastfeeding among medical pro-fessionals. Pediatrics 1982;70:912–20.

113. Schanler RJ, O'Connor KG, Lawrence RA. Pediatricians' practices and attitudes regarding breastfeeding promotion. Pediatric. 1999;103(3):E35.

114. Anonymous. Case management of a breastfeeding mother with persistent over-supply and recurrent breast infections. J Hum Lact 2000;16:221–2.

CHAPTER 8

Eyewire

A babe is fed with milk and praise.

Charles Lamb, *The First Tooth*

INFANT NUTRITION

Prepared by **Janet Sugarman Isaacs**

CHAPTER OUTLINE

- Introduction
- Assessing Newborn Health
- Infant Development
- Energy and Nutrient Needs
- Physical Growth
- Feeding in Early Infancy
- Development of Infant Feeding Skills
- Nutrition Guidance
- Common Nutritional Problems and Concerns
- Cross-Cultural Considerations
- Vegetarian Diets
- Nutrition Intervention for Risk Reduction

KEY NUTRITION CONCEPTS

1 The dynamic growth experienced in infancy is the most rapid of any age. Inadequate nutrition in infancy, however, leads to consequences that may be lifelong, harming both future growth and development.

2 Progression in feeding skills expresses important developmental steps in infancy that signal growth and nutrition status.

3 Nutrient requirements of term newborns have to be modified for preterm infants. Knowing the needs of sick and small newborns results in greater understanding of the complex nutritional needs of all newborns and infants.

4 Changing feeding practices, such as the care of infants outside the home and the early introduction of foods, are markedly affecting the nutritional status of infants.

INTRODUCTION

This chapter is about healthy *full-term infants* born at 37 weeks of gestation or later, and healthy *preterm infants* born at 34 weeks or later. These newborns are expected to have typical growth and development. The term *normal* is not used much in this chapter since its opposite, *abnormal,* is an emotionally laden term, particularly when describing babies to their parents. *Typical* is used in place of *normal* when possible.

This chapter discusses how nutrition is an important contributor to the complex development of infants. Both biological and environmental factors interact during infant growth and development. More sound research, however, is done about biological and physiological aspects, such as nutrient requirements. Animal models exist for testing a specific nutrient deficiency, but models about the interaction of biological and social interaction are less precise. They are not adequate for describing complex interactions, such as how mealtime stimulates language development and how food preferences develop during infancy. The complexity of infant development contributes to our individuality later on.

Chapter 9 discusses babies born prior to 34 weeks or who require specialty care during the first year of life, in which typical growth and development may be at risk. That chapter gives birth rates of *low birthweight infants, very low birthweight infant,* and *extremely low birthweight,* which account for a major proportion of *perinatal, neonatal* and *infant deaths,* and overall infant mortality. However, the distinction between healthy-appearing infants and those who later have health conditions is sometimes not clear until after the neonatal period, or after infancy. As a result, some topics about infants with health and nutrition risks are placed in that chapter, but may fit in either one.

FULL-TERM INFANTS Infants born between 37 and 42 weeks of gestation.

PRETERM INFANTS Infants born at or before 37 weeks of gestation.

LOW BIRTHWEIGHT INFANT (LBW) Infants weighing <2500 grams or <5 pounds 8 ounces at birth.

VERY LOW BIRTHWEIGHT INFANT (VLBW) Infants weighing <1500 grams or <3 pounds 5 ounces at birth.

EXTREMELY LOW BIRTHWEIGHT (ELBW) Infants weighing <1000 grams or 2 pounds and 3 ounces at birth.

INFANT MORTALITY Death that occurs within the first year of life.

NEONATAL DEATH Death that occurs from the day of birth through the first 28 days of life.

PERINATAL DEATH Death occurring at or after 20 weeks of gestation and through the first 28 days of life.

ASSESSING NEWBORN HEALTH

Birth is such a major event that every country measures itself by describing how newborns live or die. The importance of birthweight as an overall indicator of health is accepted worldwide. Most babies are born full-term and healthy.

Birthweight As an Outcome

The weight of a newborn is a key outcome measure of the health of the completed pregnancy. The average gestation for a full-term infant is 40 weeks, with a range from 37 to 42 weeks. Full-term infants usually weigh between 2500–3800 grams (5.5 to 8.5 pounds) and are 47–54 centimeters (18.5–21.5 inches) in length.[1] There were over 3.9 million births in the United States in 1999.[2] Over 3.6 million were full-term infants. Conversely, preterm births, regardless of birthweight, are those at 37 weeks or less. Preterm means incomplete development has taken place.[3] The importance of birthweight is demonstrated by neonatal deaths. If newborns born at the full-term weights require intensive care services, only 2% die, whereas 16% of newborns at very low weight typical of 28 weeks of gestation die.[4,5]

Infant Mortality

Worldwide neonatal mortality rates rank the United States at 26th, far worse than many other countries[2] (see Table 4.2 in Chapter 4). The reasons for the high rate of neonatal mortality are many, but the prevalence of preterm births has been identified as one important component. In 1999, 11.8% of live births in the United States were preterm births, and more than 90% of all neonatal deaths were in preterm infants.[2] In 1999 there were 27,500 infant deaths in the United States.[2] The role of race of the mother in the prevalence of preterm birth has been especially well documented. Table 8.1 shows infant deaths in the United

Table 8.1   United States neonatal mortality rates by race of the mother in 1999.[2]

	Deaths/1000 Live Births
All newborns	4.8
White	4.0
Black	9.2
Asian or Pacific Islander	3.5
American Indian or Alaska Native	4.5
Hispanic	4.0

States based on race of the mother, although this variability is unexplained by biology.[2] The basis for racial disparity and preterm birth is a major focus in federal initiatives to combat infant deaths in the United States. In spite of efforts to lower neonatal and infant deaths, rates are still too high.

Combating Infant Mortality

Efforts to improve newborn health are underway on many levels, such as national public health programs reviewed in Chapter 4. In the United States, improved access to specialized care for mothers and infants has been credited in part for a decline in the infant mortality rate.[2] This is a multi-faceted problem, however, affected by:

- Social and economic status of the local area, region, state or country reporting
- Access to health care, and related barriers such as transportation
- Interventions to save newborns
- Interventions to decrease teenage pregnancy rates
- Availability of abortion services
- Failure to prevent preterm and low birthweight births.

Resources have been concentrated on the proportion of newborns identified at risk. Some of these resources are major payers of health services, such as Medicaid and the Child Health Initiatives Program (CHIP).[6] These concepts underscore the commitment of resources for infants:

- Recognition that birthweight is important for long-term health outcomes.
- Prevention of complications and treatment for at-risk infants are investments for the future.

The emphasis on prevention is seen in various programs. The *EPSDT* Program is a major source of preventive and routine care for infants in low-income families. Immunizations during infancy are another example of a prevention approach.

Nutrition is included in some national prevention programs. The Special Supplemental Food Program for Women, Infants and Children (WIC) and the Centers for Disease Control (CDC) collaborate to track growth parameters as a part of the Nutrition Surveillance Program. Another example is called "Bright Futures."[6] Starting in 1990, the U.S. government funded this program to promote and improve the health, education and well-being of infants, children, adolescents, families and communities. Nutrition is just one component; the program includes guidelines

for various ages and recommendations about common issues and concerns, and provides tools for practitioners. Bright Futures is an example of a comprehensive approach to health supervision by a collaboration of government and professional groups.[6]

Standard Newborn Assessment Tests

The period of infancy is a continuation of the rapid growth and development of the fetus that took place in utero. Adequate birthweight is an indicator of fetal maturation, the adequacy of the placenta, and maternal factors such as weight gain.[7] Newborn health status is assessed by a number of indicators of growth and development taken right after birth. Indicators are birthweight, length, and head circumference, in the context of gestational age, such as AGA, LGA, and SGA[6,7] (see Chapter 4). These and the designation of *IUGR* are based on a clinical examination of the newborn and sometimes of the placenta.[8] The designation "small for gestational age," which is sometimes called "small for dates," is the same as "intrauterine growth retardation" or "intrauterine growth restriction" in reflecting undergrowth of the fetus without identifying a specific cause. A specific cause is identified if known instead of using such general terms. Multiple births are an example of a known cause. Findings of LGA, SGA, IUGR and low placental weight make interpreting future growth with the standard growth charts more difficult.[3,7] Although weight at birth is a reflection of the intrauterine environment and health history of the newborn, it may or may not have set the stage for a growth pattern overall.[6,7] Risks for *failure to thrive* and other slow growth patterns are statistically correlated with low birthweight.[6]

The newborn's health is initially assessed by rating systems such as the *APGAR* score.[3] Five elements comprise the APGAR scale:

- Color
- Heart rate

EPSDT   The Early Periodic Screening, Detection, and Treatment Program is a part of Medicaid and provides routine checkups for low-income families.

INTRAUTERINE GROWTH RETARDATION (IUGR)   Fetal undergrowth from any cause, resulting in a disproportionality in weight, length, or weight-for-length percentiles for gestational age. Sometimes called *intrauterine growth restriction*.

FAILURE TO THRIVE (FTT)   Condition of inadequate weight or height gain thought to result from a caloric deficit, whether or not the cause can be identified as a health problem.

APGAR   A rating score for newborns which is assigned at one minute and at five minutes after birth with the lowest score being zero and highest score ten.

- Respiratory effort
- Muscle tone
- Reflex irritability

The APGAR score has a maximum of 10. It is taken at one and five minutes after birth and reflects the newborn's general status and immediate adaptation to life outside the uterus. Efforts to correlate low APGAR scores with later developmental and neurological problems have not been as close as expected, but newborns with low APGAR scores generally have had significant distress during labor and delivery.[3]

INFANT DEVELOPMENT

Monitoring nutrition for infants requires an understanding of their overall development, as growth and feeding are not only nutrition outcomes, but also developmental outcomes. Full-term newborns have a wider range of abilities than previously recognized.[9] They hear and move in response to familiar sounds. Newborns demonstrate four states of arousal ranging from sleeping to fully alert, and responsiveness differs in part on their state of arousal.[9] Recognizing the state of arousal is a part of nursing successfully.

REFLEX   An automatic (unlearned) response that is triggered by a specific stimulus.

ROOTING REFLEX   Action that occurs if one cheek is touched, resulting in the infant's head turning towards that cheek and opening his mouth.

SUCKING REFLEX   Action in which an infant will suck on anything placed in the mouth, with the tongue raising and lowering.

SUCKLE   A reflexive movement of the tongue moving forward and backward; earliest feeding skill.

Organs and systems developed during gestation continue to increase in size and complexity during infancy. The newborn's central nervous system is immature; that is, the neurons in the brain are less organized compared to the older infant. As a result the newborn gives inconsistent or subtle cues of hunger, or other needs, compared to the cues given later. The fact that newborns can *root, suckle,* and coordinate swallowing and breathing within hours of birth shows that feeding is directed by reflexes and the central nervous system.[9,10] Newborn *reflexes* are protective for them. These reflexes fade as they are replaced by purposeful movements during the first few months of life.[9] Table 8.2 lists major reflexes in newborn.[9,10]

Motor Development

There are several models for describing infant development, but none provide a complete description and explanation of the rapid advances of intellectual, motor, and sensory skills achieved during infancy.[9,11] Illustration 8.1 depicts motor development during the first fifteen months.[7] Motor development is concerned with voluntary muscle control. It is a great source of pride for parents when their baby first rolls over or sits up. The development of muscle control is top down, meaning head control is the start and last comes lower legs.[10] Muscle development is also from central to peripheral, meaning the infant learns to control the shoulder and arm muscles before muscles in the hands.[10,12]

The infant's stage of motor development affects her nutrition needs and vice versa. Motor development influences both the ability of the infant to feed and the amount of calories expended in the activity.[13] An example of how motor development affects feeding is the ability to sit in a high chair. Only when an infant has achieved the motor development of head control and sitting balance and certain reflexes have disappeared can oral feeding with a spoon be achieved.[12] The development of motor skills slowly increases the caloric needs of infants over time as related to physical activity.[13] Infants who crawl expend more calories in physical activity than younger infants who cannot roll over.

Table 8.2   Major reflexes found in newborns.

Name	Response	Significance
Babinski	A baby's toes fan out when the sole of the foot is stroked from heel to toe	Perhaps a remnant of evolution
Blink	A baby's eyes close in response to bright light or loud noise	.Protects the eyes
Moro	A baby throws its arms out and then inward (as if embracing) in response to loud noise or when its head falls.	May help a baby cling to its mother
Palmar	A baby grasps an object placed in the palm of its hand.	Precursor to voluntary grasping
Rooting	When a baby's cheek is stroked, it turns its head toward the cheek that was stroked and opens its mouth.	Helps a baby find the nipple
Stepping	A baby who is held upright by an adult and is then moved forward begins to step rhythmically.	Precursor to voluntary walking
Sucking	A baby sucks when an object is placed in its mouth.	Permits feeding
Withdrawal	A baby withdraws its foot when the sole is pricked with a pin.	Protects a baby from unpleasant stimulation

(SOURCE: Kail, Robert V. and John C. Cavanaugh. *Human Development: A Lifespan View.* 2nd ed. (2000). Belmont CA: Wadsworth/Thompson Learning, page 99.)

Illustration 8.1 Gross motor skills.

SOURCE: Kail, Robert V. and John C. Cavanaugh. *Human Development: A Lifespan View.* 2nd ed. (2000). Belmont CA: Wadsworth/Thomson Learning, page 99.

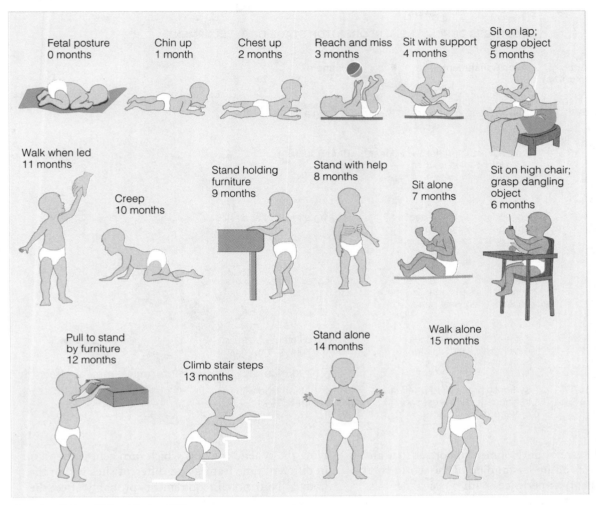

Fetal posture
0 months

Chin up
1 month

Chest up
2 months

Reach and miss
3 months

Sit with support
4 months

Sit on lap;
grasp object
5 months

Walk when led
11 months

Creep
10 months

Stand holding
furniture
9 months

Stand with help
8 months

Sit alone
7 months

Sit on high chair;
grasp dangling
object
6 months

Pull to stand
by furniture
12 months

Climb stair steps
13 months

Stand alone
14 months

Walk alone
15 months

Based on Shirley, 1931 and Bayley, 1969.

Critical Periods

The concept of critical period is based on a fixed time period during which certain behaviors emerge. This theory of development suggests that there is a time period, or window of development, in which certain skills must be learned in order for subsequent learning to occur.[11] Such a concept is one basis for *early intervention programs* for infants. A critical period for the development of oral feeding may explain some later feeding problems in infancy.[12] In the typical healthy newborn, the mouth is a source of pleasure and exploring, an important form of early learning. When a newborn has a prolonged period of respiratory support, for example, the baby may not associate mouth sensations with pleasure, but rather with discomfort. Under such circumstances the critical period for associating mouth sensations with pleasure and exploration may have been missed. After discharge, such an infant may be a reluctant feeder and have difficulty learning to enjoy food from a spoon.

Cognitive Development

The concept of biological and environmental systems interacting is seen in Illustration 8.2 showing *sensorimotor* development.[7] These skills influence feeding in important ways. For example, the stage when infants are very sensitive to food texture is also when their speech skills are emerging.[9] One broad theory of cognitive development of infants stresses sensorimotor learning. Piaget's

EARLY INTERVENTION PROGRAM Educational intervention for the development of children from birth up to three years of age.

SENSORIMOTOR An early learning system in which the infant's senses and motor skills provide input to the central nervous system.

Illustration 8.2 Sensorimotor stage of development.

SOURCE: Kail, Robert V. and John C. Cavanaugh. *Human Development: A Lifespan View.* 2nd ed. (2000). Belmont CA: Wadsworth/Thomson Learning, page 128.

SUBSTAGES DURING THE SENSORIMOTOR STAGE OF DEVELOPMENT

Substage	Age (months)	Accomplishment	Example	
1	0–1	Reflexes become coordinated.	Sucking a nipple	
2	1–4	Primary circular reactions appear —an infant's first learned reactions to the world.	Thumb sucking	
3	4–8	Secondary circular reactions emerge, allowing infants to explore the world of objects.	Shaking a toy to hear a rattle	
4	8–12	Means–end sequencing of schemes is seen, marking the onset of intentional behavior.	Moving an obstacle to reach a toy	
5	12–18	Tertiary circular reactions develop, allowing children to experiment.	Shaking different toys to hear the sounds they make	
6	18–24	Symbolic processing is revealed in language, gestures, and pretend play.	Eating pretend food with a pretend fork	

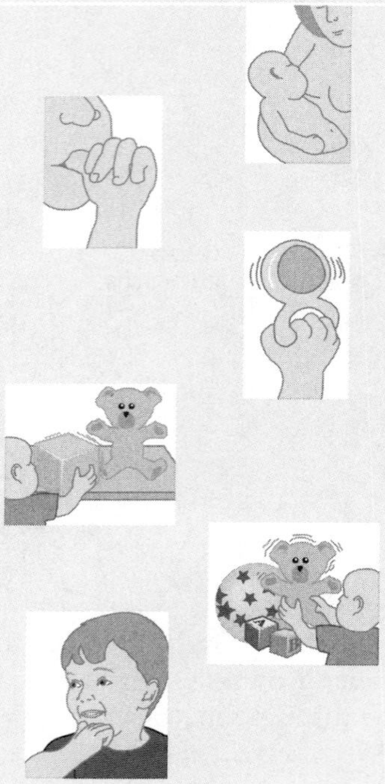

sensorimotor stage of development theorizes that an infant constructs an understanding of the world by co-ordinating sensory experiences with physical actions.[9] Piaget's theory is based on the maturation processes of various parts of the brain. Accordingly, learning can be stimulated in infants by recognizing their learning abilities at each age. Selecting toys and environments that are matched to the level of development is essential for the child's progress.[9] The suggestion is that learning occurs as an interplay of the environment and genetic makeup of the infant, which also supports the importance of early intervention programs for infants.[14] Human brain growth after birth is an area of active research, and the interaction of the environment with the infant is now seen as helping to structure the nervous system long-term.[14] The latest research suggests that access to adequate calories and protein may not be sufficient for maximizing brain maturation if the social and emotional growth of the infant is not stimulated simultaneously.[14]

Genetics and Development

Infant development theories are likely to emerge from genetic research. Mapping the human genome is clari-fying the genetic controls which turn genes on and off in different time frames and different sites within the body.[15] Nutritional requirements probably affect the interaction of genetics and environment in specific ways at specific times, and maybe in specific organs. An example of the precision required is the role of folic acid in the formation of the spinal cord early in gestation.[15] Various models are addressing various aspects of infant development as greater understanding of the brain emerges. Some developmental disabilities diagnosed in young children are hypothesized to result from faulty control of the timing of regulatory genes.[16]

Digestion and Development

Now good digestion wait on appetite, and health on both.

William Shakespeare, *Macbeth*

A healthy digestive system is necessary for successful feeding. Parents may worry about GI problems in part because of misinformation about nutrition in infancy.[17] For example, an infant with soft, loose stools may be considered by the parents to have diarrhea if they do not know that this is typical for breastfed infants.[18] Another common example is parents being concerned that their infant's GI discomfort may interfere with weight gain, even though growth usually progresses well. It takes up to six months for the infant GI tract to mature, and the process varies considerably from one individual to another. Some infants have GI problems that look like feeding problems for a long time.

During the intrauterine period in the third trimester, amniotic fluid swallowed by the fetus stimulates the lining of the intestine to grow and mature.[17] The healthy newborn's digestive system is then sufficiently mature to digest fats, protein, and simple sugars and to absorb fats and amino acids. Although healthy newborns do not have the same levels of digestive enzymes or stomach-emptying as older infants, the gut is functional at birth.[17]

After birth and through the early infancy period, the coordination of peristalsis within the GI tract improves. Maturation of peristalsis and rate of passage are associated with some forms of GI discomfort in infants.[15] Infants often have conditions that reflect the immaturity of the gut, such as colic, *gastroesophageal reflux,* unexplained diarrhea, and constipation.[3,19] Such conditions do not interfere with the ability of the intestinal villa to absorb nutrients, and typically do not hinder growth. Other factors influence the rate of passage through the colon and the GI discomfort seen in infants. These include:[3]

- *Osmolarity* of foods or liquids which affects how much water is in the intestine
- Colon bacterial flora
- Water and fluid balance in the body

Another link between GI function and development is demonstrated in the digestion of fats. Fat digestion requires lipase (a fat-digesting enzyme), as well as gall bladder, pancreas, and liver maturity. As fat-chain length increases, absorption efficacy decreases.[18] Fat malabsorption may result in changes in the colon bacteria of the large intestine, resulting in gas and abdominal pain. Infant formulas attempt to mimic breast milk's fat composition for ease of digestion. Formulas with a high percentage of medium-chain fatty acids are more expensive than standard formulas, however, and are more likely to be selected for sick infants or those with malabsorption.

Parenting

> *A babe in the house is a well-spring of pleasure.*
>
> Martin Farquhar Tupper, *On Education*

Even though the newborn can breast- or bottle-feed near birth, skills of new parents develop slowly. The ability of parents to recognize and respond to infant cues of hunger and satiety improve over time. New parents have to learn the temperament of their infant. Temperament has a biological basis and includes the infant's style or patterns of behavior.[9] Temperament includes the infant's emotional reactions to new situations, activity level, and sociability. The fit between the infant's temperament and the parents can increase or decrease feeding problems. For example, new parents may take a while to recognize that their newborn is more comfortable nursing in one noise level in the home than in another. Infants who are six months of age or older are better able to let parents know their needs and temperament. Conflicts in temperament may escalate as time progresses and become a factor in failure to thrive or other growth and feeding problems in later years.[8]

● ENERGY AND NUTRIENT NEEDS

The Dietary Guidelines for Americans address the needs of children age two and older, and not infants. Some recommendations are for older infants, such as guidance about sugar intake in beverages including juices. Some older infants are offered juice and juice-based drinks so frequently that such drinks become a source of sugar and excess calories. Excess juice for an infant is more than 4 fl oz in a day.[6]

Caloric Needs

The caloric needs of typical infants are higher per pound of body weight than at any other time of life. The range in caloric requirements for individual infants is broad, ranging from 80–120 cal per kg, (2.2 lb) body weight.[13] The average caloric need of infants in the first six months of life is 108 cal per kg body weight, based on growth in breastfed infants.[13] From 6 to 12 months of age, the average caloric need is 98 cal/kg.[13] Factors that account for the range of calorie needs of infants include the following.

> **GASTROESOPHAGEAL REFLUX (GER)** Movement of the stomach contents backwards into the esophagus, due to stomach muscle contractions. The condition may require treatment depending on its duration and degree. Also known as *gastroesophageal reflux disease (GERD)*.
>
> **OSMOLARITY** Measure of the number of particles in a solution, which predicts the tendency of the particles to move from high to low concentration. Osmolarity is a factor in many systems such as in fluid and electrolyte balance.

- Weight
- Growth rate
- Sleep/wake cycle
- Temperature and climate
- Physical activity

- Metabolic response to food
- Health status and recovery from illness

Current recommendations for infants of 108 and 98 cal are considered about 15% too high based on new study results.[13] However too few energy expenditure studies have been done on infants to reach a consensus about changing energy requirements.

Protein Needs

Recommended protein intake from birth up to six months averages 2.2 g of protein per kg of body weight, and from 6 to 12 months the need is for 1.6 g of protein per kg.[13] Protein needs of individual infants vary with the same factors listed for calorie needs. Protein needs are influenced more directly by body composition than calorie needs, since metabolically active muscles use more protein. The infants examined for determining protein requirements were 3-4 months old.

SHORT-CHAIN FATS   Carbon molecules that provide fatty acids less than 6 carbons long, as products of energy generation from fat breakdown inside cells. Short-chain fatty acids are not usually found in foods.

MEDIUM-CHAIN FATS   Carbon molecules that provide fatty acids with 6–10 carbons, again not typically found in foods.

LONG-CHAIN FATS   Carbon molecules that provide fatty acids with 12 or more carbons, which are commonly found in foods.

Most young infants who breastfeed or consume the recommended amounts of infant formula meet protein needs without added foods. Infants may exceed their protein needs based on the RDA when they consume more formula than recommended for age and when protein sources such as baby cereal are added to infant formula. Inadequate or excessive protein intake can result for infants who are offered formula that is not made correctly, such as when more or less water is used in preparation of powder or concentrates. The essential amino acids required by infants are the same over the first year of life, except for very low birthweight babies shortly after birth.[18]

Fats

There is no specific recommended intake level of fats for infants. Fat restriction is not recommended. The recommended level of 30% of calories from fat for healthy Americans is too low for infants. Breast milk provides 55% of its calories from fat.[19] The main source of fat in most infant diets is breast milk or formula. The percentage of fat in the diet drops after the infant has baby foods on a spoon added since most baby foods are low in fat. Infants need fat, which is a concentrated source of calories to support their high need for calories. Fat requirements in infancy are complicated by the differences in digestion and transport of fats based on fatty acid chain length.[20] *Short-* and *medium-chain fats* such as those in breast milk are more readily utilized than *longer chain fats,* such as in some infant formulas. Long-chain fatty acids are the most common type in food, but are more difficult for young infants to utilize. Examples of long-chain fatty acids are C16–C18 and include palmitic (C16:0), stearic (C18:1), and linoleic (C18:2) acids.[20]

Infants use fats to supply energy to the liver, brain, and muscles including the heart. The fact that infants have high caloric needs compared to older children means that they use fats more frequently for generating energy. Young infants cannot tolerate fasting because they use up both carbohydrate and fat energy sources so quickly, which is partially why they cannot sleep through the night. In rare cases, some infants cannot metabolize fats due to a genetic condition that blocks specific enzymes needed to generate energy. Such infants may get sick suddenly; a few have been identified with this rare condition of fat metabolism only after dying of what appeared to be Sudden Infant Death Syndrome (SIDS).[21]

Fats in food provide the two essential fatty acids, linoleic and alpha-linolenic acid. Essential fatty acids are substrates for hormones, steroids, endocrine, and neuroactive compounds in the developing brain.[20] The long-chain polyunsaturated fatty acids docosahexaenoic acid (DHA) and arachidonic acid (AA) are derived from the essential fatty acids. DHA and AA have been studied in eye and brain development and function.[22] Higher levels of these long-chain fatty acids were documented in breastfed compared to formula-fed infants, resulting in concern about essential fatty acids and neurological development. Fat composition of formulas and mental development is an area of active research in infants, particularly preterm infants.[22] Essential fatty acids are precursors of phospholipids, prostoglandins, and many other hormone-like compounds.[22] Cholesterol intake should not be limited in infancy because infants have a high need for it and its related metabolites in gonad and brain development.[23]

Metabolic Rate, Calories, Fats, and Protein—How Do They All Tie Together?

Growth can be simplified as a process of increasing stored protein and fats in organs, muscles, and tissues. Metabolic rate is the reason calories, fats, and protein are so important in infancy. The metabolic rate of infants is the highest of any period after birth, followed by the rate in adolescents.[11] The higher rate is primarily related to infants' rapid growth rate and also the high proportion of infant weight that is made up of muscle.

The usual body fuel for metabolism is glucose. When sufficient glucose is available, growth is likely to proceed. When glucose from carbohydrates is limited, amino acids will be converted into glucose and therefore made unavailable for growth. The conversion of amino acids into glucose is a more dynamic process in infancy compared to adults. The breakdown of amino acids for use as energy occurs during illness in adults, but can occur daily in fast-growing infants. Circulating amino acids in the blood from ingested foods will be used for glucose production, and if these are not sufficient, the body will release amino acids from muscles. This process of breaking down body protein to generate energy is catabolism. If catabolism goes on too long, it will slow or stop growth in infants. The precise site of all this metabolic activity is inside organs such as the liver, and in *mitochondria* within cells. If carbohydrates are not provided in sufficient amounts, growth will plateau as ingested protein and fats will be used for meeting energy needs. Although fatty acids can be converted into ketones to produce energy, they do not yield glucose.

Other Nutrients

The Dietary Reference Intakes (DRIs) provide recommendations for 0–6 and for 7–12 month old infants.[24,25] These are listed on the inside front cover of this text. Not all DRIs are listed for infants, although they are determined for older people. For example, iron and zinc have sufficient experimental support for recommended dietary allowances and estimated average requirements.[24] Some vitamins and minerals for infants up to 12 months, such as phosphorus, silicon, and vitamin B_6 are listed as not determined.[24] Recommendations for such nutrients are based on adequate intake levels, which are approximations or estimates when there is not sufficient evidence for establishing recommended dietary allowances.[24,25] Healthy infants born after 32 weeks of gestation are included in these recommendations, so these and other nutrient guidelines are appropriate for full-term and healthy premature babies.[24,25] Sick and recovering infants who are born with very low birthweight or with health complications were not included in the research that supports these recommendations, so their nutrient requirements are less clear. Key nutrients that may be limiting in infancy are fluoride, vitamin D, and iron, which are discussed later in the section on iron-deficiency anemia.

FLUORIDE　DRI for fluoride is 0.1 mg daily for infants less than six months of age, and 0.5 mg daily for 7–12 month olds.[24] Fluoride is incorporated into forming teeth, including those not yet erupted. If an infant does not meet the DRI for fluoride, dental

caries in early childhood are more frequent. If an infant has more fluoride than recommended, tooth discoloration may result later. Community water fluoridation is safe for breastfeeding women and for infants.[6] Most infants who live in areas with fluoridated water do not require another source of fluoride. Fluoride is low in breast milk.[18] In areas in which fluoridated water is not available, prescribed vitamins containing fluoride are recommended for breastfed infants. If families routinely purchase bottled water, they should select water that has fluoride added.[6]

VITAMIN D　Vitamin D, or preformed forms of vitamin D such as cholecalciferol, is required for bone mineralization.[24] DRI for infants up to 12 months is 5 mcg each day.[24,25] Sunlight may be a sufficient source for meeting the vitamin D requirement for infants who are taken outside in clothing that allows direct exposure of the skin to sunlight. Families are, however, encouraged to apply ultraviolet light skin-protective lotions on their children's exposed skin when outside. The higher levels of skin protective creams, though, block out vitamin D also. Families who have infants with limited exposure to sunlight, such as those with dark skin or those living in areas with limited sunny days, may need a vitamin D supplement. Commonly prescribed vitamin supplements for breastfed infants contain vitamin D.[6] Standard infant formulas also are supplemented with vitamin D.

MITOCHONDRIA　Intracellular unit in which fatty acid breakdown takes place and many enzyme systems for energy production inside cells are regulated.

SODIUM　Sodium is a major component of extracellular fluid and therefore an important regulator of fluid balance. Estimated sodium minimum requirements are 120 mg for 0–5 months and 200 mg for 6–11 months.[13] Breast milk was used as the basis for setting the minimum sodium requirements for infants, and infant formula is supplemented with sodium to match the amount in breast milk. Typical infants do not have difficulty maintaining body fluids and electrolytes, even though they may not show thirst as a separate signal from hunger. Young infants do not sweat as much as older children, so losses from sweating are not major losses for infants.[13] During illnesses such as diarrhea or vomiting, the risk of dehydration includes extra loss of sodium. Infants do not need salt added to foods to maintain adequate sodium intake. Many foods contain sodium even though salt is not an added ingredient.[27] Examples are commercially prepared baby foods, such as carrots and beets.

LEAD　Although lead is not a nutrient, it can be associated with nutrient intake of iron and other minerals during infancy. Blood lead elevation can be toxic to the developing brain, interfere with calcium and iron absorption, and bring about slowed growth and shorter stature.[28,29] Infants may inadvertently be exposed to environmental sources of lead. Lead may be a contaminant in water used in making infant formula or in the home, particularly if the house was built before 1950. Screening for lead poisoning is recommended starting at 9 to 12 months of age.[6] If siblings have been found to have lead poisoning, screening for lead may be started at 6 months. Infancy is not the peak age for lead poisoning, but infants can be exposed if their parents work with lead-containing products. For example if the father is a truck driver who uses leaded gasoline, his work clothes may have lead dust on them. If these clothes are mixed with the rest of the household laundry, or children play in the laundry room where lead dust has settled, any infant in the home may become exposed.

FIBER   Although there are dietary fiber recommendations for toddlers and children, we have no dietary fiber recommendations for infants.[30] Commercial and home-made baby foods are generally not significant sources of dietary fiber, since preparation methods reduce dietary fiber.[27] However, fiber-containing foods such as fruits, vegetables, and grains are appropriate foods for infants. High-fiber foods can be offered in infancy and are recommended to establish good food habits such as accepting a variety of fruits and vegetables. At older ages, foods which are high in dietary fiber are recommended for preventing constipation. In infancy certain types of juices with high osmolarity may be helpful for constipation management, rather than high-dietary fiber foods that require chewing. Both juices and dietary fiber impact the infant's bowel pattern by bringing more water into the lower intestine to relieve constipation.[3]

PHYSICAL GROWTH

Why Is It Important to Track Infant Growth?

Tracking growth in length and in weight helps identify health problems early, preventing or minimizing slowing of a growth rate. Parents understand that a sign of health is growth of their baby. By school age,

most homes have a wall proudly displaying many height measurements over time. Healthy newborns double their birthweight by 4–6 months and triple it by one year.[1] Growth reflects nutritional adequacy, health status, economics and other environmental influences on the family. There is a wide range of normally, however, and healthy babies may follow different patterns of growth. Often healthy infants have short periods when their weight gain is slower or faster than at other times. Slight variations in growth rate can result from illness, teething, inappropriate feeding position or family disruption. The overall growth pattern is important and each assessment is compared to the whole picture. Table 8.3 shows typical growth rates during the first year of life.

Accurate assessment of growth and interpretation of growth rates are important components of health care for infants (Illustration 8.3). Accuracy requires calibrated scales, a recumbent-length measurement board with an attached right-angled headpiece, and a nonstretch tape for measurement of head circumference. Table 8.4 shows how to avoid common errors that interfere with accurate measurements. Makers of measuring equipment recommend checking their accuracy periodically, such as once a month. Calibration of measuring equipment is carried out by using standard weights (or lengths) to confirm accuracy and precision over the range that the equipment measures.

Standard measurement techniques should be used; these require practice and consistency. Equipment needed to measure infant growth is different from equipment for measuring children and adults. The scale bed must be long enough to allow the infant to lie down. Length is measured with the infant lying down with a head and foot board at right angles to the firm surface. Positioning the baby quickly is a skill needed for accurate measurement of recumbent length. In weighing, clothing, hair ornaments, the amount the baby jiggles the scale, and the state of health of the infant are examples of factors that could add error to measurements.

Interpretation of Growth Data

The Centers for Disease Control (CDC) 2000 infant growth charts are improved in representing diversity.[1] These growth charts are based on five national sur-

Table 8.3   Typical gains in weight and height for age in infancy.[31]

AGE	WEIGHT GAIN GRAMS	WEIGHT GAIN GRAMS (POUNDS)	LENGTH GAIN mm	LENGTH GAIN mm (INCHES)
	Per Day	**Per Month**	**Per Day**	**Per Month**
0–3 months	20–30	600–900 (1.3–2)	1	30 (1.2)
3–6 months	15–21	450–630 (1–1.4)	0.68	20 (0.8)
6–12 months	10–13	300–390 (0.7–0.9)	0.47	14 (0.6)

Illustrations 8.3   Infant being measured on length board and scale.

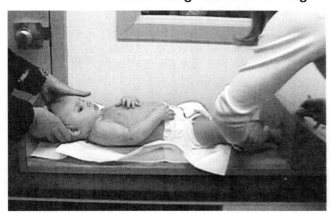
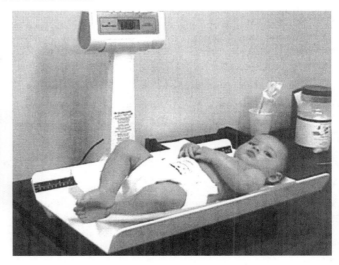

veys and represent a larger sample of infants than that used for previous growth charts. Growth data from infants with birthweights <1500 grams are excluded from the 2000 growth charts.[1] The infant growth charts are based on infants weighed nude on calibrated scales. The charts take into account differences in the growth patterns for formula-fed compared to breastfed infants, regardless of race or ethnic background.[1] Various growth charts are provided in Appendix A. Growth charts for 0–36 month olds consist of a prepared graph for each gender, showing:

- Weight for age
- Length for age
- Weight for length
- Head circumference for age

The purpose of assessing growth status multiple times is for assessment of growth progress. The more times a baby has been measured and the growth plotted, the more likely the growth trend will be clear, in spite of minor errors. Multiple measures can identify a change in weight or length gain and the need for intervention.

Growth is so fast that it may be easier to determine growth problems during infancy than later. Every month in infancy there is an increase in both weight and length, which is not expected in older children. Warning signs of growth difficulties are lack of weight or height gain, plateau in weight, length, or head circumference for more than one month, or a drop in weight without regain within a few weeks.[7] Head circumference increases as a result of brain growth. If head circumference is not increasing typically, neither weight nor height increases are likely to track on standard growth percentiles. In the rare circumstance that an infant has a rapid increase in head circumference, this is not a sign of nutrition or growth, but may signal a condition that requires im-

mediate attention to protect brain development.[3] The rate of weight gain during infancy is not necessarily predictive of future growth patterns after infancy, nor a risk for long-term overweight, compared to the weight gain pattern later in childhood.[32] Nevertheless, an infant that is gaining weight faster than expected may be at risk for overweight in the short term, and needs interventions based on feeding assessment and interviews with the parents.[33] Examples of interventions are working with parents to look for signs of

Table 8.4
Measuring growth accurately in infants.

To Avoid Measurement Errors
- Use measuring equipment that was calibrated recently.
- Confirm that the scale is on zero before starting.
- Make sure the infant is not holding or wearing anything that adds weight or length.
- Confirm the position of the infant for length measurements:

    Head position—the infant's eyes are looking straight up and the head is in midline, touching the head board. Neither hips nor knees are bent.
    Heel is measured with foot flat against the foot board.

- Head circumference measure is at the widest part of the head.

To Avoid Growth Plotting Errors
- Calculate the age accurately in months after confirming the date of birth.
- Confirm plotting on the metric scale if kilograms were measured, not the pound scale.
- Confirm that the plotted weight and length are marked well enough to read easily without being so large as to change percentiles.

hunger after trying other soothing methods, such as swaddling and rocking, and for formula-fed infants, to reduce the volume in each bottle, adding a water bottle if they still see hunger signs.

FEEDING IN EARLY INFANCY

Food is the first enjoyment of life.

Lin Yutan, *The Importance of Living*

Breast Milk and Formula

The American Academy of Pediatrics and the American Dietetic Association recommend exclusive breastfeeding for the first six months of life and continuation of breastfeeding for the second six months as optimum nutrition in infancy.[34,35] Infants who were born prematurely benefit from breastfeeding, too. Encouragement of breastfeeding right after birth, well before the mother's milk supply is visible to her, is an example of a birthing practice that is endorsed.[34] Other recommended practices include teaching safe handling by refrigeration, freezing, and heating of expressed human milk as an important support service for families with a premature infant.[36]

Breast milk nutrient composition is detailed in Chapter 6. For young infants less than 4–6 months old, no other liquids are recommended.[33] If water is offered, some families may inadvertently allow water to replace adequate amounts of breast milk or formula. Volume of formula for young infants before

Table 8.5 Typical daily volumes for young infants not being breastfed.

Age of Infant	Typical Intake of Formula per Day (24 hours)
Birth to one month	16—20 fl oz per day, 8–12 feedings/day, 1–2.5 fl oz per feeding
One to two months	18–26 fl oz per day, 8–10 feedings/day, 2–4 fl oz per feeding
Two to three months	22–30 fl oz per day, 6–8 feedings/day, 3–5 fl oz per feeding
Three to four months	24–32 fl oz per day, 4–6 feedings/day, 4–8 fl oz per feeding

adding food on a spoon is shown in Table 8.5. The growth rate and health status of an infant are better indicators of the adequacy of the baby's intake than the volume of breast milk or formula. Infant formulas for full-term newborns are typically 20 cal/fl oz when prepared as directed. Commercially prepared formula for infants who were born prematurely have either 22 or 24 cal/fl oz when prepared. Some health providers recommend further increasing the caloric density of formula for some preterm infants, but such recommendations are not appropriate for most infants. Most infants can be quite flexible in accepting formula lukewarm or cold, and in accepting changes in formula brands.

Table 8.6 gives an overview of commercially available infant formulas and how they are modified

Table 8.6 How infant formulas are modified compared to breast milk.

Macronutrients	Breast milk	Cow's milk based formula	Soybean based formula
Protein	7% of calories	9–12%	11–13%
Carbohydrates	38% of calories	41–43%	39–45%
Fats	55% of calories	48–50%	45–49%

Other ways infant formulas are modified compared to breast milk

Calorie level	Increase in calories from 20 calories/fl oz to 22 or 24 calories/fl oz (for preterm infants)
Form of protein	Formula protein is broken down to short amino acids fragments or to single amino acids
Type of sugar	Lactose from cow's milk is replaced by other sugars, such as sucrose
Type of fat	Medium-chain fatty acids partially substituted for longer chain fatty acids
Allergy/intolerance	Replacement of milk-based protein with protein from soybeans or smaller amino acid fragments or single amino acids
Micronutrients	Increased calcium and phosphorus concentration for preterm infants
Thickness	Added rice

compared to breast milk.[37] Some formulas have been developed for common conditions for healthy infants, such as gastroesophageal reflux (GER) or frequent diarrhea. The specialty formula market appears to be growing, to address health concerns about infants, but also lifestyle factors. Examples of newer commercial formulas are follow-up formulas that are marketed for late infancy. Most infant formula does not require a prescription for purchase. Physician and other health care provider recommendations compete with many other sources of information for parents who are selecting infant formula. Internet parent chat groups, WIC, and social marketing are examples of the wide variety of sources that offer advice to parents about selecting infant formulas.

Cow's Milk During Infancy

The American Academy of Pediatrics recommends that whole cow's milk, skim milk, and reduced fat milks not be used in infancy.[38] Iron-deficiency anemia has been linked to early introduction of whole cow's milk. Low iron availability may come about as a result of gastrointestinal blood loss, low absorption from binding with other minerals (calcium and phosphorus) or from the lack of other iron-rich foods in the diet.[38] Studies on infants who were 7.5 months of age confirmed earlier findings that blood loss with whole cow's milk is more likely if the infant had been breastfed earlier rather than fed infant formula.[39] In 1997, 39% of infants were fed whole cow's milk at or before age 12 months.[39] The high cost of infant formulas may result in families selecting cow's milk for older infants who are not breastfed.

DEVELOPMENT OF INFANT FEEDING SKILLS

Infants are born with reflexes that prepare them to feed successfully. As noted earlier, these reflexes include rooting, mouthing, head turning, gagging, swallowing, and coordinating breathing and swallowing.[9] Infants are also born with food-intake regulation mechanisms that adjust over time with development of the infant.[40] In early infancy, self-regulation of feeding is mediated by the pleasure of the sensation. Fullness is more comfortable than hunger. Inherent preferences are in place for sweetness, which is also a pleasurable sensation.[41] After the first 4–6 weeks, reflexes fade and infants learn to purposely signal wants and needs. However it is not until much later, about age 3 that children can verbalize that they are hungry. In between the reflexes fading and the child speaking, appetite and food intake are regulated by biological and environmental factors interacting with one another.

Table 8.7 shows infant developmental milestones and readiness for feeding skills.[9,10,12] The interaction of biology and environment prevails here, too. For example, depression in a caregiver may be an underestimated variable in the development of infant feeding. Maternal depression may bring about a lower level of interaction between the parent and infant during feeding, reducing the number or volume of feedings, and increasing the risk of slower weight gain.[42] Media influences and changes in social practices also affect how babies are fed. Examples are cultural and ethnic perceptions of breastfeeding and the availability of quality childcare for infants.

Several models help assess readiness for a breastfed infant to start being fed from a spoon in the 4–6 month period. The developmental model is based on looking for signs of readiness such as being able to move the tongue from side to side, without moving the head.[10,12] The infant must be able to keep her head upright and sit with little support before initiating spoon-feeding. Outdated models are a chronological-age model, in which only the age of the infant is considered, as well as a model based only on cues from the infant. For example, the mother looks for cues that the infant is dissatisfied with a full-liquid diet, so spoon-feeding is started as early as two months or as late as eight months.

Most infants adapt to a variety of feeding regimens, and various feeding practices can be healthy for them. The ability of the parents to read the infant's cues of hunger, satiation, tiredness, and discomfort influence the feeding skill progression. The cues infants give may include:

- Watching the food being opened in anticipation of eating
- Tight fists or reaching for the spoon as a sign of hunger

Connections
Infant Formulas

- Is infant formula a food or a drug? The Food and Drug Administration Center for Food Safety Applied Nutrition regulates the composition, labeling, and inspection of infant formula, both imported and domestic forms.

- Infant formula is a $3.5 billion industry overall in the United States. The average consumer spends $1,000 a year on baby formula, according to PBM products, maker of Babymil®.

Connections
Baby Senses

- Babies have taste buds all over the insides of their mouths, not just on their tongues, so they are more sensitive to taste than adults and children.

- Babies have a strong sense of smell, enabling them to recognize their mothers by scent.

Table 8.7 Development of infant feeding skills.

Chronological Age	Developmental Milestone	Feeding Skills
Birth to 1 month	Vision is blurry; hears clearly Head is oversized for muscle strength of the neck and upper body.	Suckling and sucking reflexes Frequent feedings of >8–12 per 24 hr Only thin liquids tolerated.
1–3 months	Cannot separate movement of tongue from head movements Head control emerges Smiles and laughs Puts hands together	Volume increases up to 6–8 fl oz per feeding, so number of feeding per day drops to 4–8 per 24 hr Sucking pattern allows thin liquids to be easily swallowed Learns to recognize bottle (if bottle-fed)
4–6 months	Able to move tongue from side to side. Working on sitting balance with stable sitting emerging Drooling is uncontrolled Disappearance of newborn reflexes allows more voluntary movements Teething and eruption of upper and lower central incisors	Interest in munching, biting, and new tastes Cannot easily swallow lumpy foods, but pureed foods swallowed 6–8 fl oz per feeding and 4–5 feedings per day (may be variable if breastfeeding). Holds bottle (if bottle-fed)
7–9 months	Hand use emerges, with pincer grasp and ability to release Stable independent sitting Crawling on hands and knees Starting to use sounds, may say "mama" and "dada"	Self-feeding with hands emerges Munching and biting emerges Indicates hunger and fullness clearly Prefers bottle but little loss from a held open cup
10–12 months	Can pull to stand, standing alone emerges Enjoys making sounds as if words Can pick up small objects, such as a raisin Can bang toys together with two hands Has consistent routines about bedtime, diaper changing Usually does not drool any more	Likes self-feeding with hands Spoon self-feeding emerges Drinks from an open cup as well as from a bottle Uses upper and lower lip to clear food off a spoon Enjoys chopped or easily chewed food or foods with lumps Sitting position for eating Enjoys table foods even if some baby foods still used

- Showing irritation if the feeding pace is too slow or if the feeder temporarily stops
- Starting to play with the food or spoon as the infant begins to get full
- Slowing the pace of eating, or turning away as they are wanting to end the meal
- Stopping eating or spitting out food when not hungry

Infants relate positive and pleasurable attributes to satisfaction of their hunger as part of a successful feeding experience. If there are long episodes of pain from gastroesophageal reflux or constipation, these can become the basis for later feeding problems, as the association of eating and pleasure is replaced by an association of eating and discomfort.[43] An infant who makes the association between eating and discomfort is likely to be seen as an irritable baby. This may set up a cycle of the infant being difficult to calm and the parent being frustrated. If this cycle is not re-placed by the more positive association of eating and pleasure, the feeding difficulty in infancy may later be characterized by pickiness, food refusals, and difficult mealtime behavior in an older child. A negative association of pain and eating may persist, and appear as a behavioral problem at meal times.

The Importance of Infant Feeding Position

Positioning infants for feeding with a bottle and for eating from a spoon are important because improper positioning is associated with choking, discomfort while eating, and ear infections.[10] The semi-upright feeding position as exemplified in car seats or infant carriers is recommended for the first few months.[10] Unsafe feeding positions, such as propping a bottle or placing the baby on a pillow, increase the risk for choking and overfeeding. The recommended sleeping position for

Case Study 8.1
Baby Samantha Does Not Like to Eat

Samantha is a healthy 8 month old girl who lives with her mother Kathy, her father, and older sister, who is almost 3 years. Both parents now work full-time and both children attend daycare full-time. Kathy nursed Samantha exclusively before she returned to work, and built up a supply of frozen breast milk. She nurses her twice per day now, early in the morning and before she goes to sleep. Samantha gets breast milk offered in bottles at daycare. Samantha is reported by the daycare staff to be a good baby. However, when Kathy picks her up after work, Samantha wants to be held and will not sit in her highchair or eat dinner. She cries if she is not held. Samantha's sister wants to eat as soon at they get home. Kathy has so much to do at home after work, she finds it difficult to hold Samantha at such a busy time. Kathy thinks that Samantha must be hungry and that she would be less irritable if she ate her dinner.

Intervention

Kathy is encouraging an upset baby to eat, which is not a good idea. Samantha at 8 months of age is still learning self-regulation and is used to nursing more often. She is learning to calm herself in new ways other than being held and nursing, but this takes time. Kathy may interpret Samantha's signs of irritability as signs of hunger, but they may be signs of needing quiet comfort and soothing. It is likely that Kathy can slightly change her routine to accommodate Samantha's needs. She could bring a snack for the older child with her from her work and briefly stop at a park on the way home, especially in nice weather. Then Kathy can hold Samantha quietly while her sister has a snack, away from home where Kathy feels she has so many other things to do. This gives mother and baby the quiet time that Samantha needs before she can show an interest in eating dinner. Her older sister may associate eating with returning home from daycare, but it may be that she also wants attention from her mother more than she wants a snack.

young infants is lying on the back without elevating the head on a pillow. This position is not recommended for feeding, which reinforces why feeding an infant in bed with a bottle is not generally recommended.[3,19]

Spoon-feeding also has a recommended infant feeding position. The infant can better control his mouth and head in a seated position with good support for the back and feet. The person offering the spoon should sit directly in front of the infant and make eye contact without requiring the baby to turn his head.[10,12] A highchair is an appropriate feeding chair when the infant can sit without assistance. The infant should be kept in a sitting position by use of a seat belt so that the hips and legs are at 90 degrees. This position assists the infant's balance and digestion. If the infant is sliding out from under the tray of the highchair with the hips forward, the stomach is under more pressure and spitting up is more likely.

Some apparently healthy infants show resistance to learning feeding skills or react to the introduction of foods in an unusual manner. These problems in early feeding experiences are sometimes warning signs of more general health or developmental problems. They may signal emerging problems that cannot be diagnosed until later. Families who call attention to early feeding problems may assist their infants in the long run by having the problem recognized earlier. For example, some infants who start and stop feeding frequently, but then do not feed for several minutes in a row, may later have heart problems found. The coordination of eating and breathing may have been the basis for the starting and stopping. Some infants are very reluctant feeders and are later diagnosed with a milk protein intolerance.

> **WEANING** Discontinuation of breastfeeding or bottle feeding and substitution of food for breast milk or infant formula.

Preparing for Drinking from a Cup

The process of *weaning* starts in infancy, and usually is completed in toddlerhood. The recommended age for weaning the infant from the breast or from a bottle to drinking from a cup is from 12–24 months.[12] Breastfed infants may be transitioned to drinking from a cup without ever having any liquids in a bottle. If breastfeeding is continued as recommended for the first year of life, introducing a cup for water and for juices after six months is recommended, near the time that foods are offered on a spoon. By the time weaning from breastfeeding is planned, the one year old infant will be skilled enough at drinking from a cup to meet her fluid needs without a bottle. Infants who are not exclusively breastfed, or are breastfed for less than 12 months need to have additional fluids offered in a bottle since their ability to meet their fluid needs by drinking from a cup are not sufficiently developed. Developmental readiness for a cup begins at 6–8 months.[12] Eight month old infants enjoy trying to mimic drinking from open cups that they see in their home. However the ability to elevate the tongue and control the liquid emerges later, at closer to a year. The 10–12 month infant enjoys drinking from a held cup and trying to hold her own cup, even though the breast or bottle is the main feeding method. Infants are likely to decrease total intake of calories from breast milk or infant formula if a cup is used, since they are less efficient in the mouth skills needed. At first the typical portion size of fluid from a cup is one to two ounces. The infant who is weaned too soon may plateau in weight because of decreased total calories. The drop in total fluids consumed may result in constipation. Changing from a bottle to a covered sippy cup with a small spout is not the same developmental step as weaning to an open cup.[12] The mouth skills in controlling liquids with the tongue are more advanced with an open cup. The skills learned in drinking from an open cup also encourage speech development.

HYPOALLERGENIC   Foods or products that have a low risk of developing food or other allergies.

> They say fingers were made before forks and hands before knives.
>
> Jonathan Swift

Food Texture and Development

Weaning is not complete until the nutritional value of breast milk is provided from foods. Infants advance from swallowing only fluids to pureed soupy foods at 4–6 months.[12] Before that they can only move liquids from the front to the back of the mouth. The mouth is exquisitely sensitive to texture. If food with soft lumps is presented too soon, it causes an unpleasant sensation of choking. When infants are 4–6 months, they can move their tongue from side to side. When they

are 6–8 months, they are ready for foods with a lumpy but soft texture to elicit munching and jaw movements.[12] These movements simulate chewing. By 8–10 months infants are able to chew and swallow soft mashed foods without choking. It is important to offer infants foods that do not require much chewing, since infants do not develop mature chewing skills until they are toddlers. Infants are likely to choke if offered foods that require chewing or large pieces of foods.

What Infants Eat

Infants begin with food offered on a spoon in a small portion size of 1–2 tablespoons for a meal, with one or two meals per day. The purpose of offering food on a spoon to infants at 4–6 months is to stimulate mouth muscle development, not for nutritional needs, which are met from breast milk. The first food generally recommended for infants at 4–6 months is baby cereal, such as dry rice cereal mixed with water or breast milk. Rice cereal is a common first food since it is easily digested and *hypoallergenic*. When to add baby cereal or other food to an infant's diet is determined not only by developmental milestones as recommended, but by other reasons, such as these:

- Some parents add baby foods because they think this will make the baby sleep longer. This practice is neither recommended nor effective for most infants. This common belief may result in introduction of baby cereal before the infant has developed the skills to eat from a spoon.
- Some families are instructed by pediatricians to add dry rice cereal as part of the treatment for gastrointestinal problems, as it tends to thicken infant formula.

Fruits and vegetables, such as pears, applesauce, or carrots are also sometimes first foods for infants. What are considered healthy first foods for infants varies in different cultures and ethnic groups. Regardless of what foods are offered first, the timing and spacing of new foods can be used to identify any negative reactions. Common recommendations for parents of 6 month olds are to add only one new food at a time and to offer it over two or three days. There are specific recommendations regarding the timing and spacing of foods known to trigger

Connections
Baby Foods

- The first Gerber baby foods were sold in the late 1920s and included peas, carrots, prunes, spinach, and beef-vegetable soup.

- Organic baby foods comprise a fast growing component of the market. Sales of organic baby foods, such as the Heinz brand Earth's Best®, represent up to 10% of total baby food sales in the United States, and 40% in Britain.

food allergies in families with histories of this problem (discussed later in the Food Allergy section).

Commercial baby foods are not required by infants. Parents can prepare baby foods at home using a blender or food processor, or by mashing with a fork. Care must be taken, however, to avoid contamination of home-prepared baby food by bacteria on food or from unsanitary storage methods. The nutrient content of home-prepared baby foods can vary widely depending on how it is prepared and stored. Adding salt and sugar to home-prepared baby foods are examples of variables that can decrease nutritional quality. The advantage of home-prepared baby foods is that a wider variety of foods may be introduced that are likely to be a part of the diet later. Additionally, money unspent on commercial baby foods may be significant savings for some families.

Commercial baby foods are commonly selected because of their sanitation and convenience. Several are listed in Table 8.8.[27] Moreover, families who pack food for daycare or who travel with infants find commercial baby foods convenient. Parents have a lot of choices to make in selecting baby foods. Selection should be based on the nutritional needs of the infant, not on what is available in local stores and the eating habits of the adult shoppers. Examples of baby foods that may reflect shopper's selections rather than baby needs are fruits with added tapioca or baby food desserts and snack foods. They are not recommended for most infants. Jar sizes of baby foods are based on industry standards, not necessarily recommended portion sizes. Portion sizes for infants should be based on appetite. Finishing an opened jar of baby food may encourage overeating if parents do not pay attention to signs from the infant.

Many foods eaten by other family members are appropriate foods for infants who are 9–12 months of age. Examples are regular applesauce, yogurt, soft cooked green beans, mashed potatoes, cooked Cream of Wheat, and Cheerios.

Table 8.8 Commercial baby foods.[27]

Type	Portion Sizes	Target Age/Skills
Single-ingredient pureed fruits and vegetables, 25–70 cal/jar	71 g (2.5 oz) jar	4–6 months, introduced on a spoon, with portion size of 2–3 tbs/meal
Powdered cereals, 60 cal/serving	Dry: ½ oz or 4 tbs/serving, mixed with water, juice, breast milk, formula, or other liquids	4–8 months
Jarred cereal mixed with fruits, 90 cal/jar	"Wet:" 113 g (4 oz.) jar	4–8 months
Juices, fruit, and vegetables, 60–100 cal/bottle with added vitamin C, calcium	4 and 6 fl oz bottles	4–10 months
Fruits and vegetables, pureed textures, added ingredients, such as tapioca and mixtures	113 g (4 oz) jar, two jars per meal	4–8 months, no munching nor food intolerances
Meat mixtures, containing 3–4 g protein/jar, 50–70 cal/jar	113 g (4 oz) jar, one jar per meal	6–9 months, no munching or food intolerances
Desserts, 0–2 g protein/jar and 80–100 cal/jar	113 g (4 oz) jar	6–9 months, no munching
Fruits and vegetables with mixed textures, 0–4 g protein/jar, 70–160 cal/jar	170 gm (6 oz) jar, one jar or 12 tbs/meal	9–12 months, side to side tongue movements and munching
Meat-based dinners, with mixed textures, 3–5 g protein/jar and 90–130 cal/jar	170 g (6 oz) jar, one jar or 12 tbs/meal	9–12 months, side to side tongue movements and munching
Meats with textures, 10–11 g protein/jar and 70–100 cal/jar	71 g (2.5 oz) jars	9–12 months, side to side tongue movements and munching
High-texture baby foods. Diced fruits, vegetables, and meats. Dinners have pieces as lumps.	Dices are 71 or 128 g jars (2.5 or 4.5 oz). Dinners are 170 g (6 oz) jars	Finger foods (to be picked up), requires self-feeding with hands and spoon, chewing
Baked products, such as zwieback toast and biter biscuits	One zwieback toast is 7 gm. One biter biscuit is 11 gm	10–12 months. Requires biting and munching, limited chewing

Photo Disc

Case Study 8.2
Paul and His Baby Food

Paul is 7.5 months old and is fed infant formula and jarred baby foods. He is offered two or three different foods at both his lunch and dinner. He opens his mouth and likes to eat at first, but then he starts reaching to put his hands in the food and smear it on the highchair tray. His mother thinks he is enjoying his meal, but his Grandmother thinks that his hands should be held so that he cannot play in the food. His father thinks that he needs regular foods, not baby foods, and that he smears the baby foods because he does not like it. He wants to offer him some foods from his plate. Each adult tries to feed Paul in a different way, and no one is sure what is best for Paul.

Intervention

The three adults are each a little right and a little wrong about how Paul should be fed at almost 8 months of age. Paul's father may be correct in noticing that Paul may be able to handle more texture than provided in baby foods. Commercial baby foods are not the only option. Some infants eat better when challenged to move their tongues and jaws in new ways. He may enjoy fork-mashed foods such as banana, potatoes, regular oatmeal, and canned fruits. He cannot safely eat foods that require chewing at this age. However he may be ready for foods that will stimulate his ability to munch on foods. His mother is correct in letting him have the experiences of touching his food to explore it. Many babies teach themselves new skills by such activities. If Paul resisted touching his food with his hands that might be a sign of immature development of fine motor skills. Too much focus on being neat, such as restricting his hands, is not recommended. However, his grandmother may be right to catch a negative behavior early. If he is learning to get attention by playing rather than eating during meal time, he may continue to do this as an older child. Many toddlers use not eating as a way to gain attention at meal time, and this behavior may interfere with development of a consistent meal and snack routine in the home. The baby is likely to adjust to the different feeding styles of his caretakers, even though a consistent meal-time routine is recommended. Continuing to have all three adults enjoy caring for Paul is best for him.

Water

Breast milk or formula generally provides adequate water for healthy infants for the first 4–6 months. Infant drooling generally does not increase the need for water. In hot, humid climates infants may increase their need for water, but water should not replace breast milk or formula. Added water can be used to meet fluid needs, but not calories and nutrient needs. All forms of fluids contribute to meeting the infant's water needs. These include breast milk, formula, juices, and baby foods. Often parents are reluctant to say they offered their infants sips from their own glasses containing soft drinks or drinks containing caffeine or alcohol. This may be important information to include in a food intake record,

especially if the contents are not recommended for babies. The replacement of an infant formula with a less nutritionally rich alternative such as juice, "sports drinks," cola, or tea has been found to be a contributor to lower quality diets for infants.[19]

Water needs of infants are a concern since dehydration is such a common response to illness in infancy. The infant has limited ability to signal thirst, especially when sick. Vomiting and diarrhea result in dehydration more rapidly in infants than in older children, with symptoms that are more difficult to interpret.[44] The loss of electrolytes is a part of dehydration.[6] Replacement of electrolytes has been the basis for a variety of over-the-counter fluid replacement products, such as Pedialyte®, "sports drinks", and Gatorade® that are marketed to parents. These products contain some glucose (dextrose) along with sodium, potassium, and water. The amount of glucose provides significantly lower calories than breast milk or formula, usually 3 cal per fl oz compared to 20 cal per fl oz. Such liquids can be overused, and may result in weight loss even in healthy infants.

How Much is Enough?

A wide variation in food intake for healthy infants can be seen. Parent reports of infant feeding behavior changes as the infant/parent interaction matures from early to late infancy. During early infancy while the sleep/wake cycle of the infant is irregular, it is common for new parents to interpret all discomfort as signs of hunger. The infant's ability to self-calm is a developmental step that plays out differently with different temperaments and parenting styles.[7] The infant who is quite sensitive to what is happening around him is likely to be viewed as irritable and hungry if he cries frequently. In contrast, the infant who sleeps through usual household noises and is less reactive to her immediate environment is likely to be offered food fewer times per day. As a result of different responses to temperament, a pattern of excessive or inadequate food and formula intakes may result.

The following is an example of excessive intake for a 3-month old not being breastfed.

> Total formula intake: 33 fl oz: seven bottles per day, ranging from 3–5 fl oz per bottle, offered at 7:30 AM, 11:00 AM, 12:45 PM, 2:30 PM, 5:30 PM, 8:45 PM, and 11:30 PM. One jar of baby food applesauce fed on a spoon at 9:00 AM.

This is over-feeding because excessive formula is being offered, along with premature offering of spooned food. The frequency of the bottles being offered suggest that the parents are interpreting her signs of discomfort as hunger when she may have other needs, such as being held, changed, or calmed by movement or touch. Overfeeding is less likely with breastfeeding.

In the first few months, the oral need to suck is easily confused with hunger by new parents. The typical forward and backward tongue movements of the infant's first attempts to eat from a spoon may seem to be a rejection of food by new parents.[12] The infant appears to spit out the food, but this is a sign of learning to swallow and not necessarily a taste preference. The same food that appears to be rejected will be accepted as the infant learns to move the food from the front to the back of the mouth.[12] It may appear to a parent that the infant does not like a food if he appears to choke. This choking response is more likely based on the position of the spoon on the tongue. The mouth is very sensitive, particularly towards the back. If the bowl of the spoon is too far back, it will cause a gagging reaction, regardless of the taste of the food.

How Infants Learn Food Preferences

Infants learn what foods are good to eat. Many factors contribute to how sensitive infants are to different tastes. Breastfed infants may be exposed to a wider variety of tastes within breast milk than infants offered formula.[45] The different foods that the breastfeeding mother eats may result in some flavor compounds being transmitted to the nursing infant. Studies on infants in the age range of 4–7 months showed that acceptance of new foods was more rapid than acceptance of new foods after the first year of life.[40] In the 1920–1930s, a pediatrician conducted studies on the self-selection of food by weaning infants.[46] Her classic studies were interpreted by later generations as supporting the concept that infants and children instinctively select a well-balanced diet. However, these studies were subject to misperception because careful attention was not paid to the original methods, in which only nutritious and unsweetened foods were available.[47]

How much selection of foods by infants is based on biological factors as compared to psychological factors is still controversial. Food preferences of infants are largely learned, but genetics, food availability, and parents' preferences affect them. Food preferences developed in infancy set the stage for lifelong food habits. The development of trust and security for an infant are crucial, but this need not be linked to overfeeding the infant.[14] Recognition of an infant's specific needs and responding to them appropriately is important. For example, a tired baby needs a nap, and offering food to a tired baby is not meeting her needs at that time.

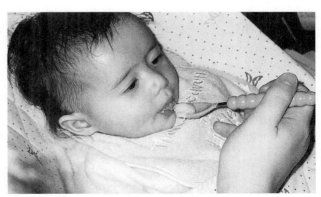

Illustration 8.4 Infant reaching with her tongue for a spoon.

Billy: A Hard-to-Feed Infant

Infants who refuse to accept new foods but have the mouth skills to eat them are difficult to feed. Billy, at 7 months, does not accept any foods with lumps, is difficult to get to sleep, waking up frequently during the night, and is so irritable that he does not stop crying when shown a new toy.

POSSIBLE PHYSIOLOGICAL BASIS   Billy needs to be taken for a medical check-up. An infant who is so difficult to feed and to get to sleep may have untreated GER, which causes discomfort when eating and sleeping. Assessment and treatment may prevent this feeding problem from interfering with Billy's growth.

POSSIBLE DEVELOPMENTAL BASIS   Billy may be at risk for developmental delay. If the infant refuses new foods, shows little interest in toys, and is unable to self-calm enough to sleep, but does not have any physical health problem, he may have delayed development. Billy would benefit from a full assessment of his overall development and referral for early intervention services.

POSSIBLE PARENTING BASIS   Billy's parents may need help in reading his cues about hunger, pain, or need for stimulation from his parents. If he is offered only a limited variety of foods with little interaction during the meal, he may learn to refuse to eat as a method to gain attention. For example, 10-month old infants enjoy throwing food on the floor just to have some one bring them more. He may enjoy the sound of banging a spoon on a high chair more than eating. If he is getting lots of bottles of formula and juice, he may not be hungry at mealtime. The infant who learns to get attention by not eating is likely to manipulate the behavior of adults even more successfully as a toddler. Referral for early intervention evaluation would be appropriate since parent education is a part of the service. Suggestions for Billy's feeding and sleep schedule and emotional support for his parents may improve their ability to read Billy's cues of hunger compared to tiredness.

NUTRITION GUIDANCE

Nutrition guidance materials have been developed for parents from many sources, such as the WIC program, makers of infant foods, and professionals such as those in the Bright Future in Practice initiative.[6] The need for nutrition education was demonstrated in a study of mothers and pregnant women which showed that misunderstanding about infant nutrition was common.[17] Infant feeding recommendations from nutrition education materials are sampled in Table 8.9.

Inappropriate and Unsafe Food Choices

New parents may inadvertently select foods for infants based on their own likes and dislikes, rather than on the infant's needs. Such choices are problematic when they increase the risk of choking. Examples of unsafe foods for infants are:

- Popcorn
- Peanuts
- Raisins, whole grapes
- Uncut stringy meats

Table 8.9   Infant feeding recommendations.

Topics	Nutrition Education Sample Content
Appropriate use of infant formula (if not breastfeeding)	Mixing instructions for diluting concentrate formula, keeping formula sanitary by refrigeration and monitoring how long offered formula is left out. Feeding positions for the infant and bottle, and burping the infant during feedings.
Baby food and sanitation	Serving sizes for infants of different ages, refrigeration and sanitation for opened jars of baby food, problems from mixing different baby food.
Preparing baby foods at home	Avoiding spices, salt, and pepper in baby foods. Using safe food-handling techniques in preparing and storing servings.
Prevention of dental caries	Recommendations for bedtime and naptime to avoid sugary liquids pooling in the mouth. Identifying liquids that may promote dental caries.
Feeding position	Feeding positions for starting food on a spoon. How to tell if the infant highchair is safe for feeding.
Signs of hunger and fullness	Identifying early signs compared to later signs of hunger. How infants of various ages communicate at mealtime. Reinforcing and rewarding infant signs of hunger and fullness.
Preventing accidents and injury	Checking temperature of baby foods and liquids. Use of appropriate car seats and safety belts.
Spitting up—when to be concerned	Typical feeding behaviors in young infants. Signs of overfeeding. Spitting up and signs of illness. Discussing signs and symptoms with health providers.

- Gum and gummy textured candies
- Hard candy, jelly beans
- Hot dog pieces
- Hard raw fruits or vegetables, such as apples, green beans

Some foods present a choking risk for infants because of their lower chewing skills. Under-chewed pieces of food can obstruct the infant's airway since voluntary coughing and clearing the throat is a skill not yet learned.[10] Moreover, the infant may not be able to clear food from the roof of the mouth. A sticky food such as peanut butter against the hard palate may fall to the back of the mouth and present a risk of choking. A food that does not easily stay together, such as potato chips, also can cause choking. A chip breaks apart in the mouth, but little pieces may stay crunchy. Small pieces may present to the back of the mouth before the infant can use the tongue to move the pieces to the sides and initiate munching.

Infants and Exercise

Infants differ in how active they are. The exercise and fitness benefits for adults do not apply to infants. There are no recommendations to increase physical activity for healthy infants. Providing a stimulating environment is recommended, so infants can explore and move as a part of their developmental milestones. The American Academy on Pediatrics Committee on Sports Medicine policy statement recommends that structured infant exercise programs should not be promoted as being therapeutically beneficial for healthy infants.[48] Infants do not have the strength or reflexes to protect themselves, and their bones are more easily broken than those of older children and adults.

Supplements for Infants

Specific supplements are recommended for breastfed infants in the United States and Canada, under certain circumstances.

- Fluoride supplements are recommended if the family lives in a place that does not provide fluoridated water.
- Elemental iron (at 3 mg per kg body weight of the infant) may be prescribed if the mother was anemic during pregnancy.[6]
- Vitamin B_{12} may be prescribed if the mother is a vegan who eats no meats.[6]
- Vitamin D supplement may be needed if the infant is not exposed to adequate amounts of sunlight.
- If breast milk is the only form of nutrition after 4–6 months, a baby vitamin such as Tri vi

Flor® (Vitamins A, D, C and fluoride) or Poly Vi-Flor® may be prescribed.

Supplements may also be prescribed for infants who were born early at low birthweight. They may need vitamin A and E and iron, because of low stores of these nutrients accumulated late in pregnancy. A liquid multivitamin/mineral with fluoride is a common prescription for the healthy premature baby, with or without breastfeeding.[6]

COMMON NUTRITIONAL PROBLEMS AND CONCERNS

Families attribute many common health conditions to what their infants eat, or don't eat. Common nutrition concerns are failure to thrive, colic, iron-deficiency anemia, constipation, dental caries, and food allergies. Although parents tend to overestimate the association between eating and common concerns, it is good advice for parents to discuss their concerns with the infant's pediatrician and nutrition expert. Suffering an illness or just starting to become sick generally decreases an infant's appetite. It is also common for an infant with constipation to eat less than usual.

Failure to Thrive

Failure to thrive (FTT) is a diagnosis that can be made during infancy or later. There are various terms used to refine FTT, such as *organic* (meaning a diagnosed medical illness is the basis), and *nonorganic* (meaning without a medical diagnosis), and mixed type.[49] Although growth failure may be brought about by a variety of medical and social conditions, FTT is primarily used to describe conditions in which a calorie deficit is suspected.[8] FTT is an emotionally loaded diagnosis for parents, as the term implies someone failed. Examples of nonorganic or environmental factors are maternal depression, mental illness, alcohol or drug abuse in the home, or feeding delegated to siblings or others unable to respond to the infant, and overdilution of the formula. The relationship of FTT to poverty has been well documented.[8] Examples of organic reasons for FTT commonly found are untreated GER, chronic illnesses such as ear infection or respiratory illness, and *developmental disabilities.* (The connection between FTT and developmental disabilities is further discussed in

ORGANIC FAILURE TO THRIVE Inadequate weight or height gain resulting from a health problem, such as iron-deficiency anemia or a cardiac or genetic disease.

NONORGANIC FAILURE TO THRIVE Inadequate weight or height gain without an identifiable biological cause, so that an environmental cause is suspected.

DEVELOPMENTAL DISABILITIES General term used to group specific diagnoses together that limit daily living and functioning and occur before age 21.

Chapter 9). If there is a medical basis for expecting that the infant will not fit standard growth projections, FTT is not an appropriate term to use. For example, growth for an infant born with IUGR should be based on this medical history and related testing near birth. If this infant at 11 months of age is taken to a new health care setting, FTT may be suspected because of the infant's small size, unless the true cause, IUGR was revealed.

Table 8.10 is an example of an assessment to determine if FTT is present. The assessment of FTT depends on tracking growth. Once FTT is suspected, review of medical records often indicates that growth measurements have been taken in a variety of health care settings with different equipment and personnel, at times when the infant was both well and sick. These records may produce an irregular growth pattern that is difficult to interpret.

FTT Nutrition Intervention

FTT may be a basis for referral to a registered dietitian. Correction of FTT usually is not as simple as just feeding the baby, but increasing caloric and protein intake is the first step.[8] The registered dietitian's role is to assess the growth and nutritional adequacy, establish a care plan, and provide follow-up as part of a team approach. She may work with other specialists concerning medical or psychological aspects. Nutrition interventions may establish caloric and protein intake goals and a feeding schedule to assure adequate nutrition is being provided. Other interventions may include:

COLIC   A condition marked by a sudden onset of irritability, fussiness, or crying in a young infant between 2 weeks and 3 months of age who is otherwise growing and healthy.[33]

- Gain agreement from the caregivers about how and when intake and weight monitoring will be done.
- Enroll the infant in an early intervention program in the local area.
- Arrange for transportation or solve other barriers to follow-up care.
- Assess social supports to assure a constant supply of food and formula (if used).
- Assist the family in advocating for the infant within the health care delivery system, such as locating a local pediatrician and getting prescriptions filled.

FTT is a reason social service agencies become involved with families. Most new parents handle stress without hurting their infants, but a few react in ways that result in infants presenting with FTT or worse. After investigation, FTT may be determined to be a form of child abuse, as a result of neglect. Some in-

Table 8.10   Complete nutrition assessment of an infant to rule out failure to thrive.

Components:
- Review birth records with attention to weight, length, head circumference, fetal or maternal risk factors such as rate of weight gain during pregnancy, newborn screening results, AGPAR scores, and physical exams after birth.
- Review and interpret growth records from all available sources such as primary care, WIC, and emergency room visits.
- Interpret current growth measurements of the infant, including an indication of body composition, such as fat skinfold measurements.
- Review of family structure, education, and social supports with attention to access to food and formula (if not breastfed).
- Analyze and interpret current food and fluid intakes as reported by the primary caregiver(s)
- Review the complete physical assessment and medical history to rule out a biological basis for FTT.
- Observe infant feeding by the primary caregiver and interpret parent-child interactions, feeding duration, and the feeding skills of the infant.

fants diagnosed with FTT become at risk for child abuse and need foster care if their home situation further deteriorates. About one-fourth of all children entering foster care are infants.[50] Depression in the caregiver, financial stress, and low education of mothers have been identified as risk factors for infants who are physically abused.[42] Mandatory reporting rules for child abuse apply to all health providers, including registered dietitians.

Colic

Colic is the sudden onset of irritability, fussiness, or crying in a young infant.[51] Parents usually think that the infant has abdominal pain. Episodes may have a pattern of onset at the same time of day, for about the same duration every day, with all symptoms disappearing by the third or fourth month. The association of colic symptoms, gastrointestinal upset, and infant feeding practices has been studied, but no definitive cause has been shown.[51] The response to colic is often to change baby formulas if the infant is not breastfeeding, although the change in formulas usually does not change the pattern of colic. Recommendations to relieve colic may include rocking, swaddling, bathing, or other ways of calming the infant, positioning the baby well for eating, or burping to relieve gas. A randomized multicenter trial with a commonly prescribed medication, simethicone (which is thought to accelerate the passage of gas through the intestine), showed no more pain relief than a placebo.[51] One theory about why infants have colic points to the mother's diet while breastfeeding, particularly her consumption of milk or specific foods such as onions. The origin of colic still needs more research.

Iron-Deficiency Anemia

Infant iron deficiency is less common than iron deficiency in toddlers. Iron reserves in full-term infants reflect the prenatal iron stores of the mother.[52] Women with iron-deficiency anemia during pregnancy pass on less iron to the fetus, a condition that may increase risk of anemia during infancy. Infants who have iron deficiency may be exposed to other risk factors to their overall development, such as low birthweight, elevated lead levels, or generalized undernutrition.[52] Family income at or below the poverty level is also a risk for iron deficiency.[52,53] Research on poor developmental outcome in infants who have iron-deficiency anemia, however, has not clarified a mechanism that explains the association. There are many theories about the role of iron in central nervous system development.[15] Treatment of iron deficiency anemia in infancy is generally by prescribed oral elemental iron administered as a liquid.[6]

Breastfed infants may be prescribed elemental oral iron and also receive iron through iron-fortified baby cereal after 4–6 months of age. For infants who are not breastfed, a usual source of iron for formula-fed babies is iron-fortified infant formula. Iron from this source improves iron status measured during the first year and is well accepted.[53] In a randomized study of healthy infants, those who received iron-fortified infant formula had by the end of the first year significantly improved biochemical measures of iron status compared to those who received infant formula without added iron. However, there were no differences in the developmental scores of the two groups by 15 months of age.[53] Other studies have also showed that correction of the laboratory indicators of iron-deficiency anemia may not correlate with overall developmental outcome measures. Such studies show that prevention of iron-deficiency anemia in infants is important when selecting iron-fortified infant formula.

The level of iron in iron-fortified formula has been 15 mg per liter or 11.5 mg per quart based on the RDA of 6 mg of iron for infants up to six months and 10 mg from 6 to 12 months.[13] New infant formulas with a lower level of supplementation are also marketed in part as a result of gastrointestinal side effects that have been attributed to iron added to formula. The "low iron" formula has 8 mg iron per liter or 4.5 mg of iron per quart. Some manufacturers are not recommending the low iron formula beyond 4 months of age since if would not meet the RDA for iron.

Constipation and Diarrhea

The maturity of the gut is the underlying basis for constipation and diarrhea during early infancy. Diarrhea and constipation may be attributed to dietary components such as breast milk or use of an iron supplement. Parents think that diarrhea and constipation are related to the infant's diet and want to change the diet or feeding plan to lower GI upset. In fact, diarrhea can result from viral and bacterial infections, food intolerance such a lactose intolerance, or changes in fluid intake.[44] Typically young infants have more stools per day than older infants, and have them timed closely after oral intake.[54] The number of stools varies widely from two per day to six per day, and decrease as the infant matures.[54] Parents of breastfed infants generally do not have concerns about constipation as the infant's stool is generally soft. Infants fed soybean-based infant formulas may have more constipation than those fed a cow's milk-based formula. Recommendations for avoiding constipation are to assure that the infant is getting sufficient fluids and to avoid medications unless prescribed for the infant. Some parents use prune or other juices that have a laxative effect for an older infant, but there is a risk of creating a fluid imbalance and subsequent diarrhea.[44] Foods with high dietary fiber are generally not recommended for infants with constipation since many sources such as whole wheat crackers or apples with peels present a choking hazard for infants.

The cause of diarrhea during infancy may or may not be identifiable. Diarrhea in an infant may become a serious problem if the infant becomes dehydrated or less responsive.[44] Most infants have one or two days of loose stools without weight loss or signs of illness, such as after getting immunizations. General recommendations are to continue to feed the usual diet during diarrhea. Breast milk does not cause diarrhea. Continuing breast milk or infant formula through a brief illness assures adequate fluids to prevent dehydration.

Prevention of Baby Bottle Caries and Ear Infections

Baby bottle caries are found in children older than one year, but are initiated by feeding practices during infancy. Infants have high oral needs, which means they love to suck, to explore by putting things in their mouths. They derive comfort from sucking and may relax or fall asleep while sucking. The use of a bottle containing formula, juice, or other high-carbohydrate foods to calm a baby enough to sleep may set them up for dental caries.[55] During sleep the infant swallows less, allowing the contents of the bottle to pool in the mouth. These pools of formula or juice create a rich environment for the bacteria that cause tooth decay to proliferate, increasing the risk for tooth decay.

Risk for ear infections is also correlated with excessive use of a baby bottle as a bedtime practice, as a result of the feeding position.[55] The shorter and more vertical tubes in the ears of infants are under different pressure during the process of sucking from a bottle.[3] If the infant is feeding by lying down while drinking, the liquid does not fully drain from the ear tubes. The build-up of liquid in the tubes increases the risk of ear infections. In a study of over 200 infants, feeding practices were identified that were linked with frequent ear infections. Pacifiers and bottle-feeding were correlated with greater prevalence of ear infections.[55] Infants who were breastfed did not have as high a rate of ear infections.[55]

Good feeding practices to limit baby bottle caries and ear infections related to baby bottles are:

- Limit the use of a bottle as part of a bedtime ritual.
- Offer juices in a cup, not a bottle.
- Put only water in a bottle if offered to sleep.
- Examine and clean emerging baby teeth to prevent caries developing.

Food Allergies and Intolerances

Allergy testing is not generally ordered for infants whose immune systems are still developing. The prevalence of true food allergies is higher in younger than in older children. About 6–8% of children under four years of age have allergies that started in infancy.[56] An infant may develop a food allergy to the protein in a cow milk-based formula over time. Often such a problem follows a gastrointestinal illness. When the infant is well, protein is broken down during digestion so that absorption in the small intestine is of groups of two or three amino acids linked together. After an illness, small patches of irritated or inflamed intestinal lining may allow protein fragments of larger lengths of amino acids to be absorbed. Such fragments are hypothesized to trigger a reaction as if a foreign protein had invaded, setting up a local immune or inflammatory response.[56] This absorption of intact protein fragments is the basis for allergic reactions. When this happens with cow milk protein, it is likely that soy-based formulas will also cause the same allergic reaction.

The most common allergic reactions are respiratory and skin symptoms, such as wheezing or skin rashes.[57] Food allergies are confirmed by specific laboratory tests after infancy.[56] True allergies can present as an array of

HYDROLYZED PROTEIN FORMULA Formula that contains enzymatically digested protein, or single amino acids, rather than protein as it naturally occurs in foods.

LACTOSE   A form of sugar or carbohydrate composed of galactose and glucose.

reactions building up over time, so that it may take several years for the initiating cause to be identified.

Food intolerances are frequently suspected in infants. Families may consider skin rashes, upper airway congestion, diarrhea, and other forms of GI upset to be food allergies, but often they are not. Food intolerances are tested by laboratory tests and sometimes by elimination diets.[56] As in the case of food allergies, infants are not usually given all the tests for food intolerance, but are treated for symptoms. The infant with suspected protein intolerance may be changed to a specialized formula composed of *hydrolyzed protein*.[57] Since the protein of a hydrolyzed formula is already broken down, it does not trigger the same response as intact protein fragments. A family with a known allergy or intolerance may lower the risk of their infant by breastfeeding, and by postponing into the second or third year introduction of allergy-causing foods, such as wheat, eggs, and peanut butter.

How Parents Respond to Suspected Food Allergies or Intolerance

If a food intolerance is suspected but not confirmed, it is likely that the family will continue to avoid the offending food. The list of foods that can cause allergic symptoms in some infants includes foods of high nutrition value, such as wheat, shellfish, eggs, tomato and citrus, peanuts, other kinds of nuts, and milk.[57] It is important that families do not restrict such foods from an infant's diet unless required. If many foods are being avoided, there may be consequences such as decreasing the nutritional adequacy of the diet and reinforcing behaviors of rejecting foods and limiting variety. Allergy and intolerance symptoms are more common in response to non-food items such as grasses and dust, so many different sources of symptoms must be considered.

Lactose Intolerance

Lactose intolerance is a food intolerance in infancy characterized by cramps, nausea, and pain, and alternating diarrhea and constipation. Infants who are breastfeeding may develop lactose intolerance, as breast milk has lactose in it.[19] Gastrointestinal infections may temporarily cause lactose intolerance, since the irritated area of the intestine interferes with lactase production.[44] The ability to digest lactose generally returns shortly after the illness subsides. *Lactose* is in all dairy products, so it is in cow milk-based infant formulas. Symptoms of lactose intolerance are upsetting to parents, so many formula changes may be tried. Pediatricians may refer sick infants for definitive testing by a breath hydrogen test, but it is more common to substitute with a formula that does not have lactose.[44]

Lactose-free infant formulas are soy bean-based or lactose-free cow's-milk based. Lactose intolerance is less common during infancy than at older ages in groups that are susceptible to it. Heating and other forms of preparing dairy products break down lactose. As a result, variable amounts of lactose occur in yogurt, cheese, and ice cream. An infant who was fed a lactose-free formula is likely to be able to eat dairy products later. Since dairy products are such an important source of calcium, introducing foods with low lactose are recommended for older infants who appeared to be lactose intolerant when younger.

CROSS-CULTURAL CONSIDERATIONS

Commercial baby foods reflect the bias of the dominant American culture. There is no ethnic diversity in baby foods: no collards or Mexican beans. Many successful avenues to nourish a healthy infant are available, and room ought to be made for different cultural patterns in the development of feeding practices. Some cultural practices are clearly unsafe and must be discouraged, such as a mother prechewing meats for a baby. Cultural practices that support the development of competence in parents can be encouraged, even if not part of the dominant culture. Examples of practices that may reflect cultural choices are swaddling an infant, or having an infant sleep in the parent's bed or in a certain temperature room. Practices based on family traditions may be forms of social support for new parents. Only if new parents have not considered the safety of the infant or have little knowledge of other, safer alternatives should cultural practices be discouraged. For example, it may be a cultural practice to offer meats to adults but not to infants. The basis for this practice used to be the cost of the meats, but new parents may not face that barrier. They should be informed that older infants may safely eat meats that are cut up or soft-cooked to avoid causing choking. Some cultures consider meat-based soups as infant foods.

Cultural considerations may impact the willingness of the family to participate in assistance programs such as WIC or early intervention programs. The dignity of the family unit, including extended relatives, has to be considered in educating the family about using various infant feeding practices and programs. Food-based cultural patterns may be part of a religious tradition, so sensitivity to the family unit would include such practices.

VEGETARIAN DIETS

Various food choices may be included in the general term "vegetarian diet"; usually they are self-defined by the consumer. Consumption of a vegetarian diet does not necessarily put an infant at risk. A vegetarian diet may be selected for many different reasons, such as religious beliefs or health concerns. Studies have found that infants who receive well-planned vegetarian diets grow normally.[58] The most restrictive diets, vegan and macrobiotic diets, have been associated with slower growth rates in infancy, particularly if infants do not receive enough breast milk.[58,59] Breastfed vegan infants are recommended to have supplements of nutrients such as vitamin K, vitamin D, vitamin B_{12}, iron, fluoride, and possibly zinc.[59] The composition of the breast milk from vegan mothers may differ in small ways from standard breast milk.[59] An example is in the ratio of types of fat, although the total fat in breast milk from vegan mothers is the same. Impacts of these differences on the health of infants are generally not known.

Vegetarian diets range from adequate to inadequate, depending on the degree to which the diet is restricted, just as diets for omnivores range from adequate to inadequate. The family following a lacto-ovo-vegetarian diet consumes fruits, vegetables, starches, milk, and eggs, but other animal protein is avoided.[58] The infant in this family usually can meet nutritional needs from foods offered on a spoon as well as other infants. Either food sources, such as fortified infant cereals and soy milk, or supplements can be used for assuring adequate vitamin and mineral intake. Vegetarian families who avoid all products of animal origin, including milk and eggs, require carefully selected fortified foods or a higher degree of supplementation for their infant.[59] They may include foods such as iron-fortified infant cereal with wheat germ, tofu, juice supplemented with vitamin C and calcium, and vegetables such as winter squash or kidney beans.[59] Such choices are implemented by families making their own baby foods, which requires careful attention to home sanitation and storage methods. Potentially limiting nutrients are protein, iron, zinc and vitamin B_{12}.[58] Periodic assessments of dietary intake, growth, and health status can be used to monitor the infant fed a vegetarian diet. Vegetarian infants have similar risk for developing food allergies from soy products and nuts as others.

Families with infants who follow vegetarian diets have two common difficulties:

- Infants may not be able to consume the volume of solids or liquids needed to obtain adequate calories, protein, and some minerals since nonanimal products have lower levels of these needed nutrients.
- Alternative food products developed for the vegetarian market are not of high nutritional quality or in portion sizes for infants.

For example, various beans and peas may have the needed calcium and iron appropriate for an older infant. However, the portion sizes are not realistic for 10 month olds. For cow's milk alternatives, fortified soy-based formulas are usually acceptable, but rice-based drinks in the diet for infants are generally of lower nutritional quality. The education of the family who selects a vegetarian diet for infants and their willingness to use fortified foods or nutritional supplements are important factors for successful implementation of the diet.

NUTRITION INTERVENTION FOR RISK REDUCTION

Early Head Start Program is an example of a federal program that is focused on prevention and reducing risks to infant development from their environment.[6] The Early Head Start Program was developed to work with infants and their families, especially new families at risk due to drug abuse, infants with disabilities, or teen-age mothers. Nutrition services are among a wide range of services typically offered in an Early Head Start Program. Other services may include home-based early childhood education, case management, mental health support services, as well as health and socialization services. Services for teen mothers typically include adult education services, vocational training, parent training, and substance abuse services. The person who is eligible for Early Head Start must have an income below the federal poverty level, and must be pregnant at the time of entering the program or have a child one year of age or younger. The Early Head Start program would assist the family in coordinating WIC participation with food stamps, routine well-baby visits, and day care, as needed.

GALACTOSEMIA   A rare genetic condition of carbohydrate metabolism in which a blocked or inactive enzyme does not allow breakdown of galactose, causing serious illness in infancy.

HYPOTHYROIDISM   Condition in which thyroid hormone is not produced in sufficient quantities, interfering with growth and mental development if untreated in infants.

MEDIUM CHAIN FATTY ACID OXIDATION DISORDER   Condition in which breakdown of fatty acids for generation of energy inside the cell is blocked.

Model Program: Newborn Screening and Expanded Newborn Screening

In the United States and many other countries, all newborns are screened for rare conditions that may cause disability or death. Such screening from a small dried blood spot was initiated in the 1960s after early treatment of phenylketonuria (PKU) was shown to prevent later mental retardation in young children.[60] Most states screen for three to six different conditions, such as PKU, *galactosemia, hypothyroidism,* and sickle cell disease.[60] New technology has resulted in expanded newborn screening, so some states are now testing for as many as 30 different conditions from the same dried blood test.[60]

Many of the disorders that can be detected by expanded newborn screening are treated by diet. They are disorders of protein, carbohydrate, and fat metabolism. Dietary treatment avoids the substance that has a metabolic block and replaces other dietary components that are usually provided in the foods that are avoided. An example of a condition that can be screened in expanded newborn screening is a disorder of fat metabolism called a *medium chain fatty acid oxidation disorder.*[39] It may cause coma and death if undiagnosed, but the dietary treatment is easy in a baby that is identified early. Treatment involves prevention of fasting, especially during illness, and for some infants carbohydrate-containing foods may be increased. Parents are instructed to wake up the infant during the night to feed them to avoid fasting, even if the baby is able to sleep through the night. There are no symptoms when the infant with a medium chain fatty acid disorder is well. Other conditions identified from expanded screening have shown that specific vitamins act as enzyme cofactors. Expanded infant screening for genetic disorders is likely to continue to expand nutrition knowledge overall.

WIC

WIC, officially the Special Supplemental Food Program for Women, Infants and Children, was envisioned in the 1960s as part of the "war on poverty" initiative. WIC provides nutrition education and access to specific foods for eligible participants. Eligibility is based on household income of no more than 185% of the federal poverty level and the presence of a nutrition or health risk. Nutrition risks for infants include:

- Mother had a nutrition risk during pregnancy.
- Growth assessment shows underweight
- Iron status as shown by hemoblobin or hematocrit results are low.
- Diet-based risks such as inadequate dietary pattern.

WIC serves 45% of all infants born in the United States.[61] The WIC Program's effectiveness has been demonstrated in improving infant feeding practices,

Case Study 8.3
Baby Derrick and Tina, a Young New Mother

Photo Disc

Derrick is a 5 month old boy who was born at 39 weeks. He has gotten his routine immunizations and has not been sick. He lives with his mother Tina, who is studying for her high school equivalency test, and his maternal grandmother, who works outside the home. He was nursed for the first two months of his life, and then Tina tried giving him a bottle to assure herself that he was getting enough. She was surprised that he liked the bottle so well and drank 3 fl oz. Derrick became less satisfied with nursing over the next two weeks. By the time he was three months old, he was no longer breastfeeding at all. At 5 months of age, he takes between 30–36 fl oz of Similac with iron per day, which Tina purchases. Tina likes that her mother can help her feed Derrick. She is still concerned that he may not be getting enough, so she tried giving him some baby rice cereal on a spoon at 4.5 months of age. He spit it back out and cried. His grandmother thinks he needs cereal in the bottle, but his mother does not know if that is right or not. She had a routine pediatrician visit and was told Derrick was growing well, and that he would eat from a spoon when he was ready. Tina is still concerned that he is not getting enough to eat.

Intervention

This case example shows the need for the WIC program and its nutrition education component. As a young single parent, Tina may have been eligible for the WIC program as a pregnant teen or as a postpartum mother. Derrick may be eligible depending on his nutritional risks, as shown on a blood iron test. Tina would benefit from nutrition education about how infants progress in their feeding skills and show signs of readiness for food on a spoon and signs of hunger. This knowledge may help Tina gain confidence in making decisions about feeding Derrick. Although it is too late for her decision about breastfeeding, WIC education may impact her breastfeeding decision if she had another child later. Tina is likely to receive education from the WIC program at a level appropriate for her education. Then she may be better equipped to ask questions at Derrick's pediatrician visits. Derrick's growth will be monitored at the WIC visits; that may reassure her that he is growing and healthy.

such as promoting breastfeeding and supporting breastfeeding mothers, and supplying iron-fortified formula for infants who are not breastfed. The WIC program also functions as a gateway program to other health services for low-income families with infants, such as immunizations, a regular source of medical care and social services. Each state administers its own WIC program, but the eligibility criteria and regulations are the same nationwide.[61] Mothers of breastfed infants are provided vouchers for milk, cheese, juices, fortified cereals, legumes, and peanut butter. Families of infants not breastfed are provided vouchers for iron-fortified infant formula for the first six months, and baby cereal and juices are added for the second half of infancy. The amount of formula is limited to provide the equivalent of 26 fl oz per day or 8 lb of powdered formula per month.[61] The formula selection is limited to iron-fortified cow's-milk and soy-based formulas. The amount of infant formula provided generally requires purchase of additional infant formula for older infants. Nutrition assistance and education are important parts of WIC services.

Resources

American Academy of Pediatrics

Reliable and credible sources of pediatric medical expertise in position papers, with sections for consumers and health care providers.
Available from: www.pediatrics.org

American Dietetic Association

Consumer and provider information includes child nutrition and health and access to credible resources.
Available from: www.eatright.org

Food Allergy Network

Reliable and scientifically based information for families with diagnosed allergies. It routinely includes recipes.
Available from: www.foodallergy.org

Health/Infants

Consumer and marketing information about infant growth and development based on information available to mass media.
Available from: www.cnn.com/health/indepth.health/infants

National Center for Growth Statistics (source of growth charts)

Site for obtaining growth charts and guidelines for their use.
Available from: www.cdc.gov/nchs/

Nutrition/WIC (part of United States Department of Agriculture)

Nutrition education materials and information for low-income families.
Available from: www.nal.usda.gov/fnic/Wicdbase/

References

1. National Center for Health Statistics: NCHS growth curves for children 0–19 years. U.S. Vital and Health Statistics, Health Resources Administration U.S. Government Printing Office 2000.

2. U.S. Vital and Health Statistics. Births: final data for 1999. National Vital Statistics Report 2001;49;1–99.

3. Blackman JA. Medical aspects of developmental disabilities in children birth to three. Gaithersburg, MD: Aspen Publications; 1997: 55–58; 234–237.

4. Richardson, DK, Gray JE, Gortmaker SL, et al. Declining severity adjusted mortality: evidence of improving neonatal intensive care. Pediatrics 1998;102:893–899.

5. Morse SB, Haywood JL, Goldenberg RL, et al. Estimation of neonatal outcome and perinatal therapy use. Pediatrics 2000; 105:1046–50.

6. Story M, Holt K, Sofka D, eds. Bright Futures in practice: nutrition. National Center for Education and Child Health, Arlington, VA 2000:25–56, 266–70.

7. Beaver PK. Gestational age and birth weight. In Lowdermilk DL, Perry SE, and Bobak, IM, eds. Maternity and women's health care. 6th ed. St. Louis, MO: Mosby; 1997: pp 1015–42.

8. Maggioni A, Litchitz F. Nutritional management of failure to thrive. Pediatric Clinics of North America 1995; 42: 791–810.

9. Kail RV, Cavanaugh JC. Human development: A lifespan view. 2nd ed. Belmont, CA: Wadsworth/Thompson Learning; 2000: pp 83–121.

10. Cech D, Martin ST. Functional movement development across the life span. Philadelphia: W B Saunder; 1995: pp 76–86.

11. Santrock JW. Life-span development. 7th ed. New York, NY: McGraw Hill; 1999: pp 124–9.

12. Colangelo CA. Biomechanical frame of reference. In: Kramer P, Hinojosa J. Frames of reference for pediatric occupational therapy. Philadelphia: Williams and Wilkins; 1993: pp 233–56.

13. National Research Council. Recommended dietary allowances. 10th ed. Washington, DC. National Academy Press; 1989.

14. Ramey CT, Ramey SL. Right from birth, building your child's foundation for life. New York: Goddard Press; 1999.

15. Perry SE. Genetics, conception and fetal development. In Lowdermilk DL, Perry SE, Bobak IM, eds. Maternity and women's health care 6th ed. St Louis, MO: Mosby; 1997: 101–35.

16. Willard HF, Hendrich BD. Breaking the silence in Rett syndrome. Nature Genetics; 1999:23:127–8.

17. Hobbie C, Baker S, Bayerl C. Parental understanding of basic infant nutrition: misinformed feeding choices. J Pediatr Health Care 2000;14:26–31.

18. Anderson DM. Nutrition for the low-birth weight infant. In: Mahan LK, Escott-Stump S, eds. Krause's food, nutrition and diet therapy. 10th ed. Philadelphia PA:W.B. Saunders; 2000: pp 214–38.

19. Akers SM, Groh-Wargo SL. Normal nutrition during infancy. In: Samour PQ, Helm KK, Lang CE, eds. Handbook of pediatric nutrition. 2nd ed. Gaithersburg, MD: Aspen: 1999: pp 65–98.

20. Jones PH, Kubow S. Lipids, sterols and their metabolites. In: Shils ME, et al. eds. Modern nutrition in health and disease. 9th ed. Baltimore MD:Williams and Wilkins; 1999: pp 67–94.

21. Winter SC, Buist NR. Clinical treatment guide to inborn errors of metabolism. J Rare Diseases 1998;4:18–46.

22. Uauy RP, Peirano D, Hoffman P, et al. Role of essential fatty acids in the function of the developing nervous system. Lipids 1996;31:Supplement167–617.

23. Elias ER, Irons MB, Hurley AD, et al. Clinical effects of cholesterol supplementation in six patients with Smith-Lemli-Opitz syndrome. Am. J Med Genetics 1997:68: 305–10.

24. Trumbo P, Yates AA, Schlicker SA, et al. Dietary reference intakes; vitamin A, vitamin K, arsenic, boron, chromium, copper, iodine, iron, manganese, molybdenum, nickel, silicon, vanadium and zinc. J Am Diet Assoc 2001;101:294–301.

25. Yates AA, Schlicker SA, Suitor CW. Dietary reference intakes; the new basis for recommendations for calcium and related nutrients, B vitamins and choline. J Am Diet Assoc 1998; 98:699–706.

26. United States Department of Agriculture and United States Department of Health and Human Services. Dietary guidelines for Americans, 2000. 5th ed. Home and Garden Bulletin No 232.

27. Gerber Products Company. Nutrient values. Fremont, MI 2000.

28. Ballew C, Khan LJ, Kaufman R, et al. Blood lead concentration and children's anthropometric dimensions in the Third National Health and Nutrition Examination Study (NHANES III), 1988–1994. J Pediatr 1999;134:623–30.

29. McLaren DS. Clinical manifestations of human vitamin and mineral disorders: a resume. In: Shils ME, et al., eds. Modern nutrition in health and disease. 9th ed. Baltimore MD:Williams and Wilkins;1999: pp 485–503.

30. Hampl, JS, Betts NM, Benes BA. The "age+5" rule: comparisons of dietary fiber intake among 1–4 year old children. J Am Diet Assoc 1998;98:1418–23.

31. Fomon S, Haschke F, Ziegler E, et al.

Body composition of reference children from birth to age 10 years. Am J Clin Nutr 1982;35:1169–75.

32. Dietz WH. Health consequences of obesity in youth: childhood predictors of adult disease. Pediatrics 1998;101:518S–25S.

33. Georgia Department of Human Resources, Division of Public Health, Office of Nutrition. Nutrition. Guidelines for practice 1997:Section IIIa–e.

34. American Academy of Pediatrics Work Group on Breastfeeding. Breastfeeding and the use of human milk. Pediatrics 1997;100:1035–39.

35. American Dietetic Association. Position of the American Dietetic Association: promotion and support of breast-feeding. J Am Diet Assoc 1997;97:662.

36. Trachtenbarg DE, and Golemon TB. Care of the premature infant: part 1. monitoring growth and development. Amer Family Physician 1998:57;2123–30.

37. American Dietetic Association. Pediatric manual of clinical dietetics. CP Williams, ed. 1998:appendix 1;563–95.

38. American Academy of Pediatrics. Committee on Nutrition (RE9251). Policy Statement. The use of whole cow's milk in infancy. Pediatrics 1992;89:1105–09.

39. Ziegler EE, Jiang T, Romero E, et al. Cow's milk and intestinal blood loss in late infancy. J Pediatrics 1999;135:720–6.

40. Birch LL, Gunder L, Grimm-Thomas K, et al. Infants' consumption of a new food enhances acceptance of similar foods. Appetite 1998;30:283–95.

41. Capaldi ED, Powley TL, eds. Taste, experience and feeding. Washington, DC: American Psychological Association; 1990:75–93.

42. Cadzow SP, Armstrong KL, Fraser JA. Stressed parents with infants: reassessing physical abuse risk factors. Child Abuse & Neglect 1999;23:845–53.

43. Shaffer SE. Gastroesophageal reflux. In: Altschuler SM, Liacouras CA, eds. Clinical pediatric gastroenterology. Philadelphia PA: Harcourt Brace and Co.;1998: pp 181–6.

44. Baldassano RN, Cochran WJ. Diarrhea. In: Altschuler SM, Liacouras CA, eds. Clinical pediatric gastroenterology. Philadelphia PA: Harcourt Brace and Co.;1998: pp 9–18.

45. Mennella JA, Beauchamg GK. Maternal diet alters the sensory qualities of human milk and the nurslings' behavior. Pediatrics 1991;88:737–44.

46. Davis CM. Self selection of diet by newly weaned infants: an experimental study. Am. J Diseases of Children 1928;36:651–79.

47. Story M, Brown JE. Do young children instinctively know what to eat? New England J Med 1998;103–6.

48. American Academy of Pediatrics. Policy Statement (RE8132). Infant Exercise programs. Pediatrics 1988;82:800. Reaffirmed 11/94.

49. Bithoney WG, Dubowitz H, Egan H. failure to thrive/growth deficiency. Pediatrics in Review 1992;13:453–9.

50. Needell B, Barth RP. Infants entering foster care compared to other infants using birth status. Child Abuse & Neglect 1998;22:1179–87.

51. Metcalf T, Irons TG, Sher L, et al. Simethicone in the treatment of infant colic: a randomized, placebo-controlled multicenter trial. Pediatrics 1994;94:29–34.

52. Recommendations to prevent and control iron deficiency in the United States. MNWR April 3, 1998, vol 47.

53. Moffatt MEK, Longstaffe S, Besant J, et al. Prevention of iron deficiency and psychomotor decline in high risk infants through use of iron-fortified infant formula: a randomized clinical trial. J. Pediatics 1994;125:527–34.

54. Wenner WJ. Constipation and encopresis. In: Altschuler SM, Liacouras CA, eds. Clinical pediatric gastroenterology. Philadelphia PA: Harcourt Brace and Co.;1998: pp 165–8.

55. Jackson JM, Mourino AP. Pacifier use and otitis media in infants twelve months of age or younger. Pediatric Dentistry 1999;21:255–60.

56. Sicherer SH, Morrow EH, Sampson HA. Dose-response in double-blind, placebo-controlled oral food challenges in children with atopic dermatitis. J Allergy and Clinical Immunology 2000;105:582–6.

57. Christie L. Food hypersensitivities. In: Samour PQ, Kelm KK, and Lang CE, eds. Handbook of pediatric nutrition. 2nd ed. Gaithersburg, MD: Aspen; 1999: pp 149–72.

58. Sanders T. Vegetarian diets and children. Pediatric clinics of North America 1995;42(4):955–65.

59. Mangels AR, Messina V. Considerations in planning vegan diets: infants. J Am Diet Assoc 2001;101:670–7.

60. Chance DH, Naylor EW. Expansion of newborn screening programs using automated tandem mass spectrometry. Mental Retardation and Developmental Disabilities Research Reviews 1999;5:150–4.

61. United State Department of Agriculture website www.usda.fns accessed on 10/26/2000.

CHAPTER 9

Photo Disc

Man eats to live, he does not live to eat.

—Abraham ibn Ezra, Spanish poet and scientist (1092–1161)

INFANT NUTRITION:

Conditions and Interventions

Prepared by **Janet Sugarman Isaacs**

CHAPTER OUTLINE

- Introduction
- Infants at Risk
- Energy and Nutrient Needs
- Growth
- Common Nutrition Problems
- Preterm Infants and Infants with Special Health Care Needs
- Severe Preterm Birth
- Infants with Chronic Illness
- Feeding Problems
- Nutrition Interventions
- Nutrition Services

KEY NUTRITION CONCEPTS

1 Infants who were born preterm or who were sick early in life require adjustments in assessing how they are meeting their nutrition requirements, and growing and developing.

2 Early nutrition services and other interventions can improve long-term health and growth for infants born with a variety of conditions.

3 The number of infants requiring specialized nutrition and health care is increasing due to the improved survival rates of small and sick newborns.

INTRODUCTION

Most infants are born healthy and then grow and develop in the usual manner. This chapter addresses the nutritional needs of infants who have health problems before or shortly after birth, so are at risk for health or developmental problems. Most infants have minor illnesses within the first year of life that do not interfere with growth and development. However infants who were sick or small as neonates are likely to have conditions that may change the course of growth or development. *Children with Special Health Care Needs* is a broad term that includes the infants discussed in this chapter. As in Chapter 8, this chapter models sensitive communication with families by avoiding the word *normal* and, by implication, *abnormal* when referring to infants with special health care needs. Similarly, the designation *normal growth* is replaced by the phrase *standard growth* when referring to the CDC growth charts. Language such as "She is below normal on the growth charts" is replaced with "family friendly" language, such as "She is the weight of a typical 6 month old." Most families use the word *premature* (or *prematurity*) comfortably, but *preterm birth* is the conventional usage in maternity care. Both terms refer to infants born before 37 weeks of gestation, and both are used in this chapter.

CHILDREN WITH SPECIAL HEALTH CARE NEEDS   A federal category of services for infants, children, and adolescents with, or at risk for, physical or developmental disability, or with a chronic medical condition caused by or associated with genetic/metabolic disorders, birth defects, prematurity, trauma, infection, or perinatal exposure to drugs.

DOWN SYNDROME   Condition in which three copies of chromosome 21 occur, resulting in lower muscle strength, lower intelligence, and greater risk for overweight.

SEIZURES   Condition in which electrical nerve transmission in the brain is disrupted, resulting in periods of loss of function that vary in severity.

INFANTS AT RISK

The overall U.S. infant mortality rate decreased 39% between 1980 and 1995.[1] It appears, however, that the health care system has been more successful at saving ill infants than preventing chronic conditions. The number of infants requiring nutritional services is increasing in large measure because of advances in neonatal intensive care. Small preterm infants who did not survive in the past are now being "saved." The smallest living newborns, who weigh 501–600 grams (1 pound 2 ounces to 1 pound 5 ounces), have a 31% chance of survival at birth.[2] This birthweight range is about 23 weeks of gestation, in the second trimester of pregnancy. Infants with birthweights of 901–1000 grams (2 pounds to 2 pounds 3 ounces) are in the 29

weeks range of gestation and have an 88% change of survival.[2] Infants with genetic disorders, malformations, or birth complications have also benefited from advances in treatment, and are less likely to die in infancy. However, they are much more likely to have chronic conditions, with increased need for medical, nutritional, and educational services later.

Regardless of what condition is involved, these nutrition questions are likely:

- How is the baby growing?
- Is the diet providing all required nutrients?
- How is the infant being fed?

In-depth nutrition assessments make sure nutrition is not limiting an infant's growth and development. Such assessments are needed by three main groups of infants.

- Infants born before 34 weeks of gestation. Preterm infants are born at less than 37 weeks of gestation, but generally only those born before 34 weeks of gestation have higher nutritional risks.
- Infants born with consequences of abnormal development during pregnancy, such as infants born with heart malformations as a result of the heart not forming correctly or exposure to toxins during gestation. Exposure to alcohol during gestation may interfere with brain formation and result in permanent changes in the brain function. This second category would include infants with genetic syndromes, such as *Down syndrome*.
- Infants at risk for chronic health problems. Risks may come from the treatment needed to save their lives, or from the home environment that the baby enters. Examples of conditions which increase risks are *seizures* or cocaine

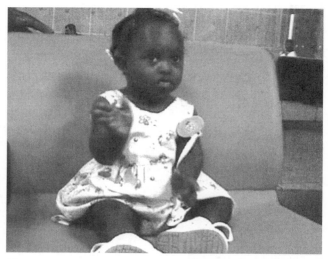

Illustration 9.1   **Infant girl with Down syndrome, after her heart surgery.**

withdrawal symptoms. Long-term consequences, such as later learning problems may
not be known for years.

Families of Infants with Special Health Care Needs

Every parent's wish is that his or her baby will be
healthy. When parents find out their newborn has
medical problems, they grieve for the loss of the perfect child of their dreams. The emotional impact of
having a sick newborn overwhelms many parents,
and providers of services for these families must be
sensitive to parents' emotional needs. Coping styles of
various family members vary, even if they are well
prepared. It may take over a year for some family
members to understand how the baby is doing and to
adjust to the special needs of the child. For conditions
with long-term consequences, the first year may not
be long enough for parents to see that their infant is
developing differently from other babies.

ENERGY AND NUTRIENT NEEDS

Nutrient requirements for infants with health conditions are based on the recommendations for healthy
infants. Specific nutrients may be adjusted higher or
lower based on the health condition involved. Adjusting the diet to changing conditions and close monitoring of growth and development may result in
changing recommendations quite frequently in the
first year. Scientific frontiers of medicine, genetics,
nutrition, and technology interact in caring for sick
infants. Nutrition requirements are not known for
every condition, and individuals respond at their own
pace of growth and development, so many nutrition
recommendations are based on the best judgement
under the circumstances.

Energy Needs

Caloric needs may be the same as for healthy infants
and based on the RDA for infants of 108 calories per
kilogram body weight.[3] Some conditions in newborns
that have caloric needs based on the RDA are cleft lip
and palate or phenylketonuria (PKU). The more common situation is that caloric needs are increased.
Caloric needs of sick infants can be estimated with
measurements such as *indirect calorimetry*.[4] Machines
that measure indirect calorimetry are often available
in intensive care nurseries. Estimated energy needs
can change, however, with medications, activity,
health conditions, and growth. Such estimates show

that sick and small infants vary in their caloric needs
more widely than expected, and that caloric deficits
in preterm infants may be more common than previously known.[5] Extra calories are needed in circumstances such as the following:

- Infections
- Fever
- Difficulty breathing
- Temperature regulation
- Recovery from surgery and complications

Infants who were born preterm at less than 34
weeks of gestation particularly need higher energy. The
American Academy of Pediatrics suggests that premature infants need 120 cal/kg.[6] Intakes for recovering
premature babies may be even higher, with a wide
range. The European Society for Gastroenterology and
Nutrition gives a range from 95–165 cal/kg.[7] The
amount of calories needed to gain 15 grams per day is
recommended.[8] Infants who were born with VLBW or
ELBW may still be weak and have difficulty feeding
when they come home. Their higher calorie needs may
be more difficult to provide in the small volume they
can consume. Over time, the recovering infant may
increase intake so that even 180 cal/kg per day may
be reported.

Some infants need less energy than the RDA.
These are infants born with smaller muscles or lower
activity as a result of the inability to move certain
muscles. An infant who has Down syndrome or one
with repaired spina bifida needs fewer calories than
the RDA of 108 cal/kg body weight.[3] Too many calories would interfere with their efforts to crawl; their
weak muscles would have more body weight to move.

Protein Requirements

As noted with energy needs, protein requirements of
infants with special health care needs may be higher,
lower, or the same as other infants. Protein requirements based on the
RDA of 2.2 grams of
protein per kilogram
body weight are rec-
ommended if the

INDIRECT CALORIMETRY Measurement
of energy requirements based on oxygen
consumption and carbon dioxide
production.

condition does not impact growth or digestion.[3]
Protein recommendations are sufficient if total calories are high enough to meet energy needs. The concept of protein sparing is important in fast-growing
infants. If enough energy is available from glucose
generated from foods containing fats and carbohydrates, then the amino acid generated from protein-
containing foods are spared, that is, the amino acids
are available for growth. If glucose from foods is not
sufficient for meeting energy needs, however, amino
acids from digestion of protein foods will be used to

meet energy needs, and therefore less will be available for growth. When total caloric intake is low, protein-rich foods become an energy source. In this circumstance, providing the RDA for protein may be inadequate protein intake, and result in slow growth. With preterm infants a sign of inadequate protein may be slow head growth, which is an indicator of brain growth.[9]

Conditions which could slow growth may require higher protein levels than the RDA for protein for infants. Higher protein recommendations are common in early infancy for conditions such as recovery from surgery or LBW. Protein intakes of 3.0–3.5 g/kg are appropriate for premature and recovering infants.[8] For recovering from some complications of ELBW, high protein intakes, as much as 4 g/kg, appear safe with adequate fluids, and without kidney problems.[8] The importance of protein to growing neonates is hard to overemphasize. Protein deficits in preterm infants have been shown even when high protein is provided, depending on how small and sick they are and how soon after birth the infant is being fed.[5]

Generally protein recommendations lower than the RDA are unusual in infancy. Infants with lower muscle activity as a result of smaller sized muscles generally need lower protein. One example is Down syndrome. Conditions that lower physical activity and movement are often not identified until the infant is old enough to be moving around, closer to the end of the first year. During infancy muscle tone is known to change over time, so muscle coordination and movement problems are usually confirmed later.

FORM OF PROTEIN   Many illnesses interfere with the functioning of the GI tract and digestion, even though newborns are born with intact enzymes for protein digestion. Protein and fat digestion depend on liver and pancreatic enzymes for intestinal absorption. However, many conditions associated with preterm birth and illness stress the liver and reduce its ability to function, causing changes in protein and fat digestion. Sick infants may require forms of protein in which amino acids are in short chains, such as in hydrolyzed protein, or single amino acids. Other examples of those needing protein that has been broken down are infants with metabolic disorders.[10] Naturally occurring pro-

MCT OIL   A liquid form of dietary fat used to boost calories; composed of medium-chain triglycerides.

CATCH-UP GROWTH   Period of time shortly after a slow growth period when the rate of weight and height gains is likely to be faster than expected for age and gender.

HYPOCALCEMIA   Condition in which body pools of calcium are unbalanced, and low levels are measured in blood as a part of a generalized reaction to illnesses.

tein has to be limited, and partially replaced by mixtures of specific amino acids. An example is phenylketonuria. Naturally occurring protein from meats and dairy products contain too much of the amino acid phenylalanine that must be limited.

Fats

Infants need a high-fat diet compared to older people, since fats provide energy. Up to 54% of calories from fats may be recommended.[9] The need for calories provided by fats are especially important in sick or recovering infants. Low-fat diets are generally not recommended for infants. Conditions which require limiting fats for infants are uncommon. One example is very sick infants who required heart-lung bypass machines as a part of major surgery. They have recommendations for avoiding high-fat diets after infancy, without a specific level of fat restriction.[8] Fats are more difficult to absorb for infants with VLBW or ELBW since they require pancreatic and liver enzymes.[9] These enzyme systems may also be impaired in sick infants. Naturally occurring long-chain fats in breast milk may be supplemented with shorter chain fatty acids for sick infants. The medium-chain triglycerides do not require bile for absorption, so they are preferred.[9] *MCT oil* can be added to ensure calories from fats are available. Additionally, the essential fatty acids, alpha-linolenic and linoleic acid, as well as docosahexaenoic acid (DHA) and arachidonic acid (AA) are provided in breast milk, breast milk fortifier, or special formulas.[11]

Vitamins and Minerals

DRIs for vitamins and minerals are appropriate for many infants with health conditions since recommendations are set with a safety margin.[12] However DRIs are based on growth of typical infants, not those in which *catch-up growth* is required.[5] Vitamin and mineral requirements are impacted by various health conditions, particularly those that involve digestion. Prescribed medications may increase the turnover for specific vitamins.[13] Some infants with special health care needs have restrictions in volume consumed or activity that increase or decrease needs for specific vitamins or minerals. For example, limited volume of liquids may rule out vitamin-rich juices in the diet of infants with breathing problems.

High-potency vitamins and mineral supplements are usually prescribed for sick or recovering infants. Calcium is a potentially limiting nutrient in sick infants since calcium imbalance and *hypocalcemia* occur with a variety of conditions.[11] Iron, B_{12}, vitamin D, and fluoride are limiting in some specific situations.[11] Even after infants who were LBW or VLBW are eating well, vitamin and mineral requirements may be higher than the RDA, depending on specific health conditions. After preterm birth and discharge home, copper, zinc and vitamin D deficiency are rare deficiencies, but may be checked by blood tests.[14] The early signs of rickets as seen by X-ray are considered a sign of needing more vitamin D, above the 400 IU daily recommendation for term and preterm infants.[14]

In order to meet the higher requirements of specific vitamins and minerals for VLBW infants, products have been developed that can be added to breast milk. Such products are expensive and generally used under specific conditions, such as when an infant can only tolerate a low volume per feeding. These products are classified as Human Milk Fortifiers and used primarily in neonatal intensive care units.[14] They are intended to bridge the gap between breast milk and the extra needs of a VLBW infant.[8] Major ingredients are vitamins A, D, and C, and minerals such as calcium, phosphorus, sodium, and chloride.[9] Iron is not in human milk fortifiers, and is prescribed as needed.[9]

Another product that also provides vitamin and mineral intake in concentrated small amounts of liquids is formula for premature infants. Such formulas are supplemented with extra calcium, phosphorus, copper, and zinc compared to standard formulas.[8] (The high levels of vitamins and minerals in premature infant formulas are shown in Table 9.2 later in this chapter). For some conditions, vitamins are used not only as dietary components, but as pharmaceuticals. An example is a condition in which vitamin B_{12} is injected as a part of therapy for a rare genetic disorder of protein metabolism.[10]

GROWTH

Growth in infancy is usually a reassuring sign that sufficient nutrients are provided. CDC 2000 growth charts are a good starting point for monitoring the growth of infants with risks for various health conditions.[15] The first goal is to maintain growth for age and gender. Later, this may be modified if there is a growth pattern typical for a specific condition, but not during the first year. A steady accretion of weight or height is a sign of adequate growth, even if gains are not at the typical rate. Plateaus in weight or height, or weight gain followed by weight loss are signs of inadequate growth. Growth may be assessed reliably using each infant as her own control regardless of health conditions. As noted in Chapter 8, the methods of assessing growth require consistency and accuracy in order to make sure growth is interpreted correctly. Errors such as confusing pounds and kilograms in plotting interfere with interpretation no matter what growth chart is being used.

Usually providing sufficient calories and nutrients results in good growth, but not in all cases. Sometimes slow growth is a symptom of an underlying condition, rather than a sign of inadequate nutrition. For example, infants who are born with genetic forms of kidney disease are short even when adequate nutrition services are provided in the first year. Refinements in the usual methods and interpretation of growth are needed in conditions known to influence growth and development. These include:

- Using growth charts for specific diagnoses, such as Down syndrome growth charts.[16] (A list of specialty growth charts is in Chapter 11.)
- Biochemical indicators of tissues stores such as iron or protein, and of electrolytes such as potassium and sodium.
- Indicators of body composition, such as body fat measurements, can be used to show calories are not limiting growth if fat stores are good.
- Special attention to indicators of brain growth, such as measuring head circumference, may be helpful to explain short stature or other unusual growth patterns.
- Using treatment guidelines or published protocols so that diagnosis-specific weight-gain recommendations replace standard growth projections.
- Medication side effects that change weight gain, appetite, or body composition can explain rapid changes in weight.

Growth in Preterm Infants

CDC growth charts include the growth of preterm infants, unless they were born quite early, with birth weights under 2500 grams.[15] These are LBW infants. Infants who had gestational ages from 35–37 weeks would probably fit well on the standard growth charts. Some full-term infants weigh as little as LBW infants, and are considered at risk based on their low weight. They should be plotted on the standard growth charts. Many LBW infants do not fit well on the standard growth charts shortly after birth. The body composition of infants born preterm is not the same as that of term infants, in part because the third trimester is when fat is usually added at the most rapid rate.[11] In utero body fat is added the last 12–14 weeks.[9] When born early, this may not happen in the same time frame. In fact, body fat build-up is a late sign of recovery from prematurity. Body composition at various gestational ages is used to adjust growth expectations based on age. Treatment of the infant's medical condition may also impact growth expectations, for instance, fluid accumulation may artificially increase weight. As a result of such considerations, VLBW and ELBW infants are not represented by the standard growth charts.[15] The growth charts intended for premature infants are called the "IHDP Growth Charts"; these are based on the *Infant Health and Development Program (IHDP)*.[17] This large research program created four charts, two for girls and two for boys, at two different birthweights. One set called "LBW Premature" is for infants with birthweights in the range of 1501–2500 grams; the other is "VLBW Premature," for infants with very low birthweight, who had birthweights equal to or less than 1500 grams and were less than 38 weeks of gestation.[17] Each chart has:

- Head circumference for gestation-adjusted age
- Weight for gestation-adjusted age
- Length for gestation-adjusted age

These growth charts for prematurity can be used into toddlerhood, although standard growth charts often replace them if the child is growing well.

CORRECTION FOR GESTATIONAL AGE

The gestational age at birth is used in determining what is a good growth rate for a neonate. There are several methods for determining the gestational age of a newborn, such as determination based on the last menstrual period reported by the mother, ultrasound examination of the fetus during the pregnancy, and physical examination and rating of the newborn. The IHDP charts are based on gestation-adjusted age, which is a calculation of chronological age for preterm delivery.[17] Gestational age at birth is subtracted from 40 weeks (the length of a full-term pregnancy) with number of weeks reported as months (two weeks being 0.5 months). The result in months is then subtracted from the infant's current chronological age. The gestation-adjusted age in months is plotted on the IHDP chart. For example if an infant was born at 30 weeks gestation, or 2.5 months early, her gestation-adjusted age at 3 months old (after birth) is two weeks, or 0.5 months. This age of 0.5 months would be used in plotting her growth on the IHDP growth chart as part of assessing her growth and development.

INFANT HEALTH AND DEVELOPMENT PROGRAM (IHDP)   Growth charts with percentiles for VLBW (<1500 gram birthweight) and LBW (<2500 gram birthweight)

Apples don't fall from a pear tree.

French saying

Does Intrauterine Growth Predict Growth Outside?

The answer is yes, no, and maybe! Fetal monitoring during pregnancy and in-depth knowledge about the development of various organ systems provides a clear pattern of growth at various gestational ages. However many factors during and after pregnancy are known to affect growth rate. In summary, these are:

- Intrauterine environment, particularly the adequacy of the placenta in delivering nutrients, the presence of toxins such as viruses, alcohol, or maternal medications, or the depletion of a needed substance, such as folic acid.
- Fetal origin for errors in cell migration or formation of organs, whether or not a cause is known. Various specific nutrients, such as vitamin A, have been implicated.
- Unknown factors which cause preterm birth, such as environmental toxins in air pollution.

Knowledge from studying the human genome is likely to identify additional factors influencing fetal development.

Research is emerging which shows some infants were born prematurely due to conditions originating during the intrauterine period.[18] As discussed in Chapter 8, *SGA* and *intrauterine growth retardation* are terms used to describe infants who are smaller than expected at birth. SGA is the more general term since it is based on the population of infants of the same gestational age.[19] Both predict higher medical risks and need for close growth monitoring.

If the intrauterine insult was early in gestation, body weight, length, and head size (brain size) are impacted.[19] There has been a change in the number and size of fetal cells. The abnormal fetal growth pattern may persist in spite of adequate medical and nutrition support after birth. Examples of conditions causing early insults are infants born after cociane and alcohol exposure, which are associated with IUGR and preterm birth.[20] Later exposure in the second or early part of the third trimester may result in preservation of head size and body length, but low weight.[19,20] Some genetic conditions characterized by small size are not diagnosed until childhood, but have IUGR noted in the medical history.[19]

Intrauterine growth may not predict growth for some infants whose birth removes them from toxins within the intrauterine environment. Examples are maternal uncontrolled diabetes, smoking, phenylketonuria, or maternal seizures treated by medications. In such cases the rate of growth after birth may be improved and normalized during the first year of life.[19] In general, the earlier the exposure to the toxin, the worse the effects on later growth, but there is quite a bit of variability.[20] Sometimes marijuana, alcohol, tobacco, and crack/cocaine have been used at various times during pregnancy, so growth impacts may be based on amount, timing, and interactions of toxins.[20] The most important risks for later growth may be neonates born with smaller head circumferences.[20]

A large category that changes the rate of intrauterine growth is early medical treatment. If the intrauterine growth was fine, preterm birth or its complications may slow or plateau growth early in infancy. This may mask whether or not typical growth goals are appropriate all during infancy. Various studies have measured whether or not recovery from early growth problems occurs. A group of ELBW infants that were followed into adolescents and did not have major disabilities were shorter in stature and lower in weight than others of the same age and gender, suggesting growth is impacted long-term.[21] Another study of infants who were short for age and were provided nutritional supplements for two years found that they were still short for age at 11 and 12 years.[22] Whether or not adequate nutrition is provided early in life of course depends on how *adequate nutrition* is defined. Advocates of increasing current nutritional recommendations for preterm infants expect growth problems could be lessened if higher levels of nutrition are provided, regardless of other factors.[5]

Conditions which impact growth may be time limited. The treatment course may result in slowing of growth, but then the rate of growth may increase. Catch-up growth may be seen during recovery in many infants, resulting in changing growth interpretation. Examples of conditions in which growth rate may change are those treated by surgical intervention, such as heart conditions, respiratory illnesses that resolve with medications, or other conditions in which weakness resolves. Usually increased access to adequate nutrition improves growth. Only close monitoring over time may show signs of catch-up growth early, such as an increase in fat stores or length.

Catch-up growth for preterm babies allows use of the standard growth charts eventually. The amount of time needed for catch-up growth for premature infants differs based on gestational age at birth and subsequent complications. Clinical conventions are to provide one year for catch-up growth for infants born 32 weeks or later, and three years for catch-up growth for VLBW or ELBW infants. It is likely such infants when children will have growth percentiles in the lower end of the normal range.

Interpretation of Growth

Hospital discharge after preterm birth may be based on a pattern of weight gain, such as 20–30 grams per day.[14] Strong emphasis is placed on growth as a sign of improving health in monitoring small and sick infants after discharge, but various complications make this difficult to show. Growth rate changes are closely associated with the frequency of illness, hospitalizations, and medical history.[11] Conditions acquired as a result of preterm birth that make growth difficult to interpret include:

- Symptoms related to intestinal absorption that can temporarily or permanently change nutrient requirements.
- *Microcephaly* (small head size) or *macrocephaly* (large head size) compared to other growth indicators may be a sign that growth may be impacted as a result of neurological consequences. Both large and small head size in infancy can affect muscle mass, body composition, and subsequent growth.
- Variable rates of recovery are seen for infants with the same diagnoses. Individual variation is hard to quantify even in infants of the same gestational age. Infants are as different from each other as the rest of us, but these differences are hard to see soon after birth.

COMMON NUTRITION PROBLEMS

Infants who were small or sick near birth may have major nutrition problems in growing and feeding. Infancy is such a vulnerable time of life that most health conditions occurring this early interfere with growth and development. Over time, most of these problems resolve, although some become chronic and a few result in death. Nutrition plays an important part in preventing illness, maintaining health, and treating conditions in infancy. Nutrition tends to become more important over time to maintain growth if conditions are chronic.

Table 9.1 shows nutrition problems in infants with special health care needs. Nutrition assessment documents these concerns, and nutrition services intervene

> **MICROCEPHALY** Small head size for age and gender as measured by centimeters (or inches) of head circumference.
>
> **MACROCEPHALY** Large head size for age and gender as measured by centimeters (or inches) of head circumference.

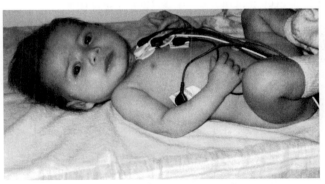

Illustration 9.2 Baby girl wearing a heart monitor at home.

Table 9.1   Nutrition concerns in infants with special health care needs.

<u>**Growth**</u>	Slow rate of weight gain
	Fast rate of weight gain
	Slow rate of gain in length
	Disproportionate rate of weight to height gain
	Unusual growth pattern with plateau in weight or length gain
	Altered body composition that decreases or increases muscle size or activity
	Altered brain size that decreases or increases muscle size or activity
	Altered size of organs or skeleton, such as an enlarged liver or shortened leg length
<u>**Nutritional adequacy**</u>	Calorie needs are higher or lower
	Nutrient requirements higher or lower overall
	Specific nutrients, such as protein or sodium, are required in higher or lower amounts
	Vitamins, minerals, or co-factors (carnitine) are required in higher or lower amounts
<u>**Feeding**</u>	Disruption of the delivery of nutrients as a result of :
	• Structure or functioning of the mouth or oral cavity
	• Structure or functioning of the GI tract, including diarrhea, vomiting, and constipation
	• Appetite suppression by constipation or medications
	• Interaction of the infant with the parent is disrupting, such as infant cues being so subtle that parent responses are delayed
	• Posture or position promotes or interferes during meal times
	• Timing of nursing, meals, and snacks throughout the day
	• Inappropriate food choices or methods of preparation
	• Interruptions in adequate shelter for feeding and sleeping
	• Instructions were unclear or too complicated for the parent to follow

based on the assessment. These concerns are later described for specific groups, such as severe preterm birth, chronic illnesses, and acquired conditions.

Nutrition Risks and Development

Many health conditions change the rate of development of the infant. *Developmental delay* describes the interaction of a chronic condition with development. The terms *Children with Special Health Care Needs* and *developmental delay* are general terms used to allow nutritional, medical, and developmental services to be provided for infants.

Developmental delay is used to describe a wide range of symptoms that reflect slow development. Symptoms that relate to nutrition are common. These symptoms include infants who are growing slower than expected for age, or have difficulty in feeding, such as re-

> **DEVELOPMENTAL DELAY**   Conditions represented by at least a 25% delay by standard evaluation in one or more areas of development, such as gross or fine motor, cognitive, communication, social or emotional development.
>
> **AUTISM**   Condition of deficits in communication and social interaction with onset generally before age 3 years, in which mealtime behavior and eating problems occur along with other behavioral and sensory problems.

fusing food from a spoon by 8 months of age. An example is a two month old girl who does not breastfeed for more than a few minutes per side. At first this may appear to be a problem of breastfeeding position or frequency. By four months, weight gain is slower than expected, so now growth concerns and feeding concerns are interacting; it is not clear if they are separate or related problems. These concerns are sufficient for providing an evaluation for eligibility for intervention services. Several months later, after various services have been put in place, it may still be unclear if these nutrition problems are from development, a health condition such as a heart murmur, or the interaction of both. In any case, the girl fits the category of a child with a special health care need. This allows the family to benefit from nutritional, medical, and developmental interventions and a specific diagnosis is not required. Infants generally are not old enough to have a specific diagnosis related to development, such as mental retardation or *autism.*

Down syndrome is an example of a condition in which developmental delay is noted in infancy. Down syndrome prevalence is about 9 cases per 10,000 live births.[23] Nutrition concerns with infants who have Down syndrome are feeding difficulties related to weak muscles in the face and overall, high risk of overweight, and constipation.[19] Heart and intestinal conditions are more common in infants with Down syndrome, so their nutrient needs may be increased if surgery is required. This is also an example of a chronic condition in which nutrition problems increase over time if prevention and maintaining health are not addressed. Growth requires close monitoring to identify and prevent overweight starting in infancy. Infants with Down syndrome love to suck and have things in their mouths so much that it is easy to overfeed them. Development of movement occurs at a slower rate, with lower physical activity, which also can contribute to overweight. Giving parents their own copy of Down syndrome growth charts for infants is recommended after the diagnosis is confirmed.[16] It may be helpful in recognizing typical growth and preventing overweight early. These special growth charts are available from places that serve children with Down syndrome, such as developmental or genetics clinics in major medical centers.[16]

Not all children with developmental delay in infancy have developmental disabilities later. For example, an infant with breathing problems may be slower to grow and to crawl as a result of his higher caloric needs during the first year of life. Such an infant may show developmental delay in motor skills, but by age three he will have improved health overall and have caught up to others in motor skills. He would not have a developmental disability. Other examples are infants from high-risk pregnancies, such as those with LGA as a result of maternal diabetes. Many require short stays in intensive care units for glucose regulation; some may have long-term risks for their development.

Some infants with developmental delays continue to have slower development over time. After infancy, when standard testing and evaluation can be performed the term *developmental delay* may be replaced with a more specific type of medical or developmental diagnosis. Developmental disabilities in toddlers and children and their nutritional aspects are discussed in Chapters 11 and 13.

PRETERM INFANTS AND INFANTS WITH SPECIAL HEALTH CARE NEEDS

Many infants who were born preterm grow and develop within months of birth similarly to term infants. In 1998, about 60,000 infants with birthweights of 2000–2499 grams (4 pounds 6 ounces to 5 pounds 8 ounces) at 32–35 weeks of gestation were born preterm.[24] They probably did not require intensive care services, but were monitored briefly to confirm they were nursing and sent home within a week after birth. These infants are in the LBW category and considered at risk for public health and educational programs within federal and state regulations. The underlying concepts are:

- LBW is a risk factor for later health and developmental concerns.
- Identifying risk factors early will benefit infant growth and development in the long-term.
- Interventions are more effective if problems are identified early.
- Prevention of secondary complications may result from identifying problems early.

Within the United States, some state public health tracking systems use birth records to identify infants at risk.[25] Health providers consider infants with low birthweight, including VLBW and ELBW infants, to need closer monitoring than other infants.[14] Under such guidelines, services for Children with Special Health Care Needs include all infants with low birthweight.[25] When programs use birthweight to trigger closer monitoring, they will also pick up infants who were not preterm but were small for gestational age (SGA). Infants who are SGA may be born later than 34 weeks. The WIC program is an example of a program that uses birthweight as a nutrition risk for providing services.[26] Other state systems and programs require evaluations or referrals of infants to document factors that require close monitoring. Under these guidelines low birthweight is not an automatic trigger for services, so not all low birthweight infants are within the Children with Special Health Care Needs category.

SEVERE PRETERM BIRTH

Prevalence of preterm birth in the United States is high at 11.6% of all live births.[27] LBW infants were 7.6% of all births, almost 300,000 infants in 1998.[27] VLBW babies were 1.4% of all births, over 55,000 newborns in 1998.[27] VLBW yearly incidence is about the same as the population of a small city, such as Iowa City, Iowa. Infants with birthweights near 1500 grams (3 pounds 4 ounces) have gestational ages from 28–32 weeks, and a survival rate of almost 90%.[24] Each infant required immediate intensive care hospitalization and if they survived, continued to have high nutrition needs through all of infancy. ELBW infants have gestational ages ranging from 23–28 weeks, so they have skipped the third trimester of pregnancy in utero. In spite of advances in their care, disability such as delayed development is a common outcome of ELBW.[28] Some outcome studies are demonstrating lifelong consequences of low birth weight, such as on later employment as adults.[29]

Nutrition problems as a result of VLBW and ELBW preterm birth are discussed based on the time of their presentation. The initial problem after birth is that the newborn cannot nurse like a full-term infant, and most require respiratory support to breathe. Getting adequate nutrients into the preterm infant requires *nutrition support,* usually first *parenteral feeding* and then *enteral feeding* methods.[11] Feeding problems of preterm infants are discussed later in this chapter.

An underlying concept in providing adequate nutrition is that newborns have high metabolic rates, as discussed in Chapter 8. If sufficient calories are not provided to meet the need for glucose in the body, the infant metabolizes ingested protein and fats in order to generate glucose. When an infant is small, sick, or must have surgery, the need for sufficient calories and nutrients may be higher.[4] Infants use fat stores and protein in tissues and muscles for calories sooner than adults do to meet glucose needs. Providing sufficient nutrition to meet requirements and preserve ingested protein and calories for growth is the goal, but it may be difficult and take more time than expected in sick and recovering infants.

NUTRITION SUPPORT   Provision of nutrients by methods other than eating regular foods or drinking regular beverages, such as directly accessing the stomach by tube or placing nutrients into the bloodstream.

TOTAL PARENTERAL NUTRITION (TPN) OR PARENTERAL NUTRITION (PN) Delivery of nutrients directly to the bloodstream.

ENTERAL FEEDING   Method of delivering nutrients directly to the digestive system, in contrast to methods which bypass the digestive system.

NECROTIZING ENTEROCOLITIS (NEC) Condition with inflammation or damage to a section of the intestine, with a grading from mild to severe.

ORAL-GASTRIC FEEDING (OG)   A form of enteral nutrition support for delivering nutrition by tube placement from the mouth to the stomach.

TRANSPYLORIC FEEDING (TP)   Form of enteral nutrition support for delivering nutrition by tube placement from the nose or mouth into the upper part of the small intestine.

GASTROSTOMY FEEDING   Form of enteral nutrition support for delivering nutrition by tube placement directly into the stomach, bypassing the mouth through a surgical procedure which creates an opening through the abdominal wall and stomach.

JEJUNOSTOMY FEEDING   Form of enteral nutrition support for delivering nutrition by tube placement directly into the upper part of the small intestine.

Delivering Nutrients

Babies are not generally released from the hospital until they are able to eat by mouth. Oral feeding, which uses the gut to keep it healthy, is the goal. If not stimulated, the absorptive surface of the small intestine decreases.[11] Immediately after birth, however, the crucial role of providing nutrients to support life is what is important, not the route by which they are delivered. Most VLBW, ELBW, and sick preterm in-

fants in neonatal intensive care units require nutrition support by parenteral nutrition (PN) in which nutrients are delivered to the bloodstream.[30] Fluids, vitamins, minerals, fats, amino acids and energy (as glucose or dextrose) are customized for the specific weight and medical condition.[11]

How Sick Babies Are Fed

Gastrointestinal upset is a response to many conditions in newborns, whether the intestines are the initial problem or not. VLBW, ELBW, and sick infants are especially vulnerable to problems related to the GI tract and its movement. Such problems directly affect how nutrition is provided and the composition of the diet. For example, if a newborn gets an infection, an early sign may be inflammation of the intestine. As a response, the method of feeding the infant has to be adjusted. Inflamed or damaged areas may slow or interrupt typical intestinal muscle movements, resulting in signs of increasing illness.[11] Blood loss from the intestines is a sign of *necrotizing enterocolitis,* a serious condition in the neonate. When this occurs, oral feeding is stopped and replaced by parenteral nutrition.

Many GI conditions interfere with feeding infants, as discussed in the previous chapter. Examples are gastroesophogeal reflux, constipation, spitting up, and vomiting. In small and sick newborns, these GI conditions may represent slow or uncoordinated movements of the intestinal muscles.[31] These conditions do not rule out enteral feeding, which stimulates the intestines and keeps them healthy. Breastfeeding such a sick infant may be attempted, but often enteral feeding methods that do not depend on the ability of the newborn to suckle, or his or her level of fatigue, are required. Feeding methods are selected based on the length of time before it is expected the baby can nurse or feed without help. Gavage feedings are slow feedings by tube that go into the stomach through the mouth or nose. Infants who are too weak to breastfeed may be offered the comfort of the breast or pacifier along with gavage feeding, as a form of non-nutritive stimulation. Since newborns breathe only through the nose, *oral-gastric (OG) feeding* is typically used.[9] Other enteral methods are *transpyloric feeding, gastrostomy feeding* and *jejunostomy feeding.*[9] These methods are used when nutrition support is expected to be needed for several months, such as when surgeries are required over time.[32]

VLBW and ELBW infants need nutrition services to maximize their growth. The provision of adequate nutrition is difficult, but important to prevent undernutrition from being superimposed on top of the infant's other medical problems. Conditions complicate growth assessment and recommendations so infants need continued nutrition services include the following:[33]

- Muscle coordination problems that stem from damage to parts of the brain controlling movement, which may later be identified as a condition such as cerebral palsy.
- Developmental delay that stems from damage to the brain overall or the thinking brain, which may later be identified as a learning problems or mental retardation.
- Breathing problems such as chronic lung disease or asthma.
- Damage to other organs such as the liver, kidneys, or intestines as a result of infection or as a consequence of treatment.

Each of these has impacts on growth and feeding which make the infant eligible for additional nutrition services.

What to Feed Preterm Infants

Breast milk is the recommended source of nutrition for preterm infants.[34] Colostrum and breast milk are made even when the mother delivers very early. Preterm human milk has increased protein content compared to term milk.[35] Hospital protocols and policies for having mothers pump and freeze breast milk for later use by their preterm infants are highly recommended.[34] Staff training to encourage new mothers at home to rest enough and pump enough to stimulate breast milk production is also recommended.[36]

Barriers to breastfeeding small and sick newborns are partially based on their abilities, how sick they are, and on the care system. Promoting breast milk for preterm infants is recommended as hospital policy by the American Academy of Pediatrics, but hospitals differ in implementation.[34] An example of promoting breastfeeding in mothers of preterm infants is a bank of pooled breast milk.[5] When an infant is monitored by a lot of machinery, positioning the infant may be difficult enough to interfere with nursing, but breast milk can still be provided. Sedation and medications may change the responsiveness of the infant so they cannot coordinate sucking and swallowing safely for nursing.

Medical conditions in the infant may undermine breastfeeding. Neonates have the coordination to safely nurse at about 37 weeks of gestation. Prior to that age, they may benefit from being put to the breast to stimulate non-nutritive sucking, which does not deliver milk to be swallowed. The few conditions in which human milk is unsafe for preterm or sick infants are when breast milk contains medications, street drugs, viruses, or other infective agents, or when the infant has a GI tract malformation or inborn errors of metabolism.[35]

Breast milk that has been supplemented usually can meet the needs of preterm and sick infants. The main consideration is that higher calorie and protein requirements of preterm infants must be delivered in small volume. Depending on the birthweight and medical condition, breast milk may be insufficient in nutrients unless supplemented by human milk fortifier and/or other sources of calories, such as MCT oil. If not fed modified breast milk or nursing, the infant's source of nutrition may be cow's milk- or soy bean-based formulas.[35] Whey as the predominant form of protein from cow's milk is recommended since its amino acids profile is closer to that of human milk.[8]

Infant formulas for preterm infants are available for home use after hospital discharge, if breast milk is not available. They provide the higher calories and nutrient levels that small infants need compared to term infants.[37] Standard formula that is 20 calories per fluid ounce can also be used for preterm infants, modified in a manner similar to breast milk to boost calories and nutrients. High-calorie formulas, such as 28 cal/fl oz may be appropriate for some infants. Such high-calorie formulas are not routinely used since they have high osmolarity, which may impact fluid and electrolyte balance. Table 9.2 shows a comparison of premature and standard formulas.[37,38] If the infant easily fatigues or is too weak to suck enough volume, 22 or 24 cal/fl oz formulas may be recommended.[9] The sources to add extra calories and nutrients are selected based on the infant's GI tolerance and volume requirements. They may include MCT oil, polycose, rice baby cereal, and rarely human milk fortifier.[37] Routine nutrition assessment

Table 9.2 Selected nutrient composition of term and preterm formulas.

Nutrients	20 Cal/Fl Oz	22 Cal/Fl Oz	24 Cal/Fl Oz
Protein	2.1 g	2.8 g	3 g
Linoleic acid	860 mg	950 mg	1060 mg
Vitamin A	300 IU	450 IU	1250 IU
Vitamin D	60 IU	80 IU	270 IU
Vitamin E	2 IU	4 IU	6.3 IU
Thiamin (B_1)	80 mcg	200 mcg	200 mcg
Riboflavin(B_2)	140 mcg	200 mcg	300 mcg
Vitamin B_6	60 mcg	100 mcg	150 mcg
VitaminB_{12}	0.3 mcg	0.3 mcg	0.25 mcg
Niacin	1000 mcg	2000 mcg	4000 mcg
Folic acid	16 mcg	26 mcg	35 mcg
Panothenic acid	500 mcg	850 mcg	1200 mcg
Biotin	3 mcg	6 mcg	4 mcg
Vitamin C	12 mg	16 mg	20 mg
Inositol	6 mg	30 mg	17 mg
Calcium	78 mg	120 mg	165 mg
Copper	75 mcg	120 mcg	125 mcg

of the infant's growth tracks how effective the diet is in providing adequate nutrients and calories.

Preterm Infants and Feeding

VLBW or ELBW infants usually progress at their own rate in developing feeding skills. The goal is the same as for all infants, to achieve good nutritional status, as indicated by growth and feeding skills progression. Most families enjoy feeding their infants and avoid long-term feeding problems. Infants who were born preterm, however, often are hard-to-feed babies. There are several reasons why an infant may be difficult to feed.[35]

- Fatigue. The levels of arousal of recovering infants may make it difficult to find sufficient times when the infant has an interest in feeding.
- Low tolerance of volume. Abdominal distention can result in changes in breathing and heart rate, so that the infant stops feeding.
- "Disorganized feeding" may result from the infant having experienced defensive and unpleasant reactions, so anything coming to the mouth causes a stress reaction, rather than a pleasurable reaction.[9]

Regardless of the associated conditions, certain feeding characteristics of preterm infants are distinct from those of term infants, as shown in Table 9.3. Most recovering infants improve in their feeding abilities with time. Anxiety decreases as parents become more comfortable caring for their infant at home. The underlying reflexes that associate pleasure with feeding re-emerge and the interaction of the infant and the feeder becomes more consistent. There is a lot of room for hope in the feeding process after discharge.

CONGENITAL ANOMALY Condition evident in a newborn that is diagnosed at or near birth, usually as a genetic or chronic condition, such as spina bifida or cleft lip and palate.

Table 9.3
Preterm and term infant feeding differences.

Preterm Infant	Term Infant
Central nervous system does not signal hunger	Signals hunger; has supportive newborn feeding reflexes
Unstable feeding position, such as a forward head position	Stable and facilitating feeding position from newborn reflexes
Oral hypersensitivity	Readily accepts food by mouth

Major advances in understanding nutritional needs of preterm babies have come about from working with smaller and smaller infants. Table 9.4 is an example of a typical diet for a premature baby, who was not breastfed and had a three-month hospital stay after birth. The infant has gastroesophageal reflux and prescribed medications which are included in his feeding instructions. This example shows that the provision of an adequate diet is an important part of the infant's growth and development and his recovery from preterm birth complications.

INFANTS WITH CHRONIC ILLNESS

Infants who are not premature but who require neonatal intensive care represent an important group at risk for chronic illness. Examples are infants born to mothers with poorly controlled diabetes, or infants with breathing difficulties shortly after birth. LGA infants may require intensive care based on glucose regulation or birth difficulties due to their large size. About half of babies within neonatal intensive care units have normal birth weights.[39] Such babies have lower mortality than LBW infants, but have a higher rate of *congenital anomalies* (22%) and require a higher rate of rehospitalization.[39] These infants require more nutrition services than typical infants

Table 9.4   Example diet for a VLBW infant at 8 months of age, with a corrected age of 4.5 months (weight 4.5 kg 5–25% on VLBW Premature Boys growth chart).

Food and Formula	Feeding Instructions
Five feedings per day of formula, each 5 fl oz, High-calorie formula (24 cal/fl oz) With 2 tbs rice cereal added	Provide support for semi-reclining feeding position during and up to 30 minutes after feeding
Medications for stomach added to 2 fl oz of 50% diluted apple juice two times per day	Encourage use of pacifier for comfort between feedings
Two meals of pureed baby foods fed with spoon, total intake one 2 oz jar	Offer bottles every 3 hours except overnight if no signs of hunger before then
Liquid vitamin/mineral supplement	Keep scheduled appointments for WIC, weigh in at MD office, bringing diet log book

Case Study 9.1
Premature Birth in an At-risk Family

Ababy named Eric was born at 30 weeks of gestation, appropriate for gestational age, at 1.4 kilograms (3 pounds). He tested positive for cocaine as did his mother. Eric received routine intensive care services, which included ruling out sepsis, and was given head ultrasound studies. By 33 weeks he was fed OG and had only transient respiratory difficulties. Prior to discharge at 37 weeks he appeared to be developing normally. He drank 22 fl oz formula/day at the rate of about 1.5–2 fl oz/feeding with 10 feedings per day. He was placed in foster care with an experienced foster mother who had two older children. His custody was reconsidered when his biological mother expressed interest when he was about 9 months of age. Shortly after that her parental rights were terminated based on criminal charges. His foster mother reported that he was a colicky baby, and he had at least three ear infections during his first year of life. He was enrolled in an early intervention program based on his prematurity and intrauterine drug exposure. Eric's initial developmental testing was within normal limits at six months.

His foster mother expressed concern about his periods of periodic crying in which he accepted no soothing, nor a bottle. Eric was diagnosed with GER and slow gastric emptying, and treated medically starting at eight months. He was slow to accept foods on a spoon, with gagging and spitting up. His growth on the IHDP LBW chart was near the 50th percentile for weight and height. His head circumference was at the fifth percentile. Eric did not sit up unassisted until 8.5 months, but he turned over from stomach to back and vice versa easily. He was sent to a genetic specialist since he appeared to have some facial features consistent with fetal alcohol syndrome, such as low set ears, wide nasal bridge, and thin upper lip. The diagnosis was not confirmed but reported as a possibility, after re-evaluation at about three years. If this diagnosis was confirmed, standard growth charts would not fit him, since short stature and low weight are part of the diagnosis, even when adequate nutrition is provided. In infancy Eric's growth was within normal limits after correction for prematurity, with the same trend of a lower head circumference. His foster mother expressed an interest in adoption when he was almost one year old. She was pleased that he needed no medication and was growing well. His early intervention services were continued based on his at-risk status, since no specific 25% delay was documented at one year. (He was later diagnosed with mixed developmental delay based on cognitive and speech delays at 34 months of age.) Eric was adopted by the foster family.

since their growth and feeding development requires close monitoring and intervention.

This category of high-risk infants comprises those with diagnoses that are included in the Birth Defects Monitoring Program (BDMP) and diagnoses tracked by the *Infant Mortality Attributable to Birth Defects (IMBD)*.[40] The Centers for Disease Control publish prevalence based on states and hospitals that participate voluntarily in surveillance programs. The United States keeps prevalence data not just for infants, but also for children of various ages. IMBD accounts for approximately 2 deaths per 1000 live births.[40] Whereas infant mortality is decreasing in the United States, the proportion of infants who died from IMBD is increasing.[40] The major type of birth defect associated with death are first heart malformations and then central nervous system defects[40] Examples of central nervous system congenital anomalies are spina bifida and *anencephaly*. These conditions occur early in the first trimester of pregnancy and may be decreased by prepregnancy folic acid prevention efforts[41]

IMBD   Infant mortality attributable to birth defects. Category used in tracking infant deaths in which specific diagnoses have a high mortality.

ANENCEPHALY   Condition initiated early in gestation of the central nervous system in which the brain is not formed correctly, resulting in neonatal death.

Infants with Spina Bifida

Babies with spina bifida have a prevalence at birth of about 4 cases per 10,000 live births during 1983–1990, a period before implementation of programs to increase folic acid consumption and supplementation.[41] Their nutrition problems during infancy are higher nutrient needs for wound healing after surgery for closure of the opening in the back, lower calorie needs due to lower activity of large muscle groups that are paralyzed, and difficulties related to feeding as a result of constipation.

Infants with congenital anomalies, genetic syndromes, and malformations fit in the Children with Special Health Care Needs category, so they are eligible for a wide range of medical, nutritional, and educational services to maximize their growth and development.[25] All babies born with such conditions have risks to maintaining good nutrition status as they receive treatment. The nutrition concerns are growth, the adequacy of the diet in providing required nutrients, and feeding development as shown in Table 9.1. Nutrition services range from temporary to long-term, and are as diverse as the many different types of conditions

DIAPHRAGMATIC HERNIA   Displacement of the intestines up into the lung area due to incomplete formation of the diaphragm in utero.

TRACHEOESOPHAGEAL ATRESIA   Incomplete connection between the esophagus and the stomach in utero, resulting in a shortened esophagus.

CLEFT LIP AND PALATE   Condition in which the upper lip and roof of the mouth are not formed completely and are surgically corrected, resulting in feeding, speaking, and hearing difficulties in childhood.

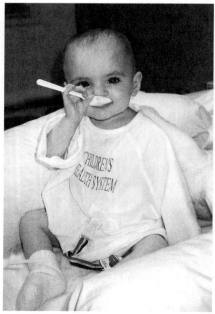

Illustration 9.3   Infant in hospital being encouraged to eat with a spoon

involved. Several examples of disorders with minor and major nutritional consequences follow.

Major nutritional impacts are exemplified by disorders that involve the GI tract. Infants with *diaphragmatic hernia* or *tracheoesophageal atresia* cannot safely eat by mouth and require nutrition support and several surgeries during infancy.[32] Diaphragmatic hernia occurs in one in 4000 live births as a result of failure of the diaphram to form completely. Tracheoesophageal atresia occurs in one in 4500 live births as a result of an error in development of the trachea. These examples of conditions treated by neonatal surgery and intensive care have been credited with lower infant mortality for such congenital anomalies.[39] Both conditions change the motility of the GI tract, so sufficient calories, nutrients to maintain growth, and oral feeding are important parts of the treatment plan.[32] Such infants miss the windows of development when oral feeding is pleasurable and may have residual feeding problems, such as disliking eating by mouth well into early childhood. Financing such intensive health care and maintaining the child's normal social and emotional development are major issues for the family. The families of such infants have many specialty health care providers, and their complex financial and emotional reactions that can also impact the infant's development. Both of these conditions eventually result in children being able to eat like every one else.

A more common example of needed nutrition services for congenital anomalies are those for infants with *cleft lip and palate.* Major feeding difficulties occur before and after corrective surgeries, which sometimes interfere with growth in infancy and as young children.[42] Assistance in feeding by registered dietitians as part of a team approach is needed, since hearing, speech and language problems are associated with cleft lip and palate. Positions for eating are adapted and use of special feeding devices for the infant's cleft are needed.[42] Cleft lip and palate may occur alone or as a part of various rare genetic conditions, so growth and feeding problems should also be assessed after corrective surgery.[43]

Disorders that do not involve the GI tract may not increase nutritional requirements at all. Nutrition services are usually provided if growth or feeding is affected. Infants with orthopedic deformities such as club feet would be eligible for services as infants with special health care needs, but they probably would not need nutrition services, even though they may have a higher risk of overweight if their mobility is restricted for many months.

Infants with Genetic Disorders

Infants diagnosed with genetic disorders near birth are a small subset of infants with congenital anomalies or chronic conditions. They also fit in the category of infants with special health care needs. The number of genetic disorders that can be identified in newborns is increasing rapidly, particularly through the expanded infant screening programs described in the previous chapter. The nutritional implications of expanded genetic screening and diagnosis is that more newborns require special diets immediately. Infants with rare genetics conditions such as galactosemia or *maple syrup urine disease* need the diet started within days of birth—waiting even one week can result in irreversible brain damage or death.[10] The infants who are identified in newborn screening with metabolic or genetic conditions usually require special formulas. Genetic centers or inborn errors of metabolism clinics are notified when a newborn screening result needs follow-up. Immediate action is taken to locate the family and confirm the diagnosis. In such circumstances, newborn screening results in early diagnosis and avoids a costly stay in the hospital intensive care unit. For example an infant with galactosemia, if not picked up by the initial abnormal screening result, is likely to have been hospitalized with possible sepsis or liver problems by the time a second screen is to be collected. If the baby with galactosemia receives supportive measures such as sugar solutions, and then soy-based infant formulas without galactose, recovery is usually rapid. Some of the disorders picked up by newborn screening do not make the baby sick in early infancy, but later. It is difficult for parents of a healthy-appearing newborn to be told that a special diet is needed to prevent illness later. An example of a condition that can be picked up by newborn screening before illness is cystic fibrosis.

Genetic tests for infants are becoming more common overall. Genetic tests for Down syndrome, *Fragile X syndrome*, and many disorders are based on blood tests. This concept of increased use of genetic tests is exemplified by the condition called *DiGeorge syndrome.* Any infant with a heart defect may have this test ordered. DiGeorge sydrome is a condition in which a small piece of chromosome 22 is deleted.[44] Recent incidence estimates place it at one in every 4000 births.[44] This makes DiGeorge relatively common for a genetic condition: more common that PKU or cystic fibrosis, and second only to Down syndrome as a cause of mental retardation. Infants with DiGeorge syndrome may have a wide range of conditions, impacting the heart, immune system, calcium balance, and later speech and learning problems. Only when the genetic probe became available was it understood that three separate disorders involved the same deletion. As a result, its incidence was under reported before the probe was available. Nutrition services may be required based on short stature, heart malformations, heart surgery, and resulting feeding problems. See Case Study 9.2.

FEEDING PROBLEMS

Infants who were born preterm or have chronic health problems tend to be more irritable and less able to signal their wants and needs compared to healthy infants.[35] Feeding difficulties are reported in 40–45% of families with VLBW infants.[35] Children with developmental disabilities have more frequent feeding problems, as high as 70%, that may or may not be identified in infancy.[6] Feeding problems in infants may be related to GER, especially if neurological problems are suspected.[31] As noted earlier, the GI tract is sometimes reactive even if the health condition is not related to the GI tract. For example, a baby with a heart problem may have GER also, so formula with hydrolyzed protein may be recommended to prevent more GI discomfort.

All infants may refuse foods or have digestive difficulties once in a while. Infants who are born with cocaine-exposure, severe preterm, or breathing problems have periods of refusing foods, too. It can be difficult to tell if feeding problems are typical or more severe, and starting to interfere with growth. Table 9.5 gives signs of feeding difficulties that can be caught early. By the time feeding problems require interventions, to protect later growth and development, families and infants may be frustrated from their feeding experiences. Infants who are difficult to feed are at risk for failure to thrive (FTT), and child abuse and neglect.[35]

> **MAPLE SYRUP URINE DISEASE**   Rare genetic condition of protein metabolism in which breakdown by-products build up in blood and urine, causing coma and death if untreated.
>
> **FRAGILE X SYNDROME**   Condition mainly identified in males characterized by mental retardation and physical signs.
>
> **DIGEORGE SYNDROME**   Condition in which chromosome 22 has a small deletion, resulting in a wide range of heart, speech, and learning difficulties.

Do Infant Feeding Guidelines Apply?

Infant feeding guidelines, such as those within Bright Futures in Practice or the WIC program that were discussed in the previous chapter, may or may not apply to infants with special health care needs.[45,26] Routine guidelines such as feeding an infant in response to signs of hunger are appropriate in conditions such as infants with spina bifida, or PKU, or an infant whose mother has uncontrolled diabetes. Infants who are too weak to nurse, however, or who have medication side effects that make them sleepy would not benefit from such a recommendation. They may give such weak hunger signals that they are missed, resulting in inadequate calories and nutrients being consumed. In some circumstances, routine

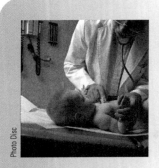

Photo Disc

Case Study 9.2
Noah's Cardiac and Genetic Condition

A baby named Noah required open heart surgery, which resulted in a diagnosis of DiGeorge syndrome. His mother was successful at expressing her milk and maintaining her breastmilk supply by using the breast pump supplied by the hospital during the hospitalization. Pumping and freezing the breast milk for the baby was important to the family. Noah was a reluctant eater by breast or bottle, with intake usually only one or two ounces per feeding, when allowed to feed. He was too weak to feed throughout most of the hospitalization.

At discharge the family was referred to a local early intervention program, WIC, Supplemental Social Insurance (SSI), and the State Program for Children with Special Needs. Specialty clinic and local follow-up appointments were made. Feeding difficulties concerned the family, and both a lactation consultant and a registered dietitian were involved at home. Noah nursed frequently, but briefly, due to fatigue. Growth was slower than expected, but the cardiologist did not think his slow weight gain was a result of the cardiac problem.

Noah tolerated only small volumes, even after recommendations from the lactation consultant were implemented. Breast milk offered in a bottle so that its caloric density could be increased by added rice cereal and MCT was recommended. The family perceived this recommendation as undermining the mother's effort to breastfeed, and they were unwilling to try it. Noah liked to nurse but for such a short time that the richer hind-milk may not have been available. Over time the family offered food on a spoon and continued nursing. Weight and height gain did not fit the expected rate, but appeared to be fairly consistent.

The family increasingly enjoyed parenting, and thought their son was a beautiful infant. They did not contact WIC, nor the early intervention program. They regarded their baby as getting better after surgery. They considered his small size a result of heart surgery. Many that they met assumed he was a premature baby, but his small size was not as much a matter of concern to his parents as it was to health providers.

This case example demonstrates inadequate growth rate, feeding problems, and questionable adequacy of the diet. The genetic syndrome, heart condition, nutritional inadequacy of the baby's diet, and the impact of stress and coping on the mother's milk supply all could have explained his slow growth. Growth expectations are unclear as there is not a growth chart for this genetic syndrome, nor for infants with cardiac anomalies. The standard growth chart is the only one available, but it may not be appropriate to predict future growth.

Important parts of his growth assessment were that his fat stores were good, showing he had access to enough calories, and that his head circumference percentile was low, which suggested that his brain was not growing at the typical rate. This could have been due to neurological damage during surgery or afterwards, or to the underlying genetic syndrome. It is probably not due to inadequate nutrition in early infancy, since the body tries to preserve brain growth, but there is no way to rule that out. Noah's feeding problems were subtle signs of his developmental delay, although this is only clear in hindsight.

nutrition recommendations may be inappropriate or even harmful. For example, if a family is facing surgery for their infant with a heart defect, weaning at the usual time frame may be not recommended. The recovering infant may require the bottle to deliver sufficient volume of concentrated infant formula. Infants undergoing repeated hospitalizations may also have plateaus or periods of regression in development, so exclusively nursing or bottle-feeding may be appropriate in this circumstance.

Table 9.5 Signs of feeding problems in infants.

In Early Infancy (under six months of age)

- Baby has a weak suck and cannot make a seal on nipple, breast milk or formula runs out of the mouth on whatever side is lower, with obvious fatigue after a few minutes of sucking.

- Baby appears to be hungry all the time due to low volume consumed per feeding, and/or time between feedings does not appear to increase from one month to the next.

- Extended feeding times are seen with the baby napping during the feeding in spite of efforts to keep the baby interested in the feeding.

- The mother is not sure that the baby is swallowing, although appearing to suck

In Later Infancy (over six months of age)

- The baby cannot maintain good head control while being fed from a spoon.

- The baby resists spoon feeding by not opening her mouth when food is offered.

- The baby drinks from a bottle but does not accept baby foods after trying repeatedly.

- The baby resists anything in the mouth except a bottle, breast, nipple, or pacifier.

- The baby does not explore the mouth with fingers or try to mouth toys.

- The baby resists lumpy and textured foods; may turn face away or push food away.

- The baby does not give signs to the parents that clearly indicate hunger or fullness.

Following infant feeding guidelines may result in recognizing early signs of feeding difficulties that are important clues to an infant's underlying condition.[46] For example, an infant who was SGA at birth may nurse well and show rapid weight gain in the first few months. These positive signs may clarify that intrauterine causes may have limited nutrients before birth, but that catch-up growth is now taking place. Another infant who was SGA at birth may have difficulty nursing, lose milk from the mouth, and spit up after nursing. The infant may need extra support to latch onto the breast. These feeding difficulties when described with other aspects of development can clarify the infant's health problems. It may take a year to document delayed development that clarifies a diagnosis such as fetal alcohol syndrome in a newborn with SGA.

Infant feeding guidelines for term infants are appropriate for many preterm infants if they were healthy and had gestational ages such as 35 weeks. Preterm infants who were VLBW or ELBW need infant feeding guidelines based on their adjusted gestational age. As an example, the recommendation for adding food on a spoon at 4–6 months would be adjusted to 6–8 months for an infant who was born at 32 weeks of gestation. Even with this adjustment, feeding problems are common since preterm infants may be extra sensitive.[35,36] Signs of feeding readiness consider characteristics such as resisting spoon-feeding and inconsistent patterns in sleeping.[35] Parents may have difficulty distinguishing hunger from tiredness. As a result of the emphasis on weight gain as a sign of readiness to go home, some families report that they are feeding the infant all the time. This is an example of having difficulty determining signs of hunger. The emphasis on weight gain and catch-up growth inadvertently results in overfeeding and signs of GI discomfort, such as spitting up. Some preterm infants by late infancy have learned to get attention by devises such as refusing to eat, dropping food off the highchair, or throwing a cup.

NUTRITION INTERVENTIONS

When feeding problems are identified in infancy, interventions are required to assure growth and development. Interventions may include any or all of these:[46]

- Assess growth more frequently or more in-depth, such as by measuring body fat stores to identify a change in rate of weight or length gain. This would include head growth measurement.
- Monitor the infant's intake of all liquids and foods by a diet analysis to document that enough calories and nutrients are being consumed. The infant's intake may be variable due to illness, congestion, or medications that lower the appetite.
- Change the frequency or volume of feedings as needed to meet calorie and nutrient needs.
- Adjust the timing of nursing, snacks, or meals as needed to fit medication or sleeping schedules.
- Assess and support as needed the feeding position of the infant. This may be important if the infant cannot sit without support.

- Change the diet composition to improve the nutrient density, so that the infant has to expend less effort to meet their calorie or nutrient needs.
- Provide parent education or support services as needed so that the feeding environment is positive and low in stress.
- Observe the interaction of the infant and mother (or whoever is routinely feeding the infant) at home or in a developmental program to make sure that signs of hunger and comfort result in a positive feeding experience for the pair.
- Adjust routine nutrition guidelines to the developmental abilities of the infant even if different from the chronological age or gestation-corrected age.

Often attempts to improve the feeding experience are successful in meeting the infant's calorie and nutrient needs. However, when calorie and nutrient needs are higher than usual, additional steps are needed to make sure that the diet is enriched. Table 9.6 shows some of the special formula that may be used by infants who have feeding problems or chronic conditions that increase their nutrient requirements.[37]

NUTRITION SERVICES

Infants who were born preterm or with special health care needs have access to more nutrition services than other infants. A variety of state and federal programs include nutrition services for infants.[25,45,47] The following programs are sources of nutrition services or finances to pay for nutrition services:

- Federal disability programs
- Individuals with Disabilities Education Act (IDEA) Part C
- Early Head Start
- WIC
- State funding from the MCH Block grant

Infants with disabling conditions are eligible for Supplemental Social Insurance (SSI), a federal program within the Social Security Administration.[47] SSI provides the family with a disability check and access to heath insurance if their income meets federal guidelines. If an infant had a brain hemorrhage, it may not be clear for years if the child will get diagnosed with cerebral palsy, but the brain damage would make the infant qualify for SSI if the family applied. Infants with genetic disorders are eligible for SSI in most cases. For example, a newborn with Down's syndrome whose families meet the income requirement would qualify for SSI for life. Although the family could then use the funds to buy nutritional supplements, there is no provision that SSI has to cover any specific service.

Nutrition services are part of educational programs in IDEA, including services for children zero through two years of age in Part C.[47] Each state varies in resources available. The Early Intervention Team may visit the infant at home to assess him and provide parent education. The team may provide case management services to assist families in coordinating appointments. Some services based in early intervention centers offer parents group support and advocacy programs. The criteria to receive early intervention services are set by each state, but they require assessment and documentation of at least a 25% delay in an area of child development.[47] Most LBW infants are eligible. Nutrition services generally include growth assessment and interventions to bring about progress in feeding skills. Measuring the adequacy of the diet, need for supplements, or special feeding supports are examples of other services.

Table 9.6   Examples of infant formula for special needs.

Condition	Example of Special Infant Formula
Pulmonary problems such as bronchiopulmonary dysplasia or cardiac defect	Breast milk or standard infant formula with polycose and MCT oil to provide 28 cal/fl oz (high calories in a low volume)
Phenylketonuria (genetic disorder of protein metabolism)	Mixture of amino acids, carbohydrates, fats, vitamins, and minerals without the amino acid phenylalanine
Maple syrup urine disease (genetic disorder of protein metabolism)	Mixture of amino acids, carbohydrates, fats, vitamins, and minerals without the amino acids leucine, isoleucine, and valine
VLBW infant who required surgery after necrotizing enterocolitis	Mixture of amino acids, carbohydrates, fats, vitamins, and minerals
Gastroesophageal reflux and swallowing problem.	Standard infant formula with baby rice cereal (increased thickness is to lower risk of choking and vomiting)
Chronic renal failure (hereditary kidney disease)	Concentrated natural protein, fats, carbohydrates providing 40 cal/fl oz

Each Early Head Start program has to have as 10% of caseload those with special health care needs. This requirement could be met by including mothers who have mental retardation or mental illness, or infants who were preterm or had been diagnosed with a chronic condition. Infants with a nutrition-related diagnosis would be eligible, such as PKU or a cardiac problem. Early Head Start staff would be provided specific instructions to make sure the infant is provided her special diet.

Each state has to designate a portion of the federal Maternal and Child Health (MCH) Block grant for children with special health care needs.[25] Services differ state to state, but all provide care for infants with chronic conditions. Examples of how nutrition services are provided include:

- Specialty clinic services, such as having a nutrition consultant attend a cystic fibrosis clinic

- Contractual services for providing special formulas or therapy for groups of patients who need more nutrition care than usually provided
- Visiting at schools or programs to conduct nutrition assessments or coordinate follow-up recommendations, such as making sure that a specific diet is being offered or that meal-time behavior is being monitored
- Transporting teams of specialists to rural or isolated areas for direct care
- Development and distribution of nutrition education materials for staff training

Also every state has a program to identify and advocate for children with special needs funded by MCH.[25,47] An example is the Developmental Disabilities Council.

Resources

Tufts Nutrition and Health

This site (newsletter) has extensive information for parents about credible nutrition information sources as well as a section on special diets and resources.
Available from: www.healthletter.tufts.edu

National Association of Developmental Disabilities Councils

This website includes public policy and advocacy resources for providers and parents all over the United States concerning various disabilities, including services for infants.
Available from: www.naddc.org

National Early Childhood Technical Assistance Center

This website provides publications and other resources about programs for infants and children. Staff training material are also available for those who are working with infants and young children.
Available from: www.nectus.unc.edu

Emory University Pediatrics Department

This website includes information on preterm births and risks for providers and parents, with sections on nutrition in various health conditions.
Available from: www.emory.edu/PEDS/

United Cerebral Palsy Associations

This website provides service sites in the United States.
Available from: www.ucpa.org

References

1. Trends in infant mortality attributable to birth defects—United States 1980–1995. MMWR September 25, 1998/47(37);773–8.

2. Cooper TR, Bereth CL, Adams, JM, et al. Actuarial survival in the premature infant less than 30 weeks' gestation. Pediatrics 1998;101;975–8.

3. National Research Council. Recommended dietary allowances. 10th Ed. Washington D.C. National Academy Press; 1989.

4. Stallings VA. Resting energy expenditure. In: Altschuler SM, Liacouras CA, eds. Clinical pediatric gastroenterology. Philadelphia: Harcourt Brace and Co.; 1998: pp 607–11.

5. Embleton NE, Pang N, Cooke RJ. Postnatal malnutrition and growth retardation: an inevitable consequence of current recommendations in preterm infants? Pediatrics 2001:107;270–3.

6. American Academy of Pediatrics, Committee on Nutrition: Nutritional needs of preterm infants. In: Kleinman RE, ed. Pediatric nutrition handbook 4th ed. Elk Grove, IL: American Academy of Pediatrics; 1998: p 55.

7. Committee on Nutrition of the Preterm Infant. European Society of Paediatic Gastroenterology and Nutrition in Nutrition and feeding of the preterm infant. Oxford, UK: Blackwell Scientific Publications; 1987.

8. Hall RT, Carroll RE. Infant feeding. Pediatrics in Review 2000;21:191–200.

9. Townsend SF, Johnson CB, Hay WW. Enteral nutrition. In: Merenstein GB, Gardner SL, eds. Handbook of neonatal intensive care 4th ed. St. Louis: Mosby 1998: 275–99.

10. Elsas, LJ, Acosta, PB. Nutritional support of inherited metabolic disease. In: Shils ME, Olson JA, Shike M, eds. Modern nutrition in health and disease 9th ed. Philadelphia: Williams and Wilkins; 1999:1003–56.

11. Anderson DM. Nutrition for the low-birth weight infant. In: Mahan LK, Escott-Stump S, eds. Krause's food, nutrition and diet therapy 10th ed. Philadelphia:W.B. Saunders; 2000; pp 214–38.

12. Trumbo P, Yates AA, Schlicker SA, et al. Dietary reference intakes; vitamin A, vitamin K, arsenic, boron, chromium, copper, iodine, iron, manganese, molybdenum, nickel, silicon, vanadium and zinc. J Am Diet Assoc 2001;101:294–301.

13. American Academy of Pediatrics. Alternative routes of drug administration—advantages and disadvantages (subject review). Pediatrics 1997;100:143–52.

14. Trachtenbarg DE, Golemon TB. Care of the premature infant: part 1. monitoring growth and development. Amer Family Physician 1998;57:2123–30.

15. National Center for Health Statistics: NCHS growth curves for children 0–19 years. U.S. Vital and Health Statistics, Health Resources Administration U.S. Government Printing Office, 2000.

16. Growth references: third trimester to adulthood 2nd ed. Clinton SC: Greenwood Genetic Center; 1998.

17. The Infant Health and Development Program: Enhancing the outcomes of low-birth-weight, premature infants JAMA 1990:263(22):3035–42.

18. Jones, KL. Smith's recognizable patterns of human malformation 5th ed. Philadelphia: W.B. Sanders Co. 1997: 677–81.

19. Blackman, JA. Medical aspects of developmental disabilities in children birth to three. Gaithersburg, MD: Aspen Publications; 1997: pp 55–8,234–7

20. Eyler FD, Behnke M, Conlon M, et al. Birth outcome from a prospective, matched study of prenatal crack/cocaine use: I. interactive and dose effects on health and growth. Pediatrics 1998;101: 229–37.

21. Peralta-Carcelen M, Jackson DA, Goran M I, et al. Growth of adolescents who were born at extremely low birth weight without major disability. Pediatrics 2000;136:633–40.

22. Walker SP, Grantham-Mcgregor SM, Powell CA, et al. Effects of growth restriction in early childhood on growth, IQ, and cognition at age 11 to 12 years and the benefits of nutritional supplementation and psychosocial stimulation. J Pediatrics 2000;137:36–41.

23. Down syndrome prevalence at birth—United States 1983–1990. MMWR August 26, 1994:43(33):617–22.

24. Lepley CJ, Gardner SL, Lubchanco LO. Initial nursery care. In: Merenstein GB, Gardner SL, eds. Handbook of neonatal intensive care 4th ed. 1998: pp 70–99.

25. Website for Maternal and Child Health Bureau, Health and Human Services available at http://mchb.hrsa.gov/html/drte.html, accessed January 12, 2001.

26. Website of WIC Program, United States Department of Agriculture available at http://fns.usda.gov/wic/, accessed January 12, 2001.

27. United States. 1998 National Vital Statistics Report Table 45. Number and percent of low birthweight and number of live births by birthweight, by age and race and Hispanic origin of mother. 2000: pp 48; 75–8.

28. Wood, NS, Marlow N, Costeloe K, et al. Neurological and developmental disability after extremely preterm birth. N Eng J of Med 2000;343:378–84.

29. Strauss R. Adult functional outcome of those born small for gestational age: twenty-six year follow-up of the 1970 British birth cohort. JAMA 2000;283: 625–32.

30. Kilbride HW, Bendorf K, Wheeler R. Total parenteral nutrition. In: Merenstein GB, Gardner SL, eds. Handbook of neonatal intensive care 4th ed. 1998: pp 300–16.

31. Shaffer SE. Gastroesophageal reflux. In: Altschuler SM, Liacouras CA, eds. Clinical pediatric gastroenterology. Philadelphia: Harcourt Brace and Co.;1998: pp 181–6.

32. Holland RM, Price FN, Bensard DD. Pediatric surgery. In: Merenstein GB, Gardner SL. Handbook of neonatal intensive care 4th ed. St. Louis: Mosby;1998: pp 625–46.

33. Baer MT, Harris AB. Pediatric nutrition assessment: identifying children at risk. J Am Diet Assoc1997;97:S107–S115.

34. American Academy on Pediatrics. Breastfeeding and the use of human milk (RE9729) Pediatrics 1997;100:1035–9.

35. Gardner S, Snell BJ, Lawrence RA. Breastfeeding the neonate with special needs. In Merenstein GB, Gardner S. Handbook of neonatal intensive care 4th ed. St. Louis: Mosby; 1998: pp 333–66.

36. Elliott S, Reimer C. Postdischarge telephone follow-up program for breastfeeding preterm infants discharged from a special care nursery. Neonat Net 1998;17:41–5.

37. Williams, CP, ed. Pediatric manual of clinical dietetics. American Dietetic Association; 1998: pp 561–95.

38. Mead-Johnson Nutritionals website available at http://www.meadjohnson. com/products/index.html/, accessed February 12, 2001.

39. Gray JE, McCormick MC, Richardson DK, et al. Normal birth weight intensive care unit survivors: outcome assessment. Pediatrics1996;97:832–8.

40. Trends in infant mortality attributable to birth defects—United States 1980–1995. MMWR September 25, 1998/47(37);773–8.

41. Prevalence of spina bifida at birth—United States 1983–1990: a comparison of two surveillance systems. MMWR April 19, 1996/45(SS-2);15–26.

42. Lee J, Nunn J, Wright C. Height and weight achievement in cleft lip and palate. Arch Dis Child 1997;76:70–72.

43. Milerad J, Larson O, Hagberg C, et al. Associated malformations in infants with cleft lip and palate:a prospective, population-based study. Pediatrics 1997;100: 180–6.

44. Moss EM, Batshaw ML, Solot CB, et al. Psychoeducational profile of the 22q11.2 microdeletion: A complex pattern. J Pediatrics 1999;134:193–8.

45. Story M, Holt K, Sofka D, Eds. Bright futures in practice: nutrition. Arlington, VA: National Center for Education and Child Health; 2000:266–70.

46. Stevenson RD. Feeding and nutrition in children with developmental disabilities. Pediatric Annuals1995;24:255–60.

47. General Information about Disabilities. National Information Center for Children and Youth with Disabilities, website available at http://www.nichcy.org/general.htm, accessed February 1, 2001.

CHAPTER *10*

Photo Disc

Enough is as good as a feast

Proverbs

TODDLER AND PRESCHOOLER NUTRITION

Prepared by **Nancy H. Wooldridge**

CHAPTER OUTLINE

- Introduction
- Tracking Toddler and Preschooler Health
- Normal Growth and Development
- Physiological and Cognitive Development
- Energy and Nutrient Needs
- Common Nutrition Problems
- Prevention of Nutrition-Related Disorders
- Dietary and Physical Activity Recommendations
- Nutrition Intervention for Risk-Reduction
- Public Food and Nutrition Programs.

KEY NUTRITION CONCEPTS

1 Children continue to grow and develop physically, cognitively, and emotionally during the toddler and preschool age years, adding many new skills rapidly with time.

2 Learning to enjoy new foods and developing feeding skills are important components of this period of increasing independence and exploration.

3 Children have an innate ability to self-regulate food intake. Parents and caretakers need to provide children nutritious foods and let children decide how much to eat.

4 Parents and caretakers have tremendous influence on children's development of appropriate eating, physical activity, and other health behaviors and habits formed during the toddler and preschool years. These lessons are mainly transferred by example.

241

INTRODUCTION

This chapter describes the growth and development of toddlers and preschool age children and their relationships to nutrition and the establishment of eating patterns. Growth during the toddler and preschool age years is slower than in infancy but steady. This slowing of **growth velocity** is reflected in a decreased appetite; yet, adequate calories and nutrients are needed to meet young children's nutritional needs. The eating and health habits established at this early stage of life may impact food habits and subsequent health later in life. The development of new skills and increasing independence mark the toddler and preschool stages. Learning about and accepting new foods, developing feeding skills, and establishing healthy food preferences and eating habits are important aspects of this stage of development.

GROWTH VELOCITY   The rate of growth over time.

TODDLERS   Children between the ages of one and three years.

GROSS MOTOR SKILLS   Development and use of large muscle groups as exhibited by walking alone, running, walking up stairs, riding a tricycle, hopping, and skipping.

FINE MOTOR SKILLS   Development and use of smaller muscle groups demonstrated by stacking objects, scribbling, and copying a circle or square.

PRESCHOOL AGE CHILDREN   Children between the ages of three and five years, who are not yet attending kindergarten.

Definitions of the Life Cycle Stage

Toddlers are generally defined as children between the ages of one and three years. This stage of development is characterized by a rapid increase in **gross** and **fine motor skills** with subsequent increases in independence, exploration of the environment, and language skills. *Preschool age children* are between three and five years of age. Characteristics of this stage of development include increasing autonomy; experiencing broader social circumstances, such as attending preschool or staying with friends and relatives; increasing language skills; and expanding their ability to control behavior.

Importance of Nutrition

Adequate intake of energy and nutrients is necessary for toddlers and preschool age children to achieve their full growth and developmental potential. Undernutrition during these years impairs children's cognitive development as well as their ability to explore their environments.[1] Long-term effects of undernutrition, such as failure to thrive and cognitive impairment, may be prevented or reduced with adequate nutrition and environmental support.

TRACKING TODDLER AND PRESCHOOLER HEALTH

Today, 71 million Americans are under 18 years of age.[2] Of these, 13% or 9.2 million are growing up with an array of disadvantages that can impact their quality of life and their overall well-being. According to the *Kids Count Data Book 1999*, these disadvantages include: child is not living with two parents (32%); household head is high school dropout (19%); family income is below the poverty line (21%); child is living with parent(s) who do not have steady, full-time employment (28%); family is receiving welfare benefits (12%); and child does not have health insurance (15%). When a child lives with more than one of these risk factors, the impact on the child's well-being is even greater. For the 9.2 million children living with four or more of these risk factors, disparities exist among different racial groups. About 30% of African American and 25% of all Hispanic children are in this high-risk category, while only 6% of all white children are growing up with this many disadvantages. When evaluating young children's nutritional status and offering nutrition education to parents, it is important to consider children's home environments. Establishing healthy eating habits may not be high on a family's priority list when the home environment is one of poverty and food insecurity.

Disparities in nutrition status indicators in this age group exist among races. For example:

- 8% of low-income children under age five years are growth retarded, but up to 15% of low-income African American children are.[3]
- 17% of Mexican American children and 10% of African American children aged one to two years have *iron deficiency anemia* compared to 8% of white children.[3] See page 93 for a definition of iron deficiency anemia.

Healthy People 2010

Healthy People 2010, objectives for the nation for improvements in health status by the year 2010, includes a number of objectives that directly relate to toddlers and preschoolers.[3] These are listed in Table 10.1.

NORMAL GROWTH AND DEVELOPMENT

An infant's birthweight triples in the first 12 months of life, but growth velocity slows thereafter until the adolescent growth spurt. On average, toddlers gain 8 ounces/month and 1 cm of height/month, while preschoolers gain 2 kg (4.4 pounds) and 7 cm (2.75

Table 10.1   Healthy People 2010 Objectives related to toddlers and preschool age children.[3]

Objective 19-4	Reduce growth retardation among low-income children under age five years from 8% to 5%.	
Objective 19-5	Increase the proportion of persons aged two years and older who consume at least two daily servings of fruit from 28% to 75%.	
Objective 19-6	Increase the proportion of persons aged two years and older who consume at least three daily servings of vegetables, with at least one-third being dark green or deep yellow vegetables from 3% to 50%.	
Objective 19-7	Increase the proportion of persons aged two years and older who consume at least six daily servings of grain products, with at least three being whole grains from 7% to 50%.	
Objective 19-8	Increase the proportion of persons aged two years and older who consume less than 10% of calories from saturated fat from 36% to 75%.	
Objective 19-9	Increase the proportion of persons aged two years and older who consume no more than 30% of calories from fat from 33% to 75%.	
Objective 19-10	Increase the proportion of persons aged two years and older who consume 2400 mg or less of sodium daily from 21% to 65%.	
Objective 19-11	Increase the proportion of persons aged two years and older who meet dietary recommendations for calcium from 46% to 75%.	
Objective 19-12	Reduce iron deficiency among young children and females of childbearing age from 4% and 11% to 1% and 7%.	

in) per year.[4] This decrease in rate of growth is accompanied by a reduced appetite and food intake in toddlers and preschoolers. A common complaint of parents of children this age is that their children have much less appetite and a lower interest in food or eating compared to their appetite and food intake during infancy. Parents need to be reassured that a decrease in appetite is part of normal growth and development for children in this age group.

In order to monitor a child's physical growth, it is important that children are accurately weighed and measured at periodic intervals. Toddlers less than two years of age should be weighed without clothing or a diaper. The *recumbent length* of toddlers should be measured on a length board with a fixed headboard and moveable footboard. Proper measurement of recumbent length requires two adults, one at the child's head making sure the crown of the head is placed firmly against the headboard, and the other making sure that the child's legs are fully extended and placing the footboard at the child's heels. Proper positioning for measurement of a child's recumbent length is shown in Illustration 10.1. Preschool age children should be weighed and measured without shoes and in lightweight clothing. Calibrated scales should be utilized and a height board should be used for measuring *stature*. Illustrations 10.2 and 10.3 further demonstrate the proper techniques for weighing and measuring young children. It is important that both weight and height be plotted on the appropriate growth charts, such as the 2000 CDC growth charts discussed next.

The 2000 CDC Growth Charts[5]

A full set of the newly released "CDC Growth Charts: United States" can be found in Appendix A.

Illustration 10.4 shows an example of one of the charts, which depicts the growth of a healthy child. Charts are gender specific and are available for birth to 36 months and 2 to 20 years. The health care professional can plot and monitor weight-for-age, length- or stature-for-age, head circumference-for-age, weight-for-length, weight-for-stature, and body mass index-for-age utilizing these growth charts. Children's growth will usually "track" within a fairly steady percentile range. It is important to monitor a child's growth over time and to identify any deviations in growth. It is the pattern of growth that is important to assess rather than any one single measurement. A

RECUMBENT LENGTH   Measurement of length while the child is laying down. Recumbent length is used to measure toddlers <24 months of age, and those between 24 and 36 months who are unable to stand unassisted.

STATURE   Standing height.

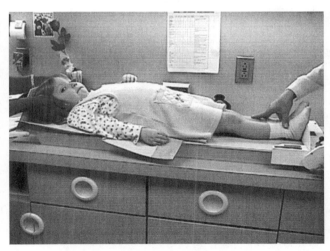

Illustration 10.1   Measuring the recumbent length of a toddler is a two-person job!

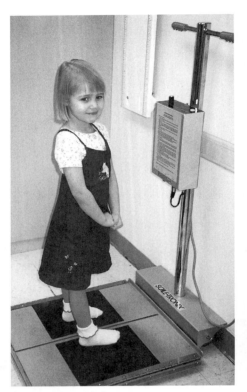

Illustration 10.2
Young child being weighed.

Illustration 10.3   Measuring the stature of a preschool age child.

weight measurement without a length or stature measurement doesn't indicate how appropriate the weight is for the child's length or stature.

Body mass index, or BMI, provides a guideline for assessing underweight and overweight in children and adults. Body mass index is predictive of body fat. A BMI of 85th percentile or greater indicates risk of being overweight, and a BMI of 95th percentile or greater indicates overweight.[8] A BMI <5th percentile indicates underweight.

Growth charts visually aid parents by demonstrating the expected slowing of the growth velocity during the toddler and preschool stage of development. Although the curves for weight-for-age and length- or stature-for-age continue to increase during the toddler and preschool age years, the curve is not as steep as during the first year of life.

> **BODY MASS INDEX**   An index that correlates with total body fat content or percent body fat, and is an acceptable measure of adiposity or body fatness in children and adults.[6, 7] It is calculated by dividing weight in kilograms by the square of height in meters (kg/m^2).

Common Problems with Measuring and Plotting Growth Data

Growth in young children that is measured or plotted incorrectly can lead to errors in health status assessment. Standard procedures should be followed, calibrated and appropriate equipment should be utilized, and plotting should be double checked, including checking the age of the child, to avoid such errors.

PHYSIOLOGICAL AND COGNITIVE DEVELOPMENT

Toddlers

An explosion in the development of new skills happens during the toddler years. Most children begin to walk independently about their first birthday. At first the walking is more like a "toddle" with a wide-based gait.[4] After practicing for several months, the toddler achieves greater steadiness and soon will be able to stop, turn, and stoop without falling over. Gross motor skills, such as sitting on a small chair and climbing on furniture, develop rapidly at this age, and with practice, great improvements in balance and agility take place. At about 15 months, children can crawl up stairs; by about 18 months, they can run stiffly. Most toddlers can walk up and down stairs one step at a time by 24 months, and jump in place. At about 30 months, children have advanced to going up stairs alternating their feet. By 36 months of age, children are ready for tricycles.

Children act increasingly more mobile and independent with improvements in gross motor skills and show fascination with these newfound skills and a readiness to put them into practice and develop new skills. However, toddlers have no sense of dangerous situations. At this age, children are especially vulnerable to accidental injuries and ingestion of harmful substances. In fact, the leading cause of death among young children is unintentional injuries.[3] Parents and caregivers have to constantly watch over toddlers, preferably in environments made "child safe."

COGNITIVE DEVELOPMENT IN TODDLERS   With the toddler's newly acquired physical skills, exploring the environment accelerates and exerting their newfound independence becomes very important to them. Toddlers now have the power to control the distance between themselves and their parents. *Nelson's*

Illustration 10.4

Birth to 36 months: Girls Length-for-age and Weight-for-age percentiles.[5]

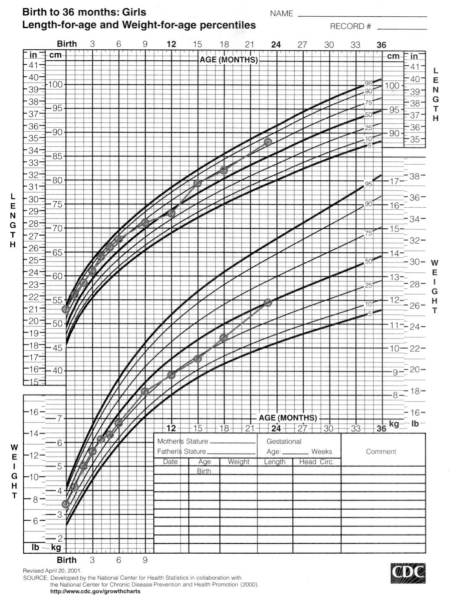

Birth to 36 months: Girls
Length-for-age and Weight-for-age percentiles

NAME _____

RECORD # _____

Revised April 20, 2001.
SOURCE: Developed by the National Center for Health Statistics in collaboration with
the National Center for Chronic Disease Prevention and Health Promotion (2000).
http://www.cdc.gov/growthcharts

CDC

Textbook of Pediatrics[4] describes how toddlers often "orbit' around their parents, like planets, moving away, looking back, moving farther, and then returning.

From a socialization standpoint, the child moves from being primarily self-centered to being more interactive. The toddler now possesses the ability to explore the environment and to develop new relationships. Fears, such as separation, darkness, loud sounds, wind, rain, and lightning commonly emerge during this period as the child learns to deal with changes in the environment. Children develop rituals in their daily activities in an attempt to deal with these fears.

Social development also involves imitating others, such as parents, caretakers, siblings, and peers, during this time. The child in this stage begins to learn about cultural customs of the family, including those related to meals and food.

Dramatic development of language skills occurs from 18 to 24 months. Once a child realizes that words can stand for things, the child's vocabulary erupts from 10 to 15 words at 18 months to 100 or more words at two years of age. The toddler will soon begin to combine words to make simple sentences. By 36 months, the child uses three-word sentences.[4]

An important social change for toddlers is increased determination to express their own will. This expression often comes in the form of negativism and the beginning of temper tantrums, which give this stage of development its label of "the terrible twos." With an increase in motor development coupled with an increasing quest for independence, the toddler tries to do more and more things, pushing his or her capabilities to the limit. Thus the toddler can become easily frustrated and negative. The child seeks more independence and at the same time needs the parents and caretakers for security and reassurance. Toddler behavior uncannily parallels the same type of behavior commonly seen in adolescents!

DEVELOPMENT OF FEEDING SKILLS IN TODDLERS

Many babies begin to wean from the bottle at about 9 to 10 months of age when their solid food intake increases and they learn to drink from a cup.[9] Parents need to pay attention to cues of readiness for weaning, such as disinterest in breastfeeding or bottle feeding. The time it takes to wean is variable and depends on both the child and the mother. Weaning will be easier for those babies who adapt well to change. Weaning is a sign of the toddler's growing independence and is usually complete by 12 to 14 months of age, although the age varies from child to child.

Gross and fine motor development during the toddler years enhances children's ability to chew foods of different textures and to self-feed. Between 12 to 18 months, toddlers are able to move the tongue from side to side (or laterally) and learn to chew food with rotary, rather than just up and down movements. Toddlers can now handle chopped or soft table food.

At about 12 months, children have a refined pincer grasp that enables them to pick up small objects, such as cooked peas and carrots, and put them into their mouths. Children will be able to use a spoon around this age but not very well. From 18 to 24 months, toddlers are able to use the tongue to clean the lips and have well-developed rotary chew movements. Now the toddler can handle meats, raw fruits and vegetables, and multiple textures of food.

A strong need for independence in self-feeding emerges during the toddler age. "I do it" and "No, no, no!" are commonly heard phrases in households where toddlers reside. As toddlers busily practice their newly found skills, they become easily distracted. Parents need to realize that their toddler's sometimes-fierce independence is part of normal growth and development and represents an ongoing process of separation from dependency on the parents and caretakers.

Increasing fine motor and visual motor coordination skills allow toddlers to use cups and spoons more effectively. Although toddlers' skill with a spoon increases during the second year, they prefer to eat with their hands. Initial attempts at self-feeding are inevitably messy, as Illustration 10.5 depicts, but represent an important stage of development. It is important that parents and caretakers keep distractions during meal times, such as television, to a minimum, and allow their toddlers to practice self-feeding skills, and to experience new foods and textures. The child derives pleasure in self-feeding and exploring new tastes. Learning to self-feed allows the child to develop mastery of an important part of everyday life.

Adult supervision of eating is imperative because of the high risk of choking on foods at this age. Toddlers should always be seated during meals and snacks, preferably in a high chair or booster seat with the family, and not allowed to "eat on the run." Foods that may cause choking, such as hard candy, popcorn, nuts, whole grapes, and hot dogs, should not be served to children less than two years of age.[9]

FEEDING BEHAVIORS OF TODDLERS   The toddler's need for rituals, a hallmark of this stage of development, may be linked to the development of food jags.

Illustration 10.5
Toddler enjoying mealtime!

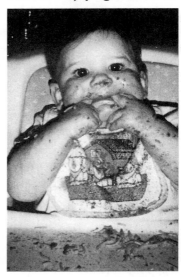

Many toddlers demonstrate strong food preferences and dislikes. They can go through prolonged periods of refusing a particular food or foods they previously liked. The intensity of the refusal or the negative attitude toward a particular food will be influenced by the child's temperament (see section on Temperament Differences on page 249). To circumvent food jags, parents can serve new foods along with familiar foods. New foods are better accepted if they are served when the child is hungry and also if she sees other members of the family eating these foods. Eventually, toddlers' natural curiosity will get the best of them. Toddlers are great imitators, which includes imitating the eating behavior of others.

Mealtime is an opportunity for toddlers to practice newly acquired language and social skills and to develop a positive self-image. It is not the time for battles over food or "force feedings." Family mealtime provides an opportunity for parents and caretakers to model healthy eating behaviors for the young child.

APPETITE AND FOOD INTAKE IN TODDLERS
Parents need to be reminded that toddlers naturally have a decreased interest in food because of slowing growth, and a corresponding decrease in appetite. Besides, with all of their newfound gross and fine motor skills, they have places to go and new environments to explore! It is a part of normal growth and development for toddlers to have a decreased interest in food and to be easily distracted at mealtime.

Toddlers need toddler-sized portions. One rule of thumb for serving size is 1 tablespoon of food per year of age. So, a serving for a two-year-old child would be about 2 tablespoons. It is better to give the child a small portion and allow him to ask for more than to serve large portions. Parents often overestimate portion sizes needed by their young child, which may contribute to labeling the child as a "picky" eater. Because toddlers can't eat a large amount of food at one time, snacks are important in meeting the child's nutritional needs. It is important that toddlers not be allowed to "graze" throughout the day on sweetened beverages and foods such as cookies and chips. These foods can "kill" their limited appetite for basic foods at meal and snack times. In considering the toddler's need for rituals and limit setting, parents and caretakers need to establish regular but flexible meal and snack times, allowing enough time between meals and snacks for the toddler to get hungry.

Preschool Age Children

Preschool age children continue to expand their gross and fine motor capabilities. At age four, the child can hop, jump on one foot, and climb well. The child can ride a tricycle, or a bicycle with training wheels, and can throw a ball overhand.[4]

COGNITIVE DEVELOPMENT OF PRESCHOOL AGE CHILDREN
Magical thinking and egocentrism characterize the preschool period.[4] Egocentrism does not mean that the child is selfish but rather that the child is not able to accept another's point of view. The child is beginning to interact with a widening circle of adults and peers. During the preschool years, children gradually move from primarily relying on external behavioral limits, such as those demanded by parents and caregivers, to learning to limit behavior internally. This transition is a prerequisite to functioning in a school classroom.[4] Also during this time, children's play starts to become more cooperative, such as building a tower of blocks together. Toward the end of the preschool years, children move to more organized group play, such as playing tag or "house."

Control is a central issue for preschool children. They will test their parents' limits and still resort to temper tantrums to get their way. Temper tantrums generally peak between the ages of two and four years.[4] The child's challenge is to separate, and the parent's challenge is to appropriately set limits and at the same time to let go, another parallel with adolescence. Parents need to strike an appropriate balance for setting limits. Too tightly controlled limits can undermine the child's sense of initiative and cause them to act out whereas loose limits can cause the child to feel anxious and that no one is in control.

Language develops rapidly during the preschool years and is an important indicator of both cognitive and emotional development. Between ages two and five, children's vocabularies increase from 50 to 100 words to more than 2000 words, and their language progresses from two- to three-word sentences to complete sentences.[4]

DEVELOPMENT OF FEEDING SKILLS IN PRESCHOOL AGE CHILDREN
The preschool age child can use a fork and a spoon and uses a cup well. Cutting and spreading with a knife may need some refinement. Children should be seated comfortably at the table for all meals and snacks. Eating is not as messy a process during the preschool years as it was during toddlerhood. Spills still do occur, but they are not intentional. Foods that cause choking in young children should be modified to make them safer, such as cutting grapes in half lengthwise and cutting hot dogs in quarters lengthwise and then cutting into small bites. Adult supervision during mealtime is still important.

FEEDING BEHAVIORS OF PRESCHOOL AGE CHILDREN
As during the toddler years, parents of preschool age children need to be reminded that the child's rate of growth continues to be relatively slow with a relatively small appetite and food intake. Growth occurs in "spurts" during childhood. Appetite and food intake increase in advance of a growth spurt, causing children to add some weight that will be used for the upcoming spurt in height.

Preschool age children want to be helpful and to please their parents and caretakers. This characteristic makes the preschool years a good time to teach children about foods, food selection, and preparation by involving them in simple food-related activities. For instance, outings to a farmers' market can introduce children to a variety of fresh vegetables and fruit. Allowing children to be involved in meal-related activities such as those listed in Table 10.2 can be quite instructive. Families of preschool age children need to continue to be encouraged to eat together, like the family in Illustration 10.6.

INNATE ABILITY TO CONTROL ENERGY INTAKE
An important principle of nutrition for young children, and one with direct application to child feeding, is children's ability to self-regulate food intake. If allowed to decide when to eat and when to stop eating without outside interference, children eat as much as they need.[11,12] Children have an innate ability to adjust their caloric intake to meet caloric needs. The preschool age child's intake may fluctuate widely from meal to meal and day to day. But over a week's time, the young child's intake remains relatively stable.[13] Parents who try to interfere with the child's ability to self-regulate intake by forcing the child to "clean her plate" or using food as a reward are asking the child to overeat or undereat.

Table 10.2   Meal preparation activities for young children.[10]

Two-year-olds:

Wipe tabletops	Snap green beans
Scrub vegetables	Wash salad greens
Tear lettuce or greens	Play with utensils
Break cauliflower	Bring ingredients from one place to another

Three-year-olds:

Wrap potatoes for baking	Shake liquids in covered container
Knead and shape yeast dough	
Pour liquids	Spread soft spreads
Mix ingredients	Place things in trash

Four-year-olds:

Peel oranges or hard-cooked eggs	Mash bananas using fork
Move hands to form round shape	Set table
Cut parsley or green onions with dull scissors	

Five- to six-year olds:

Measure ingredients	Use an eggbeater
Cut with blunt knife	

Although children can self-regulate caloric intake, no inborn mechanisms direct them to select and consume a well-balanced diet.[14] Children learn healthful eating habits.[15] Parents give up some control over what their preschool child eats if the child spends more time away from home in a child care center or with extended family members. Preschool children continue to learn about food and food habits by observing their parents, caretakers, peers, and siblings and begin to be influenced by what they see on television. Their own food habits and food preferences are established at this time.

APPETITE AND FOOD INTAKE OF PRESCHOOLERS

Parents of preschool age children often describe their children's appetite as being "picky." One reason a child may want the same foods all of the time is because familiar foods may be comforting to the child. Another reason is that the child may be trying to exert control over this aspect of his life. The child's eating and food selection can easily become a battleground between parent and child; this scenario should be avoided. Some practical suggestions for parents and caretakers of children this age include serving child-sized portions and serving the food in an attractive way. Children often do not like their foods to touch or to be mixed together, such as in casseroles or salads. They typically do not like strongly flavored veg-

Connections
The Pudding Study

Lunch at the daycare center the day of the study was a bit unusual—it started with pudding. Preschoolers were offered a serving of pudding that contained either 150 or 40 calories. Both servings looked and tasted the same. The amount of pudding each child consumed was secretly recorded, as was their food intake duing the rest of lunch. The experiment was repeated on adults.

Children compensated almost perfectly for the calories in the different puddings. Those given the higher calorie pudding ate less at lunch, and those receiving the low calorie pudding ate more. The adults, however, didn't fare so well. Their caloric intakes for the rest of lunch bore no relationship to calories consumed in the pudding. The study suggests that children are more sensitive to calorie intake on a short-term basis than adults.

SOURCE: Birch L L Presentation at the National Conference on Nutrition Education Researcy, Chicago, September 1986.

Illustration 10.6   Sharing family meals is an important aspect of development in young children.

etables and other foods, or spicy foods at this young age. Just as with toddlers, parents of preschool age children should not allow their child to eat and drink indiscriminately between meals and snacks. This behavior often "kills" his appetite at mealtime. Children should not be forced to stay at the table until they have eaten a certain amount of food as determined by the parent.

Temperament Differences

Better is a dinner of herbs where love is, than a fatted ox and hatred with it.

Proverbs 15:17

Temperament is defined as the behavioral style of the child, or the "how" of behavior. This definition was derived from analysis of data from the New York Longitudinal Study, begun in 1956 by Chess and Thomas.[16] These investigators defined three temperamental clusters, the "easy" child (about 40% of children), the "difficult" child (10%), and the "slow-to-warm-up" child (15%). The remaining children, classified as "intermediate-low" or "intermediate-high," demonstrated a mixture of behaviors but gravitated to one end of the spectrum.[16]

Children's temperaments affect feeding and mealtime behavior. The "easy" child is regular in function, adapts easily to regular schedules, and tries and accepts new foods readily. The "difficult" child on the other hand is characterized by irregularity in function and slow adaptability. This child is more reluctant to accept new foods and can be negative about them. The "slow-to-warm-up" child exhibits slow adaptability and negative responses to many new foods with mild intensity. With repeated exposures to new foods, this child can learn to accept them over time with limited complaining.[16]

The "goodness of fit" between the temperaments of the child and the parent or caretaker can influence feeding and eating experiences.[16] A mismatch can result in conflict over eating and food. Parents and caretakers need to be aware of the temperament of the child when attempting to meet nutritional needs. The "difficult" or "slow-to-warm-up" child may pose special challenges that need to be addressed by gradually exposing the child to new foods and not hurrying him to accept them.[16]

Food Preference Development, Appetite, and Satiety

Food preference development and regulation of food intake have been studied extensively by Leann Birch and associates.[17,18] It is clear that children's food pref-

erences do determine what foods they consume. Children naturally prefer sweet and slightly salty tastes and generally reject sour and bitter tastes. These preferences appear to be unlearned and present in the newborn period. Children eat foods that are familiar to them, a fact that emphasizes the importance environment plays in the development of food preferences. Children tend to reject new foods but may learn to accept a new food with repeated exposures to it. It may, however, take eight to ten exposures to a new food before it is accepted. Children who are raised in an environment where all members of the family eat a variety of foods are more likely to eat a variety of foods.

Children also appear to have preferences for foods that are energy-dense due to high levels of sugar and fat.[17,18] This preference may develop because children associate eating energy-dense foods with pleasant feelings of satiety, or because these types of foods may be associated with special social occasions such as birthday parties. The context in which foods are offered to a child influences the child's food preferences. Foods served on a limited basis but used as a reward become highly desirable. Restricting a young child's access to a palatable food may actually promote the desirability and intake of that food.[19] Coercing or forcing children to eat foods can have a long-term negative impact on their preference for that food.[17,18]

> **PRELOADS** Beverages or food such as yogurt in which the energy/macronutrient content has been varied by the use of various carbohydrate and fat sources. The preload is given before a meal or snack and subsequent intake is monitored. This study design has been employed by Birch et al. in their studies of appetite, satiety, and food preferences in young children.[18]

APPETITE, SATIETY Children's energy intake regulation has been studied by giving children *preload*s of food or beverage of varying energy content followed by self-selected meals. In one such study, children aged three to five years were given either a low-energy preload beverage made with aspartame (Nutrasweet), a low-calorie sugar substitute, or a high-energy preload beverage made with sucrose. Fat and protein content of the preloads did not differ. Children were then allowed to self-select their lunches. Children who had the low-calorie beverage before lunch consumed more calories at lunch while those who had the higher calorie beverage consumed fewer calories. These results indicate that young children are able to adjust caloric intake based on caloric need.[18,20] Similar studies were conducted in two to five year olds using foods with dietary fat or olestra, a nonenergy fat substitute. Results indicate that children compensated for the lower level of calories in food when olestra was substituted for dietary fat.

The preloading protocol just described was also used to study children's responsiveness to caloric content of foods in the presence or absence of common feeding advice from adults. In one group, teachers were trained to minimize their control over how much the children ate. In the other group, teachers were trained to focus the children on external factors to control their intake, such as rewarding the children for finishing the portions served to them or encouraging them to eat because "it was time to eat." Results of this investigation show that when the adults focused the children on external cues for eating, children lost their ability to regulate food intake based on calories. It appears children's innate ability to regulate caloric intake can be altered by child-feeding practices that focus on external cues rather than the child's own hunger and satiety signals.[20]

The effects of portion size on children's intakes were compared between classes of three-year-old and five-year-old children. The children were served either a small, medium, or large portion of macaroni and cheese along with standard amounts of other foods in their usual lunchtime setting. Analysis of amount of food eaten showed that portion size did not affect the younger children's intakes; their intakes remained constant despite the amount of food served to them. In contrast, the five-year-old children's intakes increased significantly with the larger portion sizes. The researchers conclude that by five years of age, children are influenced by the size of portions served to them, another external factor that influences intake.[21] These investigators also raise the question as to what effect large portion sizes have on overeating and, consequently, on the development of childhood obesity.

Another study of five-year-old girls and their parents looked at the effects of parents' restriction of palatable foods on their children's consumption of these foods. After a self-selected standard lunch, these five-year-old girls were given free access to snack foods, such as ice cream, potato chips, fruit-chew candy, and chocolate bars. The daughters of parents who reported restricting access to snack foods indicated to the investigators that they ate "too much" of the snack foods and also reported negative emotions about eating the snack foods. Parents' restriction of foods actually promoted the consumption of these foods by their young daughters and, of even more concern, the daughters reported feeling badly about eating these "forbidden" foods.[22]

Satter describes the optimal "feeding relationship" as one in which parents and caretakers are responsible for what children are offered to eat and the environment in which the food is served, while children are responsible for how much they eat or even whether they eat at a particular meal or snack. If this feeding relationship is respected, feeding and potential weight problems can be prevented.[23] Parenting includes influencing what is served to children and the environment in which it is served, at home and in child-care settings as well.

Table 10.3 Practical applications of child feeding research.[18]

- Parents should respond appropriately to children's hunger and satiety signals.

- Parents should focus on the long-term goal of developing healthy self-controls of eating in children, and look beyond their concerns regarding composition and quantity of foods children consume or fears that children may eat too much and become overweight.

- Parents should not attempt to control children's food intakes by attaching contingencies ("No dessert until you finish your rutabagas.") and coercive practices ("Clean your plate. Children in Bangladesh are starving.").

- Parents should be cautioned not to severely restrict "junk foods," foods high in fat and sugar, as that may make these foods even more desirable to the child.

- Parental influence should be positively focused on the child developing food preferences and selection patterns of a variety of foods consistent with a healthy diet. Parental modeling of eating a varied diet at family mealtime will have a strong influence on children.

- Children have an unlearned preference for sweet and slightly salty tastes; they tend to dislike bitter, sour, and spicy foods.

- Children tend to be wary of new foods and tastes, and it may take repeated exposures to new foods before they are accepted.

- Children need to be served appropriate child-sized servings of food.

- Child feeding experiences should take place in secure, happy, and positive environments with or under adult supervision.

- Children should never be forced to eat anything.

What implications does all of this research have on child feeding practices? Based on the results of these studies, it appears that by late preschool age, children are more responsive to external cues than their innate ability to self-regulate intake. Table 10.3 sums up the practical applications of Birch's work.[18] The importance of appropriate parenting skills in helping children learn to self-regulate food intake and possibly avoid problems with obesity is echoed by a panel of obesity experts.[24]

ENERGY AND NUTRIENT NEEDS

New Dietary Reference Intakes (DRIs) are being developed through 2002. Readers can check to see whether new DRIs for calories, protein, and other nutrients have appeared by logging onto the Web and searching the key term "Dietary Reference Intakes" or checking the National Academy Press Web site (see

Nutrition Benefits of Human Milk:

1. Isomotic - human milk has similar ion concentration to plasma, therefore no add'l water is needed.
2. Low protein content - easy on newborn kidneys
3. Whey protein forms easily digestible curds.
4. High in essential fatty acids, cholesterol, DHAS (helps in develop. of CNS)
5. Minerals are protein-bound, which helps in digestibility.

Immunological Benefits of Human Milk:

1. T- lymphocytes
2. B- lymphocytes
3. neutrophils
4. macrophages
5. epithelial cells
6. secretory IgA - protects GI
7. growth factors/hormones - mature the GI
   i.e. insulin
8. enhances immune resp. to immunizations

Anatomy of Mammary Gland

Alveoli - functional unit
  • each alveolus is composed of a cluster of secretory cells with a duct in the center.

Alveolus   (contract by oxytocin)
                  myoepithelial cells

                  epithelial cells

duct (progesterone elongates, duplicates)

Lactogenesis - 3 stages
Production, Secretion, Ejection

Five pathways for milk secretion:
1. Exocytosis of milk protein & lactose
   in secretory vesicles derived from golgi.
2. Milk fat secretion via the milk fat
   globule
3. Secretion of ions & water across cell
   membrane
4. Pinocytosis- exocytosis of immunoglobulins
5. Paracellular path for plasma components
   and leukocytes.

Lactose
    synthesized from glucose and galactose → (from maternal blood supply)
    in the secretory cells.

$$\text{glucose} + \text{galactose} \xrightarrow[\text{($\ominus$ progesterone)}]{\text{lactoalbumin}} \text{Lactose}$$
(mat. blood supp.)

Exocytosis allows secretion to ducts of:
    Lactose
    Calcium
    Phosphate
    Citrate
    Proteins

Milk proteins enter via 4 transcellular routes
    (casein, whey, Ig's, milk fat globule Ps,
    free aas)
This production is stimulated by:
    prolactin, cortisol, insulin

Case Study 10.1
Making Meal Time Pleasant

Photo Disc.

Lindsey, a 24-month-old little girl, lives with her parents. She stays at a child-care center during the week while both of her parents are at work. On the weekends, her parents enjoy their time with Lindsey, although a lot of their time is spent running errands and catching up on household chores. Partly to appease Lindsey, her parents allow her to have as much of her favorite beverage, apple juice, from a "sippy" cup as she wants between meals. Lindsey also has free access to snacks such as crackers, slices of cheese, and cookies. When the family sits down to have a meal together, Lindsey plays with her food and usually doesn't eat much. She tells her parents that she doesn't like the food being served and wants "something else." She soon becomes fussy and wants to get down from her booster seat. To try and keep her at the table with them, her parents turn on the television or play Lindsey's favorite cartoons on the VCR. If that does not quiet Lindsey, her mother offers to prepare another food item of Lindsey's choice. Mealtime has become an unpleasant experience for the family.

Lindsey's parents need to stop allowing Lindsey free access to the apple juice and snack foods between meals. Lindsey needs snacks, and foods such as the cheese and crackers are appropriate snack foods. But she should not be allowed to "graze" throughout the day. It is no wonder that she is not hungry at mealtime! If she is allowed to build up an appetite for her meals and snacks, she will be more interested in eating and in the food being served. Lindsey's parents have the right idea in trying to promote the family eating together, but they need to eliminate the distraction of watching TV and cartoons during mealtime. And Lindsey's mother needs to stop preparing special items for Lindsey. Lindsey's parents can serve her favorite basic foods along with a small portion of a new food. By employing these suggestions, the family's mealtimes should become much more pleasant.

Resource section at the end of this chapter). DRI tables that have already been published are provided on the inside front cover of this book.

Energy Needs

Energy needs of toddlers and preschool age children reflect the slowing of the growth velocity of children in this age group.[25] The RDA for energy for one- to three-year-olds is 102 calories/kg body weight, and for four- to six-year-olds, 90 calories/kg body weight. The decrease in the RDA for energy with age is a reflection of the decreased growth rate. As with other aspects of growth and development, variability of activity level is seen in young children, which impacts their energy needs.

The current RDAs for energy are based on dietary intakes associated with normal growth in groups of well-nourished children before the 1970s. The estimations take into consideration the amount of energy intake needed to balance energy expenditure of an individual who "has a body size and composition and level of physical activity consistent with long-term good health."[25] The actual recommendations are 5%

greater than reported intakes to compensate for underestimation in reporting dietary intakes.

Recent research suggests the RDAs for calories for toddlers and preschoolers may be too high. Studies using the doubly labeled water method of measuring energy requirements show that the actual energy needs of infants and young children are lower than the previous estimates by up to 25%.[26] These lower calorie values may be related to decreased energy expenditure through play activity in today's young children. The implications of lowering the RDA for energy are great since many government programs, such as the Head Start Program and Food Stamps, use the RDA for standard setting.[26]

Protein

Based on the current RDAs, the protein needs of one- to six-year-old children are approximately 1.2 grams of protein/kg body weight.[25] This recommendation is easily met with the typical American diet as well as with vegetarian diets. Adequate energy intake to meet an individual child's needs has a protein-sparing effect; that is, with adequate energy intake, protein is

DIETARY REFERENCE INTAKES (DRIs)
Quantitative estimates of nutrient intakes, used as reference values for assessing the diets of healthy people. DRIs include Recommended Dietary Allowances (RDAs), Adequate Intakes (AI), Tolerable Upper Intake Level (UL), and Estimated Average Requirement (EAR).

RECOMMENDED DIETARY ALLOWANCES (RDAs)   The average daily dietary intake levels sufficient to meet the nutrient requirements of nearly all (97% to 98%) healthy individuals in a population group. RDAs serve as goals for individuals.

HEMOGLOBIN   A protein that is the oxygen-carrying component of red blood cells. A decrease in hemoglobin concentration in red blood cells is a late indicator of iron deficiency.

HEMATOCRIT   An indicator of the proportion of whole blood occupied by red blood cells. A decrease in hematocrit is a late indicator of iron deficiency.

ANEMIA   A reduction below normal in the number of red blood cells per cubic mm in the quantity of hemoglobin, or in the volume of packed red cells per 100 ml of blood. This reduction occurs when the balance between blood loss and blood production is disturbed.

used for growth and tissue repair rather than for energy. Ingestion of high-quality protein, such as milk and other animal products, lowers the amount of total protein needed in the diet to provide the essential amino acids.

Vitamins and Minerals

Dietary Reference Intakes (DRIs) and Recommended Dietary Allowances (RDAs) for vitamins and minerals have been established for the toddler and pre-school age child. Analysis of data from NHANES I, II, and III and Continuing Survey of Food Intake by Individuals (CSFII) indicate that children's average intakes of most nutrients meet or exceed the recommendations.[27,28] Most children from birth to five years are meeting the targeted levels of consumption of most nutrients, except for iron, calcium, and zinc. The RDAs and Adequate Intakes for these key nutrients are listed in Table 10.4.

COMMON NUTRITION PROBLEMS

Iron Deficiency Anemia

Iron deficiency and iron deficiency anemia are prevalent nutrition problems among young children in the

United States. A rapid growth rate coupled with frequently inadequate intake of dietary iron places toddlers, especially 9 to 18 month olds, at the highest risk for iron deficiency.[31] According to the third National Health and Nutrition Examination Survey (NHANES III), 9% of toddlers aged one to two years are iron deficient; of these 3% have iron deficiency anemia.[32] In numbers, these percentages translate to approximately 700,000 toddlers with iron deficiency, and of these 240,000 have iron deficiency anemia.[32] The full impact of this nutrition problem is profound. Iron deficiency anemia in young children appears to cause long-term delays in cognitive development and behavioral disturbances.[1,31]

Table 10.5 depicts the progressing signs of iron deficiency. Iron deficiency can be defined as absent bone marrow iron stores, an increase in hemoglobin concentration of <1.0 g/dL after treatment with iron, or other abnormal lab values, such as serum ferritin concentration.[31] The definition of iron deficiency anemia is less than the fifth percentile of the distribution of *hemoglobin* concentration or *hematocrit* in a healthy reference population. Age and sex-specific cutoff values for anemia are derived from NHANES III data. For children one to two years of age, the diagnosis of anemia would be made if the hemoglobin concentration is <11.0 g/dL and hematocrit <32.9%. For children aged two to five years, a hemoglobin value <11.1 g/dL or hematocrit <33.0% is diagnostic of iron deficiency anemia.

Not all anemias are due to iron deficiency. Other causes of *anemia* include other nutritional deficiencies such as folate or vitamin B$_{12}$, chronic inflammation, or recent or current infection.[31]

One Healthy People 2010 objective is to reduce iron deficiency in children aged one to two years from 9% to 5% and in children aged three to four years from 4% to 1%.[3] Part of reaching this goal will mean

Table 10.4   Dietary Reference Intakes for key nutrients for toddlers and preschoolers.[25,29,30]

AGE	RECOMMENDED DIETARY ALLOWANCES		ADEQUATE INTAKE
	Iron (mg/d)	Zinc (mg/d)	Calcium (mg/d)
1–3 years	7	3	500
4–8 years	10	5	800

Table 10.5   Progression of iron deficiency.

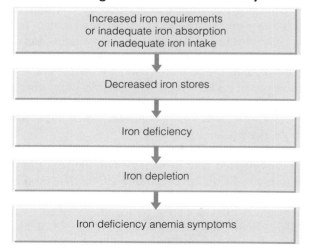

Increased iron requirements or inadequate iron absorption or inadequate iron intake

Decreased iron stores

Iron deficiency

Iron depletion

Iron deficiency anemia symptoms

reducing or eliminating disparities in iron deficiency by race and family income level. The prevalence of iron deficiency is higher in African Americans than in white children (10% versus 8% for children aged one to two years) and is highest in Mexican American children (17% of children aged one to two years).[3] Children of families with incomes ≤130% of the poverty threshold have a higher incidence of iron deficiency than those with a higher income (12% versus 7%).

PREVENTING IRON DEFICIENCY   The Centers for Disease Control have published recommendations for preventing iron deficiency in the United States.[31] It is recommended that children one to five years of age drink no more than 24 ounces of cow's milk, goat's milk, or soy milk each day because of the low iron content of these milks. Larger intakes may displace high-iron foods. For detecting iron deficiency, it is recommended that children at high risk for iron deficiency, such as low-income children and migrant and recently arrived refugee children, be tested for iron deficiency between the ages of 9 and 12 months, six months later, and then annually from ages two to five years. For children who are not at high risk for iron deficiency, selective screening of children at risk only is recommended. Children at risk include those who have a low-iron diet, consume more than 24 ounces of milk per day, have a limited access to food because of poverty or neglect, and who have special health care needs, such as an inborn error of metabolism or chronic illness.

NUTRITION INTERVENTION FOR IRON DEFICIENCY ANEMIA   Treatment of iron deficiency anemia includes supplementation with iron drops at a dose of 3 mg/kg per day, counseling of parents or caretakers about diets that prevent iron deficiency, and repeat screening in four weeks. An increase of >1 g/dL in hemoglobin concentration, or >3% in hematocrit, within four weeks of initiation of treatment confirms the diagnosis of iron deficiency. If the anemia is responsive to treatment, dietary counseling should be reinforced and the iron treatment should be continued for two months. At that time, the hemoglobin and hematocrit should be rechecked, and the child should be reassessed in six months. If the hemoglobin and hematocrit do not increase after four weeks of iron treatment, further diagnostic tests are needed. Iron status will not improve with iron supplements if the cause of the anemia is not directly related to a need for iron.[31]

Dental Caries

Approximately one in five children aged two to four years has decay in the primary or permanent teeth.[3]

A primary cause of dental decay is habitual use of a bottle with milk or fruit juice at bedtime or throughout the day. Prolonged exposure of the teeth to these fluids can produce *baby bottle tooth decay*. Upper front teeth are most severely affected by decay, which is where fluids pool when toddlers fall asleep while drinking from a bottle.

> **BABY BOTTLE TOOTH DECAY**   Dental caries in young children caused by being put to bed with a bottle or allowed to suck from a bottle for extended periods of time. Also called "baby or nursing bottle dental caries."

Toddlers with baby bottle tooth decay are at increased risk for caries in the permanent teeth.[33] The incidence of baby bottle tooth decay is highest among Hispanic, American Indian, and Alaska Native children, and among children whose parents have less than a high school education.[3]

Food sources of carbohydrates such as milk and fruit juice can have direct effects on dental caries development because *streptococcus mutans,* the main type of bacteria that cause tooth decay, uses carbohydrates for food. Bacteria present in the mouth excrete acid that causes the tooth decay.[33] Consequently, the more often and longer teeth are exposed to carbohydrates, the more the environment in the mouth is conducive to the development of tooth decay. Foods containing carbohydrates that stick to the surface of the teeth, such as sticky candy like caramel, are strong caries promoters. Rinsing the mouth with water or brushing teeth to get rid of the carbohydrate stuck to teeth reduces caries formation. Young children allowed to "graze" or indiscriminately eat or drink throughout the day likely expose their teeth to carbohydrates longer, which encourages bacteria proliferation and tooth decay. Crunchy foods such as carrot sticks and apple slices, when age appropriate, are good choices for snacks because they are less likely to promote tooth decay than sticky candies.

FLUORIDE   Children need a source of fluoride in the diet. If the water supply is not adequately fluoridated, then a fluoride supplement is recommended. The American Dental Association, the American Academy of Pediatrics, and the American Academy of Pediatric Dentistry have devised a fluoride supplementation schedule, which is based on the child's age and the fluoride content of the local water supply.[34] Children ages six months to three years need 0.25 mg of fluoride per day if their local water supply has <0.3 ppm of fluoride. Children three to six years of age need 0.5 mg fluoride per day if their water supply has <0.3 ppm, but only 0.25 mg fluoride per day if the local water has 0.3 to 0.6 ppm of fluoride.[34] Excessive fluoride supplementation, consumption of toothpaste with fluoride, and natural water supplies high in fluoride can cause *fluorosis.*

Although otherwise harmless, fluoriosis produces permanent staining of the enamel of teeth, particularly permanent teeth. Because of the risk of fluorosis, fluoride supplements are only available by prescription. Few foods contain much fluoride but fluoridated water used in beverages and food preparation does provide fluoride.

Constipation

Constipation, or hard and dry stools associated with painful bowel movements, is a common problem of young children. Sometimes "stool holding" develops when the child does not completely empty the rectum, which can lead to chronic overdistension so that eventually the child is retaining a large fecal mass.[35] Then having a bowel movement can become painful to the child, which leads to more "stool holding," and a viscous cycle ensues. A pediatrician should manage the treatment of "stool holding."

Diets providing adequate fiber for age or about 7–10 grams for young children (see fiber recommendations on page 259) guard against constipation. Some of the best food sources of dietary fiber for toddlers and preschoolers are whole grain breads and cereals, legumes, and fruits and vegetables appropriate for age. Too much fiber should be avoided, however. Young children easily develop diarrhea from high amounts, and high-fiber foods may displace other energy-dense foods.

Lead Poisoning

Approximately 4.4% of children aged one to five years have high blood lead levels, exceeding 10 mcg/dL.[3] Young children are particularly at risk for developing high levels of lead because in exploring their environment, they enjoy putting things into their mouths. Depending on their surroundings, some of these objects may be high in lead.

FLUOROSIS Permanent white or brownish staining of the enamel of teeth caused by excessive ingestion of fluoride before teeth have erupted.[33]

FOOD SECURITY Access at all times to a sufficient supply of safe, nutritious foods.

High blood lead levels affect the functioning of many tissues in the body, including the brain, blood, and kidneys. Low-level exposure to lead is associated with decreases in IQ and behavioral problems, and elevated blood lead levels may decrease growth in young children.[36]

Children living in housing built before 1950 are at increased risk of high lead levels because lead-based paint may have been used on these houses.

Lead-based paint chips taste sweet, tempting children to consume them. As the age of housing decreases nationally, so does the incidence of high lead levels in children.[3] Lead can enter the food supply through lead-soldered water pipes, contaminated water supplies, and from certain canned goods from other countries that contain lead-solder seals. Nonfood items containing lead include dirt, lead weights, and other objects. Lead is also found in ceramic glazes, pewter, and in some folk remedies. One study of a small group of toddlers, 18 to 36 months of age, who lived in an urban area with potentially high lead levels, found a correlation between the amount of lead wiped from the children's hands and the level of lead in their food.[37] These results emphasize the importance of hand washing when preparing and consuming food. Damage caused by lead exposure may begin during pregnancy as lead is transported across the placenta to the fetus.

Iron deficiency in preschool age children increases the risk of elevated lead levels.[38] With iron deficiency and a reduction in iron availability, binding receptors for iron are made available for lead transport. Also, with iron deficiency, iron uptake receptors in the small intestine are available for lead absorption. Adequate dietary calcium intake appears to protect against high blood lead levels by decreasing absorption of lead.

The CDC[39] recommends universal screening of children living in neighborhoods where the risk of lead exposure is widespread, and the federal Medicaid program requires that all eligible children be screened for elevated blood lead levels. Lead-exposure risk exists in children:

- Who live in neighborhoods where more than 27% of homes were built before 1950.
- Living in poverty.
- Who have siblings or playmates with lead poisoning.

To summarize, eliminating sources of lead in the child's environment is the most important step toward eliminating elevated blood levels in children. In addition, preventing iron deficiency and promoting a well-balanced diet, which includes good sources of calcium, help to prevent this problem in young children.

Food Security

One of the Healthy People 2010 objectives is to increase *food security* among U.S. households to 94% from a baseline of 88%.[3] In 1998, nearly 10 million people, more than one-third of them children, lived in households in which at least some members experienced hunger due to the lack of food.[40] According to the Household Food Security Report, the level of households having uncertain food availability has remained relatively constant since 1995. Food insecurity is more likely to exist among American Indian or Alaska Native, African American, and Hispanic or Latino people as compared to whites, among households with children, particularly those headed by single women, and in lower-income-level households (≤130% of poverty threshold).

Food security is particularly important for young children because of their high nutrient needs for growth and development. Young children are a vulnerable group because they must depend on their parents and caretakers to supply them with adequate access to food. It appears that children who are hungry and have multiple experiences with food insufficiency are more likely to exhibit behavioral, emotional, and academic problems as compared to other children who do not experience hunger repeatedly.[41]

Food Safety

Young children are especially vulnerable to food poisoning because they can become ill from smaller doses of organisms. Key foodborne pathogens include *Campylobacter* species and *Salmonella* species, which are the most frequently reported foodborne illnesses in the United States, and the emerging pathogens, *E. coli 0157:H7* and *Listeria monocytogenes*.[3] The most common cause of salmonella food poisoning is consumption of foods containing undercooked or raw eggs, such as raw cookie dough containing eggs. Children less than 10 years of age account for a disproportionate percent of cases of E. coli. It is a serious disease and can cause bloody diarrhea and **hemolytic uremic syndrome.** Outbreaks of E. coli have been associated with ingestion of contaminated, undercooked hamburger meat, unpasteurized apple cider and juice, and unpasteurized milk. Employing proper food storage and preparation techniques at home, in child-care centers, and in retail food establishments is essential for decreasing the incidence of food poisoning in young children.

The U.S. Environmental Protection Agency is in the process of evaluating all existing standards for pesticides by 2006.[3] A major objective of this evaluation is to ensure that the current levels of pesticides in the food supply and drinking water are safe for young children.

PREVENTION OF NUTRITION-RELATED DISORDERS

The prevalence of *overweight* and *obesity* among children, adolescents, and adults in the United States has increased and represents a major public health problem. High-energy, high-fat diets coupled with sedentary lifestyles are thought to be major contributors to the increase in weight. Cardiovascular disease, a major cause of death and morbidity in the United States today, is also thought to be influenced by diets and sedentary lifestyles. Food habits, preferences, and behaviors established during the toddler and preschool ages logically influence dietary habits later in life and subsequent health status. However, this logical deduction has not yet been substantiated by scientific research in this age group, which is sorely needed.[44] In the meantime, families are encouraged to adopt dietary and exercise patterns that promote a healthy lifestyle.

Overweight and Obesity in Toddlers and Preschoolers

Approximately 22% of children younger than age five are overweight, based on the 85th percentile cutoff point for weight-for-height. Increases in the prevalence of overweight are seen in children of all ages and race and ethnic groups, but changes are greatest for older preschool aged children.[45]

During the preschool years, a decrease in body mass index (BMI), or weight-for-height squared $(wt(kg)/ht(m)^2)$, is a normal part of growth and development. BMI usually reaches its lowest point at approximately four to six years of age and then increases gradually in the period called *adiposity rebound.* Early adiposity rebound in children increases the risk of adult obesity.[46] An annual increase in BMI of three to four units may reflect a rapid increase in body fat for most children.[24]

Obesity is a multifaceted problem that is difficult to treat, making prevention the preferred approach. A committee of obesity experts has recommended that children with a BMI greater than or equal to the 85th percentile on the new CDC growth charts with

HEMOLYTIC UREMIC SYNDROME (HUS) A serious, sometimes fatal complication associated with illness caused by *E. coli 0157:H7,* which occurs primarily in children under the age of 10 years. HUS is characterized by renal failure, **hemolytic anemia,** and a severe decrease in **platelet** count.[3]

HEMOLYTIC ANEMIA Anemia caused by shortened survival of mature red blood cells and inability of the bone marrow to compensate for the decreased life span.

PLATELETS A component of the blood that plays an important role in blood coagulation.

OVERWEIGHT Body mass index at or above the 95th percentile.

OBESITY BMI-for-age greater than the 95th percentile with excess fat stores as evidenced by increased triceps skinfold measurements above the 85th percentile.[42,43]

ADIPOSITY REBOUND A normal increase in body mass index which occurs after BMI declines and reaches its lowest point at four to six years of age.[47]

complications, such as hypertension or gallbladder disease, or with a BMI greater than or equal to the 95th percentile, be evaluated and possibly treated for obesity.[24] Children less than two years of age who fall into this category should be evaluated by a pediatric obesity specialist. Obtaining a *triceps skinfold*

TRICEPS SKINFOLD A measurement of a double layer of skin and fat tissue on the back of the upper arm. It is an index of body fatness and measured by skinfold calipers. The measurement is taken on the back of the arm midway between the shoulder and the elbow.

HEART DISEASE The leading cause of death and a common cause of illness and disability in the United States. Coronary heart disease is the principal form of heart disease and is caused by buildup of cholesterol deposits in the coronary arteries, which feed the heart.

LDL CHOLESTEROL Low-density lipoprotein cholesterol, the lipid most associated with atherosclerotic disease. Diets high in saturated fat, *trans* fatty acids, and dietary cholesterol have been shown to increase LDL-cholesterol levels.

FAMILIAL HYPERLIPIDEMIA A condition that runs in families and results in high levels of serum cholesterol and other lipids.

TRANS FATTY ACIDS Fatty acids that have unusual shapes resulting from the hydrogenation of polyunsaturated fatty acids. Trans fatty acids also occur naturally in small amounts in foods such as dairy products and beef.

ATHEROSCLEROSIS A type of hardening of the arteries in which cholesterol is deposited in the arteries. These deposits narrow the coronary arteries and may reduce the flow of blood to the heart.

measurement and comparing it to standards will further validate that the child has excess fat stores rather than increased lean muscle mass. Standards for triceps skinfold measurements include those derived from data of the National Health and Nutrition Examination Survey (NHANES) I, which are age and gender specific.[48]

Maintaining weight while gaining height can be the best treatment for obese children between the ages of two and seven. This approach allows the obese child to "grow into his or her height" and to lower BMI. However, if the child is already exhibiting secondary complications of obesity, such as mild hypertension or high cholesterol or triglyceride levels, gradual weight loss may be indicated.[24] Reducing weight at this young age is tricky because sufficient nutrients must be provided for children to reach their full height potential and to remain healthy.

Prevention and Treatment of Overweight and Obesity

Prevention is the best approach for overweight and obesity. Parenting techniques, such as finding reasons to praise the child's behavior, but never using food as a reward, foster the development of healthy eating behaviors in children and help children to self-regulate food intake. When weight control for a young child is warranted, the expert committee on

obesity evaluation and treatment recommends a general approach, including family education and involvement.[24] Examples of behavior changes, which a family can incorporate into their lifestyle, include increasing physical activity, offering nutrient-dense and not calorie-dense snacks, and focusing on behavior changes rather than weight changes.

Nutrition and Prevention of Cardiovascular Disease in Toddlers and Preschoolers

Heart disease is the number one cause of death in the United States today.[3] A leading risk factor for cardiovascular disease, which includes diseases of the heart and blood vessels, is elevated levels of *LDL cholesterol.* Children with *familial hyperlipidemias* and obese children can have high levels of LDL-cholesterol. High intakes of saturated fat, *trans fatty acids,* and, to a lesser extent, dietary cholesterol elevate LDL cholesterol levels in children and adults alike. In terms of health promotion and disease prevention, the level and type of fat in children's diets will affect their risks of developing cardiovascular disease as adults.[44] Fatty streaks, which can be precursors to the buildup of fat deposits in blood vessels, have been found in the arteries of young children. Some experts believe that these streaks can represent the beginning of *atheroscleorsis* and cardiovascular disease.[49]

The National Cholesterol Education Program recommends that healthy children consume a "Step 1 diet," which includes the following:

- Less than 10% of total calories from saturated fatty acids.
- An average total fat intake of no more than 30% of total calories.
- Less than 300 mg dietary cholesterol per day.

Toddlers over the age of two years should transition to this eating pattern.[49]

Since the publication of the National Cholesterol Education Program report, concern has been raised about the effect such heart healthy diets has on the growth of young children. Fat is an important source of energy, essential fatty acids, and fat-soluble vitamins, and some researchers do not believe that the benefits of restricting fat in children's diets outweigh the possible risks of decreasing energy intake and impacting growth.[50] Both the American Heart Association and the American Academy of Pediatrics have issued statements attesting to the safety of this population approach to lowering LDL cholesterol levels in young children.[51,52] However, the concern is that some parents and caretakers who are already preparing low-fat meals for their families may reduce the fat content of their children's diet too much. Low-fat diets in children may limit calories and essential fatty acids needed for growth. For this reason, the American Heart Association recommends that diets contain no less than 15% total fat for adults. Similarly, the American Academy of Pediatrics recommends that after two years of age, children and adolescents should gradually adopt a diet that by about five years of age, contains no more than 30%, and no less than 20%, of calories from fat.

Dietary recommendations are different for children who are at increased risk of developing premature cardiovascular disease because a parent has heart disease or because of familial hyperlipidemias. These children need periodic screening for blood cholesterol levels and close follow-up. If LDL-cholesterol levels are high and not controlled by dietary changes, further restriction of total calories from saturated fat to <7% and of dietary cholesterol to no more than 200 mg per day is recommended. On this "Step 2 diet," the child needs to be closely monitored by a physician and registered dietitian.

The Children's Health Project studied the dietary intake of children ages four to ten years, some of whom had elevated LDL cholesterol levels and some of whom had normal lipid levels, at baseline and again at three months.[53] One of the goals of the study was to determine how children changed their total diet when they reduced their total fat intake. Children who were most successful at reducing the percent of calories consumed from fat in the three-month time period did change their food intake by consuming fewer servings of meats, eggs, dairy, fats and oils, and bread. They tended to increase the number of servings from lower fat choices within these same food groups, particularly in the dairy group. These children also increased their consumption of fruits, vegetables, and desserts, but the desserts tended to be low in fat, such as flavored gelatin and popsicles. Children maintained average intakes of all nutrients, except for vitamin D, in excess of two-thirds of the respective RDAs. Low baseline levels of vitamin D intake in these children indicated a need to encourage them to consume vitamin D-fortified foods such as skim or low-fat milk. Although vitamin E intake was not below two-thirds of the RDA, vitamin E intake of these children did decrease from baseline. Because good sources of vitamin E include fats and oils, the vitamin E status of children who are decreasing their dietary fat intake warrants monitoring. These investigators conclude that lowering the dietary fat intake of young children according to the National Cholesterol Education Program does not compromise their nutrient intakes.

Vitamin and Mineral Supplements

Children who consume a variety of basic foods can meet all of their nutrient needs without vitamin or mineral supplements. Eating a diet of a variety of foods is the preferred way to get needed nutrients because foods contain many other substances, such as phytochemicals, in addition to nutrients that benefit health.

The American Academy of Pediatrics recommends vitamin and mineral supplementation for children who are at high risk of developing or have one or more nutrient deficiencies.[34] Children at risk of nutrient deficiency include:

- Children from deprived families or who suffer from abuse or neglect.
- Children with anorexia or poor appetite and poor eating habits.
- Children who consume a "fad diet" or only a few types of food.
- Children who consume a vegetarian diet without dairy products.

In spite of these recommendations, data from the NHANES III indicate that children one to five year of age are major users of supplements.[54] Approximately one in two three-year-olds in the United States are given a vitamin and mineral supplement by their parents.[55] Characteristics of mothers who give their children supplements include non-Hispanic white, older, more years of education, married, have health insurance, receive care from a private health care provider, have greater household income, and took supplements themselves during pregnancy. Children most likely to receive a supplement are those at low risk of developing nutrient deficiencies; children who would most likely benefit from a supplement are less likely to receive them.

TOLERABLE UPPER INTAKE LEVELS Highest level of daily nutrient intake that is likely to pose no risk of adverse health effects to almost all individuals in the general population; gives levels of intake that may result in adverse effects if exceeded on a regular basis.

If given to children, vitamin and mineral supplement doses should not exceed the DRI or RDA for age. Parents and caretakers should be warned against giving high amounts of vitamins and minerals to children, particularly vitamins A (retinol) and D. The *Tolerable Upper Intake Levels* shown in the DRI tables should serve as a guide to excessive levels of nutrient intake from fortified foods and supplements.

Herbal Supplements

The use of herbal remedies for various disorders is increasing in the United States today as is the use of complementary and alternative medicine practices in general. Parents and caretakers who take herbs are likely to give these products to their children. Few definitive studies exist on the effectiveness of these substances in preventing disease and promoting health in adults, much less in children. In spite of the lack of scientific evidence, anecdotal reports of benefits abound. However, some reports have linked herbal preparations to adverse effects.[56] Information on herb use should be obtained during the nutrition assessment of a child to rule out herbs as a source of health problems. At the current time, no regulation of or consistency in the composition of the products can lead to uncertain results. Children given various herbs are the "test subjects" in these uncontrolled studies. Parents should be advised of the potential risks of herbal therapies and the need for close monitoring of their child if they choose to give herbs to their child. The National Institutes of Health's (NIH) National Center for Complementary and Alternative Medicine's (NCCAM) Web site is listed in the Resources section at the end of this chapter and provides reports on the safety and effectiveness of various herbal remedies and alternative medical practices.

DIETARY AND PHYSICAL ACTIVITY RECOMMENDATIONS

Children aged 2 to 11 years should achieve healthful eating habits and participate in regular physical activity to promote optimal physical and cognitive development, attain a healthful weight, and reduce the risk of chronic disease.[57]

The American Dietetic Association

Taking into consideration calorie and nutrient needs and the common nutritional problems and concerns, it is easy to understand the importance that underlies dietary recommendations for toddlers and preschoolers. A primary recommendation is that young children eat a variety of foods. This recommendation is more easily achieved if healthful food preferences and eating habits are acquired during the early years. Food preferences in conjunction with food availability form the foundation of the child's diet. Limited food selection, therefore, will influence the adequacy of the child's diet by decreasing variety. Parents and caretakers cannot expect a child to "do as I say, but not as I do." Nutrition education aimed at the adults in the child's life becomes as important if not more so than nutrition education directed to the child.

Dietary recommendations have been developed and disseminated by the federal government and pro-fessional organizations. Two sets of guidelines for young children's diets are available: the Dietary Guidelines for Americans and the Food Guide Pyramid.[10,58] Recommendations for caloric and nutrient intake are represented in the RDAs and DRIs.

Dietary Guidelines

Dietary Guidelines for Americans, 2000 (discussed in Chapter 1), emphasize that children be offered a variety of foods that includes grain products, vegetables and fruits, and low-fat dairy products.[58] The guidelines also recommend that beans, lean meats, and poultry be added as appropriate for the child. Foods high in fat and sugar, such as candy, cookies, and cakes, should be limited in children's diets. The Dietary Guidelines also emphasize the importance of parents modeling this type of diet for their children or "to do as I say and as I do." The guidelines also emphasize the importance of physical activity. Parents are advised to encourage their children to engage in at least 60 minutes per day of vigorous physical activity and to limit the time spent in sedentary activities, such as TV watching and computer game playing, that replace physical activity.[58] It is important that parents model for their children a lifestyle that includes a varied diet and regular physical activity.

Food Guide Pyramid

The USDA has developed a Food Guide Pyramid for young children (Illustration 10.7) targeted at children ages two to six years.[10] The guide encourages children to consume a variety of foods, with grains, fruits, and vegetables forming the base of the diet, and foods high in fat and sweets being consumed in limited amounts. Foods shown in the guide represent single serving portions of foods commonly eaten by young children. To make the guide "child friendly," the food groups have shorter names, and the number of servings for each group is a single number rather than a range. To emphasize the importance of physical activity, illustrations of children being active are depicted around the pyramid. In the booklet, "Tips for Using the Food Guide Pyramid for Young Children 2 to 6 Years Old," parents are advised to give two- to three-year-old children the same number of servings as recommended for four- to six-year-old children, but to give about two-thirds of a serving for the younger children. Children two to six years of age need the same number of servings from the milk group each day, two servings. The booklet also includes age appropriate meal preparation activities for young children and some simple recipes children can help prepare. A sample meal and snack plan based on

Illustration 10.7
Food Guide Pyramid for Young Children.[10]

the "Food Guide Pyramid for Young Children" can be found in Table 10.6. Illustration 10.8 depicts a chart that parents can utilize in evaluating their children's intake as compared to the Food Guide Pyramid. The source for the booklet for parents is given in the Resource section at the end of this chapter.

Recommendations for Intake of Iron, Fiber, Fat, and Calcium

Adequate iron intake is necessary to prevent iron deficiency and iron deficiency anemia in toddlers and preschoolers. Appropriate fiber intake is needed to prevent constipation and may provide long-term disease prevention. Fat is an important source of calories, essential fatty acids, and fat-soluble vitamins in young children's diets. From age two to five years, it is recommended that the fat content of children's diets be gradually transitioned to <30% of total calories from fat. Adequate calcium intake is important for children to achieve peak bone mass, and yet about 20% of children do not meet the DRIs for calcium.[3]

IRON Adequate iron intake is important in this age group to prevent iron deficiency. Good sources of di-

etary iron can be found in Chapter 1, page 25. Meats, which are good sources of iron, can be ground or chopped to make them easier for toddlers to chew. Fortified breakfast cereals and dried beans and peas are also good sources of iron.

"Toddler" milks, or iron-fortified commercial formulas for toddlers, are available. Healthy children who consume a variety of foods, and whose milk intake is less than 24 ounces daily, obtain adequate iron without these special products. Other commercial beverages being marketed to parents include formulas that were originally designed for children with illnesses or who had to be fed complete nutrition through a feeding tube. Such special products are expensive and are unnecessary for healthy children. It would be better for parents of healthy children to spend their food dollars on a variety of healthy foods rather than on these special products.

FIBER Ample dietary fiber intake has been associated with the prevention of heart disease, certain cancers, diabetes, and hypertension in adults. Whether fiber helps prevent these problems as young children become adults is not known, but it is clear that fiber in a child's diet helps prevent constipation and is part of a healthy diet. Too much fiber in a child's diet can be detrimental because high-fiber diets have the potential of reducing the energy density of the diet, which could impact growth.[59] High-fiber diets could also impact the bioavailabilty of some minerals, such as iron and calcium.

A reasonable dietary fiber intake goal for children ages 3 through 20 is the equivalent of age plus 5.[59] For example, a three-year-old child's dietary fiber goal would be 8 grams per day, while a five-year-old child's dietary fiber goal would be 10 grams per day. Including fruits, vegetables, and whole grain breads and cereal products can increase the dietary fiber intake of children. Only 45% of children ages four to six years meet the age +5 recommendation for dietary fiber.[60] Those who meet the recommendation consume more high- and low-fiber breads and cereals, fruits, vegetables, legumes, nuts, and seeds than those who don't. Children, who meet the age +5 goal tend to have lower intakes of fat and cholesterol, and higher intakes of dietary fiber, vitamins A and E, folate, magnesium, and iron than those children who had low dietary fiber intakes.

FAT An appropriate amount of fat in a young child's diet can be achieved by employing the principles of the Dietary Guidelines and the Food Guide Pyramid that promote a diet of whole grain breads and cereals, beans and peas, fruits and vegetables, low-fat dairy products after two years of age, and

Table 10.6   One day's sample meals and snacks for four- to six-year-old children.[10] (Offer two- to three-year-olds the same variety but smaller portions.)

	GRAIN	VEGGIE	FRUIT	MILK	MEAT
Breakfast					
100% fruit juice, ¾ c			1		
Toast, 1 slice	1				
Fortified cereal, 1 oz	1				
Milk, ½ c				½	
Mid-Morning Snack					
Graham crackers, 2 squares	1				
Milk, ½ c				½	
Lunch					
Meat, poultry, or fish, 2 oz					2 oz
Macaroni, ½ c	1				
Vegetable, ½ c		1			
Fruit, ½ c			1		
Milk, ½ c				½	
Mid-Afternoon Snack					
Whole grain crackers 5	1				
Peanut butter, 1 tbs					½ oz
Cold water, ½ c					
Dinner					
Meat, poultry, or fish, 2½ oz					2½ oz
Potato, 1 medium		1			
Broccoli, ½ c		1			
Cornbread, 1 small piece	1				
Milk, ½ c				½	
Total Food Group Servings	6	3	2	2	5 oz

lean meats. Foods high in fat are used sparingly, especially foods high in saturated fat and trans fatty acids. However, an appropriate amount of dietary fat is necessary to meet children's needs for calories, essential fatty acids, and fat-soluble vitamins. As discussed in Chapter 1, good sources of the essential fatty acid, linoleic acid, are peanut, canola, corn, safflower, and other vegetable oils. Flaxseed, soy, and canola oils are good sources of the essential fatty acid, alpha-linolenic acid.

It is important to include sources of fat-soluble vitamins in the diets of young children. Good sources of vitamin A include whole eggs and dairy products. Sources of vitamin D include exposure to sunlight and vitamin D-fortified milk. Corn, soybean, and safflower oils are excellent sources of vitamin E. Vitamin K is widely distributed in both animal and plant foods.

CALCIUM   Adequate calcium intake in childhood affects peak bone mass. A high peak bone mass is thought to be protective against osteoporosis and fractures later in life.[61] Yet, many children do not consume adequate calcium. An estimated 21% of children two to eight years of age consume less than their DRI for calcium.[3] The recommendations for daily calcium intake in the DRI table is 500 mg/day for children aged one to three years, and 800 mg/day for children aged four to eight years. An important aspect of adequate calcium intake in toddlers and preschoolers is the development of eating patterns that will lead to adequate calcium intake later in childhood.[61]

Dietary sources of calcium are listed in Chapter 1, page 24. Dairy products are good sources of calcium, as are canned fish with soft bones such as sardines, dark green leafy vegetables such as kale and bok choy, tofu made with calcium, and calcium-fortified foods and beverages such as calcium-fortified orange juice. Nonfat and low-fat dairy products are low in saturated fat while still providing a good source of calcium.

Fluids

Healthy toddlers and preschoolers will consume enough fluid through beverages, foods, and sips and glasses of water to meet their needs. Fluid requirements increase with fever, vomiting, diarrhea, and when children are in hot, dry, or humid environments.

Consumption of milk has decreased among young children since the late 1970s but consumption of carbonated soft drinks has increased by about the same amount. Since that time, consumption of noncitrus juices has increased almost threefold.[62]

PLAN FOR YOUR YOUNG CHILD... The Pyramid Way

Use this chart to get an idea of the foods your child eats over a week. Pencil in the foods eaten each day and pencil in the corresponding triangular shape. (For example, if a slice of toast is eaten at breakfast, write in "toast" and fill in one Grain group pyramid.) The number of pyramids shown for each food group is the number of servings to be eaten each day. At the end of the week, if you see only a few blank pyramids...keep up the good work. If you notice several blank pyramids, offer foods from the missing food groups in the days to come.

	SUNDAY	MONDAY	TUESDAY	WEDNESDAY	THURSDAY	FRIDAY	SATURDAY
Milk	△△	△△	△△	△△	△△	△△	△△
Meat	△△	△△	△△	△△	△△	△△	△△
Vegetable	△△△	△△△	△△△	△△△	△△△	△△△	△△△
Fruit	△△	△△	△△	△△	△△	△△	△△
Grain	△△△	△△△	△△△	△△△	△△△	△△△	△△△
	△△△	△△△	△△△	△△△	△△△	△△△	△△△
Breakfast							
Snack							
Lunch							
Snack							
Dinner							

Illustration 10.8 **Plan for Your Young Child . . . The Pyramid Way.**[10]

According to food consumption surveys, young children consume large amounts of sweetened beverages, including fruit juice, soft drinks, and sweetened iced tea to the detriment of the overall nutritional balance of their diet and oral health. Approximately 50% of children aged two to five years consume soft drinks.[63] Children with high consumption of regular soft drinks (more than 9 ounces per day) consume more calories and less milk and fruit juice than those with lower consumptions of regular soft drinks. Water is a good but underused "thirst quencher" for toddlers and preschoolers, as long as milk (2 cups) and fruit juice consumption (1 cup) is part of their regular diet. Parents and caretakers can offer children water to drink between meals and snacks.

Recommended vs. Actual Food Intake

Several national surveys have examined food and nutrient intakes of young children. Comparing the Household Food Consumption Survey to the Food Guide Pyramid reveals that the mean food group intakes for children ages two to five years were below minimum recommendations for all food groups, except the dairy group with intakes at or near recommendations (see Table 10.7).[64] In this study cohort, about 70% of the children did not meet the recommended servings for fruits, grains, meats, and dairy, and about 64% did not meet the recommendations for vegetables.[57,64] Only about 1% of children met the recommendations for food intake as outlined by the Food Guide Pyramid. Children who met none of the recommendations had nutrient intakes well below the RDA for vitamin B_6, calcium, iron, and zinc and had a low fiber intake.

According to the Food and Nutrient Intakes by Children report,[62] mean energy intakes of young children meet RDAs for energy.[25] Mean percentages of total energy from carbohydrate, protein, total fat, saturated fat, and cholesterol intake in the diets of toddlers and preschoolers are shown in Table 10.8.

Table 10.7   Mean daily intakes of the Food Guide Pyramid for children in the 1989–1991 CSFII.[64]

Gender/Age	Number in study	Grain	Vegetable	Fruit	Dairy	Meat
Recommended servings		6–11	3–5	2–4	2–3	5–7 oz
Males 2–5 years	429	5.2	1.9	1.5	2.0	2.8
Females 2–5 years	416	4.9	2.0	1.5	1.9	2.7

The total fat intake of about 32% of total calories is within the target range for this age child. The saturated fat intake is greater than the recommended <10% of total calories. The three- to five-year olds have sodium intakes of 2600 mg per day, which slightly exceed the sodium recommendation of 2400 mg per day. Both the toddlers and the preschool age children appear to be meeting the recommendations for fiber, and both age groups have cholesterol intakes well below the recommendation of 300 mg per day.

In general, young children consume more than enough protein and fat. Mean vitamin and mineral intakes of young children exceed the RDAs except for vitamin E and zinc.[62] In a longitudinal study of the nutrient and food intakes of preschool children aged 24 to 60 months, mean intakes of zinc, folic acid, and vitamins D and E were consistently below the RDAs and DRIs.[65] Low intakes of zinc, vitamin E, and iron were found in toddlers ages 12 to 18 months, a time of dietary transition.[66] The means for nutrient intakes often hide problems at the extremes, however. They fail to indicate the percentage of children with low nutrient intakes <66% of the RDA and DRIs, and children with high nutrient intakes that exceed the Tolerable Upper Intake Levels.

According to the Food and Nutrient Intakes by Children report, 89% of one- to two-year old children ate snacks, and for these children snacks contributed 24% of calories and 23% of total fat intake for the day. Similarly, 88% of three- to five-year-old children consumed snacks, which contributed 22% of total energy and 20% of total fat. These data indicate the important contribution of snacks to children's overall nutrient intakes.

Cross-Cultural Considerations

When working with families from various cultures, it is important to learn as much as possible about the culture's food-related beliefs and practices. Ask the parents and caretakers about their experiences with food, including foods used for special occasions. It is also helpful to know whether foods are used for home remedies or to promote certain aspects of health. Cultural beliefs influence many child feeding practices, such as what foods are best for young children, which cause digestive upsets, or which help relieve illnesses. Perhaps one of the best-known examples is the use of chicken soup to cure illnesses! It is important for the health care provider to build on the cultural practices and to reinforce those positive practices while attempting to affect change in those practices that could be harmful to the young child.

A series of booklets entitled *Ethnic and Regional Food Practices* is available from the American Dietetic Association. This series addresses food practices, customs, and holiday foods of various ethnic groups. They provide examples for incorporating traditional foods of various groups of people into dietary recommendations. For example, peanut or polyunsaturated oils are recommended to Chinese Americans for stir-frying instead of the more traditional use of lard or chicken fat. Ordering information for this series can be found in the Resource section of this chapter.

Vegetarian Diets

Young children can grow and develop normally on vegetarian or vegan diets, as long as their dietary patterns are intelligently planned. Vegetarian diets are rich in fruits, vegetables, and whole grains, the consumption of which is encouraged for the general population. However, young children in particular need some energy-dense foods to reduce the total amount of food required. The amount of vegetarian foods needed to meet nutrient needs may be more food than young children can eat. Young children

Table 10.8   Mean percentages of total calories from carbohydrate, protein, total fat, and saturated fatty acids, and cholesterol intake.[62]

Age	Carbohydrate (%)	Protein (%)	Total fat (%)	Saturated Fat Acids (%)	Cholesterol (mg/d)
1–2 years	54	15	32	13	189
3–5 years	55	14	32	12	197

need to eat several times a day to meet their energy needs because their stomachs cannot hold a lot of food at one time.

Children who are fed *vegan* and *macrobiotic diets* tend to have lower rates of growth during the first five years of life compared to children given a mixed diet.[67] Strict vegan diets, which exclude all foods of animal origin, may be deficient in vitamins B_{12} and D, zinc, and omega-3 fatty acids, and may also be low in calcium, unless fortified foods are consumed. Children on vegan diets should receive vitamin B_{12} supplements or consume fortified breakfast cereals, textured soy protein, or soy milk fortified with vitamin B_{12}. The vitamin B_{12} status of children following vegetarian and vegan diets should be monitored on a regular basis as vitamin B_{12} deficiency may cause vitamin B_{12} deficiency anemia. Iron deficiency anemia is an infrequent problem among children consuming a vegetarian diet.

Vitamin D adequacy can be met by diet or by sun exposure. Good sources of vitamin D for children include fortified soy milk, fortified breakfast cereals, and fortified margarines. Zinc is found in foods of animal origin. Plant sources of zinc include legumes, nuts, and whole grains. Vegetable products are also lacking in omega-3 fatty acids. Therefore, including a source of these fatty acids, such as canola or soybean oils, is advisable.[67]

Foods containing phytates such as unrefined cereals, may interfere with calcium absorption. So if the child's diet contains a lot of unrefined cereals, higher calcium intakes may be needed.[67] Good sources of calcium for children on strict vegetarian diets include fortified soy milk, calcium-fortified orange juice, tofu processed with calcium, and certain vegetables, such as broccoli and kale.[9] Supplements may be necessary for some children with inadequate intakes that are not remedied by dietary means.

Guidelines recommended for diets of young children who are consuming vegetarian diets have been developed and are given here:[9]

- Allow the child to eat several times a day (i.e., three meals and two to three snacks).
- Avoid serving the child bran and an excessive amount of bulky foods, such as bran muffins and raw fruits and vegetables.
- Include in the diet some sources of energy-dense foods such as cheese and avocado.
- Include enough fat (at least 30% of total calories) and a source of omega-3 fatty acids, such as canola or soybean oils.
- Include sources of vitamin B_{12}, vitamin D, and calcium in the diet, or supplement if required.

Child-Care Nutrition Standards

All child-care programs should achieve recommended standards for meeting children's nutrition and nutrition education needs in a safe, sanitary, supportive environment that promotes healthy growth and development.[68]

The American Dietetic Association

An estimated 23 million children in the United States require child care while their parents work, making foods children eat away from home a major contribution to their overall intake. Young children eat away from home often: 37% of one- to two-year-olds, and 52% of three- to five-year-olds eat away from home daily. The most common places for away-from-home meals in children five years of age and younger are fast-food restaurants, day-care centers, and friends' houses.[62]

Nutrition standards for child-care services exist and specify minimum requirements for amounts and types of foods to include in meals and snacks and food service safety procedures.[68,69] These standards also address nutrition learning experiences and education for children, staff, and parents as well as the physical and emotional environment in which meals and snacks are served. It is recommended that children in part-day programs (four to seven hours per day) receive food that provides at least one-third of their daily calorie and nutrient needs in at least one meal and two snacks or two meals and one snack. A child in a full-day program (eight hours or more) should receive foods that meet one-half to two-thirds of the child's daily needs based on the RDAs and DRIs in at least two meals and two snacks or three snacks and one meal. Food should be offered at intervals of not less than two hours and not more than three hours and should be consistent with the Dietary Guidelines for Americans.[6]

VEGAN DIET The most restrictive of vegetarian diets, allowing only plant foods.

MACROBIOTIC DIET This diet falls between semivegetarian and vegan diets and include foods such as brown rice, other grains, and vegetables, fish, dried beans, spices, and fruits.

Physical Activity Recommendations

Inactivity is thought to be a major contributor to the increasing prevalence of obesity. The Dietary Guidelines for Americans, 2000 recommend that young children engage in play activity for at least 60 minutes every day.[58] Some of the suggested activities include:

- Playing tag.
- Riding a tricycle or bicycle.
- Walking, skipping, or running.

Parents are encouraged to set a good example for their children by being physically active themselves. Parents are also encouraged to limit the time that they allow their children to watch television and play computer games.

NUTRITION INTERVENTION FOR RISK-REDUCTION

Model Program

Bright Futures is a vision, a philosophy, a set of expert guidelines, and a practical developmental approach to providing health supervision for children of all ages, from birth through adolescence.[70]

Bright Futures in Practice: Nutrition is an example of a model program for nutrition intervention for risk reduction.[9] This guide is a component of the larger project *Bright Futures Guidelines for Health Supervision of Infants, Children, and Adolescents.*[70]

The purpose of *Bright Futures* is to foster trusting relationships between the child, health professionals, the family, and the community to promote optimal health for the child.[70] *Bright Futures* guidelines are developmentally based and address the physical, mental, cognitive, and social development of infants, children, adolescents, and their families.

Many different "user-friendly" materials and tools are available from this program to assist in the implementation of the guidelines. In addition to the *Bright Futures Guidelines for Health Supervision,* three implementation guides have been published to date; one on oral health, one on general nutrition, and one on physical activity. Future publications include: implementation guides on mental health, children with special health care needs, *Bright Futures for Families,* and training modules.

Bright Futures in Practice: Nutrition is based on three critical principles:[9]

1. Nutrition must be integrated into the lives of infants, children, adolescents, and families.
2. Good nutrition requires balance.
3. An element of joy enhances nutrition, health, and well-being.

The program is based on the premise that optimal nutrition for children be approached from the development of the child and put in the context of the environment in which the child lives.[9] It emphasizes the development of healthy eating and physical activity behaviors. Nutrition supervision guidelines are given for each age group, and within each broad age group, interview questions, screening and assessment, and nutrition counseling topics are provided. It lists desired outcomes for the child and discusses the role of the family, as well as frequently asked questions. For example, by utilizing the guidelines, health care providers will be able to provide anticipatory guidance to parents of toddlers for the proper advancements of their diets based on growth and development. The implementation guide also addresses special topics related to pediatric nutrition including oral health, vegetarian eating practices, iron-deficiency anemia, and obesity. Useful information is included in the appendix such as nutrition questionnaires for the various age groups. *Bright Futures in Practice: Nutrition* is a valuable resource to anyone who is interested in promoting healthy eating and physical activity behaviors in children. Ordering information for *Bright Futures* materials is available on their Web site, which is listed in the Resource section at the end of this chapter.

PUBLIC FOOD AND NUTRITION PROGRAMS

Young children and their families can benefit from a number of federally sponsored food and nutrition programs. Four example programs are presented here.

WIC

The Special Supplemental Nutrition Program for Women, Infants, and Children,[71] previously described in Chapter 8, is administered by the Food and Nutrition Service of the U.S. Department of Agriculture (USDA). It is one of the most successful federally funded nutrition programs in the United States. Participation in WIC services improves the growth, iron status, and the quality of dietary intake of nutritionally at-risk infants and children up to age five years.[72] The WIC Program is cost effective in that each dollar invested in the program saves up to $3 in health care.[71]

As in infancy, children must live in a low-income household, 185% or less of the federal poverty level and be at "nutrition risk" to be eligible for WIC services. "Nutrition risk" means a child has a medical or dietary-based condition, which places the child at increased risk. Such conditions include iron deficiency anemia, underweight, overweight, a chronic illness, such as cystic fibrosis, or the child consumes an inadequate diet.[71]

Children receive nutrition assistance, education, and follow-up services by specially trained registered dietitians and nutritionists. Vouchers for food items such as milk, juice, eggs, cheese, peanut butter, and fortified cereals are given to eligible families. These vouchers are exchanged for the food items at authorized retailers.

WIC's Farmers' Market Nutrition Program

The Farmers' Market Nutrition Program is a special seasonal program for WIC participants. This program provides vouchers for the purchase of locally grown produce at farmers' markets. The program is designed to help low-income families increase their consumption of fresh fruits and vegetables.

Head Start and Early Head Start

Administered by the U.S. Department of Health and Human Services, Head Start and Early Head Start are comprehensive child development programs, serving children from birth to five years of age, pregnant women, and their families. Nearly one million U.S. children participate in this program. The overall goal is to increase the readiness for school of children

from low-income families. A range of individualized, culturally appropriate services are provided through Head Start and related agencies working in education and early childhood development; medical, dental, and mental health services; nutrition services; and parent education.[73] About three in four Head Start families have incomes <$12,000 annually. More specific information about Early Head Start can be found in Chapter 9.

Food Stamps

The Food Stamp Program, administered by the USDA, is designed to help adults in low-income households buy food. The monetary amount of food vouchers provided to an eligible household depends on the number of people in it and the income of the household. Income eligibility criteria for this and a number of other federal programs can be found at the USDA's Food and Nutrition Service website: http://www.fns.usda.gov/fsp. The average monthly amount of benefits received through the Food Stamp Program in 1998 was $71 per person, enough to help families and individuals pay for a portion of the food they need. Each state must develop a food stamp nutrition education plan based on federal guidance.[74] Participation in the Food Stamp Program is associated with increased intakes of a number of nutrients.[72]

Resources

American Academy of Pediatrics.
The American Academy of Pediatrics web page contains consumer information on current news topics affecting the pediatric population. Policy statements can be found on this page and consumer publications can be ordered through the Bookstore.
Available at: www.aap.org.

American Dietetic Association
Besides information for members of The American Dietetic Association, this web page contains information for consumers on various topics of interest. There is a "Daily Tip" and a feature article, plus consumer information on topics such as food safety and healthy lifestyles.
Available at: www.eatright.org.

Bright Futures
The Bright Futures publications are developmentally based guidelines for health supervision and address the physical, mental, cognitive, and social development of infants, children, adolescents, and their families. Three implementation guides have also been published to date addressing oral health, general nutrition, and physical activity.
Available at: www. brightfutures.org. Bright Futures materials may also be ordered from: National Maternal and Child Health Clearinghouse, 2070 Chain Bridge Road, Suite 450, Vienna, VA 22182-2536, www.nmchc.org, 703-356-1964.

Centers for Disease Control and Prevention, National Center for Health Statistics
CDC growth charts: United States. May 30, 2000.
The CDC web site provides background information on the recent growth chart revisions. Also, individual growth charts can be downloaded and printed from this web page.
Available at: http://www.cdc.gov/nchs/about/major/nhanes/growthcharts/charts.htm.

Diabetes Care and Education, Dietetic Practice Group of The American Dietetic Association
Ethnic and Regional Food Practices, A Series. Chicago, IL: The American Dietetic Association, 1994-1999.
This series of booklets addresses food practices, customs, and holiday foods of various ethnic groups. Booklets describing the following ethnic groups are available: Alaska Native, Chinese American, Filipino American, Hmong American, Jewish, and Navajo.

Graves DE, Suitor CW
Celebrating Diversity: Approaching Families Through Their Food. 2nd ed. Arlington, VA: National Center for Education in Maternal and Child Health, 1998.
The purpose of this publication is to assist health professionals in learning to communicate effectively with a diverse clientele. Topics covered in the book include using food to create common ground, changing food patterns, how food choices are

made, communicating with clients and families, and working within the community.

Green M, Palfrey JS. (Eds)

Bright Futures: Guidelines for Health Supervision of Infants, Children, and Adolescents. 2nd Ed. Arlington, VA: National Center for Education in Maternal and Child Health, 2000. This Bright Futures publication contains developmentally based guidelines for health supervision and address the physical, mental, cognitive, and social developments of infants, children, adolescents, and their families.

National Academy Press, Dietary Reference Intakes

The web page of the National Academy Press, the publisher for The National Academies, contains descriptions of available books. Over 2,000 books are available online free of charge. The current Dietary Reference Intakes are also available on this web page and can be read on-line free of charge. *Available at:* http://www.nap.edu

National Center for Complementary and Alternative Medicine, National Institutes of Health

The National Center for Complementary and Alternative Medicine is dedicated to science-based information on complementary and alternative healing practices. This web page provides information for consumers and practitioners as well as information about related news and events. *Available at:* http://www.nccam.nih.gov/nccam/

National Network for Child Care

This website, hosted by Iowa State University Extension, provides articles, resources, and links on a variety of topics of interest to professionals and families who care for children and youth. Topics include child development, nutrition, and health and safety. *Available at:* http://www.nncc.org/

Patrick K, Spear BA, Holt K, and Sofka D.

Bright Futures in Practice: Physical Activity. Arlington, VA: National Center for Education in Maternal and Child Health, 2001.

This colorful and user-friendly spiral bound book, is one of the implementation guides of the Bright Futures publications. Developmentally appropriate activities are presented for infants, children, and adolescents. Special issues and concerns are addressed, such as physical activity for the child with a chronic condition such as asthma or diabetes mellitus.

USDA, Center for Nutrition Policy and Promotion

Tips for Using the Food Guide Pyramid for Young Children 2 to 6 Years Old. Government Printing Office, 202/512-1800, Stock Number 001-00004665-9.

This web page provides information on USDA materials including the Dietary Guidelines for Americans and the Food Guide Pyramid. This particular document can be found by clicking on Food Guide Pyramid for Young Children. *Available at:* http://www.usda.gov/cnpp.

References

1. Center on Hunger, Poverty and Nutrition Policy. Statement on the link between nutrition and cognitive development in children. Tufts University, School of Nutrition Science and Policy; 1998.

2. Kids count data book. Baltimore: The Annie E. Casey Foundation; 1999.

3. U.S. Department of Health and Human Services. Healthy People 2010 (conference edition, in two volumes). Washington, DC: January 2000.

4. Behrman RE, Kliegman RM, Jenson HB, eds. Nelson's textbook of pediatrics, 16th ed. Philadelphia, PA: WB Saunders Co; 1999.

5. Centers for Disease Control and Prevention, National Center for Health Statistics. CDC growth charts: United States. Available from http://www.cdc.gov/nchs/about/major/nhanes/growthcharts/charts.htm. May 30, 2000.

6. Pietrobelli A, Myles FA, Alison DB, et al. Body mass index as a measure of adiposity among children and adolescents: a validation study. J Pediatr 1998;132: 204–10.

7. Dietz WH, Robinson TN. Use of the body mass index (BMI) as a measure of overweight in children and adolescents. J Pediatr 1998;132:191–3.

8. Troiano RP, Flegal KM. Overweight children and adolescents: description, epidemiology, and demographics. Pediatrics 1998;101:497–504.

9. Story M, Holt K, Sofka D, eds. Bright futures in practice: nutrition. Arlington, VA: National Center for Education in Maternal and Child Health; 2000.

10. USDA, Center for Nutrition Policy and Promotion. Tips for using the food guide pyramid for young children 2 to 6 years old. Government Printing Office, 202/512-1800, Stock Number 001-00004665-9. Available from http://www.usda.gov/cnpp.

11. Satter E. The feeding relationship: problems and interventions. J Pediatr 1990;117:181–9.

12. Birch LL. Children's food acceptance patterns. Nutr Today 1996;31:234–40.

13. Birch LL, Johnson SL, Andresen G, et al. The variability of young children's energy intake. N Eng J Med 1991;324:232–5.

14. Story M, Brown JE. Do young children instinctively know what to eat? The studies of Clara Davis revisited. New Eng J Med 1987;316:103–6.

15. Van den Bree MBM, et al. Genetic and environmental influences on eating patterns of twins ages ≥50 years. Am J Clin Nutr 1999;70:456–65.

16. Chess S, Thomas A. Dynamics of individual behavioral development. In: Levine MD, Carey WB, Crocker AC, eds. Developmental-behavioral pediatrics. Philadelphia, PA: WB Saunders Co.; 1992: pp 84–94.

17. Birch LL, Fisher JO. Development of eating behaviors among children and adolescents. Pediatrics 1998;101:539–49 (S).

18. Birch LL, Fisher JA. Appetite and eating behavior in children. Pediatr Clinic N Amer 1995;42:931–53.

19. Fisher JO, Birch LL. Restricting access to palatable foods affects children's behavioral response, food selection, and intake. Am J Clin Nutr 1999;69:1264–72.

20. Birch LL, Fisher JO. Food intake regulation in children, fat and sugar substitutes

and intake. Ann NY Acad Sci 1997;819: 194–220.

21. Rolls BJ, Engell D, Birch LL. Serving portion size influences 5-year-old but not 3-year-old children's food intake. J Amer Diet Assoc 2000;100:232–4.

22. Fisher JO, Birch LL. Parents' restrictive feeding practices are associated with young girls' negative self-evaluation of eating. J Amer Diet Assoc 2000;100:1341–6.

23. Satter E. Feeding dynamics: helping children to eat well. J Pediatr Health Care 1995;9:178–84.

24. Barlow SE, Dietz WH. Obesity evaluation and treatment: expert committee recommendations. Pediatrics [serial online] 1998;102:E3. Available from http://www. pediatrics.org/cgi/content/full/102/3/e29, July 1, 2000.

25. National Research Council. Recommended dietary allowances, 10th ed. Washington, DC: National Academy Press; 1989.

26. Cryan J, Johnson RK. Should the current recommendations for energy intake in infants and young children be lowered? Nutr Today 1997;32:69–74.

27. Kennedy E, Goldberg J. What are American children eating? Implications for public policy. Nutr Rev 1995;53: 111–26.

28. Kennedy E, Powell R. Changing eating patterns of American children: a view from 1996. J Amer Coll of Nutr 1997;16:524–9.

29. Institute of Medicine, Food and Nutrition Board. Dietary reference intakes for calcium, phosphorus, magnesium, vitamin D, and fluoride. Washington, DC: National Academy Press; 1997.

30. Institute of Medicine, Food and Nutrition Board. Dietary reference intakes for vitamin A, vitamin K, arsenic, boron, chromium, copper, iodine, iron, manganese, molybdenum, nickel, silicon, vanadium, and zinc. Washington, DC: National Academy Press; 2001.

31. Centers for Disease Control and Prevention. Recommendations to prevent and control iron deficiency in the United States. Morbidity and Mortality Weekly Report. April 3, 1998;47:No. RR-03.

32. Looker AC, Dallman PR, Carroll MD, et al. Prevalence of iron deficiency in the United States. JAMA 1997;277:973–6.

33. Casamassimo P. Bright futures in practice: oral health. Arlington, VA: National Center for Education in Maternal and Child Health; 1996.

34. American Academy of Pediatrics, Committee on Nutrition. Pediatric nutrition handbook. Elk Grove Village, IL: American Academy of Pediatrics; 1998.

35. McClung HJ, Boyne L, Heitlinger L. Constipation and dietary fiber intake in children. Pediatrics 1995;96:999–1001 (S).

36. Ballew C, Khan LK, Kaufmann R, et al. Blood lead concentration and children's anthropometric dimensions in the Third National Health and Nutrition Examination Survey (NHANES III), 1988–1994. J Pediatr 1999;134:623–30.

37. Stanek K, Manton W, Angle C, et al. Lead consumption of 18- to 36-month-old children as determined from duplicate diet collections: Nutrient intakes, blood lead levels, and effects on growth. J Amer Diet Assoc 1998;98:155–8.

38. Mushak P, Crocetti AF. Lead and nutrition. Nutr Today 1996;31:12–17.

39. Centers for Disease Control and Prevention. Screening young children for lead poisoning: guidance for state and local public health officials. Atlanta, GA: Centers for Disease Control and Prevention, 1997. Also available at http:www.cdc.gov/nceh/ programs/lead/guide/1997/guide97.htm.

40. Household food security in the United States 1995–1998, advance report—summary. Available from http://www.fns.usda. gov/oane/MENU/Published/FSP/FILES/ fsecsum.htm.

41. Kleinman RE, Murphy JM, Little M, et al. Hunger in children in the United States: potential behavioral and emotional correlates. Pediatrics [serial online] 1998; 101:E3. Available from http://www. pediatrics.org/cgi/content/full/101/1/e3, July 1, 2000.

42. Must A, Dallal GE, Dietz WH. Reference data for obesity: 85th and 95th percentiles of body mass index (wt/ht2) and triceps skinfold thickness. Am J Clin Nutr 1991;53:839–46.

43. Dwyer JT, Stone EJ, Yang M. Prevalence of marked overweight and obesity in a multiethnic pediatric population: findings from the Child and Adolescent Trial for Cardiovascular Health (CATCH) study. J Amer Diet Assoc 2000;100: 1149–56.

44. Milner JA, Allison RG. The role of dietary fat in child nutrition and development: summary of an ASNS workshop. J Nutr 1999;129:2094–105.

45. Mei Z, Scanlon KS, Grummer-Strawn LM, et al. Increasing prevalence of overweight among US low-income preschool children: the Centers for Disease Control and Prevention pediatric nutrition surveillance. Pediatrics [serial online] 1998;101: E12. Available from http://www. pediatrics.org/cgi/content/full/101/1/e12, July 1, 2000.

46. Whitaker RC, Pepe MS, Wright JA, et al. Early adiposity rebound and the risk of adult obesity. Pediatrics [serial online] 1998;101:E5. Available from http://www. pediatrics.org/cgi/content/full/101/3/e5, July 1, 2000.

47. Dietz WH, Gortmaker SL. Preventing obesity in children and adolescents. Annu Rev Public Health 2001;22:337–53.

48. Frisancho AR. New norms of upper limb fat and muscle areas for assessment of nutritional status. Am J Clin Nutr 1981;34: 2540–5.

49. National Cholesterol Education Program. Report of the expert panel on blood cholesterol in children and adolescents. Bethesda, MD: National Cholesterol Education Program; 1991.

50. Olson RD. The folly of restricting fat in the diet of children. Nutr Today 1995;30:234–45.

51. Fisher EA, Van Horn L, McGill HC. Nutrition and children: a statement for healthcare professionals from the nutrition committee, American Heart Association. Circulation 1997;95:2332–3.

52. American Academy of Pediatrics, Committee on Nutrition. Cholesterol in childhood. Pediatrics 1998;101:141–7.

53. Dixon LB, McKenzie J, Shannon BM, et al. The effect of changes in dietary fat on the food group and nutrient intake of 4- to 10-year-old children. Pediatrics 1997; 100:863–72.

54. Vital & Health Statistics—Series 11: Data from the National Health Survey. 1999;244:1–14.

55. Yu SM, Kogan MD, Gergen P. Vitamin-mineral supplement use among preschool children in the United States. Pediatrics [serial online] 1997;100:E4. Available from http://www.pediatrics.org/ cgi/content/full/100/5/e4, July 1, 2000.

56. Buck ML, Michael RS. Talking with families about herbal products. J Pediatr 2000;136:673–8.

57. American Dietetic Association. Position of The American Dietetic Association: dietary guidance for healthy children aged 2 to 11 years. J Amer Diet Assoc 1999; 99:93–101.

58. U.S. Department of Agriculture and U.S. Department of Health and Human Services. Dietary guidelines for Americans, 2000. 5th ed. Home and Garden Bulletin No. 232.

59. A summary of conference recommendations on dietary fiber in childhood: Conference on Dietary Fiber in Childhood, New York, NY, May 24, 1994. Pediatrics 1995;96:1023–8.

60. Hampl JS, Betts NM, Benes BA. The "age+5" rule: comparisons of dietary fiber intake among 4- to 10-year-old children. J Amer Diet Assoc 1998;98:1418–23.

61. American Academy of Pediatrics, Committee on Nutrition. Calcium requirements of infants, children, and adolescents. Pediatrics 1999;104:1152–7.

62. U.S. Department of Agriculture, Agricultural Research Service. 1999. Food and nutrient intakes by children 1994–96, 1998. ARS Food Surveys Research Group, available on the "Products" page at http://www.barc.usda.gov/bhnrc/ foodsurvey/home.htm, July 1, 2000.

63. Harnack L, Stang, J, Story M. Soft drink consumption among US children and adolescents: nutritional consequences. J Amer Diet Assoc 1999;99:436–41.

64. Munoz KA, Krebs-Smith SM, et al. Food intakes of US children and adolescents compared with recommendations. Pediatrics 1997;100:323–9.

65. Skinner JD, Carruth BR, Houck KS, et al. Longitudinal study of nutrient and food intakes of white preschool children aged 24 to 60 months. J Amer Diet Assoc 1999;99:1514–21.

66. Picciano MF, Smiciklas-Wright H, Birch L, et al. Nutritional guidance is needed during dietary transition in early childhood. Pediatrics 2000;106:109–14.

67. Sanders TAB. Vegetarian diets and children. Pediatr Clinic N Amer 1995;42:955–65.

68. American Dietetic Association. Position of The American Dietetic Association: nutrition standards for child-care programs. J Amer Diet Assoc 1999;99:981–6.

69. American Public Health Association, American Academy of Pediatrics. Caring for our children, national health and safety performance standards: guidelines for out-of-home child care programs. 1992.

70. Green M, Palfrey JS, eds. Bright futures: guidelines for health supervision of infants, children, and adolescents, 2nd ed. Arlington, VA: National Center for Education in Maternal and Child Health; 2000.

71. U.S. Department of Agriculture. WIC. Available from www.usda.gov.

72. Rose D, Habicht JP, Devaney B. Household participation in the Food Stamp and WIC Programs increase the nutrient intakes of preschool children. J Nutr 1998;128:548–55.

73. U.S. Department of Health and Human Services. Head Start. Available from www.dhhs.gov.

74. U.S. Department of Agriculture, Food Stamps Program. Available from http://www.fns.usda.gov/fsp.

CHAPTER *11*

Photo Disc

> The young disease, that
> must subdue at length,
> Grows with his growth,
> and strengthens with his
> strength.

Alexander Pope

TODDLER AND PRESCHOOLER NUTRITION:

Conditions and Interventions
Prepared by **Janet Sugarman Isaacs**

CHAPTER OUTLINE

- Introduction
- Who Are Children with Special Health Care Needs?
- Nutrition Needs of Toddlers and Preschoolers with Chronic Conditions
- Growth Assessment
- Feeding Problems
- Nutrition-Related Conditions
- Dietary Supplements and Herbal Remedies
- Sources of Nutrition Services

KEY NUTRITION CONCEPTS

1 Nutrition problems in young children with special health care needs are underweight, overweight, feeding difficulties, and higher nutrient needs as a result of chronic health conditions.

2 Feeding difficulties in preschoolers and toddlers appear as food refusals, picky appetites, and concerns about growth.

3 Nutrition services for toddlers and preschoolers with chronic health problems are provided in various settings, including schools and other educational programs and specialty clinics.

4 Toddlers and preschoolers at risk for chronic conditions have the same nutritional problems and concerns as other children.

INTRODUCTION

Most toddlers and preschoolers are healthy and develop as expected. This chapter discusses children who do not fit the typical pattern, *children with special health care needs* associated with a *chronic condition* or disability or children who are at risk. Often no diagnosis has been made, and yet parents, health care providers, or preschool teachers have a nagging feeling that something is not right. Perhaps the child is slow to talk or seems more picky than expected about eating. Parents often focus on issues of feeding and eating in their efforts to try to address that nagging concern. Nutritional assessments of at-risk toddlers and preschoolers and resources for nutrition services are found in many places, such as day care, preschools, public health programs, as well as medical centers. A broad service delivery system is discussed because educational, medical, nutritional, and social services target toddlers and preschoolers who may need a variety of services. This chapter covers nutrition services for young children with food allergies, breathing or *pulmonary* problems, feeding and growth problems, *developmental delays,* and especially those at risk for needing nutrition support.

CHILDREN WITH SPECIAL HEALTH CARE NEEDS   A general term for infants and children with, or at risk for, physical or developmental disabilities, or chronic medical conditions from genetic or metabolic disorders, birth defects, premature births, trauma, infection, or prenatal exposure to drugs.

CHRONIC CONDITION   Disorder of health or development that is the usual state for an individual and unlikely to change, although secondary conditions may result over time.

PULMONARY   Related to the lungs and their movement of air for exchange of carbon dioxide and oxygen.

EARLY INTERVENTION SERVICES   Federally mandated evaluation and therapy services for children in the age range birth up to age three years under the Individuals with Disabilities Education Act.

WHO ARE CHILDREN WITH SPECIAL HEALTH CARE NEEDS?

The child who does not see, hear, or walk is easily recognized as having a chronic condition. It can be difficult and expensive, however, to identify some other children with special health care needs. Criteria for labeling chronic conditions in children vary from state to state. More than 40 different federal definitions describe "disability."[1] Criteria used for identifying disabilities in adults do not fit children because the criteria are related to one's ability to work or perform household chores. Chronic condition and disability mean the same thing in referring to toddlers and preschoolers. Confusion about the prevalence of disabilities in children is partially based on which conditions are used for testing, and what services are covered under various health insurance payers. Prevalence estimates for disabilities range from 5% to 31% of children.[1,2] Whatever the number, nutrition problems are common in children with disabilities. The highest estimate is that 90% of children with disabilities have some type of nutritional problem.[3]

Toddlers and preschoolers with chronic illnesses are entitled to the same services as older people with chronic illnesses, plus additional help. They are covered by the American with Disabilities Act, the Social Security Disability Program, Supplemental Social Security Insurance Program (SSI), and services for families without health insurance coverage.[4] Additional help comes from educational regulations that ensure that all children with disabilities have a free, appropriate public education. Nutrition services are funded within education regulations in the Individuals with Disabilities Education Act (IDEA). Most children start school at age five or six years, but children at risk or who have special needs may attend well before that, as soon as the need is identified. The sooner special educational, nutritional, and health care interventions are started, the better for the overall development of the child. Parents of a typical child choose and pay to send their child to day care or a preschool program. For a child with a special health care need, nutrition and various types of therapy may be provided within a day care or educational setting, funded by state and federal resources.

Nutrition services have been expanding for young children with special health care needs. The newer area of expansion is within educational services, based on laws first passed in 1986.[4] Nutrition services can be provided to young children within special educational programs and services covered by Part C of IDEA, which serves preschoolers (age three to five years old) and children from birth up to three years old.[4] Eligibility for services does not require a specific diagnosis as is required for older children. *Early intervention services* are based on the following:[4]

- Developmental delays in one or more of the following areas: cognitive, physical, language and speech, psychosocial, or self-help skills.
- A physical or mental condition with a high probability of delay, such as Down syndrome.
- At risk medically or environmentally for substantial developmental delay if services are not provided.

A number of chronic conditions are suspected, but not obvious in the first year of life. The diagnosis often becomes clear in the toddler and preschool years, however, when such questions as whether a preterm infant will show catch-up growth or a breathing problem will resolve are answered. Also, standardized developmental screening, evaluation, and testing for these ages show more reliability than for infants. By this age range, parents allow preschoolers to play with other children, which heightens awareness of differences between their child and others. Parents who were told about possible disabilities during infancy move beyond coping by denial or disbelief and are willing to seek out services. The tendency is to resist labeling a young child with a diagnosis, so some suspected conditions are not confirmed until school age if the delay in diagnosis will not harm the child.

NUTRITION NEEDS OF TODDLERS AND PRESCHOOLERS WITH CHRONIC CONDITIONS

Toddlers or preschoolers with chronic health conditions are still at risk for the same nutrition-related problems and concerns as a typical child. Iron deficiency and inadequate calcium intake are problem nutrients in all children in this age range.[5] Consequently, every attempt should be made to meet nutritional needs and to assure normal growth and development in toddlers and preschoolers with chronic conditions. Young children with chronic conditions may have low appetites and require increased calories and protein concentrated in small portion sizes. Table 11.1 gives examples of conditions in which caloric needs may be high or low. Each individual child must be assessed to confirm caloric needs. One child may have periods of needing additional calories and protein, followed by a need for less, based on the course of the chronic condition. Underweight or overweight is a common reason a child with a chronic condition will be referred for nutrition services.[6] The energy requirements of infants with special needs, as discussed in Chapter 9, apply to toddlers and preschoolers also.

Overweight and obesity are common features of a number of chronic conditions including Down syndrome and spina bifida.[4] A preventive approach to excessive weight gain is required for these conditions. The child does not need as many calories as others due to low muscle mass, mobility, and short stature. The health status of children with chronic conditions is worsened by excessive body weight, so matching their caloric intake to their needs is an important aspect of nutritional care.

Underweight, more commonly observed in children with chronic conditions, results in part from the chronic illness and its treatment. Children with chronic illnesses may be more likely to experience weight loss from common illnesses. If already underweight, further weight loss is likely to result in loss of muscle. To increase weight and maintain growth, underweight children may need to consume concentrated sources of calories, such as greater than 30% of calories from fat. In this circumstance, the usual recommendations to monitor fat in foods for young children so that it slowly decreases as a percent of calories in the diet are not appropriate.[5] Similarly the general recommendations from the Food Guide Pyramid and the 5-a-Day program were designed with typical children in mind. Families of toddlers and preschoolers with chronic illnesses in which extra calories from fat are needed may be confused by such general nutrition recommendations.

Recommendations regarding food intake, vitamin and mineral supplementation, and mealtime behaviors should be customized to the individual child. Children who are frequently sick or have low energy levels and appetites may dislike eating foods that are hard to chew, or take a long time to eat. Pasta, applesauce, and oatmeal may be preferences in these children. Generally food likes of toddlers and preschoolers can be incorporated into healthy diet plans for children with chronic conditions. Some nutrition problems related to chronic illness may result from the children's interaction with the parents. It is age-appropriate for

CYSTIC FIBROSIS Condition in which a genetically changed Chromosome 7 interferes with all the exocrine functions in the body, but particularly pulmonary complications, causing chronic illness.

DIPLEGIA Condition in which the part of the brain controlling movement of the legs is damaged, interfering with muscle control and ambulation.

PEDIATRIC AIDS Acquired immunodeficiency syndrome in which infection-fighting abilities of the body are destroyed by a virus.

PRADER-WILLI SYNDROME Condition in which partial deletion of chromosome 15 interferes with control of appetite, muscle development, and cognition.

BRONCHOPULMONARY DYSPLASIA (BPD) Condition in which the underdeveloped lungs in a preterm infant are damaged so that breathing requires extra effort.

Table 11.1 Chronic conditions generally associated with high and low caloric needs.

Higher Caloric Need Conditions	Lower Caloric Need Conditions
Cystic fibrosis	Down syndrome
Renal disease	Spina bifida
Ambulatory children with diplegia	Nonambulatory children with diplegia
Pediatric AIDS	Prader-Willi syndrome
Bronchopulmonary dysplasia (BPD)	Nonambulatory children with short stature

children to express their food likes and dislikes, insist on their independence, and go on food jags. It can be difficult but important to distinguish between nutrition problems related to the chronic condition and those related to "growing up" in the toddler and preschool years.

The DRIs for toddlers and preschoolers provide a good starting point for setting protein, vitamin, and mineral needs for children with chronic conditions.[7] (DRI tables are located on the inside front cover of this text.) The recommendations for typical children concerning dietary fiber, prevention of lead poisoning, and iron deficiency anemia apply to children at risk or already diagnosed with special health care needs.[5]

However, some specific conditions require adjustments to the general guidelines. The "5 + age" general guidelines for dietary fiber recommendations for children are too low for some conditions and too high for others.[8] Children with spina bifida generally need high-fiber diets, unless they are prescribed constipation management medications. Low-fiber intake is appropriate for some children with cystic fibrosis. Children with sickle cell disease have more specific blood iron and lead testing; interpretation of their results may not fit with the usual guidelines. Recommending iron-rich foods to increase their iron stores may not be appropriate, if, for example, they may be getting iron from blood transfusions.

Just like parents whose children do not have special health care needs, some parents purchase nutrition supplements and over-the-counter vitamin and mineral supplements to assure themselves that the child is getting enough nutrients. Such products may be helpful for children with identified higher needs for vitamins or minerals, or for children with self-restrictive dietary intakes. In general, the need for added vitamins and minerals or complete nutritional supplements should be identified as a part of the child's nutritional assessment.

HYDROCEPHALUS   Condition in which excess fluid collects in the brain, resulting in brain damage, often treated by a shunt.

RETT SYNDROME   Condition in which a genetic change on the X chromosome results in severe neurological delays causing children to be short and thin appearing and unable to talk.

● GROWTH ASSESSMENT

Concerns about growth of children with special health care needs are common. The child with a stable weight, or who gains and then loses weight sequentially is not growing well. This abnormal pattern may be difficult to detect. Typical toddlers appear to be "thinning out" as body composition changes from babyhood into childhood. The level of activity in this age range is so much higher than in the first year of life that some parents express concern that the typical child is running off the foods they eat. The broad range of activity reflects the abilities of the child and the family's lifestyle. For children with short parents, short stature may be constitutional and unresponsive to attempts to boost nutrition. Even if improved, nutrition would not make the child taller, the growth assessment would be used to document that the child is growing well despite short stature.

Children with special health care needs often have conditions that affect growth even when adequate nutrients are provided. In such cases the 2000 CDC growth charts require interpretation based on the child's previous growth pattern.[9] If a thin and small-appearing child has adequate fat stores, adding calories may be harmful; instead, recognize the growth pattern as healthy for that child. Trying to add calories in such a case may promote overweight in the form of excess fat stores.

Specific growth charts developed for the chronic condition, when available, are preferred. For children up to age 38 months who were born LBV or VLBV, the *Infant Health and Development Program* (IHDP) growth percentiles charts are appropriate.[10] Correction for prematurity, as discussed in Chapter 9, makes them useful for preterm babies as well as toddlers. For a child born three months early, for example, the IHDP growth charts could be used at a chronological age of 41 months. Plotting on both the 2000 CDC growth chart and also the IHDP chart documents catch-up growth. When catch-up growth happens, the child's growth pattern crosses channels on the growth chart, for weight and also for length. Without catch-up growth, the same percentiles fit even as the child gains weight.

Growth patterns in children with special health care needs are also affected by some prescribed medications, particularly those that change body composition, such as steroids and growth hormone. Conditions requiring steroids include some pulmonary conditions and inflammatory conditions, such as arthritis. The use of growth hormone improves stature for a wide variety of conditions. Examples in which growth hormone is measured and may be supplemented are genetic disorders such as Down syndrome and Prader-Willi syndrome.[11,12]

The Nellhaus chart is a head circumference growth chart commonly used by special health care providers.[13] It provides head circumference percentiles from birth to age 18 years and is used to determine whether head growth falls within normal limits or indicates a neurological condition, such as *hydrocephalus.*[3] An example of the importance of head circumference measurements is *Rett syndrome.* Before 1983 girls in this country with this condition were misdiagnosed as having autism or cerebral palsy.[14]

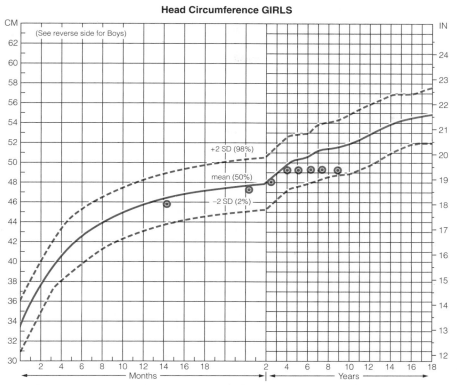

Illustration 11.1 Nellhaus head circumference growth chart plotted for girl with Rett syndrome.

Rett syndrome is a rare disorder, characterized by a reduced rate of head growth beginning in the toddler years (see Illustration 11.1).[15] Also the rate of weight and height accretion slows over time in girls with Rett syndrome.[15] Decreased rate of head growth in toddlers and preschools may be indicative of problems from infancy, such as prematurity, or consequences of infection such as *meningitis*. Some clinics plot both the standard growth chart for head circumference (up to age three years) as well as the special head circumference growth chart as part of the growth assessment.

FEEDING PROBLEMS

Children with special health care needs share many of the same developmental feeding issues as other children, like using food to control at mealtime, experiencing food jags, and self-regulating food intake, similar to those discussed in Chapter 10. However, feeding problems that are part of the underlying health condition may emerge in the toddler and preschool years on top of the usual feeding problems and require extra attention. Some feeding problems during the toddler years are typical in children who are later diagnosed with a chronic condition. Examples of such conditions include gastroesophageal reflux, asthma (pulmonary problems in general), developmental delay,

cerebral palsy, attention deficit hyperactivity disorder, and autism. These children as toddlers tend to display signs of feeding problems, such as low interest in eating, long mealtimes (over 30 minutes), preferring liquids over solids, and food refusals. Children at risk for developmental delay often prove more difficult to feed as a toddler and preschooler.[3] The child may drink liquids excessively compared to eating, or eat as if younger than his or her chronological age. Recognizing that the child needs to be treated as younger than current chronological age may be a necessary step. Offering the child food textures that she or he can eat successfully within a monotonous diet, or continuing to offer a bottle may be appropriate choices in these circumstances.

Table 11.2 shows an example of the likes and dislikes of a 2.5-year-old child. The child likes only a few foods that are not especially nutritious. Usual recommendations are to add variety to the child's diet and to assure intake of meats, milk, and vegetables. This recommendation is appropriate for a typical child, but this child's eating pattern suggests a feeding problem. The soft textures and mild tastes of preferred foods characterize a child closer to one year of age, and the foods the child dislikes require higher oral skills to eat. An evaluation of the child's overall level of functioning will likely indicate a developmental delay of the child's feeding skills.

Children with feeding problems indicative of developmental delay should be referred to an early intervention program for a developmental evaluation.

> **MENINGITIS**   Viral or bacterial infection in the central nervous system that is likely to cause a range of long-term consequences in infancy, such as mental retardation, blindness, and hearing loss.

Behavioral Feeding Problems

Every mouth prefers its own soup.

Sephardic saying

Problems in mealtime behavior and food refusals are common in children with behavioral and attention

Table 11.2   Food choices of a 2.5-year-old child with suspected developmental delay.

Likes	Dislikes
3 packets instant Cream of Wheat with added sugar and margarine (refused offered apple slices)	Hamburger meats, or any other kind of meat
Macaroni and cheese (refused offered sandwich with lettuce and bologna)	Green beans or any kind of vegetables
Banana, with peel removed (refused other cut up fruits offered)	Vegetable soup
Pudding, only chocolate	Salads of all kinds
Cheese puffs (refused corn chips)	Casseroles or any mixtures of foods
Juices of all kinds, in a sippy cup	Milk, and milk with any flavoring added

disorders. These concerns often bring parents to nutrition experts for solutions. Behavioral disorders that impact nutritional status are autism and attention-focusing problems, such as *attention deficit hyperactivity disorder (ADHD)*. ADHD may be suspected during the preschool years, but it is primarily treated during school years.[16] (ADHD is discussed further in Chapter 13.) Diagnoses of mental health and behavior problems are based on multiple observations and tests in children.[16] Table 11.3 shows the intake of a two-year-old with a feeding problem resulting in a self-restricted diet typical of autism. The child refuses to eat many foods and is rigid in what he will eat. He does not respond to feeling hungry as other two-year-old children. When he is not given foods he likes, he refuses to eat at all and has temper tantrums in which he can injure himself. His self-restricted diet is a part of the condition, which impacts how he senses everything in his environment. He prefers to drink rather than eat foods, so a high proportion of his total calories come from one type of drink. Interventions to improve the diet for this child may include a complete vitamin and mineral supplement, and adding one new food by offering it many times (15–20 times) over one or two months. Nutrition interventions should be incorporated into his overall treatment plan provided within a special education program.

NUTRITION-RELATED CONDITIONS

Children who have feeding problems related to muscle control in swallowing or control of the mouth or upper body may choke or cough while eating or refuse foods that require chewing.[3] These types of feeding problems result from conditions such as cerebral palsy or other *neuromuscular disorders* and genetic disorders such as Down syndrome. These signs of feeding and swallowing problems in toddlers or preschoolers generally appear more severe than the reactions of infants who are learning how to munch and chew foods.

The decrease in appetite expected in toddlers and preschoolers may be pronounced in children who find eating difficult and unpleasant. These feeding problems may require further study to make sure eating is safe for a child, and not related to frequent illness such as bronchitis or pneumonia. A child with *hypotonia* or *hypertonia* in the upper body may experience difficulty sitting for a meal and self-feeding with a spoon.[17] If these feeding problems are not resolved by providing therapy in early intervention programs or schools, children are likely to resist eating over time. They may then need a form of nutrition support, such as placement of a *gastrostomy*.

Most toddlers and preschoolers with chronic conditions are provided an assessment of nutritional status as a first step to determine whether more intensive levels of nutrition services are needed. The need for nutrition services are identified by answers to these sorts of questions.

- Is the child growth on track?
- Is his/her diet adequate?
- Are the child's feeding or eating skills appropriate for the child's age?
- Does the diagnosis affect nutritional needs?

Table 11.3   Dietary intake of two-year-old child with suspected autism.

Dry fruit loops cereal
10 fl oz calcium-supplemented orange juice drink
Chicken fingers from a specific fast-food restaurant
French fries
10 fl oz calcium-supplemented orange juice drink
Waverly crackers
Pringles potato chips
10 fl oz calcium-supplemented orange juice drink
oatmeal cake
10 fl oz calcium-supplemented orange juice drink

A variety of nutrition screening tools exists for assessing the nutritional status of children with chronic conditions.[17] Such tools are useful for children at risk as well as those already diagnosed with conditions such as asthma, HIV infection, allergies, and cerebral palsy. After assessment, nutrition intervention services provide methods to improve the nutritional status. Several conditions that require nutrition intervention services include failure to thrive, celiac disease, breathing problems, and muscle coordination problems.

Failure to Thrive

As discussed in Chapter 8, failure to thrive (FTT) is a condition in which a caloric deficit is suspected.[19] FTT has a slightly different basis in toddlers and preschoolers, who may have grown adequately during the first year. Their decrease in growth rate occurs at the age when appetite typically decreases and control issues at mealtime are expected, making identifying the cause of FTT more difficult. Children who have chronic illnesses or were born preterm have a higher risk of FTT as a result of abuse or *medical neglect*.[19] They have greater needs than other children, and may be more irritable and demanding, which places them at risk. FTT as a result of neglect may be suspected because growth records on the young child provide more history than for a baby. Generally FTT is suspected when a child's growth declines more than two growth percentiles, placing them near or below the lowest percentile in weight-for-age, weight-for-length, and/or BMI. FTT may result from a complex interplay of medical and environmental factors, such as the following:[4,17]

- Digestive problems such as gastrointestinal reflux or celiac disease
- Asthma or breathing problems
- Neurological conditions such as seizures
- Pediatric AIDS

Recovery from FTT can include catch-up growth, which is an acceleration in growth rate for age.[20] If calories are provided at a higher level than for a typical child of the same age, catch-up growth is likely (see Illustration 11.2). The length of time needed for catch-up growth varies, but some weight gain should be documented within a few weeks. For example, recovery from FTT for one three-year-old was a gain of 6 pounds—more weight than typical to gain in one year—within the first three months of living in a new home.

Toddler Diarrhea and Celiac Disease

Toddlers are likely to develop diarrhea. The condition is called toddler diarrhea, in which otherwise healthy growing children have diarrhea so often that their parents bring them for a checkup.[21] Testing shows no intestinal damage and normal blood levels, without FTT or weight loss. The dietary culprit is likely to be excessive intake of juices that contain sucrose or sorbitol. The diarrhea results from excess water being pulled into the intestine, so limiting juices may be recommended.[21] Limiting juices to less than 4 fluid ounces per day makes room in the diet for milk and other drinks with greater nutritional quality.[5]

Celiac disease occurs in people who are sensitive to gluten, a component of wheat, rye, barley, and oats. It has a prevalence of 1 in 3000 people with certain ethnic groups, such as those of Middle Eastern or Irish ancestry.[22] Symptoms of diarrhea and other digestive problems usually develop by two years of age. Confirmation of the condition is based on testing blood for the antibodies to gluten. Dietary management requires complete restriction of any foods with gluten. This list includes everything made with flour, such as bread and pasta, as well as foods with wheat, barley, rye, or oats as an additive.[22] The allowed foods include rice, soy, corn, and potato flours. Meats, fruits, and vegetables are not restricted, but many processed foods use wheat flour for thickening. After instituting dietary restrictions, the intestinal damage heals and the digestive symptoms disappear. The parents of preschoolers with celiac disease learn to be expert readers of food labels because intestinal damage recurs if gluten is eaten by mistake.

ATTENTION DEFICIT HYPERACTIVITY DISORDER Condition characterized by low impulsive control and short attention span, with and without a high level of overall activity.

NEUROMUSCULAR DISORDERS Conditions of the nervous system characterized by difficulty with voluntary or involuntary control of muscle movement.

HYPERTONIA Condition characterized by high muscle tone, stiffness, or spasticity.

HYPOTONIA Condition characterized by low muscle tone, floppiness, or muscle weakness.

GASTROSTOMY Form of enteral nutrition support for delivering nutrition by tube directly into the stomach, bypassing the mouth through a surgical procedure that creates an opening through the abdominal wall and stomach.

MEDICAL NEGLECT Failure of parent/caretaker to seek, obtain, and follow through with a complete diagnostic study or medical, dental, or mental health treatment for a health problem, symptom, or condition that, if untreated, could become severe enough to present a danger to the child.

Muscle Coordination Problems

Toddlers who have slow gross motor development may be at risk for cerebral palsy. Many children with neuromuscular conditions are thin-appearing and suspected to have problems with growth and feeding, which raises concerns about whether they are receiving enough nutrients.[17] Examples of children at risk for neuromuscular conditions include children who were LBW, VLBW, or ELBW infants. Later in the preschool years, specific conditions such as *spastic quadriplegia* are confirmed. One reason a specific condition may be suspected but not confirmed is that muscle tone may change over time if spastic quadriplegia is suspected.[23] Hypotonia in the legs and hypertonia in the back muscles are examples of risk factors for feeding problems.

Toddlers and preschoolers at risk for or confirmed with cerebral palsy need nutrition assessment that is particularly adjusted for assessing body composition, such as head size and fat stores.[17,24] Nutrition interventions are then based on the findings from growth assessment, such as encouraging weight gain if body fat stores are low. If the child appears thin as a result of small muscle size, not low fat stores, weight gain is not needed. Growth tracking may be based on spastic quadriplegia growth charts for some toddlers or preschoolers.[24] Part of the growth assessment for a preschooler with cerebral palsy may include an estimate of caloric needs for activity, which may be higher or lower than expected. A girl may expend higher energy in her efforts to coordinate walking while receiving physical therapy three days per week at school. Her activity may be lower if she is in wheelchair most of the time. Measuring recumbent length may be more appropriate than standing height in this example.

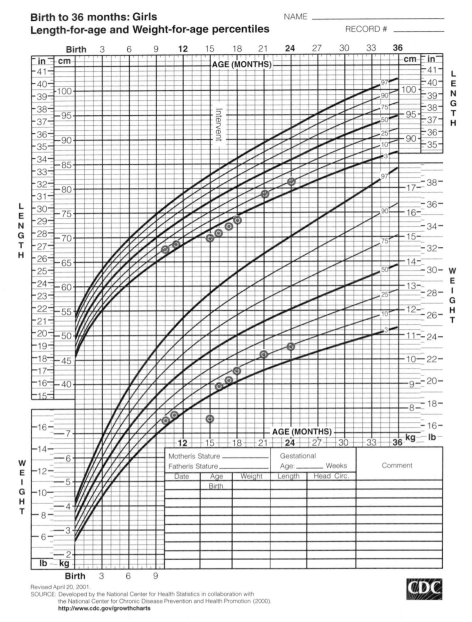

Birth to 36 months: Girls
Length-for-age and Weight-for-age percentiles

NAME _____
RECORD # _____

Revised April 20, 2001.
SOURCE: Developed by the National Center for Health Statistics in collaboration with the National Center for Chronic Disease Prevention and Health Promotion (2000).
http://www.cdc.gov/growthcharts

Illustration 11.2   Growth chart for a girl with failure to thrive before and after intervention.

SPASTIC QUADRIPLEGIA   A form of cerebral palsy in which brain damage interferes with voluntary muscle control in both arms and legs.

Feeding assessment for a child with cerebral palsy may be necessary as part of the overall nutritional assessment.[17]

The assessment may include an observation of eating to determine any restrictions in the type of foods that the child can eat, and whether coordinating muscles for chewing, swallowing, and/or using a spoon or fork are working well. Table 11.4 provides a food intake record for a four-year-old girl with spastic quadriplegia who does not walk and is receiving nutritional services for weight gain. Her meal pattern was adjusted because she tires easily while eating. She does not like to eat too much at a time, and refuses to be fed by another person, which is appropriate for her age. She can chew foods such as fresh apple, but then is too tired to eat something else. She eats a larger portion if the food is soft and does not require her to work so hard. She has gained weight at a slow rate, and her fat stores are low. The first plan is to use regular foods that are easy for her to eat to meet her nutritional needs, including cooked rather than fresh vegetables and fruits, and to avoid hard-to-chew foods, such as roast beef or corn on the cob. If she does not gain weight with the suggestions of foods in Table 11.4, she may need complete nutritional supplements to assure her nutritional needs are met within her feeding limitations.

Pulmonary Problems

Breathing conditions are examples of common problems in children with special health care needs with major nutritional consequences. Breathing problems increase nutritional needs, lower interest in eating, and can slow growth rate. Infants who were born preterm are especially likely as toddlers to have breathing problems. Up to 80% of 1000 gram infants can develop chronic lung disease.[3] Examples of pulmonary diseases or chronic lung disease are bronchopulmonary dysplasia (BPD) and *asthma*. Asthma is self reported in 58 of every 1000 children under five years of age.[25] Asthma results in more emergency room visits for children under five years at 121 visits per 10,000 people than in older children with asthma.[25] Asthma does not necessarily require nutrition services, but some children have asthma as a result of food allergies.[26] Toddlers and preschoolers with BPD have a positive long-term prognosis because new lung tissue can grow until about eight years of age.[27]

Toddlers and preschoolers with serious breathing problems generally need extra caloric intake due to the extra energy expended in breathing. Increased *work of breathing (WOB)* occurs with different pulmonary conditions and generally leads to low interest in feeding, partially as a result of tiredness.[26] Feeding difficulties have several causes in a toddler treated for BDP.[26]

- The normal progression of feeding skills is interrupted.
- Medications and their side effects contribute to high nutrition needs.
- Interrupted sleep and fatigue make hunger and fullness cues harder to interpret.

By the preschool years, impact of BPD on slowing the rate of weight gain is usually clear. Exposure to common respiratory illnesses, which are minor in typical children, can require a trip back to the hospital for some children with BPD. Increased frequency of infections adds another limitation to catch-up growth. Neither the CDC growth chart nor the IHDP preterm growth chart may be helpful in predicting the child's growth pattern, but periods of good health are usually accompanied by a increase in weight gain and appetite.

> **ASTHMA** Condition in which the lungs are unable to exchange air due to lack of expansion of air sacs. It can result in a chronic illness and sometimes unconsciousness and death if not treated.
>
> **WORK OF BREATHING (WOB)** A common term used to express extra respiratory effort in a variety of pulmonary conditions.

Dietary recommendations for toddlers with BPD are similar to those for children with weakness as shown in Table 11.4. Small frequent meals with foods that are concentrated sources of calories are needed. Easy-to-eat foods may still be recommended so that fatigue from meals is low. If the toddler with breathing problems does not gain weight as a result of the dietary recommendations in Table 11.4, the next step would be to add complete nutritional supplements to meet the higher caloric needs. The supplements such as Pediasure® are also a source of vitamins and minerals.

Developmental Delay and Evaluations

Aside from the feeding problems discussed earlier, other nutritional consequences of developmental delay in the toddler and preschool years relate to growth and nutritional requirements from a specific condition. As discussed earlier, iron deficiency and lead toxicity are risk factors for developmental problems.[5] A young child who has been sick and isolated from

Table 11.4 Meal pattern and recommended foods for an underweight girl with feeding problems as a result of weakness.

Meal Pattern: Small frequent meals and snacks to prevent tiredness at meals.	Recommended foods that are easy to chew with small portions.
Breakfast at home	**Breakfast:** ½ c oatmeal with added soft fruit, margarine, and brown sugar
Mid-morning snack	**Snacks:** 1 slice deli meat with 6 fl oz whole milk with Carnation Instant Breakfast added
Lunch (at preschool)	
Afternoon snack (at preschool)	½ soft-cooked sliced apples with added margarine
Afterschool snack	cake-type cookie (frosting allowed)
Dinner	**Dinner:** ½ c mashed potato with added margarine
Bed-time snack	3 tbs meat loaf
	3 tbs soft-cooked carrots with added margarine
	Bedtime snack: chocolate cake with frosting and 4 fl oz whole milk

other children is likely to make developmental progress after resuming a wider range of experiences. As the child gets older, standardized testing will aid in a more definitive diagnosis and subsequent appropriate recommendations for educational programs. Developmental delay is a specific diagnosis that may be replaced by *mental retardation* when the child is six or seven years old.[4]

Changes in growth rate are typical in children with developmental delay. Short stature is common and part of the unusual growth pattern that often prompts referrals for genetic testing in toddlers and preschoolers with growth or developmental issues.[28] The evaluation of growth from a genetic expert may include more in-depth analyses, such as measurements of hand and feet size and bone age.[29] Genetics syndromes can be associated with unusually fast growth also. Soto syndrome is a rare disorder in which the child is tall and large, but has delayed development.[30]

Food Allergies and Intolerance

True food allergies are estimated in 2% to 8% of children.[31] Food allergies are usually identified in toddlers and preschoolers because allergy testing in infancy is not useful due to the incomplete development of the immune system. True food allergies can result in life-threatening episodes of *anaphylaxis*.[32] Skin contact with a food that causes an allergy can result in hives, as an example of a mild reaction. Examples of food allergies that may result in anaphylaxis for some children include the following:[26,31]

- Milk
- Eggs
- Wheat
- Peanuts
- Walnuts
- Soy
- Fish

Diagnosed food allergies can greatly affect the family. For those at risk for anaphylaxis, parents and caregivers should be given instruction in emergency life-saving procedures and use of an injectible form of epinephrine.[26]

Food intolerance reactions are not generally as severe as those for food allergies. Intolerances reactions are localized. An example is a lactose intolerance, which is based in the upper small intestine. The difference between an intolerance and an allergy can be confusing.[29] For example, wheat restriction is an appropriate part of the treatment for celiac disease and also for a wheat allergy. Cow's milk intolerance, which was discussed in the chapter on infancy, rarely persists into the toddler and preschool years. However, when cow's milk intolerance does persist, symptoms in the toddler and preschool years may appear as more general allergy symptoms, such as asthma or skin rashes.[33] A high incidence of other food allergies are present in a child with confirmed cow's milk allergy, such as 35% reacting also to oranges, or 47% reacting also to soy milk.

Strict and complete avoidance of the food that causes the allergy is required. This abstinence includes all settings, such as eating nothing prepared at bake sales when the ingredients are not available. If the preschool child is on an extensively restricted diet, the quality of the diet may not met all their nutritional needs.[26] Such restrictions are also likely to result in mealtime behavioral problems. The parents may become overprotective, or the child may quickly learn to use restricted foods to get a parent's concern and attention.

MENTAL RETARDATION Substantially below-average intelligence and problems adapting to the environment, which emerge before age 18 years.

ANAPHYLAXIS Sudden onset of a reaction with mild to severe symptoms, including a decrease in ability to breathe, which may be severe enough to cause a coma.

DIETARY SUPPLEMENTS AND HERBAL REMEDIES

It is better to take food into the mouth than to take worries into the heart.

Yiddish saying

Families who are concerned that something may be wrong with their young child may be attracted to health and nutritional claims targeted and packaged for adults. The low weight of toddlers or preschoolers may not be factored in when trying various products. Other safety considerations—incidental ingredients such as caffeine—may be ignored. The family that is having difficulty finding effective treatment for a child is most at risk for inappropriate or ineffective alternative products. Parent coalitions and advocacy groups are excellent sources of networking for families, but also sources of nutritional claims for products and dietary regimens that have no scientific testing behind them. Down syndrome is a disorder for which nutritional supplementation has been marketed to parents. No specific nutrient, combination of nutrients, or herbal remedies have been shown to improve the intellectual functioning of individuals with Down syndrome.[34,35] The National Down Syndrome Society cautions parents about the ineffectiveness of nutrient and herbal supplements to discourage their use, but interest continues.[34,36] What is really being marketed is hope, which families always want and need.

Case Study 11.1
A Picky Eater

Greg is a well-groomed boy almost three years old. He has been growing as expected, but he does not talk. He can walk and move about well, but he prefers to play alone. His favorite foods are juices in his sippy cup, which he likes to carry around, macaroni and cheese, white bread without crusts, mashed potato, Honeycomb cereal, and crackers. He cries and throws food that he does not like, such as hamburgers, fruits, most vegetables, and any food combinations. He will periodically eat cheese pizza, scrambled eggs, and applesauce. His mother has tried talking to his pediatrician about his picky appetite, but the pediatrician reassured her that Greg would eat when he was hungry and not to worry. The mother is frustrated that he is so difficult to take out to eat because of the tantrums he throws in restaurants and friends' homes. He sometimes eats a large portion of a food he likes. Most of the time, he is satisfied just drinking juices all day from his sippy cup, and is rarely interested in eating when others eat. He is able to eat with a spoon, but he does not like to touch foods with his hands. He has been referred for speech therapy, but his therapy does not address his eating. His medical history shows that he was born full-term and has had three ear infections but no major illnesses.

Nutrition assessment shows that he is consuming adequate calories at 1350 calories/day or 85 calories per kilogram. His diet is excessive in vitamin C and B vitamins, with adequate protein at the RDA for his age. His sources of protein are mainly his starchy foods of bread, crackers, and dry cereal.

Interpretation and recommendations: Greg has already been identified as a child with special health care needs by the fact that he is receiving speech therapy. His feeding problem may be related to his delayed communication skills. His overall development should be evaluated to rule out developmental delay. Regardless of his evaluation results, nutrition services can be added to his speech services within his preschool by addressing his feeding problem in his Individualized Educational Plan. Nutrition services would focus on establishing a meal and snack schedule for school and home and selecting foods to replace his excess juice intake.

Constipation remedies are examples of over-the-counter products used often for children with special health care needs. Constipation is a common condition in children with various neuromuscular conditions in which muscles are weak.[3] Parents tend to try over-the-counter remedies, dietary methods, and home remedies for constipation management. The effectiveness of dietary fiber may be low when muscle weakness is an underlying problem.[3] Both overtreatment or undertreatment can get the child in trouble by worsening the constipation problem. A young child died as a result of poisoning by a laxative product administered at a higher dose than recommended.[37] Effective prescription medications for constipation management are available, but the family has to bring the problem to the attention of the provider to get the prescription.

Encouraging the family to discuss the problem with the physician before trying over-the-counter products is important.

SOURCES OF NUTRITION SERVICES

Infants and toddlers who have chronic conditions are served by a variety of resources. Registered dietitians who have training in pediatrics are qualified to provide services to toddlers and preschool children with chronic conditions.[17] Programs in which nutrition care may be accessed include the following:[38]

- State programs for children with special health care needs

Photo Disc

Case Study 11.2
Early Intervention Services for a Boy at Risk for Nutrition Support

Robert is 2.3 years in an early intervention program. He was eligible based on his preterm birth at 30 weeks gestation. His premature birth was related to exposure to an intrauterine infection. All in the family agreed that he was small, but their main concern was that he was difficult to feed. He cried and refused to eat when offered meals, even with his favorite foods. The registered dietitian who consulted at the early intervention program met with the family, assessed Robert, and reviewed his medical records. Nutrition services were first planned to boost calories to stimulate weight gain. Observing Robert being fed by his mother was part of the nutrition services. Other therapists at the early intervention center were involved in making sure that Robert was positioned well to eat, so he was sitting up without extra effort. The nutritionist and occupational therapist were concerned that Robert was so easily choking, and talked to the family about contacting his pediatrician. They faxed to the pediatrician's office their recommendation for tests to study Robert's swallowing.

Robert did not attend the early intervention program for the next three weeks. The test demonstrated he was aspirating some of his liquids into his lungs, so oral feeding was unsafe. He required a gastrostomy for feeding and had been hospitalized for surgery. His parents learned how to feed him through the gastrostomy.

When he returned to the early intervention program, his pediatrician asked the early intervention staff to monitor his weight and to reinforce the discharge feeding instruction with the family. Nutrition services provided in the early intervention program changed from working on his oral feeding to monitoring and documenting his growth as adjustments were made in his gastrostomy feeding schedule. Over the next six months, Robert gained weight. He started to be more interactive with the staff at the early intervention center and made some developmental progress in his walking and speaking. He was still a small child, but his improved nutritional status was confirmed by his adequate body fat measurements. His ability to return to eating by mouth was to be reassessed later in the year.

- Early intervention programs (age 0 up to 36 months)
- Early childhood education programs (IDEA, age 3–5 years)
- Head Start; regular program or special needs category (age 3–5 years)
- Early Head Start; regular program or special needs category (0 up to 36 months)
- WIC
- Low birthweight follow-along programs
- Child care feeding programs (USDA)

These programs are described in Chapters 8 and 10. Efforts to increase program accessibility come from state and federal governmental offices, toll-free outreach services, and Web sites. Specific outreach programs to locate toddlers and preschoolers at risk are funded in each state, under such names as "Child Find." Because every child at risk is eligible for a screening, contacting a neighborhood public school is a good starting place to locate services, even if the child is not old enough to attend the school.

Case Study 11.3
Weight Plateau

A 2.5-year-old girl, Gayle, was seen, along with her 22-year-old mother, the child's aunt, and a young cousin. She was developing normally but had not gained weight for more than four months. The pediatrician did not find a medical basis for the weight plateau. She was offered appropriate foods in appropriate serving sizes, but she mostly was throwing it on the floor or refusing it with crying tantrums. She liked to drink juices and tea from her sippy cup throughout the day. She carried the cup with her most of the time. The family was sure it was not enough for her, and they did not limit her intake because she ate little else.

Eventually the aunt mentioned how the girl used to eat well in her grandparents' house. With sensitive review of the child's social history, it became clear that the child had been living with her grandparents while the mother was incarcerated. Her mother had cared for her until eight months of age. Gayle and her mother had been living in a separate home from the grandparents until recently. Gayle had not seen her grandparents for the last several months.

Assessment and recommendation: After confirming that no nutritional or medical basis was involved, family disruption was considered the basis for the weight plateau, as a result of interfering with Gayle's hunger and appetite. The change in Gayle's eating coincided with the time of a sudden change in her life. From the child's viewpoint, her grandparents and their home were suddenly replaced by a new person and a new setting.

Gayle's mother was referred to the local early intervention program, which had parenting classes to assist her in meeting the emotional and social needs of her daughter. She was encouraged to view Gayle's eating refusals as a sign of her adjustment to her new home. Gayle's food refusals and mealtime behaviors lessened as she felt more secure with her mother in their new home. Gayle's mother met with the registered dietitian about how to reinforce Gayle's eating and how to balance the amount of juices and other fluids she drank with solid foods to gain weight.

Resources

Federal Interagency Coordinating Council Site for Families with Children with Disabilities

Identifies by state and city resources for finding local intervention programs.

Web site: www.fed-icc.org

Food Allergy

A credible source of recommendations for preventing food allergy reactions, its newsletter provides recipes to avoid foods that cause reactions.

Web site: www.foodallergy.org

The National Information Center for Children and Youth with Disabilities:

A useful site for parents and providers who are looking for intervention services for children with special needs. It is targeted mainly toward educational programs.

Web site: www.NICHCY.org

National Organization for Rare Diseases (NORD)

A credible source for parents and providers of information and resources about rare "orphan" diseases.

Web site: www.rarediseases.org

Quackwatch

Includes information on dietary supplements and products that are claimed to benefit health and nutrition.

Web site: www.quackwatch.com

References

1. Westbrook LE, Silver EJ, Stein, REK. Implications for estimates of disability in children: a comparison of definitional components. Pediatrics 1998;101:1025–30.

2. Newacheck PW, Taylor WR. Childhood chronic illness, prevalence, severity and impact. Amer J of Public Health 1992;82:364–71.

3. Blackman JA. Medical aspects of developmental disabilities in children birth to three. Gaithersburg, MD: Aspen Publications;1997:pp 113.

4. National Information Center for Children and Youth with Disabilities. General information about disabilities. Available from http://www.nichcy.org/general.htm, February 1, 2001.

5. Story M, Holt K, Sofka D, eds. Bright futures in practice: nutrition. Arlington, VA: National Center for Education in Maternal and Child Health;2000:pp 266–70.

6. Isaacs JS, Horsley J, Cialone J, et al. Children with special health care needs: a community nutrition pocket guide. American Dietetic Association, Dietitians in Developmental and Psychiatric Disorders and Pediatric Nutrition Practice Group 1997;1–2;49–54.

7. Trumbo P, Yates AA, Schlicker SA, Poos M. Dietary reference intakes: vitamin A, vitamin K, arsenic, boron, chromium, copper, iodine, iron, manganese, molybdenum, nickel, silicon, vanadium and zinc. J Amer Diet Assoc 2001;101:294–301.

8. Hampl JS, Betts NM, Benes BA. The "age+5" rule: comparisons of dietary fiber intake among 1- to 4-year old children. J Amer Diet Assoc 1998;98:1418–23.

9. National Center for Health Statistics. NCHS growth curves for children 0–19 years. U.S. Vital and Health Statistics, Health Resources Administration, U.S. Government Printing Office; 2000.

10. The Infant Health and Development Program: enhancing the outcomes of low-birthweight, premature infants. JAMA 1990;263(22):3035–42.

11. Anneren G, Gustafsson J, Sara VR, Tuvemo T. Normalized growth velocity in children with Down's syndrome during growth hormone therapy. J Intel Disabil Res 1993;371:381–7.

12. Wollmann HA, Schultz U, Grauer ML, Ranke MB. Reference values for height and weight in Prader-Willi syndrome based on 315 patients. Eur J Pediatr 1998;157:634–42.

13. Nellhaus, G. Composite international and interracial graphs. Pediatrics 1968;41:106–14.

14. Hadberg B, Aicardi J, Dias K, Ramos O. A progressive syndrome of autism, dementia, ataxia, and loss of purposeful hand use in girls: Rett syndrome. Ann Neurol 1983;471–9.

15. Moser HW, Naidu S. The discovery and study of Rett syndrome. In: Capute AJ and Accardo PJ, eds. Developmental disabilities in infancy and childhood, 2nd ed. Baltimore, MD: Paul H. Brookes; 1996:2:pp 379–86.

16. American Academy of Pediatrics clinical practice guideline: diagnosis and evaluation of the child with of attention deficit/hyperactivity disorder. Pediatrics 2000;105:1158–70.

17. Herman DR, Baer MT. Demonstrating cost-effectiveness of nutrition services for children with special health care needs: a national network. HHS, Health Resources and Services Administration; 1999.

18. Prada JA, Tsang RC. Biological mechanisms of environmentally induced causes of IUGR. Eur J Clin Nutr 1998; 52 (Suppl 1):C21–7.

19. Orelove FP, Hollahan DJ, Myles KT. Maltreatment of children with disabilities: training needs for a collaborative response. Child Abuse & Neglect 2000;24:185–94.

20. Maggioni A, Litchitz F. Nutritional management of failure to thrive. Pediatr Clinic N Amer 1995;42:791–810.

21. Toddler's diarrhea. Available from www.icondata.com/health/pedbase, December 20, 2000.

22. Celiac disease pediatric database. Available from www.icondata.com/health/pedbase, December 20, 2000.

23. Minns, RA. Neurological disorders. In: Kelnar, CJH, Savage MO, et al., eds. Growth disorders. Chapman and Hall; 1998: pp 447–70.

24. Krick J, Murphy-Miller P, Zeger S, Wright, E. Pattern of growth in children with cerebral palsy. J Amer Diet Assoc 1996;96:680–5.

25. Surveillance for asthma—United States, 1960–1995. MMWR 1998;47:SS-1.

26. Christie L. Food hypersensitivities. In: Samour, PQ, Kelm KK, and Lang CE. Handbook of pediatric nutrition, 2nd ed. Gaithersburg; MD: Aspen Publishers; 1999:pp 149–72.

27. Wooldridge, N. Pulmomary diseases. In: Samour; PQ, Kelm KK, and Lang CE. Handbook of pediatric nutrition, 2nd ed. Gaithersburg, MD: Aspen Publishers: 1999: pp 315–54.

28. Jones, KL. Smith's recognizable patterns of human malformation, 5th ed. Philadelphia: WB Sanders Co; 1997.

29. Washington State Department of Health. Genetics and your practice, 2000:34–40. Available from http://mchneighborhood.ichp.edu/wagenetics, January 5, 2001.

30. Soto Syndrome. Available from www.icondata.com/health/pedbase, December 20, 2000.

31. U.S. Food and Drug Administration. Food allergies rare but risky. Available from www.cfsan.fda.gov/dms/wh.a1rgl.html, December 10, 2000.

32. Wood R. Common myths about anaphylaxis. Food Allergy News 2000; April–May:1–5.

33. Bishop JM, Hill DJ Hosking CS. Natural history of cow milk allergy: clinical outcome. J Pediatrics 1990;116:862–7.

34. Position statement on vitamin-related therapies. National Down Syndrome Society; 1997.

35. Trissler RJ. Folic acid and Down syndrome. J Amer Diet Assoc 2000;159.

36. Statement of nutritional supplements and Piracetam for children with Down Syndrome. American College of Medical Genetics; 1996.

37. McGuire J, Kulkarni M, Baden H. Fatal hypermagnesemia in a child treated with megavitamin/megamineral therapy. J Pediatr 2000;105:318.

38. Maternal and Child Health Bureau, Health and Human Services. Available from http://mchb.hrsa.gov/html/drte.html, January 12, 2001.

CHAPTER 12

Photo Disc

Men are but children of a larger growth.

John Dryden

CHILD AND PREADOLESCENT NUTRITION

Prepared by **Nancy H. Wooldridge**

CHAPTER OUTLINE

- Introduction
- Tracking Child and Preadolescent Health
- Normal Growth and Development
- Physiological and Cognitive Development of School-Age Children
- Energy and Nutrient Needs of School-Age Children
- Common Nutrition Problems
- Prevention of Nutrition-Related Disorders in School-Age Children
- Dietary Recommendations
- Physical Activity Recommendations
- Nutrition Intervention for Risk Reduction
- Public Food and Nutrition Programs

KEY NUTRITION CONCEPTS

1 Children continue to grow and develop physically, cognitively, and emotionally during the middle childhood and preadolescent years in preparation for the physical and emotional changes of adolescence.

2 Children continue to develop eating and physical activity behaviors that affect their current and possible future states of health.

3 Although children's families continue to exert the most influence over their eating and physical activity habits, external influences, such as teachers, coaches, peers, and the media, begin to have more impact on children's health habits.

4 With increasing independence, children begin to eat more meals and snacks away from home and need to be equipped to make good food choices.

INTRODUCTION

This chapter focuses on the growth and development of school-age and preadolescent children and their relationships to nutritional status. Children continue to grow physically at a steady rate during this period, but development from a cognitive, emotional, and social standpoint is tremendous. This period in a child's life is preparation for the physical and emotional demands of the adolescent growth spurt. Having family members, teachers, and others in their lives who model healthy eating and physical activity behaviors will better equip children for making good choices during adolescence and later in life.

Definitions of the Life Cycle Stage

Middle childhood is a term that generally describes children between the ages of 5 and 10 years. This stage of growth and development is also referred to as school-age and the two terms are used interchangeably in this chapter. *Preadolescence* is generally defined as ages 9 to 11 years for girls and ages 10 to 12 years for boys. School-age is also used to describe preadolescence.

Importance of Nutrition

Adequate nutrition continues to play an important role during the school-age years in assuring that children reach their full potential for growth, development, and health. Nutrition problems can still occur during this age, such as iron-deficiency anemia, undernutrition, and dental caries. In terms of weight, both ends of the spectrum are seen during this age. The prevalence of obesity is increasing, but the beginnings of eating disorders can also be detected in some school-age and preadolescent children. Therefore, adequate nutrition and the establishment of healthy eating behaviors can help to prevent immediate health problems as well as promote a healthy lifestyle, which may reduce the risk of the child developing a chronic condition, such as obesity or cardiovascular disease, later in life.[1] Adequate nutrition, especially eating breakfast, has been associated with improved academic performance in school and reduced tardiness and absences.[2]

MIDDLE CHILDHOOD Children between the ages of 5 and 10 years; also referred to as school-age.

PREADOLESCENCE The stage of development immediately preceding adolescence; 9 to 11 years of age for girls and 10 to 12 years of age for boys.

WORKING-POOR FAMILIES Families where at least one parent worked 50 or more weeks a year and the family income was below the poverty level.

HIGH-POVERTY NEIGHBORHOODS Neighborhoods where 40% or more of the people are living in poverty.

TRACKING CHILD AND PREADOLESCENT HEALTH

Not all families benefited from the economic boom of the 1990s. The statistics regarding the environments in which many children are growing up are alarming. Approximately 22% of U.S. children live in poverty.[3] Seventeen percent of children in inner cities, who live in high poverty neighborhoods, did not have health insurance in 1998. Additional statistics include:

- 19.2 million children had no parent in the household who had a full-time, year-round job in 1998.
- 5.8 million children lived in *working-poor families.*
- 27% of families with children were headed by a single parent in 1997.
- There were 2.1 million father-only families in 1999.
- There were 8.9 million mother-only families in 1999.
- 8% of children who lived in *high-poverty neighborhoods* in inner cities did not live with either of their parents.[3]

The environment in which a child lives affects the child's health and education. In 1998, 39% of fourth grade students scored below basic reading level and 38% scored below the basic math level in 1996.[3] Lack of transportation is a significant limitation for many families. In 1997, 13% of all children, younger than 18 years, lived in a family that did not own a car or other vehicle, which was true for 50% of children living in low-income urban areas.[3] In the discussions that follow regarding nutrition during childhood, the recommendations must always be considered in the context of the individual child's environment.

Disparities in nutrition status indicators exist among the races. For example:

- African American and Mexican American girls have higher body mass index-for-age than do Caucasian girls.
- Overweight and obesity are more prevalent among African Americans and Hispanics of both sexes than whites.
- Minorities have higher percentages of total calories from dietary fat.[4]

Healthy People 2010

A number of objectives in the Healthy People 2010 document are specific to children's health and well-being. Table 12.1 lists the specific Healthy People 2010 objectives that are pertinent to a discussion of middle childhood and preadolescence.

Table 12.1 Healthy People 2010 objectives related to school-age children.[4]

Objective 19-3	Reduce the proportion of children and adolescents who are overweight or obese from 11% to 5%.
Objective 19-5	Increase the proportion of persons aged two years and older who consume at least two daily servings of fruit from 28% to 75%.
Objective 19-6	Increase the proportion of persons aged two years and older who consume at least three daily servings of vegetables, with at least one-third being dark green or deep yellow vegetables from 3% to 50%.
Objective 19-7	Increase the proportion of persons aged two years and older who consume at least six daily servings of grain products, with at least three being whole grains from 7% to 50%.
Objective 19-8	Increase the proportion of persons aged two years and older who consume less than 10% of calories from saturated fat from 36% to 75%.
Objective 19-9	Increase the proportion of persons aged two years and older who consume no more than 30% of calories from fat from 33% to 75%.
Objective 19-10	Increase the proportion of persons aged two years and older who consume 2400 mg or less of sodium daily from 21% to 65%.
Objective 19-11	Increase the proportion of persons aged two years and older who meet dietary recommendations for calcium from 46% to 75%.
Objective 19-15	Increase the proportion of children and adolescents aged 6 to 19 years whose intake of meals and snacks at schools contributes proportionally to good overall dietary quality (developmental objective).
Objective 22-8	Increase the proportion of the nation's public and private schools that require daily physical education for all students from 17% to 25%.
Objective 22-11	Increase the proportion of children and adolescents who view television two or fewer hours per day from 60% to 75%.
Objective 22-14	Increase the proportion of trips made by walking from 28% to 50%.
Objective 22-15	Increase the proportion of trips made by bicycling from 2.2% to 5.0%.

NORMAL GROWTH AND DEVELOPMENT

During the school-age years, the child's growth is steady, but the growth velocity is not as great as it was during infancy or as great as it will be during adolescence. The average annual growth during the school years is 3–3.5 kilograms (7 pounds) in weight and 6 centimeters (2.5 inches) in height.[5] Children of this age continue to have spurts of growth that usually coincide with periods of increased appetite and intake. During periods of slower growth, the child's appetite and intake will decrease. Parents should not be overly concerned with this variability in appetite and intake in their school-age child.

Periodic monitoring of growth continues to be important in order to identify any deviations in the child's growth pattern. Children should continue to be weighed on calibrated scales without shoes and in lightweight clothing. The child's stature or standing height should be measured without shoes and utilizing a height board (see Illustration 10.3). A height board consists of a nonstretchable tape on a flat surface like a wall with a moveable right-angle headboard. The child's heels should be up against the wall or flat surface and the child should be instructed to stand tall, looking straight ahead with arms by the side during the measurement. Both weight and height should be plotted on the appropriate 2000 CDC growth charts, discussed next.

The 2000 CDC Growth Charts

The "CDC Growth Charts: United States," found in Appendix A, are excellent tools for monitoring the growth of a child.[6] The growth charts, which are pertinent to the school-age child, are weight-for-age, stature-for-age, and body mass index (BMI)-for-age for boys and girls. The BMI-for-age growth charts were incorporated as part of the recent growth chart revisions and are based on data from cycles 2 and 3 of the National Health and Examination Survey (NHES) and the National Health and Nutrition Examination Surveys (NHANES) I, II, and III. However, BMI data for children greater than six years of age who participated in NHANES III, were not included as there was a known higher prevalence of overweight for these ages. Incorporating this data into the BMI-for-age growth charts would reflect an unhealthy standard.[7] Gender specific BMI-for-age greater than the 95th percentile define overweight and BMI-for-age values between the 85th and 95th percentiles identify children at risk for becoming overweight.

Illustrations 12.1 and 12.2 depict the growth of a healthy child. A chart for weight-for-stature up to a height of 122 cm or 48 in., is also available for the younger school-age child. As with the toddler and preschooler, it is the child's pattern of growth over time that is important rather than any one single measurement. The tracking of BMI-for-age is an important screening tool for overweight as well as

Illustration 12.1   2 to 20 years: Girls stature-for-age and weight-for-age percentiles.[6]

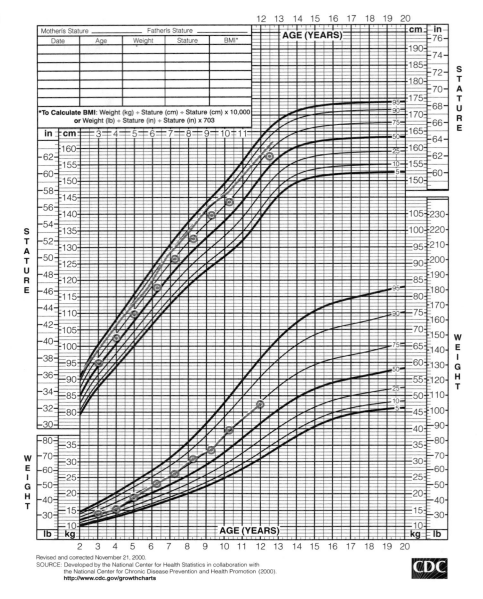

Revised and corrected November 21, 2000.
SOURCE: Developed by the National Center for Health Statistics in collaboration with the National Center for Chronic Disease Prevention and Health Promotion (2000).
http://www.cdc.gov/growthcharts

undernutrition. Making sure the correct age of the child is used when plotting on the growth charts helps to avoid errors.

PHYSIOLOGICAL AND COGNITIVE DEVELOPMENT OF SCHOOL-AGE CHILDREN

Physiological Development

During middle childhood, muscular strength, motor coordination, and stamina increase progressively.[5]

The child is able to perform more complex pattern movements, therefore affording the child opportunities to participate in activities such as dance, sports, gymnastics, and other physical activities.

Body composition and body shape remain relatively constant during school-age. During the early childhood years, percent body fat reaches a minimum of 16% in females and 13% in males. Percent body fat then increases in preparation for the adolescent growth spurt. This increase in percent body fat, which usually occurs between six to seven years of age, is called adiposity rebound and is reflected in the BMI-for-age growth charts.[8] The increase in percent body fat with puberty is earlier and greater in females

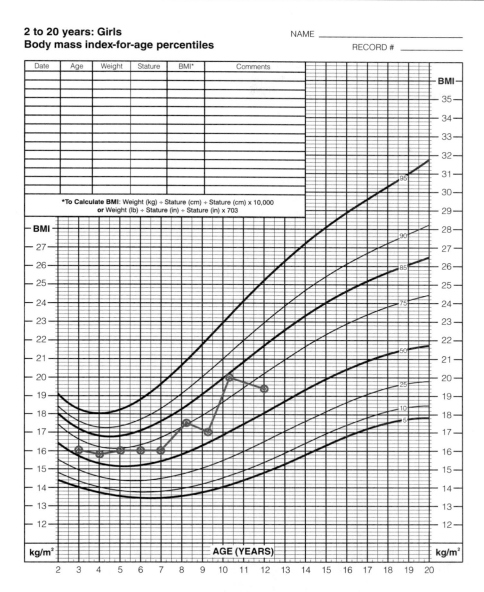

2 to 20 years: Girls
Body mass index-for-age percentiles

NAME _____

RECORD # _____

Illustration 12.2 **2 to 20 years: Girls body mass index-for-age percentiles.**[6]

Date	Age	Weight	Stature	BMI*	Comments

*To Calculate BMI: Weight (kg) ÷ Stature (cm) ÷ Stature (cm) x 10,000
or Weight (lb) ÷ Stature (in) ÷ Stature (in) x 703

AGE (YEARS)

SOURCE: Developed by the National Center for Health Statistics in collaboration with the National Center for Chronic Disease Prevention and Health Promotion (2000).
http://www.cdc.gov/growthcharts

than in males (19% for females versus 14% for males). During middle childhood, boys have more lean body mass per centimeter of height than girls. These differences in body composition become more pronounced during adolescence.[1]

With the increase in body fat, preadolescents, especially girls, may be concerned that they are becoming overweight. Parents need to be aware that an increase in body fat during this stage is part of normal growth and development. Parents need to be able to reassure their child that these changes are most likely not permanent, as well as not reinforce a preoccupation with weight and size. Boys may become concerned about developing muscle mass and need to understand

that they will not be able to increase their muscle mass until middle adolescence (see Chapter 14).[1]

Cognitive Development

The major developmental achievement during middle childhood is self-efficacy, the knowledge of what to do and the ability to do it. During the school-age years, children move from a preoperational period of development to one of "concrete operations."[5] This stage is characterized by being able to focus on several aspects of a situation at the same time; being able to have more rational cause/effect reasoning; being able to classify, reclassify, and generalize; and a

Connections
Growth Spurts

Growth during childhood occurs in irregular spurts that last an average of eight weeks and occur three to six times per year.[5] Appetite increases prior to a growth spurt and decreases after a growth spurt.

decrease in egocentrism, which allows the child the ability to see another's point of view. Schoolwork becomes increasingly complex as the child grows. School-age children also enjoy playing strategy games, which displays growing cognitive and language development.

During this stage, the child is developing a sense of self. Children become increasingly independent and are learning their roles in the family, at school, and in the community.[1] Peer relationships become increasingly important and children begin to separate from their own families by spending the night at a friend's or relative's house. More and more time is spent watching television and playing video games. Older children may be able to walk to a neighborhood store and purchase snack items. Thus, influences outside the home environment play an increasing role in all aspects of the child's life.

Development of Feeding Skills

With increased motor coordination, school-age children develop increased feeding skills. During childhood, the child masters the use of eating utensils, can be involved in simple food preparation, and can be assigned chores related to mealtime such as setting the table. By performing these tasks, the child learns to contribute to the family, which boosts developing self-esteem. The complexities of the tasks can be increased as the child grows older. At the same time, the child can be learning about different foods and some basic nutrition facts.

EATING BEHAVIORS   Parents and older siblings continue to have the most influence on a child's attitudes toward food and food choices during middle childhood and preadolescence. The eating behaviors and food preferences of parents will impact the child's food likes and dislikes. The feeding relationship between parent and child, as described in Chapter 10, still applies to the school-age child. Parents are responsible for the food environment in the home, what foods are available, and when they are served. The child is responsible for how much is eaten.[9] Parents need to continue to be positive role models for their children in terms of healthy eating behaviors. They also need to provide the necessary guidance so that their child will be able to make healthy food choices when they are away from home.

Families should try to eat together. When children are involved in school-related activities, eating together is often difficult for families to achieve because of the family members' hectic schedules. But eating together as a family should be encouraged as a goal, allowing time for conversation (see Illustration 12.3). Mealtime is not the time for excessive reprimanding and arguments.

Illustration 12.3   A family enjoying mealtime together.

One study of 9- to 14-year-old children of participants of the Nurses' Health Study II found a positive relationship between families eating dinner together and the overall quality of the children's diets.[10] Children who ate dinner with their families had higher energy intakes as well as higher intakes of nutrients such as fiber, calcium, folate, iron, and vitamins B_6, B_{12}, C, and E. These children also reported eating more fruits and vegetables, less fried foods when away from home, and drinking fewer soft drinks. The percentage of children reporting eating family dinner decreased with the age of the child. So, a higher percentage of 9-year-olds ate dinner with the family than 14-year-olds, indicating that family dinner becomes more of a challenge, as children get older.

School-age children spend more and more time away from home, which is an important part of normal growth and development. Peer influence becomes greater as the child's world expands beyond the family. The increased peer influence extends to attitudes toward foods and food choices. Children may suddenly request a new food or refuse a previous favorite food, based on recommendations from a peer.

Teachers and coaches have an increasing influence on the child's attitudes toward food and eating behaviors. Nutrition should be part of the health curriculum, and what is learned in the classroom should be reinforced by foods available in the school cafeteria. Vending machines present in school as a source of extra funding can also reinforce good nutrition with appropriate choices.

In their expanding world, children come under the influence of the media. Children want to try foods they see advertised on television. One study that analyzed the commercials aired during Saturday morning television programming found that 56.5% of all advertisements were for food.[11] Of these, 43.6% were classified in the fats, oils, and sweets food groups, which is just the opposite of recommendations of the U.S. Department of Agriculture's Food Guide Pyramid (see Chapter 1).[11,12] Fast-food establishments, with their playgrounds and give-aways, are also attractive to children.

Snacks continue to contribute significantly to a child's daily intake. During middle childhood, children cannot consume large amounts of food at one time and therefore need snacks to meet their nutrient needs. Many children prepare their own breakfasts or after school snacks. These children need to have a variety of foods available to them, be equipped with nutrition education for making their own food choices, and have some knowledge and skill in food preparation, appropriate for the child's age, assuming, of course, that the family has adequate access to food.

BODY IMAGE/EXCESSIVE DIETING Food preference development, appetite, and satiety in young children are thoroughly discussed in Chapter 10. Researchers have described the innate ability of young children to internally control their energy intake and their responsiveness to energy density. The internal controls can be altered by external factors such as child-feeding practices. Studies in 9- to 10-year-old children found that these older children were not as responsive to energy density as young preschool-age children.[13] External factors such as the time of day, the presence of other people, and the availability of good food begin to override the internal controls of hunger and satiety as children get older.

Birch and associates, who have performed extensive research in the area of the development of food preferences and appetite control in children, have also examined the relationships among children's adiposity, child-feeding practices, and children's responsiveness to energy density.[14] These researchers found that children of parents who imposed authoritarian controls on their children's eating were less likely to be responsive to energy density. In other words, these children were not able to listen to internal cues in energy regulation. In girls, regulation of energy intake was inversely related to their adiposity. Heavier girls were less likely to be able to regulate their intake based on internal cues. Parents who had difficulty controlling their own intakes seemed to impose more restrictions on their children. A study of mothers and their five-year-old daughters found that this transfer of "restrictive" eating practices may begin as early as the preschool age.[15] The more the mother is concerned with her own weight and with the risk of her daughter becoming overweight, the more likely she is to employ restrictive child-feeding practices. These researchers hypothesize that the chronic dieting and dietary restraint, which is commonly seen in adolescent girls and young women, may have their beginnings in the early regulation of energy intake and may be related to the amount of parental control exerted over the child's eating.[14]

Young girls seem to have a preoccupation with weight and size at an early age. With the normal increase in body mass index or body fatness in preadolescence, many girls and their mothers may interpret this phenomenon of normal growth and development as a sign that the child is developing a weight problem. By imposing controls and restrictions over their daughters' intakes, mothers may actually be promoting the intake of the forbidden or restricted foods.[16] Similar results were found in five-year-old girls whose parents restricted palatable snack foods.[17] Not only did parental restriction promote the consumption of these forbidden foods by the young girls, but these children reported feeling badly about eating these foods. Early "dieting" may actually be a risk factor for the development of obesity.[18] Dieting, which imposes restrictions, is similar to controlling child-feeding practices, which restrict children's intake. Both methods ignore internal cues of hunger and satiety. Not only do these types of child-feeding practices contribute to the onset of obesity and possibly a nutritionally inferior diet, but they may also be contributing to the beginnings of eating disorders. Eating disorders are discussed in more detail in Chapter 14.

ENERGY AND NUTRIENT NEEDS OF SCHOOL-AGE CHILDREN

New Dietary Reference Intakes (DRIs) are being developed through 2002. DRI tables are provided on the inside front cover of this book. Readers can check for additional DRI reports by accessing the National Academy Press Web site (see Resource section at the end of this chapter).

Energy Needs

Energy needs of school-age children reflect the slow but steady rate of growth during this stage of development.[19] The Recommended Dietary Allowance (RDA) for energy is 70 calories/kg body weight for children 7 to 10 years of age. The RDA for energy for 11- to 14-year-old children is 55 calories/kg body weight for males and 47 calories/kg body weight for females. These allowances are less energy per kilogram of body weight than required for the toddler and preschool-age child. The further decrease in the energy requirement per kilogram of body weight is a reflection of the slowing growth rate. Experts in adolescent medicine argue that it would be more appropriate to base energy requirements on sexual maturity rating rather than age during preadolescence and adolescence, since the age at which children begin their adolescent growth spurt is quite variable (see Chapter 14). Activity level also impacts an individual child's energy needs.

Protein

Based on the current RDAs, the recommended protein intake for school-age children is approximately 1 gram of protein/kg body weight for 7- to 14-year-old children, both males and females.[19] As for the young child, this recommendation can easily be met with the typical American diet as well as with vegetarian diets, provided the child's energy needs are met and that a variety of plant foods is consumed.[1] By meeting an

individual child's energy needs, protein is spared for tissue repair and growth.

Vitamins and Minerals

Dietary Reference Intakes (DRIs) and Recommended Dietary Allowances (RDAs) have been established for the school-age and preadolescent child. Analysis of data from NHANES I, II, and III and Continuing Survey of Food Intake by Individuals (CSFII) indicates that children's mean intakes of most nutrients meet or exceed the recommendations. Still, certain subsets of children do not meet their needs for key nutrients such as iron and zinc, important for growth, and calcium, needed to achieve peak bone mass.[20,21] According to NHANES III data, calcium intakes are declining in 6- to 11-year-old children.[4] The RDAs and Adequate Intakes for these key nutrients are listed in Table 12.2.

COMMON NUTRITION PROBLEMS

Iron Deficiency

Iron deficiency is not as common a problem in middle childhood as it is in the toddler age group. According to data collected during the NHANES III, 1988–1994, no more than 7% of older children were found to be iron deficient, compared to 9% of toddlers.[24]

Age- and gender-specific cutoff values for anemia are based on the 5th percentile of hemoglobin and hematocrit for age from NHANES III. For children five to under eight years of age, the diagnosis of anemia is made if the hemoglobin concentration is <11.5 g/dL and hematocrit <34.5%. For children 8 to under 12 years of age, a hemoglobin value <11.9 g/dL or hematocrit <35.4% is diagnostic of iron deficiency anemia.[25]

Dental Caries

Approximately one in two children age six to eight years have decay in their primary or permanent

Table 12.2 Dietary Reference Intakes for key nutrients for school-age children.[22,23]

AGE	RECOMMENDED DAILY ALLOWANCES		ADEQUATE INTAKE
	Iron (mg/d)	Zinc (mg/d)	Calcium (mg/d)
4–8 years	10	5	800
9–13 years	8	8	1300

teeth.[4] The amount of time that children's teeth are exposed to carbohydrates influences the risk of dental caries or tooth decay. (See explanation of carogenic process in Chapter 10.) Complex carbohydrates, such as fruits, vegetables, and grains are better choices than simple sugars, such as soft drinks and candy, in terms of oral health and nutrition. Sticky carbohydrate-containing foods, such as raisins and gummy candy, are strong caries promoters. Fats and proteins may have a protective effect on enamel. So choosing snacks that are combinations of carbohydrates, proteins, and fats may decrease the risk of developing dental caries. Having regular meal and snack times versus continual snacking throughout the day is also beneficial. Rinsing the mouth after eating or better yet, brushing the teeth regularly, will also decrease the development of cavities.[26] It is important that the school-age child continue to have a source of fluoride, either from the water supply or through supplementation. Details about fluoride supplementation are reported in Chapter 10.

During middle childhood, children lose their primary or baby teeth and begin to get their permanent teeth. If several teeth are missing, they may experience difficulty in chewing some foods such as meat. Also, orthodontic appliances, which are commonly worn by school-age children, may interfere with the child's ability to eat certain foods. Modifying food, such as chopping meat or slicing fresh fruit, can help.[1]

PREVENTION OF NUTRITION-RELATED DISORDERS IN SCHOOL-AGE CHILDREN

The prevalence of overweight among children is increasing at an alarming rate. Increases in prevalence of overweight are present in the adult population and in populations of other countries, indicating that social and environmental factors may be having an impact. Despite the increase in the prevalence of overweight, analysis of dietary data from NHANES I, II, and III indicate no corresponding increase in energy intake among children over the years. This finding suggests that physical inactivity may be a significant contributing factor to the increased prevalence of overweight.[27] The problem of increasing overweight in the United States needs to be addressed from a public health perspective.[7] Furthermore, children who are overweight are at increased risk for developing risk factors for chronic conditions, such as cardiovascular disease and type 2 diabetes mellitus.[28]

Overweight and Obesity in School-Age Children

PREVALENCE Approximately 11% of children ages 6 to 11 years are overweight with BMIs-for-age above the 95th percentile and an additional 14% considered to be at risk for overweight with BMIs between the 85th and 95th percentiles, according to the NHANES III data.[4,7] The proportion of children ages 6 to 11 years, who are at risk for becoming overweight, range from about 10% for non-Hispanic black males to approximately 17% for Mexican American females. When the NHANES III data are controlled for socioeconomic status and age, BMI levels are significantly higher for black and Mexican American girls than for white girls.[29] These ethnic differences among girls first become apparent at six to nine years of age. Black girls and boys have greater increases in BMI across age groups than whites, resulting in greater differences in BMI in the older age groups. The prevalence of overweight and obesity among participants of the Child and Adolescent Trial for Cardiovascular Health (CATCH) is comparable to the NHANES III population, with a higher prevalence among African Americans and Hispanics than whites for both genders.[30]

The prevalence of overweight among children has increased over time with the greatest increases occurring between NHANES II (1976–1980) and NHANES III (1988–1994). In fact, an intrasurvey increase of about two to six percentage points occurred for most of the gender, age, and racial-ethnic groups during the six years of the NHANES III survey.[7] As mentioned in the description of the growth charts, BMI data for children older than six years of age were not included in the revised growth charts because of the known increased prevalence of overweight for these ages in NHANES III. Inclusion of this data in the revised growth charts would reflect a heavier population and would not be a healthy standard. Further analysis shows that the heaviest, or obese, children are getting heavier.

CHARACTERISTICS OF OVERWEIGHT CHILDREN
Overweight children are usually taller, have advanced *bone ages,* and experience sexual maturity at an earlier age than their nonoverweight peers. From a psychosocial standpoint, overweight

BONE AGE Bone maturation; correlates well with stage of pubertal development.

children look older than they are, and often adults expect them to behave as if they were older. Health consequences of obesity, such as hyperlipidemia, higher concentrations of liver enzymes, hypertension, and abnormal glucose tolerance occur with increased frequency in obese children than in children of normal weight.[31] Analysis of data from the Bogalusa Heart Study, a community-based study of adverse risk factors in early life in a biracial population, confirms an increase in chronic disease risk factors with increasing BMI-for-age.[28] Increasing insulin levels show the strongest association with increasing BMI-for-age. Additionally, overweight children are more likely to have more than one chronic disease risk factor.

Type 2 diabetes mellitus is increasing in children and adolescents in the United States today with up to 85% of affected children being either overweight or obese at diagnosis.[32] According to the recommendations of a panel of experts in diabetes in children, any child who is overweight, which is defined by this group as having a BMI above the 85th percentile, and who has other risk factors, should be monitored for type 2 diabetes beginning at age 10 or at puberty. Other risk factors include a family history of type 2 diabetes, belonging to certain race and ethnic groups, including African American, Hispanic American, Asians and South Pacific Islanders, and Native Americans, and having signs of insulin resistance.[32]

It is still unclear what effect an early onset of obesity in childhood has on the risk of adult morbidity and mortality.[31] But consequences of obesity and the precursors of adult disease do occur in obese children. More studies have been performed on the relationship between obesity during adolescence and the risks of obesity in adulthood than have been performed on the relationship between childhood obesity and obesity in adulthood (see Chapter 14).[33]

PREDICTORS OF CHILDHOOD OBESITY   Dietz[8] describes critical periods in childhood for the development of obesity: gestation and early infancy, the period of adiposity rebound, and adolescence. "Adiposity rebound" (or rebound in BMI) is the normal increase in body mass index, which occurs after BMI declines and reaches its lowest point, about four to six years of age, and is reflected in the BMI-for-age growth chart. Studies suggest that the age at which adiposity rebound occurs may have a significant effect on the amount of body fat that the child will have during adolescence and into adulthood. Adolescents and adults, who as children had an early adiposity rebound, defined as beginning before 5.5 years of age, have higher BMI and *subscapular skinfold thicknesses,* than those subjects who had adiposity rebound at the average age of 6.0–6.5

SUBSCAPULAR SKINFOLD THICKNESS   A skinfold measurement that can be used with other skinfold measurements to estimate percent body fat; the measurement is taken with skinfold calipers just below the inner angle of the scapular or shoulder blade.

years or those who experienced adiposity rebound late, after seven years of age. Three possible mechanisms explain the relationship between adiposity rebound and subsequent obesity.[34] The period of adiposity rebound may be when children are beginning to express learned behaviors related to food intake and activity. Early adiposity rebound may be related to infants who were exposed to gestational diabetes and consequently have large birthweights. Although more study is needed, the conclusion is that preventive efforts need to focus on these developmental stages.[8]

Another predictor of childhood obesity is the child's home environment. Children from birth to eight years were followed over a six-year period as part of the National Longitudinal Survey of Youth.[35] The associations between the home environment and socioeconomic factors and the development of childhood obesity were examined. Maternal obesity was found to be the most significant predictor of childhood obesity followed by low family income and lower cognitive stimulation.

Parental obesity is associated with an increased risk of obesity in children.[36] In one study, parental obesity doubled the risk of adult obesity for both obese and nonobese children less than 10 years of age. An analysis of data from NHANES III indicated a higher percentage of overweight youth who had one obese parent as compared to those children who had no obese parent. The percentage of overweight youth increased further if both parents were obese.[37] The connection between parental obesity and obesity in children is likely due to genetic as well as environmental factors.[36]

EFFECTS OF TELEVISION VIEWING   Analysis of data collected during cycles II and III of the National Health Examination Survey (NHES) revealed significant associations between the time spent watching television and the prevalence of obesity in children aged 6 to 11 years and children aged 12 to 17 years.[38] Also, a dose-response effect was detected. For each additional hour of television viewed in the 12- to 17-year-old group, the prevalence of obesity increased by 2%.

A strong dose-response relationship was found between the prevalence of overweight and hours of television viewed in data analysis from the National Longitudinal Survey of Labor Market Experience, Youth Cohort (NLSY), a nationally representative sample of youths aged 10 to 15 years of age.[39] The odds of having a body mass index above the 85th percentile for age and gender are significantly greater for those youths who view more than five hours of television per day as compared to those who watch two or fewer hours of television daily. These odds remain the same when adjustments are made for confounding variables such as previous overweight of the child, maternal overweight, socioeconomic status, household structure, ethnicity, and child aptitude test scores. Approximately 33% of the youth report watching more than five hours of television per day while only 11% watch two or fewer hours of daily television, which is a Healthy People 2010 objective.[4,39]

According to NHANES III data, children aged 11 through 13 years have the highest rates of daily television viewing.[40] Children, both males and females, who watch four or more hours of television daily, have greater body fat and body mass index than those who watch less television.[40,41]

To further investigate this association, a randomized, controlled, school-based trial aimed at reducing third- and fourth-grade children's television, videotape, and video game use was conducted to assess the intervention's effects on adiposity.[42] The intervention consisted of incorporating 18 lessons of 30 to 50 minutes into the regular curriculum, which was taught by the regular third- and fourth-grade classroom teachers. These lessons were followed by a "television turnoff" during which children were encouraged to watch no television or videotapes and play no video games for 10 days. Then they were encouraged to follow a seven-hour per week television budget. After a seven-month period, children in the intervention group were found to have significant decreases in body mass index, triceps skinfold, waist circumference, and *waist-to-hip ratio* as compared to controls. Also, the children in the intervention group reported significant decreases in television viewing and meals eaten in front of the television as compared to controls. No significant differences were found between the two groups in regards to changes in high-fat food intake, moderate to vigorous physical activity, and cardiorespiratory fitness.[42] So the effect seemed to be the result of decreased television viewing.

The proposed mechanisms, by which television viewing contributes to obesity, include reduced energy expenditure by displacing physical activity and increased dietary intake by eating during viewing or as a result of food advertising.[42] Analysis of NHANES III data showed a positive correlation between intake and number of hours of television watched.[41] One study found that energy expenditure during television viewing was actually significantly lower than *resting energy expenditure* in 15 obese children and 16 normal weight children who ranged in age from 8 to 12 years.[43] Based on these findings, it is hypothesized that television viewing does contribute to the prevalence of obesity and that treatment for childhood obesity should include a reduction in the number of hours spent watching television and videos and playing video and computer games.

One of the Healthy People 2010 objectives is to increase the proportion of children and adolescents who view television two or fewer hours per day from 60% to 75%. Related data analyzed by race and ethnicity, gender, and family income level are depicted in Table 12.3.

TREATMENT OF OVERWEIGHT AND OBESITY

Recommendations of an expert committee on obesity evaluation and treatment for children are described in Chapter 10.[44] The committee recommends that children and adolescents with BMIs greater than or equal to the 95th percentile for age and gender should have an in-depth medical assessment. For children greater than seven years of age, prolonged weight maintenance is an appropriate goal as long as their BMI is between the 85th and 95th percentiles and if they do not have any secondary complications of obesity. However, for children who have a BMI between the 85th and 95th percentiles and who have a nonacute secondary complication of obesity, such as mild hypertension or hyperlipidemia, or who have a BMI at or above the 95th percentile, weight loss is recommended. Reducing sedentary behaviors and increasing physical activity are important components of obesity treatment for children.[45]

Some of the potential consequences of a weight loss program in childhood are a slowing of linear

> **WAIST-TO-HIP RATIO** The ratio of the waist circumference, measured at its narrowest, and the hip circumference, measured where it is widest. This ratio is an easy way to measure body fat distribution, with a higher ratio indicative of an abdominal fat pattern. A high waist-to-hip ratio is associated with a high risk of chronic disease.
>
> **RESTING ENERGY EXPENDITURE** The amount of energy needed by the body in a state of rest.

Table 12.3 Percentage of children and adolescents viewing television two or fewer hours per day by race/ethnicity, gender, and family income level.[4]

Children and Adolescents Aged 8 to 16 Years, 1988–1994	Television Two or Fewer Hours Per Day
Race and Ethnicity	
Mexican American	53%
Black or African American	42
White	65
Gender	
Female	64
Male	54
Family Income Level	
Poor	53
Near Poor	54
Middle/high income	64

growth and the beginnings of eating disorders. The program must ensure nutritional adequacy of the diet, a nonjudgmental approach, and attention to the child's emotional state to reduce the risks associated with weight loss in childhood.[44]

Nutrition and Prevention of Cardiovascular Disease in School-Age Children

As reported in Chapter 10, the National Cholesterol Education Program[46] recommends that healthy children consume a "Step 1 Diet," which includes the following:

- <10% of total calories from saturated fatty acids.
- An average total fat intake of no more than 30% of total calories.
- <300 mg dietary cholesterol per day.

The American Heart Association and the American Academy of Pediatrics both support these modifications of diets for healthy children. The American Heart Association recommends that diets for adults contain no less than 15% of total calories from fat.[47] The American Academy of Pediatrics recommends that after two years of age, children and adolescents gradually adopt a diet that by about five years of age, contains no more than 30% and no less than 20% of calories from fat.[48] Children who are at increased risk of developing premature cardiovascular disease by family history, require screening and follow-up and may require further modifications in their diets. Results of the Bogalusa Heart Study suggest that children who are overweight should also be screened for high lipid levels.[28] A "Step 2 Diet" further restricts saturated fat to 7% of total calories and cholesterol to no more than 200 mg per day. Children on "Step 2 Diets" require close follow-up by their physicians and registered dietitians.

Concerns have been raised about the safety and efficacy of restricting fat in children's diets.[49] Two large trials have been conducted to examine the effects of reducing total dietary fat on the growth of school-age children. The Child and Adolescent Trial for Cardiovascular Health (CATCH) studied healthy children while the Dietary Intervention Study in Children (DISC) studied children with elevated low-density lipoprotein cholesterol levels or LDL cholesterol. Both studies were successful in decreasing dietary fat intake without compromising physical growth, thus attesting to the safety and efficacy of these dietary recommendations in childhood. A discussion of these trials follows.

The Child and Adolescent Trial for Cardiovascular Health (CATCH)

CATCH was a large, randomized, controlled field trial to assess the outcomes of health behavior interventions for the primary prevention of cardiovascular disease.[50] The trial took place in public elementary schools, 56 intervention and 40 control schools, in California, Louisiana, Minnesota, and Texas, and a total of 5106 initially third-grade students from ethnically diverse backgrounds were recruited for participation in this three-year program.

The CATCH intervention consisted of a school-based component that included food service, physical education (PE), and classroom curricula, and a family-based component that included home curricula and a family fun night. The goal of the food service intervention, called Eat Smart, was to provide the children with meals that were lower in total fat (to 30% of energy), saturated fat (to 10% of energy), and sodium (600 to 1000 mg per serving). The goal of the PE intervention was to increase the amount of moderate to vigorous physical activity during PE classes at school to 40% of the PE class time. The classroom curricula focused on eating behaviors and physical activity patterns. The home curricula included activity packets that complemented the classroom curricula.

During the CATCH intervention, the intervention schools achieved a significantly greater decrease in the percentage of total energy from fat in school lunches and a significant increase in intensity of the PE classes as compared to the control schools. Students at the intervention schools reported significantly lower fat intakes and significant increases in physical activity as compared to the reports of the control students. However, blood pressure, body size, including height, weight, body mass index, and skinfold measurements, and cholesterol measurements did not differ significantly between the two groups. Growth parameters were within the normal limits for this age group. The CATCH intervention was able to modify the fat content of school lunches, increase moderate-to-vigorous physical activity in PE classes, and improve the eating and physical activity behaviors in school-age children during three school years, without having an adverse impact on growth and development.

The Dietary Intervention Study in Children (DISC)

The objective of this six-center, three-year randomized control clinical trial was to study the efficacy and safety of lowering dietary intake of fat and cholesterol in children who had elevated low-density lipoprotein cholesterol (LDL Cholesterol).[51] Prepubertal boys (n = 362) and girls (n = 301), aged 8 to 10 years, who had LDL cholesterol levels ≥80th and 98th percentiles based on U.S. Public Health Service data, were recruited for the study and randomized into an intervention group and a usual care group.[51]

The intervention consisted of individual and group sessions that focused on promoting adherence to a diet providing 28% of energy from total fat, <8% from saturated fat, up to 9% from polyunsaturated fat, and <75 mg/1000 calories per day of cholesterol (not to exceed 150 mg/d). LDL cholesterol levels and height were monitored at one year and three years as safety measures. *Serum ferritin* was monitored at three years. Secondary efficacy and safety outcomes were measured, including additional lab work, sexual maturation, and psychosocial health.

The results at three years showed a significant decrease in dietary total fat, saturated fat, and cholesterol levels in the intervention group as compared to the control group. Invention subjects also showed significant decreases in LDL cholesterol levels as compared to controls, after adjusting for baseline level and gender. No significant differences in adjusted mean height or serum ferritin levels were observed between the two groups. The DISC dietary intervention was found to be safe in maintaining adequate growth, iron stores, nutritional adequacy, and psychological well-being in preadolescent children while achieving a lowering of LDL cholesterol.[51]

Vitamin and Mineral Supplements

For most healthy children who eat a variety of foods, experts generally agree that dietary supplements beyond a daily multivitamin and mineral supplement are not necessary, let alone safe or effective.[52]

Children who are healthy and consume a diet of a variety of foods do not require a vitamin and mineral supplement to meet their nutrient needs. The American Academy of Pediatrics recommends vitamin and mineral supplementation for children who are at high risk of developing, or have one or more, nutrient deficiencies.[53] See Chapter 10 for a list of children at risk for nutrient deficiency.

If vitamin and mineral supplements are given to school-age children, the supplement should not exceed the Recommended Dietary Allowances or the Dietary Reference Intakes for age. Parents should be warned against giving toxic amounts of vitamins and

minerals, especially of the fat-soluble vitamins A (retinol) and D, and iron.

It is not clear to what extent herbal supplements are given to school-age children. Herbal supplements are used in some cultures as home remedies. It is important to obtain this information from parents and caretakers as part of the child's health history. The use of herbal supplements, botanicals, and vitamin/mineral supplements may be a more prevalent practice by parents of children with special health care needs (see Chapters 11 and 13).

DIETARY RECOMMENDATIONS

The basic dietary recommendation for school-age and preadolescent children is to eat a diet of a variety of foods, which is why it remains so important through these school years for children to have a variety of foods available to them. The available food environment will affect children's food choices. Parents

> **SERUM FERRITIN** The major iron storage protein. Serum ferritin is low in iron deficiency.

and other adult role models need to continue to model appropriate eating behaviors for children.

Dietary recommendations, as outlined by the USDA in the Dietary Guidelines for Americans and the Food Guide Pyramid, apply to school-age children as well as to other segments of the population.[12,54] Professional organizations, such as the American Dietetic Association, have also published positions on dietary guidance for healthy children, which support the federal guidelines.[55]

Recommendations for Intake of Iron, Fiber, Fat, and Calcium

Adequate iron nutrition is still important during middle childhood and preadolescence to prevent iron deficiency anemia and its consequences. According to food consumption surveys, children are not eating the recommended amounts of fiber in their diets. Children are exceeding the recommendations of total calories from fat and saturated fat. Calcium requirements increase during the preadolescent years, but calcium intake decreases with age.

IRON   Although iron deficiency is not as prevalent during the school-age years as it was during the toddler and preschool-age years, adequate intake of iron is still important. The inclusion of iron-rich foods, such as meats, fortified breakfast cereals, and dry beans and peas, in children's diets is important. A good vitamin C source, such as orange juice, will enhance the absorption of iron. See Chapter 1 for a more complete list of high-iron foods.

FIBER   As reported in Chapter 10, many health effects of *dietary fiber* have been identified, including prevention of chronic disease in adulthood, such as heart disease, certain cancers, diabetes, and hypertension. The American Health Foundation's recommendation of "age + 5" grams of dietary fiber as a reasonable goal for children ages 3 to 20 years has been endorsed by the Conference on Dietary Fiber in Childhood.[56] For example, a 10-year-old child's recommended fiber intake would be 15 grams per day using this guideline.

Only 32% of children ages 7 to 10 years meet the age + 5 recommendations for dietary fiber intake.[57] As with the four- to six-year-old group, those who meet the recommendations for dietary fiber intake consume more high- and low-fiber breads and cereals, fruits, vegetables, legumes, nuts, and seeds than those who don't.

To increase the dietary fiber in children's diets, parents and caretakers can begin by increasing the amount of fresh fruits and vegetables and whole grain breads and cereals being offered. Following the recommendations of the Food Guide Pyramid will ensure an adequate intake of dietary fiber. High-fiber fruits, such as apples with peels, have about 3 grams per serving, while fruit juices are low in fiber. High-fiber vegetables, such as fresh broccoli, have about 2.5 grams per serving. Whole grain breads, cereals, and brown rice have about 2.5 grams per serving. High-fiber cereals, such as bran, have about 8 to 10 grams per serving. Served alone, these high fiber cereals are not well accepted by young children, but can be mixed with other cereals or used in recipes for food items such as muffins. Dried beans and peas are also excellent sources of fiber, providing 4 to 7 grams of fiber per ½ cup serving.[58]

DIETARY FIBER   Complex carbohydrates and *lignins* found mainly in the plant cell wall. Dietary fiber cannot be broken down by human digestive enzymes.

LIGNIN   Noncarbohydrate polymer that contributes to dietary fiber.

FAT   Food intakes that follows the recommendations of the Dietary Guidelines for Americans and the Food Guide Pyramid provides an appropriate amount of fat for school-age and preadolescent children. Healthy diets include whole grain breads and cereals, beans and peas, fruits and vegetables, low-fat dairy products, and lean meats, fish, and poultry. Foods high in fat, especially those high in saturated fat and trans fatty acids, are used sparingly. However, an appropriate amount of dietary fat is necessary to meet

children's needs for calories, essential fatty acids, and fat soluble vitamins. As mentioned earlier, fat intakes <20% of total calories are not recommended for children.[48]

CALCIUM   The recommendations for adequate daily intakes of calcium are 800 milligrams for children aged four to eight years and 1300 milligrams for children 9 through 18 years.[22] The higher recommendation for older children is a reflection of the fact that the majority of bone formation occurs during puberty. Adequate calcium intake during this time is necessary to achieve peak bone formation, which may prevent osteoporosis later in life.[59]

Good sources of calcium are listed in Chapter 1. It is difficult to meet the higher recommendations of calcium without the inclusion of dairy products, preferably low-fat dairy products. For those individuals who are lactose intolerant, lactose-reduced dairy products are available. Calcium-fortified foods such as fruit juice and soy milk are also available. For those children whose calcium intake is inadequate, calcium supplements need to be given under the guidance of a physician or registered dietitian.

FLUIDS   In regards to fluids, it is of particular importance for school-age children to drink enough fluids to prevent dehydration during periods of exercise and during participation in sports. Preadolescent children need to be more careful about staying hydrated than do adults and adolescents for several reasons.[60] Children sweat less, and they get hotter during exercise. Some sports, such as football and hockey, require special protective gear, which may prevent the body from being able to cool off. Children should never deprive themselves of food or water in order to meet a certain weight category, such as in wrestling.

The adults who are supervising the physical activities need to make sure that children drink fluids before, during, and after exercise. The thirst mechanism does not work as well during exercise, and children may not realize that they need fluids. Cold water is the best fluid for children. However, children may be more likely to drink more fluids if they are flavored. Sports drinks, which contain 6% to 8% carbohydrate or 15 to 18 grams of carbohydrate per cup and diluted fruit juice, are appropriate for children. Children should not be given soft drinks or undiluted juice, because the carbohydrate load is too high and could cause stomach cramps, nausea, and diarrhea. Beverages that contain caffeine will further dehydrate the body.[60]

SOFT DRINKS   Approximately 32% of school-age children consume up to 8.9 ounces of soft drinks per day, 32% consume greater than or equal to 9 ounces per day, while only 36% are nonconsumers of soft drinks.[61] School-age children consume more soft drinks than preschool-age children, but not as much as adolescents, indicating an increase in consumption with age. Energy intake increases with increased consumption of nondiet soft drinks. Children with high consumption of regular soft drinks (more than 9 ounces per day) consume less milk and fruit juice than those with lower consumptions of regular soft drinks. According to analysis of NHANES III data, overweight children have a higher proportion of their energy intake from soft drinks than nonoverweight children.[27] Soft drinks can contribute significantly to children's overall calorie intake, while contributing little to the overall nutritional value of their diets and displacing more nutritious foods.

Recommended vs. Actual Food Intake

Comparing children's food intake data to the Food Guide Pyramid reveals that the mean food group intakes for children ages 2 to 19 years are below minimum recommendations for all food groups except for the dairy group with intakes at or near recommendations (see Table 12.4).[62] When the data are analyzed by racial/ethnic groups, blacks have a significantly higher number of servings of meat than do whites or Hispanics. Whites have significantly higher intakes of dairy than do blacks or Hispanics. Fruit consumption is lower for children living in households where the family income is less than 131% of poverty index. Dairy consumption is also higher with increasing income.

According to the Food and Nutrient Intakes by Children report, mean energy intakes of school-age boys meet the 1989 Recommended Dietary Allowances for energy.[19] Mean energy intake for girls is 91% of the RDAs. The composition of the children's diets can be found in Table 12.5, which depicts the mean percentages of total energy from carbohydrate, protein, total fat, and saturated fatty acids, and cholesterol intake for six- to nine-year-old and 6- to 11-year-old males and females. Fat and saturated fat intakes have decreased slightly since the NHANES III

survey, which found a total fat intake of 33.8% and a saturated fat intake of 12.6% of total calories among 6- to 11-year-old boys and a total fat intake of 33.6% and a saturated fat intake of 12.3% among 6- to 11-year-old girls.[27] Cholesterol intake is well below the recommendation of 300 mg per day.

As in the toddler and preschool-age group, mean vitamin and mineral intakes exceeded the RDAs except for vitamin E and zinc.[63] Table 12.6 depicts a further analysis of children's diets in terms of dietary fiber, sodium, and caffeine intake. Dietary fiber intake was below recommended levels for both males and females. Sodium intake for both males and females exceeded the recommendation of 2400 milligrams per day. School-age children's caffeine intake has risen dramatically for both males and females from an average daily intake of 12.7 milligrams during the preschool years. Higher caffeine consumption is seen in the 6- to 11-year-old group versus the six- to nine-year-old group, indicating an increased consumption of caffeine with age. This coincides with an increase in soft drink consumption.

According to these data, children, both boys and girls, are exceeding the recommendations for total calories from fat of 30% and from saturated fat of less than 10%. Analysis of NHANES III data shows that the percentage of energy from fat is higher for black and Mexican American girls and black boys than for white girls and boys.[29] These differences are seen by six to nine years of age in black and Mexican American girls and by 10 to 13 years of age in black boys.

Tables 12.7 and 12.8 depict the mean percentages of nutrient intake contributed by foods eaten at snacks and foods eaten away from home for one day. These figures further illustrate the important contributions that snacks and meals eaten away from home make to the total daily food intake of children, especially in terms of total calories, total fat, and saturated fat. Analysis of food consumption data indicates that snacking among children has increased over the years, and the contribution of snacks to energy intake has increased from 20% in 1977 to 25% in 1996.[64] When eating away from home, school-age children most often eat at the school cafeteria, followed by someone else's house, and fast-food restaurants.[63]

Table 12.4 Mean daily intakes of the Food Guide Pyramid for children in the 1989–1991 CSFII.[62]

Gender and Age	Number in Study	Grain	Vegetable	Fruit	Dairy	Meat
Recommendations (servings)		6–11	3–5	2–4	2–3	5–7 ounces
Males 6–11 years	599	5.7	2.4	1.3	2.2	3.8
Females 6–11 years	573	5.5	2.4	1.4	2.1	3.7

Table 12.5   Mean percentages of food energy from carbohydrate, protein, total fat, saturated fatty acids, and cholesterol intake of 6- to 11-year-old children.[63]

Gender and Age	Carbohydrate (%)	Protein (%)	Total Fat (%)	Saturated Fatty Acids (%)	Cholesterol (mg/d)
Males:					
6–9 years	55	14	33	12	225
6–11 years	55	14	33	12	232
Females:					
6–9 years	55	14	32	12	190
6–11 years	55	14	33	12	199

According to the NHANES III data, children ages 6 to 11 years obtain about 20% of their total energy intake from beverages, with milk, soft drinks, and juice drinks being the largest contributors.[27] Drinking whole milk makes significant contributions to children's overall fat and saturated fat intakes.

Cross-Cultural Considerations

Cross-cultural considerations in general are discussed in Chapter 1 and specifically for the pediatric population in Chapter 10. When data from various studies in pediatrics are reported, often differences in racial/ethnic groups are identified. One such study is the National Heart, Lung, and Blood Institute's Growth and Health Study, which is a longitudinal study of preadolescent girls designed to study the development of obesity and its later effect on cardiovascular risk factors.[65]

One aspect of the study was to note racial differences in 11 weight-related eating practices such as eating in front of the television, eating while doing homework, and skipping meals. Upon analysis, it was found that black girls are more than twice as likely to engage in the weight-related eating practices targeted by the study. Even when controlling for socioeconomic and demographic effects, black girls remain more likely to engage in these eating practices than whites. A frequent practice of these behaviors is associated with a higher energy intake than for those girls who practice these behaviors infrequently. These behaviors need

to be targeted in early nutrition education efforts with culturally appropriate education materials.

Vegetarian Diets

The nutrition basics of vegetarian diets are discussed in Chapter 10. Young children who are following vegetarian diets are usually following their parents' eating practices. Preadolescents, on the other hand, may choose to follow a vegetarian diet independently of their family, motivated by concerns about animal welfare, ecology, and the environment.[1] A vegetarian diet is a socially acceptable way to reduce total fat in the diet and is often adopted by adolescents with eating disorders (see Chapter 14).

PHYSICAL ACTIVITY RECOMMENDATIONS

Physical activity has many proven health benefits including prevention of coronary heart disease. Physical activity is one of the health behaviors that is important to establish in childhood with the hopes that this pattern will continue into adolescence and adulthood. With the increased prevalence of childhood obesity, increasing physical activity and decreasing sedentary behaviors become important factors in controlling childhood overweight.[41]

Recommendations vs. Actual

It is recommended that children engage in at least 60 minutes of physical activity every day.[54] Parents are encouraged to:

- Set a good example by being physically active themselves and joining their children in physical activity.
- Encourage children to be physically active at home, at school, and with friends.
- Limit television watching, computer games, and other inactive forms of play by alternating with periods of physical activity.

Table 12.6   Mean dietary fiber, sodium, and caffeine intake of 6- to 11-year-old children.[63]

Gender and Age	Dietary Fiber (gms)	Sodium (mg)	Caffeine (mg)
Males:			
6–9 years	13	3195	23
6–11 years	14	3264	25
Females:			
6–9 years	12	2764	19
6–11 years	12	2839	23

Table 12.7 **Mean percentages of nutrient intake contributed by foods eaten at snacks for one day.**[63]

Gender and Age	Individuals Eating Snacks (%)	Food Energy (%)	Total Fat (%)	Saturated Fatty Acids (%)
Males:				
6–9 years	83	21	19	19
6–11 years	83	21	20	20
Females:				
6–9 years	84	21	20	20
6–11 years	82	20	19	19

Additionally, physical activity and daily physical education should be encouraged at schools and during after-school care programs. But presently, only about 17% of middle and junior high schools require daily physical activity for all students.[4] Healthy People 2010 objectives include increasing the proportion of trips made by walking and by bicycling by school-age children. Currently, only about 28% of children and adolescents ages 5 to 15 years take walking trips to school less than one mile, and only about 2% of children and adolescents ages 5 to 15 years take trips to school less than two miles on a bicycle. In order for these goals to be met, communities need to assure safe places for children to walk and ride their bicycles. Bicycle safety measures, such as wearing a helmet, need to be employed. Communities can also offer youth sports and recreation programs that are developmentally appropriate and fun for all young people. To achieve such a community environment, partnerships need to be established among federal, state, and local governments, nongovernment organizations, and private entities. The Centers for Disease Control and Prevention have proposed strategies for promoting physical activity for children in family, school, and community settings.[66]

Determinants of Physical Activity

It is important to try and understand children's physical activity patterns and determinants of physical activity so that vulnerable groups can be identified and appropriate intervention programs designed. Potential determinants of physical activity behaviors among children include physiological, environmental, psychological, social, and demographic factors.[67] Childhood physical activity has been difficult to assess and to track into adulthood. Many of the studies that have been reported identify correlates of physical activity behavior rather than predictors. The determinants of childhood physical activity are probably multidimensional and interrelated. More work needs to be done in this area, but some generalities resulting from existing studies are listed here:

- Girls are less active than boys.
- Physical activity decreases with age.
- Seasonal and climate differences are seen in children's activity levels.
- Physical education in schools has decreased.

School and neighborhood safety is an important issue in promoting physical activity. Parents have direct and indirect effects on children's physical activity levels.

One cross-sectional study of 107 African American and Caucasian children ranging in age from 6.5 to 13 years found several interesting correlations:[68]

- Children from single-parent homes watched more hours of television, had less physical education exercise, but participated in more

Table 12.8 **Mean percentage of nutrient intake contributed by foods obtained and eaten away from home for one day.**[63]

Gender and Age	Individuals Eating Snacks (%)	Food Energy (%)	Total Fat (%)	Saturated Fatty Acids (%)
Males:				
6–9 years	65	26	27	28
6–11 years	65	26	28	28
Females:				
6–9 years	64	29	30	31
6–11 years	66	30	31	32

vigorous physical exercise than children from two-parent households.

- African American children had less physical education exercise than Caucasians.
- There was more sports team participation among older, yet physically immature children.
- Girls had lower physical activity levels than boys.
- There were higher physical activity levels among boys, Caucasians, physically mature children, and children from single-parent homes.

Organized Sports

Many school-age and preadolescent children participate in organized sport activities, through schools or other community organizations. An analysis of NHANES III data indicates that children who participate in team sports and exercise programs are less likely to be overweight as compared to nonparticipants.[37] The American Academy of Pediatrics (AAP) recommends that children who are involved in sports be encouraged to participate in a variety of different activities. The proper use of safety equipment, such as helmets, pads, mouth guards, and goggles, should be encouraged. The AAP warns against intensive, specialized training for children. The AAP's recommendations include:

- Monitoring the child athlete's physical condition and development on a regular basis.
- Prevention of stress or overuse injuries, with the child's physician and coach working together.
- Identifying and addressing eating disorders.
- Instructing families, coaches, and child athletes about recognizing and preventing heat injury.[69]

NUTRITION INTERVENTION FOR RISK REDUCTION

It is the position of The American Dietetic Association, the Society for Nutrition Education, and the American School Food Service Association that comprehensive school-based nutrition programs and services be provided to all the nation's elementary and secondary students. These programs and services include: effective education in foods and nutrition; a school environment that provides opportunity and reinforcement for healthful eating and physical activity; involvement of parents and the community; and screening, counseling, and referral for nutrition problems as part of school health services.[70]

Nutrition Education

Eating a healthy diet and participating in physical activity are important components of a healthy lifestyle that may prevent chronic disease in childhood and into adolescence and adulthood. School age is a prime time for learning about healthy lifestyles and incorporating them into daily behaviors. Schools can provide an appropriate environment for nutrition education and learning healthy lifestyle behaviors. Nutrition education studies have been conducted in school settings as well as outside of schools. Some of these programs have been knowledge-based nutrition education programs, with the focus on improving the knowledge, skills, and attitudes of children in regards to food and nutrition issues.[71] Other nutrition education programs have been more behaviorally focused, emphasizing disease risk reduction as well as enhancing health. The CDC has published "Guidelines for School Health Programs to Promote Lifelong Healthy Eating" and these recommendations can be found in Table 12.9.

Table 12.9 Recommendations for School Health Programs Promoting Healthy Eating.[72]

Recommendation 1. Policy: Adopt a coordinated school nutrition policy that promotes healthy eating through classroom lessons and a supportive school environment.

Recommendation 2. Curriculum for nutrition education: Implement nutrition education from preschool through secondary school as part of a sequential, comprehensive school health education curriculum designed to help students adopt healthy eating behaviors.

Recommendation 3. Instruction for students: Provide nutrition education through developmentally appropriate, culturally relevant, fun, participatory activities that involve social learning strategies.

Recommendation 4. Integration of school food service and nutrition education: Coordinate school food service with nutrition education and with other components of the comprehensive school health program to reinforce messages on healthy eating.

Recommendation 5. Training for school staff: Provide staff involved in nutrition education with adequate preservice and ongoing in-service training that focuses on teaching strategies for behavioral change.

Recommendation 6. Family and community involvement: Involve family members and the community in supporting and reinforcing nutrition education.

Recommendation 7. Program evaluation: Regularly evaluate the effectiveness of the school health program in promoting healthy eating, and change the program as appropriate to increase its effectiveness.

Nutrition Integrity in Schools

The school and community have a shared responsibility to provide all students with access to high-quality foods and nutrition services . . . Educational goals, including the nutrition goals of the National School Lunch Program and School Breakfast Program, should be supported and extended through school district policies that create an overall school environment with learning experiences that enable students to develop lifelong, healthful eating habits.[73]

The American Dietetic Association

Nutrition integrity in schools is defined as ensuring that all foods available to children in schools are consistent with the U.S. Dietary Guidelines for Americans, the Dietary Reference Intakes, and the Recommended Dietary Allowances.[19,54] School nutrition programs are vital to reinforcing healthy eating habits in school-age children. Sound nutrition policies need support of the community and school environments, and must involve students in order to be successful. Preparing community leaders for involvement in policy development is one of the nutrition integrity core concepts.[73] Training food service personnel, teachers, administrators, and parents is an integral part of this process. The school environment must be one that supports healthy eating and exercise patterns. Foods sold from vending machines and snack bars often do not support healthy eating and may undermine sound nutrition programs. However, in some schools, vending machine proceeds are important sources of revenue for underfunded schools. It is against USDA regulations to sell *competitive foods* of minimal nutritional value. However, it is not against USDA regulations to sell these foods to students at time other than meal times or in other areas of the school, outside of food service areas. Adequate time allotted for meals is another important component of a sound nutrition program. Students can be involved in a nutrition advisory council, providing feedback about menu preferences, meal environment, and serving as a communication link with other students. Model school-based programs are highlighted in Case Studies 12.1 and 12.2 as examples of what can be accomplished through nutrition education and community involvement of parents, teachers, school food service personnel, and industry.

Model Programs

The 5 A Day for Better Health program models public-private partnership to enhance community nutrition education and to affect change in eating be-

haviors. In 1991, the National Cancer Institute (NCI), acknowledging the strong association of increased fruit and vegetable consumption with decreased risk of certain cancers, launched the 5 A Day for Better Health program. NCI partnered with the Produce for Better Health Foundation, a nonprofit organization that represents the produce industry, for this campaign. Industry participants included supermarkets, suppliers, commodity groups, and food service operations.[74]

The 5 A Day program includes retail, media, community, and research components. Supermarkets provide information to consumers at the retail level. The NCI and Produce for Better Health Program work together to develop a media campaign. At the community level, health, educational, agricultural, and voluntary agencies work together, sometimes forming coalitions, to reach consumers at the local level. For the research component, NCI funded nine studies to develop, implement, and evaluate interventions in specific communities to increase the consumption of fruits and vegetables. The results of two of these funded studies, based in elementary schools, are discussed in Case Study 12.2.

COMPETITIVE FOODS Foods sold to children in food service areas during meal times that compete with the federal meal programs.

PUBLIC FOOD AND NUTRITION PROGRAMS

All children and adolescents should have access to adequate food and nutrition programs, regardless of economic status, special needs, and cultural diversity. Appropriate . . . programs include food assistance and feeding programs and nutrition education, screening, assessment, and intervention.[77]

The American Dietetic Association

Child Nutrition Programs, which have had a federal legislative basis since 1946, contribute significantly to the food intake of school-age children. The purpose of the Child Nutrition Programs is to provide nutritious meals to all children. These programs can also reinforce nutrition education in the classroom. Increasing the proportion of children and adolescents ages 6 to 19 years whose overall dietary quality is enhanced by meals and snacks at schools is addressed in one of the Healthy People 2010 objectives.[4] Child Nutrition Programs include the National School Lunch Program, School Breakfast Program, Child and Adult Care Food Program, Summer Food Service Program, Special Milk Program, Commodity Assistance for

Photo Disc

Case Study 12.1
High 5 Alabama

The purpose of this study was to evaluate the effectiveness of a school-based dietary intervention program in increasing the fruit and vegetable consumption among fourth-graders. Twenty-eight elementary schools in the Birmingham, Alabama metropolitan area were paired within three school districts based on ethnic composition and the proportion of students receiving free or reduced-price meals through the National School Lunch Program. One school in each pair was randomly assigned to an intervention group or a usual care control group. Assessments were completed at baseline (at the end of third grade), after year 1 (at the end of fourth grade), and after year 2 (at the end of fifth grade).

The intervention consisted of three components: classroom, parent, and food service. The classroom component of the intervention included 14 lessons, taught biweekly by trained curriculum coordinators with assistance from the regular classroom teachers. The parent component consisted of an overview during a kickoff meeting and completion of seven homework assignments by the parent and the child. Parents were also asked to encourage and support behavior change in their children. The food service component consisted of food service managers and workers receiving half-day training by High 5 nutritionists in purchasing, preparing, and promoting fruit and vegetables within the High 5 guidelines. Data analyzed included 24-hour recalls from the students, cafeteria observations, psychosocial measures, and parent measures.[75]

Results, found in Table 12.10, indicate that mean daily consumption of fruit and vegetables was higher at Year 1 follow-up and Year 2 follow-up for the intervention group as compared to the controls. At Year 1 follow-up, the mean daily consumption of fruits and vegetables was higher for the intervention parents as compared to control parents, but no difference was found at Year 2 follow-up. The intervention was found to be effective in subsamples suggesting that the program can be used with boys and girls, African American and European Americans, low, middle, and higher income families, and with parents of low, medium, and high educational levels. The intervention was found to be effective in changing the fruit and vegetable consumption of fourth-grade students. Future studies are recommended to enhance the effectiveness of the intervention in changing parents' consumption patterns and to test the effectiveness of the intervention when delivered by the regular classroom teachers.

Child Nutrition Programs, Special Supplemental Food Program for Women, Infants, and Children, Nutrition Education and Training Program, and the National Food Service Management Institute.[78] A description of several of these programs follows.

Table 12.10   Mean number of fruit and vegetable servings per day of intervention group versus control group, High 5 Alabama.[75]

Number of Fruit and Vegetable Servings per Day	Intervention	Control
Year 1 Follow-up	3.96	2.28
Year 2 Follow-up	3.2	2.21

The National School Lunch Program

The federal government provides financial assistance to schools participating in the National School Lunch Program (NSLP) through cash reimbursements for all lunches served, with additional cash for lunches served to needy children, and through commodities.[78,79] Schools must meet five major requirements in order to participate in the NSLP:

1. Lunches must be based on nutritional standards.
2. Children who are unable to pay for lunches must receive lunches for free or at a reduced

Case Study 12.2
5-A-Day POWER PLUS Program

A similar study was conducted in the St. Paul, Minnesota Public School District among fourth- and fifth-grade students. Twenty schools were recruited within the school district and matched into 10 pairs based on the school size, ethnic makeup, and percentage of students participating in the free or reduced-price lunch program. One school within each pair was randomly assigned to the intervention or delayed program condition. Baseline and follow-up data were collected.

The intervention had four components: behavioral curricula in the fourth and fifth grades, parental involvement/education, school food service changes, and industry involvement and support. Two different curricula were written for the fourth- and fifth-graders: "High 5" and "5 for 5". All fourth- and fifth-grade teachers participated in a one-day training session before implementing the curricula, each of which included sixteen 40- to 45-minute classroom sessions taught biweekly for eight weeks. In the fourth grade, parental involvement included assisting their student in the completion of five information/activity packets, which the students brought home to their parents. In the fifth grade, students brought home four snack packs prepared by the food service staff to be shared by the students with their families at home. The food service intervention encouraged selection and consumption of fruits and vegetables at school lunch and included a two-hour training session for food service staff before each curriculum was implemented. The industry component included support from the 72-member Minnesota 5-a-Day Coalition. Coalition members provided fruits and vegetables for the intervention as well as additional educational and incentive materials. Data analyzed included 24-hour recalls from the students, lunchroom observations, parent telephone surveys, health behavior questionnaires, and demographics.

Results indicate that the intervention increased lunchtime and daily fruit consumption, combined fruit and vegetable consumption, and lunchtime vegetable consumption among girls. This multicomponent school-based program was found to be effective in increasing the fruit and vegetable consumption among fourth- and fifth-grade students in an urban, ethnically diverse school district. Future intervention could focus on increasing vegetable consumption, especially among boys, and increasing parental involvement.[76]

price, with no discriminating between paying and nonpaying children.

3. The programs operate on a non-profit basis.
4. The programs must be accountable.
5. Schools must participate in the *commodity program.*

School lunches must provide one-third of the Recommended Dietary Allowances or Dietary Reference Intakes for the children being served and must be consistent with the most recent version of the U.S. Dietary Guidelines for Americans when analyzed over a week's time.[78,79] These programs must also meet the needs of children with disabilities and special health care needs (see Chapter 13). Although not federally mandated, schools are encouraged to allow adequate time for chil-

dren to eat their lunches. Schools receive payments from the federal government based on the number of meals served by category, paid, free, or reduced-price.

School Breakfast Program

The School Breakfast Program is similar to the National School Lunch Program and was first authorized as a pilot program in 1966. It is a voluntary federal program, but many state legislatures have mandated breakfast programs, especially

COMMODITY PROGRAM A USDA program in which food products are sent to schools for use in the Child Nutrition Programs. Commodities are usually acquired for farm price support and surplus removal reasons.[79]

in schools serving needy populations.[79] Many of the same rules apply to the School Breakfast Program as for the NSLP. School breakfasts must provide one-fourth of the Recommended Dietary Allowances or Dietary Reference Intakes and comply with the U.S. Dietary Guidelines for Americans when analyzed over a week's time. It is a special challenge for schools to allow enough time for school breakfasts before school when most of the participating children arrive at about the same time. Currently, universal breakfast programs in elementary schools are being pilot tested.

Summer Food Service Program

The Summer Food Service Program provides meals to children from needy areas when school is not in session. Schools, local government agencies, or other public and private nonprofit agencies operate these programs. The federal government provides financial assistance to these programs for providing meals in areas where 50% or more of the participating children are from families whose incomes are lower than 185% of the poverty level.[79] The Summer Food Service Program is an important source of food for many children from food-insecure families.

Nutrition Education and Training Program (NET)

The purpose of the Nutrition Education and Training Program (NET) is to provide funds to states for nutrition training programs, which link school meals to classroom education about nutrition. The NET Program was first authorized by Congress in 1977, but was only fully funded for the first two years. Since that time, funding has been insufficient and erratic. NET is the only source of federal dollars for states to use in meeting their nutrition training needs, including the proper training of food service personnel. Nutrition education and training are important components of Child Nutrition Programs, and without adequate funding, this piece of the program is absent or inadequate. Educators and health professionals can advocate to their national legislators for adequate NET Program funding.[79]

Resources

Annie E. Casey Foundation
This web site provides data on critical issues affecting at-risk children and their families. The publication Kids Count is also available on-line at this web address.
Available from: http://www.aecf.org

Bright Futures
The Bright Futues publications are developmentally based guidelines for health supervision and address the physical, mental, cognitive, and social development of infants, children, adolescents, and their families. Bright Futures materials may also be ordered from: National Maternal and Child Health Clearinghouse, 2070 Chain Bridge Road, Suite 450, Vienna, VA 22182-2536, www.nmchc.org, 703-356-1964.
Available from: http://www.brightfutures.org

Centers for Disease Control and Prevention, National Center for Health Statistics
The CDC web site provides background information on the recent growth chart revisions. Also, individual growth charts can be downloaded and printed from this web page.
Available from: http://www.cdc.gov/nchs/about/major/nhanes/growthcharts/charts.htm.

Child Nutrition Programs
The Child Nutrition Program web page provides information on all of their programs including the National School Lunch Program, School Breakfast Program, Special Milk Program, Summer Food Service Program, and Child and Adult Care Food Program.
Available from: http://www.fns.udsa.gov/cnd/

Dietary Reference Intakes: National Academy Press
Over 2,000 books are available online at this site, free of charge, including the current Dietary Reference Intakes.
Available from: http://www.nap.edu

Federal Trade Commission, Bureau of Consumer Protection
See "Promotions for Kids' Dietary Supplements Leave Sour Taste." This article explores what is known about the use of dietary supplements in the pediatric population and includes a list of pointers for parents.
Available from: http://www.ftc.gov/bcp/conline/features/kidsupp.htm.

Fit, Healthy, and Ready to Learn: A School Health Policy Guide
This in-depth guide includes policies to encourage physical activity, policies to encourage healthy eating and to discourage tobacco use. The policy to encourage healthy eating addresses nutrition education, the food service program, and other food choices at school.
Available from: http://www.nasbe.org/healthyschools/fithealthy.mgi

National Center for Complementary and Alternative Medicine, National Institutes of Health
This web page provides science based information for consumers and practitioners as well as information about related news and events.
Available from: http://www.nccam.nih.gov/nccam

Promoting Better Health for Young People Through Physical Activity and Sports
A Report to the President from the Secretary of Health and Human Services and the Secretary of Education. This report outlines ten strategies to promote health through lifelong participation in physical activity and sports.
Available from: http://www.cdc.gov/nccdphp/dash/presphysactrpt/

Team Nutrition

This web site provides information on the USDA program, Team Nutrition, that is designed to help implement the Dietary Guidelines in Child Nutrition Programs. *Available from:* http://www.fns.usda.gov/tn

The American Dietetic Association, Diabetes Care and Education Dietetic Practice Group

Ethnic and Regional Food Practices A Series. Chicago, IL: The American Dietetic Association, 1994–1999. This series of booklets addresses food practices, customs, and holiday foods of various ethnic groups.

Graves DE, Suitor CW.

Celebrating Diversity: Approaching Families Through Their Food. 2nd ed. Arlington, VA: National Center for Education in Maternal and Child Health, 1998. The purpose of this pub-

lication is to assist health professionals in learning to communicate effectively with a diverse clientele.

Green M, Palfrey JS. (eds).

Bright Futures: Guidelines for Health Supervision of Infants, Children, and Adolescents. 2nd ed. Arlington, VA: National Center for Education in Maternal and Child Health, 2000. This publication addresses the physical, mental, cognitive, and social development of infants, children, adolescents, and their families.

Patrick K, Spear BA, Holt K, and Sofka D.

Bright Futures in Practice: Physical Activity. Arlington, VA: National Center for Education in Maternal and Child Health, 2001. Developmentally appropriate activities are presented for infants, children, and adolescents.

References

1. Story M, Holt K, Sofka D, eds. Bright futures in practice: nutrition. Arlington, VA: National Center for Education in Maternal and Child Health; 2000.

2. Meyer AF, Sampson AE, Weitzman M, et al. School breakfast program and school performance. Am J Dis Child 1989;143:1234–9.

3. Kids Count Data Book. Baltimore: The Annie E. Casey Foundation; 2000.

4. U.S. Department of Health and Human Services. Healthy People 2010 (Conference Edition, in two volumes). Washington, DC; January 2000.

5. Behrman RE, Kliegman RM, Jenson HB, eds. Nelson's textbook of pediatrics, 16th ed. Philadelphia: WB Saunders Co; 1996.

6. Centers for Disease Control and Prevention, National Center for Health Statistics. CDC growth charts: United States. Available from http://www.cdc.gov/nchs/about/major/nhanes/growthcharts/charts.htm, May 30, 2000.

7. Troiano RP, Flegal KM. Overweight children and adolescents: description, epidemiology, and demographics. Pediatrics 1998;101:497–504.

8. Dietz WH. Critical periods in childhood for the development of obesity. Am J Clin Nutr 1994;59:955–9.

9. Satter E. Feeding dynamics: helping children to eat well. J Pediatr Health Care 1995;9:178–84.

10. Gillman MW, Rifas-Shiman SL, Frazier AL, et al. Family dinner and diet quality among older children and adolescents. Arch Fam Med 2000;9:235–40.

11. Kotz K, Story M. Food advertisements during children's Saturday morning television programming: are they consistent with dietary recommendations? J Amer Diet Assoc 1994;94:1296–1300.

12. U.S. Department of Agriculture. Food guide pyramid: a guide to daily food choice. Washington, DC: USDA, Human Nutrition Information Service; 1992. Home and Garden Bulletin No. 252.

13. Birch LL, Fisher JO. Food intake regulation in children, fat and sugar substitutes and intake. Ann NY Acad of Sci 1997;819:194–220.

14. Birch LL, Fisher JA. Appetite and eating behavior in children. Pediatr Clinic N Amer 1995;42:931–53.

15. Birch LL, Fisher JO. Mothers' child-feeding practices influence daughters' eating and weight. Amer J Clin Nutr 2000;71:1054–61.

16. Birch LL. Psychological influences on the childhood diet. J Nutr 1998;128:407S–10S.

17. Fisher JO, Birch LL. Parents' restrictive feeding practices are associated with young girls' negative self evaluation of eating. J Amer Diet Assoc 2000;100:1341–6.

18. Birch LL, Fisher JO. Development of eating behaviors among children and adolescents. Pediatrics 1998;101:539–49(s).

19. National Research Council. Recommended dietary allowances, 10th ed. Washington, DC: National Academy Press; 1989.

20. Kennedy E, Goldberg J. What are American children eating? Implications for public policy. Nutr Rev 1995;53:111–26.

21. Kennedy E, Powell R. Changing eating patterns of American children: a view from 1996. J Amer Coll Nutr 1997;16:524–9.

22. Institute of Medicine, Food and Nutrition Board. Dietary Reference Intakes for calcium, phosphorus, magnesium, vitamin D, and fluoride. Washington, DC: National Academy Press; 1997.

23. Institute of Medicine, Food and Nutrition Board. Dietary Reference Intakes for vitamin A, vitamin K, arsenic, boron, chromium, copper, iodine, iron, manganese, molybdenum, nickel, silicon, vanadium, and zinc. Washington, DC: National Academy Press; 2001.

24. Looker AC, Dallman PR, Carroll MD, et al. Prevalence of iron deficiency in the United States. JAMA 1997;277:973–6.

25. Centers for Disease Control and Prevention. Recommendations to prevent and control iron deficiency in the United States. Morbidity and Mortality Weekly Report. April 3, 1998;47:RR-03.

26. Casamassimo P. Bright futures in practice: oral health. Arlington, VA: National Center for Education in Maternal and Child Health; 1996.

27. Troiano RP, Briefel RR, Carroll MD, et al. Energy and fat intakes of children and adolescents in the United States: data from the National Health and Nutrition Examination Surveys. Am J Clin Nutr 2000;72 (suppl):1343S–53S.

28. Freedman DS, Dietz WH, Srinivasan SR, Berenson GS. The relation of overweight to cardiovascular risk factors among children and adolescents: The Bogalusa Heart Study. Pediatrics 1999;103:1175–82.

29. Winkleby MA, Robinson TH, Sundquist J, Kraemer HC. Ethnic variation in cardiovascular disease risk factors among children and young adults: findings from the Third National Health and Nutrition Examination Survey, 1988–1994. JAMA 1999;281:1006–13.

30. Dwyer JT, Stone EJ, Yang M. Prevalence of marked overweight and obesity in a multiethnic pediatric population: findings from the Child and Adolescent Trial for Cardiovascular Health (CATCH) study. J Amer Diet Assoc 2000;100:1149–56.

31. Dietz WH. Health consequences of obesity in youth: childhood predictors of adult disease. Pediatrics 1998;101:518S–25S.

32. American Diabetes Association. Type 2 diabetes in children and adolescents. Pediatrics 2000;105:671–80.

33. Dietz WH. Childhood weight affects adult morbidty and mortality. J Nutr 1998;128:411S–14S.

34. Dietz WH, Gortmaker SL. Preventing obesity in children and adolescents. Annu Rev Public Health 2001;22:337–53.

35. Strauss RS, Knight J. Influence of the home environment on the development of

obesity in children. Pediatrics [serial online] 1999;103:e85. Available from http://www.pediatrics.org/cgi/content/full/103/1/e85.

36. Whitaker RC, Wright JA, Pepe MS, et al. Predicting obesity in young adulthood from childhood and parental obesity. New Eng J Med 1997;337:869–73.

37. Dowda M, Ainsworth BE, Addy CL. Environmental influences, physical activity, and weight status in 8- to 16-year-olds. Arch Pediatr Adolesc Med 2001;155:711–7.

38. Dietz WH, Gortmaker SL. Do we fatten our children at the television set? Obesity and television viewing in children and adolescents. Pediatrics 1985;75:807–12.

39. Gortmaker SL, Must A, Sobol A, et al. Television viewing as a cause of increasing obesity among children in the United States, 1986–1990. Arch Pediatr Adolesc Med 1996;150:356–62.

40. Andersen RE, Crespo CJ, Bartlett SJ, et al. Relationship of physical activity and television watching with body weight and level of fatness among children: results from the Third National Health and Nutrition Examination Survey. JAMA 1998;279: 938–42.

41. Crespo CJ, Smit E, Troiano RP, et al. Television watching, energy intake, and obesity in US children. Arch Pediatr Adolesc Med 2001;155:360–5.

42. Robinson TN. Reducing children's television viewing to prevent obesity: a randomized controlled trial. JAMA 1999;282: 1561–67.

43. Klesges RC, Shelton ML, Klesges LM. Effects of television on metabolic rate: potential implications for childhood obesity. Pediatrics 1993;91:281–6.

44. Barlow SE, Dietz WH. Obesity evaluation and treatment: expert committee recommendations. Pediatrics [serial online] 1998;102:e29. Available from http://www.pediatrics.org/cgi/content/full/102/3/e29, July 1, 2000.

45. Epstein LH, Paluch RA, Gordy CC, Dorn J. Decreasing sedentary behaviors in treating pediatric obesity. Arch Pediatr Adolesc Med 2000;154:220–6.

46. National Cholesterol Education Program. Report of the expert panel on blood cholesterol in children and adolescents. Bethesda, MD: National Cholesterol Education Program; 1991.

47. Fisher EA, Van Horn L, McGill HC. Nutrition and children: a statement for healthcare professionals from the nutrition committee. American Heart Association. 1997;95:2332–33.

48. American Academy of Pediatrics, Committee on Nutrition. Cholesterol in childhood. Pediatrics 1998;101:141–7.

49. Olson R E. The dietary recommendations of the American Academy of Pediatrics. Am J Clin Nutr 1995;61:271–3.

50. Luepker RV, Perry CL, McKinlay SM, et al. for the CATCH Collaborative Group. Outcomes of a field trial to improve chil-

dren's dietary patterns and physical activity. The Child and Adolescent Trial for Cardiovascular Health (CATCH). JAMA 1996;275:768–76.

51. The Writing Group for the DISC Collaborative Research Group. Efficacy and safety of lowering dietary intake of fat and cholesterol in children with elevated low-density lipoprotein cholesterol. The Dietary Intervention Study in Children (DISC). JAMA 1995;273:11429–35.

52. Federal update. J Amer Diet Assoc 2000;100:877.

53. American Academy of Pediatrics, Committee on Nutrition. Pediatric nutrition handbook. Elk Grove Village, IL: American Academy of Pediatrics; 1998.

54. U.S. Department of Agriculture and U.S. Department of Health and Human Services. Dietary guidelines for Americans, 2000, 5th ed. Home and Garden Bulletin No. 232.

55. American Dietetic Association. Position of the American Dietetic Association: dietary guidance for healthy children aged 2 to 11 years. J Amer Diet Assoc 1999;99: 93–101.

56. A summary of conference recommendations on dietary fiber in childhood: Conference on Dietary Fiber in Childhood, New York, May 24, 1994. Pediatrics 1995;96:1023–28.

57. Hampl JS, Betts NM, Benes BA. The "age+5" rule: comparisons of dietary fiber intake among 1- to 4-year-old children. J Amer Diet Assoc 1998;98:1418–23.

58. Dwyer JT. Dietary fiber for children: how much? Pediatrics 1995;96:1019S–22S.

59. American Academy of Pediatrics, Committee on Nutrition. Calcium requirements of infants, children, and adolescents. Pediatrics 1999;104:1152–7.

60. Jennings DS, Steen SN. Play hard eat right, a parents' guide to sports nutrition for children. Minneapolis, MN: Chronimed Publishing; 1995.

61. Harnack L, Stang J, Story M. Soft drink consumption among US children and adolescents: nutritional consequences. J Amer Diet Assoc 1999;99:436–41.

62. Munoz KA, Krebs-Smith SM, Ballard-Barbash R, Cleveland LE. Food intakes of US children and adolescents compared with recommendations. Pediatrics 1997; 100:323–9.

63. U.S. Department of Agriculture, Agricultural Research Service. Food and Nutrient Intakes by Children 1994–1996. Available from htttp://www.barc.usda.gov/bhnrc/foodsurvey/home.htm (Products page), July 1, 2000.

64. Jahns L, Siega-Riz AM, Popkins BM. The increasing prevalence of snacking among US children from 1977 to 1996. J Pediatr 2001;138:493–8.

65. McNutt SW, Yuanreng HU, Schreiber GB, et al. A longitudinal study of the dietary practices of black and white girls 9

and 10 years old at enrollment: The NHLBI Growth and Health Study. J Adolesc Health 1997;20:27–37.

66. Department of Health and Human Services and Department of Education. Promoting better health for young people through physical activity and sports: a report to the president from the secretary of Health and Human Services and the secretary of Education. Silver Spring, MD: Centers for Disease Control and Prevention; 2000. Available from http://www.cdc.gov/nccdphp/dash/presphysactrpt.

67. Kohl HW, Hobbs KE. Development of physical activity behaviors among children and adolescents. Pediatrics 1998;101: 549–54.

68. Lindquist CH, Reynolds KD, Goran MI. Sociocultural determinants of physical activity among children. Prev Med 1999; 29:305–12.

69. American Academy of Pediatrics, Committee on Sports Medicine and Fitness. Intensive training and sports specialization in young athletes. Pediatrics 2000;106: 154–7.

70. American Dietetic Association. Position of ADA, SNE, and ASFSA: school-based nutrition programs and services. J Amer Diet Assoc 1995;95:367–9.

71. Lytle L. Nutrition education for school-aged children. J Nutr Ed 1995;27: 298–311.

72. Centers for Disease Control. Guidelines for school health programs to promote lifelong healthy eating. J Sch Health 1997; 67:9–26.

73. American Dietetic Association. Position of The American Dietetic Association: local support for nutrition integrity in schools. J Amer Diet Assoc 2000;100:108–11.

74. Havas S, Heimendinger J, Reynolds K. 5 a day for better health: a new research initiative. J Amer Diet Assoc 1994; 94:32–6.

75. Reynolds KD, Franklin FA, Binkley D, et al. Increasing the fruit and vegetable consumption of fourth-graders: results from the high 5 project. Prev Med 2000;30: 309–19.

76. Perry CL, Bishop DB, Taylor G, et al. Changing fruit and vegetable consumption among children: the 5-a-day power plus program in St. Paul, Minnesota. Amer J Pub Health 1998;88:603–9.

77. American Dietetic Association. Position of the American Dietetic Association: child and adolescent food and nutrition programs. J Amer Diet Assoc 1996;96:913–7.

78. 7CFR Parts 210 and 220. Child Nutrition Programs; School Meal Initiative for Healthy Children; Final Regulation. June 1995.

79. Martin J. Overview of federal child nutrition legislation. In: Martin J, Conklin MT, eds. Managing child nutrition programs leadership for excellence. Gaithersburg, MD: Aspen Publishers; 1999.

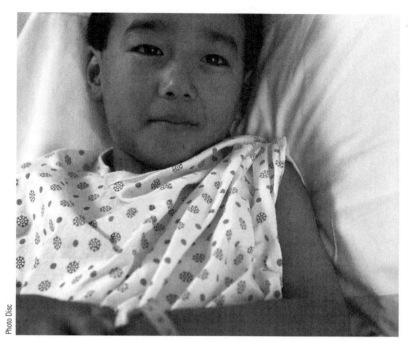

Photo Disc

Every sickness begins in the stomach.

Yemenite saying

CHILD AND PREADOLESCENT NUTRITION:

Conditions and Interventions

Prepared by **Janet Sugarman Isaacs**

CHAPTER OUTLINE

- Introduction
- "Children Are Children First"—What Does that Mean?
- Nutrition Needs of Children with Special Health Care Needs
- Growth Assessment
- Nutrition Recommendations
- Nutrition-Related Conditions
- Dietary Supplements and Herbal Remedies
- Sources of Nutrition Services

KEY NUTRITION CONCEPTS

1 Children are children first, even if they have conditions that affect their growth and nutritional requirements.

2 Children who have special health conditions receive more intensive nutrition services in schools and health care settings than other children.

3 Family meal patterns and routines affect nutrition for children, so providing adequate support for families improves the nutritional status of children with chronic health conditions.

4 Common nutrition problems in children with chronic conditions are underweight and overweight, and difficulties eating enough to meet higher nutrient requirements.

INTRODUCTION

Nutrition services need to be part of the goal to help a child reach his or her full potential. This chapter discusses the nutrition needs of children with chronic conditions, such as cystic fibrosis, diabetes mellitus, cerebral palsy, phenylketonuria (PKU), and behavioral disorders. Nutrition recommendations are based on those for children generally as discussed in Chapter 12, but they may be modified by the condition and its consequences on growth, nutrient requirements, and/or eating abilities. Other factors such as activity that increase or decrease caloric needs are also discussed. Expanding school and community resources include nutrition services for children with special health care needs or those with developmental disabilities. Advocates for those with disabilities prefer "people-first language," which this chapter models. This convention names the person first and then the condition. An example is "a girl with *Down syndrome*," rather than the "Downs girl". Advocates for those with disabilities also prefer the word *disabilities* rather than *handicapped*. (The origin of *handicap* is using a cap to beg.)

"CHILDREN ARE CHILDREN FIRST"—WHAT DOES THAT MEAN?

Children with special health care needs are children first, even if their conditions change their nutrition, medical, and social needs. As discussed in Chapter 12, children are expected to become more independent in making food choices, assisting with meal preparation, and participating at meal times with other family members. These same expectations are appropriate for children with special health care needs. For example, the child with spina bifida should be encouraged to make a salad or set the table. Modifications may be needed to help the child be successful, such as storing plates and utensils on low shelves, or lower counter heights to accommodate a wheelchair at a kitchen sink. The cognitive developmental gains of childhood and participation in meal preparation and mealtime are the same for children with chronic conditions.

This concept has been acknowledged in schools through federal legislation in the Individuals with Disabilities Education Act (IDEA). This law requires the least restrictive environment and is resulting in inclusive setting for more children with disabilities.[1] This concept of inclusion has major ramifications for how children receive all types of services, such as schools providing alternative foods as required in the main cafeteria with all the other children. As a result of inclusion, children in wheel chairs or Down syndrome spend time in regular classrooms with others of the same age. Nutritional problems related to food refusals, mealtime behavior, or a special diet are being addressed in the neighborhood school in the regular classroom as often as possible.

This same concept of treating the child as a child first is recommended at home too. A special diet is part of an overall treatment plan that incorporates normal developmental steps of childhood. Children with diabetes or PKU do not benefit from being treated in a special manner at mealtimes. As soon as possible, children are taught to take responsibility for making food choices consistent with their diet plans. Consistency and structure in the home support a child's normal development. This structure includes regular meal and snack times, and the child accepting increasing levels of responsibility in preparing foods. These approaches lower the chance of the child being overprotected or manipulating adults because of her or his illness or its treatment.

General nutrition guidelines for children, school nutrition educational materials, and nutrition prevention strategies may or may not be applicable to children with specific conditions. Many nutrition education curricula appropriately target the prevention of overweight, lowering fat, and increasing fruit and vegetable intake. These curricula provide appropriate education for the most part to children with conditions such as diabetes and Down syndrome. Conditions in which slow weight gain and underweight are common, such as cerebral palsy, may not fit such curricula.[2] Nutrition education may not address their need for high fat foods as part of high caloric needs. Another example of how general nutrition recommendations may not fit all children concerns the widely used Food Guide Pyramid.[3] A child with PKU cannot ever have protein-rich meats or dairy products.[4] When this pyramid is discussed and encouraged, the child with PKU may feel isolated and confused about whether their diet is healthy. When possible, children with special health care needs are encouraged to participate in school nutrition education programs with modifications as needed.

Train a child in the way he should go, and when he is old, he will not depart from it.

Proverbs

NUTRITION REQUIREMENTS OF CHILDREN WITH SPECIAL HEALTH CARE NEEDS

As discussed in Chapter 12, the caloric and protein needs of childhood and preadolescence are lower than in the toddler and preschool years.[5] Children with special health care needs have a wider range of nutritional needs and more variability than other children based on these factors:[6,7]

- Low caloric intake may be appropriate with small muscle size.
- High protein is needed with high protein losses, such as skin breakdown.
- High fluid volume is needed with frequent losses from vomiting or diarrhea.
- High fiber may be needed for chronic constipation management.
- Long-term use of prescribed medications may increase or decrease vitamin or mineral requirements, or change the balance of vitamins and minerals needed as a result of medication side effects.
- Routine illness is more likely to result in hospitalization or resurgence of symptoms of the underlying disorder.

Energy Needs

Children with special health care needs may need more, less, or the same caloric intakes as other children of the same age. Energy needs are amazingly complex in children, let alone those with special needs. Under ideal conditions the caloric needs are measured by indirect calorimetry, but usually they are estimated using standard calculations that cannot take into account the specific conditions involved.[8] Machines that measure indirect calorimetry are becoming available, but they still can only give estimates of energy needs at rest, without considerations of energy needed for activity and growth. Conditions that slow growth or decrease muscle size generally result in lower caloric needs.[6,9,10] Caloric needs in a child with Prader-Willi syndrome may be only 66% of the caloric needs of child the same age and gender. Other factors that change energy requirements are related to activity level and frequency of illnesses. Children with chronic condition are encouraged to participate in age-appropriate sports activities. Conditions in which activity may be especially beneficial include diabetes and children with mild cerebral palsy. The level of activity of children with chronic conditions may be higher or lower than activity in other children. Children who are very active may appear thin as a result of low caloric intake. Children with autism and attention deficit hyperactivity disorder are generally more active than other children, and/or they may sleep less time.[11,12] Such a range in level of activity is addressed as a part of a thorough nutrition assessment. Questions such as "Is the child receiving physical therapy one or three times per week?" and "How much time does the child use a walker compared to a wheelchair?" are examples of how activity can be assessed in determining caloric needs.

Protein Needs

Protein needs also can be higher, lower, or the same as those for other children, based on the condition. Healing burns and cystic fibrosis are examples of disorders with high protein needs at 150% of the RDA.[13] Conditions such as PKU and other protein-based inborn errors of metabolism require greatly reduced amounts of natural protein in the diet.[4] Children with diabetes mellitus do not have modified protein needs.[14] The importance of protein for wound healing and for maintaining a healthy immune system makes protein requirements key for various conditions with frequent illnesses or surgeries. For example a child with cerebral palsy, who is scheduled to have hip surgery, would have protein needs evaluated in a complete nutritional assessment. Higher protein may be recommended for wound healing and for skin breakdown while in a cast after surgery.

Other Nutrients

DRIs are good starting places to assess the need for vitamins and minerals in chronic conditions.[15] (DRI are located on the back cover.) As in all children, if the diet provides sufficient foods to meet the needs for protein, fats, and carbohydrates, it is likely the vitamin and mineral needs are also met. However, children with chronic conditions may have more difficulty meeting the DRI for vitamins and minerals as a result of these considerations:[7,16]

- Eating or feeding difficulties may restrict intake of foods requiring chewing, such as meats, so that certain minerals may be low in the diet.
- Prescribed medications and side effects can increase turnover for specific nutrients, raising the recommended amount needed.
- Food refusals are common with recurrent illness, so total intake may be more variable day-to-day than in other children of the same age.
- Treatment of the condition necessitates specific dietary restrictions, so that vitamins and minerals usually provided in restricted foods have to be supplemented.

Nutrients such as calcium that are low in the general population of children are also problem nutrients for children with chronic conditions.[17] The American Academy of Pediatrics statement on calcium would apply to children with chronic conditions. This statement recommends good quality food sources of calcium for all children. Food sources of calcium may avoid lead contamination that has been reported in a variety of over-the-counter calcium supplements.[18] Taking high levels of supplemental calcium does not clearly benefit children with chronic conditions.[17]

GROWTH ASSESSMENT

Children are expected to grow during middle childhood and preadolescence. CDC 2000 growth charts are a good starting place for assessing growth of any child. The growth chart with percentiles from the 5th to 95th are used most often, but those with 3rd and 97th percentile curves may be appropriate for tracking growth in some conditions.[19] As noted in Chapter 12, identification of children at risk for overweight and prevention of long-term cardiovascular risks are important aspects of growth assessments of children. Such concerns may or may not apply to children with chronic conditions, but a nutrition assessment can tailor nutrition goals to specific conditions. Families dealing with conditions that shorten life, such as cystic fibrosis, would not benefit from including overweight and its long-term risks as part of their child's nutrition assessment.

Families of children with severe disabilities who have wheelchairs may be apprehensive about growth as a goal for their child. They may have long-term concerns about caring for the child at home. Families may not want the child to grow at the usual rate when activities such as lifting from the bath tub or out of a wheelchair may become more difficult. Also children with rare degenerative conditions such as *cerebral spinal atrophy* have such major decreases in muscle size that growth may not occur.[10]

Most children with chronic conditions do grow, and assessing growth is an important component of nutrition services. If the child's condition is known to change the rate of weight or height gain, either slowing or accelerating it, the following signs need attention regardless of what growth chart is used:

- A plateau in weight.
- A pattern of gain and then weight loss.
- Not regaining weight lost during an illness.
- A pattern of unexplained and unintentional weight gain.

CEREBRAL SPINAL ATROPHY Condition in which muscle control declines over time as a result of nerve loss, causing death in childhood.

SECONDARY CONDITION Common consequence of a condition, which may or may not be preventable over time.

SEIZURES Condition in which electrical nerve transmission in the brain is disrupted, resulting in periods of loss of function that vary in severity.

SCOLIOSIS Condition in which the vertebral bones in the back show a side-to-side curve, resulting in a shorter stature than expected if the back was straight.

Growth Assessment and Interpretation in Children with Chronic Conditions

Factors that impact growth assessment and interpretation in childhood may not have been detectable earlier in younger children. These factors are the age of onset of the condition, *secondary conditions,* and activity. When the condition started may influence whether CDC growth charts are applicable. Early onset is more likely to impact growth than later onset in conditions such as *seizures.*[10] If the seizures started in middle childhood, the standard growth chart may be appropriate because the growth pattern of the child is already established. Onset of seizures in the neonatal period may reflect more severe brain damage, which slows growth rate markedly. Then the child's own growth record over time would be the best indicator of future growth.

Toddlers and preschoolers with cerebral palsy usually do not develop secondary conditions until childhood or later. *Scoliosis* is a secondary condition that interferes with accurate measurement of stature.[6] It may develop as a result of muscle incoordination and weakness in some forms of cerebral palsy in preadolescence. If a child with cerebral palsy has stature measurements that plateau or decline, it may be a result of cerebral palsy, scoliosis, or lack of adequate nutritional intake, or a combination of these three factors.[10] Nutritional interventions cannot prevent scoliosis, although nutritional consequences of its treatment may arise. Children may be provided custom-fitted back braces, so weight gain means the brace needs to be replaced. Also children with scoliosis braces may become less active because the brace restricts some types of movement. If scoliosis surgery is performed, the child may become slightly taller immediately, again showing that stature measurements have to be interpreted with care.

Body Composition and Growth

Children with special health care needs may or may not be typically proportioned in muscle size, bone structure, and fat stores. Some children in good nutritional status may plot at or below the lowest percentile on a standard growth chart for height. In fact low percentile heights are usual for a child with Down syndrome if growth is plotted on the CDC chart rather than the special growth chart for Down syndrome.[20] Short stature, low muscle tone, and low weight compared to age-matched peers are not attributed to not eating enough. They characterize the natural consequences of the *neuromuscular* changes within Down syndrome. Similarly a child with low muscle size could have a low weight and short stature. It would be unfair to assume that the child's diet is inadequate because of the low weight. A thorough assessment that includes body composition is necessary. For example, a thin-appearing child needs to have body fat stores measured before diet recommendations are made. If body fat stores are fine, adding calories is more likely to contribute to overweight.

Children with small muscle size will have lower weights than those with regular-sized muscles.[8,10] Conditions with altered muscle size may be described by terms such as *hypotonia* or *hypertonia*. Examples include cerebral palsy, Down syndrome, and spina bifida.[10] Not all muscles are affected. For example, some children with spina bifida have larger muscle size in the upper body and smaller muscle size in the lower body. Variation in size of muscles may make growth interpretation more difficult. Any assessment must address risks for overweight, which were discussed in Chapter 12, such as body mass (BMI) index and adiposity rebound. Standard interpretation may suggest a risk of overweight, but it may not accurately reflect that short stature is part of the child's condition. By standard interpretation, every child with Down syndrome or spina bifida could be overweight. For now, no established BMI tables cover specific conditions or the appropriate time for adiposity rebound.

Measuring body fat is another indicator of body composition. Skinfold fat measurements and their interpretation have to be based on consistent and repeatable standard methods.[21,22] Measurements of fat stores in children are not like measuring fat stores in adults because of the changes in body composition that come with age and growth. Calculated formulas and methods for body composition for children are not the same as adults.[8,23] Estimates of body composition for children with chronic conditions may be based on smaller sample sizes than for other children, but still such information is helpful. Identified low fat stores trigger recommendations to boost calories.

In-depth growth assessment may include head circumference measurement for all ages, with plotting and interpretation based on the Nellhaus head circumference growth chart as discussed in Chapter 9.[24] The reason head circumference is important is that children with unusually small heads have smaller brains, which is associated with short stature. Even with adequate diet and no documented eating problems, children with various genetic disorders tend to be shorter than age-matched peers.[25]

SPECIAL GROWTH CHARTS   Special growth charts have been published for a variety of genetic conditions.[20] Table 13.1 includes examples of these special growth charts. The number of children reported in such growth charts is not as large, nor as representative as the CDC 2000 growth charts. They are revised often based on new information emerging about the natural course of rare conditions. Some special growth charts are based on only the more severe forms of the condition, such as those living in residential care. Many chronic conditions do not have special growth charts because they present with a wide range of severity. Conditions without a specialty growth chart, which may or may not match the standard growth charts, include the following:

NEUROMUSCULAR   Term pertaining to the central nervous system's control of muscle coordination and movement.

JUVENILE RHEUMATOID ARTHRITIS   Condition in which joints become enlarged and painful as a result of the immune system, generally occurring in children or teens.

- *Juvenile rheumatoid arthritis*
- Cystic fibrosis
- Rett syndrome
- Spina bifida
- Seizures
- Diabetes

NUTRITION RECOMMENDATIONS

Children with chronic conditions require nutrition assessments to determine whether they are meeting their nutrient and caloric needs, whether eating problems such as food refusals or mealtime behavior is interfering with meeting nutritional needs, and whether growth is on target for their age and gender. Then nutrition interventions are provided based on the assessment. The goal is for the child to maintain good

Table 13.1
Examples of specialty growth charts.[2,20]

Conditions with Special Growth Charts	Comment
Achrondroplasia, Down Syndrome, Trisomy 13, Trisomy 18	Form of dwarfing
	Short stature, variable weight
Fragile X Syndrome	Short stature, primary in males
Prader-Willi Syndrome	Short stature, overweight
Rubinstein-Tabyi Syndrome	Short stature
Sickle Cell Disease	Short stature
Turner Syndrome	Short stature
Spastic Quadriplegia	Short stature, low weight
Marfan Syndrome	Tall stature

nutritional status, preventing nutrition-related problems from being superimposed on the primary condition.

Children with special health care needs benefit from the same nutritional recommendations as other children as discussed in Chapter 12, particularly in general areas such as dietary fiber or appropriate use of soft drinks. However, children with special health care needs may require recommendations regarding use of special formulas and nutrition support not needed for most children. Although developing feeding skills is important for toddlers and preschoolers, in childhood, abilities and/or disabilities that influence self-feeding and using utensils may be clearly recognized. More aggressive forms of providing sufficient nutrients may be required for feeding problems that have not been resolved by middle childhood and preadolescence. Forms of nutrition support common for children are enteral supplements, when oral feeding of regular foods is insufficient in quality or amount to maintain health.[26] Complete nutritional supplements are necessary to assure growth is not being limited in cases of questionable intake. Table 13.2 provides a list of commonly used complete nutritional supplements and examples of their use. Children under 10 years are generally provided a pediatric formula, but adult formulas may be used for children.

Methods of Meeting Nutritional Requirements

Children with special health care needs who cannot meet their nutrient requirements from regular foods may have complete nutritional supplements in addition to regular meals or in partial or complete replacement for meals. The first choice is that required supplements are drunk or eaten in the usual way. If this method does not work out, complete nutritional supplements can be administered by placement of a feeding *gastrostomy*.[6] Gastrostomy feeding may be required in children with kidney diseases, some forms of cancer, and severe forms of cerebral palsy and cystic fibrosis.[2,27,28] Many families experience difficulty accepting a gastrostomy for meeting nutritional requirements because feeding is such an important aspect of parenting.[29] Aside from emotional aspects, insurance coverage and financial aspects of paying for formulas fed by gastrostomies are major concerns for some families.

Children fed by gastrostomy can have many different schedules, such as eating orally during school and being fed by gastrostomy overnight. Table 13.3 gives an example of a feeding plan that includes gastrostomy feeding and oral feeding. If medications are required, they can be given through the gastrostomy also. For example, for children with pediatric AIDS who require many medications during the day, compliance with taking the drugs improved after gastros-

Table 13.2 Examples of nutritional supplements and formula for children.[26]

Formula	Comments
Pediatric versions of complete nutritional supplements, such as Pediasure®	Generally recommended for children under 10 years; can be used for gastrostomy or oral nutrition support.
Adult complete nutritional supplements, such as Ensure®	Generally 1 calorie per milliliter are recommended for children.
Enrichment of beverages, such as Carnation instant Breakfast® added to milk	Requires that milk is tolerated.
Predigested formula with amino acids and medium chain fatty acids, such as Peptamen junior®	For conditions in which intestinal absorption may be impaired.
Special formulas for inborn errors of metabolism (PKU), such as Phenex-2®.	Usually a powder that is mixed as a beverage, but other forms such as bars and capsules are available.
High-calorie booster for cystic fibrosis, such as scandishake®.	Generally 2.5 calories per milliliter to concentrate calories in small volume.

Table 13.3 Example of a feeding and eating schedule for an eight-year-old who eats by mouth and by gastrostomy.

Daily Schedule	Comments
6:30 A.M. Night feeding pump turned off	Overnight feeding by gastrostomy runs from 9:30 P.M. until 6:30 A.M., providing about 3 fl oz per hour, so no hunger in the morning is common.
7:15 A.M. Breakfast: refused	
8:00 A.M. Bus to school	
11:30 A.M. School lunch offered and about ½ is eaten: ½ chicken sandwich, all of French fries, with ⅓ of whole milk pint	Child has slow eating pace and is easily distracted by school lunchroom sounds.
3:30 P.M. After school snack at home of 4 oz pudding cup, two plain cookies, and 4 fl oz orange drink.	Mealtime behavior at home includes many attempts to leave the table, with prompting to eat from parents.
6:30 P.M. Evening meal at home: ½ cup mashed potato, 6 fl oz whole milk, refused vegetable and meat.	Parents hook up night feeding pump while the child is sleeping.
8:30 P.M. bedtime.	

tomies were placed. The parents spend less time trying to administer the medications to their children, and some children improved overall as a result of taking all of the required medications. Another example is a child who cannot safely drink liquids as a result of cerebral palsy. The child could have fluids given by gastrostomy, but eat solids foods by mouth. Children with gastrostomies can swim, bath, and do any activities they could do before the gastrostomy was placed.

Most formula fed by gastrostomy can be consumed as beverages. Even regular foods can be blended in a recipe for gastrostomy feeding for some children. Such "home brews" for gastrostomy feeding have to be carefully monitored because they are rich medium for bacterial contamination. Part of the decision-making process about use of special formulas and gastrostomy feeding often hinges on prior rejection of other feeding methods. When possible, gastrostomy feeding is planned as a temporary measure, with a return to oral eating later. For example a child who has a gastrostomy as a result of a kidney condition, may have a kidney transplant that allows removal of the gastrostomy after recovery.

Other nutrition supplements fed by gastrostomy have specific components that are unusual in beverages because they have such a strange taste. For example, formulas that contain individual amino acids generally are only accepted by those who have had them from infancy, as in the formulas for PKU. If a child required a new formula with amino acids, gastrostomy feeding would be more successful than oral feeding in most cases.

VITAMIN AND MINERAL SUPPLEMENTS FOR CHRONIC CONDITIONS

Children's complete vitamin and mineral supplements are recommended for a variety of chronic conditions to assure that the DRIs are provided. However most over-the-counter supplements are in the form of chewable tablets, so children who cannot chew well may require a liquid form of vitamins and minerals. The composition of vitamin and mineral supplements may be important because some have added ingredients not recommended for certain chronic conditions. Examples are vitamin and mineral brands with added carbohydrates, which are not allowed on a *ketogenic diet,* or those made with an artificial sweetener containing phenylalanine not recommended for children with PKU. (The ketogenic diet and PKU are discussed later in this chapter.)

The underlying diagnosis can make specific nutrients so important in the diet that they may be prescribed as pharmaceuticals. Cystic fibrosis treatment (discussed later in this chapter) requires fat-soluble vitamins supplemented as a result of the lower rate of intestinal absorption. Examples are vitamins A, D, E, and K for cystic fibrosis, or B_{12} injections for a protein-based inborn error of metabolism.[4,13] Vitamin C may be prescribed at high levels above the DRI for some children with spina bifida who have frequent bladder infections.[6] The vitamin is functioning as a medication rather than as a nutrient in this circumstance.

An excessive level of vitamins and minerals above the DRI can be a risk, especially for underweight children with various types of supplementation who may inadvertently exceed recommendations. For example, a child may be recommended Carnation Instant Breakfast from one source, another complete nutritional supplement by another provider, and take a chewable children's vitamin/mineral tablet too. Strategies to prevent both excessive and inadequate supplementation are to confirm that all medications, prescribed and over-the-counter, are identified within the nutrition and medical care plans.

Children with chronic conditions that limit activity or require medications that affect bone growth need special attention regarding calcium and vitamin D requirements.[17] Examples of such conditions are cerebral palsy and *galactosemia* in which dairy products are eliminated from the diet. Some children with these conditions are like older women with *osteoporosis* in that their bones are not strengthened by weight-bearing exercise and so calcium may move out of bones faster than it goes in. Providing additional calcium, phosphorous, and vitamin D may be recommended although the effectiveness of such supplementation is not known. Finding calcium supplements that can be taken for years or decades by children raises concerns about the lead content found in some calcium supplements.[18]

KETOGENIC DIET High-fat, low-carbohydrate meal plan in which ketones are made from metabolic pathways used in converting fat as a source of energy.

GALACTOSEMIA A rare genetic condition of carbohydrate metabolism in which a blocked or inactive enzyme does not allow breakdown of galactose. It can cause serious illness if not identified and treated soon after birth.

OSTEOPOROSIS Condition in which low density or weak bone structure leads to an increased risk of bone fracture.

Fluids

A chronic condition generally does not change the fluid requirement when the child is well. Guidelines for fluids for all children are appropriate, as discussed in Chapter 12. Special considerations for children with special health care needs are routine high fluid losses, such as children with drooling as a result of cerebral palsy.[6] Because constipation is common in children with neuromuscular disorders, adequate fluids are often stressed in those with cerebral palsy and spina bifida. Children with limited ability to talk may have more difficulty indicating thirst. Overall the child with a chronic health condition may have higher risks for dehydration when ill or as a result of prescribed medications' side effects.

NUTRITION-RELATED CONDITIONS

Children with chronic conditions may need more nutrition services than other children because of growth and eating problems. Children receive nutrition services in various settings, but state and national public health nutrition guidelines cover the practice of services for those with chronic conditions.[30] Diabetes, cerebral palsy, and PKU exemplify conditions in which nutrition services are an important component of overall management. Some conditions in younger children, discussed earlier, continue to have impacts on growth and eating during childhood, particularly feeding problems.

Eating and Feeding Problems in Children with Special Health Care Needs

Eating and feeding problems are when children have difficulty accepting foods, chewing them safely, or having sufficient foods and beverages to meet their nutritional requirements. About 70% of children with developmental delays have feeding problems, independent of whether neuromuscular problems have been identified.[6] Examples of these feeding problems include the following:

INSULIN Hormone usually produced in the pancreas to regulate movement of glucose from the blood stream into cells within organs and muscles.

- Self-feeding skills are lower than the child's chronological age, requiring assistance and supervision to assure adequate intake.
- Meals take so long or so much food is lost in the process of eating that the actual intake is lower than expected.
- The condition requires adjustment in the timing of meals and snacks at home and at school.

In children who do not have developmental delay, the impact of the chronic condition on eating may be behavioral problems at mealtimes, conflicts about control and independence about food choices, and variability in appetite. Families of children with

chronic conditions may focus on mealtimes and foods as methods to coping with their concerns about child's future. For example, families may be overprotective and restrict a child from eating at friend's homes, when such social activities may be appropriate.

CYSTIC FIBROSIS Cystic fibrosis (CF) is one of the most common lethal genetic disorders, with an incidence of one in 1500–2000 live births.[13] Its incidence is highest among Caucasians, with an incidence of one in 17,000 live births for American blacks. The CF gene is on the long arm of Chromosome 7, with many different genetic versions. The most common genetic mutation only fits 67% of the cases. CF impacts all the exocrine functions in the body, with the lung complications often causing death. Its major nutrition-related consequences are malabsorption of various nutrients, which can result in a slower rate of weight and height gain, and higher energy needs due to chronic lung infections. A multidisciplinary team, including a registered dietitian, respiratory therapist, pulmonary specialist, and social worker has been modeled in regional treatment programs.

Nutrition interventions for CF include monitoring growth, assessing dietary intake, and increasing calories and protein two to four times the usual recommendations to compensate for malabsorption. Every time a child with CF eats a meal or snack, he must take pills containing enzymes. Frequent eating and large calorie dense meals are encouraged. Gastrostomy feeding at night to boost calories is sometimes required.[19] Vitamin and mineral supplementation, particularly fat-soluble vitamins, is a part of daily management. Children with CF are at risk for developing diabetes because the pancreas is a target organ of CF damage.[13] In recent years, those diagnosed with CF live longer, as evidenced by the number of young adults living with CF.

Nutrition experts who work with children who have CF struggle to balance their high nutrient needs and frequent illnesses. Many children with CF have slow growth and are lower weight and shorter than expected. Even with nutrition support, decline in pulmonary function over time continues. Some children with CF have lung transplants if they meet strict eligibility requirements. CF is on the leading edge of gene therapy research, giving hope to families with young children.

DIABETES MELLITUS Diabetes mellitus is a disorder of *insulin* regulation in which dietary management is crucial. The hormone insulin may be underproduced, mistimed in its release, and/or peripheral tissues may become insensitive to it.[31] Both type 1 and type 2 diabetes mellitus are increasing in children for reasons that are not well understood.[14] Diabetes incidence estimates vary widely in children based on age and ethnicity. Children of Pima Indians in Arizona having a high incidence of type 2 diabetes at 22 per 1000 children compared to 7.2 cases per 100,000 in children seen in Ohio. Diabetes mellitus type 1 is more common in children than type 2. Type 1 diabetes is related to immune function and results in a lack of any insulin production. Type 2 diabetes is more common in older children, teens, and adults, and some insulin may be produced in the body.[32] Type 2 diabetes is closely associated with overweight, which may partially be managed by weight loss.[31] Consequences of poorly controlled diabetes in children are the same as in adults, and include risk of heart, eye, and kidney damage and premature death. Major changes in diabetes management for adults are based on 1994 guidelines that recommend increased monitoring of blood glucose and diet flexibility.[32] These concepts are also appropriate for children with diabetes, although some specific blood monitoring levels for adults are not appropriate for children. Children with diabetes are more likely to have both high and low blood sugars, not only high blood sugars.[14]

Treatment for diabetes is regulation of the timing and composition of meals and exercise with insulin or medications.[14] A third-grader with type 1 diabetes is likely to require an insulin injection once or twice per day, with oversight and modification of school breakfast, school lunch, and snack time, based on physical activity in school and after school activities. If the child is invited to a birthday party, the timing of meals and snack can be adjusted to allow the child to attend the party and eat most of the foods there. Common colds or foods a child refuses to eat can cause wide variation in blood sugar, contributing to irritability, sleepiness, or difficulty with schoolwork.

The trigger for diabetes type 1 in children is not yet known. One theory, now dispelled, blamed milk as a trigger for type I diabetes in children.[14] Education about all aspects of diabetes management is provided to the family and to the child. Different methods for controlling blood sugar place different emphasis on allowed and restricted foods. The method for young children may be to restrict simple sugars, but older children may be taught to count carbohydrates within meals.[26] Children with diabetes have many food choices but timing of meals and snacks and activity is important also. Parent and child education is crucial for good management, particularly for sick days. A genetic basis for diabetes is likely to be clarified as more is learned about environmental triggers and gene regulation.

SEIZURES   Epilepsy is a condition in which seizures occur. Seizures are uncontrolled electrical disturbances in the brain. Epilepsy and seizures are the same disorder. Seizures in children are a relatively common condition, with an incidence of 3.5 per 1000 children.[33] Seizure activity has a range of outward signs, from uncontrollable jerking of the whole body to mild blinking. Currently, no known nutrients bring on seizures.[34] Children who have seizures are usually treated by medications that prevent seizures. After some types of seizures, the child may have a period of semiconsciousness called a *postictal state* and appear to be sleeping but is difficult to wake. Feeding or eating during the postictal state is not recommended because the child may choke. Some children have long enough postictal states to miss meals. Then adding other eating times is needed to make up for the lost calories and nutrients.

When seizures are controlled by medications, growth usually continues at the rate typical for that child. Dietary consequences of controlled seizures are primarily related to drug-nutrient side effects, such as change in hunger or sleepiness. Some drugs should be taken without food, and others may be offered with snacks or meals. Most drugs have to be taken on a strict schedule and are not stopped without medical supervision.

Some children have uncontrolled seizures that may cause further brain damage over time. For reasons that remain unknown, seizures decrease when brain metabolism is switched from the usual fuel, glucose, to *ketones* from fat metabolism.[35] Some specialty clinics administer the *ketogenic diet* for uncontrolled seizures. The ketogenic diet severely limits carbohydrates and increases calories from fat. An example intake on a ketogenic diet is in Table 13.4. The diet is adequate in calories and protein. Vitamins and minerals have to be added as supplements because the allowed food sources of carbohydrates are not sufficient to meet vitamin and mineral requirements. The ketogenic diet may allow seizure medications to be reduced or eliminated over time. However, many difficulties lie in monitoring the reaction of the body to such a severe carbohydrate restriction, such as measuring growth, blood glucose, and ketones in urine. Growth during the time on a ketogenic diet may be different from the child's previous pattern. Some children improve in both weight and height when seizure activity declines. The ketogenic diet is so high in fat that some children gain weight faster than expected. The diet is generally recommended for two years, if it shows demonstrated effectiveness.

POSTICTAL STATE   Time after a seizure of altered consciousness appearing as a deep sleep.

KETOGENESIS   Condition in which ketones are made from metabolic pathways used in converting fat as a source of energy.

KETONES   Small two-carbon chemicals generated by breakdown of fatty acids for energy.

Table 13.4 Sample day for a ketogenic diet for a seven-year-old treated for seizures. Portion sizes are prescribed, and parents are taught to measure foods.

Breakfast
Beverage: heavy whipping cream (4 fl oz)
scrambled egg with bacon and mushrooms

Lunch
heavy whipping cream (4 fl oz)
hard-boiled egg mixed with mayonnaise
2 tablespoons green bean with butter

Afterschool snack
Carbohydrate-free multivitamin and mineral pill
sugar-free popsicle
diet soda

Dinner
heavy whipping cream (4 fl oz)
black olives (3)
sugar free gelatin topped with whipped cream
slice of full-fat ham
slice of tomato

Snack
walnuts (3)
diet soda

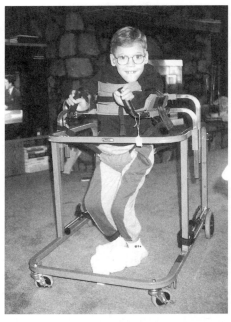

Illustration 13.1   Boy with CP in a walker.

CEREBRAL PALSY   Cerebral palsy (CP) is one of the more common conditions in children with severe disabilities. Overall incidence of cerebral palsy is about 2.5 per 1000 children.[33] *Cerebral palsy* is a general term well understood by the public; it covers a broad range of conditions resulting from brain damage. The form of cerebral palsy that presents the most nutrition problems is spastic quadriplegia, involving all limbs.[2] Most children with spastic quadriplegia appear thin, but this appearance may be a result of brain damage or muscle size. Children with cerebral palsy often have other forms of brain damage at the same time: 31%–54% have mental retardation; 30% have seizures, 10% have hearing impairment. The prevalence of CP in children born with very low birthweight has been increasing, but this trend may be a consequence of the overall increase in survivors of severe prematurity.[6] Children with CP can enjoy many activities, attend school, and later contribute to society. Persons with CP show a wide range of abilities. As indicated in the growth chart in Illustration 13.2, children with spastic quadriplegia grow, but their growth is slower than others, with or without gastrostomy feeding.[2]

Nutritional consequences of spastic quadriplegia are slow weight gain and other growth concerns, difficulty with feeding and eating, and changes in body composition. No specific vitamins or minerals are known to correct CP. Problem nutrients are likely to

ATHETOSIS Uncontrolled movements of the large muscle groups as a result of damage to the central nervous system.

be those related to bone density, calcium, and vitamin D, or nutrients needed in higher amounts as a result of medication side effects. Recommendations for caloric needs are difficult to determine, even with an in-depth growth assessment.[8] Children with small or weak muscles have lower caloric needs due to less active as a result of little voluntary muscle control. In contrast, types of cerebral palsy characterized by increased uncontrolled movement require extra calories as a result of higher activity level. *Athetosis* is an example of this less common form of CP in which increased energy needs were documented.[36] Altered body composition affects many aspects of the child's nutrition and eating abilities.[7] Eating or feeding problems may appear as spilling food, long mealtimes, fatigue at mealtime, and/or requiring assistance to eat. Difficulty in controlling muscles used in head position and sitting, and in the jaw, tongue, lips, and swallowing may contribute to feeding and eating problems.[6]

Nutrition experts who provide services for children with CP monitor their growth, and then make recommendations for food choices that fit their abilities for eating, for nutritional supplements if food and beverages are not providing sufficient nutrients, and for nutrition support if needed. Nutrition interventions may include the following:

- Working with a team approach to stimulate oral feeding.
- Working with school therapists on methods to promote healthy eating at school.
- Adjusting menus and timing of meals and

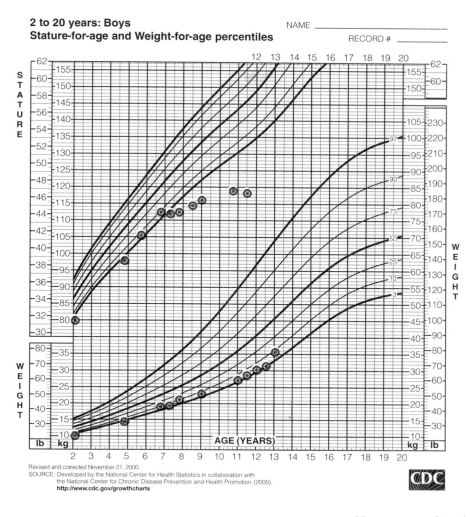

2 to 20 years: Boys
Stature-for-age and Weight-for-age percentiles

NAME _____

RECORD # _____

Revised and corrected November 21, 2000.
SOURCE: Developed by the National Center for Health Statistics in collaboration with
the National Center for Chronic Disease Prevention and Health Promotion (2000).
http://www.cdc.gov/growthcharts

Illustration 13.2 Growth chart for gastrostomy feeding for a boy with spastic quadriplegia and scoliosis.

PHENYLKETONURIA (PKU)

PKU is an inborn error of protein metabolism with a prevalence of 1 in 12,000 live births.[4] The only treatment is dietary management for life in which more than 80% of natural protein from foods and beverages need to be replaced by protein in which the amino acid phenylalanine has been removed. The enzyme that uses phenylalanine as a substrate is either not working at all or only partially active in a person with PKU. Treatment reduces this amino acid to the minimum amount needed as an essential amino acid. This strategy limits toxic breakdown products of accumulated phenylalanine that the body has difficulty clearing.

snacks at home or school for meeting nutrient needs from foods that minimize fatigue during meals.

- Assessing and adjusting the child's diet over time, such as after illness or before a planned surgery.
- Recommending adapted self-feeding utensils or other types of feeding equipment to allow the child to enjoy eating and to eat as independently as possible.

The various causes of CP all involve damage to the brain early in life, either before or after birth. The initial site of brain damage does not progress, but progression of secondary effects occurs over time. Secondary effects may include contractures, scoliosis, gastroesophageal reflux, and constipation.[6,10] The reason why many children with CP may have constipation is that coordinated muscle movements are part of bowel emptying, including the muscles in and over the intestines. Muscle coordination problems seen in the moving the arms and legs may show in muscles all over the body, including the abdominal muscles that assist in bowel evacuation.[7]

How excess phenylalanine causes mental retardation is not known. The PKU diet is required throughout life. If natural protein is consumed in too high amounts, PKU becomes a slowly degenerative disease affecting the brain at whatever age the treatment is stopped. Maternal PKU is a condition in which the fetus is damaged early in pregnancy by the high phenylalanine that crosses the placenta from the mother's blood. A woman with PKU has to continue strict dietary treatment during her pregnancy because her high level of phenylalanine would affect every pregnancy, even if the infant did not inherit her recessive gene.

When done correctly, children with PKU appear to be eating meals with less food than meals for other children. The diet is adequate in all vitamins, minerals, protein, fats and calories, but more nutrients are in liquid than in solid forms. Foods to be avoided completely are protein-rich foods such as meats, eggs, regular dairy products, peanuts, and soybeans in all forms. Allowed natural sources of protein are limited amounts of regular crackers, potato chips, rice, and potato. Many fruits and vegetables are encouraged, if offered without added sources of protein. Some foods

Illustration 13.3   A girl who does not appear to have a chronic illness, but who has PKU.

Table 13.5   Dietary recall for a five-year-old child with well-controlled PKU.

Breakfast	Dinner
2 slices low-protein bread with jelly and margarine	pickle spears (dill, 3 wedges)
6 cut-up orange pieces	1 c low protein imitation rice containing 1.5 tbs margarine
8 fl oz PKU formula	½ c grilled onions, green peppers, and mushrooms (on rice)
Lunch	1 c canned peaches in heavy syrup
½ c fruit cocktail in heavy syrup	8 fl oz PKU formula
1 c tossed salad (lettuce, tomato, celery, cucumber only) with 2 tbs Ranch dressing	**Snack**
17 French fries with ketchup	skittles candy (small snack size)
6 fl oz apple juice	4 fl oz apple juice
Afterschool snack	
½ c microwave popcorn	
8 fl oz PKU formula	

that are high in fats and sugars and generally low in natural protein, such as fried vegetables or candy canes, are safe for children with PKU. The phenylalanine-deficient protein is generally served as a liquid, called a medical food or formula. The vitamins and minerals required to meet the RDA are in the phenylalanine-deficient protein powder. If the child does not drink enough of the PKU formula, foods that the child eats to meet their vitamin, mineral, and calorie needs will elevate the blood phenylalanine. Table 13.5 is a diet recall from a child in good control of PKU. The phenylalanine-deficient protein also is available as bars and pills. It can be expensive to buy substitute low-protein alternative foods, such as low-protein pizza crusts, low-protein cheese, and low-protein baking mixes. Successful compliance requires use of low-protein foods to allow variety in the diet, such as low-protein pasta.

NEUROBEHAVIORAL   Pertains to control of behavior by the nervous system.

PSYCHOSTIMULANT   Classification of medication that acts on the brain to increase mental or emotional behavior.

ATTENTION DEFICIT HYPERACTIVITY DISORDER (ADHD)
Attention deficit disorder and attention deficit hyperactivity disorder are the most common *neurobehavioral* conditions in children. Its incidence is estimated at 3% to 5% of schoolage children.[37] Children suspected of having ADHD may have a chaotic meal and snack pattern and the inability to

stay seated for a meal. They may be given fewer opportunities to use kitchen appliances and get their own snack due to impulsiveness. Theories about specific foods or nutrients causing ADHD have not been proven scientifically, but high interest in nutrition as a cause and treatment continues. Recommended procedures to confirm ADHD include at least two sites completing observation checklists about behavior. Treatment with two approaches has been most effective. The two approaches are:

1. A structured behavioral approach that may also include mental health counseling and support, such as parenting classes.
2. A *psychostimulant* prescribed medication; examples are ritalin® or adderal®.

ADHD medications have been studied extensively regarding school performance, behavior, and side effects.[38] Nutritional concerns in ADHD include medication side effects that decrease appetite, maintaining growth while being medicated, and mealtime behavior. Low appetite as a result of treatment of ADHD is quite variable and depends on the timing of the medications compared to meals, dosage, and medication schedule, and how long the child has had to adjust to the medication. Less interference with appetite and growth is likely if the child does not take the medications during school holidays.

Regardless of the child's dosage schedule, ADHD medication peak activity is aimed for school hours, which includes school lunch. Nutrition interventions for children on psychostimulant medications are to adjust meals and snacks around the time the medication's action peaks. For example adding a large bedtime snack is a typical recommendation to make up for calories and nutrients not eaten during the day.

Monitoring weight and height carefully over time is needed to prevent growth plateaus. Education for the school's lunchroom supervisors and teachers may be helpful to deal with food refusals and mealtime behavior for the child having side effects.

DIETARY SUPPLEMENTS AND HERBAL REMEDIES

Children diagnosed and treated for chronic health conditions are often on many different medications. Families with children in a lengthy process of diagnosis, where the diagnosis does not lead to a definite treatment, and when expense, insurance coverage, and administrative problems tax their coping are at greatest risk of seeking alternative therapies, some of which are questionable in effectiveness, and perhaps even harmful. Nutritional claims for treatment abound for various chronic conditions. Families hear from one another about various micronutrients, such as magnesium, zinc, and B_6, sold with various combinations of amino acids for Down's syndrome and autism.[39] Restrictive diets, such as avoiding dairy products or gluten, have been researched for one condition, and then extrapolated for another. Sports drinks and high-protein products marketed for athletes may attract families with children who have difficulty gaining weight.

Strategies to counter unscientific nutritional claims for various products include the following:

- Recognize the benefits of supports for families, such as advocacy groups.
- Improve communication with health care providers, so that families ask more questions about nutrition claims of alternative treatments.
- Give families factual information without endorsing any claim, such as scientific literature or fact sheets, so families have some control over decision making for their children.

Including documentation in the child's records about all medications, including those supplements a family has selected, is important, even if the supplement does not interfere with the health of the child. If negative health impacts are suspected from such supplements, the nutrition professional may need to inform other providers and seek administrative or legal guidance.

SOURCES OF NUTRITION SERVICES

Most children who have chronic health problems attend regular schools, and do everything other children do. Major federal and state agencies provide resources for children with special health care needs, including greater access to nutrition services than generally available in the private sector.[40] As discussed in previous chapters, children with chronic conditions that interfere with their ability to function may be eligible for Supplemental Social Insurance (SSI). Low-income families are eligible for SSI depending on the child's condition. Examples of conditions usually qualifying for SSI are chromosomal disorders, mental retardation, and severe forms of seizures, cerebral palsy, and CF. A child with treated PKU would generally not be eligible for SSI because treatments prevent decline in learning abilities. Also, the American with Disabilities Act applies to all ages. It requires, for example, that school cafeteria lines accommodate wheelchairs.

Child Nutrition Program of the U.S. Department of Agriculture (USDA)

The Child Nutrition Program, as described in previous chapters, requires that school breakfast and lunch menus be modified for children with diagnosis-specific diets or changes in the texture of foods. Parents of children who want their child to participate in the Child Nutrition Program cannot be charged an additional fee for providing a special diet for their child. A registered dietitian or another health provider completes a prescription ordering special breakfasts or lunches. Examples of diet prescription orders are a reduced-calorie school lunch and breakfast, a pureed diet, or a nutrient-modified diet, such as a PKU diet. If a family does not want to participate in the Child Nutrition Program, preferring to pack a lunch, they can change their decision any time during the school year. Formulas administered by gastrostomy are not required to be supplied by the Child Nutrition Program.

Maternal and Child Health Block Program of the U.S. Department of Health and Human Services (HHS)

Every state has a designated portion of federal funding for children with special health care needs.[41] A wide range of services can be provided based on state planning as reported back to HHS. Nutrition services may be in specialty clinics, county health departments, or contracted for providing care, assuring resources such as formulas, foods, and nutrition education. Nutrition experts work with children in various settings, including schools, early intervention

Case Study 13.1.
Adjusting Caloric Intake for a Child with Spina Bifida

S am is a third-grader in regular classes at his public school. He uses a wheel-chair all the time and can transfer from his wheelchair to a chair by himself. He is on a toileting schedule at school with the assistance of a nurse. He participates in modified physical education as part of his physical therapy treatment. He likes to eat with his friends at school. His mother tries to make him cut back at the evening meal and has stopped buying some of his favorite snacks. He is mad at his mother since he likes his snacks after school when he is bored.

Nutrition assessment from his last visit at the spina bifida clinic at the local hospital showed that he was overweight by measuring his fat stores. Because he cannot stand, his stature was estimated by measuring his length lying down, and comparing it with his last length measurement. Standard methods could not be used to measure him, which limits the interpretation of his growth using the CDC growth chart. It showed he was at the 75% percentile in weight for his age, which is not overweight for his age. His rate of weight gain of 8 pounds per year, typical for a boy his age, is too fast for his low level of activity. His estimated calorie needs are 1100 per day due to low activity and short stature, about two-thirds of the caloric needs of others his age. Sam said he did not care about his size or being overweight. His mother was quite concerned that she would not be able to assist him if he fell or needed to be lifted.

Recommendations: The nutritionist at the clinic completed a school lunch pre-scription to reduce his caloric intake from 650 calories to 350 calories per lunch. His meal pattern was adjusted to two meals (breakfast and lunch) and two snacks per day at home, which better fits his low caloric needs. Sam was allowed to choose his favorite snack foods replacing his evening meal. Giving him choices about his snacks increased his sense of being in control and "growing up" and lowered the instances of his expressing anger at his mother about snack foods. The clinic nutritionist called the school to review his level of activity and confirm that the lunch changes were being implemented. The physical therapist at school found after-school swimming lessons and recommended them to his mother as a way to increase Sam's activity and socialization. In order to motivate Sam to pay attention to his eating and weight gain, his teacher and mother set up a monthly nonfood reward for him if he did not gain any weight. The effectiveness of the plan to cut his caloric intake and increase his activity was to be assessed at his next clinic visit when he would be weighed and fat stores measured.

programs, homes, clinics, and facilities. Also a program in every state identifies and advocates for children with special needs. An example is the Developmental Disabilities Council.

Public Schools Regulations: 504 Accommodation and IDEA

Two sets of regulations guide how schools provide nutrition services in addition to the Child Nutrition Program. Nutrition services are generally more available for younger children in schools than for older children, due in part to nutrition being written into more recent legislation. Children in regular education

have different access to services than do children eligible for special educational services.

504 ACCOMODATION   Children with special health care needs in the regular curriculum can have nutrition needs met through the 504 Accommodation of the Rehabilitation Act of 1973.[1] Examples of children who may need 504 accommodations are those with diabetes, cystic fibrosis, or a child treated for arthritis. A written plan, set in place, may provide an additional snack during the school day or additional time to eat during school lunch. The 504 accommodation procedures would be appropriate for documenting that a child with PKU needs an alternative

Case Study 13.2
Dealing with Food Allergies in School Settings

Judy was to start regular kindergarten now. When she was two years old, she was diagnosed with a peanut food allergy after many episodes of asthma and hives. Her health improved with avoidance of peanuts in all forms as a preschooler. The family carefully watched what she ate. However, at age four she had an episode of breathing difficulty that required an emergency room visit. This incident made the family quite concerned about her eating at school. She is generally not allowed to go to friends' homes to play; friends come to her house so the family can watch out for her. She has been instructed not to take any food from anyone. She has not been in day care or preschool, so starting school is a big step for the family.

Her mother met with the school staff to discuss plans for Judy at school to avoid exposure to peanuts in any form. The family did not want to participate in the school lunch program. She planned to pack Judy a lunch from home, although most children eat food provided at school. She provided a letter about the peanut allergy from her pediatrician to the school nurse, teacher, and principal at the school. They discussed snack time for Judy and her eating at the cafeteria, which periodically served food cooked in peanut oil, or containing peanuts. The snack at kindergarten is provided by parents, based on a rotation schedule. It is usually milk or juice with cookies or fruit. A plan was put in place for the teacher to check the snack foods and offer a replacement snack provided by Judy's mother if she was unsure about what was sent.

The school working-group wrote out a 504 accommodation plan for her peanut allergy. It included making sure that tables where children were eating peanut-containing foods were washed well, and posting signs in the cafeteria with Judy's picture to make sure she did not inadvertently get peanut-containing food from another child or in a food activity.

After Judy had been in school one month, the family met with the school group. In that month two episodes resulted in what may have been hives, and her family was worried that Judy was not adjusting well. At snack time Judy did not recognize some of the foods; she refused to eat snack most days. She appeared hungry after school at home. Her mother would like to send a snack to school that she knows her daughter will eat, and attend school snack time to make sure Judy is not being teased.

Assessment and recommendations: This case is an example of the careful balance between being a typical child and safely managing a health problem. Food issues at school come up with birthday parties, hands-on activities, and children swapping foods, not just snack time. How the family transfers the responsibility of avoiding peanuts to Judy's school is a key developmental step for all involved. To the extent possible the child should be treated like everyone else at school, and learn to accept a variety of snacks without peanuts. Giving the family a transition time of several months to adjust to the idea would be best. The plan may include the following: (1) send all parents a note saying no peanuts or peanut butter is to be in any snacks, (2) track Judy's acceptance of new foods month-to-month for one quarter. It is the parent's responsibility to have Judy reassessed medically every year to confirm her peanut allergy is being managed well. The medical visit would be the basis for changing plans for next year.

food during a classroom birthday or Halloween party, or when food activities take place in the classroom. This procedure gives parents a mechanism to

make sure all are aware of required dietary restrictions. It is not the responsibility of the child with PKU to inform new staff or substitutes about her need for a special diet. As a result of implementing the 504 accommodation, a child with PKU is not left out of food activities at school.

INDIVIDUALS WITH DISABILITIES EDUCATION ACT (IDEA)

Children eligible for special education are covered by regulations within the Individuals with Disabilities Education Act (IDEA).[1] It requires each child to have an individualized education plan (IEP) that may include nutrition-related goals and objectives as needed. The school staff must involve the parent in developing the IEP. For some diagnoses, it would be appropriate for a nutritionist to attend the IEP meeting to make sure the teacher, teacher aide, and other staff understand what the child needs. Nutrition, eating, and feeding problems may be a part of that plan, and apply to food offered in the classroom as well as that served in the regular school cafeteria. An example of IEP goals and objectives can be found in Table 13.6. For this particular plan, the child's education includes learning to eat by mouth with prompting and assistance. Nutritional supplements may be purchased as part of an education intervention called for in the child's IEP.

Nutrition Intervention Model Program

The Maternal and Child Health Bureau (MCH) is a part of the Department of Health and Human Services and involved in nutrition services for chronically ill children.[41] It develops and promotes model programs by funding competitive grants that emphasize training health care providers including nutrition experts. Model programs that are targeted for chil-

Table 13.6 Example of nutrition objectives in an individualized education plan for an eight-year-old boy with limited oral feeding skills.

1. In three of five trials JR will hold food on the spoon as he moves it to his mouth without hand over hand assistance of his aides during three meals per week.

2. JR will point to what he wants to eat with his left hand three trials after two prompts per meal three days each week.

3. JR will cooperate in having his gastrostomy site checked at feedings by pulling up his shirt three days in a row of each week.

dren with special health care needs are necessary because most health care providers are more comfortable caring for typical children than those with rare conditions. Families of children with special health care needs often have difficulty locating nutrition experts and other providers who are familiar with their complex nutritional, medical, and educational needs. Training programs vary in length from short, intensive courses to year-long traineeships. Topics vary from training for nutrition experts working with infants receiving intensive care services in hospitals to training counselors in nutrition-related of adolescence. Examples of such federal grant programs are the Pediatric Pulmonary Centers and Leadership Education in Neurodevelopmental Disabilities.[41] Incorporating nutrition faculty is one of the requirements of these grants. Aside from training clinical nutritionists through direct services, training may include developing family education materials, providing consultation for schools and community health care providers, and scholarly research.

Resources

Ketogenic Diet

This website is maintained by the Packard Children's Hospital at Stanford, which provides credible information for parents and providers about the ketogenic diet and places that use it.
Web site: www.stanford.edu/group/ketodiet

Exceptional Parent

This magazine is an excellent resource for a wide variety of conditions and how to work with providers and educators.
Web site: www.eparent.com

Cystic Fibrosis Foundation

This large organization has information about research, services, and policy related to cystic fibrosis, including nutritional products and recommendations.
Web site: www.cff.org

American Diabetes Association

Allows searches about diabetes in children around the world and provides professional and consumer publications.
Web Site: www.diabetes.org

National Down Syndrome Society

Includes information about its policy about nutritional products and directs parents to local resources for working with schools.
Web site: www.ndss.org

References

1. General Information about Disabilities. National Information Center for Children and Youth with Disabilities. U.S. Government Printing Office. Available from http://www.Nichcy.org/general.htm, February 1, 2001.

2. Krick J, Murphy-Miller P, Zeger S, Wright, E. Pattern of growth in children with cerebral palsy. J Am Diet Assoc 1996;96:680–685.

3. U.S. Department of Agriculture, Human Nutrition Information Service. Food Guide Pyramid; 1992; Home and Garden Bulletin No. 572.

4. Elsas LJ, Acosta, PB. Nutritional support of inherited metabolic disease. In: Shils ME, Olson JA, Shike M, Ross AC, eds. Modern nutrition in health and disease, 9th ed. Philadelphia: Williams and Wilkins; 1999:1003–56.

5. National Research Council. Recommended dietary allowances, 10th ed. Washington, DC: National Academy Press; 1989:12.

6. Blackman, JA. Medical aspects of developmental disabilities in children birth to three. Gaithersburg, MD: Aspen Publications; 1997;44–54, 88–90.

7. Stevenson, RD. Feeding and nutrition in children with developmental disabilities. Ped Annuals;1995:24:255–60.

8. Stallings VA. Resting energy expenditure. In: Altschuler SM, Liacouras CA. Clinical pediatric gastroenterology. Philadelphia: Harcourt Brace and Co.; 1998: 607–11.

9. L'Alleman D, Eiholzer U, Schlump M, et al. Cardiovascular risk factors improve during three years of growth hormone therapy in Prader-Willi syndrome. Eur J Pediatr 2000;159:836–42.

10. Minns, RA. Neurological disorders. In: Kelnar, CJH, et al., eds. Growth disorders. Chapman and Hall; 1998:447–70.

11. Schertz M, Adesman AR, Alfieri NE, Bienkowski RS. Predictors of weight loss in children with attention deficit hyperactivity disorder treated with stimulant medication. Pediatrics 1996;98:763–9.

12. Berney TP. Autism—an evolving concept. British J of Psychiatry 2000;176:20-5.

13. Wooldridge N. Pulmomary diseases. In: Samour PQ, Kelm KK, Lang CE. Handbook of pediatric nutrition, 2nd ed. Gaithersburg, MD: Aspen Publishers; 1999: 315–54.

14. Chalmers KH. Diabetes. In: Samour PQ, Kelm KK, and Lang CE. Handbook of pediatric nutrition, 2nd ed. Gaithersburg, MD; Aspen Publishers: 1999:425–51.

15. Trumbo P,Yates AA, Schlicker SA, Poos M. Dietary reference intakes; vitamin A, vitamin K, arsenic, boron, chromium, copper, iodine, iron, manganese, molybdenum, nickel, silicon, vanadium, and zinc. J Am Diet Assoc 2001;101:294–301.

16. American Academy of Pediatrics. Alternative routes of drug administration—advantages and disadvantages (subject review). Pediatrics 1997;100:143–52.

17. American Academy of Pediatrics. Calcium requirements of infants, children, and adolescents (RE9904). Pediatrics 1999; 1152–7.

18. Ross EA, Szabo NJ, Tabbett IR. Lead content of calcium supplements. J Amer Med Assoc 2000;284:1425–9.

19. National Center for Health Statistics. NCHS growth curves for children 0–19 years. U.S. Vital and Health Statistics, Health Resources Administration, U.S. Government Printing Office; 2000.

20. Growth references: third trimester to adulthood, 2nd ed. Clinton, SC: Greenwood Genetic Center; 1998.

21. Frisancho AR. Triceps skinfold and upper arm muscle size norms for assessment of nutritional status. Amer J Clin Nutr 1974;27:1052–7.

22. Fomon S, Haschke F, Ziegler E, Nelson SE. Body composition of reference children from birth to age 10 years. American J Clin Nutr 1982;35:1169–75.

23. Goran MI, Driscoll P, Johnson R, et al. Cross-calibration of body-composition techniques against dual-energy x-ray absorptiometry in young children. Am J Clin Nutr 1996;63:299–305.

24. Nellhaus G. Composite international and interracial graphs. Pediatrics 1968;41: 106–14.

25. Washington State Genetics. Genetics and your practice, 3rd ed. Available from http://mchneighborhood.ichp.edu/wagenetics/906317226.html, January 12, 2001.

26. Williams CP, ed. Pediatric manual of clinical dietetics. American Dietetic Association; 1998:479–525.

27. Steinkamp G, Von der Hardt H. Improvement of nutritional status and lung function after long-term nocturnal gastrostomy feedings in cystic fibrosis. J Pediatrics1994;121:244–9.

28. Sullivan PB. Gastrostomy feeding in the disabled child: when is the antireflux procedure required? Arch Dis Child. 1999;81: 463–4.

29. Shingadia D, Viani RM, Yogev R, et al. Gastrostomy tube insertion for improvement of adherence to highly active antiretrovial therapy in pediatric patients with human immunodeficiency virus. Pediatrics 2000;105:1–5. Available from http://www.pediatrics.org/cgi/content/full/105/6/e80, February 1, 2001.

30. Georgia Department of Human Resources, Division of Public Health, Office of Nutrition. Nutrition guidelines for practice, 1997. Section VI–VIII:1–11.

31. American Diabetes Association. Type 2 diabetes in children and adolescents. Pediatrics 2000;671–80.

32. American Diabetes Association. Nutrition recommendations and principles for people with diabetes mellitus. Diabetes Care 1998;21:532–5.

33. Lipkin PH. Epidemiology of the developmental disabilities. In: Capute AJ, Accardo PJ, ed. Developmental disabilities in infancy and childhood, 2nd ed. Baltimore, MD: Paul H. Brookes; 1996:137–56.

34. Devinsky O. A guide to understanding and living with epilepsy. Philadelphia: F. A. Davis Co.; 1994:320:220–6.

35. Tallian KB, Nahata MC, Tsao C-Y. Role of the ketogenic diet in children with intractable seizures. Ann Pharmacotherapy 1998;32:349–56.

36. Johnson RK, Goran MI, Ferrara MS, Poehlman ET. Athetosis increases resting metabolic rate in adults with cerebral palsy. J. Amer Diet Assoc 1996;96:145–8.

37. American Academy of Pediatrics Clinical Practice. Guideline: diagnosis and evaluation of the child with attention deficit/hyperactivity disorder. Pediatrics 2000;105:1158–70.

38. Schertz M, Adesman AR, Alfieri NE, Bienkowski RS. Predictors of weight loss in children with attention deficit hyperactivity disorder treated with stimulant medication. Pediatrics 1996;98:763–9.

39. Position statement on vitamin-related therapies. National Down Syndrome Society; 1997.

40. Story M, Holt K, Sofka D, eds. Bright futures in practice: nutrition. Arlington, VA: National Center for Education and Child Health; 2000:266–70.

41. Maternal and Child Health Bureau, Health and Human Services. Available from http://mchb.hrsa.gov/html/drte.html, January 12, 2001.

CHAPTER *14*

Photo Disc

The willingness of adolescents to try out new behaviors creates a unique opportunity for nutrition education and health promotion. Adolescence is an especially important time in the life cycle for nutrition education since dietary habits adopted during this period are likely to persist into adulthood.

Jamie Stang

ADOLESCENT NUTRITION

Prepared by **Jamie Stang**

CHAPTER OUTLINE

- Introduction
- Nutritional Needs in a Time of Change
- Normal Physical Growth and Development
- Normal Psychosocial Development
- Health and Eating-Related Behaviors during Adolescence
- Energy and Nutrient Requirements of Adolescents
- Nutrition Screening, Assessment, and Interventions
- Physical Activity and Sports
- Promoting Healthy Eating and Physical Activity Behaviors
- Discussion Points

KEY NUTRITION CONCEPTS

1 Nutrition needs should be determined by the degree of sexual maturation and biological maturity (biological age) instead of by chronological age.

2 Unhealthy eating behaviors common among adolescents include frequent dieting, meal skipping, use of unhealthy dieting practices, and frequent consumption of foods high in fat and sugar, such as fast foods, soft drinks, and savory snacks.

3 Concrete thinking and abstract reasoning abilities do not develop fully until late adolescence or early adulthood; therefore, education efforts need to be highly specific and based on concrete principles.

INTRODUCTION

Adolescence is defined as the period of life between 11 and 21 years of age. It is a time of profound biological, emotional, social, and cognitive changes during which a child develops into an adult. Physical, emotional, and cognitive maturity is accomplished during adolescence (Illustration 14.1). Many adults view adolescence as a tumultuous, irrational phase that children must go through. However, this view does disservice to its developmental importance. The tasks of adolescence, not unlike those experienced during the toddler years, include the development of a personal identity and a unique value system separate from parents and other family members, a struggle for personal independence accompanied by the need for economic and emotional family support, and the adjustment to a new body that has changed in shape, size, and physiological capacity. When the seemingly irrational behaviors of adolescents are reframed as essential endeavors and viewed in light of these developmental tasks, adolescence can and should be viewed as a unique, positive, and integral part of human development.

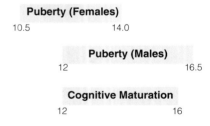

PUBERTY   The timeframe during which the body matures from that of a child to the body of a young adult.

NUTRITIONAL NEEDS IN A TIME OF CHANGE

The biological, psychosocial, and cognitive changes associated with adolescence have direct effects on

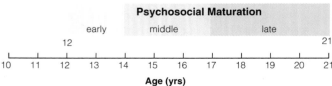

Illustration 14.1   Average ages of pubertal, cognitive, and psychosocial maturation.

SOURCE: Reprinted with permission from Johnson, RL. Adolescent growth and development. In: Hoffman A, Greydanus D, eds. Adolescent Medicine. New York: McGraw Hill; 1988.

nutritional status. The dramatic physical growth and development experienced by adolescents significantly increases their needs for energy, protein, vitamins, and minerals. However, the struggle for independence that characterizes adolescent psychosocial development often leads to the development of health-compromising eating behaviors, such as excessive dieting, meal skipping, use of unconventional nutritional and nonnutritional supplements, and the adoption of fad diets. These disparate situations create a great challenge for health care professionals. The challenging behaviors of adolescents can become opportunities for change at a time during which adult health behaviors are being formed. The search for personal identity and independence among adolescents can lead to positive, health-enhancing behaviors such as adoption of healthful eating practices, participation in competitive and noncompetitive physical activities, and an overall interest in developing a healthy lifestyle. These interests and behaviors provide a good foundation on which nutrition education can build.

This chapter provides an overview of normal biological and psychosocial growth and development among adolescents and how these experiences affect the nutrient needs and eating behaviors of teens. Common concerns related to adolescent nutrition and effective methods for educating and counseling teens are also discussed.

NORMAL PHYSICAL GROWTH AND DEVELOPMENT

Early adolescence encompasses the occurrence of *puberty,* the physical transformation of a child into a young adult. The biological changes that occur during puberty include sexual maturation, increases in height and weight, accumulation of skeletal mass, and changes in body composition. Even though the sequence of these events during puberty is consistent among adolescents, the age of onset, duration, and tempo of these events vary a great deal between and within individuals. Thus, the physical appearance of adolescents of the same chronological age covers a wide range. These variations directly affect the nutrition requirements of adolescents. A 14-year-old male who has already experienced rapid linear growth and muscular development will have noticeably different energy and nutrient needs than a 14-year-old male peer who has not yet entered puberty. For this reason, sexual maturation (or biological age) should be used to assess biological growth and development and the individual nutritional needs of adolescents rather than chronological age.

Sexual Maturation Rating (SMR), also known as "Tanner Stages," is a scale of *secondary sexual characteristics* that allows health professionals to assess the degree of pubertal maturation among adolescents, regardless of chronological age (Table 14.1). SMR is based on breast development and the appearance of pubic hair among females; and testicular and penile development and the appearance of pubic hair among males. SMR stage 1 corresponds with prepubertal growth and development, while stages 2 through 5 denote the occurrence of puberty. At SMR stage 5, sexual maturation has concluded. Sexual maturation correlates highly with linear growth, changes in weight and body composition, and hormonal changes.[1]

The onset of *menses* and changes in height relative to the development of secondary sexual characteristics that occur in females during puberty are shown in Illustration 14.2. Among females, the first signs of puberty are the development of breast buds and sparse, fine pubic hair occurring on average between 8 to 13 years of age (SMR stage 2). *Menarche* occurs two to four years after the initial development of breast buds and pubic hair, most commonly during SMR stage 4. The average age of menarche is 12.4 years, but menarche can occur as early as 9 or 10 years or as late as 17 years of age. Menarche may be delayed in highly competitive athletes or in girls who severely restrict their caloric intake to limit body fat.

Ethnic and racial differences are evident in the initiation of sexual maturation among females. Recent research suggests that African American girls experience puberty earlier than their Caucasian American peers.[2] By eight years of age, 48% of African American girls had reached SMR stage 2, compared to only 15% of Caucasian females. The average age of initial breast development is approximately 8.8 years for African American girls and 9.9 years for Caucasian females, while pubic hair growth begins at 8.7 years in African American females and about two years later in Caucasian girls. Menarche, however, occurs at about the same time: 12.2 years for African American and 12.8 for Caucasian adolescents. These findings suggest that the average length of puberty may be somewhat longer for African American females than for Caucasian females.

Onset of the linear growth spurt occurs most commonly during SMR stage 2 in females, beginning between the ages of 9.5 to 14.5 years in most females

> **SECONDARY SEXUAL CHARACTERISTICS** Physiological changes that signal puberty, including enlargement of the testes, penis, and breasts, and the development of pubic and facial hair.
>
> **MENSES** The process of menstruation.
>
> **MENARCHE** The occurrence of the first menstrual cycle.

Table 14.1. Sexual maturity rating for girls and boys.

Girls

Stage	Breast Development	Pubic Hair Growth
1	Prepubertal; nipple elevation only	Prepubertal; no pubic hair
2	Small, raised breast bud	Sparse growth of hair along labia
3	General enlargement of raising of breast and areola	Pigmentation, coarsening and curling, with an increase in amount
4	Further enlargement with projection of areola and nipple as secondary mound	Hair resembles adult type, but not spread to medial thighs
5	Mature, adult contour, with areola in same contour as breast, and only nipple projecting	Adult type and quantity, spread to medial thighs

Boys

Stage	Genital Development	Pubic Hair Growth
1	Prepubertal; no change in size or proportion of testes, scrotum, and penis from early childhood	Prepubertal; no pubic hair
2	Enlargement of scrotum and testes; reddening and change in texture in skin of scrotum; little or no penis enlargement	Sparse growth of hair at base of penis
3	Increase first in length then width of penis; growth of testes and scrotum	Darkening, coarsening and curling, increase in amount
4	Enlargement of penis with growth in breadth and development of glands; further growth of testes and scrotum, darkening of scrotal skin	Hair resembles adult type, but not spread to medial thighs
5	Adult size and shape genitalia	Adult type and quantity, spread to medial thighs

SOURCE: Reprinted with permission. Tanner, JM. Growth at adolescence. Oxford: Blackwell; 1962.

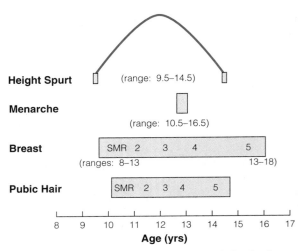

Illustration 14.2 Sequence of physiological changes during puberty in females.

SOURCE: Reprinted with permission. Tanner, JM. Growth at adolescence. Oxford: Blackwell; 1962.

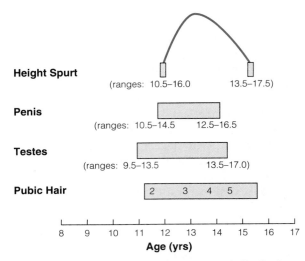

Illustration 14.3 Sequence of physiological changes during puberty in males.

SOURCE: Adapted from Tanner, JM. Growth at adolescence. Oxford: Blackwell; 1962.

(Illustration 14.2). Peak velocity in linear growth occurs during the end of SMR stage 2 and during SMR stage 3, approximately 6 to 12 months prior to menarche. As much as 15% to 25% of final adult height will be gained during puberty, with an average increase in height of 9.8 inches (25 cm).[3] During the peak of the adolescent growth spurt, females gain approximately 3.5 inches (8–9 cm) a year. The linear growth spurt lasts 24 to 26 months, ceasing by age 16 in most females. Some adolescent females experience small increments of growth past age 19 years, however. Linear growth may be delayed or slowed among females who severely restrict their caloric intake.

Enlargement of the *testes* and change in *scrotal* coloring are most often the first signs of puberty among males (Illustration 14.3), occurring between the ages of 10.5 and 14.5 years with 11.6 years being the average age. The development of pubic hair is also common during SMR stage 2. Testicular enlargement begins between the ages of 9.5 to 13.5 years in males (SMR 2 to 3) and concludes between the ages of 12.7 and 17 years. The average age of "spermarche" is approximately 14 years among males. Clearly, males show a great deal of variation in chronological age at which sexual maturation takes place.

On average, peak velocity of linear growth among males occurs during SMR stage 4, coinciding with or just following testicular development and the appearance of faint facial hair. The peak velocity of linear growth occurs at 14.4 years of age, on average.

TESTES One of two male reproductive glands located in the scrotum.

At the peak of the growth spurt, adolescent males will increase their height by 2.8 to 4.8 inches (7 to 12 cm) a year. Linear growth will continue throughout adolescence, at a progressively slower rate, ceasing about 21 years of age.

Changes in Weight, Body Composition, and Skeletal Mass

As much as 50% of ideal adult body weight is gained during adolescence. Among females, peak weight gain follows the linear growth spurt by three to six months. During the peak velocity of weight change, which occurs at an average age of 12.5 years, girls will gain approximately 18.3 pounds (8.3 kg) per year.[3] Weight gain slows around the time of menarche, but will continue into late adolescence. Adolescent females may gain as much as 14 pounds (6.3 kg) during the latter half of adolescence. Peak accumulation of muscle mass occurs around or just after the onset of menses.

Body composition changes dramatically among females during puberty, with average lean body mass falling from 80% to 74% of body weight while average body fat increases from 16% to 27% at full maturity. Females experience a 44% increase in lean body mass and a 120% increase in body fat during puberty.[4] Adolescent females gain approximately 2.5 pounds (1.14 kg) of body fat mass each year during puberty. Adolescent body fat levels peak among females between the ages of 15 to 16 years. Research by Frisch suggests that a level of 17% body fat is required for menarche to occur and that 25% body fat is required for the development and maintenance of regular ovulatory cycles.[5] Normal changes in body fat mass can be mediated by excessive physical activity and/or severe caloric restriction.

Even though the accumulation of body fat by females is obviously a normal and physiologically necessary process, adolescent females often view it negatively. Weight dissatisfaction is common among adolescent females during and immediately following puberty, leading to potentially health compromising behaviors such as excessive caloric restriction, chronic dieting, use of diet pills or laxatives, and, in some cases, the development of body image distortions and eating disorders.

Among males, peak weight gain coincides with the timing of peak linear growth and peak muscle mass accumulation.[3] During peak weight gain, adolescent males gain an average of 20 pounds (9 kg) per year. Body fat decreases in males during adolescence, resulting in an average of approximately 12% by the end of puberty.

Almost half of adult peak bone mass is accrued during adolescence. By age 18, more than 90% of adult skeletal mass has been formed.[1] A variety of factors contribute to the accretion of bone mass, including genetics, hormonal changes, weight-bearing exercise, cigarette smoking, consumption of alcohol, and dietary intake of calcium, vitamin D, protein, phosphorus, boron and iron. Because bone is comprised largely of calcium, phosphorus, and protein and because a great deal of bone mass is accrued during adolescence, adequate intakes of these nutrients are critical to support optimal bone growth and development.

NORMAL PSYCHOSOCIAL DEVELOPMENT

During adolescence, an individual develops a sense of personal identity, a moral and ethical value system,

feelings of self-esteem or self-worth, and a vision of occupational aspirations. Psychosocial development is most readily understood when it is divided into three periods: early adolescence (11 to 14 years), middle adolescence (15 to 17 years), and late adolescence (18 to 21 years). Each period of psychosocial development is marked by the mastery of new emotional, cognitive, and social skills (Table 14.2).

During early adolescence, individuals begin to experience dramatic biological changes related to puberty. The development of body image and an increased awareness of sexuality are central psychosocial tasks during this period of adolescence. The dramatic changes in body shape and size can cause a great deal of ambivalence among adolescents, leading to the development of poor body image and eating disturbances if not addressed by family or health care professionals.

Peer influence is very strong during early adolescence. Young teens, conscious of their physical appearance and social behaviors, strive to "fit in" with their peer group. The need to fit in can affect nutritional intake among adolescents. Focus groups conducted with adolescent females divide food into two main categories: junk foods and healthy foods.[6] Consumption of junk food was associated with friends, fun, weight gain, and guilt, while consumption of healthy food was associated with family, family meals, and home life. Clearly, teens express their ability and willingness to fit in with a group of peers by adopting food preferences and making food choices based on peer influences and by refuting family preferences and choices.

The wide chronological age range during which pubertal growth and development begins and proceeds can become a major source of personal dissatisfaction for many adolescents. Males considered to be "late bloomers" often feel inferior to their peers who mature earlier and may resort to the use of anabolic steroids and other supplements in an effort to

Table 14.2   Psychosocial processes and the substages of adolescent development.

Substage	Emotionally Related	Cognitively Related	Socially Related
Early adolescence	Adjustment to a new body image; adaptation to emerging sexuality	Concrete thinking; early moral concepts	Strong peer effect
Middle adolescence	Establishment of emotional separation from parents	Emergence of abstract thinking; expansion of verbal abilities and conventional morality; adjustment to increased school demands	Increased health risk behavior; heterosexual peer interests; early vocational plans
Late adolescence	Establishment of a personal sense of identity; further separation from parents	Development of abstract, complex thinking; emergence of post-conventional morality	Increased impulse control; emerging social autonomy; establishment of vocational capability

SOURCE: Reprinted with permission. Ingersoll GM. Psychological and social development. In: Textbook of adolescent medicine. Saunders Publishing; 1992.

increase linear growth and muscle development. Females who mature early have been found to have more eating problems and poorer body image than their later developing peers.[7,8] They are also more likely to initiate "grown-up" behaviors such as smoking, drinking alcohol, and engaging in sexual intercourse at an earlier age.[9] Education of young adolescents on normal variations in tempo and timing of growth and development can help to facilitate the development of a positive self-image and body image and may reduce the likelihood of early initiation of health-compromising behaviors.

Cognitively, early adolescence is a time dominated by concrete thinking, egocentrism, and impulsive behavior. Abstract reasoning abilities are not yet developed to a great extent in most adolescents, limiting their ability to understand complex health and nutrition issues. Young adolescents also lack the ability to see how their current behavior can affect their future health status or health-related behaviors.

Middle adolescence marks the development of emotional and social independence from family, especially parents. Conflicts over personal issues, including eating and physical activity behaviors, are heightened during mid-adolescence. Peer groups become more influential and their influence on food choices peaks. Physical growth and development are mostly completed during this stage. Body image issues are still of concern, especially among males who are late to mature and females. Peer acceptance remains important, and the initiation of and participation in health compromising behaviors often occurs during this stage of development. Adolescents may believe they are invincible during this stage of development.

SELF-EFFICACY The ability to make effective decisions and to take responsible action based upon one's own needs and desires.

The emergence of abstract reasoning skills occurs rapidly during middle adolescence; however, these skills may not be applied to all areas of life. Adolescents will revert to concrete thinking skills if they feel overwhelmed or experience psychosocial stress. Teens begin to understand the relationship between current health-related behaviors and future health, even though their need to "fit in" may supplant this understanding.

Late adolescence is characterized by the development of a personal identity and individual moral beliefs. Physical growth and development is largely concluded, and body image issues are less prevalent. Older teens become more confident in their ability to handle increasingly sophisticated social situations, which is accompanied by reductions in impulsive behaviors and peer pressure. Adolescents become increasingly less economically and emotionally dependent upon parents. Relationships with one individual become more influential than the need to fit in with a group of peers. Personal choice emerges.

Abstract thinking capabilities are realized during late adolescence, which assists teens in developing a sense of future goals and interests. Adolescents are now able to understand the perspectives of others and can fully perceive future consequences associated with current behaviors. This capability is especially important among adolescent females who plan to have children or who become pregnant.

HEALTH AND EATING-RELATED BEHAVIORS DURING ADOLESCENCE

Eating patterns and behaviors of adolescents are influenced by many factors, including peer influences, parental modeling, food availability, food preferences, cost, convenience, personal and cultural beliefs, mass media, and body image. Illustration 14.4 presents a conceptual model of the many factors that influence eating behaviors of adolescents. The model depicts three interacting levels of influence that impact adolescent eating behaviors: personal or individual, environmental, and macrosystems. Personal factors that influence eating behavior include attitudes, beliefs, food preferences, *self-efficacy,* and also biological changes. Environmental factors include the immediate social environment such as family, friends and peer networks, and other factors such as school, fast-food outlets, and social and cultural norms. Macrosystem factors, which include food availability, food production and distribution systems, and mass media and advertising, play a more distal and indirect role in determining food behaviors, yet can exert a powerful influence on eating behaviors. To improve the eating patterns of youth, nutrition interventions should be aimed at each of the three levels of influence.

Eating habits of adolescents are not static; they fluctuate throughout adolescence, in relation to psychosocial and cognitive development. Longitudinal data of adolescent females suggest that even though body weight percentiles track throughout adolescence, little consistency guides the intakes of the energy, nutrients, vitamins, and minerals from early to late adolescence.[10] Health professionals must therefore refrain from jumping to conclusions about the dietary habits of adolescents, (even if they have been evaluated for nutritional status at an earlier age) and take the time to assess the current dietary intake of the individual.

Adolescents lead busy lives. Many are involved in extracurricular sports or academic activities, others are employed, and many must care for younger children in a family for part of the day. These activities,

Illustration 14.4 **Conceptual model for factors influencing eating behavior of adolescents.**

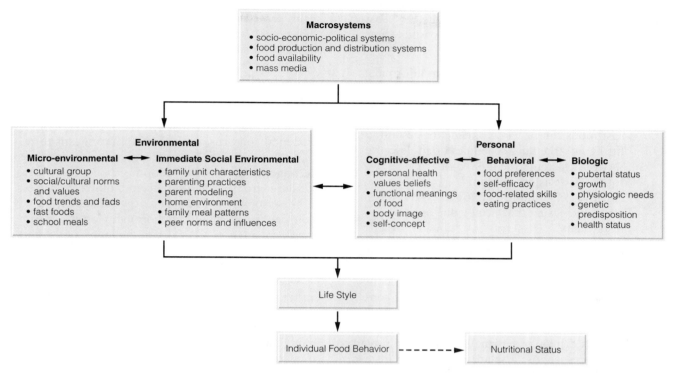

Macrosystems
- socio-economic-political systems
- food production and distribution systems
- food availability
- mass media

Environmental

Micro-environmental
- cultural group
- social/cultural norms and values
- food trends and fads
- fast foods
- school meals

Immediate Social Environmental
- family unit characteristics
- parenting practices
- parent modeling
- home environment
- family meal patterns
- peer norms and influences

Personal

Cognitive-affective
- personal health values beliefs
- functional meanings of food
- body image
- self-concept

Behavioral
- food preferences
- self-efficacy
- food-related skills
- eating practices

Biologic
- pubertal status
- growth
- physiologic needs
- genetic predisposition
- health status

Life Style

Individual Food Behavior - - - - -> Nutritional Status

SOURCE: Reprinted with permission. Story M, Alton I. Becoming a woman: nutrition in adolescence. In: Krummel DA, Kris-Etherton PM, eds. Nutrition in women's health. Gaithersburg, MD: Aspen Publishers; 1996.

combined with the increased need for social and peer contact and approval, and increasing academic demands as they proceed through school, leave little time for adolescents to sit down to eat a meal. Snacking and meal skipping are commonplace among adolescents.

Almost all adolescents consume at least one snack per day, with a range of one to seven.[11,12] One study of adolescents found that during an average week, males ate 18.2 meals and 10.9 snacks, while females ate 16.9 meals and 9.9 snacks.[6] Snacks account for 25% to 33% of daily energy intakes among adolescents.[12] National data suggest that the proportion of calories and nutrients from foods consumed as snacks has risen during the past decade.[13] Unfortunately, the food choices made by adolescents while snacking tend to be high in sugar, sodium, and fat, while relatively low in vitamins and minerals. Soft drinks are the most commonly chosen snacks for adolescent females and account for about 6% of total caloric intake.[14] This trend causes significant concern because high consumption of soft drinks increases the risk for bone fractures over an individual's lifetime.[15]

The occurrence of meal skipping increases as adolescents mature. Breakfast is the most commonly skipped meal; only 29% of adolescent females tend to eat breakfast daily.[11] Skipping breakfast can dramatically decrease intakes of energy, protein, fiber, calcium, and folate due to the absence of breakfast cereal or other nutrient-dense foods commonly consumed at breakfast. Lunch is skipped by almost one quarter of adolescents.[16,17] As with breakfast, skipping lunch reduces intakes of energy, protein, and other nutrients. Adolescents who skip meals should be counseled on convenient, portable, and healthy food choices that can be taken with them and eaten as meals or snacks.

As adolescents mature, they spend less time with family and more time with their peer group. Eating away from home becomes prevalent: female adolescents, for example, eat almost one-third of their meals away from home and the average teen eats at a fast-

Connections
Motivational Messages

As present-oriented persons, teens are generally unconcerned about how their current eating will affect their health in later years. However, they *are* concerned about their appearance, weight, and having a lot of energy. They should be told to "eat well because it will help you in what you do and what you want to become." This message contrasts with food advertising and images of health and attractiveness presented to teens by magazines, television, and other media sources.

food restaurant twice a week. Fast-food visits account for 31% of all food eaten away from home, and make up 83% of adolescent visits to restaurants.[16,18] Fast-food restaurants and food courts are favorite eating places of teens for several reasons:

- They offer a social setting with an informal, comfortable atmosphere for adolescents.
- Fast foods are relatively inexpensive and offer socially acceptable choices.
- Fast foods can be eaten outside the restaurant, fitting into the busy schedules of adolescents.
- Service is fast and the limited offerings allow for quick decision making.
- Fast-food restaurants employ many adolescents, increasing the social value of these restaurants.

Eating at fast-food restaurants has direct bearing on the nutritional status of adolescents. Many fast foods are high in fat and low in fiber and nutrients. However, specific choices increase the nutrient content of fast-food meals and decrease the fat content. Adolescents can be counseled to ask for juice or milk instead of soft drinks, order small sandwiches instead of larger choices, choose a salad as a side dish instead of fries, order grilled items as opposed to fried sandwiches, and avoid "super sizing" meals, even if it seems to offer a better economic deal.

Vegetarian Diets

The term *vegetarian* is used quite broadly and can consist of many different eating patterns. Table 14.3 lists the most common vegetarian diet patterns along with the foods most commonly excluded. Among

Table 14.3
Types of vegetarian diets and food excluded.

Type of Vegetarian Diet	Foods Excluded
Semi- or partial-vegetarian	Red meat
Lacto-ovovegetarian	Meat, poultry, fish, seafood
Lacto-vegetarian	Meat, poultry, fish, seafood, eggs
Vegan (total vegetarian)	Meat, poultry, fish, seafood, eggs, dairy products (may exclude honey)
Macrobiotic	Meat, poultry, eggs, dairy, seafood, fish (fish may be included in the diets of some macrobiotic vegetarians)

SOURCE: Reprinted with permission. Haddad E, Johnston P. Vegetarian diets and pregnant teens. In: Story M, Stang J, eds. Nutrition and the pregnant adolescent: a practical reference guide. Minneapolis, MN: Center for Leadership, Education, and Training in Maternal and Child Nutrition, University of Minnesota; 2000.

low-literacy populations, vegetarian may be thought to mean that a person eats vegetables. Therefore, health professionals should ask adolescents to define what type of vegetarian diet they consume and to elicit a complete list of foods that are avoided.

The prevalence of vegetarianism among adolescents is small—approximately 1% of adolescents report consuming a vegetarian diet.[19] Adolescents adopt vegetarian eating plans for a variety of reasons, including cultural or religious beliefs, moral or environmental concerns, health beliefs, as a means to restrict calories and/or fat intake, and as a means of exerting independence by adopting eating behaviors that differ from those of the teen's family. Regardless of the reason for consuming a vegetarian diet, the adolescent's diet should be thoroughly assessed for nutritional adequacy. As a rule, the more foods that are restricted in the diet, the more likely it is that nutritional deficiencies will result.

Vegetarian adolescents have been found to be shorter and leaner than omnivores during childhood and to enter puberty at a later age. On average, menarche occurs six months later in vegetarians than among omnivores.[20] After puberty, vegetarian adolescents are as tall or taller than omnivores and are generally leaner, although final adult height may be reached at a later age.[20,21]

Well-planned vegetarian diets can offer many health advantages to adolescents, such as a high intake of complex carbohydrates and relatively high intake of vitamins and minerals found in plant-based foods. Data suggest that vegetarian adolescents are twice as likely to consume fruits and vegetables, one-third as likely to consume sweets, and one-fourth as likely to consume salty snack foods once per day.[19] When well planned, vegetarian diets can provide adequate protein to promote growth and development among pubescent adolescents, particularly if small amounts of animal-derived foods, such as milk or cheese, are consumed at least two times per week. If vegetarian diets restrict intake of all animal-derived food products, however, careful attention must be paid to assure adequate intakes of protein, calcium, zinc, iron, and vitamins D, B_6, and B_{12}.[21] Supplements of vitamin B_{12} are often required among vegans. A suggested dietary food guide for adolescent vegetarians is listed in Table 14.4.

Adolescents who consume vegan diets must also be assessed for adequacy of total fat and essential fatty acid intakes. Docosahexaenoic acid (DHA) is derived from alpha-linolenic acid. Although it is found in soy products, flax seed, nuts, eggs, and canola oil, intake is very low in the diets of vegans. Diets that are low in fat may not supply an adequate ratio of linoleic acid to alpha-linolenic acid (5:1 to 10:1) in order to facilitate the metabolism of alpha-linolenic acid to DHA.[21]

Table 14.4   Suggested daily food guide for lacto-ovovegetarians and vegan adolescents at various intake levels.

Food Groups	Servings per Day, by Daily Caloric Intake	
	Lacto-Ovovegetarians	**Vegans**
	11+ years	**11+ years**
	(2,200–2,800 kcal)	**(2,200–2,800 kcal)**
Breads, grains, cereal	9–11	10–12
Legumes	2–3	2–3
Vegetables	4–5	1
Fruits	4	3–4
Nuts, seeds	1	4–6
Milk, yogurt, cheese	3	1
Eggs (limit 3/week)	½	3
Fats, oils (added)	4–6	4–6
Sugar (added teaspoons)	6–9	6–9

SOURCE: Data used with permission from: Haddad EH. Development of a vegetarian food guide. Am J Clin Nutr 1994; 1248S–1254S; and Story M, Holt K, Sofka, D, eds. Bright futures in practice: nutrition. Arlington, VA: National Center for Education in Maternal and Child Health; 2000.

Therefore, particular attention should be paid to sources of fat in the diets of vegans and other vegetarians with low fat intakes. Some plant sources of alpha-linolenic acid are shown in Table 14.5.

Adolescents who consume a vegetarian diet, particularly if they report doing so for health- or weight-related reasons, should be carefully assessed for the presence of eating disorders, chronic dieting, and body image disturbances. Neumark-Sztainer and colleagues have found that vegetarian adolescents are somewhat more likely to report binge eating, almost twice as likely to report frequent or chronic dieting, four times more likely to report purging, and eight times more likely to report laxative use than nonvegetarian peers.[19] These results seem to stem from the fact that many individuals who are chronic dieters or who have eating disorders adopt a vegetarian diet as a means of self-denial or self-control rather than the vegetarian diet causing these behaviors.

Table 14.5

Plant sources of alpha-linolenic acid.

Food Source	Alpha-Linolenic Acid (g)
Flax Seed, 2 Tb	4.3
Walnuts, 1 oz	1.9
Walnut oil, 1 Tb	1.5
Canola oil, 1 Tb	1.6
Soybean oil, 1 Tb	0.9
Soybeans, ½ c cooked	0.5
Tofu, ½ c	0.4

SOURCE: Reprinted with permission. Haddad E, Johnston P. Vegetarian diets and pregnant teens. In: Story M, Stang J, eds. Nutrition and the pregnant adolescent: a practical reference guide. Minneapolis, MN: Center for Leadership, Education, and Training in Maternal and Child Nutrition, University of Minnesota; 2000.

Dietary Intake and Adequacy Among Adolescents

Data on food intakes of U.S. adolescents suggest that many adolescents consume diets that do not match the Dietary Guidelines for Americans or the Food Guide Pyramid recommendations.[22,23] National data suggest that only 1% of teens consume diets that meet recommendations for all food groups in the Food Guide Pyramid, and 5% meet the recommendations for at least four food groups.[24] More than 45% of teens meet recommendation for one of the food groups, while 7% of male and 18% of female adolescents do not meet any of the Food Guide Pyramid recommendations. Inadequate consumption of dairy products, grain products, fruits and vegetables is commonplace among adolescents (Table 14.6).

Few adolescents meet the 5 A Day recommendation for fruit and vegetable consumption. Data from the CSFII suggests that 32% of males and 18% of females met this recommendation over a three-day period.[25] Krebs-Smith and colleagues found that 56.7% of male and 63.7% of female teens consumed less than one serving of fruit per day. When all vegetables were included in analyses, 9% of males and 13% of females consumed less than 1 serving of vegetables per day. However, when french fries were excluded from analyses, 20% of males and 26% of females consumed less than one serving of vegetables per day. French fries alone account for 23% of all vegetables consumed by adolescents. Data from the Minnesota Adolescent Health Survey suggest a slightly different pattern of fruit and vegetable intake, where 28% of adolescents consumed fruit and 36% consumed vegetables on a less-than-daily basis.[26] Both surveys indicate that fruit and vegetable intake is not adequate to promote optimal health and reduce risk of chronic diseases. Clearly, adolescents do not consume diets that comply with the national nutrition recommendations or provide the recommended level of intakes for all food groups.

Socioeconomic status (SES) appears to be related to food intake patterns. Consumption of fruit and

Table 14.6   Percentage of adolescents meeting the recommended number of Food Guide Pyramid servings for select food groups.

	Male (%)	**Female (%)**
Dairy products	49	22
Fruits	17	19
Vegetables	50	46
Grains	43	21

SOURCE: Data from Munoz KA, Krebs-Smith SM, Ballard-Barbash R, Cleveland LE. Food intakes of US children and adolescents compared with recommendations. Pediatrics May 1998;101(5):952–3; Pediatrics 1997;100:323–9.

fruit juices, grain products, low fat/skim milk and milk products, soft drinks, and sugars and sweets tends to increase as SES increases.[13,25] Consumption of vegetables is also positively related to SES, with the exception of starchy vegetables (white potatoes, dried beans, green peas, corn) and fried potatoes, which are negatively related to SES. Consumption of whole milk and total meat/poultry/fish intakes decreases among adolescents as SES increases.

ENERGY AND NUTRIENT REQUIREMENTS OF ADOLESCENTS

Increases in lean body mass, skeletal mass, and body fat that occur during puberty result in energy and nutrient needs that exceed those of any other point in life. Energy and nutrient requirements of adolescence correspond with the degree of physical maturation that has taken place. Unfortunately, little available data define optimal nutrient and energy intakes during adolescence. Most existing data is extrapolated from adult or child nutritional requirements. Recommended intakes of energy, protein and some other nutrients are based upon adequate growth as opposed to optimal physiological functioning. The Dietary

Reference Intakes (DRIs) provide the best estimate of nutrient requirements for adolescents (Table 14.7). It should be noted, however, that these nutrient recommendations are classified according to chronological age, as opposed to individual levels of biological development. Thus, health care professionals must use prudent professional judgment based on SMR status, and not solely on chronological age, when determining the nutrient needs of an adolescent.

Nutrient intakes of U.S. adolescents suggest that many adolescents consume inadequate amounts of vitamins and minerals; this trend is more pronounced in females than males. It is not surprising, given the fact that most adolescents do not consume diets that comply with the Food Guide Pyramid or the Dietary Guidelines for Americans. On average, adolescents consume diets inadequate in several vitamins and minerals, including folate, Vitamins A, B₆, and E, iron, zinc, and calcium.[27] Dietary fiber intake among adolescents is also low. Diets consumed by many teens exceed current recommendations for total and saturated fats, cholesterol, sodium, and sugar. Data on nutrient intakes of adolescents taken from the 1994–1995 CSFII suggest that 47% to 62% of teens consume less than 75% of the DRI for calcium, 49% to 52% consume less than 75% of the DRI for zinc, and 36% to 55% consume less than 75% of the DRI

Table 14.7   Dietary reference intakes of selected nutrients for preadolescents and adolescents.

Life Stage Group	Calcium (mg/d)	Phosphorus (mg/d)	Magnesium (mg/d)	Vitamin D (μg/d)[a,b]	Fluoride (mg/d)	Thiamin (mg/d)	Riboflavin (mg/d)	Niacin (mg/d)[c]
Males								
9–13 years	1,300*	**1,250**	**240**	5*	2*	**0.9**	**0.9**	12
14–18 years	1,300*	**1,250**	**410**	5*	3*	1.2	1.3	16
19–30 years	1,000*	**700**	**400**	5*	4*	1.2	1.3	16
Females								
9–13 years	1,300*	**1,250**	**240**	5*	2*	**0.9**	**0.9**	12
14–18 years	1,300*	**1,250**	**360**	5*	3*	1.0	1.0	14
19–30 years	1,000*	**700**	**310**	5*	3*	1.1	1.1	14
Pregnancy								
≤ 18 years	1,300*	**1,250**	**400**	5*	3*	1.4	1.4	18
19–30 years	1,000*	**700**	**350**	5*	3*	1.4	1.4	18
Lactation								
≤ 18 years	1,300*	**1,250**	**360**	5*	3*	1.4	1.6	17
19–30 years	1,000*	**700**	**310**	5*	3*	1.4	1.6	17

NOTE: This table presents Recommended Dietary Allowances (RDAs) in **bold type** and Adequate Intakes (AIs) in ordinary type followed by an asterisk (*). RDAs and AIs may both be used as goals for individual intake. RDAs are set to meet the needs of almost all (97 to 98 percent) individuals in a group. For healthy and breastfed infants, the AI is the mean intake. The AI for other life stage and gender groups is believed to cover needs of all individuals in the group, but lack of data or uncertainty in the data prevent being able to specify with confidence the percentage of individuals covered by this intake.

[a] As cholecalciferol. 1 μg cholecalciferol=40 IU vitamin D.
[b] In the absence of adequate exposure to sunlight.
[c] As niacin equivalents (NE). 1 mg of niacin = 60 mg tryptophan; 0–6 months = preformed niacin (not NE).

for vitamins A and E. Similar findings have been reported from the School Nutrition Dietary Assessment Study (SNDAS).[28] More than one-third of females consume inadequate amounts of all of these nutrients.[27]

Based on growth and development of adolescents, as well as national findings on dietary intakes of foods and nutrients, adolescent diets should be assessed for adequacy of intake of vitamins and minerals, energy, protein, carbohydrate and fiber. Nutrients of particular concern for teens are discussed in greater detail in the following sections.

Energy

Energy needs of adolescents are influenced by activity level, basal metabolic rate, and increased requirements to support pubertal growth and development. Basal metabolic rate (BMR) is closely associated with the amount of lean body mass of individuals. Because adolescent males experience greater increases in height, weight, and lean body mass, they have significantly higher caloric requirements than females. The RDAs for total calories and calories per centimeter of height by age group are listed in Table 14.8. Due to the great variability in the timing of growth and maturation among adolescents, the calculation of energy

needs based on height will provide a better estimate than the total caloric recommendation. Adolescents, especially females, may not meet the DRI for total energy intakes. Approximately 99% of teen males and 86% of teen females meet or exceed the DRI for energy.[13]

The DRI for energy is based upon the assumption of a light to moderate activity level. Therefore, adolescents who participate in sports, those who are in training to increase muscle mass, and those who are more active than average may require additional energy to meet their individual needs. Conversely, adolescents who are not physically active or those who have chronic or handicapping conditions that limit their mobility will require less energy to meet their needs. Physical activity has been found to decline throughout adolescence, with approximately one in four adolescents involved in no physical activity.[29] Therefore, caloric needs of older adolescents who have completed puberty and are less active may be significantly lower than those of younger, active, still-growing adolescents.

Physical growth and development during puberty is sensitive to energy and nutrient intakes. When energy intakes fail to meet requirements, linear growth may be retarded and sexual maturation may be delayed. The standard way to gauge adequacy of energy

Table 14.7 Dietary reference intakes of selected nutrients for preadolescents and adolescents (continued)

Life Stage Group	Vitamin B$_6$ (mg/d)	Folate (µg/d)	Vitamin B$_{12}$ (µg/d)	Pantothenic Acid (mg/d)	Biotin (µg/d)	Choline[d] (mg/d)	Vitamin C (mg/d)	Vitamin E (mg/d)	Selenium (µg/d)
Males									
9–13 years	1.0	300	1.8	4*	20*	375*	45	11	40
14–18 years	1.3	400	2.4	5*	25*	550*	75	15	55
19–30 years	1.3	400	2.4	5*	30*	550*	90	15	55
Females									
9–13 years	1.0	300	1.8	4*	20*	375*	45	11	40
14–18 years	1.2	400[e]	2.4	5*	25*	400*	65	15	55
19–30 years	1.3	400[e]	2.4	5*	30*	425*	75	15	55
Pregnancy									
≤ 18 y	1.9	600	2.6	6*	30*	450*	80	15	60
19–30 years	1.9	600	2.6	6*	30*	450*	85	15	60
Lactation									
≤18 y	2.0	500	2.8	7*	35*	550*	115	19	70
19–30 years	2.0	500	2.8	7*	35*	550*	120	19	70

SOURCE: Food and Nutrition Board, Institute of Medicine, National Academies. Dietary Reference Intakes: recommended intakes for individuals. Washington, DC: National Academy Press; 2000.

[d] Although AIs have been set for choline, there are few data to assess whether a dietary supplement of choline is needed at all stages of the life cycle, and it may be that the choline requirement can be met by endogenous synthesis at some of these stages.

[e] In view of evidence linking folate intake with neural tube defects in the fetus, it is recommended that all women capable of becoming pregnant consume 400 µg from supplements or fortified food until their pregnancy is confirmed and they enter prenatal care, which ordinarily occurs after the end of the periconceptional period—the critical time for formation of the neural tube.

Table 14.8 **Recommended caloric (Kcal) and protein intakes for adolescents.**

LIFE STAGE GROUP	CALORIES (Kcal)		PROTEIN (GRAMS)	
Age (years)	Kcal/day	Kcal/cm*	Grams/day	Grams/cm
Females				
11–14	2,200	14.0	46	0.29
15–18	2,200	13.5	44	0.27
19–24	2,200	13.4	46	0.28
Males				
11–14	2,500	15.9	45	0.29
15–18	3,000	17.0	59	0.34
19–24	2,900	16.4	58	0.33

*2.54 cm = 1 in

SOURCE: Data taken from Gong EJ, Heard FP. Diet, nutrition, and adolescence. In: Shils ME, Olson JA, Shike M, eds. Modern nutrition in health and disease, 8th ed. Philadelphia, PA: Lead & Febiger; 1994; and 1989 Recommended Daily Allowances, 10th Edition of the RDAs, Food and Nutrition Board, Commission on Life Sciences. Washington, DC: National Academy Press; 1989.

intake is to assess height, weight, and body composition. If, over time, height as well as weight-for-height continuously fall within the same percentiles when plotted on the National Center for Health Statistics growth charts, it can be assumed that energy needs are being met. If percentile of weight-for-height measurements begin to fall or rise, a thorough assessment of energy intake should be done and adjustments in energy intake should be made accordingly. The use of body fat measurements, such as triceps skinfold measurements, can provide useful information when weight-for-height does not remain consistent. Remember, however, that transient increases and decreases in body fat are commonly noted among adolescents during puberty due to the variation in timing of increases in height, weight, and accumulation of body fat and lean body mass. Repeated measurements of weight, height, and body composition over a several-month period are needed to accurately assess adequacy of growth and development.

Protein

Protein needs of adolescents are influenced by the amount of protein required for maintenance of existing lean body mass, plus allowances for the amount required to accrue additional lean body mass during the adolescent growth spurt. Because protein needs vary with the degree of growth and development, requirements based upon developmental age will be more accurate than absolute recommendations based upon chronological age. Recommended protein intakes based upon age, gender, and height are shown

in Table 14.8. Protein requirements per unit of height are highest for females at 11 to 14 years, and for males at 15 to 18 years. These peak periods of protein need correspond to the usual timing of peak height velocity. As with energy, growth is affected by protein intakes. When protein intakes are consistently inadequate, reductions in linear growth, delays in sexual maturation, and reduced accumulation of lean body mass may be seen.

Traditionally, U.S. adolescents consume more than adequate amounts of protein. National data suggest that on average teens consume about two times the recommended level of protein intake.[28] Subgroups of adolescents may be at risk for marginal or low protein intakes, however, including those from food insecure households, those who severely restrict calories, and those who consume semivegetarian or vegetarian diets, most notably vegans.

Carbohydrates

Carbohydrates provide the body's primary source of dietary energy. Carbohydrate-rich foods, such as fruit, vegetables, grains, and legumes are also the main source of dietary fiber. Absolute requirements for carbohydrate intake among adolescents have not been established. Instead, dietary recommendations suggest that 50% or more of total daily calories should come from carbohydrate, with no more than 10% of calories derived from sweeteners, such as sucrose and high fructose corn syrup. Data from a major study suggest that adolescents consume approximately 53% of their calories as carbohydrate.[28] Foods that contribute the most carbohydrate to the diets of adolescents include (in descending order) yeast bread, soft drinks, milk, ready-to-eat cereal, and foods such as cakes, cookies, quick breads, donuts, sugars, syrups, and jams.[14]

Sweeteners and added sugars provide approximately 16% of total calories to the diets of adolescents.[24] Soft drinks are a major source of added sweeteners in the diets of adolescents, accounting for more than 12% of all carbohydrate consumed.[13] (See Table 14.9.) An estimated 86% of the soft drinks consumed by teens are sweetened, nondiet soft drinks.[30]

Dietary Fiber

Dietary fiber is important for normal bowel function and may play a role in the prevention of chronic diseases such as certain cancers, coronary artery disease, and Type II diabetes mellitus. Adequate fiber intake is also thought to reduce serum cholesterol levels, moderate blood sugar levels, and reduce the risk of obe-

Table 14.9
Soft drink consumption among adolescents.

Ounces per Day	Percent (%) of adolescents
≥ 26 oz per day	22
13 – 25 oz per day	28
0.1 – 12 oz per day	32
0 oz per day	18

SOURCE: Data from Harnack L, Stang J, Story M. Soft drink consumption among US children and adolescents: nutritional consequences. J Amer Diet Assoc 1999;99:436–41.

sity. The American Academy of Pediatrics (AAP) Committee on Nutrition has recommended that dietary fiber intakes among children and adolescents should be 0.5 grams per kilogram of body weight.[31] This corresponds to average fiber intakes of 15.5 to 34.5 grams per day among 10- to 18-year-old boys, and 16.0 to 28.5 grams per day among 10- to 18-year-old females. The AAP has further recommended that fiber intake not exceed 35 grams per day, as levels above this amount may reduce the bioavailability of some minerals. More recent recommendations by the American Health Foundation suggest that average fiber requirements be based on an "age plus five" rule, where the individual's age is added to the number five to determine minimum fiber requirements.[32] A factor of 10 is added to age to determine the recommended upper limit of fiber intake. Thus, a 14-year-old would require 19–24 grams of fiber each day, in contrast to the AAP recommendation of 25 grams per day.

National data indicate that adolescent males consume 11.5 to 15.4 grams of fiber, while female adolescents consume 10.0 to 14.0 grams of fiber each day.[33,34] During adolescence, fiber intake among males increases slightly with age while it decreases with age among females. Significant sources of fiber in the diet of adolescents include breads, ready-to-eat cereal, potatoes, popcorn and related snack foods, tomatoes, and corn.[16] The low intake of fruit and vegetables among adolescents is the greatest contributing factors affecting fiber intake among adolescents. Adolescents who skip breakfast or do not routinely consume whole grain breads or ready-to-eat cereals are at the highest risk for having an inadequate consumption of fiber.

Fat

The human body requires dietary fat and essential fatty acids for normal growth and development. Current recommendations by the National Cholesterol Education Program (NCEP) suggest that children over the age of two years consume no more than 30% of calories from fat, with no more than 10% of calories derived from saturated fat.[35] Data on energy and macronutrient intakes among adolescents suggest that approximately 33% of total calories consumed are derived from fat, with 12% originating from saturated fat.[13] Approximately two-thirds of teens meet the recommendations for total fat and saturated fat. Major sources of total and saturated fat intakes among adolescents include milk, beef, cheese, margarine, and foods such as cakes, cookies, donuts, and ice cream.[14] NCEP guidelines also suggest that adolescents consume no more than 300 milligrams of dietary cholesterol per day. This recommendation is exceeded by 48% of male and 32% of female teens.[36] Significant sources of cholesterol in adolescent diets include eggs, milk, beef, poultry, and cheese.

Calcium

Achieving an adequate intake of calcium during adolescence is crucial to physical growth and development. Calcium is the main constituent of bone mass. Because about half of peak bone mass is accrued during adolescence, calcium intake is of great importance for the development of dense bone mass and the reduction of the lifetime risk of fractures and osteoporosis. Female adolescents appear to have the greatest capability to absorb calcium about the time of menarche, with calcium absorption rates decreasing from then on.[37] Calcium absorption rates in males also peak during early adolescence, a few years later than in females. Young adolescents have been found to retain up to four times as much calcium as young adults. By age 24 in females and 26 in males, calcium accretion in bone mass is almost nonexistent.[38] Clearly, an adequate intake of calcium is of paramount importance during adolescence.

The DRI for calcium for 9- to 18-year-olds is 1300 milligrams per day (Table 14.7). National data suggest that many adolescents, most notably females, do not consume the DRI for calcium. Adolescent females consume between 536 to 815 milligrams of calcium per day, while adolescent males have been found to consume about 681 to 1146 milligrams of calcium each day.[6,39–41] These levels of dietary intake are not adequate to support the development of optimal bone mass. Supplements may be warranted for adolescents who do not consume adequate calcium from dietary sources.

Milk provides the greatest amount of calcium in the diets of adolescents, followed by cheese, ice cream, and frozen yogurt.[14] Adolescents increasingly consume their calcium in the form of fortified foods. One study indicated that more than half of teens reported drinking calcium-fortified juices and 31% ate cereals fortified with calcium.[41] Other foods that may be supplemented with calcium include margarine and bread. These foods may become important sources of calcium for adolescents.

The consumption of soft drinks by adolescents may displace the consumption of more nutrient-dense beverages, including milk and fortified juices. A study by Harnack and colleagues showed an inverse relationship between the intake of carbonated beverages and the intake of milk and juice.[30] Because milk and fortified juices are significant sources of calcium in the diets of adolescents, interventions aimed at reducing consumption of soft drinks may be warranted. Such interventions are especially important in light of the growing body of evidence that suggests that carbonated beverage consumption increases the risk of bone fractures among children and adolescents.[15]

Calcium consumption drops as age increases among both male and female adolescents; however, males consume greater amounts of calcium at all ages than do females.[6,39–40] Calcium intakes among adolescents are highly correlated with energy intakes. When dietary calcium intake is adjusted for energy intake, no differences in calcium density of diets are found between males and females.[41] This fact suggests that females who restrict calories in an effort to control their weight are at particularly high risk for inadequate calcium intakes. The relationship between socioeconomic status (SES) and calcium intake is not clear. Some studies show a weak positive relationship, while others find no relationship.[6,36,42] Some variation in calcium intake follows race categories among females: Cuban, Asian, and African American females consume less calcium on average than Mexican American, Puerto Rican, and Caucasian females.

A recent study of knowledge regarding calcium and bone health revealed some interesting findings. When adolescents were questioned about their knowledge of the health benefits of calcium, 92% knew that it was needed to strengthen bones, 60% knew that it was required for "good" teeth, and 60% realized that adolescence was a critical time for the development of peak bone mass. Only 19% of teens knew that the recommended intake of milk, milk products, or fortified soy milk was four servings per day.[41] Nutrition education and interventions that target calcium consumption by older children and young teens are needed.

SERUM IRON, PLASMA FERRITIN, AND TRANSFERRING SATURATION Measures of iron status obtained from blood plasma or serum samples.

HEME IRON Iron contained within a protein portion of hemoglobin that is in the ferrous state.

NONHEME IRON Iron contained within a protein of hemoglobin that is in the ferric state.

Iron

The rapid rate of linear growth, the increase in blood volume, and the onset of menarche during adolescence increase a teen's need for iron. The DRIs for iron for male and female adolescents are shown in Table 14.7. These recommendations are based on the amount of dietary iron intake needed to maintain a suitable level of iron storage, with additional amounts of iron added to cover the rapid linear growth and onset of menstruation that occur in male and female adolescents, respectively. Note that even though DRIs are based on chronological age, the actual iron requirements of adolescents are based on sexual maturation level. Iron needs of an adolescent will be highest during the adolescent growth spurt in males, and after menarche in females.

Estimates of iron deficiency among adolescents are 2.8% to 3.5% of 11- to 14-year-old females, 4.1% of 11- to 14-year-old males, 6.0% to 7.2% of 15- to 19-year-old females and 0.6% of 15- to 19-year-old males.[33] The age-specific hemoglobin and hematocrit values used to determine iron deficiency anemia are listed in Table 14.10. Hemoglobin and hematocrit levels, although commonly used to screen for the presence of iron deficiency anemia, are actually the last serum indicators of depleted iron stores to drop. More sensitive indicators of iron stores include *serum iron, plasma ferritin, and transferring saturation.* Recent data suggests that 17% to 25% of female adolescents had at least two abnormal indices of iron status. Thus, although the prevalence of iron deficiency anemia may be relatively low among adolescents, a larger proportion may have inadequate iron stores. This finding is particularly relevant among adolescents from low SES homes, because rates of iron deficiency tend to be higher in adolescents from low-income families.

The availability of dietary iron for absorption and utilization by the body varies by its form. The two types of dietary iron are *heme iron,* which is found in animal products, and *nonheme iron,* which is found in both animal and plant-based foods. Heme iron is highly bioavailable while nonheme iron is much less so. More than 80% of the iron consumed is in the form of nonheme iron. Bioavailability of nonheme iron can be enhanced by consuming it with heme sources of iron or vitamin C. This point is particular salient for adolescents who avoid animal foods as a means of restricting calories and those who consume few animal-based foods (semivegetarian) or vegetarian diets for moral or cultural reasons.

Dietary intakes of iron range from 10.0 to 12.5 milligrams per day in females.[33,40] Data suggest that 32% of male and 83% of female teens consume less than the DRI for iron.[36] The most common dietary

Table 14.10 Hemoglobin and hematocrit cut-point values for iron deficiency anemia in adolescents.

Sex/Age[a]	Hemoglobin (≤g/dL) Less Than:	Hematocrit (<%) Less Than:
Males and Females		
8–12 years	11.9	35.4
Males		
12–15 years	12.5	37.3
15–18 years	13.3	39.7
18+ years	13.5	39.9
Females[b]		
12–15 years	11.8	35.7
15–18 years	12	35.9
18+ years	12	35.7

[a]Age and sex-specific cutoff values for anemia are based on the 5th percentile from the third National Health and Nutrition Examination Survey (NHANES III).
[b]Nonpregnant and lactating adolescents.

SOURCE: Abridged from Centers for Disease Control and Prevention. Recommendations to prevent and control iron deficiency in the United States. *MMWR* 47 (No.RR-3); 1998.

sources of iron in diets of adolescents include ready-to-eat cereal, bread, and beef.[14]

Zinc

Zinc is particularly important during adolescence because of its role in the synthesis of RNA and protein, and its role as a cofactor in over 200 enzymes. The body's need for zinc, along with its ability to retain zinc, dramatically increases during the adolescent growth spurt. Zinc is required for sexual maturation to occur. Males who are zinc deficient experience growth failure and delayed sexual development. Zinc supplementation of both male and female zinc-deficient adolescents from developing countries often initiates accelerated growth and sexual development. Data on zinc nutrition of adolescents are limited, but evidence shows that serum zinc levels decline in response to the rapid growth and hormonal changes during adolescence. Serum zinc levels indicative of mild zinc deficiency (<10.71 umol/L) have been found in 18% to 33% of female adolescents.[33]

The bioavailability of zinc from dietary sources is highly dependant upon the source of zinc. Zinc from animal sources is more bioavailable than plant-based sources. Undigestible fibers found in many plant-based sources of zinc can inhibit its absorption by the body. Zinc and iron compete for absorption, so elevated intakes of one can reduce the absorption of the

other. Adolescents who take iron supplements may be at increased risk of developing mild zinc deficiency if iron intake is more than twice as high as that of zinc.

Dietary intakes of zinc among adolescent females range from 6.6 to 7.9 milligrams per day.[33,40] National surveys suggest that 75% of males and 81% of females consume less than the DRI for zinc, with 46% of males and 59% of females consuming less than 77% of recommended intakes.[36] The top five sources of dietary zinc consumed by adolescents include beef, milk, ready-to-eat cereal, cheese, and poultry.[14] Vegetarians, particularly vegans, and teens who do not consume many animal-derived products are at highest risk for low intakes of zinc.

Folate

Folate is an integral part of DNA, RNA, and protein synthesis. Thus, adolescents have increased requirements for folate during puberty. The DRI for folate is listed in Table 14.7. Folate in the form of folic acid is twice as bioavailable than other forms of folate. For this reason, dietary folate equivalents (DFEs) are used in the DRIs. One microgram of folic acid is equivalent to approximately 2 DFEs, while 1 microgram of other forms of folate is equivalent to 1 DFE. Folic acid is the form of folate added to fortified cereals, breads, and other refined grain products.

Severe folate deficiency results in the development of megaloblastic anemia, which is rare among adolescents. Evidence, however, indicates that a significant proportion of adolescents have inadequate folate status. Twelve percent of adolescent females are mildly folate deficient, based on low serum folate levels, while 8% to 48% of female teens have been shown to have low red cell folate levels indicative of subclinical folate deficiency.[43,44] Serum folate levels drop during adolescence among females as sexual maturation proceeds, suggesting that increased folate needs during growth and development are not being met. For this reason, sexual maturation level should be used to identify folate needs as opposed to chronological age.

Poor folate status among adolescent females also presents an issue related to reproduction. Studies show that adequate intakes of folate prior to pregnancy can reduce the incidence of spina bifida and selected other congenital anomalies and may reduce the risk of Down syndrome among offspring.[45] The protective effects of folate occur early in pregnancy, often before a woman may know she is pregnant. Thus, it is imperative that all women of reproductive age (15–44 years old) consume adequate folic acid, preferably through dietary sources, or if needed, through supplements.

National data suggest that many adolescents do not consume adequate amounts of folate. Twenty-five percent of male and 43% of female adolescents consume less than the DRI for folate.[36] The top five sources of dietary folate consumed by adolescents include ready-to-eat cereal, orange juice, bread, milk, and dried beans or lentils.[14] Teens who skip breakfast or do not commonly consume orange juice and ready-to-eat cereals are at an increased risk for having a low consumption of folate.

Vitamin A

Vitamin A deficiency is rare among adolescents in the United States; however, national studies have consistently shown low dietary intakes of this vitamin. It has been reported that 52% of teen males and 62% of teen females consume less than the DRI for vitamin A, while 34% of males and 44% of females consume less than 77% of the recommended amounts.[36] The DRI for vitamin A is shown in Table 14.7. The top five dietary sources of vitamin A in the diets of adolescents are ready-to-eat cereal, milk, carrots, margarine, and cheese. Beta-carotene, a precursor of vitamin A, is most commonly consumed by teens in carrots, tomatoes, spinach and other greens, sweet potatoes, and milk.[14] The low intake of fruits, vegetables, and milk and dairy products by adolescents contributes to their less-than-optimal intake of vitamin A.

Vitamin E

Vitamin E is well known for its antioxidant properties, a role that becomes increasingly important as body mass expands during adolescence. The DRIs for vitamin E for adolescents are shown in Table 14.7. Few data are available on the vitamin E status of adolescents. National nutrition surveys suggest that dietary intakes of vitamin E are well below recommended levels, which may be indicative of poor vitamin E status. Seventy-six percent of adolescent males and females consume less than the DRI for vitamin E, with 55% consuming less than 77% of recommended amounts.[36] Among adolescents the five most commonly consumed sources of vitamin E are margarine, cakes/cookies/quick breads/donuts, salad dressings/mayonnaise, nuts/seeds, and tomatoes.[14] Increasing adolescent intakes of vitamin E through dietary sources is a challenge, given that many of the sources of vitamin E are high fat foods.

Vitamin C

Vitamin C is involved in the synthesis of collagen and other connective tissues. For this reason, vitamin C plays an important role during adolescent growth and development. Vitamin C intakes are marginally adequate within the adolescent population; however, 17% to 35% of teens consume less than 75% of the recommended amount.[27] The five most common sources of vitamin C among adolescents are orange and grapefruit juice, fruit drinks, ready-to-eat cereals, tomatoes, and white potatoes.[14] The DRIs for 9- to 13-year-old and 14- to 18-year-old adolescents are shown in Table 14.7.

Vitamin C acts as an antioxidant. Smoking increases the need for this antioxidant within the body because it consumes vitamin C in antioxidation reactions. Consequently, smoking results in reduced serum levels of vitamin C. Recommended levels of vitamin C intake are higher among smokers. Because more than one-third of adolescents consume less than 75% of the DRI for vitamin C, an even greater proportion of adolescents who smoke will likely not consume enough vitamin C to promote optimal health. On average, adolescents who use tobacco and other substances have poorer-quality diets and consume fewer fruit and vegetables, which are primary sources of vitamin C.[26]

NUTRITION SCREENING, ASSESSMENT, AND INTERVENTIONS

The American Medical Association's Guidelines for Adolescent Preventive Services (GAPS) recommend that all adolescents receive annual health guidance related to healthy dietary habits and methods to achieve a healthy weight.[46] This health guidance begins by annually screening all adolescents for indicators of nutritional risk. Common concerns that should be investigated during nutrition screening include overweight, underweight, eating disorders, hyperlipidemia, hypertension, iron deficiency anemia, food insecurity, and excessive intake of high-fat or high-sugar foods and beverages. Pregnant adolescents should also be assessed for adequacy of weight gain and compliance with prenatal vitamin-mineral supplement recommendations.

Nutrition screening should include an accurate measurement of height and weight, and calculation of BMI (body mass index). These data, plotted on age and gender appropriate National Center for Health Statistics 2000 growth charts, indicate the presence of any weight or other growth problems. Indicators of height and weight status are listed in Table 14.11.

Teens below the 5th percentile of weight-for-height or BMI-for-age are considered to be underweight and should be referred for evaluation of metabolic disorders, chronic health conditions, or eat-

Table 14.11 Indicators of height and weight status for adolescents.

Indicator	Body Size Measure	Cut-Off Values
Stunting (low height-for-age)	Height-for-age	<3rd percentile
Thinness (low BMI-for-age)	BMI-for-age	<5th percentile
At risk for overweight	BMI-for-age	≥85th percentile, but <95th percentile
Overweight	BMI-for-age	≥95th percentile

SOURCE: Compiled with permission from World Health Organization. Physical status: the use and interpretation of anthropometry. Report of a WHO expert committee. World Health Organization Technical Report Series 854, 1995; and Himes J, Dietz W. Guidelines for overweight in adolescent preventive services: recommendations from an expert committee. Am J Clin Nutr 59, 1994. In: Story M, Holt K, Sofka D, eds. Bright futures in practice: nutrition. Arlington, VA: National Center for Education in Maternal and Child Health; 2000.

ing disorders. Adolescents with a BMI above the 85th percentile but below the 95th percentile are considered to be at risk for overweight. They should be referred for a full medical evaluation to determine the presence or absence of obesity-related complications. Teenagers with a BMI greater than 95th percentile are considered to be overweight and should also be referred for a medical evaluation. Referral to a weight management program specially designed to meet the needs of adolescents may also be warranted for overweight adolescents who have completed physical growth.[47]

Nutrition screening should also include a brief dietary assessment. Food frequency questionnaires, 24-hour recalls, and food diaries or food records are all appropriate for use with adolescents. Table 14.12 lists the advantages and disadvantages of each dietary assessment method. Less formal dietary assessment questionnaires that target specific behaviors, such as consumption of savory snacks and high-sugar beverages, can also be used for initial nutrition screening. These "rate your plate" type of questionnaires can be completed quickly and may be used to determine those adolescents in need of additional dietary assessment and nutrition counseling.

Nutrition risk indicators that may warrant further nutrition assessment and counseling are listed in Table 14.13. Adolescents who have a poor quality diet characterized by an excessive intake of high-fat

Table 14.12 Strengths and limitations of various dietary assessment methods used in clinical settings.

	Strengths	Limitations	Applications
24-Hour Recall	• Does not require literacy • Relatively low respondent burden • Data may be directly entered into a dietary analysis program • May be conducted in-person or over the telephone	• Dependent on respondent's memory • Relies on self-reported information • Requires skilled staff • Time consuming • Single recall does not represent usual intake	• Appropriate for most people as it does not require literacy • Useful for the assessment of intake of a variety of nutrients and assessment of meal patterning and food group intake • Useful counseling tool
Food Frequency	• Quick, easy, and affordable • May assess current as well as past diet • In a clinical setting, may be useful as a screening tool	• Does not provide valid estimates of absolute intake of individuals • Can't assess meal patterning • May not be appropriate for some population groups	• Does not provide valid estimates of absolute intake for individuals, thus of limited usefulness in clinical settings • May be useful as a screening tool, however, further development research is needed
Food Record	• Does not rely on memory • Food portions may be measured at the time of consumption • Multiple days of records provide valid measure of intake for most nutrients	• Recording foods eaten may influence what is eaten • Requires literacy • Relies on self-reported information • Requires skilled staff • Time consuming	• Appropriate for literate and motivated population groups • Useful for the assessment of intake of a variety of nutrients and assessment of meal patterning and food group intake • Useful counseling tool
Diet History	• Able to assess usual intake in a single interview • Appropriate for most people	• Relies on memory • Time consuming (60 to 90 minutes) • Requires skilled interviewer	• Appropriate for most people as it does not require literacy • Useful for assessing intake of nutrients, meal patterning, and food group intake • Useful counseling tool

SOURCE: Used with permission. Story M, Stang J, eds. Nutrition and the pregnant adolescent: a practical reference guide. Minneapolis, MN: Center for Leadership, Education, and Training in Maternal and Child Nutrition, University of Minnesota; 2000.

Table 14.13 Key indicators of nutrition risk for adolescents.

Indicators of Nutrition Risk	Relevance	Criteria for Further Screening and Assessment
FOOD CHOICES		
Consumes fewer than two servings fruit or fruit juice per day	Fruits and vegetables provide dietary fiber and several vitamins (such as A and C) and minerals. Low intake of fruits and vegetables is associated with an increased risk of many types of cancer. In females of child-bearing age, low intake of folic acid is associated with an increased risk of giving birth to an infant with neural tube defects.	Assess the adolescent who is consuming less than one serving of fruit or fruit juice per day.
Consumes fewer than three servings of vegetables per day		Assess the adolescent who is consuming fewer than two servings of vegetables per day.
Consumes fewer than six servings of bread, cereal, pasta, rice, or other grains per day	Grain products provide complex carbohydrates, dietary fiber, vitamins, and minerals. Low intake of dietary fiber is associated with constipation and an increased risk of colon cancer.	Assess the adolescent who is consuming fewer than three servings of bread, cereal, pasta, rice, or other grains per day.
Consumes fewer than three servings of dairy products per day	Dairy products are a good source of protein, vitamins, and calcium and other minerals. Low intake of dairy products may reduce peak bone mass and contribute to later risk of osteoporosis	Assess the adolescent who is consuming fewer than two servings of dairy products per day.
		Assess the adolescent who has a milk allergy or is lactose intolerant.
		Assess the adolescent who is consuming more than 20 ounces of soft drinks per day.
Consumes fewer than two servings of meat or meat alternatives (e.g., beans, eggs, nuts, seeds) per day	Protein rich foods (e.g., meats, beans, dairy products are good sources of B vitamins, iron, and zinc. Low intake of protein-rich foods may impair growth and increase the risk of iron deficiency anemia and of delayed growth and sexual maturation. Low intake of meat or meat alternatives may indicate inadequate availability of these foods at home. Special attention should be paid to children and adolescents who follow a vegetarian diet.	Assess the adolescent who is consuming fewer than two servings of meat or meat alternatives per day or who consumes a vegan diet.
Has excessive intake of dietary fat	Excessive intake of total fat contributes to the risk of cardiovascular diseases and obesity and is associated with some cancers.	Assess the adolescent who has a family history of premature cardio-vascular disease.
		Assess the adolescent who has a body mass index (BMI) greater than or equal to the 85th percentile
EATING BEHAVIORS		
Exhibits poor appetite	A poor appetite may indicate depression, emotional stress, chronic disease, or eating disorder.	Assess the adolescent if BMI is less than the 15th percentile or if weight loss has occurred.
		Assess if irregular menses or amenorrhea have occurred for three months or more.
		Assess for organic and psychiatric disease.
Consumes food from fast-food restaurants three or more times per week	Excessive consumption of convenience foods and foods from fast-food restaurants is associated with high fat, calorie, and sodium intakes, as well as low intake of certain vitamins and minerals.	Assess the adolescent who is at-risk for overweight/obesity, or who has diabetes mellitus, hyperlipidemia, or other conditions requiring reduction in dietary fat.

Table 14.13 Key indicators of nutrition risk for adolescents (continued)

Indicators of Nutrition Risk	Relevance	Criteria for Further Screening and Assessment
Skips breakfast, lunch, or dinner/supper three or more times per week	Meal skipping is associated with a low intake of energy and essential nutrients, and, if it is a regular practice, could compromise growth and sexual development. Repeatedly skipping meals decreases the nutritional adequacy of the diet.	Assess the adolescent to ensure that meal skipping is not due to inadequate food resources or unhealthy weight loss practices.
Consumes a vegetarian diet	Vegetarian diets can provide adequate nutrients and energy to support growth and development if well planned. Vegan diets may lack calcium, iron, vitamins D and B_{12}. Low-fat vegetarian diets may be adopted by adolescents who have eating disorders.	Assess the adolescent who consumes fewer than two servings of meat alternatives per day. Assess the adolescent who consumes fewer than three servings of dairy products per day. Assess the adolescent who follows a low-fat vegetarian diet and experiences weight loss for eating disorder and adequacy of energy intake.
FOOD RESOURCES		
Has inadequate financial resources to buy food, insufficient access to food, or lack of access to cooking facilities	Poverty can result in hunger and compromised food quality and nutrition status. Inadequate dietary intake interferes with learning.	Assess the adolescent who is from a family with low income, is homeless, or is a runaway.
WEIGHT AND BODY IMAGE		
Practices unhealthy eating behaviors (e.g., chronic dieting, vomiting, and using laxatives, diuretics, or diet pills to lose weight)	Chronic dieting is associated with many health concerns (fatigue, impaired growth and sexual maturation, irritability, poor concentration, impulse to binge) and can lead to eating disorders. Frequent dieting in combination with purging is often associated with other health-compromising behaviors (substance use, suicidal behaviors). Purging is associated with serious medical complications.	Assess the adolescent for eating disorders. Assess for organic and psychiatric disease. Screen for distortion in body image and dysfunctional eating behavior; especially if adolescent desires weight loss, but BMI <85th percentile.
Is excessively concerned about body size or shape	Eating disorders are associated with significant health and psychological morbidity. Eighty-five percent of all cases of eating disorders begin during adolescence. The earlier adolescents are treated, the better their long-term prognosis.	Assess the adolescent for distorted body image and dysfunctional eating behaviors, especially if adolescent wants to lose weight, but BMI is less than the 85th percentile.
Has exhibited significant weight change in past six months	Significant weight change during the past six months may indicate stress, depression, organic disease, or an eating disorder.	Assess the adolescent to determine the cause of weight loss or weight gain (limited to too much access to food, poor appetite, meal skipping, eating disorder).
GROWTH		
Has BMI less than the 5th percentile	Thinness may indicate an eating disorder or poor nutrition.	Assess the adolescent for eating disorders. Assess for organic and psychiatric disease. Assess for inadequate food resources.
Has BMI greater than the 95th percentile	Obesity is associated with elevated cholesterol levels and elevated blood pressure. Obesity is an independent risk factor for cardiovascular disease and type 2 diabetes mellitus in adults. Overweight adolescents are more likely to be overweight adults and are at increased risk for health problems as adults.	Assess the adolescent who is overweight or at risk for becoming overweight (on the basis of present weight, weight gain patterns, family weight history).

Table 14.13 Key indicators of nutrition risk for adolescents (continued)

Indicators of Nutrition Risk	Relevance	Criteria for Further Screening and Assessment
PHYSICAL ACTIVITY		
Is physically inactive: engages in physical activity fewer than five days per week	Lack of regular physical activity is associated with overweight, fatigue, and poor muscle tone in the short term and a greater risk of heart disease in the long term. Regular physical activity reduces the risk of cardiovascular disease, hypertension, colon cancer, and type 2 diabetes mellitus. Weight-bearing physical activity is essential for normal skeletal development during adolescence. Regular physical activity is necessary for maintaining normal muscle strength, joint structure, and joint function; contributes to psychological health and well-being; and facilitates weight reduction and weight maintenance throughout life.	Assess how much time the adolescent spends watching television/videotapes and playing computer games. Assess the adolescent's definition of physical activity.
Engages in excessive physical activity	Excessive physical activity (nearly every day or more than once a day) can be unhealthy and associated with menstrual irregularity and excessive weight loss, and malnutrition.	Assess the adolescent for eating disorders.
MEDICAL CONDITIONS		
Has chronic diseases or conditions	Medical conditions (diabetes mellitus, spina bifida, renal disease, hypertension, pregnancy, HIV infection/AIDS) have significant nutritional implications.	Assess adolescent's compliance with therapeutic dietary recommendations. Refer to dietitian if appropriate.
Has hyperlipidemia	Hyperlipidemia is a major cause of atherosclerosis and cardiovascular disease in adults.	Refer adolescent to a dietitian for cardiovascular nutrition assessment.
Has iron deficiency anemia	Iron deficiency causes developmental delays and behavioral disturbances. Another consequence is increased lead absorption.	Screen adolescents if they have low iron intake, a history of iron deficiency anemia, limited access to food because of poverty or neglect, special health care needs, or extensive menstrual or other blood losses. Screen annually.
Has dental caries	Eating habits have a direct impact on oral health. Calcium and vitamin D are vital for strong bones and teeth, and vitamin C is necessary for healthy gums. Frequent consumption of carbohydrate-rich foods (e.g., lollipops, soda) that stay in the mouth longer may cause dental caries. Fluoride in water used for drinking and cooking as well as in toothpaste reduces the prevalence of dental caries.	Assess the adolescent's consumption of snacks and beverages that contain sugar, and assess snacking patterns. Assess the adolescent's access to fluoride (e.g., fluoridated water, fluoride tablets).
Is pregnant	Pregnancy increases the need for most nutrients.	Refer the adolescent to a dietitian for further assessment, education, and counseling as appropriate.
Is taking prescribed medication	Many medications interact with nutrients and can compromise nutrition status.	Assess potential interactions of prescription drugs (e.g., asthma medications, antibiotics) with nutrients.
Uses nutritional supplements	Vitamin and mineral preparations can be a healthy addition to dietary intake, especially if pregnant, lactating, or has past history of anemia; frequent use or high doses can have serious side effects. Those using other nutritional supplements for "bulking up" may be at risk for experimentation with anabolic steroids.	Inquire about type and dosage of supplement; rule out anabolic steroid use. Screen for nutrient-nutrient or drug-nutrient interactions.

Table 14.13 Key indicators of nutrition risk for adolescents (continued)

Indicators of Nutrition Risk	Relevance	Criteria for Further Screening and Assessment
LIFESTYLE		
Engages in heavy alcohol, tobacco, and other drug use	Alcohol, tobacco, and other drug use can adversely affect nutrient intake and nutrition status.	Assess the adolescent further for inadequate dietary intake of energy and nutrients.
Uses dietary supplements	Dietary supplements (e.g., vitamin and mineral preparations) can be healthy additions to a diet, especially for pregnant and lactating women and for people with a history of iron deficiency anemia; however, frequent use or high doses can have serious side effects.	Assess the adolescent for the type of supplements used and dosages.
	Adolescents who use supplements to "bulk up" may be tempted to experiment with anabolic steroids. Herbal supplements for weight loss can cause tachycardia and other side effects. They may also interact with over-the-counter prescription medications.	Assess the adolescent for use of anabolic steroids and megadoses of other supplements.

SOURCE: Adapted from Story M, Holt K, Sofka D, eds. Bright futures in practice: nutrition. Arlington, VA: National Center for Education in Maternal and Child Health; 2000.

or high-sugar foods and beverages or meal skipping should be provided with nutrition counseling that provides concrete examples of ways to improve dietary intake. Adolescents who have been found to have a nutrition-related health risk, such as hyperlipidemia, hypertension, iron deficiency anemia, overweight, or eating disorders should be referred for in-depth medical assessment and nutrition counseling. Pregnant adolescents may also benefit from in-depth nutrition assessment and counseling.

In-depth nutrition assessment should include a review of the full medical history, a review of psychosocial development, and evaluation of all laboratory data available. A complete and thorough dietary assessment should be performed, preferably using two dietary assessment methods. Most commonly, a food frequency questionnaire or a three-to-seven-day food record is combined with a 24-hour recall to provide accurate dietary intake data. Specific areas of nutrition concern can be identified during a complete nutrition assessment, and recommendations for nutrition education and counseling can be made accordingly.

Nutrition Education and Counseling

Providing nutrition education and counseling to teenagers requires a great deal of skill and a good understanding of normal adolescent physical and psychosocial development. When working with teens, it is important to treat them as individuals with unique needs and concerns. The initial component of the counseling session should involve getting to know the adolescent, including personal health or nutrition-related concerns. After establishing a rapport with the teen, the counselor should provide an overview of the events of the counseling session, including what specific nutrition topics will be discussed. Once again, the adolescent should be encouraged to add his or her own nutrition concerns to the list of topics to be discussed during the education session.

After agreeing to a list of topic areas to be covered during the nutrition education session, a complete nutrition assessment should be performed. Upon completion of the assessment, the counselor and teen should work together to establish goals for improving dietary intake and reducing nutrition risk.

It is important to involve the adolescent in decision-making processes during nutrition counseling. Allowing teens to provide input as to what aspects of their eating habits they think need to be changed and what changes they are willing to make accomplishes several important tasks during the counseling session. First, the importance of the adolescent in the decision-making process is stressed, and she or he is encouraged to become involved in personal decisions about health. Second, a good rapport established between the health professional and the adolescent may lead to greater interaction between both parties. Finally, behavior change is more likely when the adolescent has suggested ways to change, thus expressing a willingness to change.

One or two goals during a counseling session is a reasonable number to work toward. Setting too many goals reduces the probability that the adolescent can meet all of the goals and may seem overwhelming. For each goal set, several behavior change strategies should be mutually agreed upon for meeting that goal. These strategies should be concrete in nature and instigated by the teen. The adolescent and the counselor should also work together to decide how to determine when a goal is met. Frequent follow-up sessions also help to provide feedback and monitor progress toward individual goals.

PHYSICAL ACTIVITY AND SPORTS

Regular physical activity leads to many health benefits. Physical activity improves aerobic endurance and muscular strength, may reduce obesity, and builds bone mass density.[29,48] Physical activity among adolescents is consistently related to higher levels of self-esteem and self-concept and lower levels of anxiety and stress. Thus, physical activity is associated with both physiological and psychological benefits, especially during adolescence, which offers opportunities to positively influence the adoption of lifelong activity patterns. Increasing physical activity among adolescents as a goal is important because regular physical activity declines during adolescence and many American adolescents are inactive.

Physical activity is defined as any bodily movement produced by skeletal muscles that results in energy expenditure. This definition is distinguished from exercise, which is a subset of physical activity that is planned, structured, and repetitive and is done to improve or maintain physical fitness.[29] Physical fitness is a set of attributes that are either health or skill related. The International Consensus Conference on Physical Activity Guidelines for Adolescents recommends that all adolescents be physically active daily, or nearly every day, as part of play, games, sports, work, transportation, recreation, physical education, or planned exercise. Further, it is recommended that adolescents engage in three or more sessions per week of activities that last 20 minutes or more at a time and that require moderate to vigorous levels of exertion.[49] These recommendations are consistent with the surgeon general's report on physical

activity and health, which advises people of all ages to include a minimum of 30 minutes of physical activity of moderate intensity (such as brisk walking) on most, if not all days of the week. The report also acknowledges that greater health benefits can be obtained by engaging in physical activity of more vigorous intensity or of longer duration.

Despite common knowledge about the importance and benefits of physical activity, only about one-half of young people (ages 12–21 years) in the United States regularly participate in vigorous physical activity and one-fourth report no vigorous physical activity.[29] Moreover, physical activity declines steadily and dramatically during adolescence. For example, 81% of ninth-grade males report they exercised vigorously on at least three of the previous seven days, while only 67% of twelfth-grade males did so. Among females, 61% of ninth-graders exercised vigorously on at least three of the preceding seven days, which declined to only 41% among twelfth-grade girls. One explanation offered for the decline in adolescents' physical activity that occurs after the growth spurt may be related to the social demands of adolescence, changing interests, and the transition from school to work or college.

Factors Affecting Physical Activity

Individual, social, and environmental factors are associated with physical activity among adolescents. Females are less active than males, and among adolescent females, blacks are less active than whites.[48] Individual factors positively associated with physical activity among young people include confidence in one's ability to engage in exercise (i.e., self-efficacy), perceptions of physical or sports competence, having positive attitudes toward physical activity, enjoying physical activity, and perceiving positive benefits associated with physical activity (i.e., excitement, fun, adventure, staying in shape, improved appearance, weight control, improving skills). Social factors associated with engaging in physical activity are peer and family support. Environmental factors associated with physical activity are having safe and convenient places to play, sports equipment, and transportation to sports or fitness programs.

Schools offer an ideal setting for promoting physical activity through physical education classes. However, in recent years, overall daily attendance in physical education classes in grades 9–12 decreased significantly from 42% to 25%. Also, the percentage of high school students enrolled in physical education and who report being physically active for at least 20 minutes in physical education classes declined from about 81% to 70% in the 1990s.[29] Only 20% of all high school students report being physically active for

20 minutes or more in daily physical education classes.[48] Communities are essential because most physical activity among adolescents occur outside of schools.

The Centers for Disease Control and Prevention recently published Guidelines for School and Community Programs to Promote Lifelong Physical Activity Among Young People, which provides a developmental framework for comprehensive school and community physical activity programs to be used by school districts, educators, health professionals, and policymakers.[48] The guidelines include recommendations (Table 14.14) to promote lifelong physical activity including school policies, physical and social environments that encourage and enable physical activity, developmentally appropriate physical education curricula and instruction, personnel training, family and community involvement, and program evaluation.

High levels of physical activity, combined with growth and development, increase adolescents' needs for energy, protein, and select vitamins and minerals. Participation in competitive sports often means an adolescent will participate in intense training and competition during an athletic season. If the athlete competes in several sports, energy and nutrient needs will remain relatively stable throughout the year. If an athlete participates in only one sport, however, energy and nutrient needs may fluctuate based on the timing of the sports season. Therefore, adolescents must be assessed for seasonal and yearly physical activity when energy and nutrient needs are determined.

The energy and nutrient needs of adolescent athletes are largely unknown. Many of the recommendations available are based on needs of young adult athletes or are extrapolated from usual nutrient needs of adolescents. The best method of assessing the nutrient needs of athletes is to begin with general dietary needs based on Sexual Maturation Rating (SMR), adding additional allowances based upon the unique needs of the individual and the intensity of physical activity they engage in. In order to assess individual nutrient needs, health care professionals must gather information such as:

- What sport(s) does the adolescent engage in and what is the duration of the competition season?
- What is the level of competition of the adolescent? Is participation recreational, competitive, or highly competitive?
- What kind of training does the adolescent engage in? The method(s), intensity, and duration of training activities should be noted.
- Does the athlete typically sweat profusely or lose body weight during competition?

- Does the athlete follow a special diet or take supplements to improve athletic performance? Be sure to note the type, amount, and frequency of supplement use.

General energy and protein needs are shown in Table 14.8. These guidelines should provide the foundation for calculating protein and energy needs for athletes. Competitive athletes may require 500–1500 additional calories per day to meet their energy needs. Athletes and their parents should be encouraged to monitor weight stability. Any weight loss that is not transient (transient losses are often due to dehydration) signifies that the caloric intake is inadequate to support growth and development. A thorough assessment of energy and protein intakes, accompanied by measurements of body composition, should be taken when unexpected weight loss occurs. Protein should supply no more than 15% to 20% of calories in the diet. Adolescents typically consume 1.5 to 2.5 times the recommended intake of protein, so additional protein is generally not required. Exceptions would include athletes who follow vegetarian diets or restrict caloric intake. When the main sources of protein are plant-based, additional protein intake may be needed because plant-based sources of protein may be less bioavailable.

Dietary intakes of athletes should follow the Food Guide Pyramid recommendations, with the realization that the increased energy needs of athletes may require them to consume the upper limit of food group recommendations. Athletes should be encouraged to eat a pre-event meal at least two to three hours prior to exercise; eating too close to exercise may lead to indigestion and physical discomfort. Foods that are high in fat, high in protein, and high in dietary fiber should be avoided for at least four hours prior to exercise, because they take longer to digest and may cause physical discomfort during exercise. Protein and fat also displace complex carbohydrates, which are the most readily available source of energy during athletic events. Postevent meals should contain approximately 400–600 calories, and should be comprised of high-carbohydrate foods and adequate amounts of non-caffeinated fluids.

Other nutrients of concern among adolescent athletes, particularly female athletes, are iron and calcium. Teens who are athletes are at a higher risk than nonathletes for developing iron deficiency anemia, due in part to iron losses that occur during exercise.[50] Iron status should be closely monitored in athletes, especially among vegetarians and females who are postmenarcheal, in an effort to prevent iron deficiency anemia. Calcium intakes have been shown to be below the DRIs in a significant proportion of

Table 14.14
Recommendations for school health programs promoting healthy eating and physical activity.

Healthy Eating	Physical Activity
• Adopt a coordinated school nutrition policy that promotes healthy eating through classroom lessons and a supportive school environment.	• Establish policies that promote enjoyable lifelong physical activity among young people.
• Implement nutrition education from preschool through secondary school as part of a sequential, comprehensive school health education curriculum designed to help students adopt healthy eating behaviors.	• Provide physical and social environments that encourage and enable safe and enjoyable physical activity.
• Provide nutrition education through developmentally appropriate, culturally relevant, fun participatory activities that involve social learning strategies.	• Implement physical education curricula and instruction that emphasize enjoyable participation in physical activity and that help students develop the knowledge, attitudes, motor skills, behavioral skills, and confidence needed to adopt and maintain physically active lifestyles.
• Coordinate school food service with other components of the comprehensive school health program to reinforce messages on healthy eating.	• Provide extracurricular physical activity programs.
• Involve family members and the community in supporting and reinforcing nutrition education.	• Include parents and guardians in physical activity instruction and in extracurricular and community physical activity programs, and encourage them to support their children's participation in enjoyable physical activities.
• Evaluate regularly the effectiveness of the school health program in promoting healthy eating, and change the program as appropriate to increase the effectiveness.	• Provide training for education, coaching, recreation, health care, and other school and community personnel that imparts the knowledge and skills needed to effectively promote enjoyable, lifelong physical activity among young people.
	• Assess physical activity patterns among young people, counsel them about physical activity, refer them to appropriate programs, and advocate for physical activity instruction and programs for young people.
	• Provide a range of developmentally appropriate community sports and recreation programs that are attractive to all young people.
	• Regularly evaluate school and community physical activity instruction, programs, and facilities.

SOURCE: Centers for Disease Control and Prevention. Guidelines for school and community programs to promote lifelong physical activity among young people. MMWR 45;1996; and Guidelines for school health programs to promote lifelong healthy eating. MMWR 45;1996.

adolescents, especially females. Athletes' increased risk for bone fractures makes adequate calcium intake extremely importance. Recent data suggest that consumption of carbonated soft drinks by athletic females elevates their risk of developing bone fractures when compared to less active females who consume the same beverages.[16] Although the mechanism responsible for this tendency has not been identified, female adolescent athletes with low calcium consumption appear to be the highest risk group of all adolescents for bone fractures, and therefore, should make every effort to consume adequate calcium in their diets. Teen athletes who cannot or will not consume calcium from dietary sources should be counseled to take a daily calcium supplement that meets their daily needs.

PROMOTING HEALTHY EATING AND PHYSICAL ACTIVITY BEHAVIORS

Meeting the challenge of improving the nutritional health of teenagers requires the integrated efforts of teenagers, parents, educators, health care providers, schools, communities, the food industry, and policymakers all working together to create more opportunities for healthful eating.

Effective Nutrition Messages for Youth

We need to rethink how we frame our messages to youth. Years ago, Leverton[51] pointed out that too often, teenagers have been given the message that good nutrition means "eating what you don't like because it's good for you." Rather, they should be told "eat well because it will help you in what you want to do and become." Teenagers are present-oriented and tend not to be concerned about how their eating will affect them in later years. However, they are concerned about their physical appearance, achieving and maintaining a healthy weight, and having lots of energy.[52,53] Many are also interested in optimizing sports performance. Even though adolescents need to be aware of the long-term risks of an unhealthy diet and benefits of a more healthful one, focusing on the short-term benefits will have more appeal to them.

A change in the perception that healthful foods do not taste good is also needed. In a national Gallup survey, almost two-thirds (64%) of adolescents agreed a lot or somewhat with the statement that "foods that are good for you do not taste good."[54] Almost three-fourths (71%) agreed with the statement that "your favorite foods are not good for you." Getting students involved in food preparation and tasting new foods may help counter this belief. Nutrition education and intervention programs for youth need to focus not only on attitudes and acquisition of knowledge, but also on behavioral changes.[55,56] Adolescents need opportunities to learn and practice behavioral change skills. In the Gallup poll, 71% of adolescents who received nutrition information from doctors and nurses rated the advice as very useful.

Parent Involvement

Parents as well as teenagers should be targets for nutrition education because they fill the role of gate-keepers of foods and serve as role models for eating behavior. Even though parents may have little control over what their teenagers are eating outside the home, they have more control in the home environment. Teenagers tend to eat what is available and convenient. Parents can capitalize on this by stocking the kitchen with a variety of nutritious ready-to-eat foods and limiting the availability of high-sugar, high-fat foods within the home. The use of different creative settings and outlets to deliver innovative nutrition education programs to parents, including work sites, churches, community centers, libraries, supermarkets, restaurants, and parent education programs, should be explored.

School Programs

School-based programs can play important roles in promoting lifelong healthy eating and physical ac-

tivity. Efforts to promote physical activity and healthful eating should be part of a comprehensive, coordinated school health program and include school health instruction (curriculum); school physical education; school food service; health services (screening and preventive counseling); school-site health promotion programs for faculty and staff; and integrated community efforts.[48] The Centers for Disease Control and Prevention recently published two complementary reports, Guidelines for School Health Programs to Promote Lifelong Healthy Eating[57] and Guidelines for School and Community Programs to Promote Lifelong Physical Activity Among Young People,[48] which provide a developmental framework for comprehensive school nutrition and physical activity programs to be used by school districts, educators, health professionals, and policymakers. The guidelines (Table 14.14) include recommendations to promote healthy eating and lifelong physical activity including school policies, physical and social environments that encourage and enable physical activity and healthy eating, developmentally appropriate nutrition and physical education curricula and instruction, personnel training, family and community involvement, and program evaluation.

CLASSROOM NUTRITION EDUCATION   Contento and colleagues reviewed 43 school-based nutrition education programs published between 1980–1995.[58] Less than half of the studies ($n = 20$) were conducted with adolescents. Surprisingly little has been published regarding nutrition education for junior and senior high school students. Some form of nutrition education is taught in grades 7 through 9 in 92% of all public schools and 86% of grades 10 through 12.[59] The percent of schools with nutrition education requirements is substantially lower at each grade level than the percent of schools that teach nutrition education. Although 50% or more of all K–8 schools have district or state requirements for students to receive nutrition education, only 40% of schools have requirements for grades 9–10, and about 20% for grades 11–12.

By junior high, students are in the process of cognitive and social development changes that permit more advanced nutrition education concepts and activities. The ability for more abstract thinking coupled with the changing psychosocial terrain of young adolescents provides both a challenge and a unique opportunity for educators to offer new learning and teaching strategies to encourage them to make healthful food choices. Young adolescence is an ideal time to teach students how to assess their own behavior and set goals for change. As adolescents begin the social process of individuation, they become ready and eager to make their own decisions and show their individuality. Nutrition education often fails to take advantage of the social and cognitive transitions of adolescence to promote the adoption of more healthful behaviors. In addition, obstacles to implementing nutrition education programs persist, ranging from insufficient funding to teacher ambivalence to competition with other high-priority health concerns, such as HIV and substance use prevention.[57]

Because knowledge alone is inadequate when students must decide which foods to eat and how to deal with peer and social influences, as well as a widely available supply of high-fat, high-sugar foods, the focus of nutrition education and teaching methods should be on behavior change strategies and skill acquisitions to make healthful food decisions. Characteristics of teaching methods found to be most effective in school health education curricula include use of discovery learning; use of student learning stations, small work groups, and cooperative learning techniques; cross-age and peer teaching; positive approaches that emphasize the intrinsic value of good health; use of personal commitment to change and goal setting; and provision of opportunities to increase self-efficacy in modifying health behaviors.[60] Most importantly, adolescents need to be given repeated opportunities to develop, demonstrate, practice, and master the skills needed to make informed decisions and cope with social influences. Also to be effective, programs must take into account cultural factors as well as the developmental processes of adolescents.

Teacher training in basic nutrition and instructional, motivational, and behavioral change strategies increases the success of nutrition education curriculum. Training may be most effective if teachers have the opportunity to examine their own body image and assess their eating behaviors. Teacher training typically increases the time spent on teaching nutrition in the classroom.[61]

SCHOOL FOOD SERVICES The National School Lunch Program (NSLP), and School Breakfast Program (SBP) are federally sponsored nutrition programs administered by the Department of Agriculture (USDA), in conjunction with state and local education agencies. Youth from households with incomes between 130% and 185% of the poverty level receive meals at reduced rates; youth from households with incomes 130% of poverty and below receive meals free. Almost 99% of public schools participate in the NSLP and about 50% in the SBP. The National School Nutrition Dietary Assessment Study (SNDAS) funded by the USDA found that youth who participate in the NSLP have greater nutrient intakes compared with those who do not par-

ticipate.[62] The NSLP lunches provided one-third or more of the RDA for key nutrients.

The SNDAS study, coupled with accumulating research on the relationship between diet and chronic disease risk, spurred drastic changes in the national school meal programs. In 1994, Congress passed legislation that required meals served through the NLSP and SBP to comply with the Dietary Guidelines for Americans. The legislation for school meals is only for USDA-reimbursable breakfast and lunch meals. Junior and senior high school students, however, have a variety of options for lunch in which high-fat foods are easily accessible, such as in à la carte food in the cafeteria, school stores, and vending machines.

The school lunch and breakfast programs can complement and reinforce what is learned in the classroom and serve as a learning laboratory for nutrition education. The synergistic interaction between the school lunch program and classroom learning should enhance the likelihood that adolescents will adopt healthful eating practices.

NUTRITION ENVIRONMENT OF THE SCHOOL The school environment provides multiple food and nutrition activities and influences not only classroom nutrition education and school meals, but also food sold in vending machines, school stores, and snack bars; fund raising events; food rewards by teachers; corporate-sponsored nutrition education materials; and in-school advertising of food products.[63] Wolfe and Campbell found that the nutrition experiences provided in schools tend to be fragmented and multicomponent, especially when multiple people are involved in decisions about them.[64] The result can be inconsistent nutrition messages. The growing stream of commercial messages, food advertisements, and easy access to high-fat and high-sugar food products in school are at cross purposes and in direct conflict with the goals of nutrition education and may negate the efforts in the classroom and lunchroom to foster healthful eating practices.

Vending machines are commonly found in high school corridors, many of which are open throughout the school day. The majority of foods in school vending machines and school stores are high-fat or high-sugar items such as snack chips, candy, and soft drinks.[65] Healthy food choices or lower-fat alternatives are generally unavailable. In the promotion of a healthy nutrition environment, vending machines and school stores need to offer healthier choices and lower-fat alternatives.

A growing trend of commercialism and marketing in schools uses in-school advertising and corporate-sponsored education materials. A study by Consumers Union Education Services found that direct advertising in schools has mushroomed.[63] Examples include school bus advertising for soft drinks and fast-food restaurants; "free" textbook covers advertising candy, chips, and soft drinks; ads for high-sugar/high-fat products on wallboards and in hallways, in student publications such as newspapers and yearbooks, and on sports scoreboards; and product giveaways in coupons. In addition, *Channel One,* the daily news program broadcast to millions of students in grades 6 through 12 in thousands of schools, has two minutes out of each daily 12-minute program devoted to paid commercials for products that include candy bars, snack chips, and soft drinks.

Written nutrition policies provide needed guidelines for all food and nutrition activities and promotions in schools. Local and district policy initiatives can be instrumental in creating a supportive and integrated school nutrition environment with consistent health-promoting messages. Serious consideration should be given to restricting the sale of foods high in fat and sugar in schools. Corporate-sponsored education materials and programs should be carefully evaluated for nutrition accuracy, objectivity, completeness, and noncommercialism. Schools should also consider the Consumer's Union recommendation for making schools ad-free zones, where young people can pursue learning without commercial influences and messages.[66] Schools should be an environment where healthful eating behavior is normative, modeled, and reinforced. To improve the health of adolescents in the United States, schools should strive for an integrated nutritionally supportive environment.

Community Involvement in Nutritionally Supportive Environments

Promoting lifelong healthy eating and physical activity behaviors among adolescents requires attention to the multiple behavioral and environmental influences in a community. Adolescents most likely to adopt healthy behaviors receive consistent messages through multiple channels (e.g., community, home, school, and the media) and from multiple sources (e.g., parents, peers, teachers, health professionals, and the media).[29]

Most physical activity occurs outside the school setting, making community sports and recreation programs essential for promoting physical activity among young people. Healthy eating can be integrated into these efforts by providing nutritious snacks. Community coalitions or task forces can be established to assess community needs and to develop, implement, and evaluate physical activity and nutrition programs for young people. Few studies have reported on nutrition education or physical activity programs outside of the school setting.[58]

Model Nutrition Program

One example of a model community-based nutrition intervention program is the California Adolescent Nutrition and Fitness (CANfit) Program.[67] Through competitive grants, the CANfit Program supports and empowers adolescent-serving, community-based organizations to develop and implement nutrition education and physical activity programs for ethnic adolescents from low-income communities. Using a capacity-building model, the CANfit Program attempts to change the community context by improving access to healthier food choices and safe, affordable physical activity opportunities, enabling adolescents to have the decision-making skills and social support necessary for making healthy nutrition and fitness choices. Examples of CANfit grantees' projects include (1) developing a 10-week curriculum for an after-school program for African American girls that focuses on self-esteem, body image, healthy eating, cooking, and physical activity (e.g., hip hop dance); (2) developing a nutrition and physical activity program for adolescents and their parents attending Saturday Korean language schools in Los Angeles; and (3) Latino adolescents in a soccer league that worked with a local health department to train team coaches and parents in sports nutrition. Innovative programs using capacity-building models, such as the CANfit Program, can provide numerous benefits to other communities.

DISCUSSION POINTS

- Puberty may begin earlier and last longer among black girls when compared to white girls. How does that impact nutrition needs of young black females?
- French fries account for 23% of all vegetables eaten by adolescents. Which is worse: to consume vegetables that are high in fat or to consume fewer vegetables?
- More than one-third of female adolescents choose diets that are inadequate in calcium, magnesium, iron, zinc, folate, and vitamins A, B_6, and E. What types of snacks could provide these missing nutrients?
- Half of peak bone mass is accrued during adolescence. The greatest capacity to absorb calcium occurs in early adolescence, when the body can absorb almost four times more calcium than during young adulthood. However, calcium consumption drops as age increases during adolescence. What type of message would you develop to combat low calcium intakes among adolescent females?

Resources

American Medical Association Adolescent Health Online

Links to health and nutrition-related web sites specifically addressing adolescent health issues; information on key health issues that affect teens.
Web site: www.ama-assn.org/adolhlth/

Bright Futures

View and download Bright Futures publications including:
Bright futures in practice: nutrition and Bright futures in practice: physical activity.
Butts L. OK, so now you're a vegetarian: advice and 100 recipes from one vegetarian to another. New York: Broadway Books; 2000.
Pierson S. Vegetables ROCK! A complete guide for teenage vegetarians. New York: Bantam Books; 1999.
Web site: www.brightfutures.org

Centers for Disease Control and Prevention Nutrition and Physical Activity Programs

Facts on physical activity and nutrition among U.S. adults and youth, model programs, program guidelines; links to pediatric growth charts
Web site: www.cdc.gov/nccdphp/dnpa/

National Center for Education in Maternal and Child Health

Search databases on adolescent health; view and download publications and reports.
Web site: www.ncemch.org

The Vegetarian Resource Group

Information on choosing a healthy vegetarian diet; links to other resources and web sites.
Web site: www.vrg.org/nutrition/teennutrition.htm

United States Department of Agriculture Food and Nutrition Information Center

Information on Dietary Guidelines for Americans, the Food Guide Pyramid, dietary supplements, food safety, the Healthy School Meals Resource System, and other sources of nutrition information
Web site: www.nal.usda.gov/fnic

References

1. Gong EJ, Heard FP. Diet, nutrition, and adolescence. In: Shils ME, Olson JA, Shike M, eds. Modern nutrition in health and disease, 8th ed. Philadelphia: Lead & Febiger; 1994:759–69.

2. Herman-Giddens ME, Slora EJ, et al. Secondary sexual characteristics and menses in young girls seen in office practice: a study from the pediatric research in office settings network. Pediatrics 1997;99: 505–11.

3. Barnes HV. Physical growth and development during puberty. Med Clin North Am 1975;59:1305–17.

4. Frisch RE. Fatness, puberty, and fertility: The effects of nutrition and physical training on menarche and ovulation. In: Brooks-Gunn J, Peterson AC, eds. Girls at puberty: biological and psychosocial perspectives. New York: Plenum Press; 1983:29–49.

5. Frisch RE, McArthur JW. Menstrual cycles: fatness as a determinant of minimum weight for height necessary for their maintenance or onset. Science 1974;185:949–51.

6. Barr SI. Associations of social and demographic variables with calcium intakes of high school students. J Amer Diet Assoc 1994;94:260–6, 269; quiz 267–8.

7. Attie I, Brooks-Gunn J. Development of eating problems in adolescent girls: a longitudinal study. Dev Psychol 1989;25: 70–9.

8. Killen JD, Hayward C, et al. Is puberty a risk factor for eating disorders? Am J Dis Child 1992;146:323–5.

9. Wilson DM, Killen JD, et al. Timing and rate of sexual maturation and the onset of cigarette and alcohol use among teenage girls. Arch Pediatr Adolesc Med 1994;148: 789–95.

10. Cusatis DC, Chinchilli VM, et al. Longitudinal nutrient intake patterns of US adolescent women: the Penn State Young Women's Health Study. J Adolesc Health 2000;26:194–204.

11. American School Health Association, Association for the Advancement of Health Education, Society for Public Health Education. The National Adolescent Student Health Survey, a report on the health of America's youth. Oakland, CA: Third Party Publicating; 1988:1–178.

12. Bigler-Doughten S, Jenkins RM. Adolescent snacks: nutrient density and nutritional contribution to total intake. J Amer Diet Assoc 1987;87:1678–9.

13. Kennedy E, Powell R. Changing eating patterns of American children: a view from 1996. J Am Coll Nutr 1997;16:524–9.

14. Subar AF, Krebs-Smith SM, Cook A, Kahle LL. Dietary sources of nutrients among US children, 1989–1991. Pediatrics 1998;102:913–23.

15. Wyshak G. Teenaged girls, carbonated beverage consumption, and bone fractures. Arch Pediatr Adolesc Med 2000;154: 610–3.

16. Lin BH, Guthrie J, Blaylock J. The diets of America's children: influences of dining out, household characteristics, and nutrition knowledge: U.S. Department of Agriculture Economic Report Number 746 (AER-746); 1996.

17. Siega-Riz AM, Carson T, Popkin B. Three squares or mostly snacks—what do teens really eat? A sociodemographic study of meal patterns. J Adolesc Health 1998; 22:29–36.

18. Nicklas TA, Myers L, et al. Impact of breakfast consumption on nutritional adequacy of the diets of young adults in Bogalusa, Louisiana: ethnic and gender contrasts. J Amer Diet Assoc 1998;98: 1432–8.

19. Neumark-Sztainer D, Story M, Resnick MD, Blum RW. Adolescent vegetarians: a behavioral profile of a school-based population in Minnesota. Arch Pediatr Adolesc Med 1997;151:833–8.

20. Sabate J, Llorca MC, Sanchez A. Lower height of lacto-ovovegetarian girls at preadolescence: an indicator of physical maturation delay? J Amer Diet Assoc 1992; 92:1263–4.

21. Johnston P, Haddad E. Vegetarian diets and pregnant teens. In: Story M, Stang J, eds. Nutrition and the pregnant adolescent: a practical reference guide. Minneapolis, MN: Leadership, Education and Training Program in Maternal and Child Nutrition, 2000;135–45.

22. Dietary Guidelines Advisory Committee. Excerpt of the report of the Dietary Guidelines Advisory Committee on Dietary Guidelines for Americans, 2000.

23. U.S.D.A. Center for Nutrition Policy and Promotion. Food Guide Pyramid 2000.

24. Munoz KA, Krebs-Smith SM, Ballard-Barbash R, Cleveland LE. Food intakes of US children and adolescents compared with recommendations [published erratum appears in Pediatrics May 1998;101(5): 952–3]. Pediatrics 1997;100:323–9.

25. Krebs-Smith SM, Cook A, et al. Fruit and vegetable intakes of children and adolescents in the United States. Arch Pediatr Adolesc Med 1996;150:81–6.

26. Neumark-Sztainer D, Story M, Resnick MD, Blum RW. Correlates of inadequate fruit and vegetable consumption among adolescents. Prev Med 1996;25:497–505.

27. Stang J, Story MT, Harnack L, Neumark-Sztainer D. Relationships between vitamin and mineral supplement use, dietary intake, and dietary adequacy among adolescents. J Amer Diet Assoc 2000;100: 905–10.

28. Devaney BL, Gordon AR, Burghardt JA. Dietary intakes of students. Am J Clin Nutr 1995;61:205S–212S.

29. U.S. Department of Health and Human Services. Physical activity and health: a report of the surgeon general: U.S. Department of Health and Human Services, Centers for Disease Control and Prevention, National Center for Chronic Disease Prevention and Health Promotion; 1996.

30. Harnack L, Stang J, Story M. Soft drink consumption among US children and adolescents: nutritional consequences. J Amer Diet Assoc 1999;99:436–41.

31. American Academy of Pediatrics Committee on Nutrition. Carbohydrate and dietary fiber. In: Barness L, ed. Pediatric nutrition handbook, 3rd ed. Elk Grove Village, IL: American Academy of Pediatrics; 1993;100–6.

32. Williams CL, Bollella M, Wynder EL. A new recommendation for dietary fiber in childhood. Pediatrics 1995;96:985–8.

33. Donovan UM, Gibson RS. Iron and zinc status of young women aged 14 to 19 years consuming vegetarian and omnivorous diets. J Am Coll Nutr 1995;14: 463–72.

34. U.S. Department of Agriculture. Food consumption survey. Continuing survey of food intakes by individuals. Hyattsville, MD: USDA, 1985.

35. National Heart, Lung, and Blood Institute, National Cholesterol Education Program. Report of the Expert Panel on Blood Cholesterol Levels in Children and Adolescents. Bethesda, MD: National Institutes of Health; 1991.

36. Johnson R, Johnson D, et al. Characterizing nutrient intakes of adolescents by sociodemographic factors. J Adolesc Health 1994;15:149–54.

37. Weaver CM. Better bones: girls who get plenty of calcium fare well later in life. The Island Packet, June 20, 1995.

38. Hui SL, Johnston CC, Jr., Mazess RB. Bone mass in normal children and young adults. Growth 1985;49:34–43.

39. Albertson AM, Tobelmann RC, Marquart L. Estimated dietary calcium intake and food sources for adolescent females: 1980–92. J Adolesc Health 1997;20: 20–6.

40. Briefel RR, McDowell MA, Alaimo K, et al. Total energy intake of the US population: the third National Health and Nutrition Examination Survey, 1988–1991. Am J Clin Nutr 1995;62:1072S–80S.

41. Harel Z, Riggs S, Vaz R, et al. Adolescents and calcium: what they do and do not know and how much they consume. J Adolesc Health 1998;22:225–8.

42. Neumark-Sztainer D, Story M, Dixon LB, et al. Correlates of inadequate consumption of dairy products among adolescents. J Nutr Educ 1997;29:12–20.

43. Clark A, Mossholder S, Gates R. Folacin status in adolescent females. Am J Clin Nutr 1987;46:302–6.

44. Hine R. Folic acid: contemporary clinical perspective. Perspect Appl Nutr 1993;1: 3–14.

45. Institute of Medicine. Nutrition during pregnancy: part I: weight gain; part IIL: nutrient supplements. Washington, DC: National Academy Press; 1990.

46. American Medical Association. Guidelines for Adolescent Preventive Services. Chicago: American Medical Association, Department of Adolescent Health; 1992.

47. Barlow SE, Dietz WH. Obesity evaluation and treatment: Expert Committee recommendations. The Maternal and Child Health Bureau, Health Resources and Services Administration and the Department of Health and Human Services. Pediatrics 1998;102:E29.

48. Centers for Disease Control and Pre-

vention. Guidelines for school and community programs to promote lifelong physical activity among young people. Morb Mortal Wkly Rep 1997;46:1–36.

49. Sallis J, Patrick K. Physical activity guidelines for adolescents: consensus statement. Pediatric Exercise Science 1994;6: 302–14.

50. Clarkson PM. Minerals: exercise performance and supplementation in athletes. J Sports Sci 1991;9 (suppl):91–116.

51. Leverton RM. The paradox of teen-age nutrition. J Amer Diet Assoc 1968;53:13–6.

52. Olsen L. Food fight: a report on teen-aged eating habits and nutritional status. Oakland, CA: Citizen's Policy Center, 1984.

53. Story M, Resnick MD. Adolescents' views on food and nutrition. J Nutr Educ 1986;18:188.

54. American Dietetic Association IFIC, The President's Council on Physical Fitness and Sports. Food, physical activity and fun: what kids think. Washington, DC: The Gallup Organization; 1995.

55. Lytle L, Achterberg C. Changing the diet of America's children: what works and why? J Nutr Educ 1995;27:250–60.

56. Society for Nutrition Education, American Dietetic Association, ASFS. Association Joint Position of Society for Nutrition Education (SNE), The American Dietetic Association (ADA), American School Food Services Association: school-based nutrition programs and services. J Nutr Educ 1995;27:58–61.

57. Centers for Disease Control and Prevention. Guidelines for school health programs to promote lifelong healthy eating. Morb Mortal Wkly Rep 1996;45:1–37.

58. Contento L, Balch G, Bronner Y, et al. The effectiveness of nutrition education and implications for nutrition education policy, programs and research: a review of research. J Nutr Educ 1995;27: 284–418.

59. National Center for Education Statistics. Nutrition education in public elementary and secondary schools: U.S. Department of Education, Office of Educational Research and Improvement; 1996.

60. Seffrin J. Why school health education? In: Wallace HM, Patrick K, Parcel G, Igbe JB, eds. Principles and practices of student health. Oakland, CA: Third Party Publishing; 1992.

61. Contento I, Manning AD, Shannon B. Research perspective on school-aged nutrition education. J Nutr Educ 1992;24: 247–60.

62. Burghardt J, Gordon A, Chapman N, et al. The School Nutrition Dietary Assessment Study: school food service, meals offered, and dietary intakes. Princeton, NJ: Mathematica Policy Research, Inc.; 1993.

63. Consumers Union Education Services.

Captive kids: commercial pressures on kids at school. Yonkers, NY: Consumers Union of United States; 1995.

64. Wolfe WS, Campbell CC. Nutritional health of school-aged children in upstate New York: what are the problems and what can schools do? New York: Cornell University; 1991.

65. Story M, Hayes M, Kalina B. Availability of foods in high schools: is there cause for concern? J Amer Diet Assoc 1996;96:123–6.

66. Consumers Union Education Services. Selling America's kids: commercial pressures on kids of the '90s. Yonkers, NY: Consumers Union of the United States; 1990.

67. Hinkle A. Community-based nutrition interventions: reaching adolescents from low-income communities. Ann N Y Acad Sci 1997;187:83–93.

CHAPTER *15*

Photo Disc

Dramatic changes in body shape and size, heightened influence of peer perceptions, and the desire to adopt adultlike behaviors place adolescents at high risk for initiating health-compromising behaviors.

Jamie Stang

ADOLESCENT NUTRITION:

Conditions and Interventions

Prepared by **Jamie Stang**

CHAPTER OUTLINE

- Introduction
- Overweight and Obesity
- Special Concerns among Adolescent Athletes
- Adolescent Pregnancy
- Substance Use
- Dietary Supplements
- Iron Deficiency Anemia
- Hypertension
- Hyperlipidemia
- Eating Disorders
- Children and Adolescents with Chronic Health Conditions

KEY NUTRITION CONCEPTS

1 Competitive adolescent athletes require an addition 500–1500 Kcals per day to meet their energy needs. Additional protein may be required among adolescent athletes who are still growing.

3 Nutrient needs of young pregnant adolescents less than two years past menarche are greater than those of older adolescents because of continued growth and physical development.

3 Overweight adolescents are at increased risk for medical and psychosocial complications such as hypertension, hyperlipidemia, insulin resistance, type II diabetes mellitus, hypoventilation and orthopedic disorders, depression, and low self-esteem.

4 Approximately half of adolescent females and 15% of adolescent males are attempting weight loss.

INTRODUCTION

Multiple factors influence the nutritional needs and behaviors of adolescents. This chapter presents specific behaviors and nutrition concerns that affect significant numbers of adolescents, including low physical activity, participation in competitive sports, adolescent pregnancy, substance abuse, vegetarian diets, eating disorders, overweight, hypertension, and hyperlipidemia. Because overweight, sports participation, and eating disorders affect a larger group of adolescents than other conditions, they are presented in greater detail.

OVERWEIGHT AND OBESITY

The increase in the prevalence of overweight among adolescents mirrors that of adults over the past two decades. Exact reasons for this increase have not been identified. Although genetics is known to contribute to the occurrence of overweight, and having one or more overweight parent(s) increases a teen's risk of developing obesity, it alone clearly cannot account for the dramatic increase in overweight during the past two decades. Environmental factors, or interactions between genetic and environmental factors, are the most likely causes of the dramatic rise in overweight. Risk factors for the development of overweight among children and adolescents include having at least one overweight parent; coming from a low income family; being the descendant of African American, Hispanic, or American Indian/Native Alaskan parents; and being diagnosed with a chronic or disabling condition that limits mobility. Table 15.1 provides prevalence estimates of overweight among adolescents in the United States, by gender, race, and age. Inadequate levels of physical activity and consuming diets high in total calories and fat are additional risk factors among a significant proportion of adolescents. These environmental factors increase the risk of developing overweight if an adolescent is genetically predisposed to obesity.

Weight status among adolescents should be assessed by calculating body mass index (BMI). BMI is calculated by dividing a person's weight (kg) by their height^2 (m^2). BMI values are compared to age- and gender-appropriate percentiles to determine the appropriateness of the individual's weight for height. Youth with BMI values greater than the 85th but lower than the 95th percentile are considered at risk for overweight; those with BMI values above the 95th percentile are considered overweight.[1] Growth curves based on BMIs for children and adolescents are available from the National Center for Health Statistics. An example of a BMI growth curve is shown in Illustration 15.1.

Table 15.1   Prevalence of overweight by age, race, and gender.

Category	95th percentile (%)
6–11 years	
Both Sexes	10.6
Boys	
White	10.3
Black	11.9
Mexican American	17.4
Total	11.2
Girls	
White	9.2
Black	16.4
Mexican American	14.3
Total	10.0
12–17 years	
Both Sexes	10.6
Boys	
White	11.1
Black	10.7
Mexican American	14.6
Total	11.3
Girls	
White	8.5
Black	15.7
Mexican American	13.7
Total	9.8

SOURCE: Data taken from NHANES III.

The persistence of overweight from childhood throughout adulthood has not been well quantified. Research suggests that the persistence of obesity from infancy to adulthood increases with age. More than 70% of overweight adolescents can be expected to remain overweight into adulthood.[2–4] The risk of persistence of obesity from childhood into adulthood increases if at least one parent is overweight. Risk of persistence of overweight is also higher among the most overweight individuals, especially those whose weight is more than 180% of ideal weight.

Health Implications of Adolescent Overweight

A range of medical and psychosocial complications accompanies overweight among adolescents, including hypertension, dyslipidemia, insulin resistance, type II diabetes mellitus, sleep apnea and other hypoventilation disorders, orthopedic problems, body image disturbances, and lowered self-esteem.[5–8] Rates of type II diabetes mellitus are more common among overweight youth. In one study, a third of newly diagnosed diabetic patients under the age of 20 had type

Illustration 15.1   CDC Growth Charts: United States.

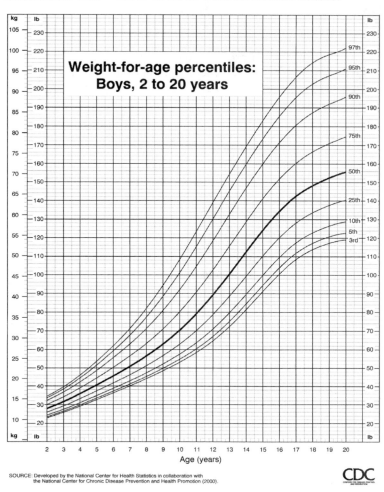

Weight-for-age percentiles:
Boys, 2 to 20 years

SOURCE: Developed by the National Center for Health Statistics in collaboration with
the National Center for Chronic Disease Prevention and Health Promotion (2000).

CDC

Developed by the National Center for Health Statistics in collaboration with the National Center for Chronic Disease Prevention and Health Promotion; 2000.

II diabetes mellitus; almost all of these patients had BMI values above the 90th percentile.[9]

All adolescents should be screened for appropriateness of weight-for-height on a yearly basis. Teens determined to be at-risk for overweight require an in-depth medical assessment to diagnose any obesity-related complications. Illustration 15.2 provides recommended screening and referral procedures for weight-for-height among adolescents.

Illustration 15.3 illustrates recommended weight goals based on BMI and age. Weight maintenance is recommended for adolescents who are at-risk for overweight and have not yet completed puberty (unless medical complications are noted). Weight loss is recommended when medical complications are present in at-risk for overweight youth, when the adolescent is determined to be overweight, and among older adolescents who have completed physical growth and devel-

opment. Guidelines for treatment of overweight are presented in Table 15.2. Weight loss should be attempted only after the adolescent and his or her family show current weight can be maintained. For severely obese adolescents, as well as those with significant medical complications, rapid weight loss may be required. Several obesity treatment centers in the United States have health professionals experienced in the management of severe obesity with complications. Staff at these centers can guide other health professionals in the treatment of such youth when necessary. The Weight-Control Information Network (WIN) can help health professionals locate specialized programs when they are required.

No specific physical activities or dietary regimens are recommended for adolescent weight management programs. Reducing or eliminating the intake of "problem" foods such as savory snacks or high-sugar beverages is often the first dietary change recommended. Health care providers generally should not recommend calorie or fat gram counting. Adolescents more readily accept replacing high-fat or high-sugar foods with healthier substitutes and monitoring portion control, resulting in a better chance of long-term behavior changes. Sedentary activities should be replaced with more strenuous activities. Referrals to community centers, athletic clubs, local parks and recreation programs, and community education programs that offer fun, noncompetitive physical activities, such as yoga, tai chi, swimming, weight lifting, and bicycling or walking clubs point adolescents in a beneficial direction.

SPECIAL CONCERNS AMONG ADOLESCENT ATHLETES

Fluids and Hydration

Fluid intake is an important issue in sports nutrition for adolescents. Young adolescents and those who are prepubertal present a particular vulnerability to heat illnesses because their bodies do not regulate body temperature as well as older adolescents. Adolescents can become so mentally and physically involved in physical activities that they do not pay attention to physiological signals of fluid loss, such as excessive sweating and thirst. Some athletes commonly assume they do not need additional fluids if they are not actively moving all of the time during exercise. Other

Illustration 15.2
Recommended overweight screening procedures.

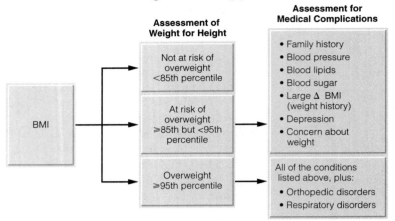

Adapted with permission from Himes JH, Dietz WH. Guidelines for overweight in adolescent preventive services: recommendations from an expert committee. The Expert Committee on Clinical Guidelines for Overweight in Adolescent Preventive Services. Am J Clin Nutr 1994;59:307–16.

factors, such as ambient temperature and humidity levels and weight of equipment (helmets, padding, etc.) worn or utilized during exercise also play a role. For instance, hockey goalies do not skate for great distances during a match, yet they may lose five or more pounds of body weight due to the weight of the padding and equipment they wear. Therefore, all athletes should be counseled to regularly consume fluids, even if they do not feel thirsty.

Athletes should consume 6–8 ounces of fluid prior to exercise, 4–6 ounces every 15–20 minutes during physical activity, and at least 8 ounces of fluid following exercise. Recommendations encourage athletes to weigh themselves periodically before and after exercise or competition to determine whether they have lost body weight. Each pound of body weight lost during an activity requires ingestion of 16 ounces of fluid following the activity to maintain proper hydration. Athletes should drink no more than 16 ounces of fluid each 30 minutes, however, to avoid potential side effects, such as nausea.

The type of fluid an athlete drinks is affected by peer pressure and mass media more than by actual physiological need. Sports drinks are very popular among teens, even those who do not participate in sports. Data on children suggest that even though water is an economical, easily available fluid, it may not provide optimal benefits for athletes who participate in physically intense events or those of great duration.[10] In such events, juice diluted at a ratio of 1:2 with water, or sports drinks that contain no more than 6% carbohydrate, may allow for better hydration and physical performance.[11] Undiluted juices, fruit drinks, carbonated beverages, and drinks that contain more than 6% carbohydrate are not recommended during exercise because they may cause gastric discomfort. Their high carbohydrate content

Illustration 15.3   Recommendations for weight goals.

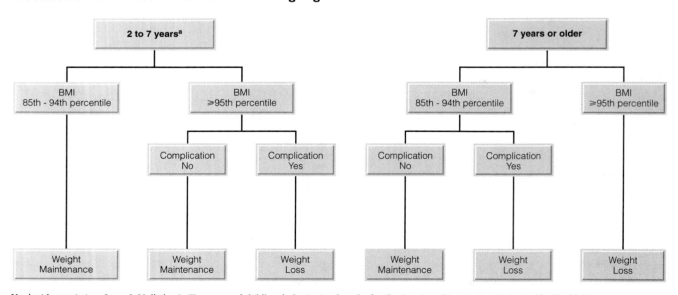

Used with permission. Stang J, Holladay L. Treatment of childhood obesity involves the family. American Dietetic Association Public Health Community Nutrition Practice Group newsletter. The Digest: Spring 2000. Adapted from Barlow SE, Dietz WH. Obesity evaluation and treatment: Expert Committee recommendations. Pediatrics 1998;102:E29.

may also delay gastric emptying. Some carbonated soft drinks contain caffeine, which promotes diuresis, reducing their effectiveness at rehydration.

Special Dietary Practices

Adolescent athletes may follow special diets or consume nutritional and nonnutritional supplements in an effort to improve physical performance and increase lean body mass. Even though data on the prevalence of supplements or special diets among adolescent athletes remain limited, data from surveys of collegiate athletes suggest that more than 10% of athletes use supplements.[12] Special diets that are noted among adolescent athletes include carbohydrate-loading regimens, high-protein diets, and "zone" diets. Distance runners and other endurance athletes traditionally used carbohydrate loading to improve the glycogen content of muscle. It involves the manipulation of training intensity and duration along with the carbohydrate content of meals to improve glycogen formation in muscle tissue. Carbohydrate loading is traditionally a weeklong process that begins with intense training one week prior to competition.[13] For the first three days of a carbohydrate-loading week, athletes choose low carbohydrate foods, but continue to exercise in an attempt to deplete muscle glycogen stores. During the three days prior to competition, athletes rest, or exercise minimally, while consuming a high-carbohydrate diet to promote glycogen formation and storage.

High-protein diets may take many forms for teen athletes. In general, athletes who follow high-protein diets may consume three to four times the recommended protein intake, accompanied by a relatively low intake of carbohydrate. High-protein diets should be discouraged among athletes for several reasons. First, many dietary protein sources are also sources of total and saturated fats, which may increase lifetime risk of coronary artery disease. Second, high protein and fat intakes delay digestion and absorption, limiting the amount of energy available for use during physical activity. Finally, more water is required for the breakdown of protein than either fat or carbohydrate due to the increased water loss that accompanies the excretion of nitrogen. This factor places an athlete at increased risk for dehydration, often accompanied by a decrease in physical performance. Adolescents should be reminded the consumption of individual amino acids or high intakes of complete proteins do not increase physical performance, and may in fact decrease physical performance.

The popularity among adults of the zone diet, which is a type of high-protein diet, has led to the adoption of this diet by adolescents. The zone theory is based upon a 40% carbohydrate, 30% fat and 30% protein intake.[14] Proponents of the zone diet suggest

Table 15.2 General guidelines for weight management therapy.

1. Early intervention is recommended, preferably before the child reaches the 95th percentile for BMI.
2. Parents should be informed of medical complications associated with childhood overweight, and all youth should be assessed for medical complications.
3. All family members and caregivers should be involved in the weight management program.
4. All family members should be assessed to determine their readiness to make behavior changes, and treatment should not begin until all family members are ready to adopt behavior changes.
5. Weight management programs should emphasize goals of improving eating and physical activity patterns as opposed to specific weight goals.
6. Programs should be skill based, with families taught to identify problem behaviors, monitor such behaviors, utilize behavior modification principles to address problem behaviors, and implement problem-solving skills when dealing with obstacles to behavior change.
7. Families should be involved in assessing current eating and activity patterns, deciding which behaviors need to be modified, setting goals for behavior changes, and determining how these goals will be achieved.
8. New behavior changes should not be instituted until previous changes have been accomplished and maintained.
9. Routine follow-up visits should be scheduled to monitor progress and prevent relapse to former eating and activity patterns.

SOURCE: Used with permission. Stang J, Holladay L. Treatment of childhood obesity involves the family. American Dietetic Association Public Health Community Nutrition Practice Group newsletter. The Digest; Spring 2000. Adapted from Barlow SE, Dietz WH. Obesity evaluation and treatment: Expert Committee recommendations. Pediatrics 1998;102:E29.

that a 40/30/30 ratio of macronutrient intakes allows the body to more efficiently burn calories, leading to loss of weight, reduction in body fat and increases in lean body mass. Even though weight loss can occur with the intake of high-protein diets due to the increased loss of body water, reductions in body fatness and increases in lean body mass have not been scientifically proven with use of this diet. Another argument supporting the 40/30/30 diet plan claims that it reduces insulin levels and insulin resistance, which proponents claim result from consuming a high-carbohydrate diet. Teens should be told that insulin levels do increase temporarily after eating a high-carbohydrate meal, in order for lean body mass and other tissues to take up carbohydrate and store it as glycogen. Carbohydrate still provides the main source of energy available for physical activity.

Dietary Supplements

Supplements reportedly used by collegiate athletes, and perhaps also by adolescents, include creatine, individual amino acids or protein powders, carnitine, dehydroepiandrosterone (DHEA), beta-hydroxy-beta-methylbutyrate, anabolic steroids, and ephedra.[12] One survey found anabolic steroid use to be at 2% among high school athletes.[15] Adolescents should be reminded that few supplements are tested on adolescents, so dosing and safety information is not available. This point is particularly important because many supplements can have detrimental health effects when used inappropriately. Experience suggests that creatine, carnitine, DHEA, and chromium picolinate are among the most widely used supplements by adolescent athletes.

CREATINE Creatine is one of the most popular supplements used by athletes, particularly among athletes who compete in strength-related sports, such as football. Creatine, formed in the liver and kidney of the human body, can be obtained in more than adequate amounts from the consumption of meat. Creatine is thought to improve anaerobic metabolism and to promote weight and lean tissue gain. Studies of creatine in adults show mixed results.[16] It appears to be of no benefit to endurance athletes, and marginally beneficial during strength-related sports. Side effects of creatine use, which seem to be dose related, include abdominal pain and cramping, nausea, diarrhea, headache, increased tendency toward muscle strains, and muscle soreness.[17] No available data document the long-term health effects related to creatine use; however, chronic use may be associated with renal damage.[18] It is not recommended for use by adolescents at this time.

CARNITINE The human body forms carnitine from the amino acids lysine and methionine. Almost all carnitine in humans resides within muscle tissue. As its main function, it metabolizes fatty acid. Animal studies suggest that it may enhance fatty acid oxidation by arterial walls.[19] Studies with athletes during athletic performance have been inconclusive. The use of carnitine seems to result in few side effects, most notably diarrhea at high doses.[18] The use of carnitine is not recommended among adolescents, however, because commercially available forms often contain impurities.

DHEA Dehydroepiandrosterone (DHEA) is a precursor of testosterone and estrogen. Naturally produced in the human body by the adrenal glands, DHEA levels fall in humans as age increases. Its reputed effects include melting fat, decreasing insulin resistance, increasing immune system function, increasing lean body mass, and decreasing risk of osteoporosis; however, no scientific evidence backs such claims.[18] As a precursor hormone to sex hormones, DHEA carries significant side effects such as irreversible gynecomastia (breast enlargement) and prostate enlargement among males and hirsuitism (facial hair) among females.[20] Adolescents primarily use this hormone to increase lean body mass, but no evidence supports this effect. Given the possibility of significant side effects of DHEA in pubescent adolescents undergoing hormonal changes, DHEA should not be used by adolescents.

CHROMIUM Studies of chromium picolinate supplementation among men, women, and athletes do not support claims that it burns fat, decreases fat deposition, increases lean body mass, and promotes weight loss. In fact, in a study of women, chromium supplementation increased body weight.[21,22] Chromium use among adolescent athletes is of concern because high doses can lead to renal failure,[23] cognitive impairment,[24] and increased risk of iron deficiency anemia.

In the highly competitive world of sports, the adolescent athlete, the coach, the trainer, and the parents easily become vulnerable to nutrition misinformation. A widespread need exists for integration of sound nutrition into team sports and training regimens for athletes. Coaches especially should be targeted for nutrition education programs because high school athletes frequently rely on coaches for nutrition information.

ADOLESCENT PREGNANCY

The birthrate among adolescent females is approximately 51 births per 1000 adolescents.[25] The majority of teens who give birth are 18–19 years old; however, pregnancies can occur as early as 11–12 years of age. The birthrate for adolescents less than 15 years old is approximately 1.0 births/1000, while birthrates for 15–17 year old and 18–19 year old teens are 30.0 births/1000 and 82 births/1000, respectively. Adolescents at risk for pregnancy include teens from low-income families, teens from unstable family situations (divorce/remarriage or unstable income), daughters of adolescent mothers, and those who demonstrate poor scholastic performance or drop out of school.[26,27] Repeat childbearing also occurs among adolescents, with 17% of adolescent mothers bearing two or more children during adolescence.[28] Among births to adolescents less than 15 years of age, 3% are repeat pregnancies.

Energy and Nutrient Needs

Many factors bear consideration when determining the energy and nutrient needs of pregnant teens, including potential for continued growth and development for the adolescent, preconceptional nutritional status, food security, and the potential for substance use. *Gynecological age* (GA) signifies the degree of physiological maturity and the potential for continued growth and development of a pregnant teen. GA is calculated by subtracting the age of menarche (first menstrual cycle) from the chronological age of a female. Adolescents with a GA of ≤2 years are likely to experience continued growth and development during pregnancy. One line of thought believes the growing adolescent competes directly with her fetus for energy and nutrients.[29,30] Therefore, the energy and nutrient needs required for adolescent growth must be added

to those required for optimal fetal development for pubescent adolescents.

Adolescents may have inadequate nutrient stores to support pregnancy at the time of conception. Females who are particularly high risk for poor preconceptional nutrition status include overweight or underweight teens, those who use substances, competitive athletes, females with eating disorders, chronic dieters, and vegetarians. Meal skipping is common among adolescent females and may lead to nutrient intakes that are inadequate to support a healthy pregnancy.

GYNECOLOGICAL AGE Defined as chronological age minus age at menarch. For example, a female with the chronological age of 14 years minus age at first menstrual cycle 12 years equals a gynecological age 2.

Warning signs of nutrition problems in pregnant adolescents are listed in Table 15.3. Nutrients most likely to be deficient in the diets of teen females include calcium, iron, zinc, magnesium, folate, and vitamins A, E, and B_6. Vegetarians may also enter pregnancy with limited intakes of vitamins D and B_{12}.

Alarmingly, 11% to 55% of pregnant adolescents report substance use.[31] Data suggest that approximately

Table 15.3 Warning signs of nutrition problems in pregnant adolescents.

Medical/Obstetric Factors	Nutrition/Dietary Factors	Psychosocial Factors
• Adolescent <16 years or gynecological age <2*	• Inadequate refrigeration or cooling facilities	• Inadequate income
• A previous pregnancy	• Lack of transportation or accessibility to grocery store	• Living alone or in an unstable family or home environment
• Closely spaced pregnancy	• Cultural or religious dietary restrictions	• Little family, partner, or peer support
• History of poor obstetric or fetal performance	• Frequent eating away from home	• Denial or failure to accept the pregnancy
• Chronic systemic disease	• Frequent snacking on low nutrient-dense foods	• Significant emotional stress or depression
• Past or present eating disorder: anorexia nervosa or bulimia nervosa	• Poor appetite	
• Underweight or overweight prior to pregnancy	• Limited, monotonous, or highly processed diet	
• Inadequate weight gain during pregnancy	• Irregular meal patterns (skipping meals)	
• Excessive weight gain during pregnancy	• History of frequent dieting	
• Persistent nausea or vomiting during pregnancy	• Exclusion of a major food group(s)	
• Iron deficiency anemia or other nutritional deficiencies	• Binge eating episodes	
• Severe infections	• Eating of nonfood substances (pica)	
• Heavy smoker	• Nontraditional dietary pattern (i.e., strict vegetarianism)	
• Alcohol or drug use	• Overuse of nutritional supplements	
	• Heavy caffeine intake	

*Gynecological age = Current age - Age at menarche
SOURCE: Reprinted with permission. Story M, Stang J, eds. Nutrition and the pregnant adolescent: a practical reference guide. Minneapolis, MN: Center for Leadership, Education, and Training in Maternal and Child Nutrition, University of Minnesota; 2000.

one-fourth of pregnant teens used tobacco, alcohol, and marijuana during the six months prior to conception, and almost half used at least one substance during pregnancy. Because substance use is associated with frequent meal skipping and overall poor quality dietary intake, females who use substances are at increased risk for poor nutrition status at the time of conception, as well as throughout the course of pregnancy. (Specific information about nutritional effects of substance use is discussed in this chapter in the section on substance use.)

The DRIs for pregnant and lactating females are included in the table on the inside front cover of this text. Energy needs of pregnant teens will vary according to the growth and development status of the teen, level of physical activity, body weight, and stage of pregnancy. Current recommendations for pregnant adult women suggest that an additional 300 calories per day are needed during the second and third trimesters of pregnancy to support fetal growth and development.[32] Younger adolescents who are likely to experience continued growth during pregnancy, and adolescents who are underweight at conception will have significantly higher energy needs than older females. Adolescents who have completed puberty and enter pregnancy overweight may require fewer additional calories. The best determination of adequate energy intake is the sufficiency of maternal weight gain.

Protein requirements increase during pregnancy in order to support the growth of the fetus and the adolescent. The additional intake of 10 grams of protein recommended throughout pregnancy brings minimum protein requirements to 55 grams per day for most teens.[32] Carbohydrate and fat should be consumed at levels that support fetal growth and optimal weight gain while also adhering to the Food Guide Pyramid.

Fluid intake is especially important during pregnancy. As blood volume increases, so does the need for water and other fluids. Adolescents should be encouraged to drink 8 to 10 cups of fluids each day, such as water, milk, and juice. Table 15.4 lists criteria for a healthy prenatal diet for teens.

Vitamin and Mineral Supplements

Vitamins and minerals that should be consumed at higher levels than those of nonpregnant adolescents include thiamin, riboflavin, niacin, folate, biotin, choline, pantothenic acid, iron, zinc, magnesium, iodine, selenium, and vitamins B_6, B_{12}, E, K, and C. A low-dose or prenatal vitamin-mineral supplement may be appropriate for pregnant teens who frequently skip meals, follow vegetarian diets, use substances, enter pregnancy with low nutrient stores,

Table 15.4   Criteria for a healthy prenatal diet.

Provides enough calories for adequate weight gain.

Is well-balanced and follows the Food Guide Pyramid.

Tastes good and is enjoyable to eat.

Spaces eating in intervals throughout the day.

Provides adequate amounts of high fiber foods.

Includes 8 cups of fluid daily.

Limits beverages that contain caffeine (2–3 servings or fewer daily).

Is moderate in fat, saturated fat, cholesterol, sugar, and sodium.

Excludes alcohol.

SOURCE: Reprinted with permission. Story M, Hermanson J. Nutrient needs during adolescence and pregnancy. In: Story M, Stang J, eds. Nutrition and the pregnant adolescent: a practical reference guide. Minneapolis, MN: Center for Leadership, Education, and Training in Maternal and Child Nutrition, University of Minnesota; 2000.

experience repeat pregnancy within two years, and have multiple gestations.[32]

Individual supplementation of 30 milligrams of elemental iron is recommended for adolescents to prevent iron deficiency anemia. Calcium supplements are recommended for pregnant teens who do not consume at least 1300 milligrams of calcium per day. Chewable vitamins and minerals may be more to the liking of adolescents than nonchewable tablets.

Pregnancy Weight Gain

Recommended weight gains for pregnant teens can be found in Table 15.5. Inadequate weight gain increases the likelihood of delivery of a premature, low birthweight, and small-for-gestational-age infant, especially among adolescents who enter pregnancy underweight.[32] Poor weight gain in the first 24 weeks of pregnancy is associated with an increased risk of delivery of a small-for-gestational-age infant, while inadequate gain after 24 weeks increases the risk of preterm delivery.[33]

Adolescents who are still growing gain 1 to 2 kilograms more weight during pregnancy than do those who have completed growth, yet they are more likely to deliver smaller infants. They should be counseled to gain the upper limit of weight gain recommendations for optimal pregnancy outcomes. Table 15.6 lists evaluation and management criteria for inadequate gestational weight gain. Excessive gestational weight gain among adolescents may signal depression and alcohol consumption.[34] Criteria for evaluation and management of excessive gestational weight gain are listed in Table 15.7.

Table 15.5   Weight gain recommendations for pregnant adolescents.

Prepregnant BMI	Total (lbs)	Trimester 1 (lbs)	Trimesters 2 & 3 (lbs/week)
Underweight	28–40	5	1.0+
Normal weight	25–35	3	1.0+
Overweight	15–25	2	0.66+
Obese	≥15	1.5	0.5+

SOURCE: Data used with permission from the National Academy of Sciences. Nutrition during pregnancy. Washington, DC: National Academy Press; 1990.

SUBSTANCE USE

The use of substances, such as tobacco, alcohol, and recreational drugs, directly affects the nutritional status of adolescents. Data on smoking rates of teens suggest that 36% of white, 31% of Hispanic, and 16% of black adolescents smoke.[35] Studies document heavy smoking in 16% of white, 7% of Hispanic, and 2% of black teens. Traditionally, male adolescents were more likely to smoke than females; however, that tendency is no longer true. Current data suggest that smoking rates are 1% to 2% higher among eighth- and tenth-grade females than among same aged males.[36] Initiation of smoking among adolescents can begin as young as nine years old, with 12–18 years of age being the highest risk age group. Use of smokeless tobacco, also known as chewing tobacco, is also seen among adolescents. The prevalence of smokeless tobacco use is thought to be at least 6.5 times higher among males than females.[35]

Alcohol intake and substance use among adolescents increases with age. A national survey suggested that by the twelfth grade, more than 80% of adolescents had tried alcohol[37] with more than one-third reporting binge drinking (drinking five or more

Table 15.6   Inadequate gestational weight gain.

Definition: <2 lbs per month after the first trimester

Evaluation

Measurement error	Inadequate food access
Excessive gain at previous visit (e.g., edema)	Homelessness
	Pica
Disordered eating	Substance abuse
Restrictive eating/dieting, meal skipping	Nausea, vomiting or heartburn
Psychosocial stress	Inadequate sleep and rest
Social isolation	High level of physical activity
Lack of partner or family support	Physically demanding job
	Gestational diabetes
Depression	Urinary ketones
Denial/rejection of pregnancy	

Management

5–6 nutrient-dense meals and snacks	Manage physical discomforts
Adequate rest and sleep	Refer to food assistance programs
Decreased physical activity	Psychosocial counseling
Stress management/relaxation techniques	Food journal

SOURCE: Used with permission. National Academy of Sciences. Nutrition during pregnancy. Washington, DC: National Academy Press; 1990.

Table 15.7
Excessive weight gain during pregnancy.

Definition: >6 lbs per month

Evaluation

Measurement error	Multiple gestation
Weight loss at previous visit	Depression
Edema	Binge eating
Smoking cessation	Psychosocial stress
Alcohol use	Social isolation
Infrequent, large meals	Emotionally based eating
High fat and/or sugar intake	Pica
Physical inactivity	

Management

Sensitive, supportive, and nonshaming manner	Alternatives to emotionally based eating
Moderate physical activity	Increased water and dietary fiber sources
Positive reinforcement for smoking cessation	Psychosocial counseling
Small, frequent meals (avoiding excess hunger and fullness)	Stress management/relaxation techniques
Decreased fat and sugar intake	Continue to gain weight at expected rate; avoid weight loss
Healthy snack and fast-food choices	Food journal
Refer to food assistance programs	Limit portions to average serving size

SOURCE: Used with permission. National Academy of Sciences. Nutrition during pregnancy. Washington, DC: National Academy Press; 1990.

alcoholic drinks during one occasion) at least one day during the past month. This survey also showed that almost one-quarter of adolescents reported using marijuana at least one time during the past month, 8% had tried cocaine, 10% had used inhalants, and 16% had used another form of recreational drug (such as LSD, PCP, speed, heroin, mushrooms, ecstasy, or methamphetamine).

Much of the data on differences in dietary intake and nutrient needs between substance users and nonusers comes from studies of adults. Data collected on adolescent substance users as part of the Minnesota Adolescent Health Survey and from high school students living in Israel provide some of the best data on dietary intakes among adolescent users. Adolescent substance users in these studies reported consuming diets that were less adequate than those of nonusers.[38,39] As a group they were more likely to skip breakfast, not eat a meal or snack at school, not eat three meals per day, report chronic dieting, and report purging than their nonsubstance using peers. All of these behaviors suggest that teens who use substances may be at high risk of poor nutritional status.

Substance use can result in depleted stores of vitamins and minerals, including thiamin, vitamin C, and iron. Chronic ingestion of alcohol and drug use can result in a reduced appetite, leading to low dietary intakes of protein, energy, vitamins A and C, thiamin, calcium, iron, and fiber.[40,41] Other adverse nutritional affects of substance use are listed in Table 15.8.

DIETARY SUPPLEMENTS

Vitamin-Mineral Supplements

Supplements may be used by adolescents for a variety of reasons, including improving health, treating iron deficiency anemia, increasing energy, building muscle, and losing weight. National data suggest that the prevalence of vitamin-mineral supplement use among teens in the United States ranges from 16% to 33%.[42,43] More than half of adolescents who report using vitamin-mineral supplements take them occasionally, with slightly less than half using them daily. Data on racial differences on adolescent supplement use are equivocal; one study suggests that white female adolescents are most likely to report using vitamin-mineral supplements, followed by white males, Mexican American females, Black females, Mexican American males and finally Black males, while another study found no differences by race. Overall, approximately 28% of female adolescents and 24% of male adolescents use vitamin-mineral supplements.

Approximately half of vitamin-mineral supplements consumed by adolescents are multivitamins without

Table 15.8   Potential effects of substance use on nutrition status.

Appetite suppression

Reduced nutrient intake

Decreased nutrient bioavailability

Increased nutrient losses/malabsorption

Altered nutrient synthesis, activation, and utilization

Impaired nutrient metabolism and absorption

Increased nutrient destruction

Higher metabolic requirements of nutrients

Inadequate weight gain/weight loss

Iron deficiency anemia

Decreased financial resources for food

SOURCE: Reprinted with permission. Alton I. Substance abuse during pregnancy. In: Story M, Stang J, eds. Nutrition and the pregnant adolescent: a practical reference guide. Minneapolis, MN: Center for Leadership, Education, and Training in Maternal and Child Nutrition, University of Minnesota; 2000.

minerals, 34% are individual vitamins or minerals, 18% are multivitamins with minerals, and 17% are iron with vitamin C tablets. Among the individual nutrient supplements used, vitamin C is the most common, followed by calcium, iron, vitamin E, and B-vitamin complex.[43] Among nutrients consumed at less than 75% of the RDA/DRI by the most adolescents are calcium and zinc, yet only one-third of supplements consumed by adolescents contain minerals. Calcium is the second most commonly used individual nutrient supplement; however, it is the nutrient most likely to be consumed at levels less than 75% of the DRI.

Adolescents who take vitamin-mineral supplements tend to consume a more nutritionally adequate diet than those who don't.[43] Supplement users obtain a smaller proportion of calories from total and saturated fats, and more from carbohydrate compared to nonsupplement users. They also consume diets that tend to be more nutrient dense, higher in folate, calcium, iron, and vitamins E, C, and A than nonusers.

Herbal Remedies

Few data shed much light on the use of nonnutritional supplements such as herbs (including herbal weight loss products) among adolescents. Experience suggests that adolescents may be likely to take herbal supplements for several reasons including weight loss, treatment of attention deficit disorder, and to increase energy and stamina. Studies are needed to determine exactly what types of herbal products are used by adolescents, because many herbs are known to have potentially dangerous side effects, and few recommendations are available to guide the use of herbs by children or adolescents.

IRON DEFICIENCY ANEMIA

Iron deficiency anemia is the most common nutritional deficiency noted among children and adolescents. Several risk factors are associated with its development among adolescents, including rapid growth, inadequate dietary intake of iron-rich foods or foods high in vitamin C, highly restrictive vegetarian diets, calorie restricted diets, meal skipping, participation in strenuous or endurance sports, and heavy menstrual bleeding.[44] The effects of iron deficiency anemia include delayed or impaired growth and development, fatigue, increased susceptibility to infection secondary to depressed immune system function, reductions in physical performance and endurance, and increased susceptibility to lead poisoning. Pregnant teens who are iron deficient in the early stages of gestation are at increased risk of preterm delivery and delivery of a low birthweight infant.

Assessment of iron deficiency anemia compares individual hemoglobin and hematocrit levels to standard reference values. Table 15.9 lists the 1998 Centers for Disease Control and Prevention criteria for determining anemia, based on age and gender. Adjustments to these values must be made for individuals who live at altitudes greater than 3000 feet and for smokers. An adjustment of +0.3 g/dL is required for adolescents who smoke.[45] Because adolescent males are not at high risk for iron deficiency anemia, they do not need to be screened unless they exhibit one or more of the risk criteria listed. All adolescent females should be screened every five years for anemia; those with one or more risk factors for anemia should be screened annually.

Treatment that follows a diagnosis of iron deficiency anemia needs to include increased dietary intake of foods rich in iron and vitamin C as well as iron supplementation. Adolescents under the age of 12 should be supplemented with 60 milligrams of elemental iron per day, and teenagers over the age of 12 should receive 60 to 120 milligrams of elemental iron per day.[45] These recommendations spark some controversy, however, given their high doses of elemental iron. Adolescents often report gastrointestinal side effects from iron supplementation, such as constipation, nausea, and cramping. These side effects can be lessened by giving smaller doses of iron more frequently throughout the day and counseling the adolescent to take the iron supplement at meal times or with food sources of vitamin C. Calcium supplements, dairy products, coffee, tea, and high-fiber foods may decrease absorption of iron supplements; these foods should be avoided within one hour of taking an iron supplement.

HYPERTENSION

Adolescents are considered hypertensive if the average of three systolic and/or diastolic blood pressure readings exceed the 95th percentile, based on age, sex, and height.[46] Blood pressure levels for the 95th percentiles for males and females are shown in Table 15.10. Classifications of blood pressure based on the average of three readings are:

- Normal blood pressure: <90th percentile
- High-normal blood pressure: ≥90th and <95th percentiles
- Hypertension: ≥95th percentile

Risk factors for hypertension among adolescents include a family history of hypertension, high dietary intake of sodium, overweight, hyperlipidemia, inactive lifestyle, and tobacco use.[46] Adolescents who display one or more of these risk factors should be routinely screened for hypertension. Nutrition counseling to decrease sodium intake, to limit fat intake to 30% or less of calories, and to consume adequate amounts of fruit, vegetables, whole grains, and low-fat dairy products should be provided when hypertension is diagnosed. Weight loss is recommended for adolescents who are hypertensive in the presence of overweight. If medications are prescribed, teens still must adhere to general dietary recommendations and should still be encouraged to reach and maintain a healthy weight for their height.

Table 15.9 Maximum hemoglobin concentration and hematocrit values for iron-deficiency anemia.

Sex/Age[a]	Hemoglobin, ≤g/dL	Hematocrit, ≤%
Males and Females		
8 to <12 years	11.9	35.4
Males		
12 to <15 years	12.5	37.3
15 to <18 years	13.3	39.7
≥18 years	13.5	39.9
Females[b]		
12 to <15 years	11.8	35.7
15 to <18 years	12	35.9
≥18 years	12	35.7

[a]Age and sex-specific cutoff values for anemia are based on the 5th percentile from the third National Health and Nutrition Examination Survey (NHANES III).
[b]Nonpregnant and lactating adolescents.
SOURCE: Data taken from Centers for Disease Control and Prevention. Recommendations to prevent and control iron deficiency in the United States. *MMWR* 47 (No. RR-3); 1998.

Table 15.10   Blood pressure levels for the 90th and 95th percentiles of blood pressure for boys and girls, ages 10 to 17 years.

		SYSTOLIC BP (mm Hg), BY HEIGHT PERCENTILE FROM STANDARD GROWTH CURVES													DIASTOLIC BP (mm Hg), BY HEIGHT PERCENTILE FROM STANDARD GROWTH CURVES													
		Boys							**Girls**							**Boys**							**Girls**						
Age	BP Percentile*	5%	10%	25%	50%	75%	90%	95%	5%	10%	25%	50%	75%	90%	95%	5%	10%	25%	50%	75%	90%	95%	5%	10%	25%	50%	75%	90%	95%
10	90th	110	112	113	115	117	118	119	112	112	114	115	116	117	118	73	74	74	75	76	77	78	73	73	73	74	75	76	76
	95th	114	115	117	119	121	122	123	116	116	117	119	120	121	122	77	78	79	80	80	81	82	74	74	75	75	76	77	78
11	90th	112	113	115	117	119	120	121	114	114	116	117	118	119	120	74	74	75	76	77	78	78	74	74	75	75	76	77	77
	95th	116	117	119	121	123	124	125	118	118	119	121	122	123	124	78	79	79	80	81	82	83	78	78	79	79	80	81	81
12	90th	115	116	117	119	121	123	123	116	116	118	119	120	121	122	75	75	76	77	78	78	79	75	75	76	76	77	78	78
	95th	119	120	121	123	125	126	127	120	120	121	123	124	125	126	79	79	80	81	82	83	83	79	79	80	80	81	82	82
13	90th	117	118	120	122	124	125	126	118	118	119	121	122	123	124	75	76	76	77	78	79	80	76	76	77	78	78	79	80
	95th	121	122	124	126	128	129	130	121	122	123	125	126	127	128	79	80	81	82	83	83	84	80	80	81	82	82	83	84
14	90th	120	121	123	125	126	128	128	119	120	121	122	124	125	126	76	76	77	78	79	80	80	77	77	78	79	79	80	81
	95th	124	125	127	128	130	132	132	123	124	125	126	128	129	130	80	81	81	82	83	84	85	81	81	82	83	83	84	85
15	90th	123	124	125	127	129	131	131	121	121	122	124	125	126	127	77	77	78	79	80	81	81	78	78	79	79	80	81	82
	95th	127	128	129	131	133	134	135	124	125	126	128	129	130	131	81	82	83	83	84	85	86	82	82	83	83	84	85	86
16	90th	125	126	128	130	132	133	134	122	122	123	125	126	127	128	79	79	80	81	82	82	83	79	79	79	80	81	82	82
	95th	129	130	132	134	136	137	138	125	126	127	128	130	131	132	83	83	84	85	86	87	87	83	83	83	84	85	86	86
17	90th	128	129	131	133	134	136	136	122	123	124	125	126	128	128	81	81	82	83	84	85	85	79	79	79	80	81	82	82
	95th	132	133	135	136	138	140	140	126	126	127	129	130	131	132	85	85	86	87	88	89	89	83	83	84	84	85	86	86

*Blood pressure percentile determined by a single movement.

SOURCE: Adapted from the National High Blood Pressure Education Program Working Group on Hypertension Control in Children and Adolescents. Update on the Task Force Report (1987) on High Blood Pressure in Children and Adolescents: A Working Group Report from the National High Blood Pressure Education Program. NIH Publications No. 97-3790. Bethesda, MD: National Heart, Lung, and Blood Institute; 1997.

HYPERLIPIDEMIA

Approximately one in four adolescents in the United States has an elevated cholesterol level.[47] Table 15.11 provides the classification criteria for elevated cholesterol levels in children and adolescents. Risk factors for hypercholesterolemia include a family history of cardiovascular disease or high blood cholesterol levels, cigarette smoking, overweight, hypertension, diabetes mellitus, and low level of physical activity. Adolescents who exhibit these risk factors should be screened to determine causes of hyperlipidemia and should be referred for treatment as required.[48] Early intervention among adolescents who have high cholesterol levels may reduce their risk of coronary artery diseases later in life.

Dietary recommendations indicate that youth over the age of five years should obtain less than 30% of their calories from fat, with no more than 10% of calories derived from saturated fat.[48,49] Dietary cholesterol intakes of 300 milligrams per day or less have also been recommended. Counseling adolescents with hyperlipidemia to follow these guidelines can be challenging, given their frequent consumption of fast foods and their preferred food choices. Suggested dietary changes should take into account the eating habits of adolescents and emphasize healthier food choices as opposed to restriction of favorite foods of teens. Health professionals can work with adolescents to make healthier choices at fast-food restaurants, to limit their portion sizes of high-fat food items such as high-fat snacks, and to consume adequate amounts of fruits, vegetables, grains, and low-fat dairy products.

Table 15.11   Classification of cholesterol levels in high-risk children and adolescents.*

	Total Cholesterol mg/dL	LDL Cholesterol mg/dL
Acceptable	<170	<110
Borderline	170–199	110–129
High	≥200	≥130

*Children and adolescents from families with hypercholesterolemia or premature cardiovascular disease.

SOURCE: Data taken from the National Institutes of Health, National Heart, Lung and Blood Institute, National Cholesterol Education Program. Report of the Expert Panel on Blood Cholesterol Levels in Children and Adolescents. Bethesda, MD: National Institutes of Health; 1991.

EATING DISORDERS

The Continuum of Eating Concerns and Disorders

Eating concerns and disorders lie on a continuum ranging from mild dissatisfaction with one's body shape to serious eating disorders such as *anorexia nervosa*, *bulimia nervosa*, and *binge eating disorder*. Along the continuum, between these endpoints, lie normative dieting behaviors and more severe disordered eating behaviors such as self-induced vomiting and binge eating (Illustration 15.4). Although engagement in anorexic behaviors and unhealthy dieting may not be frequent or intense enough to meet the formal criteria for being defined as an eating disorder, these behaviors may negatively impact health and may lead to the development of more severe eating disorders. All eating disorders present a serious public health concern in light of their prevalence and their potentially adverse effects on growth, psychosocial development, and physical health outcomes.[50]

Prevalence of Eating Disorders

An awareness of the prevalence of eating disorders is critical in effective planning for interventions aimed at their treatment and prevention. Conditions prevalent among youth warrant interventions that have the potential to reach large numbers of youth, such as community-based and school-based programs. The small percentage of the adolescent population affected by these types of conditions having severe health implications requires more intensive individual or small-group interventions. Estimates as to the prevalence of each of the eating disorders on the continuum are presented in Table 15.12.

Anorexia Nervosa

Anorexia nervosa and its impact on morbidity and mortality make it the most severe condition on the continuum of eating disorders. Among adolescent girls and young women, prevalence estimates of anorexia nervosa range from 0.2% to 1.0%.[51,52] Anorexia nervosa presents more frequently among females than among males; about 9 out of 10 individuals with anorexia nervosa are female. Only in recent years has attention been directed toward males with this condition who may not be suspected of having anorexia

nervosa and therefore may be diagnosed at later stages of the disease when treatment is more difficult.

Characteristics of anorexia nervosa include self-starvation and strong fears of being fat. An adolescent may begin with simple dieting behaviors due to social pressures to be thin or comments by others about an adolescent's weight. However, as these behaviors lead to weight loss, feelings of control, and comments from others about weight loss, anorexia nervosa develops and the condition takes on a life of its own. Diagnostic criteria for anorexia nervosa are shown in Table 15.13. Key features of anorexia nervosa are refusal to maintain body weight over a minimal normal weight for age and height; intense fear of gaining weight or becoming fat, even though underweight; a distorted body image; and amenorrhea (in females).

> **ANOREXIA NERVOSA** An eating disorder characterized by extreme weight loss, poor body image, and irrational fears of weight gain and obesity.
>
> **BULIMIA NERVOSA** An eating disorder characterized by recurrent episodes of rapid, uncontrolled eating of large amounts of food in a short period of time. Episodes of binge eating are often followed by purging.

The two types of anorexia nervosa are restricting and nonrestricting. In the restricting type, the individual does not regularly engage in binge eating or purging behaviors. The nonrestricting type exhibits regular episodes of binge eating and purging behaviors. However, both types present with a refusal to maintain a minimally normal body weight.

An estimated 10% to 15% of patients with anorexia nervosa die from their disease, although difficulties arise in assessing mortality rates from anorexia nervosa.[53] Reasons for fatality from anorexia include a weakened immune system due to undernutrition, gastric ruptures, cardiac arrhythmias, heart failure, and suicide.[54] The adolescent or the family commonly denies the condition, which delays the diagnosis and treatment, resulting in a poorer prognosis for recovery. Early recognition of possible signs of anorexia nervosa and seeking out of professional help significantly affect the time and intensity of treatment and improve chances for a successful recovery. Recovery rates are estimated at 40% to 50% for individuals with anorexia nervosa.[55]

Bulimia Nervosa

Bulimia nervosa is an eating disorder characterized by the consumption of large amounts of food with subsequent purging by self-induced vomiting, laxative or diuretic abuse, enemas, and/or obsessive exercising. Whereas anorexia nervosa is characterized by severe weight loss, bulimia nervosa may

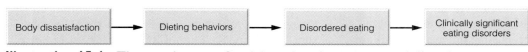

Illustration 15.4 **The continuum of weight-related concerns and disorders.**

Body dissatisfaction → Dieting behaviors → Disordered eating → Clinically significant eating disorders

Table 15.12 Estimated prevalence and brief description of weight-related concerns/disorders among adolescents.

Disorder	Estimated Prevalence
Anorexia nervosa	Approximately 0.2% to 1.0% of adolescent females and young women
Bulimia nervosa	Approximately 1% to 3% of adolescent females and young women
Binge eating disorder	Estimated 30% of weight-control population; 2% of general population
Anorexic/bulimic behaviors	Estimated 10% to 20% of adolescents although estimates vary
Dieting behaviors	Estimates vary and range from 44% of adolescent females, 15% adolescent males, to 50% to 60% of all adolescent females are attempting to lose weight
Body dissatisfaction	Estimates vary in accordance with type of measurement used and age, gender, and ethnicity of population: approximately 60% of girls and 35% of boys are not satisfied with their weight

Table 15.13 Diagnostic criteria for anorexia nervosa.

- Refusal to maintain body weight at or above a minimally normal weight for age and height (e.g., weight loss leading to maintenance of body weight less than 85% of that expected; or failure to make expected weight gain during period of growth, leading to body weight less than 85% of that expected)
- Intense fear of gaining weight or becoming fat, even though underweight
- Disturbance in the way in which one's body weight or shape is experienced, undue influence of body weight or shape on self-evaluation, or denial of the seriousness of the current low body weight
- Amenorrhea in postmenarchal women, that is, the absence of at least three consecutive menstrual cycles (A woman is considered to have amenorrhea if her menstrual periods occur only following hormone—estrogen—administration).

Restricting Type: During the episode of anorexia nervosa, the person has not regularly engaged in binge eating or purging behavior (i.e., self-induced vomiting or the misuse of laxatives, diuretics, or enemas).

Binge Eating/Purging Type: During the episode of anorexia nervosa, the person has regularly engaged in binge eating or purging behavior (i.e., self-induced vomiting or the misuse of laxatives, diuretics, or enemas).

SOURCE: Reprinted with permission. Diagnostic and statistical manual of mental disorders, 4th ed. Washington, DC: American Psychiatric Association; 1994.

show weight maintenance or extreme weight fluctuations due to alternating binges and fasts. In some individuals anorexia and bulimia nervosa overlap. Reliable estimates of bulimia nervosa range from 1.0% to 3.0%.[56] As with anorexia nervosa, about 90% of individuals with bulimia nervosa are female, probably due to the greater social pressures on women to be thin.

Diagnostic criteria for bulimia nervosa are shown in Table 15.14. Key features of bulimia nervosa include recurrent episodes of binge eating (rapid consumption of a large amount of food in a discrete period of time); a feeling of lack of control over eating behavior during the binge; self-induced vomiting; use of laxatives or diuretics; strict dieting or fasting; vigorous exercise in order to prevent weight gain; and persistent overconcern with body shape and weight. People with bulimia nervosa can be overweight, underweight, or of average weight for their height and body frame. Bulimia nervosa may be proceeded by a history of dieting or restrictive eating, which are thought to contribute to the binge-purge cycle. Mortality for bulimia nervosa appears to be lower than for anorexia nervosa. Based on a review of the existing literature in this area, Woodside has estimated that approximately 5% of pa-

BINGE EATING DISORDER An eating disorder characterized by periodic binge eating, which normally is not followed by vomiting or the use of laxatives. People must experience eating binges twice a week on average for over six months to qualify for this diagnosis.

tients die of their disease, usually due to heart failure resulting from electrolyte abnormality or suicide. Recovery rates for bulimia nervosa appear to be higher than for anorexia nervosa, with estimates of 50% to 60% for recovery.[55]

Binge Eating Disorder

Binge eating disorder (BED) is a condition in which one engages in eating large amounts of food and feels that these eating episodes are not within one's control.[57] BED is defined by recurrent episodes of binge eating at least two days a week for at least six months (Table 15.15). In addition, the person feels a subjective sense of a loss of control over binge eating, which is indicated by the presence of three of the following five criteria: eating rapidly, eating when not physically hungry, eating when alone, eating until uncomfortably full, and feeling self-disgust about bingeing. BED differs from bulimia nervosa in that binge eating is not followed by compensatory behaviors such as self-induced vomiting, as occurs in bulimia nervosa. BED is more prevalent among overweight

Table 15.14
Diagnostic criteria for bulimia nervosa.

A. *Recurrent episodes of binge eating.* An episode of binge eating is characterized by both of the following:

- eating, in a discrete period of time (e.g., within any two-hour period), an amount of food that is definitely larger than most people would eat during a similar period of time and under similar circumstances.

- a sense of lack of control over eating during the episode (e.g., a feeling that one cannot stop eating or control what or how much one is eating).

B. Recurrent inappropriate compensatory behavior in order to prevent weight gain, such as self-induced vomiting; misuse of laxatives, diuretics, enemas, or other medications; fasting; or excessive exercise.

C. The binge eating and inappropriate compensatory behaviors both occur, on average, at least twice a week for three months.

D. Self-evaluation is unduly influenced by body shape and weight.

E. The disturbance does not occur exclusively during episodes of anorexia nervosa.

Purging Type: During the current episode of bulimia nervosa, the person regularly engages in self-induced vomiting or the misuse of laxatives, diuretics, or enemas.

Nonpurging Type: During the current episode of bulimia nervosa, the person has used other inappropriate compensatory behaviors, such as fasting or excessive exercise, but has not regularly engaged in self-induced vomiting or the misuse of laxatives, diuretics, or enemas.

SOURCE: Reprinted with permission. Diagnostic and statistical manual of mental disorders, 4th ed. Washington, DC: American Psychiatric Association; 1994.

Table 15.15
Diagnostic criteria for binge eating disorder.

A. *Recurrent episodes of binge eating.* An episode of binge eating is characterized by both of the following:

- eating, in a discrete period of time (e.g., within any two-hour period), an amount of food that is definitely larger than most people would eat in a similar period of time and under similar circumstances.

- a sense of lack of control over eating during the episode (e.g., a feeling that one cannot stop eating or control what or how much one is eating).

B. The binge eating episodes are associated with three (or more) of the following:

- eating much more rapidly than normal

- eating until feeling uncomfortably full

- eating large amounts of food when not feeling physically hungry

- eating alone because of being embarrassed by how much one is eating

- feeling disgusted with oneself, depressed, or guilty after overeating

- experiencing marked distress regarding binge eating

- occuring, on average, at least two days a week for six months

C. The method of determining frequency differs from that used for bulimia nervosa; future research should address whether the preferred method of setting a frequency threshold is counting the number of days on which binges occur or counting the number of episodes of binge eating.

D. The binge eating is not associated with the regular use of inappropriate compensatory behaviors (e.g., purging, fasting, excessive exercise) and does not occur exclusively during the course of anorexia nervosa or bulimia nervosa.

SOURCE: Reprinted with permission. Diagnostic and statistical manual of mental disorders, 4th ed. Washington, DC: American Psychiatric Association; 1994.

clinical populations (30%) than among community samples (5% of females and 3% of males).[58] Studies on adolescents that assess the prevalence of binge eating include few that document prevalence rates of BED. In a college student sample, the rate of BED was 2.6%. In contrast to other weight-related conditions, significant differences were not found between male and female students. Further study of the prevalence and etiology of BED among adolescents seems critical in light of the increasing rates of obesity among youth.

Other Disordered Eating Behaviors

Some adolescents engage in anorexic or bulimic behaviors, but with less frequency or intensity than required for a formal diagnosis of an eating disorder. Behaviors typically considered in this category include self-induced vomiting, laxative use, use of diet pills, fasting or extreme dieting, binge eating, and excessive physical activity. The heterogeneity of these behaviors makes it more difficult to estimate the prevalence of anorexic and bulimic behaviors. In addition, the types of questions used to assess disordered eating behaviors may influence prevalence estimates. In a large national study of 6728 adolescents, disordered eating was reported by 13% of the girls and 7% of the boys.[59] In this study, disordered eating was assessed with the question: "Have you ever binged and purged (which is when you eat a lot of food and then make yourself throw up, vomit, or take something that makes you have diarrhea) or not?" In a large population-based study of Minnesota adolescents in grades 7 through 12, 12% of the girls reported that they had made themselves vomit for weight control purposes at least once in their lives and 2% reported having used laxatives or diuretics for weight control purposes. Rates were lower among boys: 6% reported vomiting and less than 1% reported laxative or diuretic use.[60] Binge eating (eating large amounts during a short period of time and feeling out of control while eating) was reported by 30% of the girls and 13% of the boys.[60] Based on our own research and a review of the more reliable research findings, it seems reasonable to estimate that between 10% and 20% of adolescents have engaged in anorexic or bulimic behaviors. Disordered eating behaviors such as self-induced vomiting and binge eating have serious implications for health and may be precursors to full-blown eating disorders. Therefore, interventions aimed at their prevention are essential.

Dieting Behaviors

Dieting behaviors among adolescents, and in particular among adolescent girls, tend to be alarmingly high. In a study of weight control behaviors among 459 adolescents (12–17 years of age) from four regions of the United States, current weight control behaviors were reported by 44% of the adolescent girls and 37% of the adolescent boys.[61] Based on their review of the literature, Fisher and colleagues concluded that 50% to 60% of adolescent girls consider themselves overweight and have attempted to diet.[62] Of particular concern are the increasing rates of dieting behaviors among children and young adolescents. A review of studies focusing on youth between the ages of 9 and 12 found high rates of dieting among girls (16% to 50%) and high percentages of girls who expressed a desire to be thinner (33% to 58%).[63] Estimates of dieting behaviors tend to vary across studies, in accordance with how dieting is assessed.

Dieting behaviors among youth are of concern in that they are often used by youth who are not overweight. Furthermore, unhealthful dieting behaviors in which meals are skipped, energy intake is severely restricted, or food groups are lacking are common. Dieting behaviors have been found to be associated with inadequate intakes of essential nutrients such as calcium.[64] Dieting behaviors leading adolescents to experience hunger or cravings for specific foods may place them at risk for binge eating episodes. Finally, dieting behaviors may be indicative of increased risk for the later development of eating disorders; Patton and colleagues found that the relative risk for dieters to develop an eating disorder was eight times higher than that for nondieters after a one-year period.[65] Therefore, dieting should not be viewed as a normative and acceptable behavior, in particular among children and adolescents.

Body Dissatisfaction

During adolescence, body image and self-esteem tend to be closely intertwined; therefore, body image concerns should not be viewed as acceptable and normative components of adolescence. Furthermore, body dissatisfaction is probably the main contributing factor to dieting behaviors, disordered eating behaviors, and clinical eating disorders.[66,67] Body dissatisfaction is reported by a high percentage of adolescents. In a study of 36,000 adolescents in Minnesota, Story and colleagues found that less than 40% of adolescent girls and about 65% of adolescent boys were satisfied with their weight.[68] A study by Tienboon looked at families with adolescents aged 14–15 years and reported that 41% of the girls and 14% of the boys considered themselves to be overweight.[69] Although actual weight status was directly associated with perceived weight status, a considerable number of girls who were not overweight perceived themselves as overweight.

Factors associated with increased body dissatisfaction include female gender, Caucasian ethnicity, and being overweight.[69] In working with overweight youth who express body dissatisfaction, health counselors are challenged to help them improve their body image while simultaneously working towards weight control. All adolescents, including overweight youth, should be encouraged to appreciate the positive aspects of their bodies. Overweight adolescents may need help in accepting the fact that they may never achieve the thin ideal portrayed in the media, but may strive toward a leaner and healthier body that is realistic for them. Tips for fostering a positive body image among adolescents, regardless of their weight, are shown in Table 15.16.

Etiology of Eating Disorders

The etiology of eating disorders is multifactorial, that is, many factors contribute to their onset. Some of the major contributory factors include social norms emphasizing thinness, being teased about one's weight, familial relations (e.g., chaotic lifestyles, boundaries between family members, patterns of communication), physical and sexual abuse experiences, personal body shape and size, body image, and self-esteem. These factors do not operate in isolation but interact with each other to increase the adolescent's risk for engaging in potentially harmful eating and dieting behaviors. In considering etiological issues, it is essential to realize that different etiological pathways may lead to weight-related disorders in different adolescents. For some adolescents, family issues may be major factors, while for others social norms may be the key factors leading to the onset of a condition. Furthermore, dif-

Table 15.16   **Tips for fostering a positive body image among children and adolescents.**

CHILD OR ADOLESCENT	PARENTS	HEALTH PROFESSIONAL
• Look in the mirror and focus on your positive features, not the negative ones. • Say something nice to your friends about how they look. • Think about your positive traits that are not related to appearance. • Read magazines with a critical eye, and find out what photographers and computer graphic designers do to make models look the way they do. • If you are overweight and want to lose weight, be realistic in your expectations and aim for gradual change. • Realize that everyone has a unique size and shape. • If you have questions about your size or weight, ask a health professional.	• Demonstrate healthy eating behaviors, and avoid extreme eating behaviors. • Focus on non-appearance-related traits when discussing yourself and others. • Praise your child or adolescent for academic and other successes. • Analyze media messages with your child or adolescent. • Demonstrate that you love your child or adolescent regardless of what he weighs. • If your child or adolescent is overweight, don't criticize her or his appearance—offer support instead. • Share with a health professional any concerns you have about your child's or adolescent's eating behaviors or body image.	• Discuss changes that occur during adolescence. • Assess weight concerns and body image. • If a child or adolescent has a distorted body image, explore causes and discuss potential consequences. • Discuss how the media negatively affects a child's or adolescent's body image. • Discuss the normal variation in body sizes and shapes among children and adolescents. • Educate parents, physical education instructors, and coaches about realistic and healthy body weight. • Emphasize the positive characteristics (appearance- and non-appearance-related) of children and adolescents you see. • Take extra time with an overweight child or adolescent to discuss psychosocial concerns and weight control options. • Refer children, adolescents, and parents with weight control issues to a dietitian.

SOURCE: Story M, Holt K, Sofka D, eds. Bright futures in practice: nutrition. Arlington, VA: National Center for Education in Maternal and Child Health; 2000.

ferent conditions tend to be influenced by different factors. That said, a number of factors play a major role across conditions and across individuals. Rosen and Neumark-Sztainer have grouped these potential contributory factors into socioenvironmental, personal, and behavioral domains.[70]

Socioenvironmental factors include:

- Sociocultural norms (e.g., regarding thinness, eating, food preparation, roles of women)
- Food availability (e.g., type of food, amount of food)
- Familial factors (e.g., communication patterns, parental expectations, boundaries, weight concerns and dieting behaviors of parents and siblings, family meals)
- Peer norms and behaviors (e.g., dieting behaviors, eating patterns, weight concerns)
- Abuse experiences (e.g., by family members, other adults, rape experiences)
- Media influences (e.g., images portrayed in the media, roles assigned to thin actors as compared to fat actors)

Personal factors include:

- Biological factors (e.g., genetic disposition, body mass index, age, gender, stage of development)

- Psychological factors (e.g., self-esteem, body image, drive for thinness, depression)
- Cognitive/affective factors (e.g., nutritional knowledge and attitudes, media internalization)

Behavioral factors include:

- Eating behaviors (e.g., meal patterns, fast-food consumption, nutritional variety, bingeing)
- Dieting and other weight management behaviors (e.g., dieting frequency, types of methods used, purging behaviors)
- Physical activity behaviors (e.g., TV viewing, sport involvement, daily activities)
- Coping behaviors (e.g., with dieting failures, with life frustrations)
- Skills (e.g., self-efficacy in dealing with harmful social norms, skills in food preparation, media advocacy skills)

An understanding of the etiology of eating disorders is essential to the development of effective interventions aimed at their treatment and prevention. An individual clinical setting needs to allow time to assess the factors leading to the onset of the condition for that particular adolescent. In developing prevention programs to reach larger groups of adolescents, it is more feasible to identify and address factors that may be contributing to the onset of weight-related behaviors and conditions for a broad sector of the targeted population. Although not all factors may be addressed within one intervention, it is important to be aware of the broad range of factors coming into play and the interactions between them.

Treating Eating Disorders

The complex etiology of eating disorders and their numerous potential psychosocial, physical, and behavioral consequences highlight the need for a multidisciplinary treatment approach. The health care team caring for an adolescent with an eating disorder will often include a physician, nutritionist, nurse, psychologist, and/or psychiatrist. The role of the nutritionist is paramount to the treatment of eating disorders at the stages of assessment, treatment, and maintenance. Initially an adolescent may be more willing to discuss his or her concerns with a nutritionist than with a psychologist. During treatment a major role of the nutritionist is to help the adolescent normalize eating patterns and to feel comfortable with these changes. For example, in the treatment of bulimia nervosa, some of the key goals of the nutritional care include the following:

- Establishing a regular pattern of nutritionally balanced meals and snacks
- Ensuring adequate but not excessive levels of energy intake with the goal of weight maintenance
- Ensuring adequate dietary fat and fiber intake to promote satiety
- Avoiding dieting behaviors and excessive exercise; gradually including formerly forbidden foods in the diet
- Dietary record keeping and review; stimulus control strategies to control high-risk situations; and weighing at scheduled intervals only[71]

For some adolescents, any denial of the condition or a lack of motivation for change make the work quite challenging. It is important for the nutritionist to work in close conjunction with other members of the health team to ensure that roles of different members of the team are clearly defined.

Preventing Eating Disorders

The high prevalence of eating disorders and their potentially harmful consequences point to a need for interventions aimed at their prevention. One of the most pressing current public health issues that needs to be addressed concerns the prevention of eating disorders described in previous sections and the prevention of obesity. Even the prevention of a small percentage of these conditions, at a population level, returns huge benefits in terms of reducing physical, emotional, and financial burdens.

In the development of interventions aimed at the prevention of eating disorders, it is essential to address factors that contribute to the onset of these conditions for a large proportion of the targeted population, factors that are potentially modifiable, and factors suitable for addressing within the designated setting. For example, media awareness and advocacy has been suggested as a suitable approach toward preventing eating concerns and disorders.[72] Participants may learn about how the media influences one's body image and about techniques used within the media to improve the appearance of models, and then take action towards making changes in the media. This approach is suitable in that media influences, and the internalization of media messages may contribute to weight concerns among a large sector of the adolescent population. These adolescent perceptions are potentially modifiable and suitable for addressing within clinical, community, and school-based settings where interventions may be implemented.

Any efforts toward prevention first must consider the target audience. An important question is whether to direct interventions to all adolescents or to adolescents at increased risk for eating disorders. Reasons for providing interventions for all adolescents include the high prevalence of eating concerns among adolescents, difficulties inherent in identifying and targeting high-risk individuals, and the advantages of developing positive social norms regarding eating issues within the peer group. Taking a more targeted approach offers the advantages of better use of limited resources, more intensive interventions, and interventions developed for specific high-risk groups (e.g., ballet dancers, youth with diabetes, or overweight girls). In order to be most effective in preventing eating disorders, both types of interventions seem necessary; more general approaches address the issues of the general adolescent population while more refined approaches can better meet the needs of specific high-risk groups.

Prevention interventions may be implemented within clinical, community, and school-based settings. Clinical settings provide opportunities for identifying early signs of eating disorders and working toward their prevention. Questions about body concerns and eating patterns should be a part of routine health care visits and appropriate channels be made available for discussing concerns. Key factors to be assessed in screening for eating disorders are shown in Table 15.17.

The majority of prevention programs have been implemented within school settings.[73–76] Schools provide an excellent setting for implementing prevention interventions in that they reach all adolescents; they provide a captive audience within a learning atmos-

Table 15.17   Screening elements and warning signs for individuals with eating disorders.

Screening	Warning Signs
Body image and weight history	• Distorted body image • Extreme dissatisfaction with body shape or size • Profound fear of gaining weight or becoming fat • Unexplained weight change or fluctuations greater than 10 lbs
Eating and related behaviors	• Very low caloric intake; avoidance of fatty foods • Poor appetite; frequent bloating • Difficulty eating in front of others • Chronic dieting despite not being overweight • Binge eating episodes • Self-induced vomiting; laxative or diuretic use
Meal patterns	• Fasting or frequent meal-skipping to lose weight • Erratic meal pattern with wide variations in caloric intake
Physical activity	• Participation in physical activity with weight or size requirement (e.g., gymnastics, wrestling, ballet) • Overtraining or "compulsive" attitude about physical activity
Psychosocial assessment	• Depression • Constant thoughts about food or weight • Pressure from others to be a certain shape or size • History of physical or sexual abuse or other traumatizing life event
Health history	• Secondary amenorrhea or irregular menses • Fainting episodes or frequent light-headedness • Constipation or diarrhea unexplained by other causes
Physical examination	• BMI <5th percentile • Varying heart rate, decreased blood pressure after arising suddenly • Hypothermia; cold intolerance • Loss of muscle mass • Tooth enamel demineralization

SOURCE: Used with permission. The Society for Nutrition Education. Adams LB, Shafer MB. Early manifestations of eating disorders in adolescents: defining those at risk. Journal of Nutrition Education: 20, 1988; and American Medical Association. Perkins K, Ferrari N, Rosas A, et al. You won't know unless you ask: the biopsychosocial interview for adolescents. Clinical Pediatrics: 36(2), 1997; and Guidelines for Adolescent Preventive Services (GAPS): Recommendation Monogragh, 2nd ed. Chicago, IL: American Medical Association; 1995.

phere and numerous opportunities for social interactions.[77] Ideally, a school-based intervention includes various components aimed at reaching the general population of adolescents, adolescents at increased risk for eating disorders, and school staff. Interventions should not only aim for changes in levels of personal knowledge and attitudes but also for changes in social norms (e.g., peer norms promoting thinness and dieting), policy changes (e.g., regarding tolerance levels for weight-teasing), and environmental changes (e.g., food availability within the cafeteria). Suggested components for a comprehensive school-based program for preventing eating disorders are shown in Table 15.18.

Interventions aimed at preventing eating disorders may also be implemented within other community-based settings. Neumark-Sztainer and her colleagues utilized the Girl Scout framework to reach fifth- and sixth-grade girls with an intervention aimed at promoting a positive body image and preventing unhealthy dieting.[78] The program, Free to Be Me, focused on promoting media awareness and advocacy among participants. An outline of the program is provided in Table 15.19.

Eating Disorders Among Adolescents: Summing Things Up

Eating disorders may be viewed on a continuum ranging from body dissatisfaction or dieting behaviors to clinically significant eating disorders including anorexia nervosa, bulimia nervosa, and binge eating disorder. The high prevalence of eating disorders and their potential harmful physical and psychosocial consequences indicate the need for interventions aimed at their treatment and prevention. The numerous socioenvironmental, personal, and behavioral factors leading to their onset need to be addressed in interventions. Parents, peers, educators, and health care providers play an important role in decreasing the prevalence of eating disorders.

CHILDREN AND ADOLESCENTS WITH CHRONIC HEALTH CONDITIONS

Approximately 18% of children and adolescents have a chronic condition or disability.[44] These children and adolescents are at increased risk for nutrition-related health problems because of (1) physical disorders or disabilities that may affect their ability to consume, digest, or absorb nutrients; (2) biochemical imbalances

Table 15.18   Suggested components of a comprehensive school-based program for preventing weight-related disorders.

Staff training	1. Examination of their own body image and eating behaviors 2. Knowledge about weight-related disorders 3. Skills for working with youth and program implementation
Classroom interventions for the general student body	Modula specifically aimed at preventing weight-related concerns and disorders. Options for approaches include: (1) feminist approach, (2) weight control and nutrition, and (3) promotion of life skills such as self-esteem and assertiveness
Integration of relevant material into existing curricula	Relevant information integrated into existing classes such as science, art, history, health, and physical education
Smaller and more intensive activities for high-risk students	Small group work or individual counseling for overweight students or students with excessive weight preoccupation and unhealthy dieting behaviors
Referral system within school and between school and community	Training of school staff to be alert to warning signs; referral mechanism in order to ensure that students don't get overlooked and those in need get help
Opportunities for healthy eating at school	Options for attractive nutrient-dense and low-fat foods at affordable prices in cafeteria, vending machines, and school events
Modifications in physical education and sport activities	Increased time being active in physical education classes; involvement of more students in after school sports; increased sensitivity to needs of overweight youth in physical education classes
Outreach activities to the community	By students, staff, and parents (e.g., contact with local media)

SOURCE: Adapted from Neumark-Sztainer D. School-based programs for preventing eating disturbances. Journal of School Health 1996;66.

caused by long-term medications or internal metabolic disturbances; (3) psychological stress from a chronic condition or physical disorder that may affect a child's appetite and food intake; and/or (4) environmental factors, often controlled by parents who may influence the child's access to and acceptance of food.

Nutrition reports of children and adolescents with special health care needs estimate that as many as 40% have nutrition risk factors that warrant a referral to a dietitian.[44] Common nutrition problems in children and adolescents with special health care needs include the following:

- Altered energy and nutrient needs (from inborn errors of metabolism)
- Delayed growth
- Oral-motor dysfunction
- Elimination problems
- Drug/nutrient interactions
- Appetite disturbances
- Unusual food habits (e.g., rumination)
- Dental caries, gum disease

Malnutrition has been implicated as a major factor contributing to poor growth and short stature in adolescents with a variety of diseases (e.g., chronic inflammatory bowel disease, cystic fibrosis). Factors such as inadequate intake, excessive nutrient losses,

malabsorption, and increased nutrient requirements all lead to the chronic malnourished state. Studies have shown that the energy requirements for adolescents with cystic fibrosis or inflammatory bowel disease may be as high as 30% to 50% more than the RDA for adequate growth. In addition to the increased energy needs caused by malabsorption (or in the case of adolescents with cystic fibrosis, the increased work of breathing), fever, infection, and inflammation also increase energy requirements. Whereas undernourishment is frequently seen in adolescents with chronic illnesses, obesity is common among youth with gross motor limitations or immobility. Because of limited activity, caloric requirements are lower, and the balance between intake and expenditure is often difficult, resulting in obesity.

Consideration of nutrition needs of children with chronic disabling conditions or illnesses is complex and requires specialized, individualized care by an interdisciplinary team. Assessment of nutrition status followed by nutrition intervention, when necessary, and monitoring will help ensure the health and well-being of adolescents with chronic and disabling conditions. Also, during adolescence issues of personal responsibility and independent living skills related to food purchasing and preparation may need to be addressed.

Table 15.19 **Outline of sessions in *Free to Be Me*.**

SESSION	ACTIVITIES
Body Truths I: **Body Development**	• *Let me Introduce You:* An introduction to the program and an ice-breaker activity based on the game Bingo. • *Stepping Stones:* Question-answer game in which girls learn about stages of body development. Each team advances to the next "stone" if question is answered correctly. • *Feelin' Good:* Take-home activity: Interviews with family members and friends about perceived positive traits.
Body Truths II: **Working with the Ideal Image**	• *Pin the Tail on the Timeline:* Girls attach ideal images of women throughout history to their appropriate spot on the timeline (e.g., Marilyn Monroe in the 1950s, Twiggy in the 1960s, and Tyra Banks in the 1990s). • *Sarah's Story:* Discussion of a story about a girl with poor body image who diets excessively. Discuss reasons for dieting, negative effects of skipping meals, and alternative approaches for Sarah. • *People Watching:* Take-home activity: Girls are asked to look for different body types in their community and then compare them to the body types they see in magazines.
Body Truths III: **What Else Is Out There?**	• *Looking for the Alternative (Stop-n-Go Posters):* Girls look through different teen magazines and make collages of the pictures that promote positive traits versus those that promote negative traits. Pictures are glued to either green poster board ("go") or red poster board ("stop"). • Take-home activity: Girls read "girl-friendly" magazine provided such as *New Moon* and discuss magazine with a parent.
Body Myths: **Media Madness**	• *Commercial Crazy:* Girls look at a variety of different TV commercials and look for positive and negative media messages (body types shown, messages about healthy eating versus dieting). • *What do you see on TV?:* Take-home activity: Girls and their parent(s) watch 15 to 30 minutes of TV and look for positive and negative messages in the commercials.
Take Action I: **What Can You Do to Impact the Media?**	• *Write a Letter:* Each troop writes a letter to a company that has a positive or negative impact on dieting and body image. Letters are posted on Web site for other advocates to sign. • Take-home activity: Girls develop and practice their own "girl-friendly" skit or commercial promoting positive body image and healthy eating.
Take Action II: **Spreading the Good Word**	• *Lights, Camera, Action!:* Girls perform their "girl-friendly" commercials or skits in front of their parents and troop. • Class overview.

SOURCE: Reprinted with permission. Neumark-Sztainer D, Sherwood N, Coller T, Hannan PJ. Primary prevention of disordered eating among pre-adolescent girls: feasibility and short-term impact of a community-based intervention. Journal of American Dietetic Association. In Press.

Photo Disc

Case Study 15.1
Adolescent Nutrition, Growth and Development, and Obesity

Anna is a 12-year-old female being seen in the pediatric and adolescent medicine clinic for a routine physical examination for summer camp. She reports no medical concerns. Her mother reports concern over recent weight gain. Anna's last visit was approximately five months earlier when she was seen for an upper respiratory infection.

During the current visit, Anna's height and weight are taken, and a BMI value is calculated. Her hemoglobin and hematocrit levels are checked, and a blood pressure reading is taken. The primary health care provider also takes a medical history and determines Anna's sexual maturation rating (SMR). The following data are recorded:

Age: 12 years, 5.5 months
SMR: early stage 3
Menarche: not yet occurred
Height: 58 inches (1.47 meters)
Weight: 111 pounds (50.35 kilograms)

BMI: 23.3
Hematocrit: 35.1%
Hemoglobin: 11.9 g/dL
Blood Pressure: 121/77

Anna's BMI value is charted on the National Center for Health Statistics BMI reference growth chart for girls age 2–20 years, and is compared to previous data. On her previous medical visit, Anna's BMI was calculated to be 21.4. Her previous height was 56.0 inches (1.42 meters) and her previous weight was 95 pounds (43.09 kilograms).

Questions:

1. What is the current BMI percentile for Anna, based on her age and gender?
2. What was her previous BMI percentile based on the BMI of 21.4 and age of 12.0 years ?
3. What is her current classification for height-for-weight (BMI)? Normal, at-risk for overweight, or overweight?
4. Given her current weight classification based on BMI, what are the recommendations for weight management based on current recommendations?
   a. Maintain current weight and monitor food intake, physical activity levels, weight, height, growth, and development at three- to four-month intervals.
   b. Weight loss of 1–2 pounds per week.
   c. Weight gain of 1–2 pounds per week.
   d. No concerns or actions are identified at this time.
5. What do the SMR and the absence of menarche tell you about Anna's potential for linear growth and future weight gain?
6. Using the Centers for Disease Control and Prevention (CDC) guidelines, how would you classify Anna's blood pressure readings?
7. Using the CDC guidelines, how would you classify her hemoglobin and hematocrit levels?

Answers:

1. Approximately 89th to 90th percentile.
2. Approximately 80th percentile.
3. At-risk for overweight.
4. Maintain weight and monitor every 3–4 months.
5. Because she has gained almost 17 pounds and has grown 2 inches in the past five months and she has not yet had her first menstrual cycle, she has

most likely not completed her peak growth spurt. She can be expected to gain several more inches in height in the next 6–12 months. She can be also be expected to continue to gain weight, but the rate of weight gain will slow down. Her weight gain may be excessive given her increases in height, however, so suggestions to improve her diet and increase physical activity are recommended at this time.

6. Normal systolic and diastolic per age and gender.
7. Hemoglobin is normal, but hematocrit is slightly low.

The primary health care provider refers Anna to a nutritionist at the clinic for nutrition counseling and weight management. The nutritionist conducts a 24-hour recall of food intake and also asks about physical activity and sedentary activity. The results follow.

FOOD INTAKE IN PAST 24 HOURS		ACTIVITY IN PAST 24 HOURS		SEDENTARY ACTIVITY IN PAST 24 HOURS	
Today:		**Today:**		**Today:**	
7 A.M.	8 oz apple juice at home	7:45 to 2:45	walked between classes for 1–2 minutes, six times that day	7:45 to 2:45	sat in classroom for 35–40 minutes, five times that day
10 A.M.	candy bar from school vending machine	1:00 P.M.	35 minutes of physical education (Anna played basketball for only 15 minutes due to need to rotate students on teams); physical education is offered two days per week in winter semester only	3:00 P.M.	on the computer in Internet chat rooms and e-mail for 40 minutes
12:15 P.M.	hamburger from school cafeteria and 12 oz of sweetened (nondiet) cola and 1 oz bag of potato chips from school vending machine			**Previous evening:**	
				8:00 P.M.	watched television for two hours
				10:45 P.M.	went to bed
3:00 P.M.	tortilla chips with cheese sauce (approximately 3 oz chips and ½ c cheese sauce) and 12 oz nondiet cola at home	**Previous evening:**			
		5:00 P.M.	Walked from friend's house to home (approximately 3 blocks)		
Previous evening:					
6:30 P.M.	chicken rice casserole (approximately 2 cups), 8 oz apple juice, 2 crescent rolls with 1 tbs butter				
8:30 P.M.	1 bag of butter flavored microwave popcorn (approximately 8 c)				

Questions:

1. List two nutrients that are likely to be consumed in inadequate amounts in Anna's diet.
2. In which of the Food Guide Pyramid food group(s) is/are Anna's diet inadequate?
3. List two recommendations that you would suggest to Anna to improve her food intake. (Try to use concrete examples because she is a young adolescent.)

(continued)

4. List two recommendations that you would suggest to Anna to improve her physical activity or reduce her sedentary activity. (Try to use concrete examples because she is a young adolescent.)

Answers:

1. Folic acid, calcium, iron, zinc, vitamins A, C, E, and D.
2. Fruits, vegetables, dairy products, breads/cereals/pasta.
3. Suggested responses may include: eat breakfast so as not to need a mid-morning vending machine snack; choose school lunch options such as salad bars and main dishes instead of a la carte and vending machine foods; drink low-fat milk or water at meals instead of juice or cola; limit cola to 12 oz or less, 1–2 times per week; reduce portion sizes of main dishes and supplement with vegetables and fruits if still hungry; increase fruit and vegetable consumption by choosing fresh fruit for breakfast, lunch, and dinner as well as snacks; choose baked tortilla chips and salsa instead of fried tortilla chips and cheese sauce; increase iron, calcium, zinc, and folic acid contents of diet by choosing fortified cereals at breakfast and snack times.
4. Suggested responses may include: walk to or from school if possible and safe; when possible, walk to a friend's house instead of calling on the telephone or chatting by e-mail; walk the dog each afternoon; join a local community center or health club that offers fun physical activities such as yoga classes, martial arts, dance classes, aerobic classes, swimming, or walking/hiking paths; take walks with friends or family after school or in the evenings; limit television viewing to one hour or less per day; limit Internet access to one hour or less per day; sign up for additional physical education classes at school when offered; buy or borrow a pedometer and try to walk at least 10,000 steps each day.

The nutritionist talks with Anna about ways to improve her diet by eating more healthy, convenient breakfast foods such as sandwiches, cereal, peanut butter or cheese on whole grain crackers, and string cheese for breakfast. She also suggests replacing apple juice with calcium-fortified orange juice to increase folic acid, vitamin C, and calcium intakes. The nutritionist suggests lower fat, more nutrient-dense vending machine choices that Anna can make when she eats from vending machines at school, such as animal crackers, whole grain crackers with peanut butter or cheese, yogurt, pretzels, baked chips, trail mix, dried fruit, milk, and fruit juices. Anna is encouraged to choose items from the school cafeteria for lunch instead of vending machines, especially from the salad bar. She is also given a list of low-fat, high-nutrient snacks that are easy to make or quick to eat after school. Finally, the nutritionist discusses portion sizes based on the Food Guide Pyramid with Anna and provides guidance about meeting the recommended number of servings of fruits, vegetables, whole grains, and dairy products.

The nutritionist provides suggestions on how Anna can increase physical activity and reduce sedentary activity to help in weight maintenance. She provides a list of local community centers and health clubs that offer fun classes and activities that adolescents might enjoy after school and on weekends. She suggests that Anna take daily walks with her best friend or mother, and that she limit television and computer time to no more than 2–3 hours total per day.

The nutritionist recommends that Anna come in for a follow-up appointment in 3–4 weeks to see how she is progressing with physical activity and dietary changes. She also suggests that Anna make an appointment to see her primary care provider in 3–4 months for further evaluation of her height, weight, and physical development.

Resources

American College of Sport Medicine

Web site: www.acsm.org

American Diabetes Association

Clinical practice, recommendations, nutrition, exercise, recipes, virtual grocery

Web site: www.diabetes.org

American Heart Association

An office-based practical approach to the child with hypercholesterolemia

Web site: www.americanheart.org/Scientific/pubs/hyperchol/toc.html

Nutrition tips and virtual cookbook for heart health

Web site: www.deliciousdecisions.org

Diabetes

Betschart J, Thom S. In control—A guide for teens with diabetes. Minneapolis, MN: Chronimed Publishing, 1995.

Boland, E. Teens pumping it up! 2nd ed. Sylmar, CA: Minimed Inc.; 1998.

Nutrition in the fast lane: A guide to nutrition and dietary exchange values for fast-food and casual dining. Indianapolis, IN: Franklin Publishing; 2001.

Web site: www.FastFoodFacts.com

Family Voices

Health information for children and adolescent with special health needs

Web site: www.familyvoices.org

Food Allergy Network

Educational materials and resources for youth with food allergies

Web site: www.foodallergy.org

Juvenile Diabetes Association

Web site: www.jdf.org

National Campaign to Prevent Teen Pregnancy

Web site: www.teenpregnancy.org

National Heart Lung and Blood Institute

Health Information for public: cholesterol, obesity

Health Information for professionals: NCEP ATP III, overweight and obesity

Web site: www.nhlbi.nih.gov

Weight Information Network

Publications and information on child and adolescent obesity treatment and prevention

Telephone: 1-800-946-8098

Web site: www.niddk.nih.gov//NutritionDocs.html

References

1. Barlow SE, Dietz WH. Obesity evaluation and treatment: Expert Committee recommendations. The Maternal and Child Health Bureau, Health Resources and Services Administration and the Department of Health and Human Services. Pediatrics 1998;102:E29.

2. Gortmaker SL, Must A, Perrin JM, et al. Social and economic consequences of overweight in adolescence and young adulthood. N Eng J Med 1993;329:1008–12.

3. Guo SS, Roche AF, Chumlea WC, et al. The predictive value of childhood body mass index values for overweight at age 35 y. Am J Clin Nutr 1994;59:810–9.

4. Whitaker RC, Wright JA, Pepe MS, et al. Predicting obesity in young adulthood from childhood and parental obesity. N Eng J Med 1997;337:869–73.

5. Johnson AL, Cornoni JC, Cassel JC, et al. Influence of race, sex and weight on blood pressure behavior in young adults. Am J Cardiol 1975;35:523–30.

6. Loder RT, Aronson DD, Greenfield ML. The epidemiology of bilateral slipped capital femoral epiphysis. A study of children in Michigan. J Bone Joint Surg Am 1993;75:1141–7.

7. Mallory GB, Jr., Fiser DH, Jackson R. Sleep-associated breathing disorders in morbidly obese children and adolescents. J Pediatr 1989;115:892–7.

8. Smoak CG, Burke GL, Webber LS, et al. Relation of obesity to clustering of cardiovascular disease risk factors in children and young adults. The Bogalusa Heart Study. Am J Epidemiol 1987;125:364–72.

9. Pinhas-Hamiel O, Dolan LM, Daniels SR, et al. Increased incidence of non-insulin-dependent diabetes mellitus among adolescents. J Pediatr 1996;128:608–15.

10. Boguslaw W, Bar-Or O. Effect of drink flow and NaCl on voluntary drinking and hydration in boys exercising in the heat. J Appl Physiol 1996;80:1112–7.

11. Steen SN. Nutrition for the school-age child athlete. In: Berning JR, Steen SN, eds. Nutrition for sport and exercise, 2nd ed. Gaithersburg, MD: Aspen Publishers; 1998:217–246.

12. National Collegiate Athletic Association. NCAA study of substance abuse habits of college student athletes. NCAA Committee on Medical Safeguards and Medical Aspects of Sports; 1997.

13. Coleman EJ. Carbohydrate: the master fuel. In: Berning JR, Steen SN, eds. Nutrition for sport and exercise. Gaithersburg, MD: Aspen Publishers;1998:21–44.

14. American College of Sports Medicine. Questioning 40/30/30: a guide to understanding nutrition advice. The American Dietetic Association, Women's Sports Foundation, and the Cooper Institute for Aerobics Research; 1998.

15. Schmalz K. Nutritional beliefs and practices of adolescent athletes. Journal of School Nursing 1993;9:18–22.

16. Clark JF. Creative and phosphocreative: a review of their use in exercise and sport. Journal of Athletic Training 1997;32:45–51.

17. Hultman E, Soderlund K, Timmons JA, et al. Muscle creatine loading in men. J Appl Physiol 1996;81:232–7.

18. Johnson WA, Landry GL. Nutritional supplements: fact vs. fiction. Adolesc Med State Art Rev 1998;9:501–13.

19. Dubelaar ML, Lucas CM, Hulsmann WC. Acute effect of L-carnitine on skeletal muscle force tests in dogs. Am J Physiol 1991;260:E189–93.

20. Burke ER. Nutritional ergogenic acids. In: Berning JR, Steen SN, eds. Nutrition for sport and exercise. Gaithersburg, MD: Aspen Publishers; 1998:127–8.

21. Hasten DL, Rome EP, Franks BD, Hegsted M. Effects of chromium picolinate on beginning weight training students. Int J Sport Nutr 1992;2:343–50.

22. Lefavi RG. Response (letter). Int J Sport Nutr 1993;3:120–2.

23. Wasser WG, Feldman NS, D'Agati VD. Chronic renal failure after ingestion of over-the-counter chromium picolinate. Ann Intern Med 1997;126:410.

24. Huszonek J. Over-the-counter chromium picolinate. Am J Psychiatry 1993;150:1560–1.

25. Ventura SJ, Matthews TJ, Curtin SC. Declines in teenage birthrates, 1991–1998: update of national and state trends. National vital statistics reports. Hyattsville,

MD: National Center for Health Statistics; 1999.

26. Manlove J. The influence of high school dropout and school disengagement on the risk of school-age pregnancy. J Res Adolesc 1998;8:187–220.

27. Wu LL. Effects of family instability, income, and income instability on the risk of a premarital birth. Am Soc Rev 1996;61:386–406.

28. Ventura SJ, Matthews TJ, Curtin SC. Declines in teenage birthrates, 1991–1997: national and state patterns. National vital statistics reports. Hyattsville, MD: National Center for Health Statistics; 1998.

29. Frisancho AR, Matos J, Bollettino LA. Influence of growth status and placental function on birthweight of infants born to young still-growing teenagers. Am J Clin Nutr 1984;40:801–7.

30. Naeye RL. Teenaged and pre-teenaged pregnancies: consequences of the fetal-maternal competition for nutrients. Pediatrics 1981;67:146–50.

31. Teagle SE, Brindis CD. Substance use among pregnant adolescents: a comparison of self-reported use and provider perception. J Adolesc Health 1998;22:229–38.

32. Institute of Medicine. Nutrition during pregnancy: part I: weight gain; Part II: nutrient supplements. Washington, DC: National Academy Press; 1990.

33. Scholl TO, Hediger ML. A review of the epidemiology of nutrition and adolescent pregnancy: maternal growth during pregnancy and its effect on the fetus. J Am Coll Nutr 1993;12:101–7.

34. Suiter CW. Maternal weight gain: a report of an expert work group. Arlington, VA: National Center for Education in Maternal and Child Health; 1997.

35. Centers for Disease Control and Prevention. Tobacco use among high school students—United States, 1990. Morb Mortal Wkly Rep 1991;40:617–9.

36. Johnston LD, O'Malley PM, Backman JG. National survey results on drug use from Monitoring the Future Study, 1975–1992, vol. 1. Rockville, MD: U.S. Department of Health and Human Services, Public Health Service, National Institute of Drug Abuse; 1993.

37. Kann L, Kinchen SA, Williams BI, et al. Youth risk behavior surveillance—United States, 1997. State and Local YRBSS Coordinators. J Sch Health 1998;68:355–69.

38. Isralowitz RE, Trostler N. Substance use: toward an understanding of its relation to nutrition-related attitudes and behavior among Israeli high school youth. J Adolesc Health 1996;19:184–9.

39. Neumark-Sztainer D, Story M, Toporoff E, et al. Covariations of eating behaviors with other health-related behaviors among adolescents. J Adolesc Health 1997;20:450–8.

40. Alton I. Substance use during pregnancy. In: Story M, Stang J, eds. Nutrition and the pregnant adolescent: a practical reference guide. Minneapolis, MN: Leadership, Education, and Training Program in Maternal and Child Nutrition; 2000:125–134.

41. Marion IJ, Mitchell JL. Pregnant, substance-using women. Rockville, MD: U.S. Department of Health and Human Services, Public Health Service, Substance Abuse and Mental Health Services Administration, Center for Substance Abuse Treatment; 1993.

42. Balluz LS, Kieszak SM, Philen RM, Mulinare J. Vitamin and mineral supplement use in the United States. Results from the third National Health and Nutrition Examination Survey. Arch Fam Med 2000;9:258–62.

43. Stang J, Story MT, Harnack L, Neumark-Sztainer D. Relationships between vitamin and mineral supplement use, dietary intake, and dietary adequacy among adolescents. J Am Diet Assoc 2000;100:905–10.

44. Story M, Holt K, Sofka D, eds. Bright futures in practice: nutrition. Arlington, VA: National Center for Education in Maternal and Child Health; 2000.

45. Centers for Disease Control and Prevention. Recommendations to prevent and control iron deficiency in the US MMWR 1998;47.

46. National Heart, Lung, and Blood Institute, National High Blood Pressure Education Working Group on Hypertension Control in Children and Adolescents. Update on the task force report (1987) on high blood pressure in children and adolescents: a working group report from the National High Blood Pressure Education Program. Bethesda, MD: National Institutes of Health; 1997.

47. Williams CL, Bollella M. Guidelines for screening, evaluating, and treating children with hypercholesterolemia. J of Ped 1995;9:153–62.

48. American Academy of Pediatrics CoN. Cholesterol in childhood. Pediatrics 1998;101:141–7.

49. National Heart, Lung, and Blood Institute, National Cholesterol Education Program. Report of the Expert Panel on Blood Cholesterol Levels in Children and Adolescents. Bethesda, MD: National Institutes of Health; 1991.

50. Neumark-Sztainer D. Excessive weight preoccupation: normative but not harmless. Nutr Today 1995;30:68–74.

51. Lucas AR, Beard CM, O'Fallon WM, Kurland LT. 50-year trends in the incidence of anorexia nervosa in Rochester, Minn.: a population-based study. Am J Psychiatry 1991;148:917–22.

52. Nylander I. The feeling of being fat and dieting in a school population: An epidemiological investigation. Acta Soc-Med Scand 1971;1:17–26.

53. Mott A, Lumsden D. Understanding eating disorders. Washington, DC: Tayler & Francis; 1994.

54. Kaplan A, Garfinkel P. Medical issues and eating disorders. New York: Brunner/Mazel; 1993.

55. Woodside DB. A review of anorexia nervosa and bulimia nervosa. Curr Probl Pediatr 1995;25:67–89.

56. Stein DM. The prevalence of bulimia: a review of the empirical research. J Nutr Educ 1991;23:205–13.

57. American Psychiatric Association. American psychiatric association diagnostic and statistical manual of mental disorders, 4th ed. Washington, DC: American Psychiatric Association; 1994.

58. Spitzer RL, Yanovski S, Wadden T, et al. Binge eating disorder: its further validation in a multisite study. Int J Eat Disord 1993;13:137–53.

59. Neumark-Sztainer D, Hannan PJ. Weight-related behaviors among adolescent girls and boys: results from a national survey. Arch Pediatr Adolesc Med 2000;154:569–77.

60. Neumark-Sztainer D, Story M, French SA, et al. Psychosocial concerns and health-compromising behaviors among overweight and nonoverweight adolescents. Obes Res 1997;5:237–49.

61. Neumark-Sztainer D, Rock CL, Thornquist MD, et al. Weight-control behaviors among adults and adolescents: associations with dietary intake. Prev Med 2000;30:381–91.

62. Fisher M, Golden NH, Katzman DK, et al. Eating disorders in adolescents: a background paper. J Adolesc Health 1995;16:420–37.

63. Lacher Malvey L, Neumark-Sztainer D, Story M. Prevalence of dieting and weight concerns among: a review of the literature. Under review.

64. Neumark-Sztainer D, Story M, Dixon LB, et al. Correlates of inadequate consumption of dairy products among adolescents. J Nutr Educ 1997;29:12–20.

65. Patton GC, Johnson-Sabine E, Wood K, et al. Abnormal eating attitudes in London schoolgirls—a prospective epidemiological study: outcome at twelve month follow-up. Psychol Med 1990;20:383–94.

66. French SA, Story M, Downes B, et al. Frequent dieting among adolescents: psychosocial and health behavior correlates. Am J Public Health 1995;85:695–701.

67. Neumark-Sztainer D, Butler R, Palti H. Personal and socioenvironmental predictors of disordered eating among adolescent females. J Nutr Educ 1996;28:195–201.

68. Story M, French SA, Resnick MD, Blum RW. Ethnic/racial and socioeconomic differences in dieting behaviors and body image perceptions in adolescents. Int J Eat Disord 1995;18:173–9.

69. Tienboon P, Rutishauser IH, Wahlqvist ML. Adolescents' perception of body weight and parents' weight for height status. J Adolesc Health 1994;15:263–8.

70. Rosen DS, Neumark-Sztainer D. Review of options for primary prevention of eating disturbances among adolescents. J Adolesc Health 1998;23:354–63.

71. Rock CL, Curran-Celentano J. Nutritional management of eating disorders. Psychiatr Clin North Am 1996;19:701–13.

72. Levine MP, Smolak L. Media as a context for the development of disordered eating. In: Smolak L, Levine MP, Streigel-Moore R, eds. The developmental psychopathology of eating disorders. Mahwah, NJ: Lawrence Erlbaum Associates; 1996.

73. Killen J, Barr Taylor C, Hammer L, et al. An attempt to modify unhealthful eating attitudes and weight regulation practices of young adolescent girls. Int J Eat Disord 1993;13:369–84.

74. Neumark-Sztainer D, Butler R, Palti H. Eating disturbances among adolescent girls. Evaluation of a school-based primary prevention program. J Nutr Educ 1995;27:24–31.

75. Piran N. On the move from tertiary to secondary and primary prevention: working with an elite dance school. In: Piran N, Levine MP, Steiner-Adair C, eds. Preventing eating disorders: A handbook of interventions and special challenges. Philadelphia: Brunner/Mazel; in press.

76. Smolak L, Levine MP, Schermer F. A controlled evaluation of an elementary school primary prevention program for eating problems. J Psychosom Res 1998;44:339–53.

77. Neumark-Sztainer D. School-based programs for preventing eating disturbances. J Sch Health 1996;66:64–71.

78. Neumark-Sztainer D, Sherwood N, Coller T, Hannan PJ. Primary prevention of disordered eating among pre-adolescent girls: feasibility and short-term impact of a community-based intervention. J Amer Diet Assoc; in press.

CHAPTER 16

Photo Disc

There is no dress-rehearsal for life.

Anonymous

ADULT NUTRITION

Prepared by **U. Beate Krinke**

CHAPTER OUTLINE

- Introduction
- Year 2010 Health Objectives
- Physiological Changes of Adulthood
- Maintaining a Healthy Body
- Dietary Recommendations
- Nutrient Recommendations
- Physical Activity Recommendations
- Nutrient Intervention for Risk Reduction

KEY NUTRITION CONCEPTS

1 Enjoyment of good food contributes to quality of life.

2 Food security means consistent access to safe, wholesome, and culturally acceptable food.

3 Good nutritional habits maximize peak physical and mental performance.

4 Hormones regulating reproduction also affect nutritional status.

5 Good nutrition now is an investment in healthy old age.

INTRODUCTION

Definition of Adulthood in the Lifecycle

With good luck, adulthood covers a life span of roughly 60 years. What can be expected to affect nutritional health during this time? Dividing 60 years into segments can help to address that question.

Early Adulthood: On reaching their twenties, adults have generally stopped growing. Some males grow slightly after age 20, men and women continue to develop bone density until roughly age 30, and muscle mass continues to grow as long as muscles are used. The primary tasks of adulthood include personal and career development and potentially reproduction. Challenges to good nutrition include juggling many competing demands, including work and personal schedules, travel, eating out, and exploring new relationships and community ties. Nutritional habits developed now are investments in future health.

Midlife: During their forties and fifties, most adults are reaching the peak of their career achievements. Physiologically, body composition slowly shifts somewhat, it is in tandem with hormonal shifts, but more probably due to decreased activity. On average, individuals start to gain weight after age 40. It is a good time to reassess earlier nutritional habits.

Old Age: In their sixties and beyond, adults harvest the fruits of earlier health habits. Good food and exercise habits practiced over a lifetime support continued enjoyment of sports and daily activities.

In this text, nutritional aspects of adulthood from age 20 to 64 will be covered in Chapters 16 and 17. Nutrition specifically for older adults will be covered in Chapters 18 and 19. This arbitrary division segments a natural continuum uniquely experienced by individuals. The focus of Chapter 16 is how to develop the most powerful nutrition habits possible.

Importance of Nutrition

Nutrition is about the many roles of food. Food is fuel, enjoyment, comfort, and a symbol of traditions, rituals, and celebrations. Food serves as a connection for socializing. When things go well, we take food for granted. We expect to have sufficient quantities of safe, appetizing food whenever we are hungry. During adulthood, we are often too busy to pay much attention to food.

How does food and nutrition enhance life? Lifestyle factors have a greater bearing on long life than genetics,

health care systems, and the environment.[1] Nutrition and exercise are at the top of the list of lifestyle factors; therefore individuals have some control over whether food and nutritional habits will contribute to a long and healthy life. Even though the leading causes of death at age 21 (unintentional injuries, homicide, and suicide) are not nutrition related, good nutrition throughout adulthood will reduce the risk of the leading causes of death of later adulthood, namely heart disease, cancer, stroke, and diabetes mellitus. Nutrition during adulthood supports an active lifestyle, contributes to maintenance of healthy weight, and promotes physical and mental health and well-being.

YEAR 2010 HEALTH OBJECTIVES

Since the first Healthy People 2000[2] report was generated under the guidance of the federal government in 1979, several goals have been met. Rates of heart disease, cancer, and stroke deaths have declined; fat intake (as percentage of total calories) has declined. However, rates of obesity and diabetes have increased, sugar intake has risen, and health care disparities still exist among ethnic and racial groups. Updated national health goals address these issues. Table 16.1 reflects dietary goals for disease prevention and health promotion for adults. The main focus of these goals is healthy weight maintenance.

PHYSIOLOGICAL CHANGES OF ADULTHOOD

Growth and maturation are complete by early adulthood, so that nutritional emphasis turns to maintaining physical status, continuing to build strength, and avoiding excess weight gain. Peak capacity for physical performance for most activities is reached during adulthood. Sports examples of athletes at the top of their form at age 20 or older include cycling, running, cross-country skiing, tennis (males), and soccer. The peak for endurance sports tends to be later in life than sports based on speed (for example, marathon runners are likely to be older than sprinters). However, for the average individual who participates in sport as a leisure activity, the physiological changes of early and middle adulthood neither enhance nor limit participation in, or enjoyment of, sports.

Physiological changes of adulthood differ for women and men. Women asked about their health and nutrition concerns are likely to be concerned about "not gaining weight" and about "looking good." Men are likely to say that they do not think about nutrition and health at all, but on probing will relate health concerns on strength, energy, and weight

Table 16.1 **Healthy People 2010: Health promotion goals for adults compared to current levels.**

		Population Percentage Baseline, Age 20+	
	Target	Females	Males
Increase the number who are at a healthy weight	60	45	38
Decrease the number who are obese (BMI 30+)	15	25	20
The following baseline information is for age 20–39 (Baselines for ages 40 to 59 are in parentheses) **Increase** the proportion of persons who:			
• Eat at least two daily servings of fruit	75	20 (26)	23 (28)
• Eat at least three daily servings of vegetables (At least one-third dark green or deep yellow)	50	4 (4)	3 (4)
• Eat at least six daily servings of grain products (At least three of those as whole grains)	50	4 (4)	10 (10)
• Eat less than 10% of calories from saturated fat	75	41 (42)	32 (33)
• Eat no more than 30% of calories from fat	75	38 (33)	29 (28)
The following baseline information is for age 20–49 Meet dietary recommendations (1000 mg) for calcium (level increases to 1200 mg at age 50)	75	40	64
The following baseline is for all adults, age 20+ Eat 2400 mg or less of sodium daily (Males consume more calories so they consume more sodium)	65	30	5
Increase the proportion of physician office visits made by patients with a diagnosis of cardiovascular disease, diabetes, or hyperlipidemia that includes counseling or education related to diet and nutrition	75	39	44
Increase food security and in so doing decrease hunger			
All households at or <130% of poverty	94	69*	69*
All households >130% poverty	94	94*	94*

*Adequate data to distinguish household food security levels for males and females is not available.
SOURCE: Food and Drug Administration and National Institutes of Health. Healthy People 2010: national health promotion and disease prevention objectives. Washington, DC: U.S. Department of Health and Human Services; 2000.

management. However, hormonal shifts related to reproductive capacity have the biggest impact on physiology, and therefore nutritional status and needs. Pregnancy and lactation are discussed in previous chapters. Menopause and the climacteric change in men are discussed in the following sections.

Physiological Changes in Males: Climacteric

For men, a gradual decline in testosterone levels begins about age 40 to 50. Sperm is capable of fertilizing human eggs until much later, although underweight is linked to decreased sperm production and malnutrition is linked to declining libido. Alcohol use can result in defective sperm. For the general population, alcohol intake declines with age. Alcohol ac-

counts for 5% of calorie intake between ages 20 to 34, but only 3% between ages 51 and 64.[3]

Based on cross-sectional data (see Table 16.4), males eat fewer calories as they get older. However, body weight rises slowly beginning around age 40, more likely due to declining activity levels than changing hormones.

Physiological Changes in Females: Menopause

For 40 years females have roughly 13 menstrual cycles per year (minus the ones missed during pregnancy or for other reasons such as extreme loss of body fat), totaling more than 500 menstrual cycles. After cessation of menstrual blood loss, the need for iron decreases from 18 to 8 milligrams daily.[4] Loss of estrogen leads to atrophy of tissues in the urinary tract and vagina, increased abdominal fat, and greater risk for chronic conditions such as osteoporosis and heart disease. Menopause is associated with, but not the cause of, weight gain and decreases in muscle mass.[5,6] Hormones regulate blood chemicals, that is, hormonal shifts impact nutritional status as shown in Table 16.2.

Oral contraceptives lead to an 8% to 11% increase of total serum cholesterol, with increases in LDL and triglycerides and concurrent decreases in HDL. Even though menopause negates the need for contraceptives, lack of estrogen and progesterone also leads to a rise in total cholesterol. Hormone replacement therapy (HRT, traditionally estrogen combined with progesterone, given as tablets, patch, or cream) is prescribed to maintain bone density, prevent height loss and vertebral fractures, maintain urogenital tissue structure, and to limit the risk of heart disease. HRT has also been linked to increases in blood clots and to breast cancer risk.[7] Women at low risk for heart disease do not benefit significantly from HRT.[8]

> **PHYTOESTROGEN**   A hormone-like substance found in plants, about 1/1000 to 1/2000 as potent as the human hormone, but strong enough to bind with estrogen receptors and mimic estrogen and anti-estrogen effects.

Nutritional Remedies for Symptoms of Menopause

Menopause does not lead to declining life satisfaction for women nor do most women suffer from menopause.[9] Women who do experience menopausal effects such as hot flashes, fatigue, anxiety, sleep disturbances, and memory loss are bombarded with treatment options including hormone replacement therapy, herbals, teas, foods and dietary supplements, exercise regimens, creams, and ointments. Experimental results for some treatment options have been published in reputable resources while other treatment options are at best unproven and potentially harmful.

Foods that seem to decrease menopausal symptoms due to their *phytoestrogen* content include soy, red clover, and flaxseed. Kava-kava, not a phytoestrogen, reduces anxiety and aids sleep. Phytoestrogen (e.g., lignans and isoflavones) effects depend on gut flora and hormonal milieu. Lignans, found in whole grains, beans and peas, and especially in flaxseed, are dietary precursors of phytoestrogens; gut flora convert them to usable forms. Isoflavones are especially concentrated in leguminous plants (plants that fix nitrogen in the soil, such as beans and peas). Soy and clover (red clover sprouts are eaten) contain isoflavones and have estrogenic activity. In the presence of high estrogen levels, such as before menopause, phytoestrogens link to estrogen receptors to prevent more potent human estrogens from binding with receptors. When there are low levels of estrogen, such as during or after menopause, phytoestrogens bind with estrogen receptors to supplement endogenous estrogen.

Soy, especially soy protein and soy isoflavone, are eaten to reduce unpleasant effects of menopause such as hot flashes. An epidemiological view suggests that in Asian cultures, where soy is a dietary staple, women do not report hot flashes. This factor may be a result of early exposure to soy phytoestrogens.[10] Experiments using soybeans during menopause are still relatively new, and the evidence about soy effectiveness is promising but inconsistent.[11]

Benefits of Phytoestrogens in Soy Foods

The North American Menopausal Society Recommendations state:

> Although the observed health effects in humans cannot be clearly attributed to isoflavones alone, it is clear that foods or supplements that contain isoflavones have some physiological effects.[12]

Soy isoflavones function as antioxidants and may be especially useful to women at increased risk for heart disease, which occurs at menopause.[13] Soy protein, with or without isoflavones, increases HDL cholesterol and decreases LDL and total cholesterol as well as decreasing triglycerides.[14] In fact, the Food and Drug Administration allows a health claim to be made

Table 16.2   **Hormonal shifts affect blood chemistries.**

Hormone	Triglyceride	LDL Cholesterol	HDL Cholesterol	Other Effects
Estrogen	Raises	Lowers	Raises	Slows calcium loss Increases insulin sensitivity Decreases homocysteine
Progesterone	Lowers	Raises	Lowers	......

SOURCE: Adapted from Contreras I, Parra D. Estrogen replacement therapy and the prevention of coronary heart disease in postmenopausal women. Available from www.medscape.com/ASHP/AJHP/2000/v57.n21/ajhp5721.01.cont-01.html, January 2001.

for soy: "Diets low in saturated fat and cholesterol that include 25 grams of soy protein a day may reduce the risk of heart disease." Because one consequence of menopause is an increased risk of heart disease in women, soy may be a healthful dietary addition.

Disadvantages of Adding Soy to the Diet

Soy has an estrogenic effect. Estrogen increases the risk for breast cancer. It would be unwise to suddenly add several servings of soy to the diet of someone vulnerable to breast cancer development (females with family history of breast cancer).

The taste of soy, as in soy flour and soy powder, is unappealing to many people. Soy oil, which is versatile and mild-flavored, does not contain enough phytoestrogens to impact menopausal symptoms. Tamari, or soy sauce, even in its reduced-sodium form is quite salty (540 milligrams of sodium in 1 tablespoon light soy sauce compared to 1230 milligrams in 1 tablespoon of regular soy sauce). In the small amounts used, soy sauce does not add much soy nutrition to the daily North American diet. Fermented soy products are good sources of soy protein and isoflavones but may not be widely available. Not many people are familiar with tempeh or miso, both fermented soy products.

A few ways to eat soy include the following:

- Edamame, young green soybeans, (sweet beans) are popular bar-food in Japan.
- Roasted mature soybeans look like roasted peanuts and can easily be flavored to make sweet or savory snacks (not yet an American bar-food).
- Soy protein meat analogs can substitute for meats (burgers, bacon, sausage, turkey and chicken) while providing soy nutrients.
- Soymilks do not taste like dairy milk but are used as a beverage and in cooking. Soymilk production formulas vary; try several to find your favorite.
- Processed cereals, bars, candies, and drinks and other snack foods are convenient sources of isoflavones and soy protein.

Flaxseed (see Table 16.3 for basic nutrition information) is a rich oil source that also contains high levels of antioxidants and phytoestrogens. Flaxseed can be ground, two tablespoons per day, to make nutrients accessible and help in reducing hot flashes. Also, toasted flax has a nutty flavor; Roast seeds until the kernels start to pop. (Keep the pan cover handy, it will remind you of popcorn.) Flax and flaxseed oils contain omega-3 fatty acids (alpha-linolenic acid), but

Table 16.3 Nutrient content in 2 tablespoons of ground flaxseed.

Kcal	71
Protein	3.5 gm
Carbohydrate	4.0 gm
Fiber	3.2 gm
Fat	5.3 gm
Omega 3 fatty acid	2.6 gm

Also contains significant amounts of iron, zinc, magnesium, potassium and some calcium. For fewer calories than in a slice of bread, 2 tablespoons of ground flaxseed provide the protein of one-half ounce of meat and as much fiber as in an apple. (A small apple, with skin, contains 2.9 grams of dietary fiber.)

it is the lignans (part of the fiber) in the flaxseed that are desirable for their estrogenic effects.

Other Alternatives

Estrogenic effects such as reduced hot flashes have been attributed to ginseng but experimental evidence supporting such activity is lacking.[15] Wild yams produce steroid compounds, leading to speculation about hormonal activity. They have been shown ineffective as hormone replacement therapy.[16] Kava is a relaxant and sleep enhancer; alcohol increases kava's effect. In Pacific Island cultures (Oceania) where kava is native, the root is made into a ceremonial beverage. The kava beverage first numbs the mouth, then leads to greater sociability, decreased anxiety, less fatigue, and eventually deep sleep. Therapeutic uses include muscle relaxation, sleep induction, and reduction of nervous anxiety and of restlessness. Kava extract contains kavalactones and has few side effects.[16] In reviewing cancer and diet, Willett concluded, "Moderate alcohol appears to increase endogenous estrogen levels."[17]

● MAINTAINING A HEALTHY BODY

Reproduction, weight management including maintenance of muscle mass while avoiding excess fat, and wellness are primary nutrition concerns of adulthood. Reproductive issues are discussed elsewhere in the text; the following sections focus on energy and other nutrients needed to maintain a healthy body.

Energy for Weight Management

The focus of adulthood is maintenance of physiological status rather than growth, although some males continue to gain height after age 20. Average calorie expenditures peak during late adolescence and early adulthood, and then decline roughly 20% over the course of a lifetime, although the size of this decline has not yet been clearly estimated because of limited longitudinal data.[18] In the mid-1960s, researchers at the Baltimore Longitudinal Study on Aging (BSLA) reported that caloric intake in men decreased 22%, from 2700 to 2100 calories, between age 30 and 80.[19] They suggested that the decrease was due to lowered metabolic rates as well as decreased activity levels. Subsequent research suggested that activity-related caloric expenditures are lower for women than for men, partly due to lower fat-free body mass.[20] Women's resting metabolic rate (energy expended when the body is not active) declined 2% to 4% after age 50. However, energy expenditures stayed constant as long as fat-free mass stayed constant. The Fels Longitudinal Study[21] examined weight gains over 20 years, starting at age 40, and calculated average annual increases:

> **BASAL METABOLIC RATE**   Energy expenditure at rest, 10 hours or more after eating, in a thermally neutral environment.

- Men gained 0.66 pounds per year (0.3 kg)
- Women gained 1.21 pounds per year (0.55 kg)

Determining Energy Needs

Changes in body weight are due to a complex system of interactions including gender and body size, energy intake, activity levels, health status, hormonal shifts, and individual variation.

- *Gender/body size/muscle mass:* Males in general use 5% to 10% more calories than females because they have proportionately more lean body mass. Heavier individuals need more calories.
- *Activity levels:* Activity requires energy, but more than that, exercise increases subsequent resting energy expenditures.
- *Health status:* Illness affects energy needs. For example, a fever's higher body temperatures burn more energy, 7% more per degree Fahrenheit or 13% per degree Celsius. Starvation slows energy expenditures by 20% to 30%.
- *Hormones:* The thyroid hormone thyroxin accelerates metabolism while lack of thyroid hormones leads to weight gain. Growth hormone is associated with weight gain, as is estrogen loss.
- *Individual variation:* Individuals vary by as much as 20% in their calorie expenditures during light to moderate activity.[22] In addition, the efficiency of metabolism and storage of food (the thermic effect of food) varies from person to person. Each of following approaches to estimating calorie needs tries to accommodate aspects of individual variation.

TWIN CITIES DIET MANUAL 2000   A three-step method adapted from a manual for clinical nutrition consists of estimating resting metabolic caloric needs, an activity level factor, and then adding a 10% allowance for digestion and absorption of food.[23]

1. To estimate resting metabolic needs, multiply current weight in pounds times 10 for females and times 11 calories for males. (Example, 150 pound female: $150 \times 10 = 1500$ calories per day.)
2. Additional calories to perform normal activities are derived from the following list. Calorie expenditures for specific activity levels are approximations due to variation in individual body composition, metabolism, and genetics. Estimate the individual's activity level and multiply by the resting metabolic calorie level derived in Step 1.

Activity Examples	Activity Factor
Very light activity such as sitting, standing, working at a desk	1.2
Light activity such as cleaning, strolling, shopping, child care	1.3
Moderate activity such as heavy yard work, tennis, skiing, dancing	1.4
Heavy activity such as manual labor, basketball, running, soccer	1.5
(Example: 1500 calories $\times$ 1.2 for desk job, very light activity = 1800 calories)	

3. To adjust for energy used in food metabolism, multiply the calories of Step 2 by 1.1, which adds 10% for digestion and absorption. (Example: $1800 \times 1.1 =$ **1980 calories**) The calculated caloric requirement should correspond to actual caloric intake during weight maintenance.

THE HARRIS-BENEDICT EQUATION   This two-step approach for healthy adults takes gender, age, and activity into account.[24] Physiological stress factors have also been developed for clinical use. Begin by calculating the individual's resting energy expenditure (REE), sometimes called the BEE (basal energy expenditure) or erroneously, *basal metabolic rate* (BMR). Then multiply REE by an activity factor to arrive at the desired caloric level. Metric measurements were used in the original formulas.

Step 1-a. REE for a female

655 + (9.6 × weight in pounds/2.2) + (1.8 × height inches × 2.54) − (4.7 × age) = REE

Example: Data for calculating resting energy expenditure for a 5'7" female weighing 150 pounds:

655 + (9.6 × 150/2.2) + (1.8 × 67 × 2.54) − (4.7 × 54) = REE

655 + (9.6 × 68.2) + (1.8 × 170.2) − 253.8 = REE

655 + 654.72 + 306.36 − 253.8 = **1362.3 calories**

At her 55th birthday, there would be a slight decrease in REE:

655 + 654.72 + 306.36 − 258.5 = **1357.6 calories**

Step 1-b. REE for a male

66 + (13.7 × weight in pounds/2.2) + (5 × height inches × 2.54) − (6.8 × age) = REE

Step 2. REE is multiplied by activity or injury and stress factors

Sedentary or weight maintenance	1.2
Light activity	1.3
Moderately active, infection, healing	1.5
Very active, extreme stress, burns	2.0

For the 55-year-old from step 2, a sedentary job would mean 1357.6 calories × 1.2 = 1629 calories. Comparing these two methods demonstrates why calorie calculations are called calorie estimates. Some trial and error is involved in caloric need determinations because energy expenditures are affected by complex interactions. A quick method for calorie estimations is to multiply current weight in pounds by 12–13 calories per pound (for light activity). For the 150-pound person in the example above, calorie estimates would range from 1800 to 1950 calories.

Calculations for Body Weight

Energy calculations require a known body weight. If weight is unknown, or when the desired weight is different from current weight, the Hamwi formula can be used to estimate a healthy body weight.[25,26] To adjust this formula for frame size estimate, add 10% of calculated weight for large frame or subtract 10% for small frame.

- Female: 100 pounds for the first 5 feet of height plus 5 pounds for each additional inch
- Male: 106 pounds for the first 5 feet of height plus 6 pounds for each additional inch

Alternately, the Miller method[27] more closely approximates the weights at the midpoints of weight ranges in the Metropolitan Life Tables of 1983.[25] The Miller method is not accurate for males taller than 74 inches.

- Female: 119 pounds for the first 5 feet plus 3 pounds for each additional inch
- Male: 135 pounds for the first 5 feet 3 inches plus 3 pounds for each additional inch

Other weight-for-height measures are discussed with obesity in Chapter 17.

Energy for Weight Change

It takes approximately 3500 calories to gain or lose one pound of body weight. To gain or lose a pound a week, add or subtract 500 kcal daily. These 500 calories can be a combination of intake and activity either added or subtracted. For instance, walking one mile uses roughly 100 calories, cycling five miles uses 200, and eliminating16 ounces of cola allows subtraction of 200 calories, summed to total 500 calories. Seven days of burning 300 calories and eating 200 fewer calories leads to a weight loss of approximately one pound.

We're a culture that eats everywhere and all the time. Food is everywhere. When was the last time you looked for a snack and couldn't find one?

Actual Energy Intake

Older adults eat fewer calories than younger ones. Table 16.4 shows findings from data collected by the U.S. Department of Agriculture's Continuing Survey of Food Intake by Individuals (CSFII) and by the Department of Health and Human Services in the National Health and Nutrition Examination Survey (NHANES).

Achieving Wellness: Linking Food, Nutrition, and Disease

It is not enough to know how many calories will maintain healthy weight. A more important question

Table 16.4 Comparing caloric intakes of adults, by gender, from CSFII (Continuing Survey of Food Intake by Individuals, 1994–1995) and NHANES III (National Health and Nutrition Examination Survey 1988–1994).

AGE	DAILY CALORIE INTAKE, CSFII		DAILY CALORIC INTAKE, NHANES III	
	Males	Females	Males	Females
20–29	2844	1828	3025	1957
30–39	2702	1676	2872	1883
40–49	2411	1680	2545	1764
50–59	2259	1583	2341	1629
60–69	2100	1496	2110	1578

Table 16.5   Illnesses and their modifiable risk factors.

Short-Term Illnesses	Nutritional Risk Factors
Food poisoning	Unsafe food handling practices
	Lack of hand washing, especially after toilet use
	Cross-contamination, mixing raw meat and vegetable preparation
	Lack of refrigeration, food at room temperature longer than two hours
	Inadequate temperatures in maintaining hot foods
	Contaminants such as salmonella, E. coli, mercury in some fish, etc.
Lowered immunity, Infections	Malnutrition, especially protein and calorie malnutrition
	Low antioxidant intake
	Low vitamin/mineral levels
	Low levels of zinc and excessive levels of zinc
	Dehydration

Chronic Conditions	Nutritional Risk Factors (Inactivity is associated with most)
Heart disease	Diet high in saturated fat
	More than 10% of calories from saturated fat
Atherosclerosis	High levels of trans fatty acids
	Low levels of mono-unsaturated fatty acids
	Diet high in fat, > 30% of calories from fat
	Low antioxidants, low fruit and vegetable intake
	Low intake of whole grains
	Low folic acid intake and high levels of homocysteine
	No or excess alcohol*
	Obesity (BMI ≥ 30); waist > 40" in men, > 35" in women
	Elevated plasma Apo B levels
	High levels of LDL cholesterol in men
	Low levels of HDL cholesterol in women
Cancer	Low fruit and vegetable intake
	Low level of antioxidants (especially vitamins A, C)
	Low intake of whole grains, especially fiber-rich grains
	High dietary fat intake
	Nitrosamines, burnt and charred food
	High intakes of pickled and fermented food
	Obesity
Hypertension, Stroke	High blood pressure: obesity, waist > 40" in men, > 35" in women
	High sodium
	Low potassium
	Excess alcohol
	Low levels of antioxidants
Diabetes	Obesity

* In middle aged adults, moderate alcohol reduces risk of heart disease (for men, after age 45; for women, after age 55)

is: "What should those calories look like?" Do certain foods contribute to better health? Can some foods cause illness? Dietary guidelines help to answer these questions. And to understand dietary guides or recommendations, it is useful to know about nutritional risk factors: How are our dietary habits linked to wellness and the avoidance of disease?

Much as we try, we cannot always achieve the degree of health we would like. But we can reduce the risk of contracting both short-term and chronic illnesses. Chronic conditions with modifiable risk factors include five of the 10 leading causes of death for

adults: heart disease, cancer, stroke, diabetes and hypertension. Short-term illnesses include food poisoning and infections. Table 16.5 provides a list of risk factors that can be addressed.

Nutritional risk factors are translated into dietary recommendations, such as the Dietary Guidelines for Americans and health claims on food packages as well as other public documents, including Healthy People 2010 and Diet and Health, the consensus report of the Food and Nutrition Board, National Academy of Sciences that followed the 1988 Surgeon General's Report describing nutrition and disease links.

Diet and Health

The Diet and Health summary guidelines[28] include both food and nutrition advice for disease risk reduction. They were developed for professional use and need interpretation to be applied in planning food programs, health promotions, and other nutrition services. Using this type of national consensus document in nutrition programs sets standards and facilitates consistent message development, which may decrease consumer confusion. The Healthy People 2010 goals (Table 16.1) integrate some of the recommendations from the Committee on Diet and Health:

1. Reduce total fat intakes to 30% or less of total kilocalories (kcal); see Table 16.6 for calculations. Limit saturated fat to less than 10% of total kcal, and cholesterol to no more than 300 milligrams daily; polyunsaturated fats should not exceed 10% of total kcal. *Comment:* These are still the basic recommendations of the American Heart Association, although limits on dietary cholesterol are less relevant for healthy older adults. Evidence from 128 feeding studies over 30 years suggests modest increases in dietary cholesterol lead to small, if any increases in blood cholesterol of populations.[29] McNamara calculated that six eggs per week raise serum cholesterol by 0.9 mg/dL (1994, Nutrition Close-Up, funded by the Egg Nutrition Center). Much individual variability is seen when it comes to response to any dietary intervention, and it seems that the general population is better off limiting saturated fat intake than counting cholesterol.
2. Each day, eat five or more servings of vegetables and fruits, especially dark green and deep yellow vegetables and citrus fruits; eat six or more servings from breads, cereals, or legumes. Increase of added sugars is not recommended. *Comment:* Matches Healthy People 2010 goals!
3. Maintain moderate protein intake, below twice the RDA.
4. Balance food intake and physical activity to maintain appropriate weight; avoid diets that are either excessive or severely restricted in kcal as well as large fluctuations in body weight.
5. "The committee does not recommend alcohol consumption. For those who do drink alcoholic beverages, limit consumption to the equivalent of less than 1 ounce of pure alcohol in a single day." (That's 2 cans of beer.) *Comment:* Women are more sensitive to alcohol than men. The Dietary Guidelines for Americans 2000 suggest that if you drink, do so with meals to slow absorption and keep it to one drink per day for women and two for men. (One drink: 12 oz regular beer, 5 oz wine, 1.5 oz of 80-proof distilled spirits.)
6. Limit daily intake of salt (sodium chloride or table salt) to 6 grams or less. *Comment:* 6 grams of sodium chloride correspond to 2400 milligrams sodium. "Na" is the chemical abbreviation for "sodium" seen on labels; 1 teaspoon of salt contains 2325 milligrams of Na.
7. Maintain adequate calcium intake. Potential benefits of excess intake are not documented. *Comment:* Potential for harm is now set at the tolerable upper intake level of 2000 milligrams calcium per day.[30]
8. "Avoid taking dietary supplements in excess of the RDA." *Comment:* Newer evidence about folic acid and vitamin B_{12} suggest that they are better absorbed in synthetic form during gastric illnesses; 1000 micrograms is the tolerable upper limit for folic acid, with no determined upper limit for vitamin B_{12}.[4,30]
9. Maintain an optimal intake of fluoride, available in fluoridated water.

Table 16.6 How to translate 30% of calories from fat to grams of fat.

1. Multiply total calories in the diet by 0.3 to determine calories equaling 30%.
2. Divide these 30% of calories by 9 (calories per gram of fat) to calculate the number of grams of fat that equal 30% of calories from fat.

Amount of fat that represents 30% of calories for selected calorie levels:	
Total Calorie Intake	**Fat Grams = 30%**
1200	40
1500	50
1800	60
2800	93

DIETARY RECOMMENDATIONS

How would you like your advice?

"Eat a juicy crisp apple with the peel."

or

"Eat more fiber, an apple is a good source of fiber.'

"Monounsaturated fats contribute to a lower risk for heart disease."

or

"Eating a salad with olive oil dressing makes your blood flow and gives you energy."

Governmental and private groups make food and nutrition recommendations according to their missions and goals. Three main perspectives underlie such dietary advice.

1. *Advocating for reduction of specific disease risk:* For example, the American Heart Association suggests reducing total dietary fat to decrease risk of heart disease. The National Cancer Institute (as well as the fruit and vegetable produce industry) urges us all to eat more fruits and vegetables to reduce risk of certain cancers.

2. *Ensuring adequate population intake of specific nutrients:* For example, Food and Drug Administration (FDA) regulations mandate flour and grain product fortification with B vitamins, folic acid, and iron. Folic acid was added in the late 1990s to prevent neural tube defects. Here, FDA recommendations are more than "advice," they are enacted into law. Enrichment and fortification laws are intended to improve the health status of populations.

3. *Offering guidance on what, and how much, to eat:* For example, the U.S. Department of Agriculture and U.S. Department of Health and Human Services together publish the Food Guide Pyramid and "Dietary Guidelines for Americans." On the nongovernmental side, many Dairy Councils have developed dietary guidance tools built on the basic food groups, including the cheese and milk group. Dietary guidance tools such as these can help consumers make meal plans.

Dietary Recommendations to Combat Nutritional Concerns

Dietary recommendations for the public translate and integrate documents such as Diet and Health, nutrient guidelines from the Food and Nutrition Board and the U.S. Department of Health and Human Services as well as the various health advocacy coalition positions, such as American Heart Association, National Cancer Institute, National Osteoporosis Foundation, into a consensus document. This document spells out the national health and nutrition goals and priorities, makes evident the links between eating, drinking, and health outcomes, and includes meaningful food advice.

"Dietary Guidelines for Americans, 2000" address current health issues by outlining health goals, describing diet and disease linkages, and including the Food Guide Pyramid to provide advice on what to eat for good health. Dietary Guidelines 2000 continues a series first issued in 1980 and updated every five years. The current version of health-promoting food advice reflects current nutrition science and policy. Food safety is a new topic in the 2000 guide because the number of incidents has increased. The 1995 statement about sugar changed from "choose a diet moderate in sugars" to "choose beverages and foods to moderate your intake or sugar" in order to reflect the rising intake of nonnutrient sugars, one of the contributors of obesity. Table 16.7 shows the basic components of the Dietary Guidelines 2000, including specific food-based advice.

Food Advice

Food guidance has been around since 1916, having been spurred by war and economic depression.[31] What makes food advice useful? Deciding what to eat is harder than knowing what to avoid. The message "skip ice cream" tells an audience exactly what to do: avoid eating ice cream. But the message "get no more than 30 percent of calories from fat" needs further knowledge to be translated into breakfast, lunch, snacks, or supper. In order to help individuals focus on foods to eat, a food guide needs several attributes:

- *Classify foods into relevant groups.* For example, what are the food groups for peanut butter on toast and a glass of juice?
- *Describe what constitutes a serving.* For example, do the lettuce, tomato, pickle, and ketchup on a burger count as one or more servings?
- *Identify the number of servings needed from each group to promote health.* For instance, how many grain servings will it take to meet carbohydrate and fiber recommendations?

Table 16.7   Nutrition and Your Health: Dietary Guidelines for Americans.

Aim for fitness:
  Aim for a healthy weight.
  Be physically active each day.

Build a healthy base:
  Let the pyramid guide your food choices.
  Choose a variety of grains daily, especially whole grains.
  Choose a variety of fruits and vegetables daily.
  Keep food safe to eat.

Choose sensibly:
  Choose a diet low in saturated fat and cholesterol, moderate in total fat.
  Choose beverages and foods to moderate your intake of sugars.
  Choose and prepare foods with less salt.
  If you drink alcoholic beverages, do so in moderation.

SOURCE: U.S. Department of Agriculture and U.S. Department of Health and Human Services. Home and Garden Bulletin No. 232, 5th ed. 2000.

Table 16.8 **Food Guide Pyramid recommendations for active men and women.**

	NUMBER OF SERVINGS PER kcal LEVEL		
	1500–1700 (most women)	2100–2300 (most men, active women)	2700–2900 (active men)
Food Groups			
Basic Five Food Groups			
Bread, cereal, rice, pasta (1 oz, 0.5 c cooked)	6	9	11
Vegetables (1 c leafy, 0.5 c raw, cooked)	3	4	5
Fruit (6 oz juice, 0.5 c canned, 1 med)	2	3	4
Milk, yogurt (1 c), cheese (1.5–2 oz)	2–3*	2–3*	2–3*
Meat, poultry, fish, eggs (2–3 oz); dry beans, tofu (0.5 c cooked); nuts, seeds (0.33 c)	2 (total 5 oz)	2 (total 6 oz)	3 (total 7 oz)
Fluids (1c) (amounts based on 1 ml/kcal)	6	8	10
The "Other" Group			
Fats, oils, sweets (1 tsp oil, sugar, jam)	5	7	9
Alcohol: if using, moderation is 1 drink for women and 2 drinks for men per day			

* 2 servings between age 19 and 50, 3 servings after age 50.
SOURCE: Modified from USDA/HHS Food Guide Pyramid. Available from www.usda.gov, or 202-720-2600, voice and TDD, for Braille, large print, or alternative means of communication.

The Pyramid

The Food Guide Pyramid, Table 16.8, lists food groups, serving sizes, and number of servings for various calorie levels. (See also Illustration 16.1.) Consuming the recommended number of servings from each group to meet caloric needs over a period of three days on average, and 5–10 days at most, will meet individual overall nutrient needs.[31] When all the chosen foods from each group are lean, low in fat, and contain no added sugars, calorie levels of the basic five food groups range from 1200 to 2000

Illustration 16.1 **The Food Guide Pyramid.**

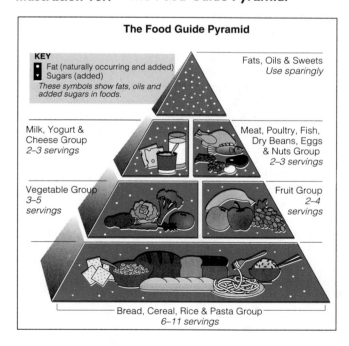

The Food Guide Pyramid

KEY
■ Fat (naturally occurring and added)
▼ Sugars (added)
These symbols show where fats, oils and added sugars in foods.

Fats, Oils & Sweets
Use sparingly

Milk, Yogurt & Cheese Group
2–3 servings

Meat, Poultry, Fish, Dry Beans, Eggs & Nuts Group
2–3 servings

Vegetable Group
3–5 servings

Fruit Group
2–4 servings

Bread, Cereal, Rice & Pasta Group
6–11 servings

"Dietary Guidelines for Americans," 2000. Available from www.usda.gov/cnpp.

WHAT COUNTS AS ONE SERVING?

The amount you eat may be more than one serving. For example, a dinner portion of spaghetti would count as two or three servings.

Bread, Cereal, Rice, and Pasta Group
1 slice of bread
½ cup of cooked rice or pasta
½ cup of cooked cereal
1 ounce of ready-to-eat cereal

Milk, Yogurt, and Cheese Group
1 cup of milk or yogurt
1½ ounces of natural cheese
2 ounces of process cheese

Vegetable Group
½ cup of chopped raw or cooked vegetables
1 cup of leafy raw vegetables

Meat, Poultry, Fish, Dry Beans, Eggs, and Nuts Group
2½ to 3 ounces of cooked lean meat, poultry, or fish
Count ½ cup of cooked beans, or 1 egg, or 2 tablespoons of peanut butter as 1 ounce of lean meat

Fruit Group
1 piece of fruit or melon wedge
¾ cup of juice
½ cup of canned fruit
¼ cup of dried fruit

Fats and Sweets
LIMIT CALORIES FROM THESE especially if you need to lose weight.

calories and 16% of the calories are provided from fats. Fats, oils, and sweets make up the "other" food group "to be used sparingly."

Food Advice Reflects Cultural Food Patterns

Food group guidance comes in many forms other than the pyramid. For example, Canada's Food Guide to Healthy Eating is a rainbow (Illustration 16.2). Inclusion of canned and frozen food pictures makes it obvious that processed foods count. The rainbow groups fruits and vegetables, lists metric and English amounts for serving sizes, and does not picture "other" foods.

The United States uses a pyramid, Canada a rainbow, and China uses a pagoda. China's pagoda highlights how food guides reflect their target audience's unique dietary needs and goals (Illustration 16.3). Following pagoda guidance ensures that individuals eat from 1800 to 2800 calories per day. Additional advice to Chinese residents includes distributing food evenly throughout the day, making full use of local food sources, and making good nutrition a lifetime commitment. The gram-weights listed with food groups show that the Chinese diet is cereal-based, includes greater quantities of vegetables than fruit, includes fish daily, and uses small portions of meat. The placement of milk/milk products and bean/bean products together in one group is also unique to the pagoda.

Dietary advice for health promotion, Japanese style, includes the following recommendations.

Make all activities pertaining to food and eating pleasurable ones:

1) use the mealtime as an occasion for family communications and
2) treasure family taste and home cooking.

Alcohol: Food, Drug, and Nutrient

Canadian nutrition recommendations are to limit alcohol calories to 5% or less of total calories or two drinks a day, whichever is less.[32] U.S. guidelines are "If you drink, do so in moderation," defined as no more than two drinks per day for males, and no more than one drink per day for females. Saving up drinks to have all 7 to 14 on the weekend is not "moderation." The Canadian guide states that more than four drinks at a time is risky.

What is a drink? Equivalent alcohol contents for "a drink" are 12 ounces of regular beer, 5 ounces of wine, 1.5 ounces of 80 proof distilled spirits. A drink contains roughly 13–15 grams of alcohol. New

Zealand gives guidelines in grams of alcohol: 20 or fewer grams of alcohol per day for women and 30 grams or less for men.

Alcohol is considered a drug for its central nervous system effects, but it is also a nutrient because it yields

Connections
Food Guide Pyramid FAQs

Sample questions from teaching the Food Guide Pyramid:

1. **Q:** *How do combination foods (pizza, spaghetti, and soup) fit into the pyramid?*
   A: Estimate the components and count separately: a wedge of pizza would be one serving of bread, a half to one vegetable (tomato and other, one serving = .5 c), and one or more ounces of meat/cheese, depending on the topping.

2. **Q:** *Does juice count as part of the fruit group? If so, how much?*
   A: Yes, 6 ounces of fruit juice counts as one serving.

3. **Q:** *Does chocolate milk count as milk?*
   A: Yes, chocolate milk is regular milk flavored with sugar and cocoa, the label should indicate that 8 ounces chocolate milk provide at least 30% of the recommended intake of calcium and of vitamin D (300 mg calcium and 100 IU of vitamin D). Chocolate and other flavored milks vary as to A and D fortification; check the ingredient panel.

4. **Q:** *Where do I put soymilk?*
   A: Soymilk, good question! When fortified with calcium, it could count in the milk group, although the calcium is less well absorbed. Eight ounces soymilk contain 6.6 grams of protein, roughly the amount in 1 ounce of meat, so the beans/nuts/meat alternate group is also a fine category.

5. **Q:** *What is the difference between serving size and portion size?*
   A: Servings are standardized amounts of food such as 1 slice bread (~1oz), 1 teaspoon butter and 6 ounces juice. A hamburger bun counts as 2 or more bread servings, depending on size. General use of the term portion is for amount of food served. For example, that 2-ounce hamburger bun is 1 portion of bun but 2 servings of bread.

Taken from a list from the Society for Nutrition Education (October 20, 2000, posted to SNEEZE by Ann Marie Dawson, a graduate student in Madison, WI). These questions show the challenges of developing an effective food guidance tool.

Illustration 16.2 **Canada's Food Guide to Healthy Eating—the rainbow.**

SOURCE: Health and Welfare Canada. Nutrition recommendations. Canada: Minister of Supply Services; 1990.

energy, that is, seven calories per gram. When developing a dietary assessment tool for adults, alcohol needs a separate line because it supplies a significant number of calories (3% to 4% of total calories) to the diet.[3,33] Males drink more alcohol than females (12 grams versus 5 grams daily for the total population, all ages). Roughly one-third of adults abstain from alcohol.[34,35] Therefore, alcohol ingestion per person is really somewhat higher than the population averages might indicate. Current data reports average alcohol consumption for alcohol per se. Caloric contributions of alcoholic beverages are higher. Based on an analysis of the first (1977–1978) Nationwide Food Consumption Survey of just those individuals who drank alcoholic beverages, males consumed 17% and females 22% of total calories from mixed drinks, wine, and beer.[36]

Fluids

Adults can survive roughly 60 days without food but only three to four days without water.[6,37] Because fluid is such an integral part of metabolism, the body has many regulatory mechanisms to decrease risk of dehydration (increase thirst, concentrate urine to re-absorb water, stop sweating, activity slowdown).

Methods to determine fluid recommendations relate to caloric intake or body weight.

The Food and Nutrition Board recommends 1 milliliter water per calorie of food ingested. Therefore, 2000 milliliter of fluid should accompany the 2000-calorie intake used as reference diet on food labels. (Average calorie intakes for 20- to 29-year-olds are 1828 for females and 3025 for males, see Table 16.9).

Community practitioners who work with metric units might appreciate the method in Chernoff's Geriatric Nutrition handbook.[38,39] She suggests 30 milliliters of fluid per kilogram body weight with a minimum of 1500 milliliters per day. For example, a 143-pound person weighing 65 kilograms would need 65 × 30 or 1950 milliliters fluid daily. A 170-pound person weighing 77 kilograms would need 77 × 30 or 2310 milliliters fluid per day.

A step-wise formula that allows adjustment for increasing or decreasing body weight and insures a 1500 milliliters minimum daily intake yields somewhat higher fluid recommendations for smaller, lighter adults:

100 ml/kg body weight for the first 10 kg
50 ml/kg body weight for the next 10 kg
15 ml/kg body weight for the remaining kilograms

Illustration 16.3
The Food Pagoda for Chinese residents.

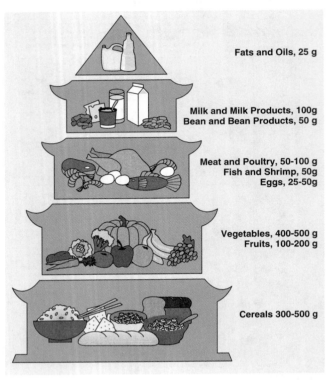

Fats and Oils, 25 g

Milk and Milk Products, 100g
Bean and Bean Products, 50 g

Meat and Poultry, 50-100 g
Fish and Shrimp, 50g
Eggs, 25-50g

Vegetables, 400-500 g
Fruits, 100-200 g

Cereals 300-500 g

Used by permission. The Chinese Nutrition Society, People's Republic of China.

Applying this formula to a 143-pound (65 kg) person leads to a recommendation of just over two liters (eight cups), and roughly 10 cups for a 170-pound (77 kg) person.

Body weight of	143 pounds	170 pounds
10 kg × 100 =	1,000 ml	1000 ml
10 kg × 50 =	500 ml	500 ml
Remaining ___kg × 15 =	45 × 15 = 675 ml	57 × 15 = 855 ml
Total =	2,175 ml	2355 ml

240 ml = 8 ounces = 1 cup; 4 cups = 1 quart, roughly equivalent to 1 liter

A simple, yet practical guide to adequate fluid intake that does not rely on numbers is the color of urine: when it is nearly colorless or pale yellow, the individual is drinking enough. (Note: Therapeutic amounts of riboflavin supplements as well as some medications and foods affect the color of urine, e.g., flavins are pigments.)

Diuretic Effects of Caffeine and Alcohol

What about fluids containing caffeine? Occasionally someone suggests that coffee, tea, or other caffeine-containing beverages not be counted as fluid intake because they have a diuretic effect on the body.

Caffeine is a stimulant that relaxes the esophageal sphincter (leads to acid reflux), has a laxative effect, and increases urine production.[40] But no evidence connects drinking coffee or tea or other caffeine-containing-fluids with greater dehydration.[41] At this time, the issue of whether to count coffee, tea, or other caffeine-containing beverages in or out of the fluid allotment is a policy issue rather than a scientific one.

Several countries address caffeine in their dietary guidelines; none address whether caffeine-containing beverages should count as fluids.[31] South Africa suggests restricting caffeine (no amounts given) as part of dietary guidance. Canada is the only country to suggest an amount for caffeine consumption: "The Canadian diet should contain no more caffeine than the equivalent of four regular cups of coffee per day."[32]

Fluids that contain alcohol have a dehydrating effect on cells, so some health professionals do not count alcoholic beverages as contributing fluids to the diet. If alcohol use is moderate, the amounts of fluid in question are necessarily small.

International Guidance Related to Fluids

We tend to take water for granted and not emphasize it in our dietary guidelines. The "Dietary Guidelines for Americans" integrate fluids into the sugar and sodium advisories, suggesting "Choose sensibly to limit your intake of beverages and foods that are high in added sugars" and "Drink water often" (p. 33).[42] Part of the sodium advice is: "Drink water freely. It is usually very low in sodium. Check the label on bottled water for sodium content" (p. 35). (For example, 8 ounces of spring water typically contain 2 milligrams of sodium; 8 ounces of municipal tap water contain 7 milligrams; and household softened water 9 milligrams of sodium. These levels vary dramatically by geographic region.) Here is how several countries address the need for adequate fluid in their dietary guidance documents.[31]

- New Zealand: Drink one to two liters per day (1000 to 2000 ml, about 8 cups) and choose from tap water and other beverages, including tea and coffee.
- South Africa: Drink at least one liter of clean, safe water or fluids.
- Switzerland: More fluids.
- Venezuela: Drink water.
- Hungary, Korea, Philippines, Singapore, and Switzerland suggest "milk every day" or "more milk" as part of their dietary recommendations. This suggestion may have more to do with increasing calcium and other nutrients than fluid intake.

Table 16.9 Selected nutrient intakes of adults, Continuing Survey of Food Intake by Individuals, CSFII, 1994–1995 data collection, compared to recommendations.

	INTAKE				
	ACTUAL				RECOMMENDED[a]
	20–29 years		40–49 years		31–50 years
Nutrient	Males	Females	Males	Females	Males/Females
Protein, g	106	68	96	64	63/50
Total fat, g	104	66	91	63	65 (DV)[b]
Saturated fat, g	36	23	31	21	20 (DV)[b]
Monounsaturated fat, g	40	25	35	24	-
Cholesterol, mg	349	226	339	225	300 or less
Fiber, g	18	12	18	14	25 (DV)
Sodium, mg	4627	2974	4061	2771	Up to 2400
Potassium, mg	3278	2240	3225	2411	3500
Vitamin A, mcg RE	1430	1149	1000	800	900/700
Carotenes, mcg RE	539	506	539	500	-
Vitamin E, mg TE	10.1	7.1	9.4	7.9	15
Vitamin C, mg	123	87	106	93	90/75
Vitamin B_6, mg	2.4	1.6	2.1	1.5	1.3/1.3
Vitamin B_{12}, mcg	6.2	3.8	6.2	4.4	2.4/2.4
Folate, mcg[c]	323	226	288	226	400/400
Calcium, mg	1005	890	717	643	1000/1000
Magnesium, mg	341	229	330	241	420/320[d]
Zinc, mg	15.2	9.7	13.2	9.4	11/8
Iron, mg	19.8	13.5	17.6	13.1	8/18
Copper, mcg	1600	1100	1500	1100	900

SOURCE: Consumption data from Wilson et al. Data tables: combined results from USDA's 1994 and 1995 Continuing Survey of Food Intakes by Individuals and 1994 and 1995 Diet and Health Knowledge Survey; 1997.

[a]Recommended intake according to DRI 2001[5] or to DV nutrient reference amount used on food labels, relevant to a 2000 kcal diet.[43]
[b]Based on a 2000 kcal diet; for other caloric levels, recommendation is no more than 30% of calories from fat and 10% from saturated fat.
[c]Synthetic, prior to 1998 grain fortification with folic acid.
[d]Nonfood sources.

NUTRIENT RECOMMENDATIONS

Nutrient recommendations are established and regularly revised by the National Academy of Sciences.[4,22] Dietary Reference Intake (DRI) values for all ages can be found on the inside cover of this text. DRIs for adults are included in Table 16.9. Recommended nutrient levels used in food labeling as daily values (DV) as well as the new Tolerable Upper Intake Levels, are included in Chapter 18, Table 18.16. The Tolerable Upper Intake Level (UL) indicates a potentially toxic limit for vitamins and minerals. Diet alone is unlikely to get anyone to these UL intake levels, eating fortified foods and taking vitamin-mineral supplements might. Intake should stay below the UL to avoid *iatrogenic* illness (when what you do to be healthy makes you sick instead).

Nutrients of Public Health Concern

Fat, fiber, folic acid, vitamin E, calcium, magnesium, plus iron for women are of public health concern be-cause intake levels do not meet recommendations (Table 16.9).

FAT An analysis of NHANES data shows that roughly 34% of the average population's caloric intake is supplied by fat, with 12% from saturated fat.[33] Figures are similar using CSFII data, with 33% of total from fat and 11% from saturated fatty acids.[44] This figure is down from 41% of calories in the late 1970s.

IATROGENIC Used in reference to disease, it is a condition secondary to a medical treatment. For example, when a person who is hospitalized to cure pneumonia gets a staphylococcus infection, that resultant staph infection is an iatrogenic disease.

FIBER As fruit, vegetable, and whole grain intakes rise to meet food guide recommendations, adequate fiber intake will result. Chapter 18 includes examples.

FOLIC ACID The evidence for folic acid in the pre-vention of neural tube defects is so strong that the

Food and Drug Administration mandated folic acid fortification of grains and grain products beginning in 1998. Some evidence also supports folic acid as a way to lower risk of heart disease and colon cancer. (More on folic acid in Chapter 18.)

VITAMIN E   This versatile antioxidant has been implicated in prostate cancer, heart disease, and cataract risk reduction. Food sources include sunflower seeds, walnuts, wheat germ, olive oil and mayonnaise, avocado, and kale. (More in Chapter 18.)

CALCIUM   Adults could meet calcium goals by adding a glass of milk, 8 ounces of yogurt, or 8 ounces of calcium-fortified orange juice to their diets daily. Eight ounces of milk contain roughly 300 milligrams of calcium, regardless of butterfat. Some milks and yogurt provide additional calcium (>30% of DV per serving) because additional milk solids have been used to give the product extra body and flavor. See Table 16.10 for strategies that optimize calcium intake.

Calcium is a public health concern for several reasons. Calcium plus vitamin D and magnesium help to develop and maintain bone density, which delays osteoporosis and reduces risk of bone fractures. Adequate calcium is a potential contributor to lowered risk of colon cancer. Milk as part of the DASH diet (see Chapter 19) has been successfully used as part of blood pressure reduction intervention.[48] Calcium's crucial role in metabolism is supported by physiological mechanisms to keep calcium in tight balance including calcium intake, amount of calcium absorbed, and calcium excreted through urine and feces. Of these factors, intake and fractional absorption of calcium account for 25% of the variance in calcium balance while losses through urine account for another 50%.[49] Because protein and sodium promote urinary calcium excretion, eating a diet high in protein and sodium (like the typical adult diet) contributes to calcium losses through urine. Other factors that decrease calcium absorption are ingestion of glucocorticoids (used to treat inflammatory diseases such as rheumatoid arthritis, asthma, and inflammatory bowel disease). As the body ingests higher levels of calcium, the percentage absorbed decreases. Conversely, as calcium intake drops, the percentage absorbed increases. A larger portion will be absorbed from a 500-milligram-per-day intake than from a 1500-milligram intake.

The tolerable upper limit for calcium is 2500 milligrams. It would be difficult to exceed this from food sources alone. For example, the basic diet without dairy products equals 400 milligrams of calcium. Four 8-ounce glasses of milk = 1200 milligrams of calcium; an ounce of cheese = 250 milligrams. These sources

Table 16.10   Optimizing calcium intake for osteoporosis prevention.

- **Vitamin D:** Adequate intake of Vitamin D (1,25-Dihydroxyvitamin D, also called calcitriol) stimulates active transport of calcium in the small intestine and the colon.
- **Sodium:** Lower levels of sodium intake lead to lower levels of urinary calcium, that is, less calcium is excreted while consuming a low-sodium diet.
- **Lower protein intakes:** Excess protein leads to greater urinary excretion of calcium. In the Nurses Health Study, women (aged 35 to 59) who ate more than 95 grams protein per day suffered more osteoporotic forearm (but not hip) fractures over 12 years.[45]
- **Caffeine:** Two to three cups of coffee daily in postmenopausal women, when consumed with less than 800 mg/day of calcium, were associated with bone loss.[46]
- **Timing:** Drink milk and take supplements with a meal because food slows intestinal transit time and allows more calcium to be absorbed from the gut. Calcium carbonate is best taken with meals; calcium citrate may be taken at any time. In addition, stomach acidity enhances absorption, therefore calcium is better absorbed when not consumed with antacids.
- **Moderation:** Consuming nutrients at recommended levels (enough but not too much) helps to synthesize and develop collagen matrix into which minerals are deposited during bone mineralization.[52]
- **Other vitamins:** Vitamin C plays a role in the development of the protein bone matrix (collagen), but has no defined role in treatment. Vitamin B_6 is a cofactor in stabilizing collagen cross-links and has been linked to preventing hip fractures. Vitamin K is required for the formation of proteins that stimulate osteoblasts to build new bone and attract osteoclasts to initiate bone reabsorption (e.g., glutamic acid residues are carboxylated to form gla-proteins; most studied is osteocalcin).[47]

add up to approximately 1850 milligrams calcium. However, supplement usage adds up more quickly.

Calcium carbonate is the cheapest form of calcium supplement. However, taking it can decrease or interfere with iron absorption. Citrate- or ascorbic acid-bound calcium supplements on the other hand, lead to greater iron absorption. Citrate and ascorbic acid calcium supplements are more easily absorbed than calcium carbonate. They usually cost more, too. Calcium and its binder are bulky, so multi-vitamin vitamin-mineral supplements tend to contain little calcium. Calcium is best absorbed when taken in divided doses.

MAGNESIUM   Poor diet, diabetes, or prolonged illness can cause magnesium deficiency, with adverse effects on bone strength and heart health. This issue tends to be problematic in older adults. See section in Chapter 18. Good magnesium sources are sunflower seeds, almonds, beans, milk, whole grains, spinach, and bananas.

IRON For women, iron needs drop from 18 to 8 milligrams per day at menopause. The benefits of sufficient iron are healthy blood cells that carry oxygen for metabolism and overall energy, drawbacks of excess iron are greater need for antioxidants. High-iron foods include red meats, fortified cereals, strawberries, and dried beans as in baked beans, bean soup, bean dip, and chili. Changing from a high-iron to a low-iron cereal during the menopausal transition can help to adjust for the dramatic shift in iron needs.

Actual Intake of Food

FOOD GROUPS AND PORTION SIZE How are Americans doing when it comes to meeting the pyramid guidelines? Several years ago, a picture of a top-heavy pyramid reflected the American plate. Fruit and vegetable servings totaled 3.5 rather than 5.0. But we are close to five servings of fruits and vegetables now, although intake from the "other" group at the tip is still high.[3] CSFII tracks the percentage of individuals who consume foods from each food group. Among 20- to 29-year-old adults, fewer than half eat fruit, but four of five eat vegetables daily. Two-thirds (67.5% of males) to three-fourths (74.5% of females) consume some milk or milk products every day. On any given day, 7% to 8% eat nuts and seeds, 11% to 13% eat beans and other legumes, 40% to 50% eat sugars and sweets, and 62% to 68% consume carbonated soft drinks.[44] However, the movement to super-size restaurant and take-out foods (e.g., fries, drinks, rolls, and cookies) has led to "portion distortion." For example, American "Chinese food" portions greatly exceed the authentic versions. Try to estimate your next fast-food meal in pyramid portions. It is not easy to think "normal."

Using NHANES data, Kant calculated that 27% of the average individual's calories are eaten as energy-dense, nutrient-poor (EDNP) foods, that is, the "other" food group.[3] When we add to that the 3% to 4% of calories alcohol provides, it turns out that we rely on 69% to 70% of our calories for the bulk of vitamins and minerals.

EATING OCCASIONS More and more people are eating away from home. For example, 62% to 72% of 20- to 29-year-olds consume more than a third of their calories away from home. Greater numbers of fast-food meals are associated with higher total and fat calories.[50]

Supplements: Vitamins and Minerals

Adults can eat nutritiously without dietary supplements. Circumstances under which dietary supplements may enhance peak functioning are childbearing (vitamins and minerals for reproduction are covered elsewhere), high-level athletic performance, illness, or when eating restricted diets, for example, a wheat-free diet due to allergies or a calorie-restricted diet for weight loss. Immune status enhancement from physiological doses (RDA, DRI levels) of supplement taken for a year or more has been found in older adults.[51,52] Younger adults might consider supplementing dietary gaps (potentially calcium, iron, magnesium, vitamin E, and folic acid, plus vitamins B_{12} and D for nonmeat eaters and vegetarians).

CSFII data reports that 42% of males and 56% of females aged 20 and older consume a vitamin or mineral supplement daily, nearly every day or "every so often."[44] Supplement use by males gradually rises with age (37% to 45% of individuals from age 20 to age 70+). Female usage rises from 51% of individuals at age 20 to a peak of 62% between ages 50 to 59 and then dropping to 54% by age 70 and older. The greatest percentage of males who take supplements uses a multivitamin. For females, the pattern shifts, a multivitamin is most popular until age 40, after which single vitamins and minerals are used more. At age 70 and older, 23.4% of women use a single vitamin or mineral and another 23.4% of women use a multivitamin. Findings in NHANES are slightly different than CSFII, showing that 40% of the U.S. population (two months and older) took at least one vitamin or mineral supplement in the previous month (35% of males and 44% of females). The most commonly found nutrients in supplement products are ascorbic acid, vitamin B_{12}, vitamin B_6, niacin, thiamin, riboflavin, vitamin E, beta carotene (vitamin A), vitamin D, and folic acid.[53]

Supplements: Herbal Products

Herbal products are often sold with other nutritional supplements. Table 16.11 lists examples of some of the more popular herbals and suggests potential uses and side effects.

Herbs and foods with special nutritive and medicinal attributes (functional foods) comprise a rapidly growing and largely unregulated market. *When in doubt, throw it out* is a food safety education message that seems appropriate for herbal products as well.

The Food and Drug Administration has banned several herbs, although they occasionally appear on the market (sassafras, yohimbe/yohimbine, comfrey, chaparral, lobelia). Others that are unsafe and/or ineffective include apricot pits (cyanide), belladonna, blue cohosh, coltsfoot, dong quai or angelica, ephedra or ma huang, eyebright, garcinia, life root, mistletoe, pokeroot (except the berries), wild yam, willowbark, and wormwood.[15] Additional information on the complex topic of herbal supplements can be found via Web sources listed at the end of the chapter.

Mmm, mushrooms, a natural example. Morels, truffles, porcini, portobello, enoki, shiitake . . . they taste so good. Yet some will kill you.

UBK, day-dreaming about edible natural things

Cross-Cultural Considerations

The health and nutrition goals outlined in Healthy People 2010 are designed to decrease health disparities among ethnic groups. Cultural groups that are part of the North American landscape can be role models for the way their traditional diets are designed. For instance, early Native Americans in the southwestern part of the United States relied on corn, beans, and squash, the three "sister-vegetables." They are a solid pyramid base. Here are a few examples of dietary patterns from various cultures:

1. Asian diets tend to be high in carbohydrate, low in fat, especially in saturated fats. The diet includes many varieties of fish. Beans, including fermented soy products, supply protein, fiber, and antioxidants. Even Americanized Vietnamese restaurants have many low-fat, high-vegetable choices on the menu. Spring rolls with plum and chili sauce are a nonfried alternative to deep-fried egg rolls. Thai dishes are based on vegetables, noodles, and fish. Fish sauce provides flavor and adds protein without fat. Peanut sauce and coconut milk are high in fat and eaten in small amounts. Peanuts provide protein.

2. Indian diets offer many vegetarian choices, are low in saturated fats and high in fiber. Extensive use of spices provides flavor without added fat.

Table 16.11
Proposed effects and potential side effects of herbal and other dietary supplement remedies.

Herb/Other Remedy	Proposed Effects	Potential Side Effects
Glucosamine-chondroitin	Slows progression of osteoarthritis and its pain	Gastrointestinal upset, fatigue, headache
Ginseng	Increases energy, normalizes blood glucose, stimulates immune function, relieves impotence in males	Insomnia, hypertension, low blood glucose, menstrual dysfunction. Interacts with estrogen and blood thinners.*
SAMe	Relieves mild depression, pain relief for arthritis	May trigger manic excitement, nausea
Garlic	Lowers blood cholesterol, relieves colds and other infections, asthma, other	Heartburn, gas, blood thinner
Cholestin	Maintains desirable blood cholesterol levels	Safety of some ingredients unknown
Echinacea	Prevents and treats colds and sore throat	Allergies to plant components
DHEA	Improves memory, mood, physical well-being	Increases risk of breast cancer
Creatine	Sport supplement (increased performance in short, high-intensity events)	Kidney disease
Saw palmetto	Improves urine flow, reduces urgency of urination in men with prostate enlargement	Nausea, abdominal pain
Ginger	Calms stomach upset, fights nausea	Central nervous system depression, heart rhythm disturbances if using very large overdoses
Ginkgo biloba	Increases mental skills, delays progression of Alzheimer's disease, increases blood flow, decreases depression	Nervousness, headache, stomachache. Interacts with blood thinners.
Shark cartilage	Treats lung cancer	Safety unknown
Saint John's wort	Relieves depression	Dry mouth, dizziness. Interacts with may drugs.
Ephedra (Ma Huang)	Promotes weight loss	Insomnia, headaches, nervousness, seizures, death
Peppermint	Treat indigestion and flatulence, spasmolytic (relaxes muscles), antibacterial agent	Safe (not for infants, if contains menthol), allergic reactions

*Blood thinners include aspirin, warfarin, coumarin.
SOURCE: Brown JE. Nutrition now, 3rd ed. Belmont, CA: West/Wadsworth Publishing Company; 1999: Table 24.4.

3. North African cuisine (for example, Amharic, Oromo, Ethiopian, Egyptian) relies heavily on beans and whole grains. Buddeena, the flat bread that feels a bit like a pancake, holds dabs of lentils, split peas, lamb, beef, and yogurt. Sharing is customary. Each person pulls off a piece of the buddeena and uses it to pick up fillings such as peas and lentils (dahl), beef, lamb, or vegetable mixtures. Imagine how much filling you can pick up with a piece of soft bread. You would be eating pyramid style, much grain and a little animal protein.

4. Caribbean dishes are made with lots of potatoes, vegetables such as spinach, fruit, and small portions of meat, often chicken.

5. Traditionally prepared Mexican food is based on corn, grains, beans, vegetables, and small amounts of meat and fat. Loads of cheese and sour cream on chips with some pickled jalapeno slices are an American restaurant invention.

When population groups migrate to another country, nutritional problems may arise because acculturation requires rebalancing old lifestyles. For instance, St. Paul, Minnesota, saw a large influx of Hmong refugees after the Vietnam War displaced them from their farms in Cambodia and Laos. Hmong (pronounced "mung") youths (but not their elders) quickly adopted American food habits, but families were no longer pursuing their old physically active lifestyles. Overweight and obesity have increased dramatically in Hmong youths since coming to Minnesota. In a study of eating habits of mostly first-generation Korean immigrants (average age 41 years, time in United States 15 years), Korean foods were consumed nearly twice as often as American foods. Consumption of Korean foods declined with adoption of American culture.[54]

Cross-Cultural Dietary Guidance

Several years ago, the Women, Infants, and Children's program (WIC) developed a set of posters (see Illustration 16.4) for use in cross-cultural nutrition education. Potential audiences were asked how they thought about food; the answers suggested developing a guide that grouped foods according to function. In other words, the guide would answer "What will this food do for me?"

1. *Staple foods:* Grains, rice, cereals, and starchy roots provide energy and bulk to the diet. They are the filling basis of the diet and, barring bad luck, are readily available.

2. *Body-building foods:* Beans, lentils, peas, nuts, seeds, eggs, milk, cheese, fish, poultry, and red meats are all high in protein and provide amino acids for tissue building and repair.

3. *Protective foods:* Vegetables and fruits contribute vitamins and minerals, antioxidants, and other phytochemicals to keep the individual healthy.

For the WIC program, grouping food according to function dealt with the milk/dairy dilemma faced by some Asian and African American audiences. Milk, cheese, calcium-fortified soymilk, or yogurt can all be placed in the protein or "body-building" group. In cross-cultural nutrition work, it is important to find

Illustration 16.4   Foods grouped according to function for cross-cultural nutrition counseling.

Photos courtesy of General Mills, Inc.

out what the other person or group thinks and believes. Go ahead. Ask.

Vegetarian Diets

How many people consider themselves to be vegetarian? Among individuals aged 18 and older, 3% to 7% consider themselves "vegetarian."[55] Fewer than 1% of the population eats no animal products, roughly 6% eat no red meat, 4% never eat eggs, 4% never eat seafood, and 3% never eat poultry. One difficult thing about polling people about vegetarianism is defining just exactly what it means to be a vegetarian. Table 16.12 gives an overview.

Individuals can achieve a high-quality diet whether or not they eat meat. Can adopting a vegetarian diet make one healthier? Often people eating vegetarian diets are health conscious in other areas of their life as well. So far, studies haven't consistently controlled for all lifestyle factors that potentially contribute to health. A well-chosen vegetarian diet is associated with decreased mortality and morbidity; "Diet is clearly a contributing factor."[56] In a relatively small study (80 women) that controlled for body composition, aerobic fitness, and psychological well-being, Nieman and colleagues found that except for zinc, eating or not eating meat makes no difference to immune status.[57] Higher intakes of zinc (3–25 grams) negatively impacted immune status.

Potential health benefits of vegetarian diets outlined in the American Dietetic Association (ADA) Position Statement[56] include:

- Disease specific benefits
    Decreased risk of mortality and symptoms of heart disease
    Lower incidence of hypertension
    Decreased risk of some colorectal cancer, in some cultures also less breast cancer
- General diet quality
    Higher intake of vegetables and fruits improves overall nutritional risk picture
    Diet tends to be lower in saturated fats and may also be lower in calories
- Environmental benefits
    Vegetable foods are lower on the food chain than meat, fish, and poultry
    Depending on packaging, transportation, food production uses fewer resources

Nutrients singled out for evaluation in vegetarian diets include protein, vitamins B_{12} and D, calcium, iron, and linolenic acid.

1. *Protein:* Eating a variety of amino acids throughout the day insures amino acid balance. Eating complementary proteins at each meal is not necessary.[58] Protein intake of vegetarians "appears to be adequate" despite lower overall protein quality and intake.[56] See Table 16.13 for nutrient content of several high-protein foods from the Food Guide Pyramid.

2. *Vitamin B_{12}:* Plant foods are not reliable sources of vitamin B_{12}; most of the B_{12} in sea vegetables and fermented soy is an inactive ana-

Table 16.12   Classification of vegetarians by food groups consumed.

Type of Vegetarian	Foods Not Eaten	Foods Eaten
Vegan	No animal foods of any kind, no honey	Grains, beans, nuts, seeds, nut butters, fruits, vegetables, sugar, molasses, oils, margarine, soda, alcohol, soy analogs (e.g., textured vegetable bacon, burgers, "meats")
Lacto-vegetarian	No meat, poultry, fish	Above plus milk and other dairy products, cheese, yogurt, butter
Lacto-ovovegetarian	No meat, poultry, fish	Above plus eggs of any sort
Vegetarian	No beef, pork, lamb, venison, buffalo, other red meat; "No red meat"	Depends on individual interpretation: fish, both fin-fish and shellfish; poultry and game birds may be allowed

Table 16.13   Examples of nonmeat protein foods and serving sizes used in the Food Guide Pyramid.

Protein Food Serving	Protein Content (grams)	Total fat (grams)	kcal
Almonds, 2 tbs whole	3.5	9.3	105
Peanuts, 2 tbs dry-roasted	4.8	9.0	106
Peanut butter, 2 tbs	8.1	16.5	191
Sesame seeds, 2 tbs	4.2	8.8	94
Egg, 1 large	6.2	5.0	74
Kidney beans, 0.5 c cooked	7.7	0.4	112
Soy milk, 1 c	6.6	4.6	79
Lentils, 0.5 c cooked	8.9	0.4	115
Tofu, firm and soft, 0.25 c	5.7	3.4	54
Cheese, 1.5 oz	10.6	14.1	171

log. Sources of B_{12} are fortified ready-to-eat breakfast cereals, soy milks, nutritional yeast, and meat analogs. (Check the label, they vary!)

3. *Vitamin D:* One good source of D is sunshine (all year in the south; during spring, summer, and fall in northern United States and Canada). Vitamin D formation is blocked by sunscreen. Another source is Vitamin D fortified milk providing 2.5 microgram (100 IU) per cup of the 10 micrograms (400 IU) needed per day. Fortified cereals, soy milk, dairy milk, and vitamin tablets are reliable sources of vitamin D.

4. *Calcium:* Absorption is variable from different foods. The calcium content in vegetables such as spinach and fruits such as rhubarb looks impressive in a food table, but doesn't reach the bones to the extent that calcium from yogurt or cheese sources does. Some dark leafy greens such as collards are low in oxalic acid so that the calcium can be absorbed. No general rule fits calcium. Fortunately, vegetarians do not tend to have a high protein intake. If they also keep sodium intakes low, their body needs less calcium.

5. *Iron:* Enriched flour and breads, beans, and fortified cereals, especially when eaten in the presence of vitamin C and not washed down with coffee or tea, can provide all the iron a healthy adult needs. An example of how to reach the 10 milligrams of iron needed daily by all adults except premenopausal women is to include two slices whole wheat bread (1.8 gm), 1 cup tomato juice (1.3 gm), 1 cup cooked chard (3.8 gm), 0.5 cup kidney beans (1.5gm), and 2 tablespoons of pumpkin seeds (2.5 gm) in the day's intake.

6. *Fats:* Most plant foods do not contain significant amounts of omega-3 (linolenic) fatty acids as fish and eggs do. It is not clear how that affects vegetarians except that the recommendation is to include good sources in the diet.[56] Good sources of omega-3 fatty acids are ground flaxseed and *linseed* oil, walnuts, walnut oil, canola and soybean oils.

The American Dietetic Association suggests that a health-promoting vegetarian diet include a variety of foods from each basic food group. Whole, unrefined foods are more nutrient-dense than highly processed ones, containing phytochemicals that were not lost in processing and about which we know too little to replace, e.g. red grapes contain reservatrol, a compound implicated in heart health and in reducing cancer risk. Real grapes contain reservatrol, manufactured imitation grape beverages do not. Consuming low-fat dairy products will help to keep the saturated fat within dietary guidelines.[56]

PHYSICAL ACTIVITY RECOMMENDATIONS

I could have danced all night!

Who can improve on the following paragraph from Healthy People 2010?

> Because the highest risk of death and disability is found among those who do no regular physical activity, engaging in any amount of physical activity is preferable to none. Physical activity should be encouraged as part of a daily routine. While moderate physical activity for at least 30 minutes is preferable, intermittent physical activity also increases caloric expenditure and may be important for those who cannot fit 30 minutes of sustained activity into their daily schedules.

In addition, resistance training (at least 30 minutes per day, two to three times per week) develops muscle mass, strength, and endurance.[59] See Table 16.14 for current activity goals for adults.

Shall we dance?

NUTRITION INTERVENTION FOR RISK REDUCTION

A Model Health-Promotion Program

Nutrition intervention occurs through the provision of food as well as by providing counseling and education to help improve nutritional status. Food pantries, community kitchens, and the 5 A Day for Better Health Campaign are all examples of nutrition interventions that mobilize individuals and communities to improve nutritional status.

LINSEED From the flax plant, *linum* is another name for flaxseed. Linseed oil is used in paints, varnishes, and inks but is also produced in food form for its rich nutrient content.

An example of a health promotion program for adults illustrating some of the concepts of nutrition intervention is Health Partner's Better Health Restaurant Challenge®. Health Partners is a large midwestern health maintenance organization (HMO).

- Health goal: Improve nutritional health by enabling HMO members to choose tasty, health-promoting, low-fat foods when eating away from home.
- Target audience: Health Partners sees its mission as serving the whole community, so the whole metropolitan community was eligible to participate in the program.
- Theoretical basis: Health behavior change depends on social support as well as on individual

goals and commitments. Health Partners wanted to build social networks that enabled healthy choices.

- Intervention strategies: Participating restaurants (sit-down and fast-food) were challenged to feature health-promoting items on the menu. Diners vote on favorite items, the HMO compiled and publicized the restaurants and the winning items. The HMO provides additional publicity for restaurants in its publications and through local media advertising, offering discount coupons for members to visit participating restaurants.

- Evaluation: Member satisfaction increased program participation from year to year, and a growing list of restaurants wished to join the program. Member surveys addressed the nutritional goal of enabling access to tasty, health-promoting menu items by asking about restaurant ordering patterns and satisfaction with featured menu items. Diner surveys revealed that 90% of diners would order low-fat items again and 50% thought low-fat items tasted better than expected.

Public Food and Nutrition Programs

One of the aims of national health goals is to minimize health disparities.[60] This challenge is a daunting one. Living in poverty is linked to poor diets and to adverse health outcomes such as increased rates of obesity.[60,61] Roughly 13% of U.S. residents live in poverty. Poverty guidelines are published annually in the Federal Register by the Department of Health and Human Services, (www.fns.usda.gov/fsp/charts/incomechart.htm). In the year 2001, a person with an annual income of $8,590 or less is living in poverty.[62] For a four-person household, the poverty threshold is $17,650. Poverty figures are derived from adding the cost of foods needed for basic dietary requirements (according to the Thrifty Food Plan) and multiplying that cost by three. Three is the factor that was applicable when the poverty index was developed. At that time, food costs made up one-third of the average household budget. The poverty threshold is used as the income level for many social service programs. Food-related public programs include the Child Nutrition Program (school lunch), which can help whole families stretch their food budget. Children can eat breakfast and/or lunch free or for reduced rates, depending on their family's income (185% or 130% of poverty). The Women, Infants, and Children (WIC) and the Mothers and Children (MAC) programs also help whole households by assisting mothers and children.

Other programs are geared to adults or family units. Programs such as the Food Stamp Program, Extension Services, Second Harvest food bank and community kitchen network, Commodity Food distributions, Fare for All, Store-to-Door home-delivered groceries, Meals-on-Wheels, and Senior Nutrition Programs help individuals gain consistent access to safe, wholesome foods that are culturally acceptable. Such access is the basis of food security, which is one of the national health goals (see Table 16.1).

Nutrition and Health Promotion

What are the most powerful nutritional habits that we can develop? The general message is to follow the principles of variety and moderation in choosing a diet that will achieve and maintain a healthy body weight when combined with activity. More specifically, nutrition educators are developing scoring systems to quantify that general message.[63] The "Diet Quality Index Revised" (DQIR) is an example of a scoring system that reflects key nutritional habits for promoting health and reducing overall disease risk. DQIR system designers recognized that no single magic bullet (e.g. "eat less fat") will make us all well. Instead, the system is designed to help track the many aspects that contribute to diet quality. The following 10 nutritional habits (for 10 points each) are used to measure diet quality:

1. Total fat = < 30% of calories
2. Saturated fat = < 10% of energy intake
3. Dietary cholesterol < 300 mg daily (if revising this tool, consider fiber here)
4. 2–4 servings fruit per day
5. 3–5 servings vegetables per day

Table 16.14 Healthy People 2010 physical activity goals and baseline activity levels.

	BASELINE LEVELS		TARGET
	Age 18 and older	Age 25–44	All adults
Engaged in NO leisure time physical activity	40%	34%	20%
Engaged in regular physical activity at least 20 minutes, three times per week	15%	15%	30%
Perform strengthening activities two or more days per week	19%	22%	30%
Perform stretching and flexibility activities	30%	34%	40%

6. 6–11 servings grain per day
7. Calcium intake to meet age and gender recommendations
8. Iron intake to meet age and gender recommendations
9. Dietary diversity score: one-half or more servings of each of 23 food component groupings including seven grain groups, seven vegetable groups, two fruit groups, and seven meat/dairy groups during a two-day period
10. Dietary moderation score composed of four items; teaspoons of added sugar with 1 teaspoon = 4 grams carbohydrate (12 ounce cola = 38 grams carbohydrate, thus contributing 9.5 teaspoons of sugar); discretionary fat grams (25 grams or less per day); sodium intake below 2400 milligrams per day; alcohol in moderation.

Knowing does not automatically lead to doing. To make health-promoting behavior changes takes knowledge of desired behaviors, skills to practice them, belief in one's own ability to carry them out, and the intention or commitment to do so.

Resources

American Dietetic Association

Provides information, advice, and links.
Web site: www.eatright.org

American Heart Association

Site includes assessments, eating advice, and definitions.
Web site: www.americanheart.org

Calorie Control Council

Lists overall diet and fitness information. (Don't look for this site to sponsor the Health-at-any-size movement.)
Web site: www.caloriecontrol.org

Centers for Disease Control and Prevention

The main site also includes morbidity and mortality data at www.cdc.gov/mmwr.
Web site: www.cdc.gov

Consumer Lab

Check out vitamin and mineral supplements. (Also use www.fda.gov.)
Web site: www.consumerlab.com

Environmental Protection Agency

Among other things, ratings for local water supplies are available.
Web site: www.epa.gov/safewater

Food and Drug Administration

Resource for labeling information, health claims on foods and supplements, fortification rules, and more.
Web site: www.nutrition.gov
U.S. government's gateway site for nutrition information.
Web site: www.fda.gov

U.S. Census Bureau

The site to go to for who we are and where we live.
Web site: www.census.gov/population/estimats/nation/intfil2-1.txt.

U.S. Department of Agriculture

Wonder about your diet? This site includes a diet self-assessment. Follow the HEI icon to USDA's interactive Healthy Eating Index. Key in what you've eaten and see how your intake stacks up against the Food Pyramid recommendations. For another view, go to www.cyberdiet.com.
Web site: www.usda.gov/cnpp

Vegetarian Resource Group

Provides information and recipes.
Web site: http://www.vrg.org

References

1. McGinnis JM, Foege WH. Actual causes of death in the United States. JAMA 1993; 270:2207–12.

2. Healthy People 2000: national health promotion and disease prevention objectives. DHHS Publication No. (PHS) 91-50212. Washington, DC: U.S. Department of Health and Human Services; 1991.

3. Kant AK. Consumption of energy-dense, nutrient-poor foods by adult Americans: nutritional and health implications. The Third National Health and Nutrition Examination Survey, 1988–1994. Am J Clin Nutr 2000;72:929–36.

4. Trumbo P, Yates AA, Schlicker S, Poos M. Dietary reference intakes: vitamin A, vitamin K, arsenic, boron, chromium, copper, iodine, iron, manganese, molybdenum, nickel, silicon, vanadium, and zinc. J Amer Diet Assoc 2001;101:294–301.

5. Crawford SL, Casey VA, Avis NE, McKinlay SM. A longitudinal study of weight and the menopause transition: results from the Massachusetts Women's Health Study. Menopause 2000;7:96–104.

6. Forbes GB. Human body composition. Growth, aging, nutrition, and activity. New York: Springer-Verlag; 1987:31.

7. Reid IR. Pharmacological management of osteoporosis in postmenopausal women: a comparative review. Drugs Aging 1999;15:349–63.

8. Col NF, Pauker SG, Goldberg RJ, et al. Individualizing therapy to prevent long-term consequences of estrogen deficiency in postmenopausal women. Arch Intern Med 1999;159:1458–66.

9. Dennerstein L, Dudley E, Guthrie J, Barrett-Connor E. Life satisfaction, symptoms, and the menopausal transition. Available from www.medscape.com/Medscape/WomensHealth/journal/2000/v05.n04/wh7254.denn-01.html, January 2001.

10. Barnes S. Phytoestrogens and breast cancer. Baillieres Clin Endocrinol Metab 1998;12:559–79.

11. Messina MJ. Legumes and soybeans: overview of their nutritional profiles and

health effects. Am J Clin Nutr 1999;70: 439S–50S.

12. The role of isoflavones in menopausal health: consensus opinion of The North American Menopause Society. Menopause 2000;7:215–29. Available from www.menopause.org.

13. Wiseman H, O'Reilly JD, Adlercreutz H, et al. Isoflavone phytoestrogens consumed in soy decrease F(2)-isoprostane concentrations and increase resistance of low-density lipoprotein to oxidation in humans. Am J Clin Nutr 2000;72:395–400.

14. Schryver T. Soy update 2000. Today's Dietitian 2000;2:2–7.

15. Foster S, Tyler VE. Tyler's honest herbal. New York: Haworth Herbal Press; 1999.

16. Robbers JE, Tyler VE. Tyler's herbs of choice. New York: Haworth Herbal Press; 1999.

17. Willett WC. Diet, nutrition and the prevention of cancer. In: Shils ME, et al., eds. Modern nutrition in health and disease. Baltimore, MD: Williams and Wilkins; 1999:1250.

18. Harper EJ. Changing perspectives on aging and energy requirements: aging and energy intakes in humans, dogs and cats. J Nutr 1998;128:2623S–26S.

19. McGandy RB, Barrows CH, Spanias A, et al. Nutrient intakes and energy expenditure in men of different ages. J Gerontol 1966;21:581–7.

20. Poehlman ET, Toth MJ, Bunyard LB, et al. Physiological predictors of increasing total and central adiposity in aging men and women. Arch Intern Med 1995;155: 2443–8.

21. Guo SS, Zeller C, Chumlea WC, Siervogel RM. Aging, body composition, and lifestyle: the Fels Longitudinal Study. Am J Clin Nutr 1999;70:405–11.

22. Food and Nutrition Board. National Academy of Sciences (Institute of Medicine) NRC, Subcommittee on the 10th Edition of the RDAs, Commission on Life Sciences. Recommended dietary allowances. Washington, DC: National Academy Press; 1989.

23. Twin Cities District Dietetic Association. Manual of clinical nutrition 2000. St. Paul, MN: Twin Cities District Dietetic Association; 1999.

24. Harris JA, Benedict FG. Standard basal metabolism constants for physiologists and clinicians: a biometric study of basal metabolism in man. Washington, DC: Carnegie Institute of Washington; 1919.

25. Roth JH. What is optimum body weight? J Amer Diet Assoc 1995;95:856–7.

26. Hamwi GJ. Therapy: changing dietary concepts. In: Danowski TS, ed. Diabetes mellitus: diagnosis and treatment. New York: American Diabetes Association, 1964:73–8.

27. Miller MA. A calculated method for

determination of ideal body weight. Nutr Support Serv 1985;5:31–3.

28. Committee on Diet and Health (Food and Nutrition Board; National Research Council). Diet and health. Implications for reducing chronic disease risk. Washington, DC: National Academy Press; 1989.

29. McNamara DJ. Dietary cholesterol and the optimal diet for reducing risk of atherosclerosis. Can J Cardiol 1995;11 Suppl G:123G–6G.

30. Yates AA, Schlicker SA, Suitor CW. Dietary Reference Intakes: the new basis for recommendations for calcium and related nutrients, B vitamins, and choline. J Amer Diet Assoc 1998;98:699–706.

31. Welch S. Nutrient standards, dietary guidelines, and food guides. In: Ziegler EE, Filer LJ, eds. Present knowledge in nutrition. Washington, DC: ILSI Press; 1996: 630–46.

32. Health and Welfare Canada. Nutrition recommendations. The report of the Scientific Review Committee. Canada: Minister of Supply Services; 1990:6.

33. McDowell MA, Briefel RR, Alaimo K, et al. Energy and macronutrient intakes of persons ages two months and over in the United States: Third National Health and Nutrition Examination Survey, Phase 1, 1988–1991. Adv Data 1994:1–24.

34. Health United States, 2000. Vol. 2001. Available from www.cdc.gov/nchs, 2000.

35. Parker DR, McPhillips JB, Derby CA, et al. High-density-lipoprotein cholesterol and types of alcoholic beverages consumed among men and women. Am J Public Health 1996;86:1022–7.

36. Windham CT, Wyse BW, Hansen RG. Alcohol consumption and nutrient density of diets in the Nationwide Food Consumption Survey. J Amer Diet Assoc 1983;82:364–70, 373.

37. Forbes GB. Longitudinal changes in adult fat-free mass: influence of body weight. Am J Clin Nutr 1999;70: 1025–31.

38. Chernoff R, ed. Geriatric nutrition: the health professional's handbook. Gaithersburg, MD: Aspen; 1991.

39. Chernoff R, ed. Geriatric nutrition: the health professional's handbook. Gaithersburg, MD: Aspen; 1999.

40. Coffee break or caffeine fix? Environmental Nutrition 2000;23:1,6.

41. Grandjean AC, Reimers KJ, Bannick KE, Haven MC. The effect of caffeinated, non-caffeinated, caloric and non-caloric beverages on hydration. J Am Coll Nutr 2000;19:591–600.

42. U.S. Department of Agriculture, U.S. Department of Health and Human Services. Nutrition and your health: dietary guidelines for Americans, 5th ed. 2000.

43. Pennington JA, Hubbard VS. Derivation of daily values used for nutrition labeling. J Amer Diet Assoc 1997;97:1407–12.

44. Wilson JW, Enns CW, Goldman KS, et al. Data tables: combined results from USDA's 1994 and 1995 Continuing Survey of Food Intakes by Individuals and 1994 and 1995 Diet and Health Knowledge Survey, 1997.

45. Feskanich D, Willett WC, Stampfer MJ, Colditz GA. Protein consumption and bone fractures in women. Am J Epidemiol 1996;143:472–9.

46. Krall EA, Dawson-Hughes B. Osteoporosis. In: Shils ME, Olson JA, Shike M, Ross AC, eds. Modern nutrition in health and disease. Philadelphia, PA: Lippincott, Williams & Wilkins; 1999: 1353–64.

47. Weber P. The role of vitamins in the prevention of osteoporosis—a brief status report. Int J Vitam Nutr Res 1999;69: 194–7.

48. Appel LJ, Moore TJ, Obarzanek E, et al. A clinical trial of the effects of dietary patterns on blood pressure. DASH Collaborative Research Group. N Eng J Med 1997;336:1117–24.

49. Optimal calcium intake. NIH Consens Statement 1994;12:1–31.

50. Jeffery RW, French SA. Epidemic obesity in the United States: are fast foods and television viewing contributing? Am J Public Health 1998;88:277–80.

51. Bogden JD, Bendich A, Kemp FW, et al. Daily micronutrient supplements enhance delayed-hypersensitivity skin test responses in older people. Am J Clin Nutr 1994;60:437–47.

52. Chandra RK. Effect of vitamin and trace-element supplementation on immune responses and infection in elderly subjects. Lancet 1992;340:1124–7.

53. Balluz LS, Kieszak SM, Philen RM, Mulinare J. Vitamin and mineral supplement use in the United States. Results from the Third National Health and Nutrition Examination Survey. Arch Fam Med 2000;9:258–62.

54. Lee S-K, Sobal J, Frongillo EA. Acculturation, food consumption, and diet-related factors among Korean Americans. J Nutr Ed 1999;31:321–30.

55. Vegetarian Resource Group. Available from www.vrg.org, 1994.

56. Messina VK, Burke KI. Position of the American Dietetic Association: vegetarian diets. J Am Diet Assoc 1997;97:1317–21.

57. Nieman DC, Butterworth DE, Henson DA, et al. Animal product intake and immune function. Vegetarian Nutr 1997;1: 5–11.

58. Young VR, Pellett PL. Plant proteins in relation to human protein and amino acid nutrition. Am J Clin Nutr 1994;59: 1203S–12S.

59. Pollock ML, Franklin BA, Balady GJ, et al. AHA Science Advisory. Resistance exercise in individuals with and without cardiovascular disease: benefits, rationale,

safety, and prescription: An advisory from the Committee on Exercise, Rehabilitation, and Prevention, Council on Clinical Cardiology, American Heart Association. Position paper endorsed by the American College of Sports Medicine. Circulation 2000;101:828–33.

60. Healthy People 2010: national health promotion and disease prevention objec-tives. Washington, DC: U.S. Department of Health and Human Services; 2000.

61. Kamimoto LA, Easton AN, Maurice E, et al. Surveillance for five health risks among older adults—United States, 1993–1997. Mor Mortal Wkly Rep CDC Surveill Summ 1999;48:89–130.

62. U.S. Department of Health and Human Services. The 2001 HHS Poverty Guidelines. Federal Register Feb. 16, 2001;66:10695–7. Available from http://aspe.hhs.gov/poverty/01poverty.htm.

63. Haines PS, Siega-Riz AM, Popkin BM. The Diet Quality Index revised: a measure-ment instrument for populations. J Amer Diet Assoc 1999;99:697–704.

CHAPTER 17

Photo Disc

> **Don't tell me what I can't eat. Tell me about the health value of food.**
>
> Margaret Meter

ADULT NUTRITION:

Conditions and Interventions

Prepared by **U. Beate Krinke**

CHAPTER OUTLINE

- Introduction
- Importance of Nutrition
- Heart Disease/Cardiovascular Diseases (CVD)
- Overweight and Obesity

KEY NUTRITION CONCEPTS

1 Small steps over a lifetime make a big impact.

2 Tastes can be trained to appreciate the health value of foods.

3 Nutrition practiced in the context of wellness is dynamic; it is evaluated by the changing relationships as well as the absolute values of health measures (e.g., weight-for-height, total to HDL cholesterol, percentage of recommended nutrient intake).

4 Optimal nutrition intervention during illness relies on good information and a support system for follow through.

5 In cases of illness (e.g., diagnosed heart disease) foods may assume extra power; they become functional foods.

6 Relevant diet counseling considers the many roles food plays in life.

INTRODUCTION

Currently more people in the United States die of lifestyle-related diseases than from any other cause. Heart disease (31.4%), cancer (23.3%), stroke (6.9%) and diabetes (2.7%) are among the 10 leading causes of death in the United States (percentages for 1997).[1] They add up to nearly two–thirds of all deaths each and every year, from causes that can, in part, be modified. Chapter 17 deals with the most common of these, heart disease. It also deals with obesity, a risk factor linking lifestyle diseases. Chapter 19 covers nutrition-related conditions that are present in younger adults but are more prevalent in old age, including stroke and hypertension. This separation simplifies discussion; real life, of course, does not draw such boundaries.

IMPORTANCE OF NUTRITION

Nutrition, health, and disease relationships are reflected in food labeling programs. In addition to nutrient content labeling, health claims on food packages emphasize the health value of food and its components. The Food and Drug Administration (FDA) has developed a series of health claims for food package labels that covers circumstances in which scientific evidence supports a food or food component/health relationship. The labeled food must naturally contain at least minimum amounts of nutrients such as vitamins A and C, calcium, iron, protein or fiber, or limited amounts of sodium. The FDA spells out the potential conditions and a standard format for making health claims; Illustration 17.1 provides an example relating to heart disease.

The FDA recognizes the following links as strong enough to allow manufacturers to make health claims on the Nutrition Facts panel of food labels.[2,3]

For heart disease:

1. Fruits, vegetables, and grain products that contain fiber, particularly soluble fiber.
2. Soluble fiber from whole grain oats and oat products (e.g., Cheerios).
3. Soluble fiber from psyllium seed husks.
4. Saturated fat and cholesterol.
5. Soybean protein (with isoflavones) substituted for animal protein.

For high blood pressure or stroke:

6. Sodium.

Illustration 17.1   A health claim for oatmeal that pertains to heart disease: "Soluble fiber from oatmeal, as part of a low-saturated fat, low-cholesterol diet, may reduce the risk of heart disease."

7. Foods that are good sources of potassium and low in sodium.

For cancer:

8. Fruits and vegetables.
9. Fiber-rich grains, fruits, and vegetables.
10. High levels of dietary fat.

For dental health and reduced risk of dental caries:

11. Sugar alcohol (sorbitol, xylitol, and mannitol).

For reduced risk of osteoporosis:

12. Calcium-rich diets.

For reduced risk of neural tube birth defects:

13. Increased folic acid.

Health claims help consumers appreciate the potential contributions food makes to disease risk reduction and therefore to health. In addition to the dietary guidance documents and tools discussed in Chapter 16, a program for health claims represents an example of public policy supporting recognized food and disease relationships. For adults, the nutrition-related conditions of greatest public health concern are heart disease and obesity.

Case Study 17.1
Using Dietary Recommendations and Nutrient Advice from Food Labels

The label of Brand X orange juice includes information that it is "100% pure squeezed orange juice," that it is a good source of potassium, and that "Diets containing foods that are good sources of potassium and low in sodium may reduce the risk of high blood pressure and stroke." Brand X also shows the American Heart Association heart check symbol with text stating "Meets American Heart Association food criteria for saturated fat and cholesterol for healthy people over age 2." The Nutrition Facts panel states that 8 ounces provide 110 calories, 450 milligrams of potassium, and 15% of the daily value of folate. Brand Y states it contains "100% orange juice" but makes no health claims. And while the Brand Y Nutrition Facts panel also states that 8 ounces provide 110 calories, it includes no information about folate.

1. You've decided to eat more fruit every day. How will you decide which orange juice to buy? Which brand is more nutritious?
2. You notice that the orange juice is pasteurized and remember that folic acid is unstable when heated. Should you buy fresh oranges instead?
3. Why might Brand X include folate information on its Nutrition Facts panel?

[Information that might help decide #2: A fresh orange provides 60 calories and 45 micrograms of folic acid; the Daily Value for folic acid used in labeling is 400 mcg.]

HEART DISEASE/CARDIOVASCULAR DISEASES (CVD)

When the heart is unable to support blood circulation to all parts of the body, an individual has heart disease. Diseases of the heart and blood vessels (cardiovascular diseases) include *atherosclerosis* (plaque buildup), *arteriosclerosis* (hardened or brittle arteries), coronary heart disease (such as congestive heart failure), ischemic heart disease (lack of oxygenated blood to the brain), heart valve malfunctions, rheumatic heart disease, and stroke. Hypertension (i.e., high blood pressure) is related to, and can be a risk factor for, heart disease but is discussed as a separate condition in Chapter 19.

Definition

Cardiovascular diseases are often grouped as "heart disease and stroke," as in Healthy People 2010[1] and the most recent revision of the American Heart Association Dietary Guidelines (statement for health professionals).[3] "Heart disease" includes atherosclerosis.

Prevalence

Mortality rates from heart disease and stroke have been falling since the 1960s.[1] Still, the American Heart Association (AHA) estimates that close to 1 million deaths (953,110) occur each year due to heart disease.[4] Heart disease is the third leading cause of death in 25- to 44-year olds, the second in 45- to 64-year olds, and the first in adults aged 65 years and older. Table 17.1 shows the annual number of deaths from heart disease by ethnic group.

ARTERIOSCLEROSIS Age-related thickening and hardening of the artery walls, much like an old rubber hose that becomes brittle or hard.

Etiology

ATHEROSCLEROSIS Beginning in childhood, normal nutrient substances such as cholesterol, fatty acid, and calcium become part of the scar tissue that naturally forms over injured arterial wall cells, creating a plaque. Eventually plaque buildup is severe enough to narrow the blood vessels and block blood circulation. Elevated blood lipids (LDL cholesterol, triglycerides)

Table 17.1 Deaths from heart disease and stroke compared to national goals.

NATIONAL GOALS TO REDUCE MORTALITY		
	Per 100,000 Population	
	Target	**Baseline, 1997**
Reduce coronary heart		
disease deaths	166	216
American Indian or Alaska Native	166	134
Asian or Pacific Islander	166	125
Black or African American	166	257
Hispanic or Latino	166	151
White	166	214
Reduce stroke deaths	48	60
American Indian or Alaska Native	48	39
Asian or Pacific Islander	48	55
Black or African American	48	82
Hispanic or Latino	48	40
White	48	60

SOURCE: Healthy People 2010: national health promotion and disease prevention objectives. Washington, DC: U.S. Department of Health and Human Services; 2000: Tables 12-15 and 12-18.

provide additional material for plaque formation. Atherosclerosis can occur in several ways:

- Formation of fatty streaks in the smooth muscle of the *artery* wall
- Formation of fibrous plaques (deposits of fats, cholesterol, collagen, muscle, and other cells) inside the artery walls, which slow blood flowing through the artery
- Calcification of fibrous plaques into lesions that eventually deteriorate, become infected, or in general weaken the artery wall.

Atherosclerosis can be slowed and partially reversed. For example, in the Ornish program, significant lifestyle changes including a very low-fat vegetarian diet that was also low in saturated fat and high in fiber, participating in regular exercise, using a stress management program, and attending group support sessions resulted in shrinking plaques and fatty streaks for adults with heart disease.[5–7]

ARTERY Blood vessel carrying oxygenated blood from the heart to the rest of the body.

OTHER POTENTIAL CAUSES OF HEART DISEASE

Cardiovascular events occur in individuals who have normal blood lipid levels, leading researchers to examine other potential causes for heart disease, including high levels of homocysteine,[3] inflammation, abnormal blood clotting factors, and apolipoprotein metabolism.[8]

Effects

The effects of heart disease can range from shortness of breath after exertion to chest pain (angina) and fi-

nally death from a heart attack (closing off of an artery prevents blood from getting to the brain and thus prevents nerve impulses from keeping the body functioning). Brittle arteries preventing adequate blood circulation result in less energy, decline in organ function, and eventual inability to perform activities of daily living. Buildup of fatty streaks, plaque, and lesions inside the blood vessels leave less room for blood flow. Consequently, the heart has to work harder to pump blood through this narrower space to reach all parts of the body, leading to higher blood pressure levels and potentially strokes. An analysis of data from more than 12,000 middle-aged men in the Seven Countries Study[9] found that increases in blood pressure (diastolic and systolic) led to increased heart disease, just as higher levels of blood cholesterol lead to atherosclerosis.

Risk Factors

Risk reduction for heart disease is most effective before damage to blood vessels and the heart occurs. Table 17.2 lists national nutrition-related objectives for decreasing heart disease and stroke. Healthy weight goals contributing to heart health are shown in Table 16.1.

National health goals are formulated after reviewing available evidence regarding the contributing risk factors. The following list contains potentially

Table 17.2 Healthy People 2010 nutrition-related health objectives to reduce heart disease and stroke among adults. (See Table 17.3 for related life expectancy gains.)

	Percentage of All Adults	
	Target	**Baseline**
Increase the proportion of adults with high blood pressure whose blood pressure is under control.	50%	18%
Increase the proportion of adults with high blood pressure who are taking action (for example, losing weight or reducing sodium intake) to help control their blood pressure.	95%	72%
Reduce the mean total blood cholesterol levels among adults.	199 mg/dL	206 mg/dL
Reduce the proportion of adults with high total blood cholesterol levels (240 mg/dL or greater).	17%	21%
Eat 2400 mg or less of sodium daily.	65%	21%

SOURCE: Healthy People 2010: national health promotion and disease prevention objectives. Washington, DC: U.S. Department of Health and Human Services; 2000: Sections 12 and 19.

modifiable and non-modifiable heart disease risk factors.[3,9–15]

Potentially Modifiable Risk Factors

Anthropometric measures

- Excess weight for height, e.g., BMI ⩾30
- Abdominal fat (waist measurement of 35 inches or more for females; 40 inches or more for males)
- Waist-to-hip ratio in older women[16]

Biological factors

- Blood pressure above 140/90 mmHg

Chemistry/laboratory values

- High-density lipoprotein (HDL) cholesterol levels of less than 40 mg/dL throughout, especially in women
- Elevated low-density lipoprotein (LDL) with 130–159 mg/dL, borderline high; 160–189 high; ⩾190 very high
- Total cholesterol levels of 200 mg/dL or more (moderate risk) or above 240 mg/dL (high risk)
- Combination of high blood triglyceride (TG) of >170 mg/dL and/or serum LDL cholesterol of >160
- Elevated plasma apolipoprotein B (atherogenic lipoprotein)
- Diabetes, especially if uncontrolled; elevated fasting plasma insulin
- Low levels of estrogen in women

Dietary intake

- High saturated fats (>10%) and total fat (>30%) calories with low percentage of poly- and monounsaturated fatty acids
- Few vegetables and fruits
- Few whole grains
- Low folic acid leading to high homocysteine levels[17]
- High cholesterol in cholesterol-sensitive individuals
- Lack of fish oils

Other modifiable risk factors for developing atherosclerosis

- Smoking cigarettes and cigars; chewing tobacco
- Lack of consistent physical activity
- Unresolved emotional stress, hostility, angry personality

Nonmodifiable Factors Associated with a Higher Risk of Atherosclerosis

- Family history of atherosclerotic heart disease
- Male gender, females after menopause
- Old age
- Geography: Japanese men incur least, European and North Americans incur most heart disease, irrespective of smoking, blood pressure, and blood cholesterol, and age

Nutritional Remedies

IMPROVING LIFE EXPECTANCY Nutrition interventions for heart disease include risk reduction to prevent damage and then managing symptoms should they develop. One reason to begin nutrition intervention in early adulthood is to enhance life expectancy. Table 17.3 shows that heart disease interventions will benefit men differently than women, and that younger individuals are more likely to gain additional lifetime than older adults.

Abnormal blood lipids are the strongest risk factors for heart disease, and the strongest evidence for atherosclerotic plaque reduction supports limiting saturated fatty acids in the diet, including trans fatty acids. However, no single action guarantees long life. Humans evolved eating varied diets and are most likely to benefit from a variety of dietary strategies to stay healthy.

REMEDIES TO CHANGE RISK FACTORS: PRIMARY PREVENTION The National Heart, Lung and Blood Institute (NHLBI) initiated the National Cholesterol Education Program (NCEP) in 1985 in order to decrease the population's average blood cholesterol, which was considered to be the most modifiable nutrition-related risk factor for heart disease. An NCEP Expert Panel on Detection, Evaluation, and Treatment of High Blood Cholesterol developed population-based strategies to reduce heart disease[10,11] through identification of cut-points for diagnosis of risk-associated blood lipid levels, development of corresponding dietary and pharmaceutical treatment suggestions, and implementation of health promotion campaigns. Population levels of cholesterol have decreased.

NCEP's dietary guidelines were made in the form of the Step I and Step II diets,[10,11] which were replaced by the AHA Dietary Guidelines released October 2000.[3] The Nutrition Committee of AHA based its report on three underlying philosophies: (1) diet and lifestyle practices can be safely followed throughout life; (2) individual intake is to be evaluated over extended time rather than a single meal; and (3) the guidelines are a population framework into which individual needs are to be integrated. The AHA dietary guidelines, summarized in Table 17.4, underscore the potential benefits of eating a varied, whole foods diet.

Table 17.3 Life expectancy gains related to cardiovascular disease.

GENERAL POPULATION	INTERVENTION	Average Time in Months	
		Males	**Females**
35-year-old men	Exercise consuming 2000 kcal/week for 30 years	6.2	NA
35-year-olds	Stop smoking cigarettes	10	8
35-year-olds	Blood pressure: reduce to 88 mm Hg	13	5
35-year-olds	Serum cholesterol: reduce to 200 mg/dL	8	10
35-year-olds	Reduce weight to ideal	7	5

HIGH-RISK GROUPS	INTERVENTION	Average Time in Months	
		Males	**Females**
35-year-olds	Blood pressure: Reduce diastolic to 88 mm Hg from >90–94	13–64	11–68
60-year-olds with hypertension	Reduce systolic by 14.3% Antihypertensive treatment	24 / 2–3	22 / N/A
35-year-olds with high cholesterol	Reduce serum cholesterol to 200 mg/dL (Low gains began @ 200-239 mg/dL chol and high gains began @ >300 mg/dL)	6–50	5–76
60-year-olds with multiple risk factors	Reduction of cholesterol by 3% to 20% through dietary program	1–5	5–29
35-year-olds who are overweight	Reduce weight to ideal (Low gains <30% above ideal and high gains >30% above ideal)	8–20	6–13

SOURCE: Adapted from Wright JC, Weinstein M. Gains in life expectancy from medical interventions—standardizing data on outcomes. N Eng J Med 1998;339:380–6; and Tsevat J, Weinstein M, et al. Expected gains in life expectancy from various coronary heart disease risk factor modifications. Circulation 1991;83:1194–201.

The AHA, health promotion planners, and public policy developers use guidelines established as consensus statements like the one in Table 17.4 to develop consistent, nonconfusing public messages. Alcohol is an example of a topic that can easily lead to public confusion. The AHA committee recommends that people do not start drinking if they do not do so now, due to the role alcohol plays in addiction, hypertension, breast cancer, weight gain, and physical abuse. Potential benefits of moderate alcohol use (wine and other forms of alcohol) are that it may increase the body's capacity to dissolve blood clots and that it raises blood HDL cholesterol, although not if liver damage is present. The effect of alcohol is more potent in women. These potential benefits and drawbacks translate roughly to "if you drink, do so in moderation" chosen as the message given in the AHA dietary guidelines.

REMEDIES TO REDUCE SIGNS OF HEART DISEASE: SECONDARY PREVENTION
The strictest and proba-bly most recognized approach to dietary treatment of heart disease is the Ornish Reversal Program.[6] It is a lifestyle program that includes a very-low-fat (fewer than 10% of calories from fat) vegetarian diet (no animal products except egg white and one cup per day nonfat milk or yogurt), no caffeine, no calorie restrictions, and 15% or more calories from protein. Other program components include the following:

- Moderate aerobic exercise (walking)
- Stress management (breathing techniques, meditation, yoga, relaxation exercises)
- No smoking
- Group support

For the experimental group reviewed in 1990, changes in diet, exercise, and stress reduction (20 experimental cases compared to 15 controls) resulted in weight loss, serum total and LDL cholesterol reduction, and stable HDL cholesterol levels. A subsequent report showed that highly motivated individuals who

Table 17.4   AHA Dietary Guidelines: Revision 2000: A statement for healthcare professionals.

GUIDELINES FOR THE GENERAL POPULATION

Achieve and maintain a healthy eating pattern that includes foods from each of the major food groups.
- Consume a variety of fruits and vegetables; choose five or more servings per day.
- Consume a variety of grain, including whole grains; choose six or more servings daily.

Achieve and maintain a healthy body weight.
- Match intake of total energy (calories) to overall energy needs.
- Achieve a level of physical activity that matches (for weight maintenance) or exceeds (for weight loss) energy intake.

Achieve and maintain a desirable blood cholesterol and lipoprotein profile.
- Limit intake of foods with high content of cholesterol-raising fatty acids.
- Limit the intake of foods high in cholesterol.
- Substitute grains and unsaturated fatty acids from fish, vegetables, legumes, and nuts.

Achieve and maintain a normal blood pressure.
- Limit salt (sodium chloride) intake.
- Maintain a healthy body weight.
- Limit alcohol intake.
- Maintain a dietary pattern that emphasizes fruits, vegetables, and low-fat dairy products and is reduced in fat.

Issues for Special Populations
- Older individuals (guidelines fit adults of all ages)
- Individuals with special medical conditions:

a) elevated LDL cholesterol or preexisting cardiovascular disease

b) diabetes mellitus and insulin resistance

c) congestive heart failure

d) kidney disease

Ancillary Lifestyle and Dietary Issues
- Smoking
- Alcohol use
- Diets with extremes of macronutrient (carbohydrate, protein, fat) intake

a) High-unsaturated-fat diets

b) Very-low-fat diets

c) High-protein diets

Issues That Merit Further Research
- Antioxidants
- B vitamins and homocysteine lowering (to reduce risk of vascular diseases)
- Soy protein and isoflavones
- Fiber supplements
- Omega-3 fatty acid supplements
- Stanol-sterol ester containing foods
- Fat substitutes
- Genetic influences on nutrient requirements and dietary response

SOURCE: Outlined from Krauss RM, et al. AHA Dietary Guidelines: revision 2000: a statement for healthcare professionals from the Nutrition Committee of the American Heart Association. Circulation 2000;102:2284–99. See www.circulation.org for full report.

follow this regimen closely (for instance, maintaining 8% of total calories from fat) could reduce heart disease and its risk factors.[5] Ornish-program supporters point to improved heart health, critics point to the difficulties of adhering to such a drastically changed and rigid lifestyle.

An example of more commonly used nutritional treatment for heart disease, based on strategies of a cardiac rehabilitation service, are shown in Table 17.5. Potential adjustments for the needs of older adults, different in focus but not in philosophy, are also listed. Ideally, dietary strategies are always matched to individual lifestyle.

HERBAL PRODUCTS IN SECONDARY PREVENTION
In addition to the areas that merit further research by the AHA (e.g., antioxidants, B vitamins, omega-3 fatty acids, and stanol esters used in margarine), sev-

eral effective herbal approaches may be taken to reduce heart disease.[18] Evidence regarding the effectiveness regarding herbal medicines has not yet reached the highest scientific level of randomized controlled double-blind trials, but studies are in progress.

1. *Hawthorn:* The leaves and flowers and/or fruits contain antioxidants (flavonoids and procyanidins), which are used in the early stages of congestive heart failure and arrhythmias. Hawthorn fruit made into jelly and candy has not been tested.
2. *Garlic:* Cloves of garlic contain allicin and antioxidants. Garlic has antibiotic properties, decreases blood cholesterol, and inhibits platelet aggregation or blood clotting. Recommended doses range from 1 clove (4 grams) to 5–20 cloves of fresh garlic per

day; chopping and drying release the active compound allicin, although special processing can result in active dried garlic for supplement use. The German Commission E approves garlic to treat hyperlipoproteinemia and to slow arteriosclerosis.

3. *Green tea extract:* Green tea and its extract contain antioxidants (polyphenols and proanthocyanidins). Consuming 10 or more cups of green tea daily is associated with decreased LDL cholesterol, decreased triglycerides, and increased HDL cholesterol.

4. *Red yeast:* Chinese records from A.D. 800 report red yeast as useful for treating diarrhea,

indigestion, for stomach health and to improve blood circulation. Red yeast is a fungus used in rice wine fermentation and is the food coloring in Peking duck. Red yeast is used to treat hyperlipoproteinemia due to its content of monacolin K, also known as the cholesterol-lowering drug lovastatin (with the same side effects). The monacolin K in red yeast blocks cholesterol formation.

As scientists learn more about the underlying mechanisms of heart disease, it will become easier to evaluate the effectiveness of various remedies, including nutritional approaches and herbal products.

Table 17.5 Treatment factors for individuals with heart disease.

Target Area	Adults (<70 years)	Adults (>70 years)
Decrease amount and type of fat	<30% of calories from fat, < 10% of calories from saturated fat	Focus on 1–2 items to make low-fat in the regular diet rather than change all things
• Use lean meats	Poultry, fish, game, bison, trimmed beef and pork	Ensure adequate protein
• Substitute saturated fatty acids with PUFA & MUFA	Oils more liquid at room temperature, e.g. corn, soybean, olive, safflower instead of palm kernel and coconut; meatless meals	Focus on oils currently using and suggest one to change if appropriate
• Decrease trans fatty acids	Give brief description of trans fatty acids and sources: margarine, vegetable shortening, cookies, pastries, and other processed fats	Same as adult, concept may be overwhelming
Reduce cholesterol intake	Body makes all it needs, a 1% reduction in blood cholesterol decreases heart disease risk by 2%	Focus on 1–2 food items; studies conflicting on role of cholesterol in older adults; liver makes less
Increase fiber, fruits, and vegetables	Use the 5 A day plan, emphasize adequate fluid intake to prevent constipation	Work with the fruits and vegetables that the individual can chew (i.e., if dentures, do they fit?)
Healthy cooking	Broil, bake, grill, steam	May not be controllable if Meals-on-Wheels; goal is to ensure adequate intake
Limit salt	Use herbs and spices; use label information to identify high-sodium processed foods like soups, sauces, condiments, and snacks	Focus on "no added salt," no salt shaker on the table
Label reading	Evaluate fat, saturated fat, sodium, fiber, whole grain ingredients	May be difficult if eyesight is poor; educate the food shopper
Exercise	Obtain doctor's approval prior to starting; emphasize the health benefits: improved blood circulation, muscle efficiency, psychological effects, tension reliever	Obtain doctor's approval prior to starting; emphasize health benefits centered around mobility and agility; emphasize walking is exercise
Maintain healthy weight	Exercise and proper diet; look at weight history to set realistic weight goals for maintenance	Strongly influenced by functional status of the individual; emphasis on adequate intake
Reduce stress	Exercise, relaxation techniques	Exercise, relaxation techniques
Quit smoking	Refer to smoking cessation program; continue the no smoking followed while in hospital, discuss anticipated weight gain	Refer to smoking cessation program; continue the no smoking followed while in hospital; discuss anticipated weight gain

SOURCE: Slight modification of chart developed by Anne Faricy Gerlach, RD MPH; includes principles used in a cardiac rehabilitation program and presented in guest lectures in public health course on "Nutrition for Adults and the Elderly," University of Minnesota.

OVERWEIGHT AND OBESITY

The second day of a diet is always easier than the first. By the second day you're off it.

Jackie Gleason

Casual conversations about food and weight tend to reveal that men and women are aware of their own weight and whether it is currently up or down from the usual. Most people are likely to have dieted at some time or other. Despite the high level of awareness of body weight and knowledge about weight maintenance, the number of overweight and obese individuals is steadily increasing. Obesity is being called an "epidemic."[19]

Overweight and obesity are often examined together because obesity is a degree of overweight. In the past, health professionals used to use the term *ideal body weight*. Ideal weights were based on survivorship of healthy adults applying for life insurance, published as the 1983 Metropolitan Life Insurance Company actuarial tables. The term *ideal body weight* has been replaced with *healthy body weight*, although no agreement has been reached (except for the BMI ranges of 18.5 to 24.9) on exactly what desirable or healthy weight should be for each individual. In general, a healthy weight is one that can be maintained through a health-promoting lifestyle. Weight becomes "overweight" when it reaches a level associated with higher risk for disease, disability, and death.

Definition

The Obesity Education Initiative of the National Institute of Health (NIH) Clinical Guidelines classifies overweight and obesity by body mass index (BMI).[20] Based on this single number expressing weight-for-height relationship, which is used as an indicator of relative proportion of fatness, obesity or a BMI of 30 or greater is roughly equivalent to being 30 or more pounds overweight. NIH classifications of body weight are:

<18.5	"thin"
18.5–24.9	"normal"
25.0–29.9	"overweight"
30.0–34.9	"obesity" Class I
35.0–39.9	"obesity" Class II
≥40.0	"obesity" Class III, extreme obesity

Calculating BMI with Nonmetric Measures

1. Multiply body weight in pounds by 704.5 (sometimes rounded to 705).
2. Divide that number by height in inches.
3. Divide that number by height once more.

For example, calculating the BMI for a 5'7" individual weighing 150 pounds:

1. 150 pounds × 704.5 = 105,675
2. 105,675 divided by 67 inches = 1577.239
3. 1577.239 divided by 67 inches = 23.54 BMI

Gaining 10 pounds would place this individual into the "overweight" category: (160 pounds × 704.5) divided by 67 and then divided by 67 again = 25.1

Calculating BMI with a Metric Formula

BMI = Body weight in kilograms / (height in meters)(height in meters)

For the 5'7" individual weighing 150 pounds:

1. 150 pounds/2.2 pounds per kilogram = 68.2 kilograms
2. 67 inches x 2.54 centimeters per inch = 170.2 centimeters (100 cm = 1 meter)

$$\frac{68.2 \text{ kg}}{(1.702 \text{ meters})(1.702 \text{ meters})} = \frac{68.2}{2.897} = 23.54$$

BMI Charts

The Dietary Guidelines for Americans 2000 include a chart listing BMI scores for selected height and weight values.[21]

Canada defines BMI levels for healthy and overweight as follows:

Healthy weight	BMI between 20 and 25
Caution	BMI between 25 and 27
Overweight	BMI of 27 or greater

For older adults, the Consulting Group of the American Dietetic Association has suggested that BMI levels of 19 to 27 be used as an acceptable and health-promoting range.[22] This range updates Nutrition Screening Initiative materials developed prior to 1999, including the Level 1 and 2 assessments that indicated a healthy BMI for older adults ranged between 22 to 27.

BMI CAUTION Body fat is more important than body weight when evaluating overweight. Although BMI approximates body fat for most individuals, a BMI value is not the same as a measurement of body fat. For instance, a heavily muscled football player who is 6 feet tall and weighs 200 pounds has a BMI of 27.2—without being fat. BMI measures don't accurately represent healthy weights for athletes and others with greater-than-average percentages of muscle mass, individuals with dense, large bones, and dehydrated individuals.

Prevalence

Obesity rates have risen rapidly in the last decade, so that currently 1 in 5 people is considered "obese."

Roughly one of two American adults is considered overweight and/or obese.[20,23] More men than women are overweight, but a greater proportion of women are classified as obese, see Table 17.6.

Obesity prevalence can vary by data collection method. For example, a report of the Behavioral Risk Factor Surveillance System using BMI values calculated from self-reported heights and weights found that 17.9% of the population is obese (17.7% of men, 18.1% of women). This survey found obesity to be increasing most rapidly in adults aged 18 to 29 years old and least in adults over 70 years old.[19] However, increasing age is still associated with weight gain, and higher rates of overweight are seen in the population over age 55 than in younger adults.[24]

Based on NHANES III measured (not self-reported) data, the population percentage currently overweight is 55% of women and 63% of men aged 25 years or older, see Table 17.6.[24] This prevalence is higher than it was 30 years ago when roughly 39% of women and 48% of men were overweight or obese.[25] Seen another way, the mean BMI of U.S. adults increased from 25.2 to 26.3 in the 20 years between NHANES I and NHANES III survey.[26] Currently, 42% of the adult population is at "normal" weight, using the NIH criterion of BMI as 18.5–24.9.[1]

Etiology

Between 1970 and 1990, average caloric intake increased by as much as 300 calories daily.[27] From the perspective of weight change, eating or not eating 100 calories per day adds up to approximately 10 pounds per year gained or lost. An excess of 3500 calories leads to a one pound weight gain. (Water weight is a different issue, dehydration and subsequent hydration can rapidly change weight by 3–5 pounds; a pint of water weighs roughly a pound.) Even though Americans ate fewer fat calories in the last few years, they ate more total calories and they exercised less. For example, two of five, or 40% of the population, participate in no leisure time physical activity (see Table 16.14). Only 15% of American adults partici-

pate in regular physical exercise (at least 20 minutes, three times per week). An abundant food supply (3600 calories available for every man, woman, and child in the United States) coupled with sedentary lifestyles contribute to the population's rising weight.

Overweight and obesity may appear to be simply a matter of intake exceeding output. But if it were so simple, the weight loss industry would hardly be so huge. Overweight and obesity are complex and chronic conditions, stemming from multiple causes. Environmental, genetic, physiological, psychological, socioeconomic, and cultural factors all play a role in the development of obesity.

Effects

The Surgeon General convened a Conference on Obesity to plan a nationwide anti-obesity campaign because excess body weight effects are a threat to the public's health.[28] Overweight, especially obesity, is associated with increased risk of hypertension, dyslipidemia, coronary artery disease, type 2 diabetes, stroke, gallbladder disease, osteoarthritis, sleep apnea, and endometrial, breast, prostate, and colon cancers.[20] Weight loss in overweight and obese persons can decrease the impact or severity of current health problems such as arthritis, elevated LDL cholesterol and triglycerides, and elevated fasting glucose.

However, obesity is not consistently linked to shorter life, although cycling weight (as in yo-yo dieting) is linked to earlier death.[29]

Nutrition-Related Risk Factors

Obesity is an evolutionary survival mechanism; it allows individuals to store energy during times of abundance. Eating more calories than the body uses leads to stores of glycogen in muscle and fat in and around other tissues and organs. Hormonal status affects body weight. Individuals taking birth-control pills or estrogen replacement therapy can verify that these lead to a small, but noticeable, weight gain. Unresolved stress increases cortisol, which in turn increases appetite and contributes to weight gain. Other risk factors for becoming obese are having obese or overweight parents, very low calorie intake during the first half of the mother's gestational period, and growing an excess number of fat cells in youth or periods of rapid weight gain. People with lower educational levels and living in poverty have higher prevalence of obesity.

Nutritional Remedies

The Healthy People 2010 target is to reduce obesity from a population average of 23% to 15% of adults.[1] A successful weight loss program includes a diet that allows lifelong personal adherence (engaging

Table 17.6 Percentage of the adult population who is overweight or obese, by gender.

Gender	Total Overweight and Obese	Overweight	Obese
Males	63%	42%	21%
Females	55%	28%	27%

SOURCE: Data from an analysis of NHANES III data by Must A et al. The disease burden associated with overweight and obesity. JAMA 1999;282:1523–9.

Case Study 17.2
Maintaining a Healthy Weight

Adam is 5'11" tall and weighs 190 pounds. He lives alone. The commute to his software development job takes about 90 minutes each day. He likes his coworkers and the work environment and is happy that his workplace provides a cafeteria so he doesn't have to bring a lunch. But he is trying to work from home occasionally to cut car expenses. His main hobby is golf; the course he plays encourages carts. He's an avid football fan. In his spare time, he is restoring an old car.

1. Calculate Adam's current BMI. How would you classify his weight status based on the NIH classifications?
2. What would it take for Adam to achieve a BMI of 24? Calculate an energy level and estimate the number of weeks it would take at that level.
3. What would you consider a "healthy weight" for Adam?
4. What are some suggestions you would discuss with Adam in order to decrease his BMI?

in the desired practices at least four days per week), that is safe (a minimum of 1200 calories each day), and that results in a calorie deficit over the long run (ability to maintain a 5% or greater loss of body weight for a year or more).[30] Compared to normal weight individuals or those who regain lost weight, individuals who successfully maintain weight loss use more behavioral strategies supporting weight loss and maintenance. These behaviors include consistently controlling caloric intake, exercising more often and more strenuously, and tracking their own weight.[31]

Weight loss programs tend to combine diet with one or more of the following approaches: physical activity (speeds up metabolism, burns calories), behavior modification (integrates weight management into everyday life), drug therapy (stimulates weight loss, likely to have unpleasant side effects), or gastric surgery (invasive, last resort action with many side effects). To "diet" means eating 250 to 500 fewer calories per day in order to lose 0.5 to 1 pound per week, or eating 1000 fewer calories each day to lose 2 pounds per week (e.g., for a large male). As a general rule, intake of 10 calories per pound current weight will result in appropriate weight loss.

Consider the following checkpoints for selecting a successful weight management program:

1. *Realistic goals:* Identify a healthy weight and a feasible rate of loss (0.5–1 pound/week), with the ability to self-monitor progress.
2. *Caloric deficit:* Develop an individualized meal plan with sufficient calories (at least 1200 for females and 1500 for males) to achieve and maintain metabolic balance.
3. *Real foods:* Meal plan is built around a variety of foods that can be readily obtained and enjoyed by the entire household.

4. *Dietary practices for a lifetime:* Weight gain does not happen overnight, neither does weight loss. The weight management plan is built around learning and practicing behaviors that can be maintained for a lifetime.
5. *Practice problem-solving techniques:* Develop strategies to anticipate and solve potential weight management problems.
6. *Stress-management:* Learn strategies other than eating to deal with stressful situations.
7. *Maintenance:* Availability of support options when goal weight is reached.
8. *Regular exercise:* Strength training builds lean body mass and aerobic exercises enhance fitness and energy; both types of exercise burn calories.
9. *Cultivate self-image:* Build self-confidence while guarding against disordered eating.
10. *Commitment:* If a weight loss approach seems too good to be true, it probably is.

Although weight loss can benefit health status, inappropriate eating habits, herbals (such as ephedra) and drugs can also be harmful to health. Using the preceding checklist can provide some assurance that the proposed weight loss approach will be helpful rather than harmful because the diet industry is not closely regulated. If weight loss were easy, we would not be spending more than $33 billion per year on the weight loss industry.

Eat smart. Play hard

Nutrition Education Promotion Campaign of USDA and American Dietetic Association

Resources

See Chapter 16.

References

1. Healthy People 2010: national health promotion and disease prevention objectives. Washington, DC: U.S. Dept. of Health and Human Services; 2000.

2. Brecher SJ, Bender MM, Wilkening VL, et al. Status of nutrition labeling, health claims, and nutrient content claims for processed foods: 1997 Food Label and Package Survey. J Amer Diet Assoc 2000;100:1057–62.

3. Krauss RM, Eckel RH, Howard B, et al. AHA Dietary Guidelines: revision 2000: a statement for healthcare professionals from the Nutrition Committee of the American Heart Association. Circulation 2000; 102: 2284–99.

4. American Heart Association. Available from www.americanheart.org, July 2000.

5. Ornish D, Scherwitz LW, Billings JH, et al. Intensive lifestyle changes for reversal of coronary heart disease. JAMA 1998;280: 2001–7.

6. Ornish D, Brown SE, Scherwitz LW, et al. Can lifestyle changes reverse coronary heart disease? Lancet 1990;336:129–33.

7. Gould KL, Ornish D, Scherwitz L, et al. Changes in myocardial perfusion abnormalities by positron emission tomography after long-term, intense risk factor modification. JAMA 1995;274:894–901.

8. Ridker PM, Hennekens CH, Buring JE, Rifai N. C-reactive protein and other markers of inflammation in the prediction of cardiovascular disease in women. N Eng J Med 2000;342:836–43.

9. Van den Hoogen PC, Feskens EJ, Nagelkerke NJ, et al. The relation between blood pressure and mortality due to coronary heart disease among men in different parts of the world. Seven Countries Study Research Group. N Eng J Med 2000;342: 1–8.

10. Report of the National Cholesterol Education Program Expert Panel on Detection, Evaluation, and Treatment of High Blood Cholesterol in Adults. The Expert Panel. Arch Intern Med 1988; 148:36–69.

11. Summary of the second report of the National Cholesterol Education Program (NCEP) Expert Panel on Detection, Evaluation, and Treatment of High Blood

Cholesterol in Adults (Adult Treatment Panel II). JAMA 1993;269:3015–23.

12. Executive summary of the third report of the National Cholesterol Education Program (NCEP) Expert Panel on Detection, Evaluation, and Treatment of High Blood Cholesterol in Adults (Adult Treatment Panel III). JAMA 2001;285: 2486–97; "At a Glance" desk reference for physicians. Available from http://www. nhlbi.nih.gov/guidelines/cholesterol/ dskref.htm.

13. www.aace.com/indexjava.htm, August 31, 2000.

14. Westerveld HT, van Lennep JE, van Lennep HW, et al. Apolipoprotein B and coronary artery disease in women: a cross-sectional study in women undergoing their first coronary angiography. Arterioscler Thromb Vasc Biol 1998;18:1101–7.

15. Lamarche B, Tchernof A, Mauriege P, et al. Fasting insulin and apolipoprotein B levels and low-density lipoprotein particle size as risk factors for ischemic heart disease. JAMA 1998;279:1955–61.

16. Folsom AR, Kushi LH, Anderson KE, et al. Associations of general and abdominal obesity with multiple health outcomes in older women: the Iowa Women's Health Study. Arch Intern Med 2000;160: 2117–28.

17. Giles WH, Croft JB, Greenlund KJ, et al. Association between total homocyst-(e)ine and the likelihood for a history of acute myocardial infarction by race and ethnicity: Results from the Third National Health and Nutrition Examination Survey. Am Heart J 2000;139:446–53. Available from www.medscape.com/mosby/ AmHeartJ/2000/v139.no3, July 2000.

18. Robbers JE, Tyler VE. Tyler's herbs of choice. New York: Haworth Herbal Press; 1999.

19. Mokdad AH, Serdula MK, Dietz WH, et al. The spread of the obesity epidemic in the United States, 1991–1998. JAMA 1999; 282:1519–22.

20. NHLBI Obesity Education Initiative Expert Panel on the Identification Evaluation, and Treatment of Overweight and Obesity in Adults. Clinical guidelines on the identification, evaluation, and treatment of

overweight and obesity in adults: the evidence report. Bethesda, MD: National Institutes of Health; 1998.

21. U.S. Department of Agriculture, U.S. Department of Health and Human Services. Nutrition and your health: dietary guidelines for Americans, 5th ed. 2000.

22. American Dietetic Association, Consultant Dietitians in Health Care Facilities. Nutrition risk assessment: form, guide, strategies and interventions. Chicago, IL: American Dietetic Association; 1999.

23. St. Jeor ST, Brunner RL, Harrington ME, et al. A classification system to evaluate weight maintainers, gainers, and losers. J Amer Diet Assoc 1997;97:481–8.

24. Must A, Spadano J, Coakley EH, et al. The disease burden associated with overweight and obesity. JAMA 1999;282: 1523–9.

25. National Heart, Lung, and Blood Institute. HeartFacts in HeartMemo. National Cholesterol Education Program newsletter; 1999.

26. Kuczmarski RJ, Flegal KM, Campbell SM, Johnson CL. Increasing prevalence of overweight among US adults: The National Health and Nutrition Examination Surveys, 1960–1991. JAMA 1994;272:205–11.

27. Flegal KM. Trends in body weight and overweight in the US population. Nutr Rev 1996;54:S97–100.

28. Satcher D. Surgeon General's Conference on Obesity, a year-long process to develop a weight reduction plan, co-sponsored by National Institutes of Health, Centers for Disease Control and Prevention, and the Office of Public Health and Science; January 8, 2001.

29. Ernsberger P, Koletsky RJ. Biomedical rationale for a wellness approach to obesity: an alternative to a focus on weight loss. J Soc Issues 1999;55:221–59.

30. Position of the American Dietetic Association: weight management. J Amer Diet Assoc 1997;97:71–4.

31. McGuire MT, Wing RR, Klem ML, Hill JO. Behavioral strategies of individuals who have maintained long-term weight losses. Obes Res 1999;7:334–41.

CHAPTER *18*

Photo Disc

Scientific evidence increasingly supports that good nutrition is essential to the health, self-sufficiency, and quality of life of older adults.

American Dietetic Association[1]

NUTRITION AND THE ELDERLY

Prepared by **U. Beate Krinke**

CHAPTER OUTLINE

KEY NUTRITION CONCEPTS

1 Eating and enjoying a varied diet contributes to mental and physical well-being.

2 Generalizations relative to health status changes with aging are unwise because "older adults" are a heterogeneous population.

3 Diseases and disabilities are *not* inevitable consequences of aging.

4 Functional status is more indicative of health in older adults than chronological age.

5 Body composition changes that occur with aging have the greatest impact on nutritional needs.

INTRODUCTION

An ounce of prevention is worth a pound of cure.

Traditional

In "normal" aging, inevitable and irreversible physical changes occur over time, and some diseases are more prevalent. The leading causes of death are heart disease, cancer, and stroke; these are linked with nutrition in complex ways. The majority of older people consider themselves to be healthy. More than anything, they want to remain healthy and independent and certainly do not want to be a burden to others.[2] Older adults feel that good nutrition and exercise are the most important health habits they can maintain in order to avoid losing autonomy and independence.[2]

Just exactly what is good nutrition for older adults? They can meet their decreasing energy requirements by choosing more nutrient-dense foods. They can drink more water to stay hydrated, even when not thirsty. Dietary factors shown to reduce incidence of the prevalent diseases are adequate vegetable, fruit, and whole grain intakes, eating fats in balance, and drinking alcohol only in moderation. Diet quality is linked to the longevity of older men and women.[3,4] Older adults can live longer and better (i.e., postponing disability and shortening the period of decreased functional capabilities at the end of life) with good health habits.[5] This chapter defines aging and provides information about the nutrient requirements, dietary recommendations, and food and nutrition programs designed to support healthy aging.

What Counts As Old?

Many chronological ages have been used as cut-points to mark the beginning of "old age." While no biological benchmark signals a person's becoming old, there are societal and governmental definitions for "old." Societal definitions label someone as "old" at a fairly young age. Turning 50 makes you eligible to join the American Association for Retired Persons (AARP). Views on "normal" retirement ages are changing, however. At age 60, many cafeterias, movie theaters, and shops give senior discounts. What counts as "old" depends on who is counting.

Governmental definitions of old age include the following categories suggested by the U.S. Census Bureau:

> **LONGEVITY** Length of life; it is a measure of life's duration in years.

- 65 to 74 years is "young old"

- 75 to 84 years is "aged"
- 85 and older is "oldest old."

The Elderly Nutrition Program, first funded under the Older Americans Act in 1972, calls "older adults" those 60 years and older. The Social Security Program identifies people 65 years old as being eligible for retirement benefits (62 years is "early retirement"). These various age cut-points are important because services and programs that support nutrition for the elderly do not use the same age limits. The arbitrarily set retirement age of 65 years is most commonly used in textbooks, and we will use that label here. Readers will also learn that many factors besides chronological age affect nutrition in the elderly.

Chronological age is a marker of where we are on life's path. Our functional status, defined as how well we accomplish the desired tasks of daily living, is more indicative of health than chronological age. Rather than ask: "How old are you?" we should ask: "What can you do?" Nonetheless, chronological age is an easy measure, so we continue to ask it. Our perception is that age is a proxy for predicting health status and functional abilities.

Food Matters: Nutrition Contributes to a Long and Healthy Life

> *We found that tomatoes were the primary food associated with higher functional status in centenarians . . . even when controlled for other factors such as illness, depression, and gastrointestinal problems.*
>
> Dr. Mary Ann Johnson, commenting on findings that elderly nuns with higher functional status also had higher blood lycopene levels[6]

Tomatoes may not be nature's perfect food, but the cumulative effects of lifelong dietary habits determine nutritional status in old age. Good nutrition throughout life contributes to optimal growth, appropriate weight, and to nutrient levels in blood and other tissues that provide disease resistance. In trying to assess the contribution good nutrition can make to longer life, the Centers for Disease Control and Prevention suggest that *longevity* depends 19% on genetics, 10% on access to high-quality health care, 20% on environmental factors such as pollution, and 51% on lifestyle factors.

Diet and exercise are estimated to be the lifestyle factors contributing most to decreased mortality, or longer life.[7] In a longitudinal study including diet monitoring, older women who ate the best diets were 30% less likely to die than the women who ate few whole grains, fruits, vegetables, low-fat dairy products, and lean meats.[4]

The role of food and nutrition often changes during aging. Besides reducing risk of disease and delaying death, diet plays a role in health and longevity by contributing to wellness. Wellness means having the energy and ability to do the things one wants to do and to feel in control of one's life. Being able to choose, purchase or prepare, and eat a satisfying diet every day, enjoying traditional foods at holidays, birthdays, and other special occasions, and having the resources to purchase desired foods on a regular basis all contribute to independence and a higher quality of life. Good nutrition, as defined by dietary guidelines covered later in this chapter, can help to "add life to years" as well as "add years to life."

A PICTURE OF THE AGING POPULATION: VITAL STATISTICS

More and more of us are growing old and older. During Roman times, fewer than 1% of the population reached age 65, but today, roughly 13% of the North American population reaches age 65 (The figure is slightly higher at 15% in Italy, the UK, Germany, and France.) Thirteen percent of the U.S. population means nearly 40 million people.

The percentage of people aged 65 and older more than tripled from 4% in 1900 to 13% in 2000 and is expected to increase to 20% by 2030.[8] When today's 21-year olds turn 65 (around 2046–2050), projections are that they will join a population of roughly 80 million people aged 65 years and older.

Persons aged 85 years and older are the fastest growing segment of the population, accounting for 2% of the population in 2000 and projected to increase to nearly 5% of the U.S. population in 2050.[8] The last White House Conference on Aging called this age wave "a demographic revolution" and predicted that it will change the twenty-first century much as information technologies revolutionized the twentieth century and the industrial revolution affected the nineteenth century.[9]

Global Population Trends: Life Expectancy and Life Span

Today, life expectancy of older adults is quite different from 1900, when average life expectancy was 47 years. The National Center for Health Statistics (NCHS) report that women born in 1995 in the United States can expect to live 79 years, whereas men can expect to live 72 years (see Table 18.1). For babies born in 2050, life expectancy is expected to increase to 84 years for women and 80 years for men.[10] Life expectancy in the United States lags behind that of many other countries, primarily because infants in the United States have higher mortality rates than those in other countries. Rates of childhood mortality, infectious and chronic diseases, death from violence, and accidents are other contributors to life expectancy.

Immunizations and other risk-reduction measures, treatment of disease, decreased infant and childhood mortality rates, clean water, and ample

Table 18.1 Examples of life expectancy for people born 1990–1995 in selected countries.

Country	Female Life Expectancy at Birth, years	Country	Male Life Expectancy at Birth, years
Japan	82.9	Japan	76.4
France	82.6	Sweden	76.2
Switzerland	81.9	Israel	75.3
Sweden	81.6	**Canada**	**75.2**
Spain	81.5	Switzerland	75.1
Canada	**81.2**	Greece	75.1
Australia	80.9	Australia	75.0
Italy	80.8	Norway	74.9
Norway	80.7	Netherlands	74.6
Netherlands	80.4	Italy	74.4
United States	**78.9**	Costa Rica	73.0
Chile	77.8	Finland	72.8
Costa Rica	77.8	**United States**	**72.5**
Kenya	57.2	Kenya	54.2
Nigeria	52.0	Nigeria	48.8

SOURCES: Healthy People 2010 (cohort born in 1995);[10] National Center for Health Statistics, 2000, cdc.gov/nchs, Table K, accessed 10-1-2000 (life expectancy cohorts from 1990–1995, including estimates from United Nations data); Health, United States, 2000.[8] Numbers do not include Tuljapurkar et al.'s[12] recalculated expectancies.

Connections
Life Expectancy vs. Life Span: The Jeanne Calment Story

In France, part of the contract when buying an apartment is that the buyer cannot move in until the current tenant decides to leave. Jeanne Calment's wish when selling her apartment to Andre-Francois Raffray was to stay until she died. Mr. Raffray agreed. He purchased Jeanne Calment's apartment in 1965 for monthly payments of $500, which he promised to make until Jeanne died, at which time he could move into the apartment. She was 90, he was 47. She lived another 32 years, to age 122. Unfortunately, he only lived 30 more years, dying at age 77. Who owned the apartment after his death? His heirs, who were obligated to continue payments for two more years until Jeanne Calment died as the oldest known living person in the world.

amounts of safe food have increased average *life expectancies,* which are getting closer to the potential human *life span.* For instance, from 1970 to 1990 life expectancy increased by more than four years largely due to fewer deaths from heart disease, stroke, and accidents among older adults.[8]

While life expectancy is rising and the population is aging, human life span remains stable around 110 to 120 years.[11] Few people currently survive to age 120. However, Jeanne Calment, a French woman who lived to age 122, is an exception. Learn more about Ms. Calment in the Connection in this section. Nonetheless, the increase in centenarians (persons living to age 100 and beyond) may add to our understanding of current limits on the human life span. Demographers are predicting that by 2030, there will be 381,000 centenarians in the United States, up from 65,000 in the year 2000.

Nutrition: A Component of Health Objectives for the Older Adult Population

Behaviors that can enhance the health of an aging population are reflected in national health objectives.

LIFE EXPECTANCY Average number of years of life remaining for persons in a population cohort or group; most commonly reported as life expectancy from birth.

LIFE SPAN Maximum number of years someone might live; human life span is projected between 110 to 120 years.

Table 18.2 identifies dietary goals related to disease prevention and to health promotion for older adults. The current set of health objectives emphasizes public education about health consequences of overweight and obesity. The greatest dietary improvement in the older adult population would be to eat more vegetables and grain, especially whole grain products.

THEORIES OF AGING

What triggers aging? Theories explaining aging grow from human desire to understand the biologic processes that determine how long and how well we live. Aging theory tries to explain the mechanisms behind loss of physical resilience, decreased resistance to disease, and other physical and mental changes that accompany aging.

Table 18.2
Healthy People 2010 goals for older adults.

PERCENTAGE OF POPULATION		Current, Age 60+	
	Target	Females	Males
Increase the number who are at a healthy weight	60	37	33
Decrease the number who are obese (BMI 30+)	15	26	21
Increase the proportion of persons who:			
• Eat at least 2 daily servings of fruit	75	35	40
• Eat at least 3 daily servings of vegetables (at least one-third dark green or deep yellow)	50	6	5
• Eat at least 6 daily servings of grain products (at least 3 of those as whole grains)	50	4	11
• Eat less than 10% of calories from saturated fat	75	47	42
• Eat no more than 30% of calories from fat	75	40	34
Baseline information is for age 50 and older			
Meet dietary recommendations (1200 mg) for calcium	75	27	35
Baseline information is for age 65+			
Increase the proportion of physician office visits made by patients with a diagnosis of cardiovascular disease, diabetes, or hyperlipidemia that include counseling or education related to diet and nutrition	75	33[a]	33[a]
Increase food security and in so doing, decrease hunger			
Households <130% of poverty, with elderly persons	94	85[a]	85[a]
Households >130% poverty, with elderly persons	94	98[a]	98[a]

[a]Adequate data to distinguish levels for males and females is not available.

SOURCE: Healthy People 2010[10]

Biologic systems are much too complex to have one theory that is robust enough to explain the mechanisms of aging.[13] Basic biologic processes involved in aging are largely determined by genetics. However, environmental factors influence expression of the genetic code by exacerbating certain traits. For example, height and weight are genetically programmed, but diet and other environmental exposures can moderate these outcomes.[14] Persons with a family history of high levels of LDL cholesterol (i.e., bad cholesterol) and early death from cardiovascular disease can moderate the outcomes suggested by their genetic programming through weight loss, a diet low in saturated fat, adequate exercise, and no smoking. Someone with a family history of type 2 diabetes can reduce their risk of developing the disease with a healthy diet, exercise, and maintenance of normal weight.

Theories of aging fall into three major categories: (1) programmed aging theories, (2) "wear and tear" theories, and (3) caloric restriction. Categories of aging theories overlap to some extent because it is clear that neither genetic programming nor environmental exposures fully account for changes related to aging.

Programmed Aging

HAYFLICK'S THEORY OF LIMITED CELL

REPLICATION   Hayflick proposed that all cells contain a genetic code that directs them to divide a certain number of times during their life span.[11] After cells divide according to their programmed limit, and barring disease or accident, cells begin to die. For example, if individual cells of a fly have a 3-day life span and replicate 15 times, a fly can live 45 days. Using this theory, Hayflick calculated the potential human life span to be in the range of 110 to 120 years, estimating that human cells replicate from 40 to 60 times. Although most human cells can regenerate (e.g., blood, liver, kidney, and skin cells reproduce themselves), not all cells have that capacity (e.g., spinal cord, nerves, and brain cells). Hayflick's theory is difficult to prove in humans because we die from age-associated chronic disease more often than from old age itself.

MOLECULAR CLOCK THEORY   Another theory of programmed aging is that of the molecular clock. *Telomeres* that cap the ends of chromosomes act as clocks, becoming a bit shorter with each cell division. Eventually loss of telomeres stops the ability of chromosomes to replicate, and they become *senescent*. Loss of chromosomal replication may produce signs of aging because new cells cannot be formed and the function of existing cells declines with time. A major thrust of current research is identification of ways to limit loss of telomeres and thus prolong cell replication.[15,16]

Wear and Tear Theories of Aging

"Wear and Tear" theories are built on the concept that things wear out with use. Mistakes in the replication of cells or buildup of damaging by-products from biological processes eventually destroy the organism. Cytotoxicity (poisoning of the cell) results when damaged cell components accumulate and become toxic to healthy cells. According to this theory, the accumulation of damaged cells and waste by-products leads to aging.

OXIDATIVE STRESS THEORY   Oxygen is an integral and versatile part of metabolic processes; it can both accept and donate electrons during chemical reactions. One of the causes of aging is thought to be oxidative stress due to the buildup of reactive (unstable) oxygen compounds. Unstable oxygen, formed normally during metabolism (e.g., hydroxyl radicals) can also damage cells by initiating reactions that break down cell membranes and modify normal metabolic processes that protect people from disease. Exposure to oxidizing agents is increased by smoking, ozone, solar radiation, and environmental pollutants. Unstable oxygen compounds are neutralized, however, when they combine with an antioxidant. This prevents them from interfering with normal cell functions. The body produces antioxidant enzymes (such as catalases, glutathione, peroxidase reductases and superoxide dismutase), but part of our need for antioxidants is met from the diet. Dietary antioxidants

Connections
Is Aging Programmable?

Experience related to Dolly the sheep, the first cloned mammal, supports theories of programmed aging. Dolly was created from single mature cells removed from a 6 year old female sheep. She has no father. Even though Dolly is physically young and healthy and has had several offspring with normal genetic material, Dolly's genetic age is that of an older sheep. Her telomeres are shorter than those of other sheep her age. Since sheep life expectancy is 13 years and Dolly was born with 6 year old mother cells, time will tell if she will live 7 or 13 years. Telomere research is an active field. Newer cloning experiments created calves with cells that were genetically younger than those of their parents.[106]

(More information available at: www.cnn.com/NATURE/9905/26/dolly.clone.02, accessed 10/4/2000)

TELOMERE   A cap-like structure that protects the end of chromosomes; it erodes during replication.

SENESCENT   Old to the point of nonfunctional.

include selenium, vitamins E and C, and other phytochemicals. Remember, these are plant substances such as beta-carotene, lycopene, flavonoids, lutein, zeaxanthin, reservatrol, and isoflavones that contribute to normal metabolism. For example, flavonoids found in grapes, apples, broccoli, and onions act as antioxidants.

RATE OF LIVING THEORY   The rate of living theory is similar to the oxidative stress theory in that it suggests that "faster" living results in faster aging. [17] For example, higher metabolic rate and energy expenditure leads to greater turnover of all body tissues. Theoretically, fast-paced living shortens lifespan, whereas living more slowly leads to a longer life. Scientists haven't adequately examined old people, including centenarians, to fully understand how this theory works.

Calorie Restriction and Longevity

Animal studies (e.g., of fruit flies, water fleas, spiders, guppies, mice, rats and other rodents) show that an energy-restricted diet that meets micro-nutrient needs can prolong healthy life. [18,19] For example, laboratory mice and rats fed calorie-restricted diets live longer and have fewer age-associated diseases than their counterparts whose diets are unrestricted. In the 1930s, McCay and colleagues suggested that delays in aging result after food restriction, due to slowed growth and development. [20] But since then, rodent studies have shown instituting caloric restrictions in mid-life, after growth and development were completed, result in longer life spans. [19,21]

RESILIENCE   Ability to bounce back, to deal with stress and recover from injury or illness.

Caloric restriction is also being examined in primates. [22,23] Since nonhuman primate life expectancy is roughly 40 years, it will be years until we learn whether life extension through caloric restriction might be applied to primates. Stay tuned.

Could calorie-restricted diets also extend human life? Experimental findings in small animals have led some individuals, such as Dr. Roy Walford of Biosphere 2, to personally adopt very low calorie diets. Walford coordinated the calorie-restricted diets of eight normal-weight people living in Biosphere 2. However, a study lasting only 6 months is too brief to determine human life span extension results from caloric restriction. Proposed anti-aging effects of caloric restriction in humans are thought to result from an evolutionary response to famine and feasting cycles. [24] The weakness in this theory is that human's need for reproductive capacity is greater than that for longevity. [11]

From an ecological view, France and Japan have lower caloric intake than the USA and both have longer life expectancies. [25] Physiologically, we know nutrition affects human longevity by moderating risks of developing chronic diseases, ameliorating certain chronic conditions, and contributing to healing in a variety of acute conditions. [10,26] These changes indirectly lower caloric intake. An example is adding vegetables to a diet in order to decrease risk of chronic disease, which also lowers the overall caloric intake. Finally, severe caloric restriction during famine or malnutrition during starvation leads to poor outcomes in human reproduction, growth, development, immune status, and healing. Caloric restriction as a means to enhance human longevity remains a very engaging theory.

PHYSIOLOGICAL CHANGES

Normal aging is associated with shifts in body composition and subsequent loss of physical *resilience*. (Aging is not all loss or decline, aging is also associated with psychosocial, personal, moral, cognitive, and spiritual development.) As scientists come to understand the human aging process, they will learn to sort through age-associated physiological changes to be able to distinguish exactly which changes are due to genetic factors and which are due to poor diets, inactivity, or other lifestyle-related factors. Physiological system changes commonly associated with healthy aging are described in Table 18.3. Disease related changes are presented in Chapter 19.

Body Composition Changes

LEAN BODY MASS (LBM) AND FAT   Individual shifts in body composition are common but neither

Table 18.3 Age-associated physiological system changes that impact nutritional health.*

Cardiovascular System
- Reduced blood vessel elasticity, stroke volume output
- Increased blood pressure

Endocrine System
- Reduced levels of estrogen, testosterone
- Decreased secretion of growth hormone
- Reduced glucose tolerance
- Decreased ability to convert provitamin D to previtamin D in skin

Gastrointestinal System
- Reduced secretion of saliva and of mucus
- Missing or poorly fitting teeth
- Dysphagia or difficulty in swallowing
- Reduced secretion of hydrochloric acid and digestive enzymes
- Slower peristalsis
- Reduced vitamin B_{12} absorption

Nervous System
- Blunted appetite regulation
- Blunted thirst regulation
- Reduced nerve conduction velocity affecting sense of smell, taste, touch, cognition
- Changed sleep as the wake cycle becomes shorter

Musculo-Skeletal System
- Reduced lean body mass (bone mass, muscle, water)
- Increased fat mass
- Decreased resting metabolic rate
- Reduced work capacity (strength)

Renal System
- Reduced number of nephrons
- Less blood flow
- Slowed glomerular filtration rate

Respiratory System
- Reduced breathing capacity
- Reduced work capacity (endurance)

*Some of these age-associated changes, such as the increase in blood pressure, are usual but not normal.

Table 18.4 Comparison of body composition of a young and an old adult.

	20 to 25 Years	70 to 75 Years
Protein/cell solids	19%	12%
Water	61%	53%
Mineral mass	6%	5%
Fat	14%	30%

Data from Chernoff, 1999[32], p. 391; based on NW Shock, *Biological Aspects of Aging*, 1962.

water, so older people have lower mineral, muscle, and water reserves to call upon when needed. At the same time, many older adults gain body fat. Compared to males in their 20s, males in their 70's have roughly 24 pounds less muscle (a decrease from 24 to 13 kg on average) and 22 pounds more fat (an increase from 15 to 25 kg on average).[31] Think about losing 24 pounds of muscle: it is roughly equivalent to the muscle mass that girls or boys gain throughout the puberty growth spurt. Over 50 years, 24 pounds of muscle slowly goes away to be replaced by 22 pounds of fat. Of course, the extra fat does provide a reserve of energy for periods of low food intake, recovery from illness or surgery, as an insulator in cold weather and to cushion falls. These shifts in body composition tend to occur even when weight is stable.[29]

WEIGHT GAIN Weight gain, although not inevitable, tends to accompany aging. Obesity is a problem for 24% of adults aged 60 years and older.[10] Reasons for age-associated gains are uncertain, but longitudinal studies are showing that lack of exercise could be a factor.[27] For example, men in the Baltimore Longitudinal Study on Aging decreased their energy expenditure by 17 to 24 calories per day after age 55 but gained weight.[33] In the Fels Longitudinal Study, men gained 0.7 pounds per year and women gained an average of 1.2 pounds as they aged.[29] Subjects were 40 to 66 years old when entering the study and were followed for up to 20 years. Weight gains were concurrent with decreases in lean body mass and increases in body fat. This overall weight and body composition shift was moderated by physical activity. For example, the two groups with moderate or high physical activity levels (as opposed to the least active group) increased lean body mass and decreased total and percentage of body fat as age increased. Physical activity effects differed by gender. In women, higher levels of physical activity were associated with higher levels of lean body mass. However, lack of estrogen seems to

LEAN BODY MASS Sum of fat-free body tissues: muscle, mineral as in bone, and water.

inevitable nor irreversible.[27] Of all physiologic changes that do occur during aging, the biggest effect on nutritional status is due to the shifts in body composition (the musculo-skeletal system, Table 18.3). Many individuals experience a decline in *lean body mass* with aging (see Table 18.4). On average, lean body mass decreases by 15% in the 50 years from the mid-twenties to the mid-seventies.[28]

The composition changes are associated with lower levels of physical activity, food intake, and hormonal changes in women.[29,30] Loss of mineral and muscle mass is also accompanied by loss of body promote fat accumulation, and total weight increased regardless of the group's activity level. (The menopausal transition is covered in Chapter 16.) Men in the highest physical activity groups in the Fels Longitudinal Study slowed their total body weight and body fat gains.

MUSCLES: USE IT OR LOSE IT   Many older people expect decreases in physiologic function with increasing age. However, physical activity contributes to staying strong, no matter the age.[28] For example, the HERITAGE Family Study compared the effects of a 20-week strength training program on older and younger men and women, finding that the training response differed by sex and by race, but *not* by age.[34] Training exercises led to increases in fat-free mass, decreases in total, subcutaneous, and visceral fat mass, and to weight loss. Weight bearing and resistance exercise increase lean muscle mass and bone density.[28,35] Since muscle tissue contains more water than fat tissue, building muscle also results in more stored water. Regular physical activity, including strengthening and flexibility exercise, contributes to maintenance of *functional status*.

> **FUNCTIONAL STATUS**   Ability to carry out the activities of daily living, including telephoning, grocery shopping, food handling and preparation, and eating.

> *I would not wish to imagine a world in which there were no games to play and no chance to satisfy the natural human impulse to run, to jump, to throw, to swim, to dance.*
>
> Sir Roger Bannister, 1989

Changing Sensual Awareness: Taste and Smell, Appetite and Thirst

TASTE AND SMELL   Although there is some argument about the extent to which aging affects the sense of taste, there is general agreement that taste and smell senses eventually decline.[36–39] Eating is a sensuous activity, involving taste buds and olfactory nerves. In healthy adults, aging is associated with a decline in ability to identify smells beginning sometime around age 55 for men and age 60 for women[40] or later.[41] Some smells are perceived more easily than others. A blunted sense of smell can lead to a blunted sense of enjoyment of food as well as decreased ability to detect spoiled or overcooked foods. Women's abilities to identify smells remain higher than men's throughout the life span.

The controversy around taste is that the number and structure of taste buds are not significantly al-

tered during aging. In addition, taste perception for sucrose does not decline with age. Bartoshuk[42] argues that taste is so important to biological survival that the body has developed "redundancy in the mediation of taste," (p. 65) meaning that several pathways (nerves and receptors) control taste mechanisms. All of the pathways would have to be damaged before the ability to identify sweet, bitter, salty, sour, or savory tastes is lost.

Disease and medications do affect taste and smell more than age itself. Roughly three of four adults have had a temporary smell loss at some time, most often due to colds, flu, or allergies.[41] Yet, age makes a difference. For example, during illness or with the use of medications, younger individuals maintained greater ability to detect salty, bitter, sour, sweet, and savory tastes than older ones.[43]

APPETITE AND THIRST   Hunger and satiety cues are also weaker in older than in younger adults. Roberts and colleagues examined the ability of 17 young men (mean age 24) and 18 older men (mean age 70) to adjust caloric intake after periods of overeating and of undereating.[44] All men were healthy and not taking medications. Food intake and weight were monitored for 10 days and then men were overfed by roughly 1000 calories, or underfed by about 800 calories for 21 days. Periods of over- or underfeeding were followed by 46 days of "ad lib" intake where all men were free to eat as much or as little as desired. After the periods of over- and underfeeding, young men adjusted their caloric intake to get back to their initial calorie intake level and weight. Older men kept overeating if they had been in the overfed group, and undereating if they had been in the underfed group. The authors suggest that older adults may need to be more conscious of food intake levels since their appetite regulating mechanism may be blunted. Whereas healthy young people adjust to cycles of more and less food intake, healthy older people's inability to adapt to these changes may lead to overweight or anorexia.

Elderly people don't seem to notice thirst as clearly as younger people do. A small study demonstrated that the thirst-regulating mechanism of older adults was less effective than that of younger individuals.[45–47] Researchers compared thirst response to fluid deprivation in a group of seven 20 to 31 year old men, and seven 67 to 75 year old men. Subjects lost 1.8% to 1.9% of body weight during 24 hours without fluids. Both groups were asked about feeling thirsty, mouth dryness, and how pleasant it would be to drink something. After fluid deprivation, the younger group reported being thirsty and having dry mouth. The older group, however, reported no change in thirst or mouth dryness. Both the older and younger groups thought that it would be pleasant to drink something after fluid deprivation. Blood measures showed that older men lost more blood volume than younger men, indicated by their plasma concentrations of sodium. Researchers also measured how much water the men drank in the hour after their 24-hour period of fluid deprivation. Older men drank less water than their younger counterparts. Younger people made up for fluid loss in 24 hours; older people did not drink enough to achieve their prior state of hydration. It appears dehydration occurs more quickly after fluid deprivation and that rehydration is less effective in older men.

NUTRITIONAL RISK FACTORS

Identifying nutritional risk factors before chronic illness occurs is basic to health promotion. Decreasing risk factors forms the basis of dietary guidance. For instance, the leading causes of death for older adults are heart disease, cancer, and stroke. In all adults, dietary risk factors that increase the likelihood of developing these diseases are consuming a high-fat diet that is also high in saturated fat, low intake of vegetables, fruits, and whole grain products, and excessive calorie intake leading to obesity. Healthy People 2010, the Food Guide Pyramid, and the Dietary Guidelines for Americans emphasize eating patterns that reduce the risk of the leading killer diseases.

Another approach to nutritional risk factor identification is to compare adequacy of current dietary intake to dietary intake recommendations such as the Recommended Dietary Allowances and the Dietary Reference Intakes. According to national intake data[48] older adults are not consuming enough protein, vitamin E, folic acid, magnesium and calcium to meet recommended nutrient levels.

A third approach is to examine a population and determine how environmental factors combine with dietary factors to predict nutritional health. This approach was used by a consortium of care providers, policy makers, and researchers to develop the Nutrition Screening Initiative (NSI).[49] The American Academy of Family Physicians, the American Dietetic Association (ADA), and the National Council on the Aging, Inc. sponsored the development of this health promotion campaign. The NSI consortium used literature review, expert discussion, and a consensus process to generate a list of warning signs of poor nutritional health in older adults. See Table 18.5 for a condensed version of the acronym they promote.[50,51]

These warning signs were integrated into a screening tool, the NSI DETERMINE checklist (Illustration 18.1), and pilot tested for use in community settings. Ten risk factors remained after testing.[52] The revised tool is used by community agencies, edu-

Table 18.5 DETERMINE: Warning signs of poor nutritional health.

Disease. Any disease, illness, or chronic condition (i.e., confusion, feeling sad or depressed, acute infections) that causes changes in the way you eat, or makes it hard for you to eat, puts your nutritional health at risk.

Eating poorly. Eating too little, too much, or the same foods day after day, or not eating fruits, vegetables, and milk products daily will cause poor nutritional health.

Tooth loss/mouth pain. It is hard to eat well with missing, loose, or rotten teeth, or dentures that do not fit well or cause mouth sores.

Economic hardship. Having less (or choosing to spend less) than $32.80 (female) to $36.60 (male) weekly for groceries makes it hard to get the foods needed to stay healthy (www.usda.gov/cnpp).

Reduced social contact. Being with people has a positive effect on morale, well-being, and eating.

Multiple medicines. The more medicines you take, the greater the chance for side effects such as change in taste, increased or decreased appetite and thirst, constipation, weakness, drowsiness, diarrhea, nausea and others. Vitamins or minerals taken in large doses can act like drugs and can cause harm.

Involuntary weight loss or gain. Losing or gaining a lot of weight when you are not trying to do so is a warning sign to discuss with your health care provider.

Needs assistance in self-care. Older people who have trouble walking, shopping, and buying and cooking food are at risk for malnutrition.

Elder years above 80. As age increases, risk of frailty and health problems also rises.

Warning signs adapted from: Nutrition Screening Initiative, a project of the American Academy of Family Physicians, the American Dietetic Association, and the National Council on the Aging, Inc., and funded by a grant from Ross Products Division, Abbott Laboratories, Inc.; dollar amounts under economic hardship inserted by author using June 2001, USDA low-cost food plan.

cators, and service providers to screen program participants and the aged public for risk of malnutrition.

The nutritional risk factors identified during the NSI process[50] are reflected in the list of risk factors identified in the ADA position on nutrition and aging.[1] Presence of any of the following conditions places older adults at greater nutritional risk:

- Hunger
- Poverty
- Inadequate food and nutrient intake
- Functional disability
- Social isolation
- Living alone
- Urban and rural demographic areas

Illustration 18.1   **DETERMINE Your Nutritional Health checklist.**

The Warning Signs of poor nutritional health are often overlooked. Use this checklist to find out if you or someone you know is at nutritional risk.

DETERMINE YOUR NUTRITIONAL HEALTH

Read the statements below. Circle the number in the yes column for those that apply to you or someone you know. For each yes answer, score the number in the box. Total your nutritional score.

	Yes
I have an illness or condition that made me change the kind and/or amount of food I eat.	2
I eat fewer than 2 meals per day.	3
I eat few fruits or vegetables, or milk products.	2
I have 3 or more drinks of beer, liquor, or wine almost every day.	2
I have tooth or mouth problems that make it hard for me to eat.	2
I don't always have enough money to buy the food I need.	4
I eat alone most of the time.	1
I take 3 or more different prescribed or over-the-counter drugs a day.	1
Without wanting to, I have lost or gained 10 pounds in the last 6 months.	2
I am not always physically able to shop, cook, and/or feed myself.	2
TOTAL	

Total your nutrition score. It it's ...

0–2   **Good!** Recheck your nutritional score in 6 months.

3–5   **You are at moderate nutritional risk.**
See what you can do to improve your eating habits and lifestyle. Your office on aging, senior nutrition program, senior citizens center, or health department can help. Recheck your nutritional score in 3 months.

6 +   **You are at high nutritional risk.**
Bring this checklist the next time you see your doctor, dietitian, or other qualified health care or social service professional. Talk with them about any problems you may have. Ask for help to improve your nutritional health.

Remember: warning signs suggest risk, but do not represent diagnosis of any condition. See Table 18.5 to learn more about the warning signs of nutritional health.

These materials developed and distributed by the Nutrition Screening Initiative, a project of: AMERICAN ACADEMY OF FAMILY PHYSICIANS, THE AMERICAN DIETETIC ASSOCIATION, and NATIONAL COUNCIL ON AGING, INC.

Used by permission.

- Depression
- Dementia
- Dependency
- Poor dentition and oral health, chewing and swallowing problems
- Presence of diet-related acute or chronic diseases or conditions
- Polypharmacy (use of multiple medications)
- Minority status
- Advanced age

Why are factors such as poverty and minority status included in a list of nutritional risk factors?

Economic security contributes to food security, one of the health goals for the nation. Lack of food security leads to food intakes below 67% of the RDA, especially in older persons, who eat fewer calories, less magnesium, calcium, zinc, and vitamins E, C, and B_6 when food-insecure.[53] While older people on average are less likely to live in poverty than are children, they are a heterogeneous population and several groups are at high risk of poverty. Minority status is related to economic status. African Americans are most likely to be poor (22–29%), then Hispanic elders (20–26%), while non-Hispanic whites (6–12%) are least likely to be poor and thus malnourished. More women (12–29%) than men (6–22%) in all race and ethnicity categories are poor. Furthermore, minority populations are more likely to be in fair or poor health while non-Hispanic whites are most likely to be in excellent or good health.

Taken individually, the risk factors identified in the DETERMINE acronym and in the ADA position statement on aging are not unique to older adults. But each is more likely to lead to nutritional problems in a frail, vulnerable population. For instance, functional disability can affect dietary intake at any age, but very old people are more likely to live alone and to have fewer resources to compensate for any type of lost function. Consequently, diet quality may decline. Fewer 65 year olds live alone than 75 and 85 year olds. More women than men live alone. Race and ethnicity affect living situations; Hispanic men and women are least likely to live alone, while white women over age 85 are most likely to live alone; 65% of older white women live alone compared to 47% of African American and 35% of Hispanic women.

A common perception is that older adults living alone eat poorly. While living alone is a nutritional risk factor, it's not clear whether the issue is eating alone or living alone. On average, meals eaten with other people last longer and supply more calories than meals eaten alone.[54] The effect of living alone starts at a younger age for men and affects nutrition more extensively than for women.[55] Women aged 75 and older eat less protein and also less sodium than those living with others; men aged 65 and older eat less protein, beta-carotene, vitamin E, phosphorus, calcium, zinc, and fiber. Men and women living alone did not consume greater amounts of any of the 23 nutrients (alcohol was not included) examined than men and women living with someone.

George: *"There's so much advice out there, how do I know what's right for me?"*
Martha: *"I know, it's all so confusing. How ARE you supposed to know what to eat?"*

DIETARY RECOMMENDATIONS

Sometimes, in all the discussions about nutritional effects on health and disease status of older adults, we forget that old age is not a disease. However, recommendations for specific nutrients do change with age, so the food on one's plate should also change over time.

Food-based Guidance: The Pyramid

The Food Guide Pyramid is easily recognized as dietary advice. Pyramid adaptations for older adults have been developed by several groups, including the American Dietetic Association,[56] the Senior Nutrition Awareness Project (SNAP) of Connecticut and Rhode Island, and Tufts University researchers (see Illustration 18.2). The American Dietetic Association pyramid for persons age 50 and over depicts fluids, but does not include vitamin-mineral supplements. In the Tufts' "elderly" pyramid, authors portray the need for fewer calories with a narrower pyramid base.[57] They also emphasize adequate fluid intake by depicting 8 glasses of water at the pyramid base, but this depiction is confusing for some older audiences who interpret the water glasses to mean that water is the only fluid that counts (personal communications, Olga Monell, nutrition provider with the Senior Nutrition Program, October, 1999). Potentially higher need for supplemental calcium, vitamins D and B$_{12}$ is flagged at the top of the pyramid. Symbols depicting fat, sugar, and fiber in foods are intended to help older adults recognize sources of those food constituents and subsequently choose a more nutrient-dense diet.

Illustration 18.2   Tufts University modified food pyramid for 70+ adults (used by permission of Tufts University)

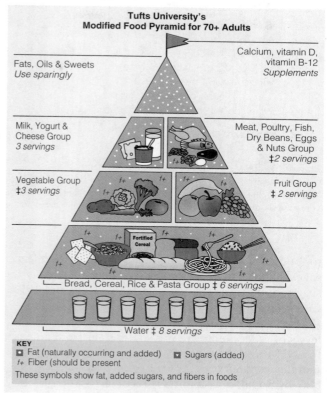

The pyramid shows food groups and corresponding serving sizes to be used for meal plan development. Table 18.6 lists food group suggestions for common calorie levels, which can be modified by choosing the more or less nutrient-dense foods from each group. For instance, citrus and berries are among the most nutrient-dense fruits. A baked potato has more vitamins and minerals per bite than french fries. Among grain-based products, whole grain breads and cereals provide fiber as well as the usual grain-nutrients. Consistently choosing nutrient-dense foods decreases caloric intake. Fortification and enrichment, while enhancing nutrient intake, complicate nutrient density calculations. For instance, unless whole grain products are also fortified with folic acid, they may actually be considered to be less nutrient-dense than their plain white, but enriched, counterparts. That is one reason variety remains a part of dietary advice. Regularly choosing a variety of foods from each group enhances overall diet quality.

Actual Food Group Intake

A picture of the eating habits of older Americans can be developed from records of consumption according

Table 18.6   Meeting food guide pyramid recommendations at selected calorie levels.

FOOD GROUPS/SERVING SIZE	NUMBER OF SERVINGS		
	1500–1700 cal	1800–2000 cal	2100–2300 cal
Water and other nonalcoholic fluid (1 cup)	6–8	7–8	8
Bread, cereal, rice, pasta (1 oz, 0.5 cup cooked)	6	7–8	8–9
Vegetables (1 cup leafy, 0.5 cup raw, cooked)	3	3	4
Fruit (6 oz juice, 0.5 cup canned, 1 med)	2	2	3
Milk or yogurt (1 cup), cheese (1.5–2 oz)	3	3	3
Meat, poultry, fish, eggs (ounces); dry beans, tofu (0.5 cup cooked); nuts, seeds (0.33 cup)	5	5	6
The "Other" Group			
Fats, oils, sweets (1 tsp oil, sugar, jam)	5	6	7
Alcohol: if using, moderation is 1 drink for women and 2 drinks for men per day			

Adapted from SNAP Senior Food Guide Pyramid, University of CT and University of RI, 1-800-595-0929, and USDA/HHS Food Guide Pyramid, (www.usda.gov or 202-720-2600, voice and TDD, for Braille, large-print or alternative means of communication). Used by permission.

to food group, by analyzing occasions for eating (e.g., snacks) and by nutrients consumed. The most recent survey of nutritional intake is USDA's 1994 and 1995 Continuing Survey of Food Intakes by Individuals (CSFII).[48] While this data may seem dated, the intake levels of older adults are not so different from an earlier NHANES (National Health and Nutrition Examination Survey),[58,59] suggesting that dietary intake by population cohorts is relatively stable. Table 18.7 compares proportions of an older and a younger population currently eating foods from each group. Greater numbers of older than younger adults eat fruit, meat, eggs, nuts and seeds, fats, and sugars daily, and fewer use carbonated and alcoholic beverages.

Eating Occasions

EATING OUT   More and more people are eating away from home, including older adults. While 62–72% of individuals in their twenties consume more than a third of their calories away from home, 39–50% of individuals in their 60s consume one-fifth (roughly 20%) of their calories away from home, and approximately 28% of people aged 70 and older consume one-eighth (13%) of their calories away from home. Food eaten away from home supplies relatively more protein and fat than carbohydrate. For example, the population aged 70 years and older obtains the following percentages of the day's calories with foods eaten away from home:[48]

Table 18.7
Daily consumption of named food groups by adults aged 70 and older, compared to younger adults.

FOOD GROUP	PERCENTAGE OF INDIVIDUALS CONSUMING IN ONE DAY			
	Aged 70+		Aged 20–29	
	Males	Females	Males	Females
Grain products	99.1	99.0	94.9	94.8
Vegetables	84.7	86.3	84.2	81.7
Fruits	70.6	71.5	42.5	45.7
Milk and milk products	86.0	82.7	67.5	74.5
Meat, poultry, fish	92.4	88.1	88.8	80.3
Eggs	27.4	18.6	17.0	17.4
Legumes (dried peas and beans)	14.6	12.0	11.0	13.3
Nuts and seeds	10.4	11.1	7.0	7.9
Fats and oils	66.9	66.0	46.2	48.5
Sugars and sweets	65.8	63.2	39.7	48.4
Beverages, alcoholic	14.9	4.6	29.5	13.8
Beverages, carbonated soft drinks	27.2	21.9	67.9	62.4

Based on CSFII data from Tables 10.1 to 10.7 in Wilson et al, 1997[48]

- 13–14% of protein
- 14–15% of fat
- 11–12% of carbohydrate

SNACKING   For older adults, snacking is less common than in other age groups. Snack foods do not appear to be nutrient-dense. Two-thirds of people aged 70 and older (65% of women, 67% of men) eat snacks that supply 12% of their food energy but only 6–10% of vitamins, minerals, and potassium consumed. The energy-nutrient contributions of snacks for this group are:

- 7% of protein consumed
- 10–11% of fat consumed
- 13–14% of carbohydrate consumed

Information about dietary intake of older adults (for example, Table 18.7 and 18.8) helps to answer questions such as: "How does Uncle Al's diet compare to that of other men his age?" and "Are there unusual patterns in Al's diet that might reflect some potential problem?" While change is good when it helps people to stay flexible, unexplained changes in diet and nutritional status need to be evaluated.

NUTRIENT RECOMMENDATIONS

Nutrient recommendations change as scientists learn more about the effect of foods on human function. Specific nutrient levels for population groups above age 51 were first established in 1997 as the Dietary Reference Intakes (DRIs).[60] The complete tables, including tolerable upper intake levels (UL) are found on the inside covers of this text. The following section will address first the macronutrient recommendations (carbohydrates, protein, and fat) and then the micronutrient recommendations relevant to older adults.

Energy Intake: A Measure of Macronutrients

The main goal for energy calculations is to maintain a healthy body weight. Decrease of physical activity and basal metabolic rate from early to late adulthood results in 20% fewer calories needed for weight maintenance.[61] Table 18.8 shows that on average, adults do eat fewer calories as they age. Women aged 60 and older eat fewer than 1500 calories and are especially vulnerable to malnutrition, despite the fact that approximately one in four is obese (BMI = 30 or more). Calorie intake and weight status is an area that illustrates the dangers of generalizing from populations to individuals or the reverse.

Table 18.8 Caloric intake comparison of younger and older adults, by gender, from CSFII (Continuing Survey of Food Intake by Individuals, 1994–1995 data collection).

AGE	ACTUAL DAILY CALORIE INTAKE		RECOMMENDED DAILY CALORIES	
	Males	Females	Males	Females
20–29	2844	1828	2900	2200
50–59	2259	1583	2300	1900
60–69	2100	1496	2300	1900
70 and older	1854	1377	2300	1900

Data from Wilson JW, Enns CW, Goldman KS, et al. Data tables: combined results from USDA's 1994 and 1995 Continuing Survey of Food Intakes by Individuals and 1994 and 1995 Diet and Health Knowledge Survey, 1997.[48]

Estimating individual calorie needs with the Harris-Benedict equation is described in Chapter 16. An easier-to-remember method that works for older adults is to multiply current weight in pounds by an activity factor to derive a daily calorie level:

10 calories per pound if sedentary or desiring weight loss
12–13 calories per pound for moderately active lifestyle
15 calories per pound for active lifestyle

It is difficult to meet vitamin and mineral needs at caloric levels below 1200–1500 but adding activity can maintain caloric needs. This is often easier to say than to do.

Nutrient Recommendations: Macro- and Micronutrients of Concern

Potentially problematic nutrients for older adults are presented in Table 18.9, listing both actual and recommended intake. Each listed nutrient plays a role in normal, healthy aging and is eaten in amounts different from the recommended level (except for dietary cholesterol and zinc). Reasons why each is of concern are presented next in the text.

Macronutrients

CARBOHYDRATE AND FIBER   Although most older adults do not eat the recommended 25 grams of fiber daily,[63] adequate carbohydrate intake of 50% to 60% of calories is generally not a problem. Of individuals aged 70 and over, males eat 51% of their calories from carbohydrate and women eat 54%.[48]

Following the Food Guide Pyramid recommendations can ensure adequate carbohydrate and fiber intake. To meet the recommended 50% to 60% of

Table 18.9   Selected nutrient intakes of older adults, Continuing Survey of Food Intake by Individuals, CSFII, 1994–1995, data collection, and relevant recommendations.

	ADULTS AGED 70 AND OLDER			
	Actual Intake		Recommended Intake[a]	
	Males	Females	Males	Females
Macronutrient				
Fiber, g	18	14	25 (DV)	25 (DV)
Protein, g	74	57	77	65
Total fat, g	69	48	65 (DV)	65 (DV)
Saturated fat, g	23	16	20 (DV)	20 (DV)
Cholesterol, mg	274	185	300 or less	300 or less
Micronutrients				
Sodium, mg	3122	2376	2400 (DV)	2400 (DV)
Vitamin A, mcg RE	1430	1149	900	700
Vitamin E, mg	8.9	6.5	15	15
Vitamin B$_{12}$, mcg	7.1	4.7	2.4	2.4
Folate, mcg	283	234	400	400
Calcium, mg	754	587	1200	1200
Magnesium, mg	286	224	420	320
Zinc, mg	11.8	8.4	11	8
Iron, mg	16.7	12.3	8	8

[a]Intake recommendations are based on Dietary Reference Intake (DRI), Daily Value (DV—the nutrient reference amount used on Nutrition Facts panel of food labels) or WW Campbell, 1996[62] (based on nitrogen balance studies for protein).

calories from carbohydrate sources, men would eat between 288 (50%) and 345 grams (60%) of carbohydrate per day and women 188 (50%) to 285 grams (60%). An example of foods containing at least 50% of calories from carbohydrate (translates to 188 grams for a female eating 1500 kcal daily) and 25 grams of fiber is presented in Table 18.10 (see Table 18.6 for serving size).

Depending on caloric intake, a range of 20–35 grams dietary fiber daily is appropriate for older adults. Males eat 18 grams and females eat 14 grams of fiber daily, 7 and 11 grams fewer than the recommended 25 grams of fiber, respectively.

Table 18.11 shows how common portions of foods eaten by older adults can supply adequate fiber. Reasons to eat adequate fiber include the strong associations between fiber intake and incidence of diverticular disease in men, non–insulin dependent diabetes in women, coronary heart disease in men, and hypertension in women and men.[64] The role of fiber in gastrointestinal problems is discussed in Chapter 19.

When increasing fiber intake, additional fluids are needed to process the fiber. Slowly adding both fiber and fluids allows the intestinal system to adapt to the additional bacterial substrate.

PROTEIN   While most adults in North America eat sufficient or even excess amounts of protein, older adults (especially those over 75 years and living alone) may be eating too little.[55] Inadequate protein intake contributes to muscle wasting (sarcopenia), a weakened immune status, and delayed wound healing. Protein guidelines for individuals aged 25 years and older are currently set at 63 grams per day for

Table 18.10   Using the pyramid groups to estimate adequate carbohydrate and fiber for daily intake.

BASIC FOOD GROUPS		
	Carbohydrate (CHO), grams	Approximate Total Fiber Content, grams
Number of Servings/ Grams of Carbohydrate per Serving		
6 servings of grain at 15 g each	90	12
2 servings of fruit at 15 g each	30	6
3 servings of milk at 12 g each	36	0
3 servings vegetable at 5 g each	15	9
Total from basic groups	**171**	**27**
OTHER CARBOHYDRATE-CONTAINING FOODS		
1 tbs sugar for coffee or tea	12	0
2 Fig Newton cookies	20	1
Total, including "Other" group	**203**	**28**

Mixed dishes like soups, sandwiches, and salads count as partial servings from their contributing food groups.

Table 18.11 **An example of fiber-containing foods that might comprise one day's intake.**

Food Item	Grams of Fiber
Oatmeal (½ cup) with wheat germ (¼ cup)	8
Banana	2
Peanut butter (2 tbs)/ whole wheat sandwich (2 sl)	6
Orange	3
Baked potato with skin	4
Green beans (½ cup)	3
Bran muffin (1 med)	2
Pear	4
Total dietary fiber	**32**

NOTE: Meat, poultry, fish, eggs, milk, sugar, and oils do not contain dietary fiber.

Table 18.12 **Protein sources and protein quality measures: Examples of protein scores used by INCAP, the Institute of Nutrition of Central America and Panama.**

Protein Source	Chemical Quality Score*	True Digestibility in Humans
Meat, eggs, milk,	100	95
Beans	80s	78
Soy protein isolate	97	94
Rice	73	88
Oats, oatmeal	63	86
Lentils	60	Not available
Corn	50	85
Wheat	44	Refined, 96; whole wheat, 86

*Amino acid score relative to amino acid score of egg, the reference protein.

Data from Torun et al, 1994 (INCAP)[68] and Shils et al, 1999,[69] based on "Energy and protein requirements: report of a Joint FAO/WHO/UNU Expert Consultation." Technical report series No. 724, Geneva: World Health Organization, 1985:119.

males and 50 grams for females.[65] Researchers are finding that older persons need higher levels of protein intake to maintain nitrogen balance.[28,62,65] Nitrogen balance studies used to determine current recommendations of 0.8 grams per kilogram (or 0.36 gram per pound) were done with young males who have proportionately more muscle mass than older adults, male or female, and are more efficient in maintaining nitrogen balance.

Nitrogen balance is also easier to achieve when the protein eaten is of high quality (such as meat, milk, eggs, see Table 18.12), is ingested with adequate calories, and when an individual is doing resistance exercises.[31] During resistance training, less nitrogen is excreted and thus nitrogen balance improves. Consuming a low-calorie diet, as many older adults do, leads to a greater need for protein in order to maintain nitrogen balance. Evans and Cyr-Campbell recalculated nitrogen-balance study data and did new laboratory tests with older adults.[28] They found that older adults on average maintain nitrogen balance eating 0.91 grams of high-quality protein per kilogram body weight per day (see Table 18.12 for quality scores of common proteins). They also calculated that 1.25 g protein/kg body weight/day would be a safe intake for 97.5% of the older adult population, recommending that older adults eat 1 to 1.25 g of protein/kg body weight daily. This translates to 0.45 to 0.57 g protein/lb body weight. This protein range echoes an earlier recommendation by Campbell:[62] "A safe protein intake for older people would be 1.0 to 1.25 gm protein per kg per day of high quality protein." (p. 196) Findings such as these do *not* support the suggestion that less lean body mass in an older person leads to smaller protein requirement for maintenance.

Leading protein sources for all adults, based on NHANES II surveys[67] are: (1) beef steak, roast, (2) hamburgers, (3) white bread, rolls, (4) whole milk, and (5) pork. The chemical protein scores of four of the five leading protein sources are 100 (see Table 18.12), but the digestibility scores of the five leading protein sources are all in the 90s (see the same table).

People eat mixed diets, so individual protein scores tell only part of the story. For example, using the human protein digestibility scale, the average American and Chinese mixed diets score 96, the rural Mexican diet scores 80, and an Indian rice and bean diet scores 78. However, for older adults who are following modified diets, who are cutting back on meat because they can't afford it or lack energy to prepare it, or who may eat too few calories, protein quality can make the difference between a good and a poor diet.

On average, older adults eat slightly more protein as percentage of total calories than adults in general (16.5% and 16.8% of calories from protein for males and females, compared to 16.0% and 16.1% percent in younger persons). But when the population is heterogeneous, average data are not enough. The following questions address protein adequacy for older individuals:

- Based on height and weight, how much protein will meet the individual's need?
- Are enough calories eaten so that protein does not have to be used for energy?
- If marginal amounts of protein are eaten, is the protein of high quality?
- Are there additional needs: wound healing, tissue repair, surgery, fracture, infection?

- Is the individual exercising? It is harder to achieve nitrogen balance while sedentary.

Until the National Academy of Sciences reviews the 1989 protein guidelines, it seems wise to use the range of 1–1.25 grams per kilogram daily recommended by Evans & Cyr-Campbell.[28] The average older woman weighs 143 pounds and the average older man weighs 170 pounds. Eating 1 gram of protein per kilogram of body weight would mean eating 65 grams/day for women and 77 grams/day for men, significantly more than the 50 grams recommended as the basis for the Daily Value food-product nutrition labels.[70] Fifty grams of protein would be adequate for an older adult weighing 50 kilograms or 110 pounds.

FATS AND CHOLESTEROL   The need for fat does not seem to change with age; high saturated fat intake continues to be a risk factor for chronic disease. Keeping the amount of saturated fat in the diet below 10% of total calories and total fat at 30 to 35% of calories is a reasonable goal for older adults. Cholesterol intake for older adults is already below recommendations for the nation (see Table 18.9). Eggs, which have high cholesterol content, are a nutrient-dense, convenient and safe food for most people (those without lipid disorders, e.g., high triglyceride and high serum cholesterol). McNamara, a professor of nutrition sciences and director of the Lipid Metabolism Laboratory at the University of Arizona, calculated that eating seven eggs per week would raise total serum cholesterol by 1.3 mg/dL.[71] Roughly two-thirds of the population is not sensitive to changes in dietary cholesterol[72] and can enjoy high-in-cholesterol foods such as shrimp, eggs, and liver for their high-quality protein and other nutrients. For individuals who are sensitive to dietary cholesterol, the public health recommendation is "no more than four egg yolks per week."

Recommendations for Fluid

Water as percentage of total body weight decreases with age, resulting in a smaller water reservoir, leaving a smaller safety margin for staying hydrated. Drinking six or more glasses of fluid per day prevents dehydration (and subsequent confusion, weakness, and altered drug metabolism) in individuals whose thirst mechanism may no longer be very sensitive.

To individualize fluid recommendations, provide 1 milliliter (ml) of fluid per calorie eaten, with a minimum of 1500 ml. For a 2000 calorie diet, that would be 2000 ml or 2 liters of fluid, roughly eight cups. Foods like stews, puddings, fruits, and vegetables contribute significant amounts of fluid to the diet but are not counted as part of the fluid allowance in healthy individuals. The Tufts (see Illustration 18.2) and Connecticut adaptations of the Food Pyramid for healthy older adults show eight glasses of water, which is adequate for a 2000-calorie diet. Individuals who need additional calories can use milk, juice, shakes, and soups as nutrient-dense fluids.

Micronutrients: Vitamins and Minerals

The nutritional health of older adults depends on modifying dietary habits that address age-associated changes in absorption and metabolism. The nutrients discussed in the following sections are of special concern for older adults because of age-associated metabolic changes or low dietary intake.

Age-Associated Changes in Metabolism: Nutrients of Concern

VITAMIN D, CALCIFEROL   Age-related metabolic changes affect vitamins D status, independent of dietary intake, primarily due to decreased ability of the skin to synthesize previtamin D_3 from its precursor, 7-dehydrocholesterol.[73–75] Declining photochemical production may be compounded by limited exposure to sunlight due to use of sunscreen or pollution, institutionalization, or being homebound. Furthermore, in northern regions (above 42° north, the latitude of Boston and Chicago), UV light is not powerful enough to synthesize vitamin D in exposed skin between November and February.[76] The sun's rays are even weaker in Edmonton, Canada (at 52° N latitude), so that previtamin D_3 is not synthesized between mid-October and mid-April. Older individuals who live in these northern latitudes are at greater risk for vitamin D deficiency because they need more vitamin D than younger ones. How far south does one go for the winter sun to convert vitamin D precursors? Tests in Los Angeles, at 34° N, showed vitamin D production in skin even in January. The sun's rays in Puerto Rico (latitude of 18° N) were even more effective.[76] Fortunately for people living in northern regions, the body stores vitamin D, so that summer sun builds a winter reserve.

Another reason that vitamin D is of concern for older adults (who use more drugs than younger persons) is that commonly used medications interfere with vitamin D metabolism. Examples include barbiturates, cholestyramine, phenytoin (Dilantin), and laxatives.

Intake recommendations for vitamin D intake are 10 mcg (400 International Units) for persons aged 51 to 70 and 15 mcg (600 IU) for individuals older than 70 years. Potentially toxic doses are those above 50 mcg (2000 IU), the tolerable upper intake level (UL) set by the National Academy of Sciences. Symptoms of toxicity are hypercalcemia (high blood calcium levels), anorexia, nausea, vomiting, general disorientation, muscular weakness, joint pains, bone demineralization, and calcification (calcium deposits) of soft tissues. Toxicity is rare from food sources. The relatively few good food sources of vitamin D are fortified cereal, milk (but not cheese), eggs, liver, salmon, tuna, catfish, and herring. Mushrooms also contain a small amount of vitamin D (USDA Provisional Table, Vitamin D, http://www.nal.usda.gov/fnic/foodcomp/). Cod liver and other fish oils contain medicinal levels of Vitamin D (about 21 mcg, or 840 IU per teaspoon). Although vitamin D intake has not been tracked in national food monitoring surveys (because technically, humans can derive vitamin D from nonfood sources, i.e., UV light), researchers have assessed individual intake. For instance, Dawson-Hughes et al.[77] found that individuals 65 years and older consumed 4.5 to 5.0 mcg (180-200 IU) on average.

For healthy aging, the critical function of vitamin D is maintenance of blood calcium levels through intestinal absorption or through bone resorption when the diet lacks sufficient calcium.[77] Osteoporosis and osteomalacia (adult rickets) prevention is covered in Chapter 19.

VITAMIN B$_{12}$ While population intake of vitamin B$_{12}$ or serum cobalamine is higher than the DRI level of 2.4 mcg (see Table 18.9, intake is 7.1 mcg for males and 4.7 mcg for females), vitamin B$_{12}$ status suggests that many older adults are unable to use B$_{12}$ efficiently. Vitamin B$_{12}$ blood levels decrease with age even in healthy adults. Compared to 18% of healthy 22–63 year olds in the Framingham Heart Study, 40% of 65–99 year olds had low serum B$_{12}$ levels. Despite adequate food intake, an estimated 30% of older adults suffer from atrophic gastritis. In atrophic gastritis, a bacterial overgrowth in the stomach leads to inflammation, and decreased secretion of hydrochloric acid and pepsin with or without decreased secretion of intrinsic factor,[78] and subsequent inability to split Vitamin B$_{12}$ from its food protein carrier. Since some of the protein-bound vitamin B$_{12}$ passing into the small intestine is split off by bacteria residing there, an individual with atrophic gastritis can still absorb a fraction of protein-bound Vitamin B$_{12}$ It takes years to develop a B$_{12}$ deficiency, but once developed, the neurological symptoms are irreversible. Symptoms include deterioration of mental function, change in personality, and loss of physical coordination.

"Food first" is usually sound advice regarding nutritional needs, but B$_{12}$ is one of the two vitamins better absorbed in synthetic or purified form. Folic acid is the other one. Synthetic, not protein-bound vitamin B$_{12}$ is found in fortified foods such as cereals and soy products. Protein-bound B12 is found in all animal products, although poultry is a surprisingly poor source (2.4% of B$_{12}$ in adult diets is derived from poultry, compared to 27.3% from beef). The leading food sources of vitamin B$_{12}$ among U.S. adults are beef, milk, fish (excluding canned tuna), and shellfish.[67]

VITAMIN A Older adults are more likely to overdose with vitamin A than to be deficient in the nutrient. They eat more than the DRI of 900 mcg for males and 700 mcg RE for females (see Table 18.9). Plasma levels and liver stores of vitamin A increase with age. This may be due to increased absorption but is more likely due to decreased clearance of Vitamin A metabolites (retinyl esters) from the blood.[79] Kidney disease further elevates serum vitamin A levels because retinol-binding protein, another vitamin A metabolite, can no longer be cleared from blood. Thus older adults are more vulnerable to vitamin A toxicity and possible liver damage than younger individuals. The Tolerable Upper Level of Intake (UL) for vitamin A is 3000 mcg (3 mg).[80] Older adults have a smaller margin of safety for vitamin A than younger adults.

The plant precursor of vitamin A, beta-carotene, will not damage the liver. Excess beta-carotene, since it is water-soluble, may give old skin a yellow-orange tint, but it will not lead to hair loss, dry skin, nausea, irritability, blurred vision, or weakness as vitamin A would. Leading food sources of vitamin A and beta-carotene among U.S. adults are carrots, ready-to-eat cereals, and milk.[67] Rich vitamin A sources are liver, milk, cheese, eggs, and fortified cereals.

IRON The need for iron decreases with aging for women after menopause. Like vitamin A, iron is stored more readily in the old than in the young; high intakes of vitamin C enhance absorption. On average, older adults eat more iron than the DRI of 8 mg (males and females, see Table 18.9). Excess iron contributes to oxidative stress, increasing the need for antioxidants to deal with oxidant overload. The UL for iron is 45 mg per day.[80]

Older adults are a heterogeneous population, and not every older person has adequate iron stores. Reasons for inadequate iron status include blood loss from disease or medication (e.g., aspirin), poor absorption due to antacid interference or decreased stomach acid secretion, and low caloric intake. Leading sources of iron in the American diet are ready-to-eat cereals, yeast bread, and beef.[67]

Low Dietary Intake: Nutrients of Concern

VITAMIN E　Also known as tocopherol, vitamin E is a potent antioxidant. It is a problematic nutrient because dietary intake is well below the recommended 15 mg or 15 IU alpha-tocopherol equivalents (TE) (see Table 18.9). Vitamin E plays a special role for the health of older adults due to its antioxidant functions, such as hindering development of cataracts[81] and heart disease.[82] Vitamin E is associated with enhanced immune function[83,84] and cognitive status.[85] The UL is 1000 mg (or IU, seen on supplement labels) alpha TE. Even though vitamin E is fat-soluble and can be stored in the body, it seems safe in doses up to 800 IU. At higher doses, vitamin E is linked to longer blood-clotting times and increased bleeding. Aspirin, anti-coagulants, and fish oil supplements also increase blood-clotting time and are incompatible with high doses of vitamin E intake. Vitamin E is thought to be more effective in its "d" (rather than "dl") form.

Leading food sources of vitamin E are salad dressings/mayonnaise, margarine, and ready-to-eat cereal. Other good sources of vitamin E are oils (especially sunflower and safflower oils), fats, whole grains, wheat germ, leafy green vegetables, tomatoes, nuts, seeds, and eggs. Food sources cannot supply the optimal dose of 200 IU vitamin E found most effective in modulating heart disease and immune status.

FOLATE, FOLIC ACID　For persons with low serum folate levels, dietary increases of folic acid (100 to 400 mcg) can lower serum homocysteine levels and subsequent risk of heart disease.[86] The DRI for males and females is 400 mcg folate (the food source) per day, with a UL of 1000 mcg of folic acid (the synthetic form). On average, older adults do not meet recommended levels (see Table 18.9). Leading food sources for U.S. adults are ready-to-eat cereals, yeast bread, and orange/grapefruit juice.[67] The folate fortification of grain products begun in the late 1990s is expected to increase older adults' average daily intake by 70 to 120 mcg; grain products are likely to remain the leading dietary sources of folic acid.

Despite intakes below recommended levels (pre-fortification levels), folate deficiency as measured by serum levels is uncommon in healthy older adults. Depletion is measured as serum folate <3 ng/ml and red blood cell folate <160 ng/ml[87] and is estimated to exist in 11% to 28% of all elderly persons.[88] Absorption of folate, like vitamin B_{12}, may be impaired by atrophic gastritis. Moreover, alcoholism is associated with folate deficiency and subsequent pernicious anemia. Folic acid deficiency can also be secondary to vitamin B_{12} deficiency, which is more common in the elderly than folic acid deficiency. Folic acid deficiency should be treated only with folic acid, or treatment may mask potential B_{12} deficiency, which will continue to cause irreversible neurological damage. Medications commonly used by older adults such as antacids, diuretics, phenytoin (Dilantin), sulfonamides, and anti-inflammatory drugs affect folate metabolism.

CALCIUM　A 1994 Consensus Conference sponsored by the National Institutes of Health[89] recommended calcium intake for women depending on estrogen status, namely 1500 mg for women aged 50 to 64 not taking supplementary estrogen and 1000 mg calcium per day for post-menopausal women taking estrogen. After age 65, it was recommended that all men and women consume 1500 mg calcium per day. This recommendation of 1500 mg was lowered in 1997, when the National Academy of Sciences set the DRI for adults aged 51 and older at 1200 mg daily, independent of gender or hormone status. Recommended levels are met by a small portion of the population; on average, calcium intake of older men and women ranges between 600 and 800 mg per day, with the lowest intake among older black women.[48,58]

Low calcium intake has been linked to colon cancer and hypertension. Study results are inconclusive, so it is too early to recommend a higher calcium intake for the purpose of decreasing colon cancer risk. Study results are stronger for the protective effects of calcium intake in the development of hypertension.[90] For example, blood pressure decreases in a subgroup of hypertensive individuals with higher calcium intake. On average, participants who adhered to the DASH (Dietary Approaches to Stop Hypertension) diet reduced their blood pressure. The diet includes 2 or more servings of low-fat dairy products (1265 mg calcium in the experimental group), 10 servings of fruits and vegetables, and reduced fat, saturated fat, and cholesterol.

The National Academy of Sciences has set the Tolerable Upper Limit for calcium at 2500 mg per day. Individuals should not consume supplements in excess of this amount. Adverse effects of excess calcium (reported at 4 grams per day) include high blood calcium levels, kidney damage, and calcium deposits in soft tissues and outside the bone matrix, such as bone spurs on the spine.[89] High calcium intake, resulting in elevated urinary excretion of calcium, could also lead to new kidney stones in individuals with a history of kidney stones.

MAGNESIUM　Of adults aged 70 and older, only 22 to 23% (male and female, respectively) meet 100% or more of their magnesium DRI. Adequate magnesium intake (see Table 18.9) is needed for bone and tooth formation, nerve activity, glucose utilization, and synthesis of fat and proteins. Magnesium deficiency can result not only from low intake, but also from malabsorption due to gastrointestinal disorders, chronic alcoholism, and diabetes. Signs of deficiency include personality changes (irritability, aggressiveness), vertigo, muscle spasms, weakness, and seizures. An indicator of the wide-ranging functions of magnesium is that it plays a part in over 300 enzyme systems.[91]

Age does not seem to affect magnesium metabolism. However, drugs commonly used by older adults such as magnesium-based antacids and cathartics may lead to magnesium overdose. The DRI is 420 mg for males and 320 mg daily for females, with a UL of 350 mg from nonfood (supplementary) sources. Signs of magnesium toxicity are diarrhea, dehydration, and impaired nerve activity. Magnesium from food sources does not result in toxicity. Leading food sources for adults are milk, yeast bread, coffee, ready-to-eat cereal, beef, and potatoes.[67]

ZINC   The role of zinc is changing from problematic nutrient because of insufficient intake to potentially problematic because of too much zinc in the diet. Few diets supplied adequate zinc before the requirement was adjusted downward to 8 mg daily for females and 11 mg for males by the 2001 DRI. Of adults aged 70 and older, only 13% of females and 16% of males ingested 100% of the 1989 guidelines for zinc (12–15 mg for females and males) (see Table 18.9 for average intakes). The UL is 40 mg; excess dietary zinc results in decreased immune response, decreases in HDL cholesterol, and copper deficiency. Leading food sources of zinc among U.S. adults are beef, ready-to-eat cereal, milk, and poultry.[67]

Healthy older individuals maintain zinc balance as well as younger ones do even though they absorb less zinc.[92] Zinc absorption is affected by conditions of the intestinal tract such as infections, surgery, muscle-wasting diseases, pancreatic insufficiency, and alcoholism, which all increase the need for zinc. Medications including diuretics, antacids, laxatives, and iron supplements also increase the need for zinc.

Of particular interest to older adults are the relationships between zinc deficiency and delayed wound healing, decreased taste acuity and immune response, and increased susceptibility to dermatitis. Zinc supplementation seems to be beneficial only when zinc status is depleted; supplementation in the absence of zinc deficiency does not improve wound healing, taste acuity, or dermatitis. Studies dealing with improved immune status following zinc supplementation are inconsistent.[92]

Micronutrient Supplements: Why, When, What?

WHY AND WHEN TO CONSIDER SUPPLEMENTS

Can older adults benefit from taking dietary supplements? Yes! Population surveys show that diets of many older persons fall short of meeting recommended nutrient levels (see Table 18.9). A boost from appropriately used supplements completes the nutritional balance of an inadequate regular diet. Dietary supplements can also decrease infection-related illness.[93,94]

Any discussion about supplements is based on the assumption that whole foods are the ideal source of nutrients and supplements are meant to boost a marginal diet. Sometimes it turns out that, as is the case with beta-carotene, the supplemental form of a nutrient does not promote health as the food form does. Furthermore, the interactions among nutrients and the composition of plants and animals making up our food supply are much too complex to replicate in supplements. However, vitamin B_{12} and folic acid are two nutrients better absorbed in a synthetic than in their protein-bound food form, but this only becomes important when normal metabolic processes fail. It is possible for healthy older adults to live well without dietary supplements.

> **USP**   United States Pharmacopeia, a nongovernmental, nonprofit organization (since 1820), establishes and maintains standards of identity, strength, quality, purity, processing, and labeling for health care products.
>
> **NF**   A uniformity standard for herbs and botanicals.

Age-associated nutritional risk factors such as poverty, isolation, chronic illness, or lack of appetite highlight circumstances when dietary intake might be inadequate and further assessment and intervention, potentially including a supplement, may be useful. Such factors include:

- Lack of appetite resulting from illness, loss of taste or smell, or depression
- Diseases or bacterial overgrowths in the gastrointestinal tract that prevent absorption
- Poor diet due to food insecurity, loss of function, dieting, or disinterest in food
- Avoidance of specific food groups such as meats, milk, or vegetables
- Contact with substances that affect absorption or metabolism: smoke, alcohol, drugs

WHAT TO TAKE

The question: "What should I take?" may not have a simple answer. Considerations guiding supplement choices are the same for older as for younger adults, and address the following five questions:

1. Does the supplement contain a balance of vitamins and minerals?
2. When all supplements and fortified foods are combined, is the dose still safe?
3. Does the supplement contain the missing nutrients?
4. Does the supplement carry a *USP* or *NF* code to assure potency and purity?

5. Is the supplement safe? ("Natural" does not mean safe; follow "buyer-beware.")

Because older adults eat fewer calories and are less resilient physically, they also have less tolerance for mistakes in dietary supplement use.

In general, multivitamin/mineral supplements should be used in physiologic rather than high-potency doses to maintain a balance of vitamins and minerals. Physiologic dose formulas unique for older adults are available with little or no iron and additional vitamin B_{12} and E. In addition, generously fortified breakfast cereals, "power" bars, and drinks count as vitamin or mineral supplements.

The supplement market is large. It can be confusing. In the United States in 1999, it was $44.5 billion, the largest supplement market in the world.[95] Most of the nearly 2500 products reported in the NHANES III survey were vitamin-mineral combinations or vitamin–single nutrient combinations.[96] The most common supplement ingredients were vitamin C, vitamin B_{12}, B_6, niacin, thiamin, riboflavin, vitamin E, beta-carotene, cholecalciferol or vitamin D, and folic acid. As a group, older adults need the vitamins B_{12}, E, D, and folic acid. The other five are superfluous and possibly unsafe (e.g., beta-carotene supplements). Balluz et al.[96] report that adults use more than 300 nonvitamin and nonmineral products, including some that have toxic effects at normal doses (e.g., blue cohosh, chaparral).

WHO TAKES SUPPLEMENTS? Current supplement users may not be the individuals who need them most. In addition, supplement users may not be choosing supplements that could help. Of the 40% to 47% of the population who use supplements,[48,96] the most likely to take them are non-Hispanic white females, as are individuals with more education and higher incomes. Overall, women are more likely than men to use supplements.[96]

WHY TAKE SUPPLEMENTS? Older people are motivated by wellness and want to take responsibility for their own health, wishing to increase their energy, improve memory, and reduce susceptibility to illness and stress.[96–98] Wisely chosen supplements contribute to good nutrition, which in turn increases energy and reduces susceptibility to disease. Table 18.13 is a summary of some of the vitamins, minerals, and other dietary supplements of special interest to older adults.

Dietary Supplements and Functional Foods

The growing area of functional foods can be a boon for older adults who can benefit from the convenience of nutrient-dense foods such as calcium-fortified orange juice, yogurt-juice beverages, fortified high-

Table 18.13 Dietary supplements and conditions for potential use by older adults.

Condition or Health Status	Dietary Supplement/Herbals
Poor appetite or dieting, leading to intake below 1200–1600 calories	Multivitamin/mineral
If underweight is a problem	Add high-calorie/protein foods/fats as oils
Vegetarian or vegan	Vitamins B_{12}, D, calcium, zinc, iron
Arthritis	Ginseng, licorice, parsley Antioxidants potentially useful (vitamins C, E)
Cataracts	Vitamin C
Constipation	Fiber (cellulose, bran, psyllium) Fluid to accompany fiber
Diarrhea	Fluid, multivitamin/mineral
Energy boosters	Evaluate total nutrient intake, adequate calories, iron if blood levels are low
Immune status enhancement	Multivitamin/mineral (DRI dosage) Vitamin E (100–400 IU daily)
Memory aids	Antioxidants, especially vitamin E Gingko biloba
Osteoporosis	Vitamins D, K, calcium, fluoride, magnesium (avoid excess K if on blood thinners)
Sleep aids	Milk and a sweet at bedtime, tryptophan
Stress reduction	Eat well and play in the sunshine, (not available in pill or tonic)

fiber cereals, yogurt with live cultures, soy milks, breakfast powders, and various fortified chews, bars, and drinks. However, regular users of functional foods, no matter what age, need to add up all intake in order to avoid potentially toxic nutrient levels (see Table 18.14 for recommended and potentially toxic intake levels). Advice that "you can't overdose on nutrients from foods" does not apply to fortified foods. High nutrient doses can act like drugs, and should be treated as such. Just as drugs, nutrient supplements interact with each other (and with medications); side effects may outweigh the desired benefits.

Eat leeks in March and wild garlic in May,
And all the year after, physicians may play.

Welsh rhyme

"Non-nutrient" Intakes: Special Interest for Older Adults

Several other "ingestibles" besides vitamins and minerals affect older adults' health status. Examples include quasi-functional foods, herbs, stimulants, and other non-vitamin or mineral food components used by older adults to promote health and ward off chronic disease:

1. Herbs and spices: rosemary, sage, and thyme, cinnamon, clove, and ginger enhance the taste of food and protect against oxidative stress.
2. Caffeine: a stimulant to which older people may become increasingly sensitive.
3. Black and green teas (Camellia sinensis): contribute fluid *and* phytochemicals, especially the antioxidant catechins and flavonols.
4. Garlic: contains allicin, which has anti-bacterial activity, and ajoene, which enhances blood thinning. Eating garlic has been linked to colon cancer protection and better heart health. Individuals taking aspirin or blood thinners shouldn't take garlic pills or eat large amounts of garlic (one or two garlic cloves are considered a small amount).
5. Herbal treatments: widely used and often effective, but herbs used as medicinal treatment can also be dangerous. Herbs should help, not hinder health! Rule #1 in using herbs is also the first rule of medicine: "Do no harm." A list of unsafe herbs is found in Chapter 16.
6. Phytochemicals: these plant-based compounds (such as reservatrol, flavonoids, carotenoids, indoles, isoflavones, lignans, and salicylates) are of special interest in aging because they have been linked to reduction of chronic disease. Eating fruits, vegetables, and whole grains automatically increases phytochemical intake.
7. Pre- and probiotics: prebiotics are nondigestible food ingredients that feed health-promoting colon bacteria, and probiotics are live health-promoting bacterial cultures such as lactobacillus acidophilus and bifidobacterium in yogurt. After a course of antibiotic treatment, biotics can re-establish intestinal bacterial life.
8. Red yeast rice: a food product long used in China; red yeast rice is sold as a food supplement to combat heart disease because it lowers mildly elevated cholesterol.
9. Hormones: DHEA (dehydroepiandrosterone) is taken to increase muscle mass and immune function, pregnenolone to enhance memory, and melatonin to enhance sleep. While there is no evidence that pregnenolone enhances human memory or improves concentration, there is evidence that the other two can work. Melatonin is particularly popular. Secretion of this bio-rhythm regulator normally decreases with age. Studies are now examining whether melatonin supplements help people with Alzheimer's syndrome sleep better.

Nutrient Recommendations: Using the Food Label

The Nutrition Facts Panel on food packages is structured to provide nutrient content information in relation to nutrient needs. The Food and Nutrition Board of the National Academies of Sciences, National Research Council worked with scientists from Canada to establish an integrated set of reference values for specific nutrients.

The recommended nutrient amounts for older adults (ages 70 years and older) are slightly different from the amounts used as references for the nutrition label and that are recommended for "all" adults (see Table 18.14). For example, older adults need more vitamin C and calcium than the reference amounts listed in the mandatory components of the label, and they need less iron and zinc than the label value. Values also differ for vitamins D, E, and B_{12}. Table 18.14 lists the levels used for food labels and the current DRI values for older adults. The UL is listed because with today's food supply, it is easy to overdose vitamins and minerals.

Older adults can still use the percentages on food labels for dietary guidance, as long as they adjust them to get more than 100% calcium, vitamins D, and C and less than 100% for iron and zinc. The vitamin B_{12} label recommendation is higher than the DRI, but poor absorption in older adults makes unsafe intake from foods unlikely. In nutrition labeling and dietary guidance, "one size does not fit all."

Table 18.14   Vitamin and mineral recommendations for nutrition labeling (Daily Reference Values or DV) compared with Dietary Reference Intakes (DRI) and Tolerable Upper Intake Levels (UL) for older adults.

MANDATORY VITAMIN AND MINERAL COMPONENTS OF THE NUTRITION LABEL

Nutrient	DV (1993, 1995) All Adults	DRI (2001) Adults over 70 Males	DRI (2001) Adults over 70 Females	UL
Vitamin A; 5 IU (= 1 RE = 1 RAE, Retinol Activity Equivalent)	5000 IU	900 RE	700 RE	3000 RE
Vitamin C, mg	60	90	75	2000
Calcium, mg	1000	1200	1200	2500
Iron, mg	18	8	8	45

VOLUNTARY VITAMIN AND MINERAL COMPONENTS OF THE NUTRITION LABEL

Nutrient	DV (1993, 1995) All Adults	Males	Females	UL
Vitamin D, 1 mcg (= 40 IU)	400 IU	15mcg	15 mcg	50 mcg
Vitamin E, mg TE (= IU)	30 IU	15	15	1000
Vitamin K, mcg	80	120	90	—
Thiamin, mg	1.5	1.2	1.1	—
Riboflavin, mg	1.7	1.3	1.1	—
Niacin, mg	20	16	14	35
Vitamin B_6, mg	2.0	1.7	1.5	100
Folic acid, mcg	400	400	400	1000[a]
Vitamin B_{12}, mcg	6.0	2.4	2.4	—
Biotin, mcg	300	30	30	—
Pantothenic acid, mg	10	5.0	5.0	—
Phosphorus, mg	1000	700	700	3000
Iodine, mcg	150	150	150	1100
Magnesium, mg	400	420	320	350[b]
Zinc, mcg	15	11	8	40
Selenium, mg	70	55	55	400
Copper, mcg	2000	900	900	10,000
Manganese, mg	2.0	2.3	1.8	11
Chromium, mcg	120	30	20	—
Molybdenum, mcg	75	45	45	2000
Chloride, mg	3400	—	—	—
Potassium, mg	3500	—	—	—
Choline, mg	—	550	425	3500
Fluoride, mg	—	4.0	3.0	10

[a]synthetic
[b]nonfood sources

SOURCE: Pennington, 1999[70]; Trumbo et al, 2001[80]

Cross-cultural Considerations in Making Dietary Recommendations

Food habits develop in cultural contexts, and we can learn about them through various ways. Travel through North America would allow us to observe cultures that make up our society. Visiting ethnic restaurants, stores, and farmers' markets can be another way to get a glimpse of cultural food diversity. Cookbooks, films, talking with individuals about their food history, and participating in ethnic celebrations are other sources of insight into food patterns of various cultures. Each new immigrant wave adds unique food traditions to the country's mix. Older adults may be stronger advocates for upholding traditional food patterns than young people. In working with them on food issues, it is useful (and interesting) to determine whether food and lifestyle habits are patterned on specific cultural considerations.

National food monitoring programs survey the population with proportionately larger samples drawn from minority groups in order to develop a balanced picture of the whole population. North America is home to rapidly growing Hispanic, Asian, Russian, and African immigrant groups. The U.S. Census tracks minority groups, but it is only completed once every 10 years. The local Area Agencies on Aging and Senior Nutrition Programs also track population trends, and are likely to have greater insight about some of the smaller ethnic population groups in their unique communities and regions.

Cultural differences are reflected in approaches to dietary guidance. For example, Chile has separate guidelines for older and younger adults. In New Zealand, older adults are encouraged to socialize at mealtimes to improve appetite. In France, South Korea, and Japan, people of all ages are encouraged to enjoy mealtimes and take pleasure in eating. Other unique guidelines are those of China, which suggest people eat 20 to 25 grams (nearly an ounce) of fish daily. Guatemala uses a bean pot as a nutritional icon, and India has developed one set of guidelines for the rich and one for the poor (listserv notes on July 6, 2000, accessed July, 2000, Eatrightmn@skypoint.com, upcoming Handbook of Nutrition and Food Assessment, CRC Press).

Communicating effectively and avoiding misinterpretation in intercultural settings is probably the most important thing a nutritionist can learn to do when working with older adults from various cultures; it requires skill in transferring information, developing and maintaining relationships, and gaining compliance.[99] Developing individual skills takes time, commitment, and practice. On the other hand, tools are easier to find. Nutrition education and guidance tools have been developed in many languages and for diverse cultures, although not typically for elders of ethnic groups. Culturally appropriate resources can be found through local extension services, cross-cultural education centers, diabetes education programs, public health agencies, and some commodity groups such as the Dairy Council. Multilingual versions of the Food Guide Pyramid can be downloaded from http://monarch.gsu.edu/nutrition. Nearly 40 languages are available, including Amharic, Russian, Somali, and Vietnamese.

PHYSICAL ACTIVITY RECOMMENDATIONS

. . . There is no segment of the population that can benefit more from exercise than the elderly.

William J. Evans[100]

Exercise is a true fountain of youth. Physical activity builds lean body mass, helps to maintain balance and flexibility, contributes to aerobic capacity and to overall fitness,[28,101] and boosts immune status.[102] Only resistance exercises—for example, weight lifting—build muscle and bone. All physical activities count toward energy expenditure and health maintenance.

Table 18.15 **Healthy People 2010 physical activity goals and baseline activity levels comparing older and younger adults.**

TYPE OF ACTIVITY	BASELINE LEVELS		TARGET
	Age 75+	Age 18–24	All Adults
Engage in *no* leisure-time physical activity	65%	31%	20%
Engage in regular physical activity at least 20 min, 3 times/wk	23%	36%	30%
Perform strengthening activities 2 or more days/wk	8%	30%	30%
Perform stretching and flexibility activities	21%	39%	40%

SOURCE: Healthy People 2010[10]

National health goals encourage increased physical activity levels for older adults (see Table 18.15). On average, older adults are less active than younger ones. Lower activity levels as well as deteriorating strength, endurance, and sense of balance are associated with, but not caused by, increasing age.

Older people benefit from exercise even more than younger people do because strength training is the only way to maintain and build muscle mass. When evaluating strength in men and women, "muscle mass (not function) is the major determinant of age- and sex-related differences in strength," independent of where muscle is located.[28] Besides becoming stronger, increased muscle mass also leads to higher caloric turnover and less likelihood of weight gain.

Age does not hinder training effects. For example, in a comparison of changes in body composition between young and old individuals during 20 weeks of an endurance training program, all individuals gained total body density and fat-free mass while losing fat mass and abdominal fat. Responses to training differed by sex and by race, but *not* by age.[34]

Exercise Guidelines

How can one predict whether or not physical activity will exacerbate existing medical conditions? Physician screening or assessment by completing a questionnaire like the one in Table 18.16 (Keep Moving—Fitness after 50 screening tool, developed by Dr. Maria Fiatarone) can identify potential problem areas.

Table 18.16   Keep moving—Fitness after 50 chart.

A. Do I get chest pains while at rest and/or during exercise?

B. If the answer to question A is "yes": Is it true that I have not had a physician diagnose these pains yet?

C. Have I ever had a heart attack?

D. If the answer to question C is "yes": Was my heart attack within the last year?

E. Do I have high blood pressure?

F. If you do not know the answer to question E, answer this: Was my last blood pressure reading more than 150/100?

G. Am I short of breath after extremely mild exertion and sometimes even at rest or at night in bed?

H. Do I have any ulcerated wounds or cuts on my feet that do not seem to heal?

I. Have I lost 10 lb or more in the past 6 months without trying and to my surprise?

J. Do I get pain in my buttocks or the back of my legs—my thighs and calves—when I walk? (This question is an attempt to identify persons who suffer from intermittent claudication. Exercise training may be extremely painful; however, it may also provide relief from pain experienced when performing lower-intensity exercise).

K. When at rest, do I frequently experience fast irregular heartbeats or, at the other extreme, very slow beats? (Although a low heart rate can be a sign of an efficient and well-conditioned heart, a very low rate can also indicate a nearly complete heart block.)

L. Am I currently being treated for any heart or circulatory condition, such as vascular disease, stroke, angina, hypertension, congestive heart failure, poor circulation to the legs, valvular heart disease, blood clots, or pulmonary disease?

M. As an adult, have I ever had a fracture of the hip, spine, or wrist?

N. Did I have a fall more than twice in the past year (no matter what the reason)? (Many older persons may have balance problems and at the initiation of a walking program will have a high chance of falling. These persons may benefit from balance training and resistance exercise before beginning a walking program.)

O. Do I have diabetes?

Questionnaire developed by Maria Fiatarone, MD, to determine which persons should be examined by a physician before beginning an exercise program.[28] Reprinted by permission.

Many types of physical activity are good, and small increments add up. Muscle mass is built through resistance or weight-bearing activities such as walking, running, and jumping, which move the body's own weight against gravity. Playing tennis and jumping rope are examples of games or training activities that incorporate weight-bearing exercise. Lifting weights is another example. Water exercises can be done as resistance training when webbed gloves or empty plastic containers are used to push against the water, or they can be done as aerobic activity. The Resources at chapter's end list several books on exercise with older adults.

Aerobic exercise builds endurance and contributes to cardiac fitness, even when mobility is limited. Nonimpact aerobics and chair aerobic classes can help people with arthritis, or hip-replacements, or those who are wheel-chair bound to get and stay strong. Elaine Costa, R.D., M.P.H., has developed the following guide for planning effective exercise sessions:

- Decide on frequency: 2–3 times per week is effective for strength training, using 8–10 different exercises with 8–12 repetitions each, the whole routine to be done in 20–30 minutes[101]
- For general health: exercise for 30 minutes on most days of the week

- Drink water when exercising (before, during, and after exercise)
- Do warm-up and cool-down activities of 5 to 10 minutes each

Individuals can take charge of their own healthy aging by developing appropriate and effective exercise habits. Simple and—occasionally challenging—fitness advice: eat good food, drink lots of water, and play hard!

NUTRITION POLICY AND INTERVENTION FOR RISK REDUCTION

Nutrition policy promotes health by combining nutrition education for individuals and population interventions. The ultimate goal of nutrition intervention is to produce a better health outcome.

Nutrition Education

The human mind, once stretched to a new idea, never goes back to its original dimension.

Oliver Wendell Holmes

Contrary to some beliefs, older people do learn and change. Someone born in 1920 has seen the invention of fast food restaurants, microwaves, television and TV dinners, and a whole host of computer-controlled kitchen appliances. To age is to grow and adapt. Learning new nutrition habits is part of aging. Nutrition education is different from education in general because its goal is changed dietary behaviors. Nutrition education consists of a set of learning experiences to facilitate voluntary adoption of nutrition-related behaviors that are conducive to health and well-being. Several requirements must be met for nutrition education, that is, behavior change, to occur (see Illustration 18.3). Think of it as the 4 C's of nutrition education.[103]

1. *Commitment:* Commitment means being motivated to adopt health-promoting behaviors and intending to adopt and maintain a new food behavior.
2. *Cognitive processing:* Understanding how a food behavior contributes to health and planning how it will fit into one's life constitutes "cognitive processing."
3. *Capability:* Acquiring the skills to practice new food behaviors is part of nutrition education. An example is learning to identify whole grain breads or to prepare vegetables when intending to adopt a high-fiber diet.
4. *Confidence:* "Nothing breeds success like success!" The best predictor that someone will practice new dietary habits is their personal confidence of being able to do so.

Educational strategies can result in better diets for individuals, but these alone may not be sufficient to enhance the dietary patterns of populations. Cultural environments either support, ignore, or punish desired behavior change. An example is a peer group that values health and fitness; group members will support each other's health-promoting behaviors.

Illustration 18.3 For essential elements to achieve and maintain individual dietary behavior change.

SOURCE: Adapted from Krinke, 2001[103]

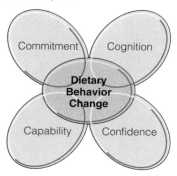

Food and nutrition policies arise from public values, beliefs, and opinions that define the cultural context in which dietary behaviors exist. Policies can be overt or unspoken. Public policies supporting the health of older adults are evident in the Social Security program, which provides financial support for post-retirement living, and the Food Stamp program, which makes grocery money available when adequate living resources are lacking.

Model Programs Exemplify Intervention Goals

NUTRITION PARTNERS Nutrition Partners is a model nutrition intervention program for free-living older adults in a large midwestern metropolitan community. Nadine Reiser, RD, project director of the Senior Dining Program, applied for and received a grant to pilot a home-visiting program to assist elders with their nutritional problems. Dietetic staff of the nutrition program make home visits in response to referrals from community agencies, physicians, and program site staff, doing dietary assessments, developing and prioritizing treatment plans together with the client, and making return visits periodically to monitor the intervention. Clients make changes appropriate to their specific lifestyle. For instance, to avoid dehydration and side effects from medications, one client decided to keep a large bottle of water in the refrigerator, refilling it each morning and drinking until it was empty at the end of the day. A daily mark on the calendar tracking refills helps to ensure that lack of thirst does not lead to dehydration.

STORE-TO-DOOR After "retiring" in the 1980s, Dr. Dave Berger surveyed his community to see where he might do some good. Many older adults told him that getting to the grocery store was impossible, and that even when they could get to the store, bringing the bags home was difficult. Winter ice and snow made things worse, for people feared falling. There had been a grocery delivery service, but it closed because profits were not meeting shareholder expectations. So Dr. Berger joined forces with his wife Fran and with friends, colleagues and a lot of volunteers to start up the non-profit Store-to-Door, a home-delivered grocery program for older people and those who are disabled. For a small fee, volunteers will shop for you. They'll buy items discount grocers offer: food of course, but also greeting cards, medicines, paper goods and cleaning supplies, although no alcohol or cigarettes. Customers can get credit for coupons. Volunteers deliver the groceries throughout the year.

After starting Store-to-Door in the upper Midwest, Dr. Berger started another in Portland, Oregon. When last heard from, he was in Ventura, California, running Shop Ahoy, his latest home-delivered grocery program!

COMMUNITY FOOD AND NUTRITION PROGRAMS

Elderly Nutrition Programs

Governmental programs for older adults include USDA's Food Stamp and Extension programs, Adult Day Services food programs, Nutrition Assistance Programs for Seniors (NAPS), meals-on-wheels and other home-delivered meal programs, and the Senior Nutrition Program of the Older Americans Act. In addition, nongovernmental home health programs provide food and nutrition services as part of a broader range of screening and assessment, nursing, and other support services. For instance, home health aides will shop for and prepare food, and clean up the kitchen afterwards. Home care services allow individuals to receive the necessary support to stay in their homes for as long as they wish. Remaining in one's home indefinitely is sometimes referred to as "aging in place." A broader definition of this concept is found in the Position of the American Dietetic Association.[1]

> Aging in place has many definitions and does not necessarily mean living in one setting or one home for a lifetime. Ideally, aging in place offers choices from a spectrum of living options and medical and supportive services customized to accommodate those who are fully active and have no impairments, those who require limited assistance, and those with more severe impairments who require care in long-term care facilities.

Other food and nutrition programs that contribute to the continuum of nutrition services include commodity foods, food pantries and soup kitchens, cooperative buying groups such as Fare For All, and screening and referral services.

Senior Nutrition Program

Congress first appropriated funds under Title VII of the Older Americans Act of 1965 to begin the Senior Nutrition Program, also called the Elderly Nutrition Program (ENP), created in 1972. The Senior Nutrition Program was created to alleviate poor nutritional intake and reduce social isolation among older adults. It was based on evidence that older adults do not eat adequately:

1. Lack of income limits ability to purchase food,

2. Lack of skills limits ability to select and prepare nourishing meals,
3. Limited mobility affects shopping and meal preparation, and
4. Feelings of isolation and loneliness decrease the incentive to eat well.

Senior Nutrition, now Title IIIC of the Older Americans Act, is a community-based nutrition program that provides meals (congregate and home-delivered), increased social contact, nutrition screening and education, information and linkages to other support programs and services, as well as volunteer opportunities. Anyone who is 60 or more years of age (and spouse regardless of age) is eligible to participate in the congregate dining program; home-delivered meal clients must be homebound and unable to prepare their own meals.

Today, there is less poverty among younger seniors, but greater nutritional and social needs among frail elderly and individuals with low incomes, chronic health conditions, limited mobility or disabilities, limited and non–English speaking ability, and among minority and isolated elders.

Senior dining sites are targeted to neighborhoods where older, frail, impoverished seniors live. Dining sites are located in community centers, senior centers, civic buildings, subsidized housing units, schools, and other accessible locations. Meals are delivered to the homes of individuals who are 60 years of age or older, homebound by reason of illness or disability, and unable to prepare meals. Services are adapted to meet each unique community's setting. For instance, meal vouchers for use at local cafes or diners are available in some small communities.

Other services to meet the needs of frail older seniors include multiple meals, weekend meals, take-home snacks, liquid supplements, nutrition screening and education, and one-to-one nutrition counseling. Also available are special diets for medical reasons and special meals for Jewish and ethnic elders. Grocery shopping assistance is an important service for frail seniors. A variety of models are used, including volunteer escorts to the supermarket, bus rides, and grocery delivery to the door. However, Title IIIC funds do not currently pay for grocery shopping assistance.

Senior or Elderly Nutrition Programs have successfully brought together millions of people to socialize and enjoy nutritious meals. In 1997, 123 million meals were served in congregate settings and 116 million meals were delivered to individuals at home. Currently, there are 15,000 dining sites nationwide.[104] Meals must meet one-third of the Recommended Dietary Allowances, the U.S. Dietary Guidelines for Americans, and provide adjustments for special ethnic dietary needs (Older Americans Act

Amendments of 2000, H.R. 782). Roughly 6% of the eligible population participates, most of whom are in their 70s and 80s. Surveys show that targeting those who need services most is successful. Increasing socialization was one of the original program goals, and continues to be one of the outcomes. Participants have 16–18% more social contacts per month than nonparticipants. A national evaluation of the senior nutrition program found that the program is targeted to those most in need.[105] Relative to the general older population, participants are older, more likely to be female, belong to an ethnic minority, live alone, and have incomes well below poverty. The nutrition program has a major impact on participants' nutritional well-being. Nutrition program meals were found to exceed the one-third RDA standard. Compared to nonparticipants, participants had up to 31% higher intake of recommended nutrients. In other words, the program is working.

The Promise of Prevention: Health Promotion

Grow old along with me, the best is yet to be!

Robert Browning

While good nutritional habits make a greater impact when started early in life,[25] sometimes individuals are not motivated to pursue these risk-reduction strategies until later in life or after experiencing a health problem. Successful strategies to reach an older audience reflect their specific needs and interests. It is time for the belief that an 80 year old is too old to learn and practice health promotion strategies to become a myth of the past.

Resources

American Association of Retired People
Information related to growing old in America: finances, politics, health issues, travel, and population statistics.
Available from: http://www.aarp.org/visit/brpromo.htm

Center for Disease Control and Prevention
Mortality and Morbidity data.
Available from: www.cdc.gov/mmwr

Florida National Policy and Resource Center on Nutrition and Aging
Aging Policy and education center
Available from: www.fiu.edu/~nutreldr/

McArdle, Katch, & Katch: *Sports, Exercise, and Nutrition,* Lippincott Williams & Wilkins, 1990.
Great information and advice on sports nutrition for adults.

National Heart, Lung and Blood Institute
Gateway site for consumers and health professionals; links and information materials about heart health and more.
Available from: www.nhlbi.nih.gov

Tufts University Center on Aging
Popular gateway to aging information.
Available from: www.navigator.tufts.edu

U.S Administration on Aging
Fact sheet, news, links to other resources.
Available from: www.aoa.dhhs.gov

References

1. Position of the American Dietetic Association: nutrition, aging, and the continuum of care. J Am Diet Assoc 2000; 100:580–95.

2. Maloney SK, Fallon B, Wittenberg CK. Executive summary. Aging and health promotion: market research for public education. Washington, DC: Office of Disease Prevention and Health Promotion, U.S. Department of Health and Human Services, PHS; 1984.

3. Nube M, Kok FJ, Vandenbroucke JP, et al. Scoring of prudent dietary habits and its relation to 25-year survival. J Am Diet Assoc 1987;87:171–5.

4. Kant AK, Schatzkin A, Graubard BI, et al. A prospective study of diet quality and mortality in women. JAMA 2000;283: 2109–15.

5. Vita AJ, Terry RB, Hubert HB, et al. Aging, health risks, and cumulative disability. N Engl J Med 1998;338:1035–41.

6. MA Johnson. Achieving 100 candles: the Georgia Centenarian Study lights the way. USDA 2000 Millenium Lecture Symposium. September 28, 2000.

7. McGinnis JM, Foege WH. Actual causes of death in the United States. JAMA 1993; 270:2207–12.

8. Health United States, 2000, www.cdc. gov/nchs, accessed January, 2000.

9. 1995 White House Conference on Aging. The road to aging policy for the 21st Century. Washington, DC: U.S. National Commission on Libraries; 1996.

10. Healthy People 2010: national health promotion and disease prevention objectives. Washington, DC: U.S. Dept. of Health and Human Services; 2000.

11. Hayflick L. How and why we age. Exp Gerontol 1998;33:639–53.

12. Tuljapurkar S, Li N, Boe C. A universal pattern of mortality decline in the G7 countries. Nature 2000;405:789–92.

13. Bernarducci MP, Owens NJ. Is there a fountain of youth? A review of current life extension strategies. Pharmacotherapy 1996;16:183–200.

14. Forbes GB. Human body composition. Growth, aging, nutrition, and activity. New York: Springer-Verlag; 1987: p 31.

15. Buys CH. Telomeres, telomerase, and cancer. N Engl J Med 2000;342:1282–3.

16. Bodnar AG, Ouellette M, Frolkis M, et al. Extension of life-span by introduction of

telomerase into normal human cells. Science 1998;279:349–52.

17. Parsons PA. The limit to human longevity: an approach through a stress theory of ageing. Mech Ageing Dev 1996;87:211–8.

18. Masoro EJ. Hormesis and the antiaging action of dietary restriction. Exp Gerontol 1998;33:61–6.

19. Weindruch R, Sohal RS. Seminars in medicine of the Beth Israel Deaconess Medical Center. Caloric intake and aging. N Engl J Med 1997;337:986–94.

20. McCay C, Crowell M, Maynard L. The effect of retarded growth upon the length of life and upon ultimate size. J Nutr 1935;10:63–79.

21. Masoro EJ. Nutrition and aging in animal models. In: Munro HN, Danford DE, eds. Nutrition, aging, and the elderly. New York: Plenum Press; 1989: pp 25–41.

22. Weindruch R. The retardation of aging by caloric restriction: studies in rodents and primates. Toxicol Pathol 1996;24:742–5.

23. Roth GS, Ingram DK, Lane MA. Calorie restriction in primates: will it work and how will we know? J Am Geriatr Soc 1993;47:896–903.

24. Holliday R. Food, reproduction and longevity. Bioessays 1989;10:125–7.

25. Weisburger JH. Approaches for chronic disease prevention based on current understanding of underlying mechanisms. Am J Clin Nutr 2000;71:1710S–4S; discussion 1715S–9S.

26. Walker AR, Walker BF. Nutritional and non-nutritional factors for 'healthy' longevity. J R Soc Health 1993;113:75–80.

27. Forbes GB. Longitudinal changes in adult fat-free mass: influence of body weight. Am J Clin Nutr 1999;70:1025–31.

28. Evans WJ, Cyr-Campbell D. Nutrition, exercise, and healthy aging. J Am Diet Assoc 1997;97:632–8.

29. Guo SS, Zeller C, Chumlea WC, et al. Aging, body composition, and lifestyle: the Fels Longitudinal Study. Am J Clin Nutr 1999;70:405–11.

30. Poehlman ET, Toth MJ, Bunyard LB, et al. Physiological predictors of increasing total and central adiposity in aging men and women. Arch Intern Med 1995;155:2443–8.

31. Carter WJ. Macronutrient requirements for elderly persons. In: Chernoff R, ed. Geriatric nutrition: the health professional's handbook. Gaithersburg, MD: Aspen; 1999.

32. Chernoff R, ed. Geriatric nutrition: the health professional's handbook. Gaithersburg, MD: Aspen; 1999.

33. Elahi VK, Elahi D, Andres R, et al. A longitudinal study of nutritional intake in men. J Gerontol 1983;38:162–80.

34. Wilmore JH, Despres JP, Stanforth PR, et al. Alterations in body weight and composition consequent to 20 weeks of endurance training: the HERITAGE Family Study. Am J Clin Nutr 1999;70:346–52.

35. Fiatarone MA, Marks EC, Ryan ND, et al. High-intensity strength training in nonagenarians. Effects on skeletal muscle. JAMA 1990;263:3029–34.

36. Kaneda H, Maeshima K, Goto N, et al. Decline in taste and odor discrimination abilities with age, and relationship between gustation and olfaction. Chem Senses 2000;25:331–7.

37. Schiffman SS. Taste and smell losses in normal aging and disease. JAMA 1997;278:1357–62.

38. McDonald RB. Influence of dietary sucrose on biological aging. Am J Clin Nutr 1995;62:284S–292S; discussion 292S–293S.

39. Cain WS, Stevens JC. Uniformity of olfactory loss in aging. Ann N Y Acad Sci 1989;561:29–38.

40. Ship JA, Pearson JD, Cruise LJ, et al. Longitudinal changes in smell identification. J Gerontol A Biol Sci Med Sci 1996;51:M86–91.

41. Wysocki CJ, Gilbert AN. National Geographic Smell Survey. Effects of age are heterogenous. Ann N Y Acad Sci 1989;561:12–28.

42. Bartoshuk LM. Taste. Robust across the age span? Ann N Y Acad Sci 1989;561:65–75.

43. Schiffman S. Changes in taste and smell: drug interactions and food preferences. Nutr Rev 1994;52(II):S11–4.

44. Roberts SB, Fuss P, Heyman MB, et al. Control of food intake in older men. JAMA1994;272:1601–6.

45. Rolls BJ. Regulation of food and fluid intake in the elderly. Ann N Y Acad Sci 1989;561:217–25.

46. Phillips PA, Johnston CI, Gray L. Disturbed fluid and electrolyte homoeostasis following dehydration in elderly people. Age Ageing 1993;22:S26–33.

47. Phillips PA, Rolls BJ, Ledingham JG, et al. Reduced thirst after water deprivation in healthy elderly men. N Engl J Med 1984;311:753–9.

48. Wilson JW, Enns CW, Goldman KS, et al. Data tables: combined results from USDA's 1994 and 1995 Continuing Survey of Food Intakes by Individuals and 1994 and 1995 Diet and Health Knowledge Survey, 1997.

49. White JV, Dwyer JT, Posner BM, et al. Nutrition screening initiative: development and implementation of the public awareness checklist and screening tools. J Am Diet Assoc 1992;92:163–7.

50. A Consensus Conference sponsored by the Nutrition Screening Initiative. Report of nutrition screening I: toward a common view, Washington, D.C., April 8–10, 1991. The Nutrition Screening Initiative.

51. White JV, Ham RJ, Lipschitz DA, et al. Consensus of the Nutrition Screening Initiative: risk factors and indicators of poor nutritional status in older Americans. J Am Diet Assoc 1991;91:783–7.

52. Posner BM, Jette AM, Smith KW, et al. Nutrition and health risks in the elderly: the nutrition screening initiative. Am J Public Health 1993;83:972–8.

53. Rose D, Oliveira V. Nutrient intakes of individuals from food-insufficient households in the United States. Am J Public Health 1997;87:1956–61.

54. De Castro JM. Social facilitation of food intake: people eat more with other people. Food Nutr News 1994;66:29–30.

55. Gerrior SA, Guthrie JF, Fox JJ, et al. How does living alone affect dietary quality? U.S. Department of Agriculture, ARS, Home Economics Research Report No. 51 in Fam Econ Nutr Rev 1995;8:44–6.

56. American Dietetic Association Foundation. Nutrition and Health for Older Americans Toolkit, 1998:content available at www.eatright.org/olderamericans/.

57. Russell RM, Rasmussen H, Lichtenstein AH. Modified Food Guide Pyramid for people over seventy years of age. J Nutr 1999;129:751–3.

58. Alaimo K, McDowell MA, Briefel RR, et al. Dietary intake of vitamins, minerals, and fiber of persons ages 2 months and over in the United States: Third National Health and Nutrition Examination Survey, Phase 1, 1988-91. Advance data from vital and health statistics. Hyattsville, MD: National Center for Health Statistics; 1994.

59. McDowell MA, Briefel RR, Alaimo K, et al. Energy and macronutrient intakes of persons ages 2 months and over in the United States: Third National Health and Nutrition Examination Survey, Phase 1, 1988–91. Adv Data 1994:1–24.

60. Yates AA, Schlicker SA, Suitor CW. Dietary Reference Intakes: the new basis for recommendations for calcium and related nutrients, B vitamins, and choline. J Am Diet Assoc 1998;98:699–706.

61. Harper EJ. Changing perspectives on aging and energy requirements: aging and energy intakes in humans, dogs and cats. J Nutr 1998;128:2623S–26S.

62. Campbell WW. Dietary protein requirements of older people: is the RDA adequate? Nutr Today 1996;31:192–7.

63. Code of Federal Regulations, Food and Drugs, Title 21, Part 101.9, Nutrition labeling of food. The Office of the Federal Register, National Archives and Records Administration. Washington, DC: U.S. Government Printing Office; 1996.

64. Fuchs CS, Giovannucci EL, Colditz GA, et al. Dietary fiber and the risk of colorectal cancer and adenoma in women. N Engl J Med 1999;340:169–76.

65. Food and Nutrition Board. National Academy of Sciences (Institute of Medicine)

NRC, Subcommittee on the 10th Edition of the RDAs, Commission on Life Sciences. Recommended dietary allowances. Washington, DC: National Academy Press; 1989.

66. Hussey GD, Klein M. A randomized, controlled trial of vitamin A in children with severe measles. N Engl J Med 1990; 323:160–4.

67. Subar AF, Krebs-Smith SM, Cook A, et al. Dietary sources of nutrients among US adults, 1989 to 1991. J Am Diet Assoc 1998;98:537–47.

68. Torun B, Menchu MT, Elias LG. Recomendaciones dieteticas diarias del INCAP. Guatemala: INCAP, 1994:[INCAP Publicacion ME/0571].

69. Shils ME, Olson JA, Shike M, et al. Modern nutrition in health and disease. Philadelphia: Lippincott, Williams & Wilkins; 1999.

70. Pennington JA, Hubbard VS. Derivation of daily values used for nutrition labeling. J Am Diet Assoc 1997;97: 1407–12.

71. McNamara DJ. Increase in egg intake minimally affects blood cholesterol levels (editorial). Washington, DC: Nutrition Close-Up Egg Nutrition Center; 1994.

72. McNamara DJ. Dietary cholesterol and the optimal diet for reducing risk of atherosclerosis. Can J Cardiol 1995;11Suppl G:123G–126G.

73. Dawson-Hughes B, Harris SS, Dallal GE. Plasma calcidiol, season, and serum parathyroid hormone concentrations in healthy elderly men and women. Am J Clin Nutr 1997;65:67–71.

74. Holick MF. McCollum Award Lecture, 1994: vitamin D—new horizons for the 21st century. Am J Clin Nutr 1994;60: 619–30.

75. MacLaughlin J, Holick MF. Aging decreases the capacity of human skin to produce vitamin D_3. J Clin Invest 1985;76: 1536–8.

76. Webb AR, Kline L, Holick MF. Influence of season and latitude on the cutaneous synthesis of vitamin D_3: exposure to winter sunlight in Boston and Edmonton will not promote vitamin D_3 synthesis in human skin. J Clin Endocrinol Metab 1988;67:373–8.

77. Dawson-Hughes B, Harris SS, Krall EA, et al. Effect of calcium and vitamin D supplementation on bone density in men and women 65 years of age or older. N Engl J Med 1997;337:670–6.

78. Saltzman JR. The aging gut. In: Chernoff R, ed. Geriatric nutrition: the health professional's handbook. Gaithersburg, MD: Aspen; 1999: pp 212–53.

79. Krasinski SD, Russell RM, Otradovec CL, et al. Relationship of vitamin A and vitamin E intake to fasting plasma retinol, retinol-binding protein, retinyl esters, carotene, alpha-tocopherol, and cholesterol among eldery people and young adults: incrased plasma retinyl esters among vitamin

A-supplement users. Am J Clin Nutr 1989; 49:112–20.

80. Trumbo P, Yates AA, Schlicker S, et al. Dietary reference intakes: vitamin A, vitamin K, arsenic, boron, chromium, copper, iodine, iron, manganese, molybdenum, nickel, silicon, vanadium, and zinc. J Am Diet Assoc 2001;101:294–301.

81. Jacques PF. The potential preventive effects of vitamins for cataract and age-related macular degeneration. Int J Vitam Nutr Res 1999;69:198–205.

82. Ames BN, Shigenaga MK, Hagen TM. Oxidants, antioxidants, and the degenerative diseases of aging. Proc Natl Acad Sci USA 1993;90:7915–22.

83. Meydani SN, Meydani M, Blumberg JB, et al. Vitamin E supplementation and in vivo immune response in healthy elderly subjects. A randomized controlled trial. JAMA1997;277:1380–6.

84. Meydani M. Effect of functional food ingredients: vitamin E modulation of cardiovascular diseases and immune status in the elderly. Am J Clin Nutr 2000;71: 1665S–8S; discussion 1674S–5S.

85. Sano M, Ernesto C, Thomas RG, et al. A controlled trial of selegiline, alpha-tocopherol, or both as treatment for Alzheimer's disease. The Alzheimer's Disease Cooperative Study. N Engl J Med 1997;336:1216–22.

86. Malinow MR, Duell PB, Hess DL, et al. Reduction of plasma homocyst(e)ine levels by breakfast cereal fortified with folic acid in patients with coronary heart disease. N Engl J Med 1998;338:1009–15.

87. Herbert V. Vitamin B-12. In: Ziegler EE, Filer LJ, eds. Present knowledge in nutrition. Washington, DC: ILSI Press; 1996: pp 191–205.

88. Keane EM, O'Broin S, Kelleher B, et al. Use of folic acid-fortified milk in the elderly population. Gerontology 1998;44: 336–9.

89. Optimal calcium intake. NIH Consens Statement 1994;12:1–31.

90. Appel LJ, Moore TJ, Obarzanek E, et al. A clinical trial of the effects of dietary patterns on blood pressure. DASH Collaborative Research Group. N Engl J Med 1997;336:1117–24.

91. Shils ME. Magnesium. In: Ziegler EE, Filer LJ, eds. Present knowledge in nutrition. Washington, DC: ILSI Press; 1996: pp 256–64.

92. Fosmire GJ. Trace metal requirements. In: Chernoff R, ed. Geriatric nutrition: the health professional's handbook. Gaithersburg, MD: Aspen; 1999: pp 84–106.

93. Girodon F, Lombard M, Galan P, et al. Effect of micronutrient supplementation on infection in institutionalized elderly subjects: a controlled trial. Ann Nutr Metab 1997;41:98–107.

94. Chandra RK. Effect of vitamin and trace-element supplementation on immune

responses and infection in elderly subjects. Lancet 1992;340:1124–7.

95. Global nutrition industry. Nutrition Business Journal 2000:10–11.

96. Balluz LS, Kieszak SM, Philen RM, et al. Vitamin and mineral supplement use in the United States. Results from the Third National Health and Nutrition Examination Survey. Arch Fam Med 2000;9: 258–62.

97. National Council for Reliable Health Information. "Prevention magazine assesses dietary supplement use". Newsletter 2000; 23:cites BA Johnson in Herbalgram 2000;48:65 and Natural Health Line, 3/10/00.

98. Eliason BC, Huebner J, Marchand L. What physicians can learn from consumers of dietary supplements. J Fam Pract 1999; 48:459–63.

99. Kittler PG, Sucher K. Food and culture in America. Belmont, CA: Wadsworth; 1998.

100. Evans WJ. Exercise training guidelines for the elderly. Med Sci Sports Exerc 1999; 31:12–7.

101. Pollock ML, Franklin BA, Balady GJ, et al. AHA Science Advisory. Resistance exercise in individuals with and without cardiovascular disease: benefits, rationale, safety, and prescription: An advisory from the Committee on Exercise, Rehabilitation, and Prevention, Council on Clinical Cardiology, American Heart Association; Position paper endorsed by the American College of Sports Medicine. Circulation 2000;101:828–33.

102. Ventrakaman JT, Fernandes G. Exercise, immunity and aging. Aging Clin Exp Res 1997;9:42–56.

103. Krinke UB. Effective nutrition education strategies to reach older adults. In: Watson RR, ed. Handbook of nutrition in the aged. Boca Raton: CRC Press; 2001: pp 319–31.

104. Administration on Aging. Executive summary: serving elders at risk: the Older Americans Act Nutrition Programs— National Evaluation of the Elderly Nutrition Program, 1993–1995: www.aoa.gov/ aoa/nutreval/execsum.html, accessed June 23, 2000.

105. Ponza M, Ohls J, Millen B. Serving elders at risk. The Older Americans Act Nutrition Programs—National Evaluation of the Elderly Nutrition Program, 1993–1995. Washington, DC: Mathematica Policy Research, Inc. and Administration on Aging; 1996.

106. Lanza RP, Cibelli JB, Blackwell C, et al. Extension of cell life-span and telomere length in animals cloned from senescent somatic cells. Science 2000;288:665–9.

107. Turturro A, Witt WW, Lewis S, et al. Growth curves and survival characteristics of the animals used in the Biomarkers of Aging Program. J Gerontol A Biol Sci Med Sci 1999;54:B492–501.

CHAPTER 19

Corbis

Knowing is not enough; we must apply.
Willing is not enough; we must do.

Goethe

NUTRITION AND THE ELDERLY:

Conditions and Interventions

Prepared by **U. Beate Krinke,** with **Lori Roth-Yousey**

KEY NUTRITION CONCEPTS

1 Multiple health problems put older adults at higher nutritional risk; interventions with the greatest potential for benefit are targeted to treat specific conditions.

2 There are nutrient thresholds: dietary nutrients that are added to treat a deficiency have little effect when the diet is adequate, and are dangerous when consumed in excess of recommendations.

3 Successful nutrition interventions complement other lifestyle choices to enhance physical and mental resilience and to stabilize physiological changes.

INTRODUCTION

Importance of Nutrition

Aging adults want to stay healthy until death. Yet, those who have chronic illnesses or age-related changes might not meet their *health* goals. Nonetheless, good food choices can limit illnesses, reduce risks, and contribute to *quality of life.* Medical nutrition therapy can be a part of treatment once diseases have occurred. For example, Congress passed a bill accepting *medical nutrition therapy* as a Medicare benefit. The current bill allows dietitians to treat diabetes and kidney diseases (www.eatright.org, 12/19/00).[1]

Foundations for the bill's passage were numerous studies that supported the link between nutritional status and health care utilization of older adults.[2-6] The Lewin Group researchers estimated that covering medical nutrition therapy for older managed care clients who had cardiovascular disease, diabetes, or renal disease would recover Medicare costs after three years and would begin to save system dollars by the fourth year.[2] For example, malnourished older patients have higher postoperative complication rates and longer hospital stays, therefore incurring greater health care costs.[7] In free-living older adults nutritional risk status was found to be the most important predictor of total number of physician visits, visits to physicians in the emergency room, and hospitalization rates.[8] Researchers are identifying specific contributions that nutrition can make to a healthier older population that uses proportionately more health care services and products than younger persons. Enjoyment of foods improves the chances of eating well, which encourages healing and recovery, thereby decreasing health care costs.

HEALTH   More than the absence of disease, health is a sense of well-being. Even individuals with a chronic condition may properly consider themselves to be healthy. For instance, a person with diabetes mellitus whose blood sugar is under control can be considered healthy.

QUALITY OF LIFE   A measure of life satisfaction that is difficult to define, especially in an aging heterogeneous population. Quality of life measures include factors such as social contacts, economic security, and functional status.

MEDICAL NUTRITION THERAPY Comprehensive nutrition services by registered dietitians to treat the nutritional aspects of acute and chronic diseases.

NUTRITION AND HEALTH

When asked, older people feel good about their health. Only 10% say their health is fair or poor, while 90% consider their health to be good, very good, or excellent.[9] This happens even when many are troubled with a chronic health problem. In contrast to older adults' perceptions, public health professionals objectively monitor health by measuring leading causes of death (mortality) and leading diagnoses of health conditions (morbidity). Heart disease and cancer are the leading causes of death for all persons aged 65 and older. Other chronic health conditions, such as arthritis and hypertension, affect almost half of older adults.[10,11]

Arthritis	49%
Hypertension	40%
Heart disease	31%
Cancers	7%

Eating patterns of older adults contribute to the incidence and course of hypertension, heart disease, and cancer and impact the limited mobility related to arthritis. For example, overeating and obesity complicate arthritis management. In turn, arthritis limits functional abilities and may limit access to food. Overall, nutritional status is a major factor in prevention, treatment, and recovery of health.

NUTRITION-RELATED CONDITIONS

Heart Disease

Heart disease (cardiovascular disease, CVD) is the leading cause of death in older adults but is potentially reversible by adopting a healthy life style. The adult risk factors and course of heart diseases have been discussed in Chapter 17. Specifics for older adults are highlighted in this section, including stroke and hypertension.

PREVALENCE   Many types of heart diseases are common problems among older adults and vary by race and gender (see Table 19.1).[11] Approximately one-third of adults aged 65 and older have some form of heart disease.[11] Of all deaths from heart disease each year, over three-fourths (78%) occur in elderly persons.

Table 19.1   Prevalence of heart diseases in adults aged 65 and older.

	Total (%)	Gender (%)		Race (%)	
		Male	Female	White	Black
Heart disease	30.8	36.2	26.9	31.5	26.1
Hypertension	40.3	34.9	44.2	39.5	53.3
Cerebrovascular diseases	7.1	7.9	6.5	7.1	*
Atherosclerosis	4.1	4.5	3.9	4.5	*

*Information not available

In adults over age 85 who are hospitalized, 53% are admitted with heart failure as the principal diagnosis.[9]

RISK FACTORS　Risk factors for cardiovascular disease in old age remain the same as in younger adulthood, except that the factors have less predictive value in old age.[12] Of adults aged 65 years and older examined in the NHANES III survey, 86% had one or more modifiable cardiovascular risk factors. These include hypertension (140 mmHg/90 mmHg), elevated LDL cholesterol (at least 130 mg/dL), and/or diabetes mellitus (physician diagnosis or fasting plasma glucose greater than 126 mg/dL).[12,13] Race is associated with risk. Older African Americans are nearly three times more likely to have one of the three cardiovascular risk factors than the average population.

Obesity increases the risk of CVD in aging adults. For example, 93% of those with a BMI over 30 (i.e., obesity) had one or more nutrition-related CVD risk factors (see Chapter 17). In the overweight category (BMI of 25 to 29.9), 87% had one or more risk factors.

OLDER ADULT NUTRITIONAL REMEDIES FOR CARDIOVASCULAR DISEASES　Assertive treatment can modify the course of heart disease at any age. Special considerations for older adults contemplating dietary changes include assessment of personal motivation regarding diet and health, functional status, and lifestyle habits. Is the individual interested in making needed diet changes? If willing to change, does the individual have the knowledge and skill to adopt new dietary behaviors? What motivated him to seek nutrition advice? What is the individual's physical and mental functional level? Is assistance needed to shop for groceries or to prepare meals? Does the individual eat alone, with family and friends, at home or in restaurants? Questions such as these impact nutritional intake and help to predict whether lifestyle changes will be adopted and maintained.

Dietary guidelines to fight cardiovascular disease in adults of all ages include those from the National Cholesterol Education Program (NCEP) and the American Heart Association (AHA). Cardiac rehabilitation programs adapting these guidelines are part of Chapter 17.

Stroke

DEFINITION　The American Heart Association defines stroke as "a cardiovascular disease that affects the blood vessels supplying blood to the brain" (www.americanheart.org). A stroke can occur in several forms, including cerebral thrombosis (a blood clot, or thrombus, blocks blood flow to the brain), cerebral embolism (an embolus, or wandering blood clot, lodges in an artery and blocks blood flow in or to the brain), or as hemorrhage (a blood vessels bursts and part of the brain goes without oxygen as a result).

PREVALENCE　Of adults aged 70 and older, 8% of females and 10% of males have had a stroke.[11] Stroke is one of leading causes of death of older adults.

ETIOLOGY　Factors that can lead to a stroke include blocked arteries, easily clotting blood cells, and weak heartbeats that are unable to keep blood circulating through the body, allowing pools of blood to form and clot. Hypertension is another stroke contributor because the force of blood may break weak vessels.

EFFECTS　Strokes deprive the brain of needed oxygen and other nutrients. Brain and nerve cells deprived of oxygen for only a few minutes can die. Brain cells do not regenerate. As a result, stroke leads to loss of function (speaking, walking, talking, and eating) for parts of the body controlled by the oxygen-deprived nerves. Quick recognition of stroke results in faster treatment and better recovery. Although brain cells do not regenerate, new nerve pathways can develop in the gray matter reservoirs of the brain. This provides hope for successful rehabilitation therapies. Nonetheless, rehabilitation from a stroke is a slow, arduous process. Learning how to feed oneself, chew, and swallow are among functions that may have to be relearned in a rehabilitation clinic.

> **CAROTID ARTERY DISEASE**　The arteries that supply blood to the brain and neck becoming damaged.
>
> **ATRIAL FIBRILLATION**　Degeneration of the heart muscle causing irregular contractions.
>
> **TRANSIENT ISCHEMIC ATTACKS (TIAs)** Temporary and insufficient blood supply to the brain.

RISK FACTORS　Gender is not a risk factor for stroke, although more women die from a stroke than men do. The following factors place an individual at higher risk for stroke:

- Family history,
- African American, Asian, and Hispanic ethnic groups,
- Having had a prior stroke,
- Long-term high blood pressure (either systolic or diastolic),
- Cigarette smoking,
- Diabetes mellitus,
- *Carotid artery disease, atrial fibrillation, transient ischemic attacks,*
- High red blood cell count,
- Sickle cell anemia,

- Living in poverty, and
- Excessive use of alcohol, use of cocaine and IV drugs.

NUTRITIONAL REMEDIES The focus of dietary advice in stroke prevention is to normalize blood pressure.[14] Other dietary goals are to reduce overweight and obesity (particularly abdominal fat) and to moderate alcohol intake.

The Dietary Approaches to Stop Hypertension, or DASH, diet (see Table 19.2) has been a promising strategy to decrease blood pressure[14] and risk of stroke in adults under age 65. It has also been shown to enhance perceptions of quality of life.[15] Theoretically, the DASH diet should also work for older adults. Other nondrug interventions have successfully lowered the blood pressure of older adults (e.g., using weight reduction and/or sodium restriction of 1800 mg per day over 30 months).[16,17]

Hypertension

DEFINITION In Western societies, blood pressure increases with age and stabilizes around age 50 to 60 for men. A blood pressure greater than or equal to 140/90 mmHg is defined as stage 1 hypertension. Although it has been suggested that older adults can tolerate higher blood pressure and may even benefit from increased blood flow to the brain, old age does not change the diagnosis criteria for high blood pressure. Higher blood pressure puts more force on potential vessel blockages and increases chances of blood vessel breakage. An individual who controls high blood pressure with medication is still considered to have hypertension.

PREVALENCE Analysis of the NHANES III data shows that 40% of older adults have hypertension.[11] This is almost two times greater than all adults (40% vs. 24%). Prevalence increases with age and differs by gender. In adults aged 75 years old and over, 64% of males and 77% of females have hypertension. Over half of individuals with hypertension (53%) control it with prescribed medications.

ETIOLOGY Family history and ethnic background increase the risk of hypertension; African Americans are most likely to have hypertension. Salt intake can also contribute to hypertension. Researchers with the Intersalt Study calculated that over time, 20% of hypertension in Western societies is attributable to salt intake.

EFFECTS Prolonged high blood pressure puts extra tension on blood vessels and organs in the body, wearing them out before the natural aging process. Damaged kidneys are a common sign of uncontrolled hypertension.

RISK FACTORS Nutritional risk factors are drinking alcohol to excess, high–saturated fat diets leading to dyslipidemia and atherosclerosis, lifestyles resulting in overweight and obesity, and a diet low in calcium.[18]

Table 19.2 The DASH diet plan for blood pressure control.

FOOD GROUP	Servings per Day	Serving Sizes of Foods within the Food Group
Grains and Grain Products Especially whole grain[a]	7–8	Breads: 1 slice or 1 oz Cereal: ½ cup cooked or dry Rice, pasta: ½ cup cooked
Vegetables Fresh, frozen, no-salt-added canned	4–5	Raw, 1 cup; cooked, ½ cup
Fruits Fresh, frozen, or canned in juice	4–5	Juice: 6 oz Fresh: 1 med piece Mixed or cut: ½ cup Juice: 6 oz Dried: ¼ cup
Dairy Foods Skim or 1% milk, fat-free dairy products	2–3	Milk: 8 oz Yogurt: 1 cup Cheese: 1½ oz
Meats, Poultry, and Fish	Up to 2	3 oz, cooked
Nuts, Seeds, Dry Beans	4–5 per week	⅓ cup or 1½ oz nuts 2 tbs or ½ oz seeds ½ cup cooked beans (legumes)
Fats and Oils[b] Select olive, canola, corn, and safflower oils	2–3	1 tsp soft margarine, oils, mayonnaise 1 tbs low-fat mayonnaise 2 tbs light salad dressing
Sweets	Up to 5 per week	1 tbs jam, jelly, syrup, or sugar ½ oz fat-free candy, jelly beans, or 12-oz sweetened beverage

[a]Whole grain is the entire edible part of wheat, corn, rice, oats, barley, and other grains. Whole grain bread has the words "whole grain" before the type of flour is listed; whole grain breakfast cereals include the word "whole" or "whole grain" before the grain name (e.g., whole grain wheat).
[b]One serving is equivalent to 5 grams of fat
SOURCE: Adapted from a 2000-calorie eating plan, The National High Blood Pressure Education Program's HeartFile, Winter 1999, National Heart, Lung, and Blood Institute and the National Dairy Council's "DASH TO THE DIET," 2000.

NUTRITIONAL REMEDIES The DASH diet described in the earlier section on stroke is an important nutrition intervention for persons with a history of hypertension. Other strategies to decrease hypertension include achieving and maintaining a healthy weight, and reducing sodium intake while maintaining adequate potassium, magnesium, and calcium intake. The "Dietary Guidelines for Americans" suggest limiting dietary sodium to 2400 mg per day. Intake of 70-year-olds (not counting salt added at the table) is 3122 mg daily for males and 2376 mg for females.[19] In the DASH-sodium study, the greatest blood pressure reduction occurred in the subjects with the strictest sodium intake limit (1500 mg a day). Blood pressure reduction occurred whether individuals were normo- or hypertensive.

Most of dietary sodium is contributed by processed foods. For example, choosing a baked potato rather than eating more highly processed potato chips limits sodium intake. Adding salt at the table increases sodium intake by roughly one-fourth.[20]

CANCERS

Definition

Cancer is a name for a group of conditions resulting from uncontrolled growth of abnormal cells. It is a complex progression through several stages: activation, initiation (injury or insult by a carcinogen), promotion (damaged DNA divides during a lag period, potentially over 10 to 30 years), progression (growth and spread), and possible remission (successful treatment or reversal). See Table 19.3.

Prevalence

Half of all new cancer cases are diagnosed in individuals aged 65 and older, with the most common sites being breast, colorectal, lung and bronchial system, and prostate.[9] In the case of prostate cancer, 80% of all cases occur in men aged 65 and older.[21] Cancers of the breast, skin, intestine, genitals, lungs, and other respiratory sites are present in 7.4% of individuals aged 65 and older (roughly 1 of 14). Diet-related cancers include cancer of the stomach, breast, prostate, and colon.

While the majority of older adults do not have cancer, it is the second leading cause of death after heart disease. Compared to the death rate from heart disease (1808 per 100,000), the death rate from cancers (listed as malignant neoplasms in some charts) is 1131 per 100,000.[11] After age 85, cancer was responsible for 10% of deaths among women and 16% among men.[22]

Table 19.3 The phases leading to transformation of a normal cell into a cancerous cell.

1. **Activation:** Certain chemicals and/or radiation can trigger a cellular change. In normal processes, an individual's body removes harmful substances; in some cases, the substance remains and attaches to the DNA within a cell.

2. **Initiation:** The changed or mutated DNA within the cell is copied. If it occurs within a specific DNA region, this makes the cell more sensitive to the harmful substances and/or radiation.

3. **Promoters:** When cells become sensitive, promoters encourage the cells to divide rapidly. If the normal sequences of DNA are damaged, a mass of abnormal cells bind together to form a mass, or tumor.

4. **Progression:** The cells continue to multiply and spread into nearby tissues. If they enter the lymph system, the abnormal cells are transported to other body organs.

5. **Reversal:** The goal of reversal is to prevent the cancer's progression, or to block any of the first four stages.

Excerpted from: "Cancer—The Intimate Enemy." In Marieb EN. Human Anatomy and Physiology. Redwood City, CA: Benjamin Cummings Publishing; 1996.

Etiology

Genetics contribute to the development of cancer; some cancers are hormone-sensitive (e.g., uterine, breast, prostate, and ovarian). As humans live longer, it becomes more likely that some mistake or insult will adversely affect the DNA replication process and ultimately cause cancer. Environmental factors can largely determine whether or not an individual develops cancer. Diet alone is estimated to account for 35% of all cancers while smoking and dietary habits together have been estimated to cause 50% or more of cancers.[20,22]

Effects

Cancer development is age-associated but not age-dependent; development is neither consistent nor linear. In a healthy, resilient individual, initiation may be repaired and subsequent cancer avoided or delayed. In someone with impaired immunity or suffering major physiological stress, initiation may proceed through promotion or progression. For older adults, cancer treatment presents a greater nutritional challenge than cancer prevention.

Cancer treatment affects all aspects of nutrition. Side effects include mouth sores and altered sense of taste and smell, which affects appetite and eating. This may include changes in intake, digestion, and absorption related to chemotherapy, surgery and radiation. For example, treatment may leave the cancer patient tired and lacking immunity to fight common diseases. In addition, ability to socialize over meals or to grocery shop may be curtailed. Mouth sores and an altered ability to smell and taste foods affects appetite. Types of cancers hinder secretion of stomach acids, and this may decrease nutrient absorption. Medications also interact with nutrient absorption. Overall, cancer treatment effects cause significant amounts of stress in older adults.

Risk Factors

Diet can refocus the genetic role in cancer development. Although dietary patterns make the biggest impact when begun early in life, older adults show greater interest in risk reduction. Vegetable and fruit consumption is the primary diet strategy for cancer risk reduction. Vegetable, fruit, and anti-oxidant intake are related to reductions in cancer initiation and progression. Extensive reviews of diet and cancer studies[23-25] conclude that high-level vegetable and fruit consumption is associated with decreased risk of cancer including oral, esophageal, stomach, colorectal, and lung cancers. Research is currently underway to examine how components in green tea hinder the body's abilities to develop cancer-promoting chemicals.

Table 19.4 offers an overview of specific types of cancers and their link to well-supported nutritional risk factors. The studies do not deal specifically with older adults but with adults of all ages. Among the more recently identified potential connections between nutrition and cancer are possible protective functions of calcium intake with colon cancer as well as the role of increasing lycopene (a flavonoid found in tomatoes) intake to decrease prostate cancer risk.

Nutritional Remedies

While undergoing cancer treatment is not the time to worry about a low-fat diet; it is a time for older adults to keep up their calorie consumption. The focus of cancer treatment is to maximize nutritional status and avoid weight loss and dehydration. This can be a challenging task. Anti-cancer medications and radiation treatments are associated with nausea, vomiting, diarrhea or constipation, fatigue, and weight loss. Finding food items that are tolerated, especially if they are calorie-dense, is one way family and friends can help medical care providers support a cancer patient's treatment.

Treating cancer is a process that tends to generate quests for alternative methods. The hope for remission and cure is a powerful motivator to try herbs, new nu-

Table 19.4 Nutritional risk factors for the most common cancers.

Type of Cancer	Nutritional Risk Factors with Convincing or Probable Evidence
Lung	Low vegetable and fruit intake
Stomach	Low vegetable and fruit intake Lack of refrigeration for food High salt intake High use of smoked, cured, and pickled foods (salt and nitrites)
Breast	Low vegetable and fruit intake Increased obesity (especially in post-menopausal women) Increased alcohol
Colon/rectum	Low vegetable and fruit intake High meat consumption (especially fat from red meats) Excess alcohol (folate may help offset detrimental effects)
Mouth/pharynx/nasopharynx	Low vegetable and fruit intake Excess alcohol, together with tobacco (smoke or chew) High salted fish intake (nasopharynx)
Liver	Excess alcohol Eating contaminated (especially aflatoxins[a]) foods
Cervix	Low vegetable and fruit intake
Esophagus	Low vegetable and fruit intake Diets generally deficient in nutrients High alcohol intake
Prostate	High meat or meat fat or dairy fat

[a]Aflatoxins are toxins produced by molds in some foods such as peanuts.
SOURCE: Adapted from Word Cancer Research Fund and American Institute for Cancer Research, Food, Nutrition, and Prevention of Cancer, p 542.

trient combinations, and complementary healing practices. Two botanicals, mayapple (*Podophyllum*) and pacific yew (*Taxus brevifolia*) are sold as prescription cancer treatments; green tea and grape seed extract are sold as cancer-preventive-herbals.[26] The following herbal products have potentially useful roles in cancer treatment but still need scientific proof regarding efficacy in broad populations:[26,27]

- Ginseng: stimulates immune function
- Garlic: relieves colds and other infections
- Saw palmetto: improves urine flow, reduces urgency in men with prostate enlargement
- Gingko biloba: increases blood flow, decreases depression
- Shark cartilage: treat lung cancer (strong community belief—no scientific evidence)

- Bitter herbs: gentian (in angostura bitters) and centaury stimulate appetite

The first rule of treatment, "Do no harm," is especially relevant during the vulnerable period of cancer treatment. A product that can be enjoyed or perhaps tolerated by a healthy person may exacerbate illness in someone with health problems. Discussing all contemplated treatments with a physician, registered dietitian, and pharmacist will help to prevent dangerous interactions.

DIABETES

Definition

Diabetes mellitus is a chronic metabolic disorder that affects all organ systems. The body cannot properly process and use carbohydrates because the pancreas is either not producing enough insulin or the insulin produced is not being effectively absorbed into the cell. Lack of adequate insulin to transport sugar into cells results in excess glucose circulating in the blood stream. Diabetes diagnosis criteria for older adults are the same as for younger adults. Symptoms of type 1 diabetes include high blood sugar, sugar in urine, excessive thirst, an increase in food intake, and unexplained weight loss. In type 2 diabetes the primary symptoms include high blood sugars and thirst. In addition, a casual (random) blood sugar of ≥200 mg/dL would be confirmed by either (1) fasting (no caloric intake for at least 8 hours) blood sugar of 126 mg/dL or greater, (2) an abnormal blood glucose (equal to or greater than 200 mg/dL following a drink of 75 grams glucose), or (3) another casual blood sugar of ≥200 mg/dL. Nondiabetic adults maintain fasting blood sugars below 110 mg/dL. Blood sugars above 109 mg/dL and below 126 mg/dL suggest impaired glucose tolerance.[13]

Two diabetes mellitus classifications relate to older adults: type 1 and type 2. Type 1 diabetes was formerly labeled insulin-dependent diabetes (IDDM) or juvenile-onset diabetes because this type often begins in youth. Type 2 was formerly referred to as non–insulin-dependent diabetes mellitus (NIDDM) or "adult-onset diabetes" because it was generally diagnosed after age 40. Older women who had impaired glucose tolerance or gestational diabetes (GDM) during pregnancy are at increased risk for developing type 2 diabetes.

Prevalence

Each day, 2200 new cases of diabetes are diagnosed.[9] Roughly 800,000 new cases are diagnosed each year, and most of those (90–95%) are type 2 diabetes. The Centers for Disease Control and Prevention estimate that there are 10.3 million individuals in the United States who have been diagnosed with diabetes, and that another 5.4 million have undiagnosed diabetes. Approximately 45% of the U.S. diabetic population are aged 65 years and older.[28]

Of adults surveyed through the National Health and Nutrition Examination Survey Epidemiologic Follow-Up Study (1971–1992), 10% developed diabetes.[29] However, 18.7% of adults aged 65 or older have diabetes, although this rate decreases to 11% of the 70 year and older population. Prevalence is higher in some ethnic groups. For example, 29.4% of Mexican American and 27.3% of African American older adults have diabetes. Using a different database, the prevalence of diabetes in Native Americans and Alaska Natives increased from 18.2% in 1990 to 22.8% in 1997 for adults aged 65 years and older.[30] Non-Hispanic black and Hispanic individuals are twice as likely to have diabetes as whites.[28] For every two people diagnosed with diabetes, it is estimated that another one to two individuals have diabetes without knowing it.

Etiology

Inheritance, environment, and lifestyle contribute to the development of diabetes. Vulnerable groups are thought to have a "thrifty" gene, but no specific genetic marker has been identified. Twin studies have supported the strong role of genetics. When one of a set of identical twins develops type 2 diabetes, the likelihood of the other twin developing it is 90–100%.[31] Diabetes has been called a disease of civilization because incidence increases with increasing industrialization. This is thought to be due to less physical activity, an increased food supply, and "westernized" diet (including greater intake of highly processed and fatty foods). The increasing prevalence of diabetes parallels the increasing obesity prevalence.

Insufficient quantities of effective insulin, insulin receptor deficiencies (too few or defective receptors), or post-receptor processing malfunctions prevent glucose derived from food (mostly from carbohydrate) from being transported into cells. Glucose remains circulating in the blood, so that diabetes is diagnosed by a higher than normal blood sugar level. High blood sugar (hyperglycemia) stimulates insulin secretion in two phases. There is a quick response after carbohydrate ingestion and then a longer, slower secretion phase that stops after blood sugar drops. Blood sugar that is not transported into cells and metabolized remains circulating in the blood, accompanied by high levels of insulin (hyperinsulinemia). Over time, excess insulin and/or glucose in the blood lead to nerve and organ damage.

Blood glucose levels above a threshold of roughly 120 mg/dL exceed the kidney's capacity for reabsorption, and subsequently glucose starts spilling into the urine. This results in sweet urine. In fact, finding urine to be sweet was an early way to diagnose diabetes. Glucose excreted in urine cannot be used for energy, so fatigue and weight loss accompanies uncontrolled hyperglycemia in type 1 diabetes. In contrast, weight gain is more common in type 2 diabetes.

Effects

In four of five older people, diabetes occurs as one of several co-existing conditions that includes heart disease, hypertension, elevated blood lipids, and obesity. Individuals with diabetes are at greater risk for heart disease and its complications. For instance, individuals with diabetes are more likely to die after suffering a myocardial infarction (MI) than those without diabetes.[31,32] Hyperhomocysteinemia, which is linked to increased risk of heart disease, is a stronger risk factor for overall mortality in individuals with type 2 diabetes, independent of CVD and other risk factors, than in individuals without diabetes.[33]

Diabetes leads to a ten-fold greater risk of amputations, macular degeneration, visual loss, cataracts, glaucoma, and neuropathies (nerve damage, pain, or tingling) of the hands and feet. Other effects of hyperglycemia may include sodium depletion and dehydration, trace mineral depletion (zinc, chromium, magnesium), insomnia, nocturia, blurred vision, increased platelet adhesiveness related to atherosclerosis, increased infection and decreased wound healing, and aggravated peripheral vascular disease.

Unlike type 1 diabetes, where hyperglycemia can lead to ketosis, type 2 diabetes is associated with hyperosmolar, hyperglycemic, nonketotic syndrome (HHNS). HHNS is characterized by weakness, fatigue, confusion, and severe dehydration.[31] Blood pressure may drop (resulting in orthostatic hypotension) which increases the potential of falls and broken bones. In HHNS, blood glucose may rise to 600 mg/dL but there will be no or only slight ketosis. HHNS can develop from taking certain medications, loss of thirst, stress as result of surgery or infection, and renal disease. HHNS is linked to high mortality; treatment with insulin and rehydration should proceed slowly to avoid further complications.

Older adults with type 2 diabetes also develop hypoglycemia, although not as often as individuals with type 1. Patients using oral hypoglycemic drugs (for example, sulfonylureas that stimulate insulin secretion) are at increased risk for hypoglycemia. Hypoglycemia is problematic for older adults because it leads to weakness, confusion, and possible falls and fractures. Alcohol and drugs such as clofibrate and salicylates (aspirin) also contribute to drops in blood sugar.

Risk Factors

Risk factors for type 1 diabetes are primarily genetic and mediated by viral and other diseases. Obesity is unlikely. Individuals with type 1 diabetes have immune and genetic bases called ideopathic diabetes. This type is usually found in individuals with African or Asian origin; risk factors for type 2 diabetes are more lifestyle related (e.g., obesity) and also include certain genetic defects of the pancreas beta-cells. Other risk factors for type 2 diabetes include:

- *Body weight and abdominal fat*: Higher body fat levels (BMI over 25, but even more so over 30) are associated with higher incidence of type 2 diabetes, especially when fat is stored in the abdominal area (waist circumference greater than 35 inches or more in females, 40 inches or more in males).
- *Race/ethnicity*: Blacks have a higher risk of developing type 2 diabetes than whites at normal or low body weight. Risks of developing diabetes are similar for blacks and whites at higher body weights.[29] NHANES data (30 to 74-year-olds) collected from 1971 to 1992 yielded risk calculations for developing type 2 diabetes relative to BMI (weight for height). Cumulative risk increases with BMI in all race and sex groups. At a BMI of 22 (i.e., desirable weight), the probability of developing diabetes was 1.87 for blacks compared to the probability of 1.76 (p <.01) for whites.[29] At BMI levels of 32 (obesity), the probability of developing diabetes was not significantly different for blacks and whites. The difference in risk may be associated with visceral adiposity (fat storage in midsection). Greater visceral adiposity impacts blacks more than whites at normal weight (BMI levels of 22) but not at levels defined as overweight or obese (BMI levels 25 or greater).

Nutritional Remedies

The goal of nutritional interventions is to achieve the best possible glycemic (blood sugar) control and reduction of cardiovascular disease risk factors "without jeopardizing the quality of life, the health, or the safety of the patient."[34] Diabetes care providers tend to agree that the following components comprise the nutritional management of both types of diabetes.[13,31,34–36]

1. Maintain as close to normal blood glucose levels as possible by balancing food intake, exercise, and insulin/medications.
2. Choose health goals similar to the USDA/HHS "Nutrition and Your Health: Dietary

Guidelines for Americans 2000";[37] these include recommendations for physical activity.

3. Individualize dietary recommendations according to the individual's ethnic group.

4. Many practitioners suggest that a stable weight (even if overweight) is better than weight cycling, since weight cycling is associated with greater risk for complications of diabetes (such as cardiovascular disease).

5. Vitamin and mineral supplementation is reviewed on an individual basis.

6. Interactions among carbohydrates, fat, fiber, and fluids make it impractical to assign useful glycemic scores to individual foods for population-wide use. The *glycemic index* of food indicates that some readily digested starches cause greater rises in blood sugar than certain simple sugars. However, the glycemic index concept for ranking response to foods is still controversial, primarily because foods are eaten in mixed diets.

Controversy is part of guideline development. The controversy in meal planning for individuals with diabetes revolves around the role of simple sugars. For example, simple sugars can lead the body to secrete endogenous insulin in individuals with type 2 diabetes. If the insulin is defective and unable to assist sugars to move into the cells to be metabolized for energy, the elevated sugar and insulin (hyperinsulinemia) circulating in the blood can damage organs and nerves. Hyperinsulinemia also promotes weight gain.[38]

Eating for blood sugar control and decreased risk of cardiovascular disease means eating high-fiber foods, oats and other whole grains, and vegetables. Fibers such as pectins and gums in citrus, apples, oats, barley, and dried beans have been shown to slow glucose absorption and lower serum cholesterol.[39] However, evidence is inconclusive about whether fiber is most effective as soluble or insoluble fiber and whether it can be supplemental or must be eaten as part of a mixed diet. Mixed evidence plus the difficulty of getting very high levels of fiber from foods (i.e., 65 grams daily compared to the 11 to 15 grams the public actually eats)[31] led the panel developing the 1994 Dietary Recommendations and the 1998 revision[40] not to suggest high fiber for blood sugar control. A mixed, high-fiber diet (eating 20–35 grams fiber per day) is part of overall dietary guidance for Americans[37] including individuals with diabetes.[40]

A high-carbohydrate diet is likely to elevate fiber intake, and it may also increase triglyceride levels. Moderate increases in carbohydrate (from 45 to 55%) have been well tolerated in adults (42–79 years, mean 62.5) with diabetes when given as breakfast cereal over a 6-month period.[41] If an individual's blood triglyceride level increases with a 3-month, higher carbohydrate meal-plan trial, this demonstrates the individual has sensitivity to a high-carbohydrate diabetic diet.[31] Carbohydrates can then be reduced to 45% of total calories while maintaining other heart-healthy features of the meal plan.

Consuming needed nutrients from a mixed diet is generally solid nutrition advice for all older adults, including those with diabetes. According to the American Diabetes Association, supplements are required to complement inadequate intake or metabolic needs. Chromium deficiency is associated with impaired glucose tolerance and elevated serum cholesterol because it is an essential nutrient required for sugar and fat metabolism.

> **GLYCEMIC INDEX**   A classification used to describe how quickly food carbohydrate appears as blood glucose; the glycemic response of various carbohydrates is compared to intake of pure glucose.
>
> **GLYCOSYLATED HEMOGLOBIN**   A laboratory test that measures how well the blood sugar level has been maintained over a prolonged period of time, also called Hemoglobin A, C.

Chromium picolinate supplements have shown some promise to reduce the need for insulin and oral hyperglycemic medications when a deficiency exists.[42] Chromium-rich foods include brewer's yeast, oysters, meats (especially liver), whole grains, skin-on potatoes, and apples. Magnesium deficiency has been linked with insulin resistance, carbohydrate intolerance, and abnormal blood lipid metabolism. Magnesium may be lost through the urine when excess glucose is being excreted, or intake may be insufficient, or intestinal absorption limited. Magnesium supplementation should be monitored closely to avoid magnesium toxicity, especially if kidney function is impaired.

In some minority populations, complementary medicine is used to treat diabetes. This is especially true in Mexican American and Native American populations. Evening primrose oil, milk thistle, fenugreek seeds, and prickly pear cactus (nopales) are some examples of such herb treatments. Alternative nutritional remedies can complement the standard nutritional therapies if they have been evaluated and found safe and effective.

Constant attention to meal planning in diabetes management has led many individuals to search for foods that will not affect blood sugar. One example is a vegetable called Jerusalem artichoke, not related to the globe artichoke or artichoke hearts seen in recipes and restaurants. Jerusalem artichokes contain inulin, a fructose polymer, which does *not* turn to insulin. Inulin may be absorbed more slowly than other starches. While Jerusalem artichokes are a perfectly fine vegetable, eating them does not lower blood sugar.

Type 2 diabetes has well-developed nutrition intervention recommendations. Cost-benefit analyses have shown that tight control of blood sugars (maintaining *glycosylated hemoglobin* below 7%) can lead to better

Photo Disc

Case Study 19.1
JT: Spiraling Out of Control?

JT, a retired computer company executive, eats out four times a week since his wife died last year. Meals at home consist of microwave dinners or supreme pizza. He belongs to a health club, which he visits three times a week. After working out and socializing, he and his friends go for beer. He developed type 2 diabetes five years ago. Last week, he visited the clinic for his annual check-up. He was measured at 5 feet 9 inches and weighed 235 pounds.

- If you were his medical doctor, what nutrition remedies would you prescribe for JT?
- What would be a reasonable weight range for JT?
- What would you ask him about his fitness routine and how would you convinced him that he needs an aggressive nutrition and fitness program?
- What sort of fluid recommendation would you make?

quality of life for the older patient with diabetes and can result in fewer complications from diabetes.[43–45]

INSULIN RESISTANCE SYNDROME (METABOLIC SYNDROME)

Definition

Also known as Syndrome X[46] insulin resistance describes a combination of atherogenic risk factors including hyperinsulinemia, obesity with an abdominal pattern of distribution, some degree of carbohydrate intolerance, hypertension, and an abnormal blood lipid pattern of increased triglyceride and decreased high-density lipoprotein (HDL) subfractions. The combination of type 2 diabetes and hypertension has also been called Syndrome X.[47]

While diabetes is diagnosed as hyperglycemia due to lack of insulin and an excess of glucose in the blood stream, insulin resistance occurs when excess insulin circulates in blood but the body loses sensitivity to insulin action. In hyperinsulinemia (high levels of insulin in blood), the pancreas continues to produce insulin in response to high blood sugars.

Prevalence

Insulin resistance is estimated to be present in 20 to 30% of healthy adults[46,48] and in almost all individuals with type 2 diabetes. Not all people with insulin resistance develop diabetes; the bodies of some are able to produce enough insulin to overcome the insulin resistance. They still have hyperinsulinemia. Obese individuals are likely to be insulin resistant.[38]

Etiology

The causes of insulin resistance are unclear. Three components, however, are associated with Syndrome X, genetic disposition, diet, and lifestyle. Familial history is a risk factor for insulin resistance, but defective genes have not been identified. A cellular defect in glucose transport and oxidation is probably due to some genetically linked messenger malfunction. A diet high in fat and simple sugars can trigger Syndrome X in rats.[49] In addition, epidemiological dietary recommendations associated with the disease triad of hypertension, diabetes, and heart disease include decreasing fat and increasing complex carbohydrates. Finally, insulin sensitivity is enhanced after exercise and with weight loss.

Effects

When the pancreas can no longer produce additional insulin to enable glucose transport, oxidation, or cell storage, hyperglycemia results. Consequently, muscle cells and fat tissues respond by making glucose available for energy, resulting in elevated triglyceride levels associated with low HDL levels (see Chapter 17).

Risk Factors

Obesity, especially abdominal or apple-shape obesity, inactivity, and family history are key risk factors for developing insulin resistance. Waist to hip ratio of 0.8 or more is a good predictor of insulin resistance.[50] Smoking and alcohol consumption of three or more drinks on a regular basis are also risk factors. A typical patient with insulin resistance is obese with a large waist and has tried weight loss repeatedly.

The cluster of risk factors for cardiovascular disease is also associated with insulin resistance syndrome:[46,48]

- Hyperinsulinemia
- High triglycerides and low HDL concentration
- Increased small, dense LDL particles
- Higher levels of fats circulating in blood after meals
- Higher concentrations of blood clotting factors such as fibrinogen

Nutritional Remedies

A meal plan providing less than 10% of total calories from saturated fat decreases cardiac risk factors. Total fat calories, emphasizing monounsaturated sources, should constitute roughly 30–35% of the meal plan. If triglycerides are elevated, a meal plan that includes up to 40% of calories from fat and 15–25% of calories from protein will help to control insulin levels better than the typical American Heart Association's low-fat, high-carbohydrate diet.[38,46] See Table 19.5.

Increasing intake of whole grain carbohydrates (whole wheat breads, oat cereals, and brown rice) and moderating total intake of carbohydrates can help to moderate blood sugar levels. Overall, for older adults, individualized dietary recommendations are similar to those for nondiabetic or nonhyperinsulinemic peers. The exception is to monitor quantity of carbohydrates and establish reasonable weight goals.

OSTEOPOROSIS

Watch it when you hug Grandma

Mary Nelsestuen, afraid that a strong hug might break her frail mother's osteoporotic ribs

Definition

Osteoporosis means "porous bone," which results from reduction in bone mass and disruption of bone architecture. Osteoporosis can develop rapidly or slowly, depending on the homeostatic mechanism in-volved. An accelerated phase of bone loss (also called Type I) occurs due to estrogen or testosterone loss. Bone mass loss is sharper for women, who are most vulnerable to bone loss in the three to five years past menopause due to estrogen decline (Type I osteoporosis).[51] Men develop osteoporosis later than women. Their testosterone levels decline roughly 40% between ages 40 to 70, leading to declining bone mass.[52] Men's bone mass losses double after surgical or hormonal castration (a form of treatment for prostate cancer).

A slow phase of bone loss, also called Type II or senile osteoporosis, occurs due to decline in calcium absorption. Osteoporosis is an age-associated disease because effects of slow bone loss are usually not seen until late adulthood.

Diagnosis of osteoporosis can be aided by assessment of clinical risk and biochemical markers and by bone densitometry measurement, for example, dual-energy X-ray absorptiometry (DEXA or DXA).[53]

Just as bones grow strong over time, they become brittle slowly. In addition, bone tissue is not all the same. Up to 50% of trabecular or spongy bone (wrist, vertebrae, and ends of long bones) may be lost, and up to 35% of cortical or compact bone (shafts of long bones) will be lost during a lifetime.[54] Average bone loss is 0.5–1% per year after approximately age 50; when bone density is below the "young normal" mean, osteoporosis is diagnosed.[55]

Prevalence

Osteoporosis is four times more common in women than men (80% compared to 20%). Men have larger bodies and denser bones than women, and thus have greater peak and total bone mass.[51] Blacks have denser bones and less osteoporosis than whites.[56]

Osteoporosis is dissimilar in men and women. Women tend to break hips and backbones, while wrist fractures are less common. Men with osteoporosis are also likely to have hip fractures but suffer fractures of or near the wrist more often than women. One out of every two women and one out of every eight men

Table 19.5 Comparing dietary approaches to insulin resistance with the American Heart Association Dietary Guidelines.

Macronurient	AHA, Krauss 2000	Syndrome X, Reaven 2000	Insulin Resistance, Coulston 1997	Low-calorie, Higher-Protein Diet, Holtmeier 2000
Carbohydrate	50–60	45	50	45
Protein	10–25	15	15–20	25
Fat	25–30	40	30–35	30

Krauss RM, Eckel RH, Howard B, et al. AHA Dietary Guidelines: revision 2000: A statement for healthcare professionals from the Nutrition Committee of the American Heart Association. Circulation 2000;102:2284–99.

over 50 will have an osteoporosis-related fracture in their lifetime.[57]

Etiology

Bone mass is gained primarily during growth periods, with peak bone density reached between ages 18 and 30.[56] Subsequently, bone mass remains stable until about age 40 to 50 for women and about age 60 for men. Inadequate building of peak bone mass coupled with significant bone loss leads to a low bone density and increased risk for fractures.

INADEQUATE BONE MASS   Although osteoporosis is seen most often in the elderly, the risk for developing osteoporosis in later years begins during childhood and adolescence. Development of osteoporosis is delayed when an individual develops bigger, denser bones during youth.[58] For example, an epidemiological study in Yugoslavia showed that higher calcium intake in youth led to higher peak bone mass, independent of exercise and other factors. Higher bone mass has also been associated with slower decline in later life.[59] More recent studies in the United States have also shown that getting enough calcium during growth spurts (between ages 11 and 17) reduces the risk of osteoporosis. In an intervention study, girls aged 11 to 12 years who received calcium supplements (500 mg calcium citrate-malate) gained an additional 1.3% bone mineral density per year compared to controls.[60] In a cross-sectional study of 18 to 31 year old women, positive correlations were found between adolescent milk intake in childhood and bone mineral density (total skeleton, spine, femoral neck, radius).[61] While there are many actions that build bones in older adults, preventing brittle bones is best done early in life.

Lack of exercise or inactivity leads to osteoporosis. Bedrest, hospitalization, and sedentary lifestyles make bones weak! Weight-bearing or resistance exercises are needed to develop bone mass because bone grows in response to pressure on the bone tissue. The more often and harder you push on the bone (not enough to break it, of course), the more the body will respond by depositing minerals (calcium, magnesium, phosphorous, fluoride, and boron) into the bone matrix. For instance, tennis players have significantly higher bone mass in their playing arm than nonathletic controls.[62] However, even the controls had more bone mass in their dominant arm than the less-used one. Exercise also stimulates growth hormone, which in turn stimulates bone development. "Use it or lose it" applies here. Lack of exercise leads to loss of bone and eventually a fragile skeleton. Fear of falling and breaking a bone keeps some older adults from getting much-needed exercise, contributing to a vicious circle.

INCREASED BONE LOSS   The skeleton acts as structural support and as a calcium reservoir for the body. Bone tissue includes jawbones and teeth. Bones and teeth contain about 99% of the calcium in an adult, roughly 2.2 to 3.3 pounds.[62] The remaining 1% of calcium is found linked with protein in blood, soft tissues, and extracellular fluids. This reservoir is needed for nerve transmission, muscle contraction, and enzyme systems such as those controlling blood clotting. Maintaining nerve transmissions takes physiologic priority over maintaining bone structure. In order to be consistently available to perform many functions, calcium is tightly regulated by hormone systems. When calcium levels in the blood fall, the body responds by secreting more parathyroid hormone (PTH). PTH acts on bone to release calcium and thus raise the blood calcium levels. Too much calcium in the blood stimulates calcitonin secretion. The hormone calcitonin slows release of stored calcium.[62] Bone mineral reserves are dissolved (resorption) and rebuilt constantly to maintain adequate calcium levels for these messenger functions of calcium.

A consistent dietary supply of bone-building minerals (i.e., calcium, magnesium, phosphorus, fluoride, boron) and vitamins (primarily D and K) coupled with regular weight-bearing exercise helps maintain the skeleton reserves and supply calcium when needed. When some portion of this build-dissolve-rebuild cycle is malfunctioning, the body's first priority is to maintain blood calcium levels for nerve, muscle, and enzyme functions. Bone loss results from inadequate nutrient levels or calcium absorption, or from excess calcium excretion.

Osteoporosis can also develop from a shortage of the mineral phosphate during bone mineralization. A balanced calcium-phosphorus ratio as provided in a varied diet allows both nutrients to be used by the bone-building cells.

Lack of sufficient phosphorus promotes release of calcium from the skeleton.[63] Although phosphorus is abundant in the food supply, some antacids bind with phosphorus, making it unusable by the body. In the absence of phosphorus, bone mineral formation is delayed until more phosphate is available. Shortage of vitamin D also delays bone mineral formation. Finally, the skin's ability to make vitamin D from the sun is less efficient with aging.

Effects

FALLS AND FRACTURES   Avoiding fractures is an important goal for older adults because fractures contribute to earlier death. Twelve percent (12%) of older women who broke a hip were dead within six months.[64] The rate for men who did not survive six months after breaking a hip was almost double (25%).[64] Death is not due to the fracture itself but to complications resulting from the break. One of these complications is impaired mobility, complicating all the activities of daily living (including eating and exercising). If an older adult has also had a stroke, impaired mobility becomes the leading cause of institutionalization in the United States. Furthermore, 50% of individuals who fall and break a bone are likely to need assistance walking, and 25% are permanently disabled.[56]

SHRINKING HEIGHT, KYPHOSIS   In contrast to hip fractures, most vertebral fractures (67%) are asymptomatic. Postmenopausal women with compression and/or a bone fracture in the spinal column have a condition known as "shrinking height," leading to dowager's hump (also known as *kyphosis*, meaning a bent upper spine). Shrinking in height is slow and usually not painful. The individual may not notice what is happening until someone else comments or until they notice that clothes no longer fit.

Risk Factors

A typical osteoporosis patient is a petite elderly white female. Brittle bones develop from a complex array of physiological factors, including nutrition and exercise. Major risk indicators are male and female glucocortcoid treatment and past history of fractures. Table 19.6 lists risk factors related to osteoporosis.

Nutritional Remedies

The first remedy is optimal diet, including intake of the recommended daily allowance of calcium for older adults, which is 1200 mg per day, not to exceed the tolerable upper limit of 2500 mg. The intake goal is to provide enough available calcium so that despite declining absorption rates, bone loss is minimized. Calcium retention increases with increasing calcium intake up to each individual's threshold. Beyond that threshold, greater intake of calcium does not lead to higher levels of calcium retention. A sample meal plan that provides the DRI for calcium in older adults is shown in Table 19.7.

In older persons, taking calcium supplements without other bone-building activity has not been shown to increase bone density. In fact, an individual at bed rest or otherwise immobilized loses bone mass rapidly. Several dietary components (protein, sodium, caffeine, vitamins) are closely linked to calcium metabolism and their intake can interfere with appropriate supplementation.

High protein intakes result in less calcium being available because protein leads to greater excretion of calcium into the urine. In the Nurses Health Study,

Table 19.6   Risk factors associated with osteoporosis in older adults.

Not Modifiable

Female, number of pregnancies, length of time between pregnancies

Age, age at which pregnancy(ies) occurred, breastfeeding

Caucasian, Asian

Thin, small-boned rather than large-boned

Family history of osteoporosis

Inadequate bone mass achieved during youth

Low body weight

Premature menopause

History of amenorrhea

Hypogonadism

Glucocorticoid fracture

Maternal history of hip fracture

Potentially Modifiable

Lack of weight-bearing exercise

Cigarette smoking

Long-term dietary phosphorus deficiency (e.g., use of phosphorus-binding antacid)

Heavy alcohol consumption

Underweight

Malnourished

Inadequate dietary calcium (<1200 mg) and vitamin D (<400/600 IU) intake

Still Controversial or Not Yet Clear

Diet high in phosphorus while low in calcium (mixed evidence)

Inadequate fluoride, boron, and magnesium in diet

Consistently high protein and/or sodium intake

Eating soy products (soymilk, textured protein, etc.) for its estrogen-like activity

women (aged 35 to 59, followed for 12 years) who ate more than 95 grams protein per day suffered more osteoporotic forearm (but not hip) fractures.[65] High sodium intake leads to higher levels of urinary calcium, that is, more calcium is excreted when higher levels of sodium are eaten. Caffeine to equal 2 to 3 cups of coffee daily (in post-menopausal women), consumed with less than 744mg/day of calcium has been associated with bone loss.[63] Along with supplementation, a number of nutritional remedies can improve calcium intake and absorption:

- Drink milk and take supplements with a meal, for food slows intestinal transit time and allows more calcium to be absorbed from the gut.

Table 19.7   A day's intake for an older adult: consuming at least 1200 mg calcium (1520 mg).

Food	Amount of Calcium mg
Oatmeal made with milk, 1 cup total	266
Banana, one medium	6
Coffee, 10 ounces with 1 ounce (2 tbs) evaporated milk	87
Turkey sandwich on whole wheat bread, lettuce, mayonnaise	54
Cheese added to sandwich, 1 ounce cheddar	148
Canned fruit cocktail, ½ cup	9
Iced tea, plain	0
Orange juice, calcium-fortified, 8 ounces	289
Roasted almonds, 2 tbs	33
Pasta with chicken, 1½ cups	54
Tomato slices, 2	2
Milk, 1%, 8 ounces	279
Sugar cookie, 1 medium	5
Chocolate milk, 8 ounces	287

Table 19.8   Amounts of Vitamin K in selected foods (DRI for ages 51 and older is 120 mcg/day for males, 90 mcg/day for females).

Food	Amount, Size	Vitamin K mcg
Kale, chopped	1 cup	547
Broccoli, cooked	1 small stalk, 5"	378
Swiss chard, chopped	1 cup	299
Pumpkin, canned	½ cup	196
Broccoli, raw	1 cup	180
Spinach, raw	1 cup	120
Cabbage, raw	1 cup shredded	102
Butterhead lettuce	1 cup chopped	67
Lettuce	½ cup shredded	59
Parsley, raw	1 tablespoon	22
Canola oil	1 tablespoon	20
Asparagus	4 small spears, ≤5"	19
Soybeans, dry roasted	¼ cup	16
Mayonnaise	1 tablespoon	11
Plums, raw	1 med, 2⅛" diameter	8
Peanut butter, smooth	2 tablespoons	3
Cucumber without skin	½ cup slices	1
Apple without skin	1 med 2¾" diameter	0.5

SOURCE: USDA provisional table, Vitamin K, http://www.nal.usda.goiv/fnic/foodcomp, accessed August 2000.

- Take calcium supplements at a different time than antacids, because stomach acidity enhances absorption.
- Consume foods that are rich in bone-building vitamins (C, D, B_6, K) at recommended levels. This will help to synthesize bone and develop the collagen matrix, into which minerals are deposited during bone mineralization.[66]
- *Vitamin C* plays a role in the development of the protein bone matrix (collagen), but has no defined role in treatment.
- *Vitamin D* (1,25-Dihydroxyvitamin D, also called calcitriol) stimulates active transport of calcium in the small intestine and the colon.
- *Vitamin B_6* is a cofactor in stabilizing collagen cross-links.
- *Vitamin K* is required for the formation of proteins that stimulate *osteoblasts* to build new bone and attract *osteoclasts* to initiate bone resorption. (see Table 19.8)

VITAMIN K CAUTION   While Vitamin K has been linked to building a stronger bone matrix, it also plays an important role in the blood clotting process. Older

OSTEOCLASTS   A bone cell that absorbs and removes unwanted tissue.

OSTEOBLASTS   Bone cells involved with bone formation; bone-building cells.

adults with a history of strokes are placed on an anti-clotting medication. Nutrition counselors advise people who are taking anti-clotting medication to maintain a stable Vitamin K intake. Big portions of broccoli and greens for a few weeks in summer could increase chances of blood clotting and this can increase the chances of causing another stroke. In addition, when a Vitamin K–containing supplement is added to the diet (e.g., taking 2–3 Viactiv™ calcium chews as per recommendation), 150% of Vitamin K is instantly provided. One can easily get too much of a good thing.

How much is too much? As yet, no tolerable upper levels are set for Vitamin K. But for fat-soluble vitamins, a general rule (although one that does not apply to Vitamin E) is to stay under 500% of the RDA or DRI.

Other Issues Impacting Nutritional Remedies

Hormones and exercise are addressed with nutritional remedies because they have a direct impact on the nutrient requirements of individuals.

HORMONES   Estrogen, testosterone, growth hormone, insulin-like growth factor-I, and parathyroid hormone increase calcium absorption rate. Hormone replacement, with or without additional calcium, has been shown to increase bone mineral density in early postmenopausal women (4.5% and 1.5% respectively).[67] Pines et al. also found that the women not receiving hormone therapy, either with or without calcium supplementation, lost bone mineral density (1.4% and 3.7% respectively).

People not taking hormone replacements may try soy (e.g., milk, textured protein, bars, soy nuts) for its estrogen-like activity. Evidence is not definitive that soy will mimic the effects of estrogen in bone development, though.

EXERCISE   While there is a relationship between calcium intake, exercise, and bone mass, it is not known whether exercise increases absorption or whether it enhances bone mineralization.

In summary, the best osteoporosis prevention strategy is exercise and adequate diet in young people when bones are first growing. For older individuals who have brittle bones, exercise supplemented with calcium and vitamin D strengthens bone mass. Medications (estrogen, selective estrogen receptor modulators, e.g., tamoxifen) can help, too, especially when people are immobilized or confined indoors. As much as possible, stay active and drink milk!

ORAL HEALTH

We can't have a healthy mouth, a great smile, or a good conversation without it. Saliva or "spit" lubricates living.

Dr. Nelson Rhodus, Director of Oral Pathology, University of Minnesota

Definition

Oral health depends on several organ systems working together: gastrointestinal secretions (saliva), the skeletal system of which teeth and jaw are a part, mucus membranes, muscles such as the tongue and jaw for chewing and swallowing, taste buds and olfactory nerves for smelling and tasting. Disturbances in oral health are associated with, but not necessarily caused by, aging.

Systemic diseases such as diabetes and hypertension have implications for oral health. For instance, the high blood sugar of diabetes accelerates periodontal disease and makes the mouth more susceptible to yeast infection (candidiasis) and *xerostomia*. Diuretic treatment for hypertension not only aids the kidney in urine excretion, but also leads to less salivary secre-

tion (makes the mouth dry). Lack of saliva for any reason provides bacteria a better environment for building plaque.

Prevalence

Certain changes in oral health are common in older adults:

- Decayed and missing teeth: nearly half of adults over 65 are edentulous (no teeth)
- Xerostomia: 41% of older adults have salivary gland dysfunction
- Candidiasis, yeast infection: roughly 83% of older adults with dentures have candidiasis; they also tend to have dysgeusia (loss of taste) and pain of the tongue (glossodynia)

Etiology and Effects

What and when we eat affects oral health. Oral health, in turn, affects what and how we eat, including our enjoyment of the process. Healthy teeth are protected by healthy enamel, but bacterial action on the breakdown products of food slowly erodes tooth enamel. For about 15 minutes after putting something into our mouth, oral bacteria feast on the food breakdown products, especially those of sucrose. Sticky foods like caramels or raisins can stay around longer, especially when they get between the teeth. Frequent eating and drinking of sugary beverages provides a continuous substrate for bacteria. The acid in carbonated beverages adds to the corrosive potential of food. Missing teeth and ill-fitting dentures can make chewing difficult and have a negative impact on eating.[68]

Saliva, which lubricates the mouth and begins the digestive process (amylase in "spit" begins starch breakdown), also helps to keep tooth enamel clean. However, saliva seems to become thicker and more viscous with age. Lack of sufficient and effective saliva, especially in the presence of gingivitis and periodontal disease, makes the oral

> **XEROSTOMIA**   Dry mouth, or xerostomia, can be a side effect of medications (especially anti-depressants), of head and neck cancer treatments, of diabetes, and also a symptom of Sjogren's syndrome, which is an autoimmune disorder for which no cure is known.

cavity more sensitive to temperature extremes and coarse textures, thus making eating painful as well as slowing absorption of nutrients from food.

Risk Factors

Nutritional factors during critical periods of growth and development influence subsequent resistance to oral disease. Better enamel is more resistant to caries, fluoride treatment or fluoridated water strengthens enamel; and a strong skeleton is likely to include strong jawbones that will securely hold the teeth.

Xerostomia and periodontal disease are most detrimental to oral health in older adults, while decayed and missing teeth and ill-fitting dentures can interfere with speech and chewing.

Nutritional Remedies

A healthy body supports a healthy immune system. Adequate calories, protein, vitamins, and minerals, as well as fluids, support a healthy oral cavity by providing materials for tissue repair and regeneration, lubrication for food, and energy for metabolism.

To keep teeth healthy:

- Rinse, brush, or chew gum after eating; do whatever you can to stimulate saliva and remove substrate for bacteria. Sugary gum is better than doing nothing, the sugar is soon gone, but the saliva continues to rinse the teeth. Gum without sugar but with xylitol (a sugar alcohol) helps to clean teeth after eating and drinking.
- If you have the habit of sipping and snacking throughout the day, stay away from the sticky stuff: caramel, raisins, and sweet rolls are all sources of sugar for bacteria. Make sips of soda, tea, or other drinks sugar free if you can't brush afterwards.
- Fight dry mouth with sips of water.

GASTROINTESTINAL DISEASES

The gastrointestinal (GI) system is roughly the length of a football field and serves so many functions that it should not be surprising to learn that, by late adulthood, it occasionally malfunctions. It seems miraculous that we so consistently eat what we like, without much thought, and our body converts that food to energy for daily living. Parts of the GI system most likely to malfunction in old age are:

LES   Lower esophageal sphincter, which is the muscle enabling closure of the junction between the esophagus and stomach.

1. The esophageal-stomach juncture: weakened muscle results in gastroesophageal reflux disease (GERD)
2. The stomach: decreased acidity leading to changes in nutrient absorption or increased acidity causes ulcers
3. The intestines: resulting in constipation, diarrhea, and some food intolerance

Often these problems are secondary to other diseases. No matter what the cause, older adults are at higher risk for some of the GI conditions discussed next, conditions which may impair their activity.

Gastroesophageal Reflux Disease (GERD)

DEFINITION   Gastroesophageal reflux disease occurs when stomach contents flow back into the esophagus.

PREVALENCE   Approximately 19 million Americans or one of five older adults have GERD.[69,70]

ETIOLOGY AND EFFECTS   It is not clear if acid in the esophagus leads to a weakened lower esophageal sphincter *(LES)* or if a weakened sphincter leads to GERD.[71] The main symptoms of GERD are heartburn and acid regurgitation. Stomach contents, which are highly acidic, spill back into the esophagus, resulting in irritation and pain. Other effects are chest pain, trouble swallowing, nausea and vomiting, and belching.

NUTRITIONAL RISK FACTORS   Alcohol in excess of seven drinks per week, obesity, and smoking are consistently linked to GERD episodes.[69,70] In addition, regular and decaffeinated coffee are associated with heartburn.[70]

NUTRITIONAL REMEDIES   The primary dietary remedy is to omit foods that are chemically or mechanically irritating. There is little consistency about which foods these are. However, general guidelines are to choose a low-fat diet and nonspicy foods. Chew thoroughly and eat slowly. Finally, to take advantage of gravity and make it difficult for the stomach acids to reflux upward, don't lay down after eating.

Constipation

DEFINITION   There are many definitions that constitute what "a normal bowel pattern" is. The range can be from two or three bowel movements per day to two or three bowel movements per week.

PREVALENCE   When two or fewer bowel movements per week is used as the definition of constipation, the prevalence of constipation was 3.8% in older adults aged 60–69 years and 6.3% in older adults aged 80 years and older.[72] In a midwestern community of people ages 65 years and older, as many as 40.1% reported some type of constipation.[73]

ETIOLOGY   The physiological causes of constipation in older adults are emerging.[74] Animal studies indicate that aging intestinal muscles respond less to triggers and that the brain-muscle transmitters are either inadequate or less responsive. Older adult's thirst mechanisms decline. Lower levels of stomach secretions and potentially less muscle strength affect peristalsis. Due to chewing problems, there is a potential for eating fewer fiber-rich foods.

EFFECTS   Persons with laxation difficulties are prone to be more anxious and focused on bowel movements as an aspect of health. Lower fiber diets can also exacerbate *diverticulitis*.

NUTRITIONAL RISK FACTORS   Risk factors for constipation include:

- Poor water intake
- Eating small amounts of food
- Medications, such as nonsteroidal anti-inflammatory drugs
- High iron mineral supplement

NUTRITIONAL REMEDIES   Remedies are to encourage increased dietary fiber in tandem with an increased fluid intake. Adding bran is a way to add bulk to help stimulate movement through the colon. It is, again, important to add water when increasing fiber.

Inflammatory Diseases

DEFINITION   Examples of inflammatory disease are osteoarthritis, rheumatoid arthritis, atrophic gastritis (typically due to *Helicobacter pylori* infection), celiac disease (*gluten* intolerance), Irritable Bowel Disease (IBS), diverticulitis (an infection in the large intestine), and asthma. Of these conditions, arthritis affects the greatest number of older individuals.

PREVALENCE   Osteoarthritis is the most common form of arthritis, affecting roughly 20 million people in the United States.[75] Prevalence of osteoarthritis at the knee is twice as common as at the hip (6% compared to 3%); the knee and hip are the two most common sites for persons aged 30 and older. Prevalence increases with age and peaks between 70 and 79 years. More men than women have osteoarthritis before age 50; after age 50, women are more often affected. (Study results are consistent about the potential benefits of estrogen treatment in osteoarthritis.)

ETIOLOGY   Cartilage loss, bone changes such as bone outgrowth, hardening of soft tissues, and inflammation all lead to tissue damage. Some wear and tear of aging also contributes to the complex disease of os-

teoarthritis. Variability in prevalence and progression of knee, hip, and hand osteoarthritis suggests that the disease may be a group of different and unique conditions. Rheumatoid arthritis is a collagen disease characterized by inflammation and increased protein turnover.

EFFECTS   Pain occurring with joint movement is common, but the cause of pain is still unexplained.[75] Osteoarthritis often leads to disability: In the words of Felson and colleagues:[75]

> The risk for disability (defined as needing help walking or climbing stairs) attributable to osteoarthritis of the knee is as great as that attributable to cardiovascular disease and greater than that due to any other medical condition in elderly persons.(p. 642)

Individuals with osteoarthritis tend to have pain, depressive symptoms, muscle weakness, and poor aerobic capacity.

NUTRITIONAL RISK FACTORS   Obesity, continuous exposure to oxidants, and possibly low vitamin D levels are risk factors for developing osteoarthritis. Low intake of vitamins C and D are risk factors for its progression. Risk factors may turn out to be different for knee or hip and hand osteoarthritis.

NUTRITIONAL REMEDIES   Weight loss is the first remedy advised. Felson et al.'s summary report[75] found that women in the Framingham study who lost 11 pounds cut their risk for knee osteoarthritis in half. However, the connection between weight loss and relief of symptoms of osteoarthritis has not yet been well studied. Evidence to date suggests weight loss helps to reduce symptoms.

> **DIVERTICULITIS**   Infected "pockets" within the large intestine.
>
> **GLUTEN**   A protein found in wheat, oats, barley, rye and triticale (all in the genus *Triticum*); gliadin is the toxic fraction of gluten.

Antioxidants   Individuals with the highest levels of vitamin C intake had significantly slower (three-fold) disease progression and less knee pain than individuals with the lowest intakes. Results of increased vitamin E and beta-carotene intake were inconsistent.

Vitamin D   Progression at higher intakes and serum levels is roughly one-third slower than at the lowest levels. The DRI for older adults is 400 to 600 IU, with a tolerable upper intake level of 2000 IU.

Flavonoids   This is a large group of phytochemicals (found in vegetables, fruits, tea, whole grains) with antioxidant and anti-inflammatory activity which help

Photo Disc

Case Study 19.2
Ms. Wetter: A Senior Suffering through a Bad Stretch

Ms. Wetter: About to turn 81, Elizabeth Wetter is 5 feet 6 inches tall and weighs 106 pounds. She has had Parkinson's disease for five years, but that is not what concerns her. Her problem is pain from arthritis and lack of energy. She saw an ad on television for a vitamin-mineral supplement with ginseng that promises "more energy." Her son also told her to take a liquid dietary supplement to "feel better." Eighteen months ago, she had successful surgery for colon cancer, which was followed by chemotherapy treatments. She is now free of cancer. After the cancer treatment she fell and broke a hip. This healed well, but serious leg pains started shortly afterwards. There seems to be no cause for the pain, and a cure is unavailable. She is no longer able to take her walks through the neighborhood or tend her prize-winning garden. She would like to weigh 118 pounds again (her "usual") and is seeking nutritional counseling to try to regain some of her energy.

- What are some of the nutritional issues faced by Ms. Wetter? (Hint: Calculate her current weight to usual body weight as a percentage.)
- How would you prioritize these in a nutritional care plan?
- Calculate her energy needs and suggest strategies she might use to regain some energy.
- What other information would you want to know in order to counsel Ms. Wetter?

---

to maintain cell membranes. Higher levels of antioxidants are required to act as free-radical scavengers in inflammatory diseases.[76]

Chondroitin and Glucosamine　These two substances naturally occur in the body as substrate for cartilage repair. Although the mechanisms by which they act are not known, most clinical trials of their use demonstrate favorable effects, with a slightly larger effect for chondroitin sulfate.[77]

SAM-e　S-adenosylmethionine may reduce pain and stiffness and possible depression; it may also cause nausea in large doses.

Capsaicin　The compound that provides the heat in peppers is used in a topical cream for pain relief. We don't know anything about the effect of eating hot peppers.

Other Treatments　Fatty acids and oils are other anti-inflammatory therapies with potential to lessen signs and symptoms in a variety of conditions including osteo- as well as rheumatic arthritis. Borage seed and evening primrose oils contain GLA or gamma-linolenic acid which plays a role in prostaglandin synthesis (GLA competes with other fatty acids, limiting

the production of inflammatory omega-6 fatty acids) and can decrease pain of arthritis after several months use.[76] Reductions in pain, morning stiffness, and swelling of rheumatoid arthritis follow intake of omega-3 fatty acids (fish oil, marine fatty acids) for at least 12 weeks.[78] A type of omega-3 fatty acid is found in vegetables such as flaxseed and purslane.

Traditional Native American medical therapy used Echinacea (three varieties) for pain relief, rheumatism, and arthritis, ginseng for asthma and rheumatism, garlic for asthma, and evening primrose oil for obesity. Today Echinacea and ginseng are marketed to boost immune function. Evening primrose oil is marketed as an antioxidant and for its role in decreasing premenstrual pain.[79]

Vegetarian Diets　Plants are rich in antioxidants that may play a role in managing inflammatory diseases. A review of the roles of vegetarian diets[80] found that it is difficult to design good experiments to answer the question: "Do vegetarians have less inflammatory disease?"

Food Allergies　Gluten has a clear role in celiac disease, which is an inflammatory disease. Food allergies have also been suggested as a cause for inflammatory diseases such as irritable bowel syndrome, Crohn's disease, and rheumatoid arthritis, but evidence from well-done studies that can point to allergens or mechanisms is not yet available.[63]

Complementary and Traditional Medications

DEFINITION   Prescription medications are medicines that a physician or other health care provider acting within the scope of their license can order. Prescription drugs, according to health care plans, do not include vitamins, herbal medicines, or over-the-counter (OTC) medicines. OTC medications are any pills, liquids, salves, creams, and supplements that are purchased at a pharmacy, discount, or food store without prescription. Most complementary medicines such as botanicals and herbs are sold as OTCs.

PREVALENCE   A national survey on prescription drug usage by older adults was completed in September 2000 by the Harvard School of Public Health.[81] When asked, "How many separate bottles of prescription drugs would you estimate you have in your medicine cabinet right now," 14% had 11 or more different prescriptions. Eighty-five percent of adults aged 65 or older were taking prescription medications regularly. When the primary respondent, aged 65 years or older, was asked how many prescription drugs he or she took on a regular basis, 59% were "heavy users" (i.e., took three or more prescriptions) and 59% took OTC medications regularly or sometimes.

ETIOLOGY   A high prevalence of ongoing health problems leaves older adults using more medications, both non- and prescription types, than younger adults.

EFFECTS   Some medications interfere with appetite, food digestion, and absorption. This leads to unintentional weight loss or gain. The expense of extra medications reduces the amount of money available for food.

MENTAL HEALTH AND COGNITIVE DISORDERS

Definition

A group of disorders called dementia or cognitive disorders includes depression, Alzheimer's disease and *memory impairment*. Often considered an expected (and dreaded) aspect of "getting old," this "dementia of aging" is being more carefully defined as knowledge grows about the mechanisms of normal aging. It is impor-

tant to note that Alzheimer's disease and dementia symptoms are not a part of normal aging.

Definitions of these conditions are:

- Depression is a brain disorder characterized by "being down in the dumps" and not wanting to socialize with others over a long period of time.
- Alzheimer's disease is when the brain atrophies and there is a slow and steady loss of sense of self and a meaningful relationship with the world. This includes memory loss, behavior and personality changes, and a decline in ability to think.
- Dementia is a clinical state in which intellectual level declines; it usually involves deterioration in memory and one or more of the other intellectual functions such as language, spatial thinking and orientation, judgment, and abstract thought.

Prevalence

Since Alzheimer's is difficult to diagnose, prevalence statistics are "approximate estimates." In 1997, the estimated prevalence of Alzheimer's Disease in the United States was 2.32 million cases (estimates range from 1.09–4.58 million).[70] The risk of developing Alzheimer's disease doubles every five years after age 65. Among people 65 years and older, as many as 3 in 100 suffer from clinical depression.[70] Table 19.9 lists prevalence data for other cognitive disorders.

Etiology

Cognitive disorders stem from brain damage including tumors, broken blood vessels, and neurological diseases such as Alzheimer's. What causes Alzheimer's disease? No one knows. We do know that aluminum cookware does not cause Alzheimer's disease.

MEMORY IMPAIRMENT   Moderate or severe impairment is when four or fewer words can be recalled from a list of 20.

Aluminum, copper, carnitine, and choline have been examined as potential causes because of the role they play in neurological functioning. Vitamin B_{12} deficiency can

Table 19.9   Percentage of persons aged 65 years and older with cognitive disorders.[22]

Age	Percentage with Moderate to Severe Memory Impairment		Percentage with Severe Depressive Symptoms	
	Males	Females	Males	Females
65–74 years	15.4	9.7	22.4	35.2
75–84 years	39.0	30.6	27.5	39.8
85+ years	37.3	35.0	22.5	23.0

lead to neurological damage and cognitive decline, but it is not an underlying factor in Alzheimer's disease.

Effects

Weight loss is a nutritional feature associated with the diagnosis of Alzheimer's disease. Short-term memory loss, difficulty in decision making, depression, and changes in smell and taste are a few of the aspects of this disease that make good nutritional habits difficult to maintain. While individuals may be able to maintain an adequate diet early in the disease, they will need nutritional assistance when the disease progresses to later stages.

Nutritional Remedies

Additional calories may be needed for increased energy expended by individuals who pace and wander. Vitamin E and selenium may be useful[70] for their antioxidant properties; choline (in soybeans and eggs) is a building block for acetylcholine but has not been proven useful in controlled research. Developing routines and eliminating distractions during meals, allowing plenty of time to eat, serving foods one course at a time, offering finger foods, and tracking fluid intake to ensure adequacy are some of the nutritional strategies in treating the disease. Safe use of kitchen tools and equipment and food safety are two food-related considerations as the disease develops.

Gingko biloba may help to enhance memory due to its antioxidant and anti-inflammatory properties.

OBESITY AND LOW BODY WEIGHT/UNDERWEIGHT

Obesity

DEFINITION   Chapter 17 covers obesity, weight guidelines, and recommended BMI values.

PREVALENCE   Obesity (BMI >30) rates are rising for adults of all ages, with 24% of men and 27% of women aged 65 years and over classified as obese.[82] Overall, the economic cost of obesity in the United States has been estimated to be nearly $100 billion (in 1995 dollars).

ETIOLOGY/EFFECTS/RISK FACTORS   Current evidence supports the notion that the body mass index associated with the lowest mortality falls within the range of 18.5–24.9 in men and women aged 30–74. However, whether age impacts the association between BMI and deaths is controversial based on the

data gathered and analyzed.[83] Based on Ernsberger's analysis,[84] morbidity and mortality is not any higher, and sometimes is lower, in older people who are at the high end of the BMI continuum. In fact, in a study of older, community-dwelling Canadians, an increased BMI was associated with lower mortality.[85]

In addition, weight cycling is associated with higher mortality, and evidence suggests that the main emphasis should be in primary prevention of obesity through lifelong weight maintenance.[86] In cases of severe obesity, gradual weight loss is warranted.

For older adults, extra weight during illness episodes, especially hospitalizations, seems to be protective. Materials developed in 1999 by the American Dietetic Association Long Term Care Task Force in conjunction with Health Care Finance Administration (HCFA) suggest a BMI range of 19–27 as an acceptable and health-promoting weight range for older adults.[87] This range updates the Nutrition Screening Initiative materials that use the BMI cut-points of 22–27 (items published before 1999).

NUTRITIONAL REMEDIES   Older obese individuals should undertake a healthy eating program based on enough nutrient-dense calories to support a slow weight loss. Age is not a factor in formulas to calculate calorie levels supporting desired activity and weight loss. Ensuring adequate nutrient intake by eating and drinking a balance of servings from basic food groups, for instance as outlined in the Food Guide Pyramid, is straightforward advice that can work for older as well as younger adults.

Low Body Weight/Underweight

DEFINITION   There is no consensus on universal definitions for underweight in the frail elderly, but a low BMI or weight in the lowest percentiles of a reference standard is a starting point. Population weight percentiles are given for older adults by decade (beginning at 50 years), gender, and ethnic racial groups.[88] An individual's current weight is often compared to "usual" body weight.

Terms such as sarcopenia (muscle loss), anorexia (no appetite), and cachexia (another way to say no appetite) are related to undernutrition in older adults and are associated with becoming frail. The National Heart, Lung, and Blood Institute (NHLBI) defined underweight as a BMI <18.5 kg/meter squared for all adults.[89] The World Health Organization (WHO) further defines levels of underweight as grades of "thinness."[90]

BMI 17.0–18.49 indicates grade 1 thinness
BMI 16.0–16.99 indicates grade 2 thinness
BMI <16.00 indicates grade 3 thinness

PREVALENCE Approximately one-third of older adults are underweight.

ETIOLOGY Underweight is not considered problematic when the individual has had a life-long low weight. However, weight cycling is problematic. In addition, unintentional weight loss is likely due to disease; when greater than 10% of total body weight in six months, it leads to mortality. Intended weight loss should be consistent with reasonable weight-for-height standards.

EFFECTS For older adults, underweight is much more serious than overweight. Being thin has been related to increased incidence of disease, but it is impossible to tell from the data whether thin precedes or follows incidences of disease.

NUTRITIONAL RISK FACTORS Protein calorie malnutrition leads to underweight. Underlying causes may be illness, poverty, or functional decline.

NUTRITIONAL REMEDIES Avoiding weight loss is desirable, but not always possible. Weight loss is most often associated with disease states. Medical nutrition therapy for a frail elderly malnourished person should occur in consultation with an experienced registered dietitian. Re-feeding will be done slowly:

- *Calories:* eat and exercise to build muscle mass, strength.
- *Protein:* 1 to 1.5 grams of protein per kilogram body weight is adequate. 1.5–2 g/kg/day is recommended for severe depletion by the American Dietetic Association's Consultant Dietitians in Health Care Facilities Practice Group. Exceptions are patients with renal or liver failure, who may need a protein restriction.
- *Water:* 1 ml per calorie, rehydrate slowly (see section on dehydration).

VITAMIN B_{12} Deficiency

Definition and Etiology

The normal process by which the body absorbs vitamin B_{12} from foods requires the following conditions. The stomach environment needs to be acidic and producing enzymes (especially pepsin) and the stomach cell lining needs to be secreting an intrinsic factor (IF). Stomach acids and enzymes split off and transfer vitamin B_{12} from foods to carrier proteins (mostly secreted by the salivary gland in the mouth). Then, the vitamin B_{12}-carrier protein complex moves into the

small intestine where the vitamin B_{12} will again be broken off and bound to the IF (that was produced in the stomach and migrated to the small intestine). The vitamin B_{12}-IF linked complex then binds to a specific site in the lining of the small intestine, is transported across, and released into blood serum to be taken to tissue cells. (See Illustration 19.1).

The two primary types of vitamin B_{12} deficiency in older adults are: (1) pernicious anemia due to lack of intrinsic factor (IF) being released from the stomach cell wall, and (2) lack of adequate dietary intake (i.e., lack of extrinsic factor) or poor absorption of vitamin B_{12}.

Pernicious Anemia

Pernicious anemia results when intrinsic factor (IF) is lacking. IF is released from the stomach wall and needed for the dietary vitamin B_{12} to be absorbed in the small intestine. Physical signs of pernicious anemia (macrocytic megaloblastic anemia) include large, undeveloped red blood cells, increased redness and swelling in the mouth (glossitis), and tongue fissures. Individuals with pernicious anemia also tend to have shrunken stomach glands and mucosa leading to a decreased enzyme and hydrochloric acid secretion. Lack of IF leads to vitamin B_{12} deficiency.

EFFECTS Effects of pernicious anemia are glossitis and irreversible neurological damage. Other symptoms may also include fatigue, anorexia, weight loss, nausea and vomiting, diarrhea or constipation, and tachycardia.[70] Pernicious anemia is often accompanied by inadequate stomach acidity (hypochlorhydria). Less stomach acid leads to impaired iron absorption.

PREVALENCE Pernicious anemia takes five to six years to develop and is rarely seen before age 35. It is found in approximately 2% of the population, most commonly in women over 65.[70]

RISK FACTORS Risk factors are a history of *Helicobacter pylori (H. pylori)* infections, decreased stomach acid production, and a family history of pernicious anemia.

NUTRITIONAL REMEDIES Vitamin B_{12} may be given orally or by injection. Injections are administered daily until lab values stabilize, then six to eight times per year throughout life.[70] In Scandinavian countries, injections are prescribed until pernicious anemia is in remission. Then, maintenance doses of oral vitamin B_{12} are used. Roughly 80% of Scandinavian therapy costs are

Illustration 19.1   Overview of vitamin B₁₂ absorption.

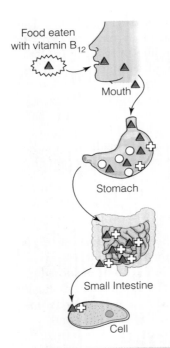

Food eaten with vitamin B₁₂

Mouth

Stomach

Small Intestine

Cell

The salivary glands in the mouth secrete enzymes and mucous to begin breaking down foods and provide substrates for binding.

The food substrate-enzyme mixture enters the stomach where acid and enzymes (especially pepsin) detach the vitamin B₁₂ from protein in foods and reattach it to binder proteins. The acid and enzymes trigger release of an intrinsic factor from the stomach cell wall (known as vitamin B₁₂ intrinsic factor, or IF).

The B₁₂-complex moves to the small intestine, where enzymes secreted by the pancreas release vitamin B₁₂ and transfer it to the IF that was secreted by the stomach. Next, the vitamin B₁₂-IF complex attaches to specific receptor sites in the intestinal cell wall, then moves across and is released into the blood plasma

COBALAMIN Another name for vitamin B₁₂. Important roles of cobalamin are fatty acid metabolism, synthesis of nucleic acid (i.e., DNA, a complex protein that controls the formation of healthy new cells), and formation of the myelin sheath that protects nerve cells.

METHYLMALONIC ACID (MMA) An intermediate product that needs vitamin B₁₂ as a coenzyme to complete the metabolic pathway for fatty acid metabolism. Vitamin B₁₂ is the only coenzyme in this reaction; when it is absent, the blood concentration of MMA rises.

KREBS CYCLE A series of metabolic reactions that produce energy from the proteins, fats, and carbohydrates that constitute food.

HOMOCYSTEINE Another intermediate product that depends on vitamin B₁₂ for complete metabolism. However, both vitamin B₁₂ and folate (another B vitamin) are coenzymes in the breakdown of certain protein components in this pathway. Thus, elevated homocysteine levels can result from vitamin B₁₂, folate, or pyridoxine deficiencies.

Key

▲ Dietary vitamin B₁₂ (extrinsic)

✚ Intrinsic Factor vitamin B₁₂

▲✚ Protein bound vitamin B₁₂ complex

○ Protein-binder complex found in stomach at normal acidity

spent on oral vitamin B₁₂ administration rather than injections. One reason why oral maintenance therapy works is that approximately 1–2% of synthetic vitamin B₁₂ (not found bound) is absorbed passively, removing the need for either intrinsic factor or carrier protein.

Foods high in vitamin B₁₂ are unlikely to maintain adequate blood levels in pernicious anemia. However, foods fortified with crystalline or synthetic vitamin B₁₂ increase the likelihood of passive absorption. Other dietary recommendations include eating proteins of high biological value and, if there is an iron deficiency, iron.

Inadequate Dietary Intake or Absorption (Protein-bound Vitamin B₁₂ Deficiency)

DEFINITION AND ETIOLOGY The more common cause of vitamin B₁₂ deficiency in older adults is related to abnormal stomach function. For instance, after prolonged inflammation, the stomach mucosa shrinks and secretes less acid and enzymes, a condition known as atrophic gastritis. Older adults with atrophic gastritis still absorb some dietary vitamin B₁₂, but they will become deficient over time. Bacterial overgrowth (usually related to infection with *H. pylori*) will also use vitamin

B₁₂ and diminish the amount available for absorption. Finally, antacid treatment in individuals neutralizes stomach acid (i.e., pH increases) and this rise in stomach pH does not allow all of the available vitamin B₁₂ to be split from the food carrier, even though the intrinsic vitamin B₁₂ factor is present. Symptoms of vitamin B₁₂ deficiency begin to appear after 3–6 years of poor absorption.[70,91]

PREVALENCE The prevalence of vitamin B₁₂ deficiency caused by poor absorption is greater in older adults than pernicious anemia. Protein-bound vitamin B₁₂ is estimated to occur in 20% of people over 69 years old (known or undiagnosed).[92] In reviewing studies, Nilsson-Ehle found prevalence rates of up to 80%, but 20–30% was typical.

Serum *cobalamin* levels are used to detect deficiency, although the cut-off limit has been set too low for the elderly.[93] It is more effective to measure the blood levels of intermediate breakdown products in pathways where vitamin B₁₂ is needed. For example, vitamin B₁₂ is needed to change *methylmalonic acid* (MMA) into a component used in the *Krebs cycle*. When adequate amounts of vitamin B₁₂ are unavailable, blood levels of MMA rise. Since B₁₂ is the only coenzyme to catalyze this reaction, a test for levels of MMA is specific to vitamin B₁₂ deficiency.

Another test is to measure blood levels of the coenzyme *homocysteine*. If vitamin B₁₂ isn't available to complete the pathway that forms DNA, then homocysteine (intermediate metabolite) levels become elevated. Since folate is also needed in this metabolic pathway, lack of either folate or vitamin B_{12} can lead to high levels of homocysteine.

Therefore, elevated MMA and homocysteine levels confirm a vitamin B_{12} deficiency. However, if MMA levels are normal but homocysteine levels are high, folic acid deficiency or another cause of increased homocysteine levels should be investigated.

EFFECTS   Vitamin B_{12} deficiency leads to irreversible neurological damage, walking and balance disturbances, and cognitive impairment (including confusion and mood changes). High levels of homocysteine are known to be a risk factor for heart and peripheral vascular disease.[94]

RISK FACTORS   Risk factors include advanced age, gastrointestinal disorders, genetic family patterns, medications, and (to some extent) inadequate food intake. An example of aging as a risk factor comes from Nilsson-Ehle; who found that 1.5% of 50-year-olds have achlorhydria and B_{12} deficiency as compared to 18% of 80-year-olds.[92] Gastrointestinal disorders that affect vitamin B_{12} absorption include atrophic gastritis (with higher risks in diabetes mellitus and autoimmune thyroid disorders), partial stomach removal, and *H. pylori* infection. Medications that suppress stomach acid secretion or impair absorption are associated with the risk of deficiency. Examples of these types of medicines are oral biguanides (e.g., metformin used to treat type 2 diabetes mellitus), modified-release potassium preparations, anesthesia, hydrogen-receptor antagonists (e.g., Cimetidine or Tagamet) and proton pump inhibitors (e.g., omeprazole/Prilosac given for gastroesophogeal reflux disease). In genetically linked cases, individuals often have antibodies for the intrinsic factor.

Vegetarians not taking a B_{12} supplement are also at risk of vitamin B_{12} deficiency. Hokin[95] studied 245 Australian Seventh Day Adventist ministers. Even with intrinsic vitamin B_{12} factor present, 53% of the group had serum cobalamin levels below the reference level for this study (171–850 picomoles per liter). Most of the subjects (70%) had a vitamin B_{12} deficiency due to a low dietary intake.

Some ethnic and gender differences may exist for developing a vitamin B_{12} deficiency. Elderly white men had the highest prevalence rates, whereas black and Asian women had the lowest rates among community-dwelling seniors (aged 60 or more) in Los Angeles.[96] In contrast, Zeitlin et al[97] concluded that there were no age- or gender-related differences in the mean levels of serum B_{12} after studying approxi-mately 440 subjects in the Bronx Aging Study (mean age = 79 years).

NUTRITIONAL REMEDIES   The DRI of vitamin B_{12} for men and women over 50 years is 2.4 mcg/day[98], and the tolerable upper intake level is not determined.[99] Good food sources include meats, shellfish (shrimp, crab, mussels), and milk and milk products. Oral pharmacological doses for vitamin B_{12}–deficient patients are 0.2–1 mg (200–1000 mcg) and are above the DRI because roughly 1–2% is absorbed through passive diffusion in the small intestine.[92] Foods high in folate and iron are also necessary in cases where folate deficiency and/or iron deficiency has been diagnosed. For older adults, a bowl of vitamin B_{12}–fortified whole grain cereal is a "power food."

FOODBORNE ILLNESS AND FOOD SAFETY

Definition

The definition of foodborne illness risk is "a biological, chemical or physical property that may cause a food to be unsafe for human consumption or the potential for a substance to cause injury under a given set of conditions."[100]

Prevalence

How widespread this problem is is basically unknown since many foodborne illnesses are not reported when individuals call it "the flu." Vulnerable groups are older adults and those with compromised immune status, such as HIV/AIDS and cancer patients.

Etiology/Effects

Poor storage, thawing, and food handling practices leading to microorganism growth are generally to blame for these illnesses.

Bacteria and viruses, especially Campylobacter, Salmonella, and Norwalk-like viruses, are among the most prevalent causes of foodborne illnesses. Signs and symptoms may appear within half an hour of eating a contaminated food or may not develop for up to three weeks. They include gastrointestinal distress, diarrhea, vomiting, and fever.

Risk Factors

The following list contains the leading practices putting an older person at risk:

- Improper holding temperatures of foods
- Poor personal hygiene

- Contaminated food preparation equipment (cutting boards, knives)
- Inadequate cooking time

Nutritional Remedies

The new "Dietary Guidelines for Americans" provide the following suggestions for remedies:

- Wash hands and surfaces often.
- Separate raw, cooked, and ready-to-eat foods while shopping, preparing, or storing.
- Cook foods—especially raw meat, poultry, fish and eggs—to a safe temperature
- Refrigerate or freeze perishable or prepared foods within two hours
- Follow the label for food safety preparation and storage instructions
- Serve hot foods hot (140°F or above) and cold foods cold (40°F or below).
- When it doubt, throw it out!

DEHYDRATION

Definition

Dehydration is the physiological state in which cells lose water and metabolic processes are hindered. Normal urination does not cause dehydration. Phillips et al.[101] defined dehydration as losing nearly 2% of initial body weight; this can occur after not drinking any fluid and eating only dry foods for 24 hours. Abnormally high serum sodium levels (>150 milliequivalents per liter) or a high ratio of blood urea nitrogen to creatinine (>25) can also be used to diagnose "significant dehydration."[102]

Dehydration can be measured as percentage of body weight lost when normal body weight is known. In a continuum of dehydration shown by Briggs and Calloway[103] (Table 19.10), originally designed for the National Aeronautics Space Administration (NASA), the indicators are somewhat different from the indicators that Gross et al. found. The NASA continuum shows how the human body responds to water losses.

Weight loss of 4% normal weight would be hard to ignore unless the individual had a compromised mental or cognitive status. Flushed skin, nausea, and apathy or lack of energy occur when 4% body weight is lost due to dehydration.

ANTI-DIURETIC HORMONE Hormone that causes the kidneys to dilute urine by absorbing more water.

Prevalence

Although the human body is quite resilient in that it can lose water up to 20% of its weight before death, dehydration is common for older people. Prevalence data is hard to gather because dehydration is usually temporary and there is no simple or commonly used standard definition for diagnosing dehydration in the elderly. For example, patients coming to a nursing home from a hospital may have IV tubes, and nursing notes may simply say, "The patient looks well-hydrated."

Etiology

Aging itself does not cause dehydration, even though the percentage of total body water shrinks from infancy to old age. But dehydration occurs more often in the elderly as a result of illness or other problems. Older people are less sensitive in detecting thirst than younger people and therefore may forget to drink.[101] Once fluids are consumed, aging kidneys may lose the ability to concentrate urine and the *anti-diuretic hormone* may become less effective. Swallowing problems, depression, or dementia may cause individuals to forget to eat or drink. Decreased mobility impairs older adults' access to water, and subsequent mobility to the bathroom. Fear of incontinence, in general, is another reason leading to decreased fluid intake and subsequent dehydration.

Table 19.10 Percent of initial weight lost due to dehydration and physiological signs.

Percent Lost	Physiological Signs
1%	Thirst (true for young people, *not* necessarily in older men or women)
4–6%	Economy of movement, flushed skin, sleepiness, apathy, nausea, tingling in arms, hands, feet, headache, heat exhaustion in fit men, increases in body temperature, pulse rate, respiratory rate
8%	Dizziness, slurred speech, weakness, confusion
12%	Cognitive signs: wakefulness, delirium
20%	Bare survival limit

So, for someone who normally weighs 160 pounds:

1% loss means weight down to 158.4 lb
4% loss means weight down to 153.6 lb
6% loss means weight down to 150.4 lb
20% loss means weight down to 128.0 lb

Adapted from Briggs and Calloway, originally NASA, 1967.[103]

Effects

In older adults, thirst and skin turgor (pinch a skin fold on the forearm, forehead, or over the breastbone and observe it fall back) are not good indicators of dehydration.[102] In the same review of 38 potential indicators of dehydration in older adults (61 to 98 years old, median age 82; n = 55, half were free-living, half were admitted to the hospital emergency department from extended care), seven signs and symptoms were strongly related to dehydration (p < .01 or p < .001), and *not* to patient age:

1. Upper body muscle weakness
2. Speech difficulty
3. Confusion
4. Dry mucous membranes in nose and mouth
5. Longitudinal tongue furrows
6. Dry tongue
7. Sunken appearance of eyes in their sockets

Although regulation of body temperature is one of the functions of the body's water compartment, fever or elevated temperature did not identify persons with dehydration status in Gross's study of individuals admitted to emergency rooms.

Dehydration increases susceptibility of developing urinary tract infection, pneumonia, and pressure ulcers; it also leads to confusion, disorientation, and dementia.[104] Because confusion and delirium are signs of—as well as risk factors for—dehydration, getting enough fluids can become a vicious circle for someone at risk for cognitive decline.

Dehydration at End-of-Life

Lack of hydration can be an issue for people with a terminal disease or who are near death. Some individuals stop eating and/or drinking hours, days, or even weeks before death. Laboratory values for blood and urine are likely to become abnormal. Treatment for dehydration at the end of life may differ from treatment of dehydration during an acute disease episode. Suggestions for treating dehydration in a dying person are to integrate four Cs:[105]

1. *Common sense:* use approaches that benefit the whole patient; ask, "what does the patient want?"
2. *Communication:* respect and acknowledge the sadness felt by friends and family
3. *Collaboration:* with the patient's permission, bring in experienced people such as hospice workers to guide treatment
4. *Caring:* listen, respond with love and compassion

Dehydration at the end of life contributes to an overall slowing down of body systems, including production of body fluids, resulting in less congestion, less edema and ascites or water retention, and less gastrointestinal action. A person experiencing dehydration at the end of life may experience slight thirst, although many dying patients are not thirsty. They may experience dry mouth that can be alleviated by sucking on ice chips or using artificial saliva. Decreased urine output can be a benefit because there is less need to go to the toilet. The most commonly reported symptoms in the last week of life of an individual with advanced progressive disease are loss of appetite, asthenia (loss of strength), dry mouth, confusion, and constipation.[106] Dehydration may also lead to increasing levels of confusion and drowsiness, which can reduce fear and anxiety related to dying.

Nutritional Remedies

The Food Guide Pyramid for older adults suggests drinking 8 glasses of fluid daily. Some health professionals suggest 1500 to 2500 ml per day, approximately 6 to 10 cups.[107] The ultimate beverage is *water*—tap or flavored, *not* sugared. Water is generally accessible, adds no calories to the diet, does provide traces of minerals needed for metabolism, and is very low in sodium, even when softened. When a water beverage provides calories, sugar or some other carbohydrate was added. Pure water doesn't provide energy, but lacking water to the point of dehydration can dramatically reduce one's energy.

Many beverages contribute nutrients in addition to providing fluid:

1. Tea, especially green tea, has been linked to cancer risk reduction; the flavonoids in tea act as antioxidants.[108]
2. Low-fat milk provides calcium, protein, riboflavin, and vitamin D and plays a role in treating hypertension and weight maintenance.
3. Regular use of cranberry juice reduces urinary tract infection in older women.[109]
4. Fruit and vegetable juices can be counted as part of the 5-a-day recommended fruit and vegetable servings.

> **Connection: Fluid**
> 8 ounces = 1 cup = 240 ml (milliliters) = 240 cc (cubic centimeters)
>
> A super-sized soda (32 ounces) equals approximately 4 cups (or 4 × 240 = 960 ml). A 2-liter bottle of soda provides a little more than 8 cups of fluid. Some foods also count as fluid. For example, soup, jello, and sherbet are considered fluids.

REHYDRATE SLOWLY   To treat dehydration in older adults, replace fluids slowly. Guidelines are to provide roughly one-fourth to one-third of the overall fluid deficit each day in the form of water or a 5% glucose solution (when the individual's blood values are stable).[110] For individuals with swallowing problems, or dysphagia, thickened liquids count as fluid.

BEREAVEMENT

Bereavement is the loss felt when someone who is personally significant dies. Losses of friends and family members happen more often in the lives of older persons. Grief, a very powerful emotion, is a natural response to bereavement. The grieving process, with its stages of shock and denial, disorganization, volatile reactions, guilt, loss and loneliness, relief, and re-establishment,[111] diverts attention from normal activities. Shopping and food preparation, eating, and drinking may get lost in the grieving process. Any loss of long-shared relationships through death, dementia, or moving brings about lack of interest in activities surrounding meal planning, preparation, shopping, and eating. People who are in mourning are vulnerable to malnutrition.

Widowhood has been shown to trigger disorganization and changes in daily routine, especially related to food preparation and eating.[112] Widowed persons who are able to enjoy mealtimes, have good appetites, have higher quality diets, and receive social support work through the grieving process with fewer health consequences. They demonstrate healthy aging, which is, in the words of Dr. Tamara Harris, Chief of Geriatric Epidemiology at the National Institutes of Health, "the ability of the individual to be resilient, to be adaptive, to be flexible, and to mobilize compensatory areas as they face adversities in all areas associated with health, disease, and decline in old age."

Resources

Fowkes WC. Prolonging Death—An American Tragedy. Long Beach, CA: The Archstone Foundation. 562-590-8655 Debates the use of life support.

Alzheimer's Disease, Education and Referral Center.
Site provides information about Alzheimer's Disease and related disorders. It is a service of the National Institute of Aging (NIA).
Available from: www.alzheimers.org

American Diabetes Association.
Site with areas for professionals and for the public; includes links.
Available from: www.diabetes.org

American Dietetic Association
Site offers food and nutrition tips, fact sheets and search to find a registered dietitian.
Available from: www.eatright.org

Choices in Dying, 1035 30th Street NE, Washington, DC 20007. 1-800-989-9455.
Information about hydration and artifical nutrition during end-of-life decision making.
Available from: www.choices.org.

Food and Drug Administration and American Association of Retired Persons (AARP).
Provides information for seniors about safe handling of foods.
Available from; http://vm.cfsan.fda.gov/~dms/seniorsd.html.

National Center for Complementary and Alternative Medicine
Site provides fact sheets for complementary medicine associated with dietary supplements, cancer prevention and treatment, and other dietary components. The Web site is supported by the National Institutes of Health (NIH).
Available from: www.nccam.nih.gov

National Institute of Diabetes and Digestive and Kidney Diseases
Site that offers health information for diabetes, metabolic illnesses, and kidney disease. Research and clinical trial information is also available at this site. The site provides online interactive diet planning interview to develop client-personalized food recommendations. A written handout worksheet can be printed.
Available from: www.niddk.nih.gov

National Institute of Mental Health.
Major depression is most prevalent in developed nations. This site provides a link for the public with a specific topic discussing depression. A further link, titled older adults, provides comprehensive information on definitions of depression and the treatment options available.
Available from: www.nimh.nih.gov

References

1. Ketch TD. MNT essential to plans for adding prescription drug benefit to Medicare. J Amer Diet Assoc 2000;100:762.

2. Sheils JF, Rubin R, Stapleton DC. The estimated costs and savings of medical nutrition therapy: the Medicare population. J Am Diet Assoc 1999;99:428–35.

3. Chima CS, Barco K, Dewitt ML, et al. Relationship of nutritional status to length of stay, hospital costs, and discharge status of patients hospitalized in the medicine service. J Amer Diet Assoc 1997;97:975–8; quiz 979–80.

4. Barents Group. The clinical and cost-effectiveness of medical nutrition therapy:

evidence and estimates of potential Medicare savings from the use of selected nutrition interventions. Washington, DC: KMPG Peat Marwick LLP; 1996.

5. Reilly JJ, Jr., Hull SF, Albert N, et al. Economic impact of malnutrition: a model system for hospitalized patients. J Parenter Enteral Nutr 1988;12:371–6.

6. Warnold I, Lundholm K. Clinical significance of preoperative nutritional status in 215 noncancer patients. Ann Surg 1984;199:299–305.

7. Sullivan DH, Sun S, Walls RC. Protein-energy undernutrition among elderly hospitalized patients: a prospective study. JAMA 1999;281:2013–9.

8. Wolinsky FD, Coe RM, Miller DK, et al. Health services utilization among the non-institutionalized elderly. J Health Soc Behav 1983;24:325–37.

9. Healthy People 2010: national health promotion and disease prevention objectives. Washington, DC: U.S. Dept. of Health and Human Services; 2000.

10. American Association of Retired Persons & Administration on Aging, U.S. Department of Health and Human Services. A profile of older Americans. Washington, DC; 1999:1-800-424-3410.

11. Desai MM, Zhang P, Hennessy CH. Surveillance for morbidity and mortality among older adults—United States, 1995–1996. Mor Mortal Wkly Rep CDC Surveill Summ 1999;48:7–25.

12. Erlinger TP, Pollack H, Appel LJ. Nutrition-related cardiovascular risk factors in older people: results from the Third National Health and Nutrition Examination Survey. J Am Geriatr Soc 2000;48:1486–9.

13. American Diabetes Association. Report of the Expert Committee on the Diagnosis and Classification of Diabetes Mellitus. Diabetes Care 2001;24:S5–S20.

14. Appel LJ, Moore TJ, Obarzanek E, et al. A clinical trial of the effects of dietary patterns on blood pressure. DASH Collaborative Research Group. N Eng J Med 1997;336:1117–24.

15. Plaisted CS, Lin PH, Ard JD, et al. The effects of dietary patterns on quality of life: a substudy of the Dietary Approaches to Stop Hypertension trial. J Amer Diet Assoc 1999;99:S84–9.

16. Whelton PK, Appel LJ, Espeland MA, et al. Sodium reduction and weight loss in the treatment of hypertension in older persons: a randomized controlled trial of non-pharmacologic interventions in the elderly (TONE). TONE Collaborative Research Group. JAMA 1998;279:839–46.

17. Sacks FM, Svetkey LP, Vollmer WM, et al. Effects on blood pressure of reduced dietary sodium and the Dietary Approaches to Stop Hypertension (DASH) diet. DASH-Sodium Collaborative Research Group. N Engl J Med 2001;344:3–10.

18. The sixth report of the Joint National Committee on prevention, detection, evaluation, and treatment of high blood pressure. Arch Intern Med 1997;157:2413–46.

19. Wilson JW, Enns CW, Goldman KS, et al. Data tables: combined results from USDA's 1994 and 1995 Continuing Survey of Food Intakes by Individuals and 1994 and 1995 Diet and Health Knowledge Survey, 1997.

20. Food and Nutrition Board. National Academy of Sciences (Institute of Medicine) NRC, Subcommittee on the 10th Edition of the RDAs, Commission on Life Sciences. Recommended dietary allowances. Washington, DC: National Academy Press; 1989.

21. Food, nutrition, and the prevention of cancer: a global perspective. Washington, DC: World Cancer Research Fund & American Institute for Cancer Research; 1997.

22. Kramarow E, Lentzner H, Rooks R, et al. Health United States, 1999. With health and aging chartbook. Hyattsville, MD: National Center for Health Statistics; 1999.

23. Steinmetz KA, Potter JD. Vegetables, fruit, and cancer prevention: a review. J Amer Diet Assoc 1996;96:1027–39.

24. Steinmetz KA, Potter JD. Vegetables, fruit, and cancer. I. Epidemiology. Cancer Causes Control 1991;2:325–57.

25. Steinmetz KA, Potter JD. Vegetables, fruit, and cancer. II. Mechanisms. Cancer Causes Control 1991;2:427–42.

26. Robbers JE, Tyler VE. Tyler's herbs of choice. New York: Haworth Herbal Press;1999.

27. Brown JE. Nutrition now. Belmont, CA: West/Wadsworth Publishing Company; 2001.

28. Centers for Disease Control and Prevention. Unrealized prevention opportunities: reducing the health and economic burden of chronic disease. Centers for Disease Control and Prevention, National Center for Chronic Disease Prevention and Health Promotion; 1997.

29. Resnick HE, Valsania P, Halter JB, et al. Differential effects of BMI on diabetes risk among black and white Americans. Diabetes Care 1998;21:1828–35.

30. Burrows NR, Geiss LS, Engelgau MM, et al. Prevalence of diabetes among Native Americans and Alaska Natives, 1990–1997: an increasing burden. Diabetes Care 2000;23:1786–90.

31. Powers MA. Handbook of diabetes medical nutrition therapy. Gaithersburg, MD: Aspen; 1996.

32. Haffner SM, Lehto S, Ronnemaa T, et al. Mortality from coronary heart disease in subjects with type 2 diabetes and in nondiabetic subjects with and without prior myocardial infarction. N Eng J Med 1998;339:229–34.

33. Hoogeveen EK, Kostense PJ, Jakobs C, et al. Hyperhomocysteinemia increases risk of death, especially in type 2 diabetes : 5-year follow-up of the Hoorn Study. Circulation 2000;101:1506–11.

34. Nuttall FQ, Chasuk RM. Nutrition and the management of type 2 diabetes. J Fam Pract 1998;47:S45–53.

35. Stanley K. Assessing the nutritional needs of the geriatric patient with diabetes. Diabetes Educ 1998;24:29–30,35–7.

36. Terpstra TL. The elderly type II diabetic: a treatment challenge. Geriatr Nurs 1998;19:253–9; quiz 259–60.

37. U.S. Department of Agriculture, U.S. Department of Health and Human Services. Nutrition and your health: dietary guidelines for Americans, 5th ed. 2000.

38. Holtmeier KB, Seim HC. The diet prescription for obesity. What works? Minn Med 2000;83:28–32.

39. Chandalia M, Garg A, Lutjohann D, et al. Beneficial effects of high dietary fiber intake in patients with type 2 diabetes mellitus. N Eng J Med 2000;342:1392–8.

40. American Diabetes Association. Nutrition recommendations and principles for people with diabetes mellitus. Diabetes Care 2001;24:S544–S547.

41. Tsihlias EB, Gibbs AL, McBurney MI, et al. Comparison of high- and low-glycemic-index breakfast cereals with monounsaturated fat in the long-term dietary management of type 2 diabetes. Am J Clin Nutr 2000;72:439–49.

42. Preuss HG, Anderson RA. Chromium update: examining recent literature 1997–1998. Curr Opin Clin Nutr Metab Care 1998;1:509–12.

43. The effect of intensive treatment of diabetes on the development and progression of long-term complications in insulin-dependent diabetes mellitus. The Diabetes Control and Complications Trial Research Group. N Eng J Med 1993;329:977–86.

44. Effect of intensive blood-glucose control with metformin on complications in overweight patients with type 2 diabetes (UKPDS 34). UK Prospective Diabetes Study (UKPDS) Group. Lancet 1998;352:854–65.

45. Intensive blood-glucose control with sulphonylureas or insulin compared with conventional treatment and risk of complications in patients with type 2 diabetes (UKPDS 33). UK Prospective Diabetes Study (UKPDS) Group. Lancet 1998;352:837–53.

46. Reaven GM, Strom TK, Fox B. Syndrome X: overcoming the silent killer that can give you a heart attack. New York: Simon & Schuster; 2000.

47. Wada Y, Tsukada M, Koizumi A. Diabetes and hypertension (syndrome X) as an increasing risk factor for cerebrovascular aging in Japan. J Anti-aging Med 1988;1: 45–52.

48. Coulston AM. Insulin resistance: its role in health and disease and implications for nutrition management. Topics in Nutrition, No. 6: Hershey Foods Corporation; 1997.

49. Barnard RJ, Roberts CK, Varon SM, et

al. Diet-induced insulin resistance precedes other aspects of the metabolic syndrome. J Appl Physiol 1998;84:1311–5.

50. Baumgartner J. "Syndrome X" (hypertension, diabetes, coronary heart disease), National Maternal Nutrition Intensive Course, Minneapolis, MN, July 10–13; 1996.

51. Looker AC, Orwoll ES, Johnston CC, Jr., et al. Prevalence of low femoral bone density in older U.S. adults from NHANES III. J Bone Miner Res 1997;12:1761–8.

52. Niewohner K. Conference on Osteoporosis. Minnesota Department of Health, Minnesota Board on Aging, Midwest Dairy Council, St. Paul, MN; September, 1998.

53. Madhok R, Allison T. Bone mineral density measurement in the management of osteoporosis: a public health perspective. In: Fordham JN, ed. Manual of bone densitometry measurements. An aid to the interpretation of bone densitometry measurements in a clinical setting. New York: Springer; 2000:1–16.

54. Faine MP. Dietary factors related to preservation of oral and skeletal bone mass in women. J Prosthet Dent 1995;73:65–72.

55. Nordin BE, Need AG, Steurer T, et al. Nutrition, osteoporosis, and aging. Ann NY Acad Sci 1998;854:336–51.

56. Chumlea WC, Guo SS. Body mass and bone mineral quality. Curr Opin Rheumatol 1999;11:307–11.

57. McGowan J. Osteoporosis research, prevention and treatment. Senate Special Committee on Aging; 1999.

58. Rubin LA, Hawker GA, Peltekova VD, et al. Determinants of peak bone mass: clinical and genetic analyses in a young female Canadian cohort. J Bone Miner Res 1999; 14:633–43.

59. Matkovic V, Kostial K, Simonovic I, et al. Bone status and fracture rates in two regions of Yugoslavia. Am J Clin Nutr 1979;32:540–9.

60. Lloyd T, Andon MB, Rollings N, et al. Calcium supplementation and bone mineral density in adolescent girls. JAMA 1993;270:841–4.

61. Teegarden D, Lyle RM, McCabe GP, et al. Dietary calcium, protein, and phosphorus are related to bone mineral density and content in young women. Am J Clin Nutr 1998;68:749–54.

62. Heaney RP. Bone biology in health and disease. In: Shils ME, Olson JA, Shike M, et al. eds. Modern nutrition in health and disease. Philadelphia: Lippincott, Williams & Wilkins; 1999: pp 1327–38.

63. Shils ME, Olson JA, Shike M, et al. Modern nutrition in health and disease. Philadelphia: Lippincott, Williams & Wilkins; 1999.

64. Economos C. Osteoporosis facts. Conference on Osteoporosis. Minnesota Department of Health, Minnesota Board on Aging, Midwest Dairy Council, St. Paul, MN; September, 1998.

65. Feskanich D, Willett WC, Stampfer MJ, et al. Protein consumption and bone fractures in women. Am J Epidemiol 1996;143:472–9.

66. Weber P. The role of vitamins in the prevention of osteoporosis—a brief status report. Int J Vitam Nutr Res 1999;69: 194–7.

67. Pines A, Katchman H, Villa Y, et al. The effect of various hormonal preparations and calcium supplementation on bone mass in early menopause. Is there a predictive value for the initial bone density and body weight? J Intern Med 1999;246: 357–61.

68. Appollonio I, Carabellese C, Frattola A, et al. Influence of dental status on dietary intake and survival in community-dwelling elderly subjects. Age Ageing 1997;26:445–56.

69. Locke GR, III, Talley NJ, Fett SL, et al. Risk factors associated with symptoms of gastroesophageal reflux. Am J Med 1999;106:642–9.

70. Brookmeyer R, Gray S, Kawas C. Projections of Alzheimer's Disease in the United States and the public health impact of delaying disease onset. Am J PH 1998; 88:1337–1342.

71. Castell DO. Gastroesophageal reflux and abnormal esophageal pressures: cause or effect? J Clin Gastroenterol 2000;30:3.

72. Harari D, Gurwitz JH, Avorn J, et al. Bowel habit in relation to age and gender. Findings from the National Health Interview Survey and clinical implications. Arch Intern Med 1996;156:315–20.

73. Talley NJ, Fleming KC, Evans JM, et al. Constipation in an elderly community: a study of prevalence and potential risk factors. Am J Gastroenterol 1996;91:19–25.

74. Camilleri M, Lee JS, Viramontes B, et al. Insights into the pathophysiology and mechanisms of constipation, irritable bowel syndrome, and diverticulosis in older people. J Am Geriatr Soc 2000;48:1142–50.

75. Felson DT, Lawrence RC, Dieppe PA, et al. Osteoarthritis: new insights. Part 1: the disease and its risk factors. Ann Intern Med 2000;133:635–46.

76. Galperin C, German BJ, Gershwin ME. Nutrition and diet in rheumatic diseases. In: Shils ME, Olson JA, Shike M, et al. eds. Modern nutrition in health and disease. Philadelphia, PA: Lippincott, Williams & Wilkins; 1999: pp 1339–51.

77. Felson DT, Lawrence RC, Hochberg MC, et al. Osteoarthritis: new insights. Part 2: treatment approaches. Ann Intern Med 2000;133:726–37.

78. Kremer JM. n-3 fatty acid supplements in rheumatoid arthritis. Am J Clin Nutr 2000;71:349S–51S.

79. Borchers AT, Keen CL, Stern JS, et al. Inflammation and Native American medicine: the role of botanicals. Am J Clin Nutr 2000;72:339–47.

80. Johnston PK. Nutritional implications of vegetarian diets. In: Shils ME, Olson JA, Shike M, et al. eds. Modern nutrition in health and disease. Philadelphia, PA: Lippincott, Williams & Wilkins; 1999: pp 1755–67.

81. Health Desk. A Partnership of the Kaiser Family Foundation and The NewsHour with Jim Lehrer. National survey on prescription drugs. The Henry J. Kaiser Family Foundation, 2400 Sand Hill Road, Menlo Park, CA 94025, accessed 1/15/01 from website address: www.kff.org, 2000.

82. Healthy People 2010: http://www.health. gov/healthypeople/document/html/volume2/ 19nutrition.html.

83. Stevens J. Impact of age on associations between weight and mortality. Nutr Rev 2000;58:129–37.

84. Ernsberger P, Koletsky RJ. Biomedical rationale for a wellness approach to obesity: an alternative to a focus on weight loss. J Soc Issues 1999;55:221–59.

85. Ostbye T, Steenhuis R, Wolfson C, et al. Predictors of five-year mortality in older Canadians: the Canadian Study of Health and Aging. J Am Geriatr Soc 1999;47: 1249–54.

86. Hegman K. Weight loss and mortality in adults. Abstract and commentary for Pamuk ER, Williamson DF, Serdula MK, et al. Weight loss and subsequent death in a cohort of US adults. Ann Intern Med 1993;119(7-pt2):744–8. In: ACP Journal Club, May–June, 1994: pp 120–81.

87. The American Dietetic Association Long Term Care Task Force. Nutrition risk assessment form, guides, strategies and interventions. Chicago, IL: American Dietetic Association; 1999.

88. Kuczmarski MF, Kuczmarski RJ, Najjar M. Descriptive anthropometric reference data for older Americans. J Amer Diet Assoc 2000;100:59–66.

89. Clinical guidelines on the identification, evaluation, and treatment of overweight and obesity in adults—the evidence report. National Institutes of Health. Obes Res 1998;6Suppl2:51S–209S.

90. Physical status: the use and interpretation of anthropometry. Report of a WHO Expert Committee. Geneva: World Health Organization; 1995.

91. Kasper H. Vitamin absorption in the elderly. Int J Vitam Nutr Res 1999;69: 169–72.

92. Norberg B. Turn of tide for oral vitamin B12 treatment. J Intern Med 1999;246: 237–8.

93. Nilsson-Ehle H. Age-related changes in cobalamin (vitamin B12) handling. Implications for therapy. Drugs Aging 1998;12:277–92.

94. Lindenbaum J, Rosenberg IH, Wilson PW, et al. Prevalence of cobalamin deficiency in the Framingham elderly population. Am J Clin Nutr 1994;60:2–11.

95. Hokin BD, Butler T. Cyanocobalamin

(vitamin B-12) status in Seventh-Day Adventist ministers in Australia. Am J Clin Nutr 1999;70:576S–578S.

96. Carmel R, Green R, Jacobsen DW, et al. Serum cobalamin, homocysteine, and methylmalonic acid concentrations in a multiethnic elderly population: ethnic and sex differences in cobalamin and metabolite abnormalities. Am J Clin Nutr 1999;70:904–10.

97. Zeitlin A, Frishman WH, Chang CJ. The association of vitamin B12 and folate blood levels with mortality and cardiovascular morbidity incidence in the old old: the Bronx aging study. Am J Ther 1997;4:275–81.

98. Yates AA, Schlicker SA, Suitor CW. Dietary Reference Intakes: the new basis for recommendations for calcium and related nutrients, B vitamins, and choline. J Amer Diet Assoc 1998;98:699–706.

99. Trumbo P, Yates AA, Schlicker S, et al. Dietary Reference Intakes: vitamin A, vitamin K, arsenic, boron, chromium, copper, iodine, iron, manganese, molybdenum, nickel, silicon, vanadium, and zinc. J Am Diet Assoc 2001;101:294–301.

100. Position of the American Dietetic Association: food and water safety. J Amer Diet Assoc 1997;97:184–9.

101. Phillips PA, Rolls BJ, Ledingham JG, et al. Reduced thirst after water deprivation in healthy elderly men. N Eng J Med 1984;311:753–9.

102. Gross CR, Lindquist RD, Woolley AC, et al. Clinical indicators of dehydration severity in elderly patients. J Emerg Med 1992;10:267–74.

103. Briggs GM, Calloway DH. Bogert's nutrition and physical fitness. New York: Holt, Rinehart and Winston; 1984.

104. Chidester JC, Spangler AA. Fluid intake in the institutionalized elderly. J Amer Diet Assoc 1997;97:23–8; quiz 29–30.

105. Fordyce M. Dehydration near the end of life. Ann Long-Term Care 2000;8:accessed 5/31/2000, www.mmhc.com/nhm/articles/NHM0005/fordyce.html.

106. Conill C, Verger E, Henriquez I, et al. Symptom prevalence in the last week of life. J Pain Symptom Manage 1997;14:328–31.

107. Chernoff R, ed. Geriatric nutrition: the health professional's handbook. Gaithersburg, MD: Aspen; 1999.

108. Bushman JL. Green tea and cancer in humans: a review of the literature. Nutr Cancer 1998;31:151–9.

109. Fleet JC. New support for a folk remedy: cranberry juice reduces bacteriuria and pyuria in elderly women. Nutr Rev 1994;52:168–78.

110. Carter WJ. Macronutrient requirements for elderly persons. In: Chernoff R, ed. Geriatric nutrition: the health professional's handbook. Gaithersburg, MD: Aspen; 1999.

111. Leming MR, Dickinson GE. Understanding dying, death, and bereavement. Fort Worth, TX: Holt, Rinehart, and Winston; 1990.

112. Rosenbloom CA, Whittington FJ. The effects of bereavement on eating behaviors and nutrient intakes in elderly widowed persons. J Gerontol 1993;48:S223–9.

CDC GROWTH CHARTS

CDC Growth Charts: United States

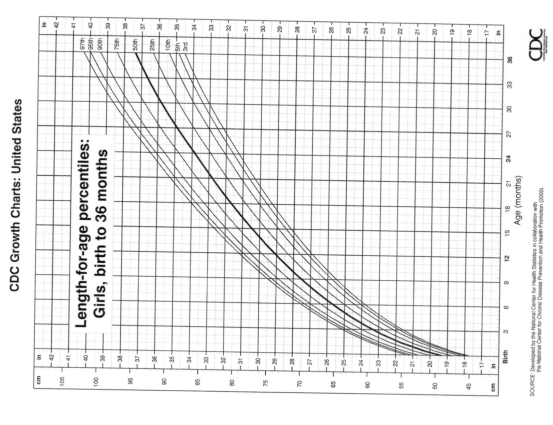

Length-for-age percentiles: Girls, birth to 36 months

SOURCE: Developed by the National Center for Health Statistics in collaboration with the National Center for Chronic Disease Prevention and Health Promotion (2000).

CDC Growth Charts: United States

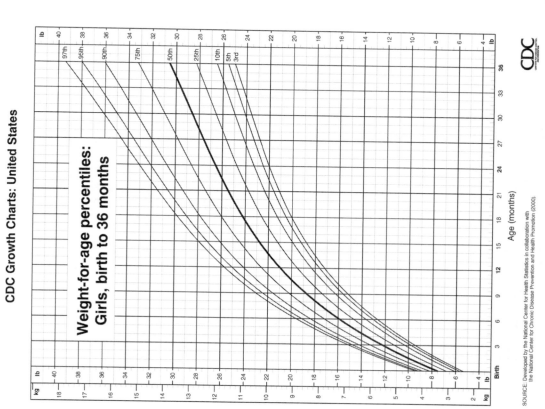

Weight-for-age percentiles: Girls, birth to 36 months

SOURCE: Developed by the National Center for Health Statistics in collaboration with the National Center for Chronic Disease Prevention and Health Promotion (2000).

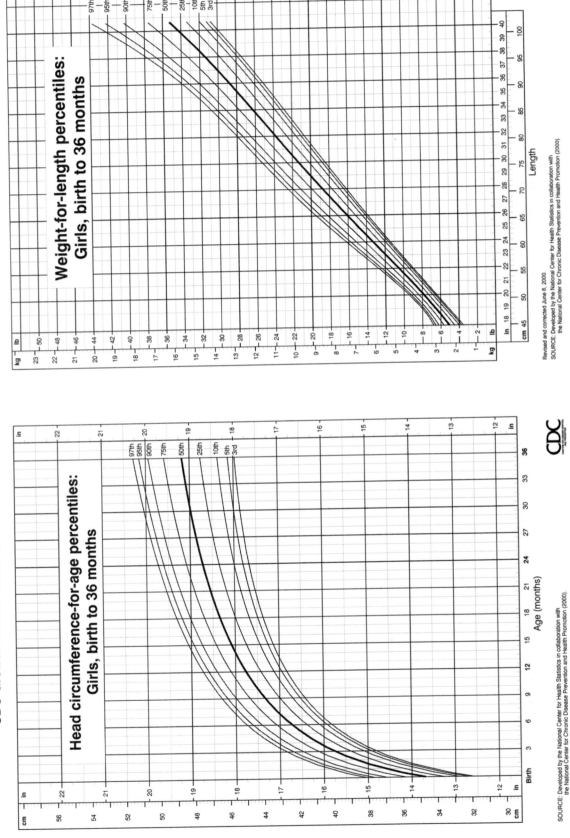

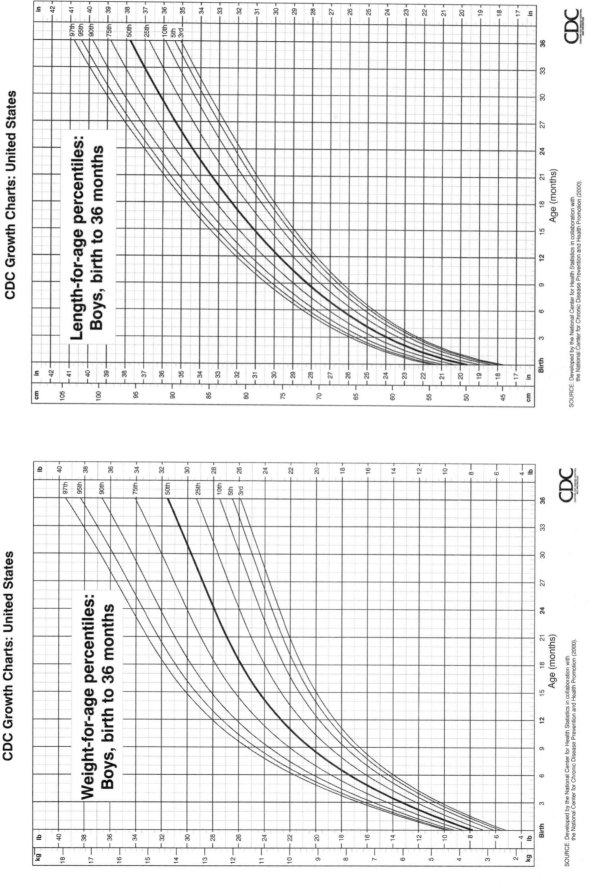

CDC Growth Charts: United States

Length-for-age percentiles: Boys, birth to 36 months

SOURCE: Developed by the National Center for Health Statistics in collaboration with the National Center for Chronic Disease Prevention and Health Promotion (2000).

CDC Growth Charts: United States

Weight-for-age percentiles: Boys, birth to 36 months

SOURCE: Developed by the National Center for Health Statistics in collaboration with the National Center for Chronic Disease Prevention and Health Promotion (2000).

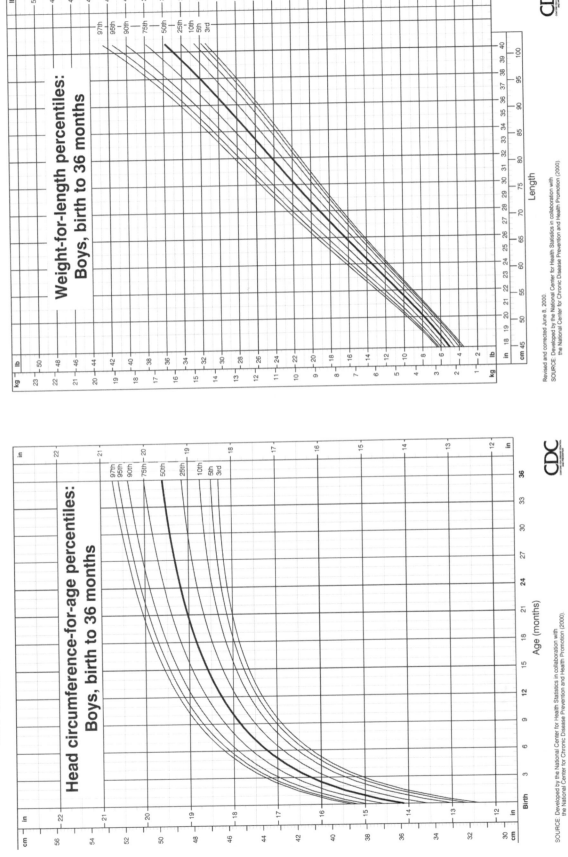

CDC Growth Charts: United States

Weight-for-length percentiles: Boys, birth to 36 months

Revised and corrected June 8, 2000.
SOURCE: Developed by the National Center for Health Statistics in collaboration with the National Center for Chronic Disease Prevention and Health Promotion (2000).

CDC Growth Charts: United States

Head circumference-for-age percentiles: Boys, birth to 36 months

SOURCE: Developed by the National Center for Health Statistics in collaboration with the National Center for Chronic Disease Prevention and Health Promotion (2000).

CDC Growth Charts: United States

Stature-for-age percentiles: Girls, 2 to 20 years

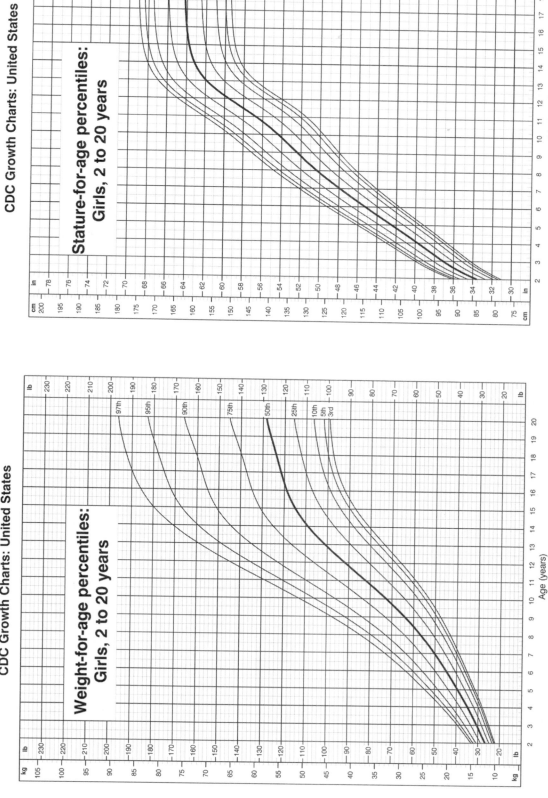

SOURCE: Developed by the National Center for Health Statistics in collaboration with the National Center for Chronic Disease Prevention and Health Promotion (2000).

CDC Growth Charts: United States

Weight-for-age percentiles: Girls, 2 to 20 years

SOURCE: Developed by the National Center for Health Statistics in collaboration with the National Center for Chronic Disease Prevention and Health Promotion (2000).

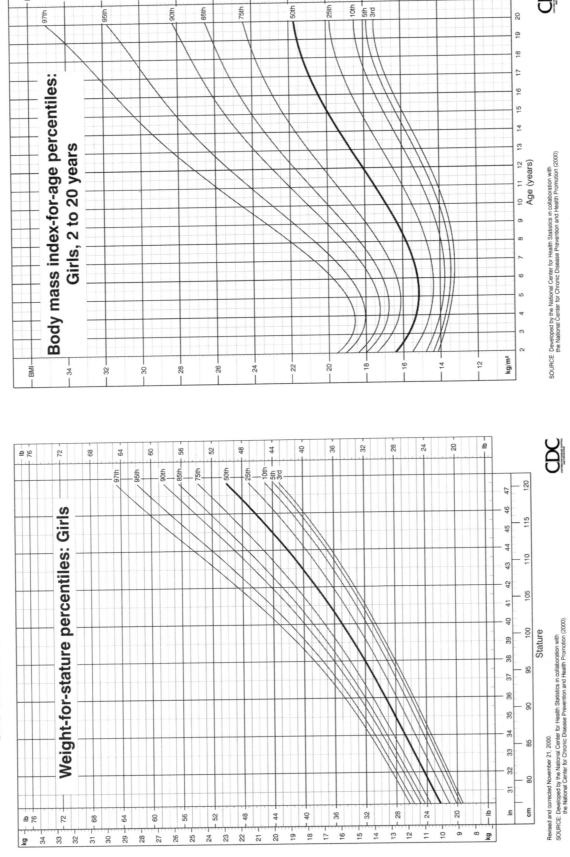

CDC Growth Charts: United States

Body mass index-for-age percentiles:
Girls, 2 to 20 years

SOURCE: Developed by the National Center for Health Statistics in collaboration with
the National Center for Chronic Disease Prevention and Health Promotion (2000).

CDC Growth Charts: United States

Weight-for-stature percentiles: Girls

Revised and corrected November 21, 2000.
SOURCE: Developed by the National Center for Health Statistics in collaboration with
the National Center for Chronic Disease Prevention and Health Promotion (2000).

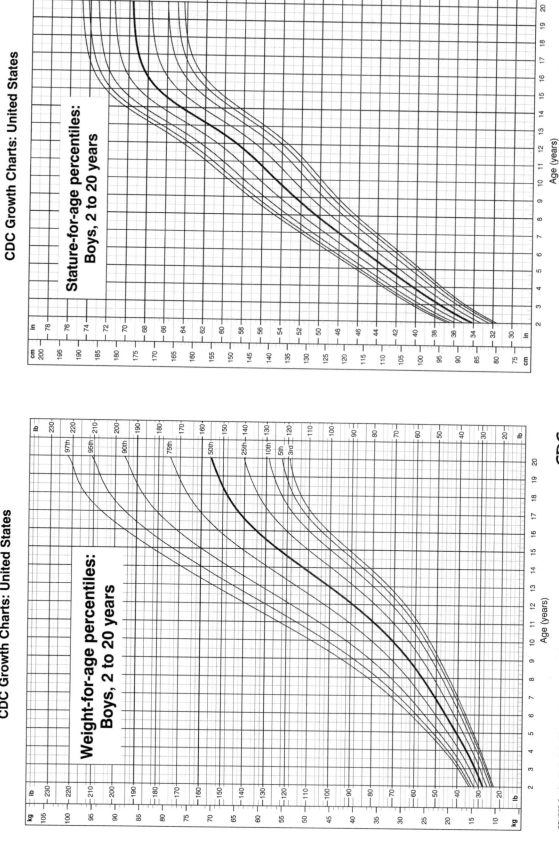

CDC Growth Charts: United States

**Stature-for-age percentiles:
Boys, 2 to 20 years**

CDC Growth Charts: United States

**Weight-for-age percentiles:
Boys, 2 to 20 years**

SOURCE: Developed by the National Center for Health Statistics in collaboration with
the National Center for Chronic Disease Prevention and Health Promotion (2000).

SOURCE: Developed by the National Center for Health Statistics in collaboration with
the National Center for Chronic Disease Prevention and Health Promotion (2000).

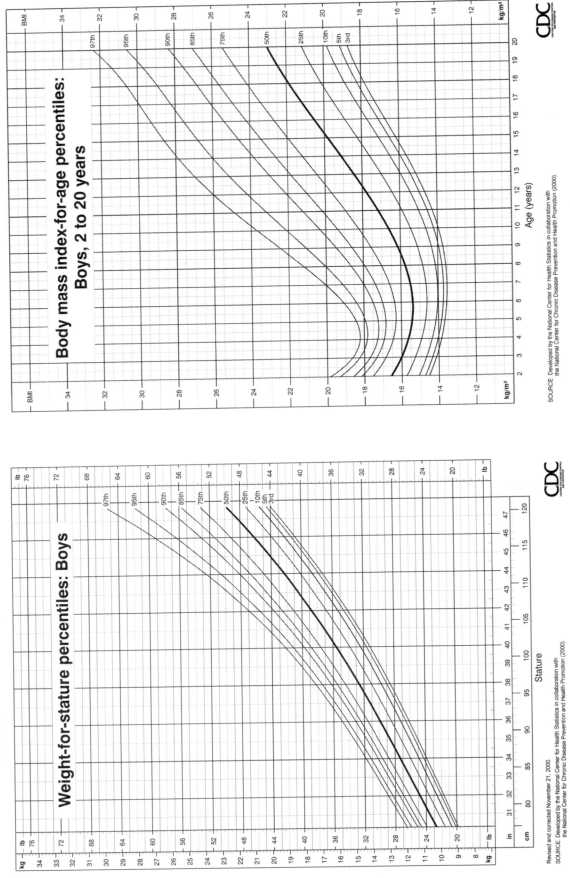

CDC Growth Charts: United States

Body mass index-for-age percentiles: Boys, 2 to 20 years

SOURCE: Developed by the National Center for Health Statistics in collaboration with the National Center for Chronic Disease Prevention and Health Promotion (2000).

CDC Growth Charts: United States

Weight-for-stature percentiles: Boys

Revised and corrected November 21, 2000.
SOURCE: Developed by the National Center for Health Statistics in collaboration with the National Center for Chronic Disease Prevention and Health Promotion (2000).

BODY MASS INDEX (BMI)

Height	18	19	20	21	22	23	24	25	26	27	28	29	30	31	32	33	34	35	36	37	38	39	40
											Body Weight (pounds)												
4'10"	86	91	96	100	105	110	115	119	124	129	134	138	143	148	153	158	162	167	172	177	181	186	191
4'11"	89	94	99	104	109	114	119	124	128	133	138	143	148	153	158	163	168	173	178	183	188	193	198
5'0"	92	97	102	107	112	118	123	128	133	138	143	148	153	158	163	168	174	179	184	189	194	199	204
5'1"	95	100	106	111	116	122	127	132	137	143	148	153	158	164	169	174	180	185	190	195	201	206	211
5'2"	98	104	109	115	120	126	131	136	142	147	153	158	164	169	175	180	186	191	196	202	207	213	218
5'3"	102	107	113	118	124	130	135	141	146	152	158	163	169	175	180	186	191	197	203	208	214	220	225
5'4"	105	110	116	122	128	134	140	145	151	157	163	169	174	180	186	192	197	204	209	215	221	227	232
5'5"	108	114	120	126	132	138	144	150	156	162	168	174	180	186	192	198	204	210	216	222	228	234	240
5'6"	112	118	124	130	136	142	148	155	161	167	173	179	186	192	198	204	210	216	223	229	235	241	247
5'7"	115	121	127	134	140	146	153	159	166	172	178	185	191	198	204	211	217	223	230	236	242	249	255
5'8"	118	125	131	138	144	151	158	164	171	177	184	190	197	203	210	216	223	230	236	243	249	256	262
5'9"	122	128	135	142	149	155	162	169	176	182	189	196	203	209	216	223	230	236	243	250	257	263	270
5'10"	126	132	139	146	153	160	167	174	181	188	195	202	209	216	222	229	236	243	250	257	264	271	278
5'11"	129	136	143	150	157	165	172	179	186	193	200	208	215	222	229	236	243	250	257	265	272	279	286
6'0"	132	140	147	154	162	169	177	184	191	199	206	213	221	228	235	242	250	258	265	272	279	287	294
6'1"	136	144	151	159	166	174	182	189	197	204	212	219	227	235	242	250	257	265	272	280	288	295	302
6'2"	141	148	155	163	171	179	186	194	202	210	218	225	233	241	249	256	264	272	280	287	295	303	311
6'3"	144	152	160	168	176	184	192	200	208	216	224	232	240	248	256	264	272	279	287	295	303	311	319
6'4"	148	156	164	172	180	189	197	205	213	221	230	238	246	254	263	271	279	287	295	304	312	320	328
6'5"	151	160	168	176	185	193	202	210	218	227	235	244	252	261	269	277	286	294	303	311	319	328	336
6'6"	155	164	172	181	190	198	207	216	224	233	241	250	259	267	276	284	293	302	310	319	328	336	345

Underweight (<18.5)	Healthy Weight (18.5–24.9)	Overweight (25–29.9)	Obese (≥30)

Find your height along the left-hand column and look across the row until you find the number that is closest to your weight. The number at the top of that column identifies your BMI. The area shaded in green represents healthy weight ranges. The figure below presents silhouettes of various BMI.

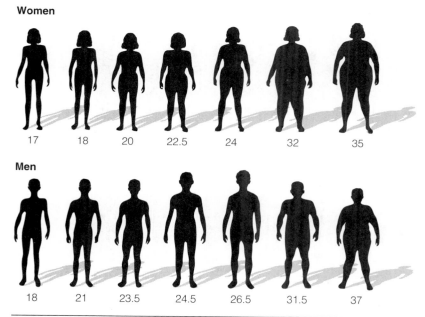

Women

17   18   20   22.5   24   32   35

Men

18   21   23.5   24.5   26.5   31.5   37

SOURCE: Reprinted from material of the Dietitians of Canada.

NUTRIENT INTAKES OF ADULTS AGED 70 AND OLDER

From CSFII[48]

NUTRIENT	MALES	FEMALES
Energy, kcal	1854	1377
Protein, gm	74	57
Total fat, gm	69	48
Saturated fatty acids, gm	23	16
Monounsaturated fatty acids, gm	27	18
Polyunsaturated fatty acids, gm	14	10
Cholesterol, mg	274	185
Total carbohydrate, gm	233	184
Dietary Fiber, gm	18	14
Vitamin A, mcg RE	1430.0	1149.0
Carotenes, mcg RE	648.0	602.0
Vitamin E, mg alpha TE	9.0	6.0
Vitamin C, mg	99.0	96.0
Thiamin, mg	1.6	1.2
Riboflavin, mg	2.0	1.5
Niacin, mg	22.2	17.5
Vitamin B6, mg	2.0	1.5
Folate, mcg	283.0	234.0
Calcium, mg	754.0	587.0
Phosphorus, mg	1211.0	912.0
Magnesium, mg	286.0	224.0
Iron, mg	16.7	12.3
Zinc, mg	11.8	8.4
Copper, mg	1.3	1.0
Sodium, mg	3122.0	2376.0
Potassium, mg	2825.0	2824.0

GLOSSARY

ADIPOSITY REBOUND   A normal increase in body mass index which occurs after BMI declines and reaches its lowest point at four to six years of age.

AIDS   Acquired Immunodeficiency Syndrome.

ALLERGY   Hypersensitivity to a physical or chemical agent.

ALVEOLUS   A rounded or oblong shaped cavity present in the breast.

AMENORRHEA   Absence of menstrual cycle.

AMINO ACIDS   The "building blocks" of protein. They are nitrogen-containing constituents of food.

AMNIOTIC FLUID   The fluid contained in the amniotic sac that surrounds the fetus in the uterus.

AMYLOPHAGIA   Compulsive consumption of laundry starch or cornstarch.

ANAPHYLAXIS   Sudden onset of a reaction with mild to severe symptoms, including a decrease in ability to breathe, which may be severe enough to cause a coma.

ANEMIA   A reduction below normal in the number of red blood cells per cu mm, in the quantity of hemoglobin, or in the volume of packed red cells per 100 ml of blood. This reduction occurs when the balance between blood loss and blood production is disturbed.

ANENCEPHALY   Condition initiated early in gestation of the central nervous system in which the brain is not formed correctly, resulting in neonatal death.

ANOREXIA NERVOSA   (anorexia = poor appetite, nervosa = mental disorder) A primarily psychological disorder characterized by extreme underweight, malnutrition, amenorrhea, low bone density, irrational fear of weight gain, restricted food intake, hyperactivity, and disturbances in body image.

ANOVULATORY CYCLES   Menstrual cycles in which ovulation does not occur.

ANTHROPOMETRY   The science of measuring the human body and its parts.

ANTI-DIURETIC HORMONE   Causes the kidneys to dilute urine by absorbing more water.

ANTIOXIDANTS   Chemical substance that prevent or repair damage to cells caused by exposure to oxidizing agents such as oxygen, ozone, and smoke, and to other oxidizing agents normally produced in the body. Many different antioxidants are found in foods and some are made by the body.

APGAR   A rating score for newborns with the lowest score of zero and highest score of ten which is assigned at one minute and five minutes after birth.

APPROPRIATE FOR GESTATIONAL AGE (AGA)   Birth weight is between 10th and 90th percentile for gestational age.

ARTERIOSCLEROSIS   Age-related thickening and hardening of the artery walls, much like an old rubber hose that becomes brittle or hard.

ARTERY   Blood vessel carrying oxygenated blood from the heart to the rest of the body.

ASTHMA   Condition in which the lungs are unable to exchange air due to lack of expansion of air sacs. It can result in a chronic illness and sometimes unconsciousness and death if not treated.

ATHEROSCLEROSIS   A type of hardening of the arteries in which cholesterol is deposited in the arteries. These deposits narrow the coronary arteries and may reduce the flow of blood.

ATHETOSIS   Uncontrolled movements of the large muscle groups as a result of damage to the central nervous system.

ATTENTION DEFICIT HYPERACTIVITY DISORDER (ADHD)   Condition characterized by low impulsive control and short attention span, with and without a high level of overall activity.

AUTISM   Condition of deficits in communication and social interaction with onset generally before age 3 years, in which mealtime behavior and eating problems often occur.

BABY BOTTLE TOOTH DECAY   Dental caries in young children caused by being put to bed with a bottle or allowed to suck from a bottle at will for extended periods of time. Also called "baby or nursing bottle dental caries.

BASAL METABOLIC RATE   Energy expenditure at rest, 10 hours or more after eating, in a thermally neutral environment.

BINGE EATING DISORDER   An eating disorder characterized by periodic binge eating, which normally is not followed by vomiting or the use of laxatives. People must experience eating binges twice a week on average for over six months to qualify for this diagnosis.

B-LYMPHOCYTES   White blood cells that are responsible for producing immunoglobulins.

BODY MASS INDEX (BMI)   Weight in kg/height in m². BMIs <19.8 are considered underweight, those between 19.8–25

correspond to normal weight, 25–30 is considered overweight, and BMIs of 30 and higher represent obesity.

BONE AGE   Bone maturation, correlates well with stage of pubertal development.

BRONCHOPULMONARY DYSPLASIA (BPD)   Condition in which the underdeveloped lungs in a preterm infant are damaged so that breathing requires extra effort.

BULIMIA NERVOSA: (BULIMIA = OX HUNGER)   A disorder characterized by repeated bouts of uncontrolled, rapid ingestion of large quantities of food (binge eating) followed by self-induced vomiting, laxatives or diuretic use, fasting, or vigorous exercise in order to prevent weight gain. Binge eating is often followed by feelings of disgust and guilt. Menstrual cycle abnormalities may accompany this disorder.

CALORIE   A unit of measure of the amount of energy supplied by food. Also know as the "kilocalorie," or the large Calorie.

CAROTENEMIA   A condition caused by ingestion of high amounts of carotenoids (or carotenes) from plant foods. High levels of carotenoids causes the skin to turn yellowish-orange.

CATCH UP GROWTH   Period of time shortly after a slow growth period when the rate of weight and height gains is likely to increase faster than expected for age and gender.

CELIAC DISEASE   Malabsorption with fatty stools (steatorrhea) due to an inherited sensitivity to the gliadin portion of gluten in wheat, rye, barley, and oats. It is often responsible for iron, folate, and zinc deficiencies. Also called Celiac sprue and non-tropical sprue.

CEREBRAL PALSY   A group of disorders characterized by impaired muscle activity and coordination present at birth or developed during early childhood.

CEREBRAL SPINAL ATROPHY   Condition in which muscle control declines over time as a result of nerve loss, causing death in childhood.

CHILDREN WITH SPECIAL HEALTH CARE NEEDS   A federal category of services for infants, children and adolescents with, or at risk for, physical or developmental disability, or with a chronic medical condition caused by or associated with genetic/metabolic disorders, birth defects, prematurity, trauma, infection, or prenatal exposure to drugs.

CHOLESTEROL   A fat-soluble, colorless liquid found in animals but not plants.

CHRONIC CONDITION   Disorder of health or development that is the usual state for an individual and unlikely to change, although secondary conditions may result over time.

CLEFT LIP AND PALATE Condition in which the upper lip and roof of the mouth are not formed completely and are surgically corrected, resulting in feeding, speaking, and hearing difficulties in childhood.

COBALAMIN Is another name for vitamin B$_{12}$. Important roles of cobalamin are: fatty acid metabolism, synthesis of nucleic acid (i.e., DNA, a complex protein that controls the formation of healthy new cells) and formation of the myelin sheath that protects nerve cells.

COGNITIVE FUNCTION The process of thinking.

COLIC A condition with a sudden onset of irritability, fussiness or crying in a young infant between 2 weeks and 3 months of age who is otherwise growing and healthy.

COLOSTRUM The milk produced in the first 2–3 days after the baby is born. Colostrum is higher in protein and lower in lactose than milk produced after a milk supply is established.

COMMODITY PROGRAM A USDA program in which food products are sent to schools for use in the Child Nutrition Programs. Commodities are usually acquired for farm price support and surplus removal reasons.

COMPETITIVE FOODS Foods—sold to children in food service areas during meal times—that compete with the federal meal programs.

CONGENITAL ABNORMALITY A structural, functional, or metabolic abnormality present at birth. They may be caused by environmental or genetic factors, or a combination of the two. Also called birth defects and congenital anomalies. Structural abnormalities are generally referred to as congenital malformations, and metabolic abnormalities as inborn errors of metabolism.

CONGENITAL ANOMALY Condition evident in a newborn that is diagnosed at or near birth, usually as a genetic or chronic condition, such as spina bifida or cleft lip and palate.

CONSTIPATION A condition characterized by hard stools that are difficult to pass.

CORPUS LUTEUM (Corpus = body, luteum = yellow.) A tissue about ½ inch in diameter formed from the follicle which contained the ovum prior to its release. It produces estrogen and progesterone. The "yellow body" derivation comes from the accumulation of lipid precursors of these hormones in the corpus luteum.

CYSTIC FIBROSIS Condition in which a genetically changed Chromosome 7 interferes with all the exocrine functions in the body, but particularly pulmonary complications, causing chronic illness.

DEVELOPMENT Progression of the physical and mental capabilities of an organism through growth and differentiation of organs and tissues, and integration of functions.

DEVELOPMENTAL DELAY Conditions represented by at least a 25% delay by standard evaluation in one or more area of development, such as gross or fine motor, cognitive, communication, social or emotional development.

DEVELOPMENTAL DISABILITIES General term used to group specific diagnoses together that limit daily living and functioning, and occur before age 21 years.

DIABETES MELLITUS Condition in which glucose regulation in the body is disturbed, requiring treatment to prevent blood sugar from being too high or low.

DIAPHRAGMATIC HERNIA Displacement of the intestines up into to the lung area due to incomplete formation of the diaphragm in utero.

DIARRHEA A condition characterized by an increase in the frequency and fluidity of stool relative to the usual pattern.

DIETARY FIBER Complex carbohydrates and lignins found mainly in plant cell wall. Dietary fiber cannot be broken down by human digestive enzymes.

DIETARY REFERENCE INTAKES (DRIs) Quantitative estimates of nutrient intakes which are used as reference values for assessing the diets of healthy people. DRIs include Recommended Dietary Allowances (RDAs), Adequate Intakes (AI), Tolerable Upper Intake Level (UL), and Estimated Average Requirement (EAR).

DiGEORGE SYNDROME Condition in which chromosome 22 has a small deletion, resulting in a wide range of heart, speech, and learning difficulties.

DIPLEGIA Condition in which the part of the brain controlling movement of the legs is damaged, interfering with muscle control and ambulation.

DIVERTICULITIS Infected "pockets" within the large intestine.

DOULA An individual who surrounds, interacts with, and aids the mother at any time within the period that includes pregnancy, birth, and lactation; may be a relative, friend, or neighbor and is usually but not necessarily female. One who gives psychological encouragement and physical assistance to a new mother.

DOWN SYNDROME Condition in which three copies of chromosome 21 occur, resulting in lower muscle strength, lower intelligence and greater risk for overweight.

DYSMENORRHEA Painful menstruation due to abdominal cramps, back pain, headache, and/or other symptoms.

EARLY INTERVENTION PROGRAM Educational intervention for the development of children from birth up to 3 years of age.

EARLY INTERVENTION SERVICES Federally mandated evaluation and therapy services for children in the age range birth up to age three years under the Individuals with Disabilities Education Act.

EDEMA Swelling (usually of the legs and feet but can also extend throughout the body) due to an accumulation of extracellular fluid.

EIGHT (8) OUNCES—1 cup = 240 ml (milliliters) = 240 cc (cubic centimeters) A super-sized soda (32 ounces) equals approximately 4 cups (or 4 × 240 = 960 ml). A 2-liter bottle of soda provides a little more than 8 cups of fluid. Some foods also count as fluid. For example, soup, Jell-O, and sherbet are considered fluids.

EMBRYO The developing organism from conception through eight weeks.

ENDOCRINE A system of ductless glands that produce secretions that affect body functions. Such glands include the thyroid, adrenal glands, ovaries, and testes.

ENDOMETRIOSIS A disease characterized by the presence of endometrial tissue in abnormal locations such as deep within the uterine wall, in the ovary, or in other sites within the body. The condition is quite painful and associated with abnormal menstrual cycles and infertility in 30–40% of affected women.

ENDOTHELIUM The layer of flat cells lining blood and lymph vessels. These cells produce a variety of proteins that play a role in blood pressure regulation and body fluid distribution.

ENTERAL FEEDING Method of delivering nutrients directly into the gastrointestinal system. The delivery can be by mouth, or through a tube that is placed into the stomach or intestines.

EPIDIDYMIS Tissues on top of the testes that store sperm.

EPITHELIAL CELLS Cells that line the surface of the body.

EPSDT The Early Periodic Screening, Detection, and Treatment Program is a part of the Medicaid program and provides routine check-ups for low-income families.

ESOPHAGEAL ATRESIA Incomplete connection between the esophagus and the stomach in utero, resulting in a shortened esophagus.

ESTROGEN The name of a group of hormones made by the ovaries, testes, fat cells, corpus luteum, and placenta (during pregnancy). Compounds with similar actions are also produced by some plants. Estradiol is the most active and abundant form of estrogen in humans. Estrogen performs many functions, among them is stimulation of the release of GnRH in the follicular phase and its inhibition in the luteal phase, and the thickening of the uterine wall during the menstrual cycle.

Exocytosis   Release of cellular contents of a vesicle when the membrane of that vesicle merges with the plasma membrane.

Exposure Index   The average infant milk intake per kilogram body weight per day × (the milk to plasma ratio divided by the rate of drug clearance) × 100. It is indicative of the amount of the drug in the breast milk that the infant ingests and is expressed as a percentage of the therapeutic (or equivalent) dose for the infant.

Extremely Low Birth Weight (ELBW)   Infants weighing <1000 grams or 2 pounds and 3 ounces at birth.

Failure to Thrive (FTT)   Condition of inadequate weight and/or height gain thought to result from a caloric deficit, whether or not the cause can be identified as a health problem.

Familial Hyperlipidemia   A condition which runs in families and results in high levels of serum cholesterol and other lipids.

Fecundity   Biological ability to bear children.

Fertility   Actual production of children. Word best applies to specific vital statistic rates, but is commonly taken to mean the ability to bear children.

Fetal Origins Hypothesis   The theory that exposures to adverse nutritional and other conditions during critical or sensitive periods of growth and development can permanently affect body structures and functions. Such changes may predispose individuals to cardiovascular diseases, type 2 diabetes, hypertension, and other disorders later in life. Also called "metabolic programming" and the "Barker Hypothesis."

Fetus   The developing organism from eight weeks after conception to the moment of birth.

Fine Motor Skills   Development and use of smaller muscle groups demonstrated by stacking objects, scribbling, and copying a circle or square.

Fluorosis   Permanent white or brownish staining of the enamel of teeth caused by excessive ingestion of fluoride before teeth have erupted.

Follicle Stimulating Hormone (FSH)   A hormone produced in the pituitary gland that stimulates the maturation of ova and sperm.

Food Allergy   Adverse reaction to a substance in food that sets up an immune system reaction, and may be expressed in several different body systems over time, such as breathing and skin reactions.

Food Allergy (hypersensitivity)   Abnormal or exaggerated immunologic response, usually immunoglobulin E (IgE) mediated, to a specific food protein.

Food Intolerance   An adverse reaction to a substance in food that is not a reaction of the immune system.

Food Security   Access at all times to a sufficient supply of safe, nutritious food.

Fragile X Syndrome   Condition mainly identified in males characterized by mental retardation and physical signs.

Full-Term Infants   Infants born between 37 and 42 weeks of gestation.

Functional Status   Ability to carry out the activities of daily living, including telephoning, grocery shopping, food handling and preparation, and eating.

Galactosemia   A rare genetic condition of carbohydrate metabolism in which a blocked or inactive enzyme does not allow breakdown of galactose, causing serious illness in infancy.

Gastroesophageal Reflux (GER)   or gastroesophageal reflux disease (GERD) Movement of the stomach contents backwards into the base of the esophagus.

Gastrostomy   Form of enteral nutrition support for delivering nutrition by tube directly into the stomach, bypassing the mouth through a surgical procedure that creates an opening through the abdominal wall and stomach.

Gastrostomy Feeding   Form of enteral nutrition support for delivering nutrition by tube placement directly into the stomach, bypassing the mouth through a surgical procedure which creates an opening through the abdominal wall and stomach.

Geophagia   Compulsive consumption of clay or dirt.

Gestational Diabetes   Carbohydrate intolerance with onset or first recognition in pregnancy.

Gliadin   The toxic fraction of gluten.

Gluten   A protein found in wheat, oats, barley, rye and triticale (all in the genus Triticum).

Glycemic Index   A classification used to describe how quickly food carbohydrate appears as blood glucose; the glycemic response of various carbohydrates is compared to intake of pure glucose.

Glycosylated Hemoglobin   A laboratory test that measures how well the blood sugar level has been maintained over a prolonged period of time, also called Hemoglobin A, C.

Golgi   Groups of intracellular vesicles where processing and packaging of secreted proteins takes place.

Gonadotropin Releasing Hormone (GnRH)   A hormone produced in the hypothalamus that stimulates the release of FSH and LH.

Gravida   Number of pregnancies a woman has experienced.

Gross Motor Skills   Development and use of large muscle groups as exhibited by walking alone, running, walking up stairs, riding a tricycle, hopping, and skipping.

Growth   Increase in an organism's size through cell multiplication (hyperplasia) and enlargement of cell size (hypertrophy).

Growth Velocity   The rate of growth over time.

Gynecological Age   Defined as chronological age minus age at menarch. For example, a female with the chronological age of 14 years minus age at first menstrual cycle 12 years equals a gynecological age 2.

Health   More than the absence of disease, health is a sense of well-being. An individual with a chronic condition may consider themselves to be healthy. For instance, a person with diabetes mellitus whose blood sugar is under control properly can be considered healthy.

Heart Disease   The leading cause of death and a common cause of illness and disability in the U.S. Coronary heart disease is the principal form of heart disease and is caused by buildup of cholesterol deposits in the coronary arteries, which feed the heart.

Hematocrit   An indicator of the proportion of whole blood occupied by red blood cells. A decrease in hematocrit is a late indicator of iron deficiency.

Heme Iron   Iron contained within a protein portion of hemoglobin that is in the ferrous state.

Hemoglobin   A protein which is the oxygen carrying component of red blood cells. A decrease in hemoglobin concentration in red blood cells is a late indicator of iron deficiency.

Hemolytic Anemia   Anemia caused by shortened survival of mature red blood cells and inability of the bone marrow to compensate for the decreased life span.

Hemolytic Uremic Syndrome (HUS)   A serious, sometimes fatal complication associated with illness caused by E. coli 0157:H7, which occurs primarily in children under the age of 10 years. HUS is characterized by renal failure, hemolytic anemia, and a severe decrease in platelet count. (HP 2010)

High-poverty Neighborhoods   Neighborhoods where 40% or more of the people are living in poverty.

HIV   Human Immunodeficiency Virus.

Homeostasis   Constancy of the internal environment. The balance of fluids, nutrients, gases, temperature, and other conditions needed to ensure ongoing, proper functioning of cells, and therefore all parts of the body.

Homocysteine   is another intermediate product that depends on vitamin B_{12} for complete metabolism. However, both vitamin B_{12} and folate (another B vitamin) are coenzymes in the breakdown of certain protein components in this pathway. Thus, elevated homocysteine levels can result from vitamin B_{12}, folate or pyridoxine deficiencies.

HYDROCEPHALUS Condition in which excess fluid collects in the brain, resulting in brain damage, often treated by a shunt.

HYDROLYZED PROTEIN FORMULA Formula that contains enzymatically digested protein, or single amino acids, rather than protein as it naturally occurs in foods.

HYPERBILIRUBINEMIA Elevated blood levels of bilirubin, a yellow pigment that is a byproduct from the breakdown of fetal hemoglobin.

HYPERTONIA Condition characterized by high muscle tone, stiffness, or spasticity.

HYPOALLERGENIC Foods or products that are associated with a low risk of developing food or other allergies.

HYPOCALCEMIA Condition in which body pools of calcium are unbalanced, and low levels are measured in blood as a part of a generalized reaction to illnesses.

HYPOGONADISM Atrophy or reduced development of testes or ovaries. Results in immature development of secondary sexual characteristics.

HYPOTHYROIDISM Condition in which thyroid hormone is not produced in sufficient quantities, interfering with growth and mental development if untreated in infants.

HYPOTONIA Condition characterized by low muscle tone, floppiness, or muscle weakness.

IATROGENIC Used in reference to disease, it is a condition secondary to a medical treatment. For example, when a person who is hospitalized to cure pneumonia gets a staphylococcus infection, that resultant staph infection is an iatrogenic disease.

IMMUNOGLOBULIN A specific protein that is produced by blood cells to fight infection.

IMMUNOLOGICAL Having to do with the immune system and its functions in protecting the body from bacterial, viral, fungal, or other infections, and from foreign proteins—or proteins that differ from other proteins normally found in the body.

IN UTERO Period of gestation or intrauterine life.

INDIRECT CALORIMETRY Measurement of energy requirements based on oxygen consumption and carbon dioxide production.

INFANT HEALTH AND DEVELOPMENT PROGRAM (IHDP) Growth charts with percentiles for VLBW (<1500 gram birth weight) and LBW (<2500 gram birth weight).

INFANT MORTALITY Death that occurs within the first year of life.

INFANT MORTALITY ATTRIBUTABLE TO BIRTH DEFECTS (IMBD) Category used in tracking infant deaths in which specific diagnoses have a high mortality.

INFANT MORTALITY RATE Number of deaths occurring within the first year of life per 1,000 life births.

INFECUNDITY Biological inability to bear children after one year of unprotected intercourse.

INFERTILITY Absence of production of children. Commonly used to mean a biological inability to bear children.

INORGANIC FAILURE TO THRIVE Inadequate weight and/or height gain without an identifiable biological cause, so that an environmental cause is suspected.

INSULIN Hormone usually produced in the pancreas to regulate movement of glucose from the blood stream into cells within organs and muscles.

INSULIN RESISTANCE A condition in which cell membranes have reduced sensitivity to insulin so that more insulin than normal is required to transport a given amount of glucose into cells. It is characterized by elevated levels of serum insulin, glucose, and triglycerides, and increased blood pressure.

INSULIN Hormone produced in the pancreas.

INTRAUTERINE GROWTH RETARDATION (IUGR) (OR INTRAUTERINE GROWTH RESTRICTION) Fetal undergrowth from any cause, resulting in a disproportionality in weight, length, and/or weight for length percentiles for gestational age.

IRON DEFICIENCY A condition marked by depleted iron stores. It is characterized by weakness, fatigue, short attention span, poor appetite, increased susceptibility to infection, and irritability.

IRON DEFICIENCY ANEMIA A condition often marked by low hemoglobin level. It is characterized by the signs of iron deficiency plus paleness, exhaustion, and a rapid heart rate.

JEJUNOSTOMY FEEDING Form of enteral nutrition support for delivering nutrition by tube placement directly into the upper part of the small intestine.

JUVENILE RHEUMATOID ARTHRITIS Condition in which joints become enlarged and painful as a result of the immune system, generally occurring in children or teens.

KERNICTERUS OR BILIRUBIN ENCEPHALOPATHY The end result of very high untreated bilirubin levels. Excessive bilirubin in the system is deposited in the brain, permanently destroying brain cells.

KETOGENESIS Condition in which ketones are made from metabolic pathways used in converting fat as a source of energy.

KETOGENIC DIET High-fat, low-carbohydrate meal plan in which ketones are made from metabolic pathways used in converting fat as a source of energy.

KETONES Small two-carbon chemicals generated by breakdown of fatty acids for energy.

KLINEFELTER'S SYNDROME A congenital abnormality in which testes are small and firm, legs abnormally long, and intelligence generally subnormal.

KREBS CYCLE A series of metabolic reactions that produce energy from the proteins, fats, and carbohydrates that constitute food.

L. MONOCYTOGENES, OR LISTERIA A food borne bacterial infection that can lead to preterm delivery and stillbirth in pregnant women. Listeria infection is commonly associated with the ingestion of soft cheeses, unpasteurized milk, ready-to-eat deli meats, and hot dogs.

LACTATION CONSULTANT Health care professional whose scope of practice is focused upon providing education and management to prevent and solve breastfeeding problems and to encourage a social environment that effectively supports the breastfeeding mother/infant dyad. Those who successfully complete the International Board of Lactation Consultant Examiners (IBLCE) certification process are entitled to use the IBCLC (International Board Certified Lactation Consultant) after their names (http://www.iblce.org/).

LACTIFEROUS SINUSES Larger ducts for storage of milk behind the nipple.

LACTOGENESIS Human milk production.

LACTOSE A form of sugar or carbohydrate composed of galactose and glucose.

LARGE FOR GESTATIONAL AGE (LGA) Birth weight is greater than 90th percentile weight for gestational age.

LDL-CHOLESTEROL Low density lipoprotein cholesterol, the lipid which is most associated with atherosclerotic disease. Diets high in saturated fat, trans fatty acids, and dietary cholesterol have been shown to increase LDL-cholesterol levels.

LE LECHE LEAGUE International, nonprofit, nonsectarian organization dedicated to providing education, information, support, and encouragement to women who want to breastfeed. Founded in 1956 by seven women who had learned about successful breastfeeding while nursing their own babies, currently approximately 7,100 accredited lay leaders facilitate more than 3,000 monthly mother-to-mother breastfeeding support group meetings around the world (http://www. lalecheleague.org)

LEAN BODY MASS Sum of fat free body tissues: muscle, mineral as in bone, and water.

LES Lower esophageal sphincter which is the muscle enabling closure of the junction between the esophagus and stomach.

LIFE EXPECTANCY Average number of years of life remaining for persons in a population cohort or group; most commonly reported as life expectancy from birth.

LIFE SPAN Maximum number of years someone might live; human life span is projected between 110 to 120 years.

LIGNIN Noncarbohydrate polymer which contributes to dietary fiber.

LINSEED From the flax plant, *linum* is another name for flaxseed. Linseed oil is used in paints, varnishes, and inks but is also produced in food form for its rich nutrient content.

LOBES Rounded structures of the mammary gland.

LONG CHAIN FATS Carbon molecules that provide fatty acids with 12 or more carbons, which are commonly found in foods. Examples are palmitic (C16:0) stearic (C18:1) linoleic (C18:2) and linolenic (C18:3) acids.

LONGEVITY Length of life; it is a measure of life's duration in years.

LOW BIRTH WEIGHT (LBW) Infants weighing less than 2500 grams (5 pound 8 ounces) at birth.

LUTEINIZING HORMONE (LH) A hormone produced in the pituitary gland that stimulates the secretion of estrogen, progesterone, and testosterone; and the growth of the corpus luteum.

MACROBIOTIC DIET This diet falls between semivegetarian and vegan diets and include foods such as brown rice, other grains, and vegetables, fish, dried beans, spices, and fruits.

MACROCEPHALY Large head size for age and gender as measured by centimeters (or inches) of head circumference.

MACROPHAGES A white blood cell that acts mainly through phagocytosis.

MALNUTRITION Poor nutrition resulting from an excess or lack of calories or nutrients.

MAMMARY GLAND The source of milk for offspring, also commonly called the breast. The presence of mammary glands is a characteristic of mammals.

MAPLE SYRUP URINE DISEASE Rare genetic condition of protein metabolism in which breakdown by-products build up in blood and urine, causing coma and death if untreated.

MCT OIL A liquid form of dietary fat used to boost calories composed of medium chain triglycerides.

MECONIUM Dark green mucilaginous material in the intestine of the full-term fetus.

MEDIAN CHAIN FATTY ACID OXIDATION DISORDER Condition in which breakdown of fatty acids for generation of energy inside the cell is blocked.

MEDICAL NEGLECT Failure of parent/caretaker to seek, obtain, and follow through with a complete diagnostic study or medical, dental, or mental health treatment for a health problem, symptom, or condition that, if untreated, could become severe enough to present a danger to the child.

MEDICAL NUTRITION THERAPY Comprehensive nutrition services by registered dietitians to treat the nutritional aspects of acute and chronic diseases.

MEDICINAL HERBS Plants used to prevent or remedy illness.

MEDIUM-CHAIN FATS Carbon molecules that provide fatty acids with 6–10 carbons, not typically found in foods.

MENARCHE The occurrence of the first menstrual cycle.

MENINGITIS Viral or bacterial infection in the central nervous system that is likely to cause a range of long-term consequences in infancy, such as mental retardation, blindness, and hearing loss.

MENOPAUSE Cessation of menses for at least 12 months. Menses decrease and then stop because the ovaries stop producing estrogen, progesterone, and testosterone. Menopause occurs over a period of years; average age in U.S. is 51.

MENSES The process of menstruation.

MENSTRUAL CYCLES An approximately 4 week interval in which hormones direct a buildup of blood and nutrient stores within the wall of the uterus, and ovum maturation and release. If the ovum is fertilized by a sperm then the stored blood and nutrients are used to support the growth of the fertilized ovum. If fertilization does not occur, stored blood and nutrients are released from the uterine wall over a period of 3 to 7 days. The period of blood flow is called the "menses." Also called the menstrual period.

MENTAL RETARDATION Substantially below-average intelligence and problems adapting to the environment, which emerge before age 18 years.

METABOLISM The chemical changes that take place in the body.

METHYLMALONIC ACID (MMA) An intermediate product that needs vitamin B_{12} as a coenzyme to complete the metabolic pathway for fatty acid metabolism. vitamin B_{12} is the only coenzyme in this reaction; when it is absent, the blood concentration of MMA rises.

MICROCEPHALY Small head size for age and gender as measured by centimeters (or inches) of head circumference.

MIDDLE CHILDHOOD Children between the ages of 5 and 10 years; also referred to as school-age.

MILK TO PLASMA DRUG CONCENTRATION RATIO (M/P RATIO) The concentration of the drug in the milk versus the concentration in maternal plasma. Since the ratio varies over time, a time averaged ratio provides more meaningful information than data obtained at a single time point.

MISCARRIAGE Generally defined as the loss of a conceptus in the first 20 weeks of pregnancy. Also called "spontaneous abortion."

MITOCHONDRIA Intracellular unit in which fatty acid breakdown takes place and many enzyme systems for energy production inside cells are regulated.

MODERATE OR SEVERE MEMORY IMPAIRMENT When four or fewer words can be recalled from a list of 20.

MORBIDITY Illness, disease.

MORTALITY Death, e.g., "mortality data" indicate how many individuals died in a given period.

MYOEPITHELIAL CELLS Specialized cells that line the alveoli and can contract to cause milk to be secreted into the duct.

NASOGASTRIC FEEDING Form of enteral nutrition support for delivering nutrition by tube placement from the nose to the stomach.

NECROTIZING ENTEROCOLITIS (NEC) Condition with inflammation or damage to a section of the intestine, with a grading from mild to severe.

NEONATAL DEATH Death occurs from the day of birth through the first 28 days of life.

NEURAL TUBE DEFECTS Spina bifida and other malformation of the neural tube. Defects result from incomplete formation of the neural tube during the first month after conception.

NEURAL TUBE DEFECTS (NTDs) Malformations of the spinal cord and brain. There are three major types of NTDs: 1)Spina bifida: spinal cord fails to close, leaving a gap where spinal fluid collects during pregnancy. Paralysis below the gap in the spinal cord occurs in severe cases; 2) Anencephaly: the absence of the brain or spinal cord; 3) Encephalocele: the protrusion of the brain through the skull.

NEUROBEHAVIORAL Pertains to control of behavior by the nervous system.

NEUROMUSCULAR Term pertaining to control of the central nervous system on muscle coordination and movement.

NEUROMUSCULAR DISORDERS Conditions of the nervous system characterized by difficulty with voluntary or involuntary control of muscle movement.

NEUTROPHILS- Class of white blood cells that are involved in the protection against infection.

NF A uniformity standard for herbs and botanicals.

NONHEME IRON   Iron contained within a protein portion of hemoglobin that is in the ferric state.

NUTRIENTS   Chemical substances found in food that are used by the body for growth and health.

NUTRITION SUPPORT   Provision of nutrients by methods other than eating regular foods or drinking regular beverages, such as directing accessing the stomach by tube or placing nutrients into the blood stream.

OBESITY   Body mass index (BMI)-for-age >95th percentile with excess fat stores as evidenced by increased triceps skinfold measurements of >85th percentile.

ORAL-GASTRIC FEEDING (OG)   A form of enteral nutrition support for delivering nutrition by tube placement from the mouth to the stomach.

ORGANIC FAILURE TO THRIVE   Inadequate weight and/or height gain resulting from a health problem such as iron deficiency anemia, or a cardiac or genetic disease.

OSMOLARITY   Measure of the number of particles in a solution which predicts the tendency of the particles to move from high to low concentration. Osmolarity is a factor in many systems such as in fluid and electrolyte balance.

OSTEOBLASTS   Bone cells involved with bone formation, bone-building cells.

OSTEOCLASTS   A bone cell that absorbs and removes unwanted tissue.

OSTEOPOROSIS   Condition in which low density or weak bone structure leads to an increased risk of bone fracture.

OVA   Eggs of the female produced and stored within the ovaries. Singular is ovum.

OVERWEIGHT   Body mass index (BMI) at or above the 95th percentile.

OXYTOCIN   A hormone produced during letdown that causes milk to be ejected into the ducts.

PAGOPHAGIA   Compulsive consumption of ice or freezer frost.

PARITY   The number of previous deliveries experienced by a woman; nulliparous = no previous deliveries, primiparous = one previous delivery, multiparous = 2 or more previous deliveries. Women who have delivered infants are considered to be "parous."

PEDIATRIC AIDS   Acquired immunodeficiency syndrome in which infection-fighting abilities of the body are destroyed by a virus.

PELVIC INFLAMMATORY DISEASE   A general term applied to infections of the cervix, uterus, fallopian tubes, or ovaries. Occurs predominately in young women and is generally caused by infection with a sexually transmitted disease such as gonorrhea or chlamydia, or with intrauterine device (IUD) use.

PERCENT DAILY VALUE (%DV)   Daily Values represent scientifically agreed-upon standards of daily intake of nutrients from the diet developed for use on nutrition labels. The "% DV" listed on nutrition labels represent the percentages of the standards obtained from one serving of the food product. Table 1.5 lists DV standard amounts for nutrients that are mandatory or voluntary components of nutrition labels.

PERICONCEPTIONAL PERIOD   Around the time of conception, generally defined as the month before and the month after conception.

PERINATAL DEATH   Death occurring at or after 20 weeks of gestation and through the first 28 days of life.

PHENYLKETONURIA (PKU)   An inherited error in phenylalanine metabolism most commonly caused by a deficiency of phenylalanine hydroxylase which converts the essential amino acid phenylalanine to the nonessential amino acid tyrosine.

PHYTOCHEMICALS   ("Phyto" = plants) Chemical substances in plants, some of which affect body processes in humans that may benefit health.

PHYTOESTROGEN   A hormone-like substance found in plants, about 1/1000 to 1/2000 as potent as the human hormone, but strong enough to bind with estrogen receptors and mimic estrogen and anti-estrogen effects.

PICA   An eating disorder characterized by the compulsion to eat substances that are not food.

PINOCYTOSIS   Process of taking fluid filled vesicles into cells.

PLACENTA   A disk-shaped organ of nutrient and gas interchange between mother and fetus. At term, the placenta weighs about 15% of the weight of the fetus.

PLATELETS   A component of the blood which plays an important role in blood coagulation.

POLYCYSTIC OVARY SYNDROME (PCOS)   (polycysts = many cysts—abnormal sacs with membranous linings) A condition in females characterized by insulin resistance, high blood insulin and testosterone levels, obesity, menstrual dysfunction, amenorrhea, infertility, hirsutism (excess body hair), and acne.

POSTICTAL STATE   Time after a seizure of altered consciousness appearing as a deep sleep.

PRADER-WILLI SYNDROME   Condition in which partial deletion of chromosome 15 interferes with control of appetite, muscle development, and cognition.

PREADOLESCENCE   The stage of development immediately preceding adolescence; 9 to 11 years of age for girls and 10 to 12 years of age for boys.

PRELOADS   Beverages or food such as yogurt in which the energy/macronutrient content has been varied by the use of various carbohydrate and fat sources. The preload is given before a meal or snack and subsequent intake is monitored. This study design has been employed by Birch et al. in their studies of appetite, satiety, and food preferences in young children.

PREMENSTRUAL SYNDROME   (Premenstrual= the period of time preceding menstrual bleeding; syndrome = a constellation of symptoms) A condition occurring among women of reproductive age that includes a group of physical, psychological, and behavioral symptoms with onset in the luteal phase and subsiding with menstrual bleeding. Also called premenstrual dysphoric disorder (PMDD).

PRESCHOOL AGE CHILDREN   Children between the ages of 3 and 5 years, who are not yet attending kindergarten.

PRETERM INFANTS   Infants born at or before 37 weeks of gestation.

PRIMARY MALNUTRITION   Malnutrition that directly results form inadequate or excessive dietary intake of energy or nutrients.

PROGESTERONE   A hormone produced by the ovaries and placenta that primarily functions to prepare the uterus for a fertilized ovum and to maintain a pregnancy; it's "progestational." Progesterone also prompts the buildup of the uterine lining during the menstrual cycle, helps to stimulate cell division of fertilized ova, and inhibits the action of testosterone. Progestin, progestogen, gestagen are names that refer to similar hormones.

PROGRAMMING   The process by which exposure to adverse nutritional or other conditions during sensitive periods of growth and development produces long-term effects on body structures, functions, and disease risk.

PROLACTIN   A hormone necessary for milk production.

PROSTACYCLIN   A potent inhibitor of platelet aggregation and a powerful vasodilator and blood pressure reducer derived from n-3 fatty acids. Prostacyclin I2 levels are generally low in women with preeclampsia.

PROSTAGLANDINS   A group of physiologically active substances made from the fatty acid arachidonic acid. They are present in many tissues and perform such functions as the constriction or dilation of blood vessels, and stimulation of smooth muscles and the uterus. (They are not produced solely in the prostate gland as once was thought.)

PSYCHOSTIMULANT   Classification of medication that acts on the brain to increase mental or emotional behavior.

PUBERTY   The stage in life during which humans become biologically capable of reproduction. In girls, puberty generally occurs somewhere between the ages of 9 to 16, and in boys between 11 and 17 years.

PULMONARY Related to the lungs and their movement of air for exchange of carbon dioxide and oxygen.

QUALITY OF LIFE A measure of life satisfaction that is difficult to define, especially in an aging heterogeneous population. Quality of life measures include factors such as social contacts, economic security, and functional status.

RECOMMENDED DIETARY ALLOWANCES (RDAs) The average daily dietary intake levels that are sufficient to meet the nutrient requirements of nearly all (97% to 98%) healthy individuals in a population group. RDAs serve as goals for individuals.

RECUMBENT LENGTH Measurement of length while the child is laying down. Recumbent length is used to measure toddlers < 24 months of age, and those between 24 and 36 months who are unable to stand unassisted.

REFLEX An automatic (unlearned) response that is triggered by a specific stimulus.

REGISTERED DIETITIAN An individual who has acquired knowledge and skills necessary to pass a national registration examination and participates in continuing, professional education.

RESILIENCE Ability to bounce back, to deal with stress and recover from injury or illness.

RESTING ENERGY EXPENDITURE The amount of energy needed by the body in a state of rest.

RETT SYNDROME A genetic change on the X chromosome that results in severe neurological delays causing children to be short and thin appearing and unable to talk.

ROOTING REFLEX Action that occurs if one cheek is touched, resulting in the infant's head turning towards that cheek and opening his mouth.

SCOLIOSIS Condition in which the vertebral bones in the back show a side-to-side curve, resulting in a shorter stature than expected if the back was straight.

SCROTUM A muscular sac of the male reproductive organ containing the testes and the spermatic cord.

SECONDARY CONDITION Common consequence of a condition which may or may not be preventable over time.

SECONDARY MALNUTRITION Malnutrition that results from a condition (eg. disease, surgical procedure, medication use) rather than primarily dietary intake.

SECONDARY SEXUAL CHARACTERISTICS Physiological changes that signal puberty, including enlargement of the testes, penis, and breasts, and the development of pubic and facial hair.

SECRETORY CELLS Cells in the acinus that are responsible for secreting milk components into the ducts.

SECRETORY IMMUNOGLOBULIN A A protein found in secretions that protect the body's mucosal surfaces from in infections. The mode of action may be by reducing the binding of a microorganism with cells lining the digestive tract. It is present in human colostrum but not transferred across the placenta.

SEIZURES Condition in which electrical nerve transmission in the brain is disrupted, resulting in periods of loss of function that vary in severity.

SELF-EFFICACY The ability to make effective decisions and to take responsible action based upon one's own needs and desires.

SEMEN The penile ejaculate containing a mixture of sperm and secretions from the testes, prostate, and other glands. It is rich is zinc, fructose, and other nutrients. Also called seminal fluid.

SENESCENT Old to the point of non-functional.

SENSORIMOTOR An early learning system in which the infant's senses and motor skills provide input to his central nervous system.

SERUM FERRITIN The major iron storage protein. Serum ferritin is low in iron deficiency.

SERUM IRON, PLASMA FERRITIN, AND TRANSFERRING SATURATION Measures of iron status obtained from blood plasma or serum samples.

SEX HORMONE BINDING GLOBULIN A protein that binds with the "sex hormones" testosterone and estrogen. Also called steroid hormone binding globulin as testosterone and estrogen are produced from cholesterol and are therefore considered to be "steroid hormones." These hormones are inactive when bound to SHBG, but are available for use when needed. Low levels of SHBG are related to increased availability of testosterone and estrogen in the body.

SGA SMALL FOR GESTATIONAL AGE Infant birth weight is <10 percentile of weight for gestational age.

SHORT-CHAIN FATS Carbon molecules that provide fatty acids less than 6 carbons long, as products of energy generation from fat breakdown inside cells. Short chain fatty acids are not usually in foods.

SHOULDER DYSTOCIA Blockage or difficulty of delivery due to obstruction of the birth canal by the infant's shoulders.

SMALL FOR GESTATIONAL AGE (SGA) Birth weight is less than 10th percentile weight for gestational age.

SOCIAL MARKETING "Combines the principles of commercial marketing with health education to promote a socially beneficial idea, practice or product" (Bryant, 1992).

SPASTIC QUADRIPLEGIA Condition in which brain damage interferes with voluntary muscle control in both arms and legs; a form of cerebral palsy.

SPINA BIFIDA Condition initiated early in gestation of the central nervous system in which the spinal cord and related structures are malformed, resulting in paralysis in the body below the lesion. A type of myelomeningocele condition.

STATURE Standing height.

STEROID HORMONES Hormones such as progesterone and estrogen produced primarily from cholesterol.

SUBFERTILITY Reduced level of fertility characterized by unusually long time to conception (over 12 months) or repeated, early pregnancy losses.

SUBSCAPULAR SKINFOLD THICKNESS A skinfold measurement that can be used with other skinfold measurements to estimate percent body fat; the measurement is taken with skinfold calipers just below the inner angle of the scapular or shoulder blade.

SUCKING REFLEX Action in which an infant will suck on anything placed in mouth; with the tongue movement raising and lowering.

SUCKLE A reflexive movement of the tongue moving forward and backward; earliest feeding skill.

T. GONDII, OR TOXOPLASMOSIS A parasitic infection that can impair fetal brain development. The source of the infection is often hands contaminated with soil or cat litter box contents, or raw or partially cooked pork, lamb, or venison.

TELOMERE A cap-like structure that protects the end of chromosomes; it erodes during replication.

TERATOGENIC Exposures that produce malformations in embryos or fetuses.

TESTES Male reproductive glands located in the scrotum.

TESTICLE (PL., TESTES) One of two male reproductive glands located in the scrotum.

TESTOSTERONE The most potent, naturally occurring male sex hormone formed in greatest quantity by the testes. It stimulates the maturation of male sex organs and sperm, the formation of muscle tissue, and serves other functions.

THROMBOXANE The parent of a group of thromboxanes derived from the n-6 fatty acid arachidonic acid. Thromboxane increases platelet aggregation and constricts blood vessels causing blood pressure to increase. Thromboxane A2 is generally low in women with preeclampsia.

T-LYMPHOCYTE A white blood cell that is active in fighting infection. (May also be called T-cell; the t in T-cell stands for thymus). These cells coordinate the immune system by secreting hormones that act on other cells.

TODDLERS   Children between the ages of 1 and 3 years.

TOLERABLE UPPER INTAKE LEVELS   Highest level of daily nutrient intake that is likely to pose no risks of adverse health effects to almost all individuals in the general population; gives levels of intake which may result in adverse effects if exceeded on a regular basis.

TOTAL PARENTERAL NUTRITION (TPN) OR PARENTERAL NUTRITION (PN)   Delivery of nutrients directly to the blood stream.

TRANS FATTY ACIDS   Fatty acids which have unusual shapes, which result from the hydrogenation of polyunsaturated fatty acids. Trans fatty acids also occur naturally in small amounts in foods such as dairy products and beef.

TRANSPYLORIC FEEDING (TP)   Form of enteral nutrition support for delivering nutrition by tube placement from the nose or mouth into the upper part of the small intestine.

TRICEPS SKINFOLD   A measurement of a double layer of skin and fat tissue on the back of the upper arm. It is an index of body fatness and measured by skinfold calipers. The measurement is taken on the back of the arm midway between the shoulder and the elbow.

UNITED STATES PHARMACOPEIA (USP)   a non-governmental, non-profit organization (since 1820) establishes and maintains standards of identity, strength, quality, purity, processing and labeling for health care products.

VEGAN DIET   The most restrictive of vegetarian diets as only plant foods are allowed.

VENOUS THROMBOEMBOLISM   Blood clot in a vein.

VERY LOW BIRTH WEIGHT INFANT (VLBW)   Infants weighing <1500 grams or <3 pounds 5 ounces at birth.

VESICLE   Small liquid-containing sac

WAIST-TO-HIP RATIO   The ratio of the waist circumference, measured at its narrowest, and the hip circumference, measured where it is widest. This ratio is an easy way to measure body fat distribution, with a higher ratio indicative of an abdominal fat pattern. A high waist-to-hip ratio is associated with a high risk of chronic disease.

WEANING   Discontinuation of breast-feeding or bottle feeding and substitution of food for breast-milk or infant formula.

WORK OF BREATHING (WOB)   A common term used to express extra respiratory effort in a variety of pulmonary conditions.

WORKING-POOR FAMILIES   Families where at least one parent worked 50 or more weeks a year and the family income was below the poverty level.

XEROSTOMIA   Dry mouth or xerostomia can be a side effect of medications (especially anti-depressants), of head and neck cancer treatments, of diabetes, and also a symptom of Sjogren's syndrome, which is an auto-immune disorder for which no cure is known.

INDEX